Remote Sensing of the Asian Seas

Vittorio Barale · Martin Gade
Editors

Remote Sensing of the Asian Seas

Springer

Editors
Vittorio Barale
European Commission
Joint Research Centre
Ispra, Italy

Martin Gade
Institute of Oceanography
University of Hamburg
Hamburg, Germany

ISBN 978-3-030-06789-2 ISBN 978-3-319-94067-0 (eBook)
https://doi.org/10.1007/978-3-319-94067-0

Softcover re-print of the Hardcover 1st edition 2019

Cover illustration: Pictures were taken from Chapter 3: Bai et al., Fig. 1; Chapter 8: Mitnik and Dubina, Fig. 8b; Chapter 7: Eriksen et al., Fig. 2; Chapter 19: Gade et al., Fig. 2; Chapter 17: He and Chen, Fig. 7.

This Springer imprint is published by the registered company Springer Nature Switzerland AG
The registered company address is: Gewerbestrasse 11, 6330 Cham, Switzerland

To Ching Shih
Shrewd businesswoman and pirate
Champion of the Asian Seas sailing traders

Preface

The Asian Seas have hosted sea routes known and sailed since ancient times—especially on the southern courses that were a counterpart of the Silk Road trade network, complementing classical land itineraries—and provided all-important connections between east and west, between the mighty Asian civilizations and the Middle East, as well as early European cultures. This historical role of the Asian Seas is now continuing with renewed interest and new endeavors in both the marine, scientific, as well as the maritime, economic, sector.

In broad terms, the expression Silk Road identified a web of commercial routes, extending primarily over land from Chang'ans (now Xian) in the east to the Mediterranean Sea in the west, established during the Han Dynasty of China in the second century BC. At the time, this Road linked even older, if shorter, arteries (e.g., the Persian Royal Road opened by the Achaemenid Empire five to three centuries BC) and ultimately connected the Chinese Empire in the east with the Roman Empire in the west. Because silk was the major product that traveled along this corridor, it was named the Silk Road in 1877 by the German geographer Ferdinand von Richthofen. However, as this was not a single thoroughfare between east and west, and included also sea legs along the way, the term Silk Routes has become increasingly favored by historians.

One of the reasons this connection had so much importance was the range of merchandises traded, silk and spices in the first place, and other material goods as well. But just as important were the sundry cultural exchanges taking place: ideas, philosophy, science, technology, architecture, language, and even religions passed along. It can be argued that the Silk Routes, and hence the parallel Asian Seas' links, have been a significant factor in the advance of great civilizations, such as those of China, India, Persia, Arabia, Egypt, Greece, and even ancient Rome. Disease also traveled along the same pathways—as evidenced in the spread of the bubonic plague that reached Constantinople in the sixth century AD and decimated the Byzantine Empire—as well as awkward commodities such as slaves.

The commercial network was used regularly to the end of the fourteenth century AD, when the Ottoman Empire boycotted trade with the west and closed the routes. The shut down of the main Silk Road over land forced merchants to take to the sea to ply their trade, even more so than ever before, thus contributing to initiate the Age of Discovery, in the fifteenth and sixteenth centuries AD, which led to worldwide interactions and the beginnings of a global community.

The enhancement of a maritime route along the Asian Seas gave rise to a new breed of entrepreneurs, who developed and maintained the sea links along coastal regions between Asia and the other continents. One of the most successful, and famous, was Ching Shih, a shrewd businesswoman who had at her service, at the height of her power, over 1800 ships and an estimated 80,000 men and women. Indeed, the nature of her business was a bit awkward, and rather controversial as well (she was one of the most feared pirates of all times), but still she embodies in a way the romantic and audacious ideal of the Asian Seas sailing traders, who contributed to mold the ancient world. Born in humble beginnings as Shi Xiang Gu, sometime around 1775, she started her career working in a floating brothel in Guangzhou. In 1801, she married Zhèng Yi (a.k.a. Cheng I), a notorious Chinese pirate with a large fleet at his command, the Red Flag Fleet, made up of about 300 ships and somewhere between 20,000 and 40,000 men. When he died, in 1807, Gu became to be known as Ching Shih, which simply means "Cheng's widow," and succeeded him at the helm of the family business. With the help of Chang Pao, a former protégé of her late husband and second in command, she continued to run the organization and also to expand and govern her outfit into something that went beyond just a consortium of pillaging pirates, essentially setting up her own government of sorts, to establish laws, a rigorous code of conduct for her people, as well as taxes. She dominated the China Seas, regulating maritime traffic along profitable but dangerous sea lanes, and guaranteeing safe passage to those merchants who wanted to pay; she controlled nearly the entire Guangdong province directly; and she held a vast spy network within the Qing Dynasty. The Chinese government raised a fleet to attack her, and even enlisted the aid of the British and Portuguese navies, with little or no success. Given the unlikelihood of a military victory, in the end, the Empire offered her and most of her organization amnesty. So in 1810, Ching Shih retired, at age 35, with her untouched dignity and all her ill-gotten loot, becoming an extremely wealthy member of the aristocracy, a mother and a grandmother, and opening a gambling house in Guangzhou, which she managed until her death in 1844.

The heritage Ching Shih left behind from the time of her rule—including continuing piracy activities in the region—has penetrated popular culture, and still evokes the adventurous life of the Asian Seas' maritime raiders. Today, the concept of a modern Maritime Silk Road, proposed by the People Republic of China and readily embraced by several other nations along the Asian Seas routes, provides a legacy to the historical perspective of the Silk Routes and to the legendary figures that roamed the Asian Seas, while it brings new demands for cooperation, at the diplomatic, economic, as well as scientific level. The Chinese strategic initiative aims to increase investments and foster cooperation across the historic sea routes, and in particular to develop maritime infrastructures, especially ports. As such, it also links with the existing "string of pearls", i.e., China's network of maritime facilities in the Pacific and Indian Oceans, focusing even more attention onto the marginal and semi-enclosed basins surrounding the Asian continent.

The present Volume, which spotlights the issues, peculiarities, and special challenges posed by the Asian Seas, is dedicated to the extraordinary, albeit controversial, figure of Ching Shih. Rather than on pillaging shipborne raids, however, it is on modern expeditions, using Earth-orbiting satellites to collect data about marine environment and maritime activities alike, that the present review concentrates. The following compilation of topical papers covers virtually the entire remote sensing potential to explore the Asian near-coastal water bodies. It addresses the assessment of surface parameters by means of both passive and active techniques, covering the use of reflected visible and near-infrared sunlight, of surface emissions at thermal infrared or microwave frequencies, or again of the return of transmitted impulses of microwave radiation. The peer-reviewed papers gathered in this Volume provide a summary of the most recent results achieved in these fields, offering an unprecedented insight into some of the dynamical and bio-geo-chemical complexity of the Asian Seas.

The Table of Contents of this book—which targets researchers working in the Earth and Marine Sciences, but also teachers, as well as students, in the same fields, particularly in Asia—includes a brief overview of the Asian marginal and enclosed seas, followed by three introductory papers on Visible, Infrared, and Microwave (passive and active) remote sensing applications, providing examples centered on Asian waters. In-depth assessments of the main Asian marginal basins, in the Arctic, Pacific, and Indian Ocean, are grouped in three dedicated sections. A fourth section will deal with other enclosed, or semi-enclosed, basins. Several examples are provided of multisensor techniques, which exploit the synergies of complementary sensors and enhance the value of their combined views. In the section on enclosed water bodies, the Caspian Sea and Aral Sea, as well as two freshwater lakes, Balkhash and Baikal, are discussed, while Mediterranean Sea and Black sea are not present, having been covered already by several papers appearing in previous Volumes[1] of this Series. A truly international group of scientists collaborated to cover the wide spectrum of environmental themes, and the diverse techniques, dealt with in the current survey. Their fundamental input, their patience, and their endurance in resisting the Editors' constant harassments are acknowledged by the inclusion of all authors' and reviewers' names and affiliations in the list of Contributors. Sincere thanks are due to all of them.

Ispra, Italy
Hamburg, Germany
January 2018

Vittorio Barale
Martin Gade

[1] Barale V, Gade M, Eds. (2008) Remote Sensing of the European Seas. Springer Science+Business Media BV, 514 pp; and Barale V, Gade M, Eds. (2014) Remote Sensing of the African Seas. Springer, New York, Heidelberg, Dordrecht, London, 428 pp.

Acknowledgements

This volume has been prepared and published with the support of the Joint Research Centre (JRC), the European Commission's in-house science service.

As the Commission's in-house science service, the Joint Research Centre mission is to provide EU policies with independent, evidence-based scientific and technical support throughout the whole policy cycle. Working in close cooperation with policy Directorates-General, the JRC addresses key societal challenges, while stimulating innovation through developing new methods, tools and standards, and sharing its know-how with the EU Member States, the scientific community and its international partners.

Contents

Authors and Reviewers

Ekaterina Abramova St. Petersburg University, St. Petersburg, Russia

Faiza Al-Yamani Kuwait Institute for Scientific Research, Kuwait

Werner Alpers Institut für Meereskunde, Universität Hamburg, Hamburg, Germany

Roy Armstrong Universidad de Puerto Rico, San Juan, Puerto Rico

Merete Badger Department of Wind Energy, Technical University of Denmark, Roskilde, Denmark

Yan Bai State Key Laboratory of Satellite Ocean Environment Dynamics, Second Institute of Oceanography, State Oceanic Administration, Hangzhou, China

A. N. Balchand Nansen Environmental Research Centre (India), Kochi, India

Thomas Ballatore Lake Basin Action Network, Moriyama, Japan

Vittorio Barale European Commission, Joint Research Centre, Ispra (VA), Italy

A. Baum Leibniz Center for Tropical Marine Ecology, Bremen, Germany

M. Bergé-Nguyen Laboratoire d'Etudes en Géophysique et Océanographie Spatiales (LEGOS), Toulouse, France

Astrid Bracher Alfred-Wegener-Institut für Polar- und Meeresforschung, Bremerhaven, Germany

Vittorio Brando Institute of Atmospheric Sciences and Climate, National Research Council of Italy, Roma, Italy

Colin Brown Ryan Institute, National University of Ireland, Galway, Ireland

Hans Burchard Leibniz-Institut für Ostseeforschung Warnemünde, Rostock, Germany

Valborg Byfield National Oceanography Center, Southampton, UK

Connor Cahalane National Centre for Geocomputation, Maynooth University, Ireland

Fabrice Chane-Ming Université de la Réunion, CNRS-Météo France-Université, La Réunion, France

Paul Chang National Ocean and Atmosphere Administration (NOAA), Washington DC, USA

Xiaoyan Chen State Key Laboratory of Satellite Ocean Environment Dynamics, Second Institute of Oceanography, State Oceanic Administration, Hangzhou, China

Yongcun Cheng Old Dominion University, Norfolk, VA, USA

Po-Kuan Chiang Center for Space and Remote Sensing Research, National Central University, Taoyuan City, Taiwan

Byung-Hun Choe Satellite Geophysics Laboratory, School of Earth and Environmental Sciences, Seoul National University, Seoul, South Korea

Jean-François Cretaux Laboratoire d'Etudes en Géophysique et Océanographie Spatiales (LEGOS), Toulouse, France

José da Silva CIIMAR, Universidade do Porto, Rua dos Bragas 289, 4050-123 Porto, Portugal

Wolfgang Dierking Alfred-Wegener Institut, Bremerhaven, Germany

Roland Doerffer Helmholtz-Zentrum Geesthacht Centre for Materials and Coastal Research, Geesthacht, Germany

Vyacheslav A. Dubina V.I. Il'ichev Pacific Oceanological Institute FEB RAS, Vladivostok, Russia

Guy Duchossois GEO Secretariat, Paris, France

Celine Dupouy Institut de Recherche pour le développement, Marseille, France

Naoto Ebuchi Institute of Low Temperature Science, Hokkaido University, Sapporo, Japan

William Emery University of Colorado, Boulder, CO, USA

T. Eriksen Institute for the Protection and Security of the Citizen, Joint Research Centre of the EC, Ispra (VA), Italy

Leif Eriksson Chalmers University, Gothenburg, Sweden

Xiangbo Feng University of Reading, Reading, UK

Ralph Foster Applied Physics Laboratory, University of Washington, Seattle, WA, USA

K. Fukudome Research Institute for Applied Mechanics, Kyushu University, Kasuga, Japan

Martin Gade Institut für Meereskunde, Universität Hamburg, Hamburg, Germany

Luis Garcia-Weil Departamento de Física, Universidad de Las Palmas de Gran Canaria (ULPGC), Las Palmas, Spain

Mary Swapna George Nansen Environmental Research Centre (India), Kochi, India

Pierre Gernez Département des Sciences de la Vie, Université de Nantes, France

Monika Gerth Leibniz-Institut für Ostseeforschung Warnemünde, Rostock, Germany

Anna I. Ginzburg P. P. Shirshov Institute of Oceanology, Moscow, Russia

Rafael Gonçalves-Araujo Alfred-Wegener-Institut für Polar- und Meeresforschung, Potsdam, Germany

Jim Gower Fisheries and Oceans Canada, Ottawa, Canada

Harm Greidanus European Commission, Joint Research Centre, Ispra (VA), Italy

Grant Gunn Michigan State University, East Lansing, MI, USA

Klaus-Werner Gurgel Institut für Meereskunde, Universität Hamburg, Hamburg, Germany

Nicholas M. J. Hall LEGOS, Université de Toulouse, CNES, CNRS, IRD, UPS Toulouse, France

Faisal Hamzah Institute for Marine Research and Observation, Agency for Marine and Fisheries Research and Development, Ministry of Marine Affairs and Fisheries, Bali, Indonesia

Xianqiang He State Key Laboratory of Satellite Ocean Environment Dynamics, Second Institute of Oceanography, State Oceanic Administration, Hangzhou, China

Birgit Heim Alfred-Wegener-Institut für Polar- und Meeresforschung, Potsdam, Germany

Sebastian Hellman Alfred-Wegener-Institut für Polar- und Meeresforschung, Potsdam, Germany

Benjamin Holt Jet Propulsion Laboratory, NASA, Pasadena, CA, USA

Jing-Shan Hong Central Weather Bureau, Taipei, Taiwan

Peter Hoogeboom Faculty of Civil Engineering and Geoscience, Delft University of Technology, Delft, The Netherlands

Nicolas Höpffner European Commission, Joint Research Centre, Ispra (VA), Italy

Ching-Yuang Huang Department of Atmospheric Sciences, National Central University, Taoyuan City, Taiwan

Kaoru Ichikawa Research Institute for Applied Mechanics, Kyushu University, Kasuga, Japan

Joji Ishizaka Institute for Space-Earth Environmental Research (ISEE), Nagoya University, Nagoya, Japan

S. Jacobsen German Aerospace Center (DLR), Maritime Safety and Security Lab, Bremen, Germany

Ch. V. C. Jayaram Nansen Environmental Research Centre (India), Kochi, India

Lianghong Jiang, State Key Laboratory of Satellite Ocean Environment Dynamics, Second Institute of Oceanography, State Oceanic Administration, Hangzhou, China

S. Jolivet Meteobooking, Bossonnens, Switzerland

Ajith K. Joseph Nansen Environmental Research Centre (India), Kochi, India

Bennet Juhls GEOMAR Helmholtz-Zentrum für Ozeanforschung Kiel, Germany

Seppo Kaitala Finnish Environment Institute, Marine Research Centre, Helsinki, Finland

Jochen Kämpf Flinders University, Adelaide, South Australia

Vladimir Karaev Institute of Applied Physics, Russian Academy of Sciences, Nizhny Novgorod, Russia

Svetlana Karimova Université de Liège, GHER—AGO Department, Liège, Belgium

Kristina Katsaros NorthWest Research Associates, Redmond, WA, USA

Stefan Kern Integrated Climate Data Center—ICDC, Center for Earth System Research and Sustainability, Universität Hamburg, Hamburg, Germany

Duk-Jin Kim Satellite Geophysics Laboratory, School of Earth and Environmental Sciences, Seoul National University, Seoul, South Korea

Andrey G. Kostianoy P. P. Shirshov Institute of Oceanology RAS, Moscow, Russia

Alexei Kouraev Laboratoire d'Etudes en Géophysique et Océanographie Spatiales (LEGOS), Toulouse, France

Alexandra Kraberg Alfred-Wegener-Institut für Polar- und Meeresforschung, Potsdam, Germany

Adam Krezel University of Gdansk, Gdansk, Poland

Raj Kumar Indian Space Research Organisation (ISRO), Ahmedabad, India

Tiit Kutser Estonian Marine Institute, University of Tartu, Tartu, Estonia

Olga Yu. Lavrova Space Research Institute RAS, Moscow, Russia

Sergey A. Lebedev P.P. Shirshov Institute of Oceanology RAS, Moscow, Russia

Yoon-Kyung Lee Korea Ocean Satellite Centre, Korea Ocean R&D Institute, Ansan, Korea

Xiaofeng Li National Oceanic and Atmospheric Administration (NOAA), College Park, MD, USA

XiaoMing Li Institute of Remote Sensing and Digital Earth, Chinese Academy of Sciences, Beijing, China

Yuei-An Liou Center for Space and Remote Sensing Research, National Central University, Taoyuan City, Taiwan

Ji-Chyun Liu ChienHsin University of Science and Technology, Tao-yuan, Taiwan

Feodor Martynov St. Petersburg University, St. Petersburg, Russia

Bernhard Mayer Institut für Meereskunde, Universität Hamburg, Hamburg, Germany

Carolin Meier Institut für Meereskunde, Universität Hamburg, Hamburg, Germany

Walt Meier National Snow & Ice Data Center, University of Colorado Boulder, USA

Christian Melsheimer Institut für Umweltphysik, Universität Bremen, Bremen, Germany

Nandini Menon Nansen Environmental Research Centre (India), Kochi, India

Peter Miller Plymouth Marine Laboratory, Plymouth, UK

Leonid M. Mitnik V.I. Il'ichev Pacific Oceanological Institute FEB RAS, Vladivostok, Russia

Marina I. Mityagina Space Research Institute RAS, Moscow, Russia

A. Morimoto Research Institute for Applied Mechanics, Kyushu University, Kasuga, Japan

Ari Murdimanto Bachelors of Faculty of Geography, University of Gadjah Mada, Jogjakarta, Indonesia

Nicholas Murray Centre for Ecosystem Science, University of New South Wales, Sydney, Australia

Archana Nair Nansen Environmental Research Centre (India), Kochi, India

Dwiyoga Nugroho Agency for Marine and Fisheries Research and Development (AMFRD), Ministry of Marine Affairs and Industry, Jacarta, Indonesia

Paul Overduin Alfred-Wegener-Institut für Polar- und Meeresforschung, Potsdam, Germany

Lasse Pettersson Nansen Environmental and Remote Sensing Center, Bergen, Norway

William Plant Applied Physics Laboratory, University of Washington, Seattle, WA, USA

Thomas Pohlmann Institut für Meereskunde, Universität Hamburg, Hamburg, Germany

Igor G. Polikarpov Environment and Life Sciences Center, Kuwait Institute for Scientific Research, Salmiya, Kuwait

Dmitry Pozdnyakov Nansen International Environmental and Remote Sensing Center, St. Petersburg, Russia

Mutiara Rachmat Putri Study Program of Oceanography, Institute Technology Bandung, Bandung, Indonesia

Dionysios Raitsos Plymouth Marine Laboratory, Plymouth, UK

Mini Raman Marine Ecosystems Division, EPSA, Space Applications Centre (ISRO), Ahmedabad, India

Bernardinus Realino Agency for Marine and Fisheries Research and Development (AMFRD), Ministry of Marine Affairs and Industry, Jacarta, Indonesia

Olivier Regniers i-Sea (SAS), Bordeaux, France

F. Rémy Laboratoire d'Etudes en Géophysique et Océanographie Spatiales (LEGOS), Toulouse, France

YongZheng Ren Institute of Remote Sensing and Digital Earth, Chinese Academy of Sciences, Beijing, China

Ernesto Rodriguez National Aeronautics and Space Administration (NASA), Pasadena, USA

Roland Romeiser University of Miami, Miami, FL, USA

Rüdiger Röttgers Helmholtz-Zentrum Geesthacht Centre for Materials and Coastal Research, Geesthacht, Germany

Angelo Rubino Università Ca' Foscari di Venezia, Venezia, Italy

Joo-Hyung Ryu Korea Ocean Satellite Centre, Korea Ocean R&D Institute, Ansan, Korea

Maria Saburova Kuwait Institute for Scientific Research, Kuwait

J. Samiaji University of Riau, Pekanbaru, Riau, Indonesia

C. Santamaria European Commission, Joint Research Centre, Ispra (VA), Italy

R. K. Sarangi SAC(ISRO), Ahmedabad, India

Agus Setiawan Agency for Marine and Fisheries Research and Development (AMFRD), Ministry of Marine Affairs and Industry, Jacarta, Indonesia

Weizeng Shao Marine Acoustics and Remote Sensing Laboratory, Zhejiang Ocean University, Zhejiang, China

Nickolay A. Sheremet P. P. Shirshov Institute of Oceanology RAS, Moscow, Russia

M. N. Shimaraev Limnological Institute, Siberian Branch of Russian Academy of Sciences, Irkutsk, Russia

Herbert Siegel Leibniz-Institut für Ostseeforschung Warnemünde, Rostock, Germany

Stine Skrunes The Arctic University of Norway, Tromsø, Norway

A. Smith Nansen Environmental Research Centre (India), Kochi, India

Dmitry M. Soloviev Remote Sensing Department, Marine Hydrophysical Institute of RAS, Sevastopol, Russian Federation

Gunnar Spreen Institut für Umweltphysik, Universität Bremen, Bremen, Germany

Sergey Stanichny Marine Hydrophysical Institute, National Academy of Sciences of Ukraine, Sevastopol, Russia

Kerstin Stelzer Brockmann Consult, Geesthacht, Germany

Ad Stoffelen Koninklijk Nederlands Meteorologisch Instituut (KNMI), de Bilt, The Netherlands

Iris Stottmeister Leibniz-Institut für Ostseeforschung Warnemünde, Rostock, Germany

A. Ya. Suknev Great Baikal Trail Buryatiya, Ulan-Ude, Russia

S. Syam Nansen Environmental Research Centre (India), Kochi, India

Xianshang Tian-Kunze Institut für Meereskunde, Universität Hamburg, Hamburg, Germany

Iis Triyulianti Institute for Marine Research and Observation, Agency for Marine and Fisheries Research and Development, Ministry of Marine Affairs and Fisheries, Bali, Indonesia

Daphne van der Wal NIOZ Royal Netherlands Institute for Sea Research, Department of Estuarine and Delta Systems, Yerseke, The Netherlands

Michele Vespe Norwegian Defence Research Establishment (FFI), Kjeller, Norway

Vasiliy Vlasenko School of Earth, Ocean and Environmental Sciences, University of Plymouth, Plymouth, UK

Anna Wendleder Deutsches Zentrum für Luft- und Raumfahrt, DFD, Oberpfaffenhofen, Germany

Sascha Willmes Universität Trier, Trier, Germany

K. Yamada Keimyung University, Korea

Xiaofeng Yang Institute of Remote Sensing and Digital Earth, Chinese Academy of Sciences, Beijing, China

J. H. Yoon Research Institute for Applied Mechanics, Kyushu University, Kasuga, Japan

Y. Yoshikawa Research Institute for Applied Mechanics, Kyushu University, Kasuga, Japan

E. A. Zakharova Laboratoire d'Etudes en Géophysique et Océanographie Spatiales (LEGOS), Toulouse, France

Zhongxiang Zhao Applied Physics Laboratory, University of Washington, Seattle, WA, USA

Giuseppe Zibordi European Commission, Joint Research Centre, Ispra (VA), Italy

Juhong Zou National Space Ocean Application Service (NSOAS), Beijing, China

About the Editors

Vittorio Barale (right) graduated in Physics in 1977, from the University of Milan, and then received both his M.S., in 1982, and Ph.D., in 1986, while at the Scripps Institution of Oceanography, from the University of California at San Diego. Since 1990, he has been with the Joint Research Centre of the European Commission. His research focuses on the ecological assessment of marginal and enclosed seas, and their coastal margins in particular, using optical remote sensing.

Martin Gade (left) received his first degree in Physics in 1992 and then his doctoral degree in Geosciences in 1996, both from the University of Hamburg, Germany. Since 1991, he has been with the Remote Sensing Unit of the Institute of Oceanography, at the University of Hamburg. His research focuses on air–sea interactions and coastal processes, and their remote sensing using active microwave techniques, as well as on laboratory and field experiments with marine surface films.

About the Editors

Acronyms

1D	One-dimensional
2D	Two-dimensional (variational data assimilation)
3D	Three-dimensional (variational data assimilation)
4D	Four-dimensional (variational data assimilation)
a.s.l.	Above Sea Level
AABW	Antarctic Bottom Water
AAIW	Antarctic Intermediate Water
AATSR	Advanced Along-Track Scanning Radiometer
$aCDOM_{440}$	Absorption of CDOM at 440 nm
ACF	Auto Correlation Function
ADCP	Acoustic Doppler Current Profiler
ADEOS	ADvanced Earth Observing Satellite
AIS	Automatic Identification System
ALI	Advanced Land Imager
ALOS	Advanced Land Observing Satellite
AM	*Ante meridiem* (morning hours)
AMI	Active Microwave Instrument
AMJ	April, May, June
AMSR	Advanced Microwave Scanning Radiometer
AMSR-E	AMSR for Earth Observing System
AO	Announcement of Opportunity
AOI	Area of Interest
AS	Arabian Sea
ASAR	Advanced Synthetic Aperture Radar
ASCAT	Advanced SCATterometer
ASF	Alaska Satellite Facility
ASI	Artist Sea Ice (ENVISAT)
ASTER	Advanced Spaceborne Thermal Emission and Reflection Radiometer
ASW	Arabian Sea Water

ATS	Application Technology Satellite
ATSR	Along-Track Scanning Radiometer
AVHRR	Advanced Very High Resolution Radiometer
AVISO	Archiving, Validation and Interpretation of Satellite Oceanographic (data) (Centre, of France)
AVNIR	Advanced Visible and Near-Infrared Radiometer
AWI	Alfred-Wegener-Institut
bbp	Particulate Backscattering Coefficient
bbp/MFLH	Ratio of Particulate Backscattering Coefficient to MFLH
BBW	Bengal Bay Water
BIX	Biological Index
BoB	Bay of Bengal
BP	British Petrol
BPLS	Badan Penanggulangan Lumpur Sidoarjo
BS	Bohai Sea
C2R	Case 2 (waters) Regional processor
CBERS	China-Brazil Earth Resources Satellite
CCD	Charge-Coupled device
CCMP	Cross-Calibrated, Multi-Platform
CDOM	Coloured Dissolved Organic Matter
CDR	Climate Data Record
CDW	Circumpolar Deep Water
CEOS	Committee on Earth Observation Satellites
CFAR	Constant False Alarm Rate
CFOSAT	Chinese-French Ocean SATellite
Chl-*a*	Chlorophyll-*a*
CLS	Collecte Localisation Satellites
CMFRI	Central Marine Fisheries Research Institute
CMODIS	Chinese Moderate Resolution Imaging Spectroradiometer
CN	Coastal Northern Zone
CNES	Centre National d'Études Spatiales
CNRS	Centre National de la Récherche Scientifique
CNSA	China National Space Administration
COCTS	Chinese Ocean Colour and Temperature Scanner
COMPIRA	Coastal and Ocean Measurement mission with Precise and Innovative Radar Altimeter
COMS	Communication, Ocean and Meteorological Satellite
COSMIC	Constellation Observing System for Meteorology, Ionosphere, and Climate
CoW	Common Water
CS	Coastal South Zone
CSEOF	Cyclostationary Empirical Orthogonal Function
CSK	Cosmo-SkyMed
CTD	Conductivity-Temperature-Depth sensor
CTOH	Center for Topographic studies of the Ocean and Hydrosphere

CTZ	Critical Transition Zone
CW	Coastal Western Zone
CZCS	Coastal Zone Color Scanner
CZI	Coastal Zone Imager
D20	20 °C isotherm depth
DAAC	Distributed Active Archive Center
DAC	Dynamic Atmospheric Correction
DBM	Digitized Bathymetry Map
DEM	Digital Elevation Model
DJF	December, January, February
DLR	Deutsches Zentrum für Luft- and Raumfahrt
DMSP	Defense Meteorological Satellite Program
DO	Dissolved Oxygen
DOC	Dissolved Organic Carbon
DOM	Dissolved Organic Matter
DUACS	Developing Use of Altimetry for Climate Studies
EACC	East African Coastal Current
EC	Equatorial Current
EC	European Commission
ECMWF	European Centre for Medium-range Weather Forecasts
ECS	East China Sea
ECV	Essential Climate Variable
EICC	East Indian Coastal Current
ENSO	El Niño/Southern Oscillation
ENVI	Environment for Visualizing Images
ENVISAT	ENVIronmental SATellite
EO	Earth Observation
EOF	Empirical Orthogonal Function
EO-GPOD	EO Grid Processing On-Demand
EOS	Earth Observing System
ERA	ECMWF Re-Analysis
Erdas	Earth Resources Data Analysis System
ERGB	Enhanced Red-Green-Blue
ERS-1/2	European Remote Sensing Satellite-1/2
ESA	European Space Agency
ESC	East Sakhalin Current
ETC	Extratropical Cyclone
ETM	Estuarine Turbidity Maximum
ETM+	Enhanced Thematic Mapper Plus
EU	European Union
EUMETSAT	European Organisation for the Exploitation of Meteorological Satellites
EW	Extra-Wide
FD	Freeman-Durden decomposition
fDOM	Fluorescent Dissolved Organic Matter

FEB RAS	Far Eastern Branch Russian Academy of Sciences
FFI	Forsvarets Forskningsinstitutt
FFT	Fast Fourier Transform
FIM	Fall Inter-Monsoon
FLH	Fluorescence Line Height
FR	Full resolution and reduced resolution (RR)
FVCOM	Finite Volume Coastal Ocean Model
FY	FengYun (series of Chinese polar weather satellites)
GC	Ground Check
GCM	Global Circulation Model of the atmosphere
GCOM-W	Global Change Observation Mission—Water
GCOS	Global Climate Observing System
GCP	Ground Control Point
GDRI	Groupement De Recherche International
GEONET	GNSS Earth Observation Network System
GES DISC	Goddard Earth Sciences Data and Information Services Center
GEWEX	Global Energy and Water Exchanges (project)
GF	*Gao-Fen* (chinese: high resolution)
GFO	Geosat Follow-On
GHRSST	Group for High Resolution SST
GIA	Glacial Isostatic Adjustment
GIOVANNI	Geospatial Interactive Online Visualization and Analysis Infrastructure
GIS	Geographic Information System
GLI	Global Imager
GMF	Geophysical Model Function
GMIS	Global Marine Information System
GMS	Geostationary Meteorological Satellite
GNSS	Global Navigation Satellite System
GNSS-R	Global Navigation Satellite System-Reflectometry
GOCE	Gravity Field and Steady-State Ocean Circulation Explorer
GOCI	Geostationary Ocean Color Imager
GODAE	Global Ocean Data Assimilation Project
GPCC	Global Precipitation Climatology Centre
GPCP	Global Precipitation Climatology Project
GPM	Global Precipitation Measurements
GPS	Global Positioning System
GR	Gradient Ratio
GRACE	Gravity Recovery and Climate Experiment
GRDH	Ground Range Detected High resolution
GRDM	Ground Range Detected Medium resolution
GW	Gravity Waves
H	Horizontal
HAB	Harmful Algae Bloom
HAMSOM	Hamburg Shelf Ocean Model

HF	High Frequency
HH	Horizontal-Horizontal
HICO	Hyperspectral Imager for the Coastal Ocean
HIX	Humification IndeX
HOCI	Hybrid Ocean Colour Index
HY	*Hai-Yang* (chinese: ocean)
IB	Inverse Barometer
ICESat	Ice Cloud and Land Elevation Satellite
IEEE	Institute of Electrical and Electronic Engineers
IEM	Integral Equation Method
IEW	Indian Equatorial Water
IFREMER	Institut Francais de Récherche pour Exploitation de la Mer
IMO	International Maritime Organization
IMRO	Institute for Marine Research and Observation
INCOSPAR	Indian National COmmittee on SPAce Research of the Coastal Environment
IndoNACE	Indonesian seas Numerical Assessment
INSTANT	International Nusantara Stratification and Transport
IO	Indian Ocean
IOCCG	International Ocean Colour Coordinating Group
IOD	Indian Ocean Dipole
IOVWST	International Ocean Vector Winds Science Team
IR	Infrared
IRS	Indian Remote Sensing Satellite
IRT	Image Reconstruction Technique
ISMR	Indian Summer Monsoon Rainfall
ISO	International Organization for Standardization
ISRO	Indian Space Research Organization
ISS	International Space Station
ITCZ	Intertropical Convergence Zone
IUW	Indonesian Upper Water
IVI	Importance Value Index
IW	Interferometric Wide Swath
IW	Internal Wave
IWWG	International Winds Working Group
J1,2,3	Jason 1,2,3
Japan-E	East coasts of Japan
JAS	July, August, September
JAXA	Japanese Aerospace eXploration Agency
JERS	Japanese Earth Resources Satellite
JFM	January, February, March
JGOFS	Joint Global Ocean Flux Study
JJAS	June, July, August, September
JONSWAP	Joint North Sea Wave Project
JPL	Jet Propulsion Laboratory

JRC	Joint Research Centre
KARI	Korean Aerospace Research Institute
KE	Kuroshio Extension
KISR	Kuwait Institute for Scientific Research
KNMI	Koninklijk Nederlands Meteorologisch Instituut
KOMPSAT	KOrea Multi-Purpose SATellite
LAIW	Large Amplitude Internal Wave
lat	Latitude
LEGOS	Laboratorie Etudes en Geophysique et Oceanographie Spatiales
LEO	Low Earth Orbit
LG	Local Gradient
LH	Lakshadweep high
LiDAR	Light Detection and Ranging
LL	Lakshadweep low
lon	Longitude
LS	Lower Stratosphere
LTAN	Local Time of equator Acending Node
MAM	March, April, May
MASIE-NH	Multisensor Analyzed Sea Ice Extent—Northern Hemisphere
MCC	Maximum Cross Correlation
MCI	Maximum Chlorophyll Index
MCS	Mesoscale Complex System
MCSST	Multi-Channel Sea Surface Temperature
MDC	Mahalanobid Distance Classification
MERIS	MEdium Resolution Imaging Spectrometer
MeSAA	Mechanistic semi-analytic algorithm
Meso-NH	Mesoscale Non-Hydrostatic Model
MetOp	Meteorology Operational Programme
MFLH	Modified Fluorescence Line Height
MICOM	Miami Isopycnic Coordinate Ocean Model
MID	Maritime Identification Digits
MIZ	Marginal Ice Zone
MLE	Maximum Likelihood Estimator (wind inversion residual)
MLLL	Mean Lowest Low Level
MMAF	Ministry for Marine Affairs and Fisheries
MMSI	Maritime Mobile Service Identity
MODIS	MODerate resolution Imaging Spectroradiometer
MOS	Modular Optoelectronic Scanner
MOS	Marine Observation Satellite
MPA	Marine Protected Area
MPI	Max Planck Institut
MPIOM	MPI Ocean Model
MSG	Meteosat Second Generation
MSL	Mean Sea Level
MSMR	Multi-frequency Scanning Microwave Radiometer

MSR	Microwave Scanning Radiometer
MW	Microwave
MWP	Mean Wave Period
MWR	Microwave Radiometer
NADW	North Atlantic Deep Water
NASA	National Aeronautics and Space Administration
NASDA	National Space Development Agency
NCAR	National Center for Atmospheric Research
NCEP	National Center for Environmental Prediction
NDBC	National Data Buoy Center
NDVI	Normalized Difference Vegetation Index
NDZ	Nitrate Depleted Zones
NEM	Northeast Monsoon
nFLH	Normalized Fluorescence Line Height
NGDC	National Geophysical Data Centre (of NOAA)
NIC	National Ice Center
NIO	North Indian Ocean
NIR	Near Infrared
NJRC	Nagano Japan Radio Co. Ltd.
NOAA	National Oceanic and Atmospheric Administration
NOWPAP	Northwest Pacific Action Plan
NPCW	North Pacific Central Waters
NPGO	North Pacific Gyre Oscillation
NPP	National Polar-orbiting Operational Environmental
NRCS	Normalized Radar Cross Section
NRT	Near Real Time
NSCAT	NASA SCATterometer
NSIDC	National Snow and Ice Data Center
NSPO	National Space Program Organization
NWP	Numerical Weather Prediction
OC2	Ocean Chlorophyll 2 (algorithm)
OC4L	Ocean Colour 4 L (algorithm for Chl-a)
OC4Mev6	Ocean Colour 4 MERIS version 6 9algorithm for Chl-a)
OC-CCI	Ocean Colour—Climate Change Initiative
OCI	Ocean Color Imager
OCM	Ocean Colour Monitor
OCTS	Ocean Color and Temperature Scanner
OGCMs	Ocean General Circulation Models
OISST	Optimally Interpolated Sea Surface Temperature
OLCI	Ocean Land Colour Instrument
OLI	Operational Land Imager (on Landsat 8)
OND	October, November, December
ONWJ	Offshore North West Java
OPS	Optical Sensor
OS	Southern Open Region

OSCAR	Observing Systems Capability Analysis and Review (WMO)
OSCAT	OceanSat SCATterometer
OSMI	Ocean Scanning Multispectral Imager
OSSE	Observing System Simulation Experiment
OSVW	Ocean Surface Vector Winds
PALSAR	Phased Array type L-band Synthetic Aperture Radar
PAR	Photosynthetically Active Radiation
pCO_2	Partial pressure of Carbon Dioxide
PCQM	Point Centre Quarter (or Quadrant Method)
PCTS	Principle Component Time Series
PDF	Probability Density Function
PDO	Pacific Decadal Oscillation
PEW	Pacific Equatorial Water
PFSM	Parameterized First-guess Spectrum Method
PM	Pierson-Moscowitz
PM	*Post meridiem* (afternoon/evening hours)
POC	Particulate Organic Carbon
PODAAC	Physical Oceanography Distributed Active Archive Center
POLDER	POLarization and Directionality of Earth's Reflectances
PORSEC	Pan-Ocean Remote Sensing Conference
PR	Polarization Ratio
PRE	Pearl River Estuary
PS/CIW	Pacific Subarctic/California Intermediate Water
PSSM	Polynya Signature Simulation Method
PSU	Practical Salinity Unit
PSUW	Pacific Subarctic Upper Water
PSW	Pacific Subarctic Water
PWV	Precipitable Water Vapor
QBO	Quasi-Biennial Oscillation
QPF	Quantitative Precipitation Forecast
QuikSCAT	Quick Scatterometer
RA	Radar Altimeter
RAMSES	Radiation Measurements Sensor, Enhanced Spectral Resolution
RAR	Real Aperture Radar
RAS	Russian Academy of Sciences
RC	Relative Closure
RD	Relative Density
RF	Relative Frequency
RFI	Radio Frequency Interference
RIAM	Research Institute for Applied Mechanics
RISAT	Radar Imaging SATellite
RLSBO	Side-Looking Real Aperture Radar (in Russian)
RMS	Root Mean Square
RMSE	Root Mean Square Error
RO	Radio Occultation

ROI	Region of Interest
ROPME	Regional Organization for the Protection of the Marine Environment
RR	Reduced Resolution
RS/PGIW	Red Sea/Persian Gulf Intermediate Waters
RTK	Real-time Kinematic
RTVK	Radio and TV Komplex
S-1	Sentinel-1
SAC-C/D	Scientific Application Satellite-C/D
SAF	Satellite Application Facility (of EUMETSAT)
SAMIR	SAtellite MIcrowave Radiometer
SAR	Synthetic Aperture Radar
SARAL	Satellite with Argos and Altika
SC	Somali Current
ScatSat	Scatterometer Satellite
SCS	South China Sea
SCS-E	East of South China Sea
SCS-W	West of South China Sea
SD	Standard Deviation
SD	Secchi Depth
SDE	Standard Deviation of Error
SEA	South East Aral
SeaDAS	SeaWiFS Data Analysis System
SEAS	South-Eastern Arabian Sea
Seasat	SEAfaring SATellite
SeaWiFS	Sea-viewing Wide Field-of-View Sensor
SEC	South Equatorial Current
SECC	South Equatorial Counter Current
SEE	Standard Error of Estimate
SFMR	Stepped-Frequency Microwave Radiometer
SGD	Submarine Groundwater Discharge
SHH	Sea Hydrometeorology and Hydrochemistry
SIA	Sea Ice Area
SIC	Sea Ice Concentration
SICW	South Indian Central Water
SIE	Sea Ice Extent
SIM	Spring Inter-Monsoon
SIO	Shirshov Institute of Oceanology
SIR-A/B/C	Spaceborne Imaging Radar -A/B/C
SIRAL	SAR Interferometer Radar Altimeter
SIT	Sea Ice Thickness
SLA	Sea Level Anomaly
SMC	Summer Monsoon Current
SMIW	Surface Manifestations of Internal Waves
SMOS	Soil Moisture and Ocean Salinity

SNR	Signal-to-Noise Ratio
SoJ	Sea of Japan
SOLAS	Safety of Life at Sea
SPARC	Stratospheric Processes and their Role in Climate
SPCW	South Pacific Central Waters
SPM	Suspended Particulate Matter
SPOT	Satellite Pour l'Observation de la Terre
SPRA	Semi Parametric Retrieval Algorithm Scheme
SRAL	SAR Radar Altimeter
SRTM	Shuttle Radar Topography Mission
SS	Standard Stripmap
SSALTO	Segment Sol multi-missions dALTimetrie
SSDH	Sea Surface Dynamic Height
SSDHA	Sea Surface Dynamic Height temporal Anomaly
SSH	Sea Surface Height
SSM/I	Special Sensor Microwave Imager
SSMIS	Special Sensor Microwave Imager Sounder
SSPP	Satellite System Preparatory Project
SSS	Sea Surface Salinity
SST	Sea Surface Temperature
S-VISSR	Stretched-Visible Infrared Spin Scan Radiometer
SWA	South-West Aral
SWC	Soya Warm Current
SWH	Significant Wave Height
SWIR	Short Wave Infrared
SWM	South-West Monsoon
SWOT	Surface Water and Ocean Topography
T/P	TOPEX/Poseidon
TC	Triple Collocation
TC	Tropical Cyclone
TD	Tropical Depression
TDX	TanDEM-X
TIR	Thermal Infrared
TIROS	Television and Infrared Observation Satellite
TIT	Thermal Ice Thickness
TM	Thematic Mapper
TMI	TRMM Microwave Imager
TOMS	Total Ozone Mapping Spectrometer
TOPEX	TOPography Experiment
TPH	Total Petroleum Hydrocarbon
TRMM	Tropical Rainfall Measuring Mission
TSIT	Thin Sea Ice Thickness
TSM	Total Suspended Matter
TSX	TerraSAR-X
TWC	Tsushima Warm Current

T-XVII	TRANSDRIFT Expedition XVII
UAE	United Arab Emirates
UAV	Unmanned Aerial Vehicle
UI_{wind}	Wind-Based Upwelling Index
UN	United Nations
UNEP	UN Environmental Program
UNESCO	UN Educational, Scientific and Cultural Organization
US	United States
USSR	Union of Soviet Socialistic Republics
UT	Upper Troposphere
UTC	Universal Time Coordinated
V	Vertical
VC	Virtual Constellation
VGPM	Vertically Generalized Production Model
VHF	Very High Frequency
VHSR	Very High Spatial Resolution
VIIRS	Visible Infrared Imager Radiometer Suite
VIRS	Visible Infrared Scanner
VTIR	Visible and Thermal Infrared Radiometer
VV	Vertical-Vertical
WBC	Western Boundary Current
WICC	West India Coastal Current
WMC	Winter Monsoon Current
WMO	World Meteorological Organisation
WVC	Wind Vector Cell
YRE	Yalu River Estuary
YS	Yellow Sea

Part I
Introduction to Remote Sensing of the Asian Seas

The Asian Marginal and Enclosed Seas: An Overview

Vittorio Barale

Abstract An astonishing variety of marginal basins ranging from polar to equatorial regions, and a few sizeable enclosed basins, can all be encompassed among the Asian Seas. The Arctic Ocean near-coastal zones off Siberia, the semi-enclosed basins of the Pacific Ocean western rim, the coastal seas of the northernmost Indian Ocean, the Red Sea as well as the easternmost reaches of the Mediterranean Sea and Black sea, the Caspian Sea, the remnants of the Aral Sea and other brackish or freshwater lakes, such as Lake Balkhash and Lake Baykal, exhibit a multiplicity of environmental features and processes. The diversity of the Asian seascapes, which range from permafrost cliffs to mangrove forests, and include wide gulfs and island groups, shallow estuaries, ship-congested straits, and some of the most crowded coastal regions in the world, where millions rely on fish for much of their protein, is an appropriate match to the continent's variety of political, economic and social systems. Understanding the peculiarities of such a large collection of marine and coastal zones, managing and exploiting the wealth of resources that the Asian Seas have to offer, requires integrated observation systems, among which orbital remote sensing must play an essential role.

Keywords Asian Seas · Asian seascapes · Marginal seas · Enclosed seas

1 Introduction

An astonishing variety of marginal basins, ranging from polar to equatorial regions, and a few sizeable enclosed basins, including some of the most troubled water bodies on the planet, can all be encompassed among the Asian Seas. These waters have been coasted since ancient times, from the primeval eastward migrations of early humans, to the ages of exploration and trading between east and west, and to the current interest for new endeavours in several maritime sectors.

V. Barale (✉)
European Commission, Joint Research Centre, Ispra, Italy
e-mail: Vittorio.BARALE@ec.europa.eu

V. Barale and M. Gade (eds.), *Remote Sensing of the Asian Seas*,
https://doi.org/10.1007/978-3-319-94067-0_1

Fig. 1 Main geographic features of the Asian continent and surrounding ocean basins. The colour coding shows bathymetry (blue tones, darker means deeper waters) over water, and vegetated to desert areas (green to yellow and brown tones) over land

The Asian Seas comprise the marginal sections of three major oceans, the Arctic Ocean, the Pacific Ocean and the Indian Ocean, as well a number of totally enclosed water bodies (Fig. 1). The diversity of the Asian coastlines and seascapes, which extend from the permafrost cliffs of near-polar reaches to the mangrove forests of near-equatorial regions, and which include wide basins and island chains, shallow estuaries, ship-congested straits, and some of the most crowded coastal regions in the world, where millions rely on fish for much of the proteins in their diet, is an appropriate match to the continent's variety of political, economic and social systems. The near-coastal basins surrounding the Asian continental landmass encompass a wide spectrum of environmental traits, spanning most of the Earth's climatic zones, and ranging from unspoiled conditions to hotspots heavily impacted by human activities. Due to their vast size, great depth and efficient water circulation, the open oceans are relatively unaffected by anthropogenic pollution (with the notable exception of plastics, and micro-plastics in particular), in comparison with near-coastal areas. Hence, degradation is generally limited to marginal seas, while most of the Arctic Ocean, Pacific Ocean and Indian Ocean ecosystems remain

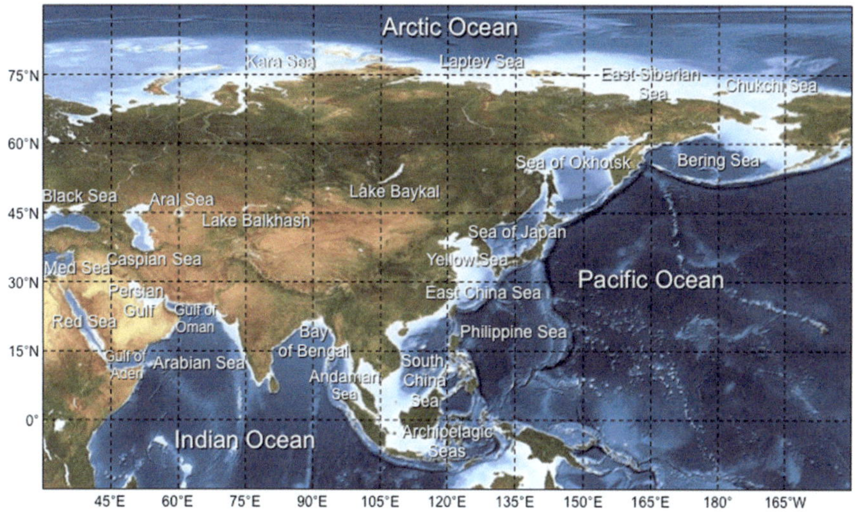

Fig. 2 Marginal and enclosed basins surrounding the Asian continent. The colour coding shows bathymetry (blue tones, darker means deeper waters) over water, and vegetated to desert areas (green to yellow and brown tones) over land

undisturbed, even though with notable exceptions (e.g. that of fish stocks decline, the reasons for which are not fully understood, but possibly include overfishing and climate change). The state of the semi-enclosed or enclosed seas, which have limited or no exchanges with the adjacent oceans, is extremely varied and strongly influenced by local processes.

The Asian Seas (Fig. 2) include, in the north, the Kara Sea, Laptev Sea and East Siberian Sea, bounded and separated by island chains, over the broad shelf extending from the Asian continental landmass into the Arctic Ocean. The Chukchi Sea and the Bering Sea, respectively north and south of the Bering Strait, provide a buffer between Arctic Ocean and Pacific Ocean. East of the Asian continent, the Sea of Okhotsk, the Sea of Japan, the Yellow Sea and the East China Sea, as well as the South China Sea, the Archipelagic Seas, and the Philippine Sea, edge on the northwest Pacific basin. South of the continent, separated by the Indian sub-continent, the Bay of Bengal and the Andaman Sea, to the east, and the Arabian Sea, to the west, constitute the northern limits of the Indian Ocean. The Arabian Sea connects to the semi-enclosed basins of the Persian Gulf, through the Gulf of Oman, and of the Red Sea, through the Gulf of Aden. Further west, the eastern Mediterranean Sea and the southern Black Sea are still bordering the Asian continent—although being commonly considered water bodies pertaining to the European (sometimes African, for the Mediterranean) geographic setting. Finally, at the divide with the European continent, the totally enclosed Caspian Sea, the remnants of the Aral Sea and, toward the center of the continent, a number of large freshwater lakes, among which Lake Balkhash and Lake Baykal.

In the following, the main features of the great oceans surrounding the Asian continent will be recalled, focusing in particular on the key oceanographic traits of their marginal basins, in order to place into perspective environmental peculiarities shown and special challenges posed by each. The review, based on published information,[1] will cover all near-coastal and land-locked basins mentioned above, with a variable degree of detail, according to current oceanographic knowledge.

2 Arctic Ocean Marginal Basins

The Arctic Ocean, connecting the marginal basins north of the Asian continent (i.e. Kara Sea, Laptev Sea, East Siberian Sea, and Chukchi Sea), is the smallest of the Earth's oceanic divisions. The basin occupies the polar region of the northern hemisphere, roughly bounded by the Arctic Circle at 66° 33′ 46.7″N, and is almost entirely surrounded by the landmasses of the European, Asian and North American continents (Fig. 3). It has about the same size of the Antarctic continent, located in the corresponding polar region of the southern hemisphere. The Lomonosov Ridge, extending between Greenland and New Siberian Islands, divides it into two main parts: the Amerasian Basin, further subdivided by the Alpha Ridge (parallel to the Lomosonov Ridge) into Makarov Basin and Canada Basin, bordering onto Canada and Alaska; and the Eurasian Basin, further subdivided by the Gakkel Ridge (also parallel to the Lomosonov Ridge) into Amundsen Basin and Nansen Basin, bordering onto Siberia.

There are different theories about the origin of the Arctic basin. There is general agreement that the Eurasian part formed as a result of spreading at the Gakkel Ridge, during the past 40–60 million years, but there is no consensus concerning the Amerasian part. Present seismic activity along the Gakkel Ridge (as well as its continuity with the mid-Atlantic Ridge) suggests it is the modern Arctic Ocean spreading axis. The Alpha Ridge has been interpreted to be a subduction zone related to the expansion of the Atlantic Basin. Geology of the area surrounding the Arctic Ocean has been used as evidence that the ocean existed during most of the Phanerozoic, i.e. the last 541 million years. An alternative idea is that the basin may be younger, and originated during the Mesozoic, from 252 to 66 million years ago. The Arctic Ocean would have been much warmer than it is at present, in the Mesozoic, and would have obtained its present day ice cover only during the middle Cenozoic, from 66 million years ago to present.

The Arctic Ocean occupies about 2.8% of the Earth's surface, representing an area of approximately 14,056,000 km^2, with an average depth of 1038 m. Its deepest point is Litke Deep in the Eurasian Basin, at 5450 m. The Arctic Ocean is

---

[1]Please note the deliberate choice of not introducing references in the text, so as to avoid lengthy listings of names and dates, which would have disrupted its readability. In order to facilitate access of readers to the appropriate literature, a Bibliography has been appended to this introduction, organized in geographic sections that mirror those of the present review.

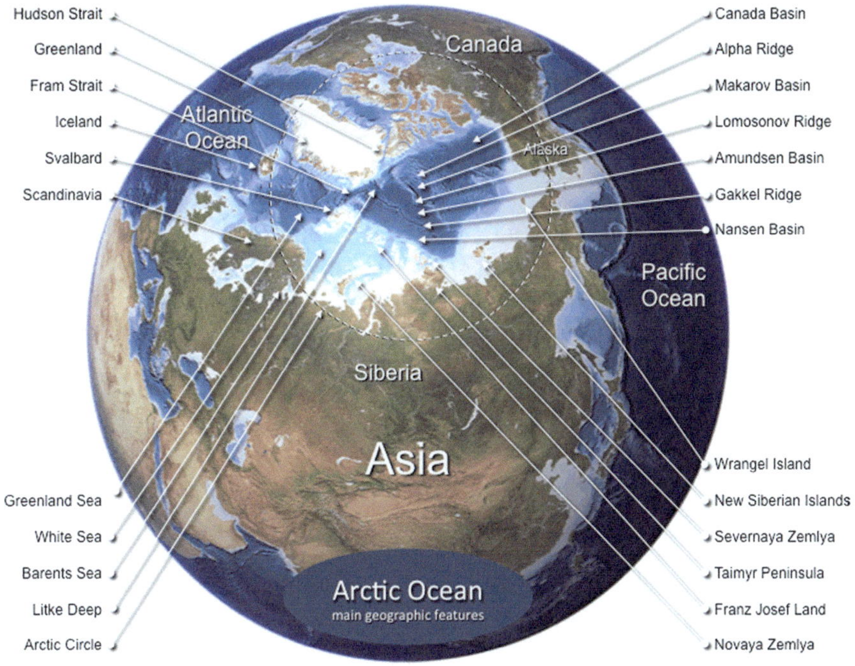

Fig. 3 Arctic Ocean main geographic features. The dashed circle represents the approximate position of the Arctic Circle, which delimits the oceanic basin. Colour coding as in Fig. 1

partly covered by sea ice (0.6–4.0 m thick) throughout the year, and almost completely covered from October to June. Seasonal variations of this ice cover, in extent and thickness, range from a maximum in April and a minimum in September. The summer shrinking of the ice cap is on the order of 50%. Since the 1980s, the mean extent of the ice has been decreasing at a rate of about 3% per decade. Sea ice can move and rotate over large areas due to wind and currents, piling up in zones of compression to form pack ice. Both surface temperature and salinity vary seasonally, as the ice cover melts and freezes again. Surface temperature is fairly constant, near the freezing point of seawater (which must reach −1.8 °C before freezing occurs). Surface salinity is the lowest of all oceans, due to low evaporation, large river water input, and limited exchange with other oceanic waters.

The uppermost section of the Arctic Ocean water column, up to 50 m deep and relatively stable, consists of the Polar Mixed Layer—fed by the freshwater input of the large Siberian and Canadian rivers, such as the Ob, Yenisei, Lena, and Mackenzie—which floats on the often warmer, but saltier, and therefore denser, oceanic water below. The underlying water column consists of three main water masses—i.e. Arctic Surface Water, at about 150–200 m of depth; Atlantic Water, in the depth range of 150–900 m; and Arctic Deep Water below 900 m of depth—resulting from the complex interplay of the inflow from both Pacific Ocean (about 0.8 Sv) and Atlantic Ocean (about 9.0 Sv) with the extreme climatic conditions of the polar region. The production of sea ice (together with the formation of dense water on the continental shelves by freezing and brine rejection) creates water dense enough to sink through the Polar Mixed Layer and feed the lower layers.

The inflow from the Pacific Ocean through the Bering Strait is cooled over the Chukchi Sea shelf by winter winds. As sea ice is formed, salt is released into the remaining non-frozen water, which becomes very dense and sinks below the cold, relatively fresh Polar Mixed Layer, to form the Pacific Halocline. Similarly, the inflow from the Atlantic Ocean that enters over the Siberian Shelf undergoes intense mixing and cooling, contributing to the formation of an Atlantic Halocline. This layer of Arctic Surface Water acts as a buffer between the frozen or near-freezing surface and the additional—salty, but much warmer—waters that come in from the Atlantic Ocean. Without the Halocline, the main Atlantic inflow would reach deep into the Arctic Ocean and melt the existing sea ice.

The Atlantic Water layer is formed by the inflow from the North Atlantic Current, cooling and sinking after entering the Eurasian Basin through Barents Sea and Greenland Sea, east and west of the Svalbard archipelago. This inflow circles the Arctic basin counter-clockwise in the Arctic Ocean Boundary Current. The Atlantic Water has about the same salinity as deeper waters, but is much warmer (up to 3 °C). This water mass is actually warmer than the surface water, and remains submerged only due to the predominant contribution to density due to its high salinity. When it reaches the Amerasian Basin, it is pushed by strong winds into the inner anticyclonic Beaufort Gyre. The northernmost component of this Gyre and the Transpolar Drift Stream, both the results of large-scale atmospheric circulation, form the return outflow towards the Atlantic Ocean. Thus, the waters of Atlantic (and to a lesser extent also of Pacific) origin exit through the Fram Strait between Svalbard and Greenland. The size of the Fram Strait, about 2700 m deep and 350 km wide, is what allows for both inflow and outflow on the Atlantic side of the Arctic Ocean. The Coriolis force concentrates the outflow into the East Greenland Current, on the western side, and the inflow into the Norwegian Current, on the eastern side. Pacific water also exits along the west coast of Greenland and through the Hudson Strait.

The Arctic Deep Water layer is composed of some of the densest water in the World Ocean and has two main sources: sinking Arctic shelf waters and Greenland Sea deep waters. This water mass is critically important because of its outflow, which contributes to the formation of the North Atlantic Deep Water (NADW). The overturning of this water in the ocean Conveyor Belt system plays a key role in maintaining global circulation and in regulating the Earth's climate.

The **Kara Sea** is a marginal basin of the Arctic Ocean (Fig. 4) extending mostly on the western Siberian Shelf, and separated from the Barents Sea to the west by Novaya Zemlya, and from the Laptev Sea to the east by the Severnaya Zemlya archipelago. Its northern limit is marked by a line running from Franz Josef Land to the northernmost point of Severnaya Zemlya. It has an area of about 926,000 km^2 and a mean depth of 131 m.

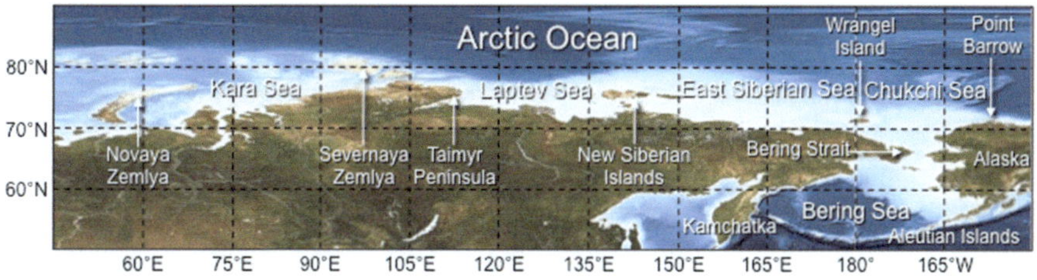

Fig. 4 Asian marginal basins of the Arctc Ocean, and Bering Sea. Colour coding as in Fig. 1

The Kara Sea owes its name to the Kara River flowing into Baydaratskaya Bay, which played an important role in the Russian conquest of northern Siberia. It receives a large amount of fresh water also from the Ob, Yenisei, Pyasina, and Taimyra rivers, so its salinity is quite variable. Compared to the Barents Sea, which receives relatively warm currents from the Atlantic Ocean, the Kara Sea is much colder, remaining frozen for over nine months a year.

The Kara Sea is important as a fishing ground, although it is ice-free only for less than three months a year. Significant discoveries of both oil and gas have been made, but have not been exploited yet. Much concern has been raised about radioactive contamination from nuclear waste, dumped at sea in Soviet Union times (primarily decommissioned naval nuclear reactors, as well as solid and liquid wastes), and the effect this may have on coastal and marine ecosystems.

In spite (or perhaps because) of this, the sizeable Kara Sea Islands section of the Great Arctic Nature Reserve—the largest nature reserve of Russia, created in 1993—extends for about 4000 km^2, and includes sites that fully represent the natural and biological diversity of eastern Arctic Ocean islands. Nearby, sites on the Franz Josef Land and Novaya Zemlya are also registered as natural sanctuaries, within the Russian Arctic National Park.

The **Laptev Sea**, named after the Russian explorers Dmitry and Khariton Laptev, is a marginal sea of the Arctic Ocean (Fig. 4) that occupies the central Siberian Shelf, between the Taimyr Peninsula and Severnaya Zemlya to the west (separating it from the Kara Sea), and the New Siberian Islands to the east (separating it from the East Siberian Sea). Its northern boundary extends from the northernmost points of the Severnaya Zemlya archipelago to that of the New Siberian Islands.

The Laptev Sea has an area of around 700,000 km^2 and a mean depth of 578 m. Its climate is one of the most severe among the Arctic seas, with temperatures below 0 °C over more than 9 months per year, low water salinity (due to the outflow of various rivers, primarily the Lena) and low depths (more than 50% of the sea rests on a continental shelf with depths below 50 m, while in the north the bottom sharply drops to the ocean floor with depths of the order of 1000 m).

The Laptev Sea is frozen for most of the year, though mostly ice-free in summer, and is a major source of Arctic sea ice, generating an outflow larger than that of all other Asian marginal seas combined. Below the surface layer (either frozen or formed by freshwater from river runoff or ice thawing) a warmer (up to 1.5 °C) intermediate halocline layer is fed by Atlantic waters. An even deeper, colder layer (at about −0.8 °C) lies below. The main surface circulation is cyclonic.

A large number of islands are located in the western part of the Laptev Sea and in the river deltas, many of which contain well-preserved mammoth remains. Both flora and fauna are scarce in the surrounding region, owing to the harsh climate. Coastal settlements are few and small. Major human activities in the area are mining and navigation on the Northern Sea Route, while fishing and hunting are practiced but have no commercial significance. Water pollution is relatively low.

The **East Siberian Sea** is a marginal basin of the Arctic Ocean (Fig. 4) located on the eastern Siberian Shelf, and bounded to the west by the New Siberian Islands (beyond which is the Laptev Sea) and to the east by Wrangel Island (beyond which is the Chukchi Sea). The basin, ideally limited to the north by a line connecting the northernmost islets of the western and eastern archipelagos, has an area of around 987,000 km^2 and a mean depth of 58 m.

The East Siberian Sea has a severe climate (average temperature in January is −30 °C, while in July 0–1 °C at sea and 2–3 °C on the coast), so that it freezes over between October and July. Sea ice melts only in August and September. The ice fields are stationary near the coast, with thickness of 2 m by the end of winter. Further out at sea, drifting ice has a thickness of 2–3 m. Winter winds may shift this ice northward, creating polynyas in the sea interior.

Surface temperature decreases northward (in winter, from −0.2 °C near river deltas, to −1.8 °C in the open sea; in summer, from 7–8 °C along the coast, to 2–3 °C at sea). Surface salinity increases from southwest to northeast (in winter, from 4–5 g/kg near river deltas, to 31–32 g/kg out at sea; in summer, it decreases by 5 g/kg due to snow melting), and decreases from the seafloor to the surface. Surface currents are directed from west to east, rather constant and weak.

The sea shores were inhabited for thousands of years by indigenous tribes, engaged in fishing, hunting and reindeer husbandry. Today, major industrial activities in the area are linked to mining and navigation on the Northern Sea Route, while commercial fishing is poorly developed. In the 1950s, labour camps of the Soviet Union's Gulag system operated on the coast, each hosting about 10,000 inmates engaged in the local mines and construction works.

The **Chukchi Sea** is a marginal basin of the Arctic Ocean (Fig. 4) at the easternmost reaches of the Siberian Shelf, comprised between Wrangel Island in the west (separating it from the East Siberian Sea) and Point Barrow, Alaska, in the east (separating it from the Beaufort Sea). The Bering Strait forms its southernmost limit and connects it to the Bering Sea and the Pacific Ocean. The basin has an approximate area of 620,000 km^2 and an average depth of 80 m.

The Chukchi Sea is fed from the south by the Pacific water throughflow, via the Bering Strait. In most of the sea, the flow field responds rapidly to the local wind, with high spatial coherence over the basin scale. The mean flow is northwards, opposed to the mean wind, but presumably forced by a sea-level slope between the Pacific and the Arctic. The inflow of nutrient-rich waters from the south, feeds one of the most abundant marine ecosystems in the world.

Temperature and salinity records show a strong annual cycle of freezing, salinization, freshening and warming, with sizable interannual variability. The largest seasonal variability is seen in the east, where warm, fresh waters escape from the buoyant, coastally trapped Alaskan Coastal Current into the interior. In the west, the seasonal Siberian Coastal Current provides a source of cold, fresh waters and a flow field less linked to the local wind.

The sea owes its name to the native Chukchi people, who are based along its western shorelines and traditionally engaged in near-coastal fishing, whaling and hunting. The 80 km Chukchi Corridor, on the shelf off northwest Alaska, has been reserved as a passageway for one of the largest marine mammal migrations in the world, concerning various marine mammal species such as the Pacific walrus, bowhead whale, beluga whale, and bearded seals.

3 Pacific Ocean Marginal Basins

The Pacific Ocean, which borders with a string of marginal seas the eastern Asian landmass (Fig. 5), is the largest of the Earth's oceanic divisions. It is twice as large as the second largest, the Atlantic Ocean, and it is larger than the landmass of all continents combined. The basin began to form about 750 million years ago, at the breakup of the supercontinent Rodinia in the Neoproterozoic era, although it is generally referred to as the Panthalassic Ocean until the breakup of Pangea, the supercontinent that existed during the late Paleozoic and early Mesozoic eras, about 200 million years ago. At present, the oldest Pacific Ocean floor is only 180 million years old, with older crust having been already subducted. Indeed, the Pacific is the only ocean almost totally bounded by subduction zones (off east Asia and west of the Americas, with the exception of Australian and Antarctic coasts). The inner oceanic basin is circled by the Ring of Fire—the world's foremost belt of explosive volcanism, associated with a nearly continuous series of trenches, volcanic arcs, chains and belts, or tectonic plate edges in general—named after the several hundreds active volcanoes that sit above the various subduction zones.

The Pacific Ocean covers about 32.4% of the Earth's surface, with an area of approximately 165,250,000 km^2, and an average depth of 4280 m. Its deepest point, still uncertain but so far measured at 11,034 m below the surface, is the Challenger Deep, located in the Mariana Trench. The basin extends in latitude from the Arctic (from which it is separated by the Bering Strait) to the Antarctic (where the boundary with the Southern Ocean has been set at 60°S). It stretches through 135° of latitude (without considering the conventional Southern Ocean boundary), for 15,500 km, while its greatest longitudinal extent measures 19,300 km, along latitude 5°N, between the coast of Colombia, in South America, and the Malay Peninsula, in Asia. Surface salinity ranges from minima of 32 g/kg in the far north, as less evaporation takes place in these frigid areas, and of 34 g/kg near the equator, because of abundant precipitation throughout the year, to a maximum of 37 g/kg in

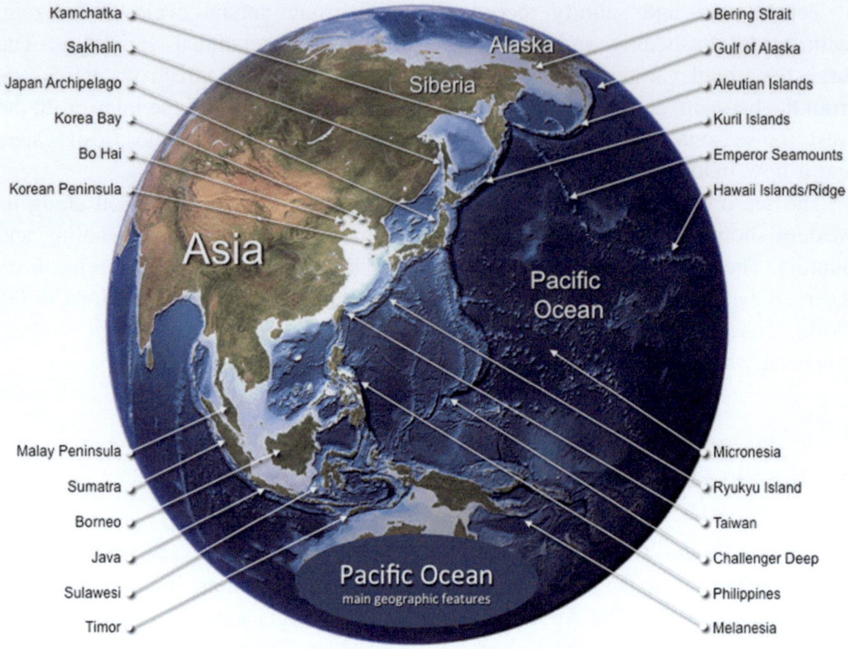

Fig. 5 Western Pacific Ocean main geographic features. Colour coding as in Fig. 1

the southeast. Surface water temperatures vary from −1.4 °C, close to the freezing point of seawater, in polar regions, to about 30 °C, near the equator.

The schematic hydrographic structure of the Pacific Ocean water column, based on combined temperature and salinity characteristics, comprises first the shallow (depths to 500 m) Upper Waters—i.e. the Pacific Equatorial Water (PEW), found in the equatorial region 20°N to 10°S, and extending for the entire longitudinal width of the basin; the (eastern and western) North Pacific Central Waters (NPCW), and (eastern and western) South Pacific Central Waters (SPCW), formed respectively in the northern and southern sub-tropical convergence zones, and extending northward and southward of the PEW, up to 50°N and 40°S respectively; and in the north the Pacific Subarctic Upper Water (PSUW). Below these Central Waters, at intermediate depths ranging from 500 to 1500 m, but closer to the surface northward of NPCW and southward of SPCW, are the Pacific (Subarctic/California) Intermediate Water (PS/CIW), formed in the northern basin by mixing processes still poorly understood, and the Antarctic Intermediate Water (AAIW), formed in the southern basin, at the Antarctic Polar Front. Still below, from 1500 m to the bottom, the northern Pacific Subarctic Water (PSW) and southern Circumpolar Deep Water (CDW), or Common Water (CoW), are the deepest water masses in the Pacific Ocean. Note that the Arctic Ocean is not a significant source of deep and bottom water for the North Pacific, because the Bering Strait (and the Aleutian Ridge) serves as partial barrier to any major flow from the Chukchi Sea (and the Bering

Sea). Instead, CDW/CoW originates at the subarctic convergence in the south western basin, covers the entire South Pacific seafloor below 2000 m, and extends across the equator to nearly 10°N. It is the largest of oceanic water masses, essentially generated by mixing of other waters (mainly North Atlantic Deep Water, NADW, and Antarctic Bottom Water, AABW) in the Southern Ocean, with subsequent injection by the Antarctic Circumpolar Current into the Indian and Pacific Ocean.

The general thermohaline circulation of the Pacific Ocean, part of the Great Ocean Conveyor Belt system, is composed by a prevalently northward flow in the deeper layer, from the southwestern (near-polar) area to a broad northeastern region, and by a prevalent southward flow in the surface layer, from the northeast to the western equatorial area and into the Indian Ocean. At the surface, the wind-driven circulation pattern consists of two main basin-wide gyres, circulating in an anticyclonic direction, i.e. clockwise in the North Pacific and counter-clockwise in the South Pacific. In between these oceanic gyres, as the trade winds drive waters of the inter-tropical region from east to west, lies what is known as the equatorial current system, composed of the westward North Equatorial Current and South Equatorial Current, separated by an eastward Equatorial Countercurrent.

In the northern gyre, the North Equatorial Current feeds the Kuroshio, flowing northward off eastern Asia. The latter is joined by the southward Oyashio (part of the counter-clockwise circulation in the Gulf of Alaska and Bering Sea, which is formed by the Alaska Current and Kamchatka Current) into the North Pacific Current, which flows eastward and then branches into the northward Alaska Current and the southward California Current closing the clockwise circulation pattern.

In the southern gyre, the South Equatorial Current moves westward and then southward, feeding the East Australia Current and the southward flow into the South Pacific Current, which eventually joins the Antarctic Circumpolar Current (circling the Earth around Antarctica). As it approaches South America, the latter branches eastward around Cape Horn, feeding the Falklands Current in the South Atlantic Ocean, and northward into the Humboldt Current (or Peru Current, further north), which flows northward and ultimately rejoins the equatorial system.

Countless oceanic islands, as well as seamounts and guyots, dot the Pacific Ocean floor, mostly between the Tropics of Cancer and Capricorn. The northernmost archipelago is that of Hawaii, associated with the Hawaiian Ridge. To the south(east), the vast area of Polynesia; further west, north of the equator, the numerous small islands of Micronesia, nearly all coralline, and south of the equator, the larger islands of Melanesia. The Aleutian Islands, in the northern ocean basin, and the Kuril Islands, the Japan archipelago, the Ryukyu Islands, Taiwan, the Philippines and the group of continental islands of Indonesia, in the western ocean basin, enclose a string of marginal seas. From north to south, these include the Bering Sea, the Sea of Okhotsk, the Sea of Japan, the Yellow Sea, the East China Sea and the South China Sea, the Archipelagic Seas and the Philippine Sea. It is of great oceanographic significance that most of the main rivers of eastern Asia—including the Amur (Heilong Jiang), the White River (Hai He) system, the Yellow River (Huang He), the Yangtze (Chang Jiang), the Pearl River (Zhu Jiang) system, including the Xi Jiang, the Red River (Yuan Jiang, or Sông Cái) and the Mekong—all enter the Pacific indirectly, by way of these marginal seas.

The **Bering Sea**—the name of which celebrates Vitus Bering, Danish explorer, and officer in the Russian Navy, who was the first to explore in depth the region, in 1728—is the northernmost marginal basin of the Pacific Ocean (Fig. 4). It is bordered to the east by Alaska; to the west by Siberia and the Kamchatka Peninsula; to the south by the Alaska Peninsula and the Aleutian Islands; to the north by the Bering Strait, which connects it to the Chukchi Sea and the Arctic Ocean.

The Bering Sea covers about 2,000,000 km^2, comprising a southwestern deep area, i.e. the Aleutian, Commander and Bowers Basins, and a northeastern shallower continental shelf. Ice forms annually, from October to June, with the ice edge roughly parallel to (but slightly poleward of) the shelf break. The ice edge position varies from year to year, and the ice cover includes open water polynyas, where sea-air exchanges can increases surface salinities to more than 35 g/kg.

The extended shelf break between the Aleutian Basin, over 3000 m deep, and the Bering Shelf is the main driver of the massive primary productivity of the Bering Sea. The upwelling of nutrient-rich deep waters, mixing up the slope with shelf waters, sustains persistent plankton blooming. Further, seasonal melting of sea ice causes stratification and hydrographic effects that also influence productivity, and contribute to trigger, in part, a vigorous spring bloom.

Freshwater and nutrient input is also due to the Anadyr River, entering the sea from the west, and to the Yukon River, from the east. The northward flow of nutrient-rich surface waters, through the Bering Strait, feeds yet another exuberant marine ecosystem in the Chukchi Sea. The basin's high productivity supports extremely profitable fisheries, such as those of crabs, salmon, pollock and groundfish. The basin has great potential for both oil and natural gas extraction.

The **Sea of Okhotsk**, which owes its name to the very first Russian base in the Far East, the present Okhotsk urban-type settlement, is a marginal sea of the western Pacific Ocean (Fig. 6), lying between the Siberian coast, to the west and north; the Kamchatka Peninsula, to the east; the Kuril Islands, to the southeast; the island of Hokkaido, to the south; and the island of Sakhalin, to the west. It has an area of 1,583,000 km^2, a mean depth of 859 m, and a maximum depth of 3372 m.

The seabed slopes from north to south, forming a continental shelf along the northern and western margins, and a continental slope in the remaining area (about 70% of the total). Surface water temperature ranges from freezing point to 8–12 °C, while surface salinity can be as low as 32 g/kg, or less (due to the freshwater input from many rivers, and primarily the Amur). Deeper waters have a mean temperature ranging from −1.8 to −1 °C, and salinity of up to 34 g/kg.

Ice cover appears at the end of October, and reaches its greatest extent in March. In the coastal areas it extends to the shore, while in the open sea ice floes have extent and thickness depending on factors such as location, time of year, currents, and of course seawater temperature. The ice vanishes in June, with the exception of the Sakhalin gulfs and the region around the Shantar Islands, where ice floes are not uncommon in July and may occur even in August.

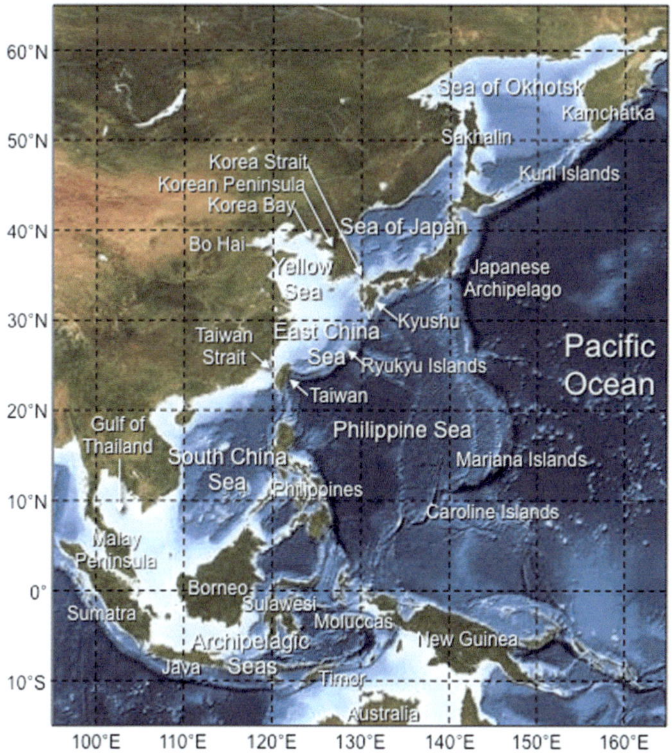

Fig. 6 Asian marginal basins of the western Pacific Ocean. Colour coding as in Fig. 1

The Sea of Okhotsk is connected to the Sea of Japan, north and south of Sakhalin, and to the North Pacific Ocean, through the Kuril island chain. In general, local currents flow clockwise around the Kuril Islands: in the northern part of the straits the flow is into the sea, but in the southern part it is out to the ocean. Relatively warmer inflows from these sources greatly affect temperature in the basin. The general water circulation in the basin is cyclonic (counterclockwise).

The Sea of Okhotsk is considered one of the richest north temperate marine ecosystems in the world, and one of the most biologically productive. Intense mixing of waters in the straits, large river runoff, and upwelling of deep, nutrient-rich ocean waters are all favorable to marine life. It is home to an abundance of fish, bird and marine mammal life. Known as a "bird lover's paradise," the Okhotsk Sea's shores serve as a haven for large colonies of several species.

In the late 1800s, American and European whalers operated extensively in the Sea of Okhotsk. In the 1900s, it was the scene of several US Navy operations, until it became a strategic basin for the Soviet Pacific Fleet, and later on for the Russian Navy. Recently, several oil and natural gas deposits have been identified on the shelf running along the coast. The local fishing industry is crucial to the economies of far eastern Russia, Kamchatka Peninsula, Sakhalin Island, and Japan.

The **Sea of Japan** (or East Sea) is a semi-enclosed basin parted from the Pacific Ocean by the Japanese archipelago, to the east and south; bounded by the Russian mainland and Sakhalin island, to the north; by the Korean Peninsula, to the west (Fig. 6). It has an area of about 978,000 km^2, a mean depth of 1752 m, and a maximum depth of 3742 m. It extends about 2255 km from north to south, up to 1070 km from east to west, and has hardly any prominent island, cape or bay.

The Sea of Japan is connected to the outer ocean by several straits, which are all rather shallow (about 100 m or less), thus hindering water exchanges with adjacent basins. This separation causes tidal ranges to be also very small. The water balance is mostly determined by the inflow and outflow through the straits. Few rivers discharge into the sea and their total contribution to the water exchange is just around 1%. The main surface circulation is cyclonic (counterclockwise).

The Sea of Japan is set in the region of the East Asian extratropical monsoon. This location, combined with water exchange through the straits, determines its hydrologic regime. A through flow, connected with the transit of warm currents from south to north, has a significant impact on the southern subtropical area. In winter, the prevailing winds are northwesterly. In summer, winds are generally weaker and variable, with a predominant southerly or southwesterly component.

The northwesterly winds bring storms and typhoons, with waves reaching 8–10 m. Tsunami waves have also been recorded. The cold and dry continental air produces strong evaporation, and enhances convection in the water column. The northern part of the sea freezes for about 4–5 months a year. Ice cover is continuous only in the inner bays, and forms floating patches in the open sea. The southwesterly winds blow warm, humid air from the ocean onto the Asian mainland.

The water column stratifies according to seasonal and regional factors. Surface temperature is mostly affected by atmospheric processes in the north, and by the currents in the south. In winter, values can be as low as 0 °C in the north (with a significant difference, up to 5–6 °C, between the colder west and warmer east), and as high as 10–14 °C in the south. In summer, temperatures rise to 18–20 °C in the north, and to 25–27 °C in the south. This east-west difference drops to 1–2 °C.

The hydrological isolation of the Sea of Japan also results in slightly lower average water salinity (34.1 g/kg), in comparison to the outer ocean. In winter, the highest salinity (34.5 g/kg) is observed in the south, where evaporation dominates over precipitation. Thawing of ice in spring reduces salinity (31.5 g/kg) in the north, while values remain high (34.6–34.7 g/kg) in the south, partly because of the inflow of salty water through the southern straits.

The basin's physical isolation is reflected also in the flora and fauna. Seawater presents an elevated concentration of dissolved oxygen, which results in a high biological productivity. Therefore, fishing is the dominant economic activity in the region. The intensity of ship traffic across the sea has been moderate, mostly owing to political issues (e.g. the Japan-Russia dispute over the Kuril islands), but it is steadily increasing as a result of the growth of East Asian economies.

The **Yellow Sea** constitutes the northernmost part of the East China Sea, a marginal basin of the western Pacific Ocean (Fig. 6). It is bounded north and west by the Chinese coast, and east by the Korean Peninsula. Its innermost northwest bay is the Bo Hai, while its northeast extension is the Korea Bay. The seas owes its name to the hue of its surface waters, due to the particles brought in by sand storms from the Gobi Desert, and by the Hai He and Huang He runoff.

The basin—a flooded section of continental shelf, formed after the last ice age, as the sea rose 120 m to its current level—extends for about 960 km from north to south and about 700 km from east to west. It has an area of nearly 380,000 km^2, and depth gradually increasing from north to south, with an average of only 44 m, and a maximum of 152 m. The shorelines and the sea floor are dominated by sand and silt deposits, brought by the rivers in the Bo Hai and the Korea Bay.

The local climate is characterized by cold, dry winters and warm, wet summers. From November to March, a strong northerly monsoon prevails, with severe blizzards in the Bo Hai, where seawater gets close to the freezing point and sea ice may form (and hinder navigation). Typhoons occur in summer, between June and October. Summer temperatures range between 22 and 28 °C. Average salinity is relatively low, ranging from 30 g/kg in the north to 33–34 g/kg in the south.

In spring and summer, the upper layer is warmed up by the sun and diluted by the rivers' freshwater input, while deeper layers remain cold and saline. Water circulation follows a cyclonic, counterclockwise pattern, fed by the Tsushima Current, which diverges from the Kuroshio Current system west of the Japanese islands and flows northward into the East China Sea and Yellow Sea. Along the continent, southward currents prevail, especially in the winter monsoon period.

A high nutrient load, mostly from river runoff, feeds extensive phytoplankton blooms (especially blue-green algae, recurring in summer). The sea is rich in seaweed (predominantly kelp) and shellfish. Species abundance increases southward and point to the high productivity of the basin, which accounts for the large fishing industry. Marine mammals, sea turtles, and larger fish drastically decreased in recent times not only due to pollution, but also mainly due to direct harvesting.

In the southern basin, a 10 km-wide belt of intertidal mudflats, consisting of highly productive sediments with a rich benthic fauna, is (or used to be) of great importance for migratory birds, making this the single most important site on northward migration in the entire East Asian–Australasian Flyway, with at least two million birds, of more than 35 species, passing through on a northward migration, and about half that number on a southward migration.

The Yellow Sea, however, is considered among the most degraded marine regions on Earth. Pollution is widespread, the quality of pelagic and benthic habitats is declining, the invasion of alien species is having a detrimental effect, and harmful algal blooms frequently occur. Loss of coastal habitats, due to land reclamation, has resulted in the destruction of more than 60% of the tidal wetlands in just 50 years. Decline in biodiversity, fisheries and ecosystem services is extensive.

The **East China Sea** is a marginal sea of the western Pacific Ocean (sometimes considered, with the South China Sea, as a part of the China Sea proper), comprised within the Asian continent to the west, and Kyushu, the Ryukyu Islands and Taiwan, to the east and south (Fig. 6). It opens onto the Yellow Sea to the north, while it connects with the Sea of Japan, via the Korea Strait, in the northeast, and with the South China Sea, via the Taiwan Strait, in the southwest.

The basin has an area of about 750,100 km^2, and is generally shallow, with an average depth of only 350 m, as three-fourths of this sea is less than 200 m deep. The Okinawa Trough, its deepest section, runs alongside the Ryukyu island chain and has a maximum depth of 2717 m. The western section is a continuation of the shelf extending from the Yellow Sea to the South China Sea. A large number of islands and shoals dot both the eastern section and the continental border.

The shallow shelf areas are covered with sediments from the bordering landmasses, deposited mainly by the Yangtze (Chang Jiang), and other rivers on the northern continental margin of the basin and in the Yellow Sea. Coarser sand deposits occur also farther out, and in some areas the sediments are up to 1.5 km thick. The Ryukyus include several volcanic islands, while earthquake epicenters are found along the island arc itself and the axis of the Okinawa Trough.

The East China Sea climate is dominated by monsoon winds, due to differential heating between land and water. In winter, the Asian landmass is much colder than the ocean, chiefly on the Tibet Plateau, and warmer in summer. Northerly winter winds bring cold, dry air from the continent. Summer heating over Asia reverses the predominant winds, which become southeasterly and bring warm, moist air from the western Pacific, producing a rainy season and recurrent typhoons.

The monsoon system influences also water circulation of the Kuroshio, flowing northward from the warm North Equatorial Current. The eastern inflow into the East China Sea through the Ryukyu island chain forms the Tsushima Current, which moves north into the Sea of Japan, while the main part of the current diverts eastward back out into the Pacific south of Kyushu, and flows east of Japan. Some of these warm waters also move northward via the Luzon and Taiwan straits.

Strengthened by the southeasterly winds, the Kuroshio is at its widest and fastest in summer, when its axis is displaced well into the East China Sea. Surface temperatures vary from 30 °C in the south to 25 °C in the north. In winter, northerly winds impact this circulation pattern, and the northward Kuroshio is reduced in strength, while southward coastal currents are strengthened. This brings in colder water, with temperatures ranging from 5 °C in the north to 23 °C in the south.

The East China Sea is a highly productive region, where fisheries and harvesting of living resources are well developed. Local ship traffic is intense, and is increased by through traffic from the South China Sea to many North Pacific ports. Oil and natural gas deposits have been discovered under the continental shelf. This has led to disputes between bordering countries over control of areas with hydrocarbon reserves, such as deep trenches, straits, rocks, and uninhabited islands.

The **South China Sea** is a marginal sea of the western Pacific Ocean (sometimes considered, with the East China Sea, as a part of the China Sea proper), bounded by the Asian continental landmass to the west and to the north; by Taiwan and the Philippines to the east; by Borneo to the southeast, and by the Gulf of Thailand and the coast of the Malay Peninsula to the southwest (Fig. 6). The basin has an area of about 3,685,000 km^2, with a mean depth of 1212 m.

Due north, it opens on the East China Sea, via the Taiwan Strait, about 160 km wide and 70 m deep; due east on the Philippine Sea, via the Luzon Strait, about 320 km wide and 2600 m deep, and on the Sulu Sea, beyond the Palawan island; and due south on the Java Sea, via the Karimata (and Gaspar) Strait, about 200 km wide and only 50 m deep. A western connection to the Indian Ocean is provided by the long Strait of Malacca, 31 km wide at its narrowest and about 30 m deep.

The major feature of the East China Sea is a deep basin in the central and eastern parts, with mean depth of 4300 m and maximum depth of 5016 m. Along the northwest side of the basin is a broad, shallow shelf, which includes the Gulf of Tonkin. On the southwest side, the Sunda shelf, one of the largest in the world, which includes all of the Gulf of Thailand. On the eastern side, the continental shelf falls off sharply near the Philippine islands and forms the Palawan Trough.

The climate is tropical and controlled by monsoons. In winter, winds blow chiefly from the northeast; in summer, from the southwest. Summer typhoons are frequent. Monsoons control surface currents as well, and water exchange with adjacent basins. In summer the inflow is from the south, via the Karimata (and Gasper) Strait. Near the mainland the flow is to the northeast, moving out via the Taiwan and Luzon Straits. In winter the flow is generally to the southwest.

The near-surface waters are relatively warm (up to 29 °C in the summer) because of the low latitude and a tendency for the equatorial currents to feed warm water into the area. In (early) summer, wind from the southwest not only moves the surface water to the northeast, but causes it to be displaced off the mainland coast, driving coastal upwelling of colder, nutrient-rich water. In winter, surface temperatures are cooler, ranging from 21 °C in the north to 27 °C in the south.

The major rivers draining into the South China Sea are the tributaries of the Pearl River (Zhu Jiang) system, the Red River (Yuan Jiang, or Song Cai) and the Mekong River. The wet summer season causes these rivers to triple sometimes their annual average flow, and even more in the case of the Red River. The extensive runoff of nutrient-laden river waters, as well as coastal upwelling in several areas, all contribute to the rich abundance of marine life in the South China Sea.

The sea is heavily fished, and is the main source of animal protein for millions in Southeast Asia. The lucrative fisheries, crucial for local food security, are coupled to huge oil and gas reserves believed to lie beneath the seabed, and to some of the world's main shipping lanes, hosting one-third of global traffic. Thus, the South China Sea carries great strategic importance, and the basin and its uninhabited islands are subject to competing claims of sovereignty by several countries.

The **Archipelagic Seas** comprise a group of semi-enclosed basins, gulfs, and straits of Southeast Asia, between the Pacific Ocean and the Indian Ocean, and connected to both (Fig. 6). These Seas are parceled out, in the east-west direction, by the Greater Sunda Islands, from Sumatra to Borneo, Sulawesi, the Moluccas and New Guinea; and they are bounded, to the north by the Philippine archipelago, to the south by the Lesser Sunda Islands, between Java and Timor.

Progressing north to south, and then west to east, the main basins in this group are the Sulu Sea, Celebes Sea, Molucca Sea, Java Sea, Flores Sea, Banda Sea,

Arafura Sea, and Timor Sea. Other minor basins, located between the scattered islands of Indonesia, are the Ceram Sea, Bali Sea, Halmahera Sea, and Savu Sea, to which the Gulf of Tomini and the Gulf of Boni, respectively north and south of Sulawesi, and the Makassar Strait, west of Sulawesi, should also be added.

These tectonically active basins extend between, and partially overlap, the Sunda Shelf of the Asian continent, to the northwest, and the (Arafura and) Sahul Shelf of the Australian continent, to the southeast. Deep hollows, some reaching almost 5000 m of depth, are set in the intermediate basins. The Karimata Strait connects the Seas to the South China Sea, in the northwest. Countless other straits and passages open onto both the Pacific Ocean and the Indian Ocean.

The region has a tropical to equatorial climate, which is generally warm and humid year round, influenced by the monsoon regime, but with no significant division between seasons. The Archipelagic Seas have complex oceanographic features, the main of which being the Indonesian Throughflow that transports water from the Pacific Ocean (mostly through the Makassar Strait) to the Indian Ocean (mostly via the Timor Current, flowing south-west between Timor and Australia).

Many of the islands bounding these Seas are covered by intact rain forest and are home to a number of endemic plants and animals. The islands are also the native habitat of plants long coveted for their use as spices. The tropical setting and warm, clear waters permit these Seas to harbor about 580 of the world's 793 species of reef-building corals, which grow as some of the most bio-diverse coral reefs in the world, and an impressive array of marine life.

The Archipelagic Seas offer some of the richest marine fisheries in the world, albeit under increasing pressure from illegal, unreported and unregulated fishing activities. Moreover, although hazardous to navigation, these basins provide important sea routes for regional trade, and have been the site of successful exploration for oil and natural gas. The Seas are also an ever more popular tourist destination, in particular for scuba diving and luxury ocean cruising.

The **Philippine Sea** is a marginal sea located east and northeast of the Philippine archipelago (Fig. 6). It is bordered by Taiwan and the Philippines to the west; by the Ryukyu Islands, and the Japanese islands of Honshu, Shikoku, and Kyushu, to the north; by the northern Mariana Islands on the east, and the western Caroline Islands on the south. It connects to the west with the East and South China Sea, and to the east with the Pacific Ocean.

The seafloor is formed into a structural basin by a series of geologic faults and fracture zones. Island arcs, extended ridges due to plate tectonic activity, enclose almost completely the basin, which has an estimated area of about 5,000,000 km^2. Depths around 6000 m are common, while deep trenches are located east of the island arcs, the deepest of which, at 10,539 m, is the Philippine Trench. Numerous seamounts, often volcanic, and capped with coral, rise from the basin floor.

The warm Pacific North Equatorial Current flows westward across the southern part of the sea. On meeting the Philippines, the current divides; part swings north to form the Kuroshio, of which some returns to the ocean as the Kuroshio Countercurrent, while part swings south as the Mindanao Current and then the

Pacific North Equatorial Countercurrent. These currents, together with areas adjacent to reefs, ridges, and seamounts, are the sites of rich fishing grounds.

A tropical marine climate, usually experienced by islands and coastal areas 10° to 20° north or south of the equator, is typical of the region. There are two main seasons, driven respectively by a warm and wet southwest monsoon from June through October, and by a cold and dry northeast monsoon from November to March. A hot and dry spring, from March to May, constitutes a third season. Typhoons, which become particularly strong in September, originate in this region.

The Philippine Sea hosts an exotic marine ecosystem, harboring about 500 species of hard and soft corals in coastal waters, and roughly 20% of the shellfish species known worldwide. However, uncontrolled deforestation in watershed areas, subsequent soil erosion, as well as air and water pollution cause severe environment damages in near-coastal regions. Increasing water pollution has particular impacts on coastal mangrove swamps, which are important fish breeding grounds.

4 Indian Ocean Marginal Basins

The Indian Ocean, connecting the marginal basins south of the Asian continent, is the third largest of the Earth's oceanic divisions, after the Pacific and Atlantic Oceans (Fig. 7). It is the youngest of the major oceans, but its origin and evolution are the most complicated. Its formation is a consequence of the breakup, about 150 million years ago, of the southern supercontinent Gondwana, followed by the movement to the northeast of the Indian subcontinent (started about 125 million years ago), which began colliding with Eurasia about 50 million years ago; by the western movement of Africa; and by the separation of Australia from Antarctica some 53 million years ago. By 36 million years ago, the Indian Ocean had taken on its present configuration.

The Indian Ocean takes up about 14.4% of the Earth's surface, representing an area of approximately 73,440,000 km^2, with an average depth of 3890 m. Its deepest point, more than 8047 m below the surface, is the Diamantina Deep, located southwest of Perth, Western Australia. The basin extends in latitude from the Tropic of Cancer to the Antarctic (where the boundary with the Southern Ocean has been set at 60°S), and reaches a longitudinal width of about 10,000 km, between the coasts of Africa and Australia. Surface salinity ranges between 32 g/kg (in areas of monsoon rains, at high latitudes and along coasts impacted by large rivers) to 37 g/kg (in the Arabian Sea and in a belt between southern Africa and southwestern Australia). Surface water temperatures vary between 22 and 28 °C, northward of the Tropic of Capricorn, but drop rather quickly southward.

In the schematic representation of the Indian Ocean water column, based on combined temperature and salinity characteristics, Upper Waters, formed by subduction at the tropics, comprise Bengal Bay Water (BBW), Arabian Sea Water (ASW), Indian Equatorial Water (IEW), Indonesian Upper Water (IUW), and South Indian Central Water (SICW) (surface to 500 m). Below this, Intermediate Waters

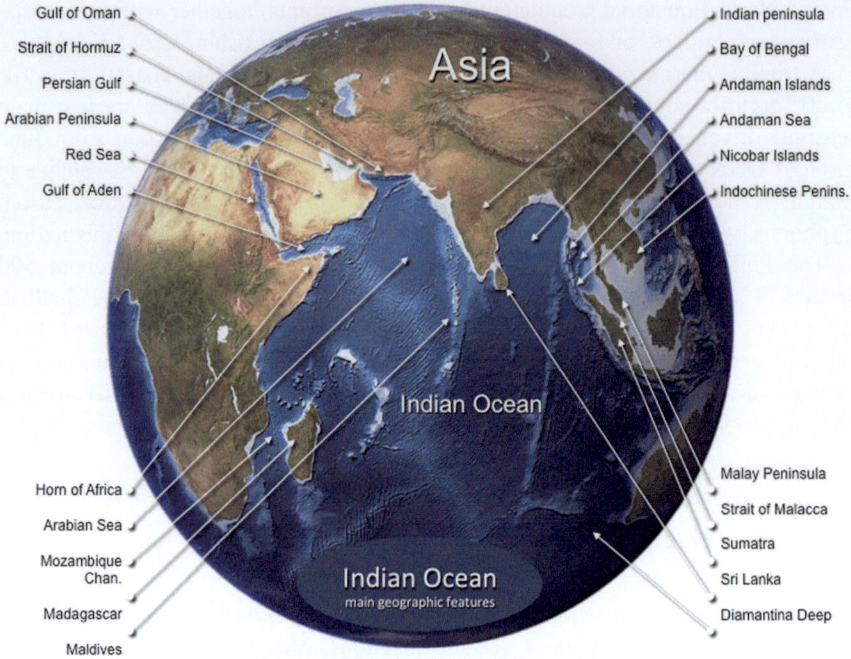

Fig. 7 Northern Indian Ocean main geographic features. Colour coding as in Fig. 1

include the Antarctic Intermediate Water (AAIW) from the south, formed by deep convection east and west of the southern tip of the American continent and spread into all oceans with the Antarctic Circumpolar Current, and by Indonesian, Red Sea and Persian Gulf Intermediate Waters (RS/PGIW) in the north (500–1500 m). Still deeper in the water column, Deep and Abyssal Waters—basically just Circumpolar Deep Water (CDW), or Common Water (CoW), a mixture of North Atlantic Deep Water (NADW) and Antarctic Bottom Water (AABW), formed by deep convection in the Southern Ocean, primarily the Weddell Sea and Ross Sea—occupy the deepest basins (1500 m to bottom).

The general thermohaline circulation of the Indian Ocean, as part of the Great Ocean Conveyor Belt system, is composed by a prevalent northeastward flow in lower layers (primarily in the western basins), and by a prevalent southwestward flow in the upper layers. At the surface, the wind-driven circulation pattern consists of variable currents, dominated by the monsoon wind reversals, north of the equator, and of a main basin-wide gyre, circulating in an anticyclonic direction, i.e. counter-clockwise, south of the equator. In the northern basin, during the winter monsoon (blowing from the northeast to the southwest), the Indian North Equatorial Current and the Northeast Monsoon Current, part of the cyclonic surface circulation in the Arabian Sea, feed a southward Somali Current off east Africa. This joins with the northern component of the Indian South Equatorial Current (the East African Coastal Current) into the Indian Equatorial Countercurrent flowing eastward, away from the African continent, and ultimately feeding also the cyclonic pattern in the Bay of Bengal. The flow reverses during the summer monsoon (blowing from the southwest to the northeast), as the northern component of the Indian South Equatorial Current (the East African Coastal Current) feeds a northward Somali Current (recirculating also in a small anticyclonic gyre off east Africa). This continues northward, eastward and then southward into the Southwest Monsoon Current, which closes the anticyclonic circulation in the Arabian Sea and feeds an analogous pattern in the Bay of Bengal.

In the southern oceanic gyre, the southward component of the Indian South Equatorial Current, in part fed by the Indonesian Throughflow from the Pacific Ocean, splits west of Madagascar, towards the Mozambique Channel, and east of Madagascar, as the East Madagascar Current, only to re-join southward to form the complex Agulhas Current system. The Agulhas Return Current retroflects eastward, south of the African continent, into the South Indian Current that eventually joins the Antarctic Circumpolar Current (which circles the Earth around Antarctica). Further east, the mainly seasonal—stronger in (austral) summer than in winter—northward flowing West Australian Current—partly running under the near-coastal, shallow, southward flowing Leeuwin Current, weakest in (austral) summer months—closes the loop between the South Indian Current and the complex equatorial current system.

The northern section of the Indian Ocean can be divided into two main groups of marginal basins, separated by the Indian sub-continent (and the Maldives Islands chain), i.e. the Bay of Bengal and the Andaman Sea, to the east; and the Arabian Sea, connected to the semi-enclosed basins of the Persian Gulf, through the Gulf of Oman, and of the Red Sea, through the Gulf of Aden, to the west. The key oceanographic features of these marginal basins are dominated by major, and seasonal, surface current systems (the Monsoon Currents, or Drifts, i.e. the seasonally varying current regime that characterizes the tropical regions of the northern Indian Ocean). These, in fact, influence physical as well as ecological processes of the marginal basins off the southern Asian continental landmass, while affecting at the same time the entire world's climate.

Also of great oceanographic impact are the river inflows, in the Arabian sea (the Indus and Narmada) and mainly in the Bay of Bengal (the Ganges, or Ganga—Known as Padma in its last section, where it joins the Jamuna and the Meghna, respectively the largest and second largest distributaries of the Brahmaputra, before entering the estuary that empties into the Bay of Bengal—as well as the Irrawaddy, Godavari, Mahanadi, Krishna and Kaveri). In the Bay of Bengal, fresher waters enter also with the Indonesian Throughflow, via the Andaman Sea, while in the Arabian Sea saltier waters enter from both Red Sea and Persian Gulf, via the Gulf of Aden and the Gulf of Oman respectively.

The **Bay of Bengal** is comprised within the Indian sub-continent, to the west (and northwest), and the Indochinese Peninsula, to the east (and northeast). On its southeastern reaches it borders the Andaman and Nicobar Islands chain, which separates it from the Andaman Sea. To the south, it opens onto the Indian Ocean

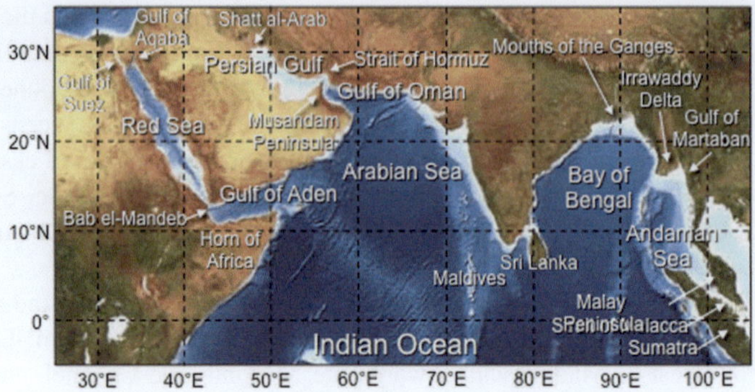

Fig. 8 Asian marginal basins of the northern Indian Ocean. Colour coding as in Fig. 1

proper. Conventionally, its southern boundary extends from the southern end of Sri Lanka in the west to the northern tip of Sumatra in the east (Fig. 8).

This vast embayment, formed as the Indian sub-continent collided with Asia, within roughly the past 50 million years, is considered to be the largest bay in the world. Its total area is on the order of 2,172,000 km^2, its average depth is 2600 m, and its maximum depth 4694 m. A number of large rivers—the Mahanadi, Godavari, Krishna, and Kaveri in the west, and primarily the Ganges (Ganga), Brahmaputra and others (local names may vary) in the north—flow into the basin.

The northern section of the Bay of Bengal is bordered by a wide continental shelf that narrows to the south, with a continental slope cut by canyons lined with the main rivers. Marine sediments are dominated by riverine terrigenous deposits coming from the Indian sub-continent and the Himalayas. The sediment fan off the Mouths of the Ganges is the widest and thickest in the world. The deep floor of the Bay's southern section is a vast, southward sloping abyssal plain.

Climate is dominated by the monsoons. In winter, November to April, a continental high-pressure system, north of the bay, produces northeasterly winds (the northeast monsoon). In summer, June to September, a low-pressure system over the continent drives air from the ocean, and rain-bearing southwesterly winds (the southwest monsoon) prevail. Cyclones, extreme tropical storms, occur in spring (May) and in fall (October), before and after the onset of monsoon rains.

The influence of the monsoons is seen in the reversal of the surface circulation and in the hydrographical conditions of surface waters as far as 10–20°S. Surface currents change direction with the season, the northeast monsoon giving them a clockwise circulation, the southeast monsoon a counterclockwise circulation. Because rainfall and river input far exceed evaporation, the bay exhibits a net water gain annually. Sea level varies considerably due to storm surges.

A unique feature of the bay is the extreme variability of its physical properties. Temperature in offshore areas, however, is commonly warm and markedly uniform in all seasons, increasing somewhat toward the south. Surface salinity, normally around 33 to 34 g/kg, can decrease to nearly half that level during the fall transition period, when river discharge is highest, extending well south in the bay. Weak upwelling occurs in the northeast during the northeast monsoon.

The whole bay hosts a distinct tropical marine ecosystem. River drainage, and a profusion of coastal marshes, wetlands and mangroves increase productivity of nearshore fish species, sustaining important local fisheries. Oil and natural gas discoveries have been made, notably off some of the main river deltas. The principal trade routes for large tankers, en route from the Persian Gulf to the Strait of Malacca, pass south of the Bay of Bengal.

The **Andaman Sea** is the easternmost marginal sea of the Indian Ocean, southeast of the Bay of Bengal, bounded in the east by the Malay Peninsula and in the west by the Andaman Islands (from which it takes its name) and Nicobar Islands. To the north, the Myanmar (Burma) coastline, with the large delta of the Irrawaddy, and the Gulf of Martaban; to the south, the island of Sumatra. At its southeastern reaches, the Strait of Malacca, parting Sumatra and Malay Peninsula (Fig. 8).

The basin measures about 1200 km, north-south, and 645 km, east-west, with a total area of 798,000 km^2. Its average depth is around 1000 m. The northern and eastern sections are less than 180 m deep, in part due to the vast quantities of silt deposited by the Irrawaddy River through its delta. The western and central sections are 900 to 3000 m deep. Less than 5% of the basin is deeper than 3000 m, but depths can exceed 4400 m just east of the Andaman-Nicobar Ridge.

The monsoon regime of Southeast Asia governs both climate and oceanographic properties of the Andaman Sea. Air temperature is stable over the year, at 26–27 °C, while sea surface temperatures range from 26–28 °C in winter to 29 °C in summer. Water temperature is constant at 4.8 °C at depths of 1600 m and below. In winter, the sea receives little rainfall or runoff, and surface salinity is rather high. In summer, the huge runoff causes much lower surface salinity in the north.

In the south, salinity values can vary from 30.0–33.0 g/kg in winter to 31.5–32.5 g/kg in summer. In the north, they can decrease locally to 20–25 g/kg, due to the inflow of fresh water from the Irrawaddy River during the summer monsoon. The oceanic flow changes direction twice during the year. Surface currents are south-easterly and easterly in winter, but turn south-westerly and westerly in summer. The semidiurnal tides have amplitudes in excess of 7 m.

The coastal areas are lined with mangrove forests and seagrass meadows, ever decreasing due to human activities. Most coastlines, particularly that of the Malay Peninsula, are rich in coral reefs and offshore islands with spectacular topography. The marine fauna includes several vulnerable species, such as the dugong (*Dugong dugon*), and the Irrawaddy dolphin (*Orcaella brevirostris*). The Andaman Sea also forms part of a major shipping route, via the Strait of Malacca.

This funnel-shaped waterway, named after the trading port of Melaka (formerly Malacca) on the Malay coast, connects the Andaman Sea and the South China Sea. It is 800 km long, and has a width of about 250 km in the north, narrowing to only 65 km in the south, for a total area of about 65,000 km^2. In the northwest, near the Andaman basin, water depth is on the order of 200 m. In the southeast, depths are usually around 27 m and rarely exceed 37 m.

Throughout the year, the current flows northwest through the Strait of Malacca. Numerous islets, reefs and sand ridges, hinder the passage at the southern entrance of the channel, which is also gradually silting along both sides, near the mouths of large rivers. As the main link between the Indian Ocean and the Pacific Ocean, the Strait is one of the most heavily traveled waterways in the world. Piracy, a historical recurrence in the Strait since the 14th century, is still on the rise.

The **Arabian Sea**, the northwestern part of the Indian Ocean, is a marginal basin comprised within the Arabian Peninsula and the Horn of Africa, to the northwest, and the Indian sub-continent, to the northeast. To the south, it opens onto the Indian Ocean, at the latitudes of the Maldives. To the north, the Gulf of Oman connects it with the Persian Gulf, via the Strait of Hormuz. To the west, the Gulf of Aden connects it with the Red Sea, via the Bab el-Mandeb (Fig. 8).

The Arabian basin was formed 50 million years ago, due to the collision of the Indian sub-continent with Asia. Its total area is on the order of 3,862,000 km^2, its average depth is 2734 m, and its maximum depth 4652 m. The Indus and the Narmada are the main rivers flowing into this Sea. The narrow continental shelf is cut by a deep submarine canyon created by the Indus River, which also has deposited an abyssal cone of thick sediments in the abyssal plain of the basin.

The Arabian Sea has a monsoon climate. In winter, northeasterly winds cause cooling and mixing. Surface temperature minima of 24–25 °C occur in the central basin, while surface salinities in excess of 36 g/kg have been recorded in this dry season. In summer, southwesterly winds induce coastal upwelling along the western continental margins. Surface temperature increases from west to east, with maxima of 28 °C. Surface salinities lower than 35 g/kg recur in this rainy season.

Shifts in surface circulation occur, over periods of several weeks, due to wind stress changes associated with the monsoon reversal: winds from the northeast result in westward currents, while winds from the southwest result in eastward currents. During winter, the surface flow is directed westward, from the Indian (and Indonesian) region towards the African area. During summer, the trend reverses, with eastward flow from the African area, to the Indian (and Indonesian) region.

Thus, the Northeast (Winter) Monsoon Current is westward from January to March, and the Southwest (Summer) Monsoon Current is eastward from April to November. As the currents shift toward the northeast, offshore Ekman transport moves warmer surface waters deep into the Arabian Sea, and causes the upwelling of cooler waters along the coast. In the water column, distinct water masses can be distinguished, in particular the Red Sea and Persian Gulf highly saline inflows.

The fertilizing effect of the monsoon-driven upwelling, which raises the surface levels of nutrients, produces an abundance of fish life throughout the western Arabian Sea. Extensive small-scale fishing takes place in the region. However, a mass mortality of fish also occurs periodically. This is attributed to a subsurface water layer of tropical origin, rich in nutrients, but poor in oxygen, brought to the surface by strong upwelling and resulting in fish death from lack of oxygen.

Sediments near the upwelling areas consist of fine muds, with a high organic component containing hydrogen sulfide. These hold much fish remains and form a true fish cemetery. Authigenic ferromanganese nodules, polymetallic sulfides, and oil and natural gas deposits have been discovered on the Arabian Sea's continental shelves. Some of the world's busiest shipping lanes are located in the basin, due to its strategic location connecting to the Red Sea and Persian Gulf.

The **Persian Gulf** (sometimes referred to as the Arab Gulf) is a shallow marginal sea of the Indian Ocean that lies between the Arabian Peninsula, to the west, and the western Asian mainland (of southwestern Iran), to the east. The river delta of the Shatt al-Arab, which carries the waters of the Tigris, Euphrates (and Karun) rivers, forms the northwest shoreline. In the southeast, the basin is connected to the Gulf of Oman, and to the open Arabian Sea, through the Strait of Hormuz (Fig. 8).

This inland basin is geologically very young, having been formed around 15,000 years ago. It has an area of about 241,000 km^2, with a north-south length of 990 km, and an east-west width varying from a maximum of 340 km to a minimum of 55 km in the Strait of Hormuz. Its waters are overall very shallow, rarely deeper than 90 m, although depths exceeding 110 m can be found at isolated localities in the southeast, and have an average depth around 50 m.

The Gulf of Oman, sometimes referred to as the Sea of Oman, is really an elongated strait, or an arm of the Arabian Sea, and not an actual gulf. It is 560 km long and 320 km wide, and 50–200 m deep, extending between the southwestern part of the Arabian Peninsula and the Asian mainland, and being separated from the Persian Gulf by the Musandam Peninsula. Its narrowest section in the north, the Strait of Hormuz, is 55–95 km wide, and only 50 m deep.

The region has a harsh climate. Temperatures are generally high, though winters may be cool in the northwest. Evaporation far exceeds freshwater inflow, due to low rainfall (mainly in winter, in the northeast) and coastal runoff. Dust carried by desert storms can affect significantly solar radiation. Surface temperatures range from 16 to 32 °C in the northwest, to 24–32 °C in the southeast. Surface salinities range from 38–41 g/kg in the northwest, to 37–38 g/kg in the southeast.

The circulation pattern in the Persian Gulf is cyclonic (counterclockwise), with surface waters entering from the Arabian Sea, undergoing intense evaporation, becoming denser and sinking, to flow out again beneath the inflow, in an anti-estuarine pattern. The wind affects local currents and sometimes reverses them. Onshore winds can cause surges in excess of 2 m, particularly in the south, flooding the coastal salt flats known as sebkhas. The tidal range varies from 1 to 3 m.

The Persian Gulf supports a rich flora and fauna, even though high temperatures and salinities lead to a lessening in the diversity typical of the Indian Ocean. The basin has extensive reefs (mostly rocky, but also coral), important fishing grounds (now impacted by large river dams, which lower nutrients inflow), and abundant pearl oysters (with traditional pearl-fishing waning since the advent of cultivated pearls in the 1930s), though its ecology has been marred by pollution.

The Persian Gulf region has acquired considerable strategic significance as the world's largest source of crude oil. Large gas finds have also been made. The Persian Gulf nations produce about 25% of the world's oil, and hold nearly 66% of the world's proven crude oil reserves, plus about 35% of the world's natural gas reserves. Exploration remains active, as new reserves are constantly being discovered, both on land and offshore, while related industries dominate in the area.

The **Red Sea** is a (semi) enclosed basin, bounded by northeastern Africa, to the west, and the Arabian Peninsula, to the east. The elongated and narrow-shaped basin extends between the Mediterranean Sea, to the north-west, and the Arabian Sea, to the south-east. At the northern end, it separates into the Gulf of Aqaba and the Gulf of Suez, which is connected to the Mediterranean Sea via the Suez Canal. At the southern end, it connects to the Gulf of Aden, via the Bab el-Mandeb (Fig. 8).

The formation of the Red Sea begun in the Eocene, when the African and Asian continental plates started to move apart, giving birth to an infant ocean, and accelerated during the Oligocene. The basin acquired its present shape over the past 4–5 million years, by slow seafloor spreading, a fact that makes it a geologically recent opening and one of the youngest oceanic zones on Earth. Today, the basin continues to widen at a rate of 1–2 cm *per* year.

The basin has a total area of about 440,000 km^2. It extends from north to south for about 1900 km, while its width from east to west ranges from 306 to 26 km. It has wide, shallow continental shelves, and a mean depth of only 524 m. A central deep trench, with depth around 3000 m, stretches for almost the entire axis of the basin. The seafloor rises in the southern region toward the Bab el-Mandeb, which is just 137 m deep at the Hanish Sill.

The Red Sea climate is characterized by two distinct seasons, respectively under the influence of the northeast monsoon, in winter, and the southwest monsoon, in summer. In this arid region comprised of desert or semi-desert areas, among the hottest in the world, extremely high temperatures characterize the weather, while rainfall is extremely low. High evaporation rate and low freshwater input render the Red Sea one of the hottest and saltiest bodies of seawater in the world.

Surface temperature ranges from 26 °C in the north to 30 °C in the south, in summer, with variations of only 2 °C, in winter. In general, values declines toward Bab el-Mandeb, due to exchanges with the Gulf of Aden, and toward the northern gulfs. Surface salinity is also high, due to low precipitation, high evaporation, and lack of river outflow. It ranges between 36 and 39 g/kg, and is usually lower in the southern region, again due to the exchanges with the Gulf of Aden.

Surface currents are weak and variable, and essentially wind-driven. In winter, the prevailing winds drive surface waters north, whereas in summer the mean flow is reversed. The overall surface drift to the north, where denser water is formed by evaporation, is balanced by a southward deeper outflow. Thus, the general circulation follows an anti-estuarine pattern, with fresher (nutrient-rich) surface water entering from Gulf of Aden, above an outflowing layer of saltier, denser water.

The Red Sea has low productivity, due to a strong nutrient limitation, but presents a rich and diverse ecosystem, in part due to the 2000 km of coral reefs, and other particular coastal habitats, such as mangrove forests, seagrass beds, salt marshes and saltpans. The basin constitutes a major shipping route for oil tankers and other vessels travelling through the Suez Canal. Fisheries and, more recently, tourism also started to play an important role in the local economy.

5 Enclosed Asian Basins

A number of semi-enclosed and enclosed basins dot the westernmost reaches of the Asian continent, as well as its interior (Fig. 1). The terminal Asian lands, in Asia Minor, and the Middle East are bounded by the eastern Mediterranean Sea and the Black Sea. Although commonly identified as European Seas (sometimes African, for the Mediterranean), these marginal basins are included in the present review for reason of completeness. The Mediterranean was commonly thought to be a remnant of the ancient Tethys Ocean. It is now known to be a structurally younger ocean basin known as Neotethys, formed during the Late Triassic and Early Jurassic rifting of the African and Eurasian plates. The Black Sea basin was formed during the Miocene orogenesis, which uplifted the mountain ranges that divided the ancient Tethys Ocean into several brackish basins, the remnants of which include also the Azov Sea, as well as the other enclosed basins of western Asia. These comprise the Caspian Sea and the remnants of the Aral Sea, which present—or presented, in the Aral Sea case—characteristics common to both seas and lakes. Further east, well into the interior of the continental mass, a score of semi-saline or freshwater lakes includes Lake Balkhash and, more prominently, Lake Baikal, some of the most peculiar water bodies on the Earth's surface.

The **Mediterranean Sea**, bounded to the north by the coast of southern Europe, to the south by that of northern Africa, and to the east by Asia Minor and the Middle East, has an area of 2,700,000 km^2 and depth averaging roughly 1500 m, with a maximum of 5267 m. In the west, it connects to the Atlantic Ocean, via the Strait of Gibraltar; in the northeast, to the Black Sea, via the Dardanelles and the Bosphorus; in the southeast, to the Red Sea, via the Suez Canal (Fig. 9).

The Mediterranean Sea is a concentration basin, where evaporation greatly exceeds precipitation and freshwater inflow from just a few large rivers (the Nile, Rhone, Po and Ebro). Its mild climate is characterized by windy, cold, wet winters and by relatively calm, hot, dry summers. Surface temperatures can vary from 14–16 °C in winter, to 20–26 °C in summer. Surface salinity differs from about 36 g/kg in the west, due to the Atlantic inflow, to 38–39 g/kg in the east.

The general circulation follows a cyclonic pattern. Oceanic waters flow in from the Atlantic through the Strait of Gibraltar, in the surface layer. This incoming water warms and becomes saltier as it travels east in the southern basin. It sinks in the Levantine basin (as well as in other areas of deep and bottom waters formation), due to winter cooling, moves back westward and ultimately spills over the sill of the Strait of Gibraltar, in the bottom layer, and out into the Atlantic.

A substantial amount of vertical mixing is provided by strong regional (winter) wind regimes. The vertical mixing ventilates the deepest parts of the basin and fertilizes surface waters. On average, the Mediterranean Sea is poor in nutrients, with consequent low phytoplankton biomass and low primary production. By contrast, a rich biodiversity characterizes the ecosystem: the fauna and flora of the Mediterranean Sea are among the richest in the world.

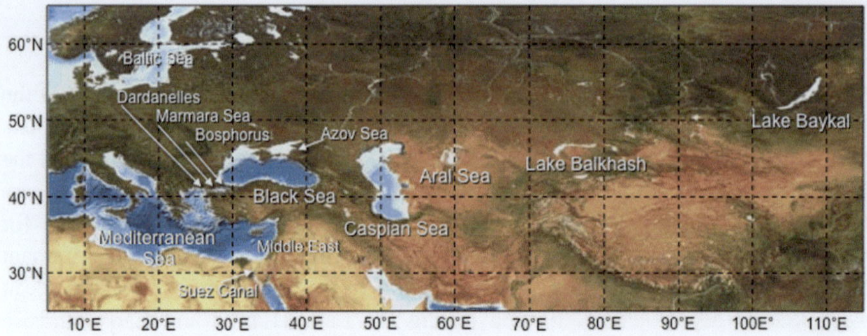

Fig. 9 Asian enclosed basins. Colour coding as in Fig. 1

The environmental state of the Mediterranean is generally good, but coastal areas are subject to various problems, due to an ever-increasing resident population (150 million people), and tourists (also about 150 million per year). Oil transport and pollution are also a growing concern, since of the 2000 cargos that sail in the Mediterranean at any given time, about 250–300 are oil tankers, transporting 400 million tons/year of petroleum products (i.e. 30% of the world traffic).

The landlocked **Black Sea**, the world's largest meromictic basin, is sited between Eastern Europe and Asia Minor. In the south-west, it connects, via the Bosphorus to the Marmara Sea and then, via the Dardanelles, to the Mediterranean Sea. In the north-east, it is also connected to the shallow Azov Sea, via the Kerch Strait (Fig. 9). The basin has a total area of about 423,000 km^2, an average depth around 1240 m, and a maximum depth of 2212 m.

A major part of the Black Sea experiences a Mediterranean climate, with warm, wet winters and hot, dry summers. Sea surface temperature varies seasonally from 8 to 30 °C, while deep sea temperature is about 8.5 °C. Salinity, regulated by the large input of freshwater from rainfall and a vast catchment area, the low evaporation rate and the limited water exchange with more saline basins, ranges between 18.0–18.5 g/kg at the surface and 22.5 g/kg in deep waters.

The deep influx of dense water from the Mediterranean Sea, over the narrow, shallow sills separating the two basins, is balanced by a surface outflow of fresher surface water from the Black Sea into the Marmara Sea. The upper layer circulation is characterized by a basin-wide cyclonic gyre, i.e. the Rim Current that encircles the entire Black Sea (but that can divide into two man gyres in winter), and by mesoscale eddies capable of advecting coastal waters into the open sea.

About 90% of the basin's volume, below 150 m of depth, is occupied by anoxic waters, with a high concentration of hydrogen sulphide. This is due to topography and fluvial inputs, which result in a positive water balance and strong stratification. The surface oxygenated waters support a rich ecosystem, in spite of eutrophication, proliferation of alien species and overfishing. In the deep anoxic waters, only protozoa, bacteria and invertebrate species are known to exist.

Various types of wetlands, providing productive spawning and feeding grounds for various species and refuge for migratory waterfowl are found in the Black Sea. They are mainly located at the mouth of large rivers, such as the Danube, Dnster, Bug, Dnper, Don, Kuban, Rioni, Sakarya and Kizilirmak. The same rivers are the main source of the excess nutrients discharged in the Black Sea, triggering extensive eutrophication and decline of water quality.

The totally landlocked **Caspian Sea** is the largest inland water body on Earth. It is bordered by the Caucasus and Elburz Mountains, from west to south, and by the Caspian depression and the central Asian steppes and deserts, from north to east. Although it has no outlet, the basin is connected to the Sea of Azov by the Manych Canal and the Volga-Don Canal, ultimately providing links with the Black Sea, Baltic Sea and White Sea, through a network of inland waterways (Fig. 9).

The Caspian Sea is often listed as the world's largest endorheic lake, by both area (371,000 km^2) and volume, though it is not a freshwater lake. The basin has an average depth of 184 m, with minima (averaging only 5 m) in the northern part, intermediate depths (788 m) in the central part, and a maximum (1025 m) in the southern part. Currently, the surface elevation is rather stable at about 28 m below global sea level, but historically has been undergoing large variations.

In winter, water temperature varies from −0.5 °C in the north, to 3–7 °C in the central basin, and to 8–10 °C in the south. In summer, it ranges around 24–26 °C, reaching 29 °C in the south. The average salinity is 12.7–12.8 g/kg, about one third the salinity of seawater (due to the freshwater input from several rivers, and in particular from the Volga, which accounts for about 80% of the inflow), with maximum of 13.2 g/kg in the east and minimum of 0.1–0.2 g/kg in the north.

The Caspian Sea crosses several climatic zones: continental, in the north; temperate, in the west; subtropical and wet, in the southwest; harsh and dry in the east. The shallow northern basin freezes from November to March, when ice thickness can reach up to 2 m. The Caspian Sea is home to a number of ecosystems, but in general flora and fauna are relatively poor. Endemic species include the endangered Caspian Seal (*Phoca caspica*) and the famous native sturgeons.

Underlying the Caspian region are some of the world's largest oil and natural gas reserves. Discharges from offshore drilling, and inflows from the highly polluted Volga river, all contribute to the degradation of water quality. While riparian countries dispute over control of and access to mineral resources, fishing grounds and major waterways, decades of environmental mismanagement have led to severe pollution problems, and to a significant impact on human health.

The **Aral Sea**, formerly the fourth largest lake in the world, with an area of 68,000 km^2, has been steadily shrinking since the 1960s, after the two major rivers that fed it, the Amu Darya and the Syr Darya, were diverted by Soviet irrigation projects. By 1997, it had declined to 10% of its original size, splitting into four lakes—the North Aral Sea, the eastern and western basins of the once far larger South Aral Sea, and one smaller intermediate lake (Fig. 9).

In an effort to save and replenish the North Aral Sea, a dam project was completed in 2005. In 2008, the water level in this lake section had risen by 12 m, compared to 2003, with deeper areas reaching a maximum depth of 42 m. Salinity dropped, and fish returned in sufficient numbers for some fishing to be viable. However, by 2009, the southeastern lake had disappeared and the southwestern lake had retreated to a thin strip at the western edge of the former southern sea.

Today the Aral Sea is a ghost basin, morphed into the Aralkum Desert, and considered "one of the planet's worst environmental disasters". The once-prosperous lake-related economic activities have completely disappeared. The region is heavily polluted, with consequent serious public health problems. The local climate has worsened considerably. The Memory of the World Register by UNESCO now lists the Aral Sea as an "environmental tragedy."

Riparian countries adopted already in 1994 an Aral Sea Basin Program. But efforts are being done only to (partly) restore the North Aral Sea. The 13 km Kok-Aral dam, completed in 2005 by Kazakhstan and financed by the World Bank, allows water of the Syr Darya to accumulate, and helps restore delta and riverine wetland ecosystems of the northern basin. The South Aral Sea, half of which lies in Uzbekistan, has been largely abandoned to its fate.

Excess water from the North Aral Sea periodically flows into the dried-up South Aral Sea through a sluice in the dike. Opening a channel between improved North and desiccated South, or redirecting the Amu Darya residual flow to the western basin, may salvage fisheries there, while relieving the flooding of the eastern basin. Here, attempts to mitigate desertification by planting vegetation on the exposed seabed are underway, but oil exploration has also been started.

Lake Balkhash is one of the largest semi-saline lakes in Asia (and the 15th largest in the world), at the heart of an endorheic basin in the vast Balkhash-Alakol depression (shared with other minor lakes, remnants of an ancient sea which once covered the entire depression), in Central Asia (Fig. 9). Seven rivers drain into the lake, the primary of which being the Ili, flowing from the mountains of China's Xinjiang region. The lake currently covers an area of about 16,400 km^2.

The area and volume of the lake vary due to both long-term and short-term fluctuations. The basin is about 600 km long, and is hydrographically divided into two different parts, eastern and western, by the Saryesik Peninsula. The width is 9–19 km in the eastern part, deeper and saltier, and up to 74 km in the western part, shallower and fresher. The average depth of the lake is almost 6 m (16 in the eastern basin), while the maximum depth (again in the eastern part) is 26 m.

The lake is elevated about 340 m above sea level, and its climate is continental. The windy, dry climate and high summer temperatures result in a high evaporation rate. The prevailing winds induce a steady clockwise circulation (at least in the western part). Surface water temperature varies from 0 °C in winter to 28 °C in summer, with annual means of 9 °C in the east and 10 °C in the west. Average salinity is about 3 g/kg. The lake freezes every year from November to April.

Lake Balkhash used to have rich flora and fauna, but in the 1970s biodiversity began to decline due to deterioration of water quality. Abundant reeds in the southern part of the lake, especially in the delta of Ili River, served as a haven for birds and other animals. Changes in the water level led to the degradation of the delta, reducing wetlands and riparian forests. Land development, use of pesticides, overgrazing and deforestation also contributed to the decrease in biodiversity.

The economic importance of the lake is mostly in its fishing. Hydroelectric power is generated on the Ili River, while river waters are drawn out of the upstream Kapchagay Reservoir for irrigation. There are serious concerns about the diversion of water from rivers that feed the lake, especially in the view of repeating the environmental disaster at the Aral Sea. Other factors affecting local ecology are the emissions due to mining and metallurgical processes.

Lake Baikal, located in the central part of southern Siberia, is a rift lake—created by the Baikal Rift Zone, where the Earth's crust is slowly pulling apart—reputed to be the largest in the world, holding 22–23% of the planet's freshwater, and one of the oldest, with an estimated age of 20–25 million years (Fig. 9). Owing to its age and long-standing isolation, it is called the "Galapagos of Russia", home to more endemic species of flora and fauna than any other lake on Earth.

At 636 km long and 79 km wide, Lake Baikal is the 7th largest lake in the world, with a total surface area of 31,722 km^2, but it also the world's deepest lake, with an average depth of 744 m, and a maximum depth of 1642 m, which makes it the world's largest lake by volume (23,615.39 km^3). The bottom of the lake is 1186.5 m below sea level, but below this are 7 km of sediments, placing the rift floor 8–11 km below the surface: the deepest continental rift on Earth.

In geological terms, the rift is young and active (widening $\sim$2 cm per year). The fault zone is also seismically active; hydrothermal vents with water up to 50 °C occur in the area, and severe earthquakes occur occasionally. The lake is fed by as many as 330 inflowing rivers. The main ones draining directly into it are the Selenga, the Barguzin, the Upper Angara, the Turka, the Sarma, and the Snezhnaya. It is drained through a single outlet, the Angara, a tributary of the Yenisei.

The local climate is milder than that of the surrounding region. Nevertheless, the entire lake surface freezes from January to May–June. The surface water temperature in August rises to 10–12 °C, and can reach 20 °C in shallow areas. In the upper 300 m, temperature increases with depth in winter, and decreases with depth in summer, with homeothermic conditions in June and November. In spite of the great depth, the lake's waters are well mixed and oxygenated, and very clear.

The lake is rich in biodiversity. It hosts more than 1000 species of plants and 2500 species of animals, but the actual figures for both groups are believed to be significantly higher. More than 80% of the species are endemic. The only true freshwater seal species—the Baikal seal, or nerpa (*Pusa sibirica*)—lives in the basin, constituting one of only three entirely freshwater seal populations in the world (the other two being subspecies or colonies of regular saltwater seals).

Lake Baikal was declared a UNESCO World Heritage Site in 1996. It is also home to the indigenous Buryat tribes, which reside on the eastern shores. Local economic activities include fisheries, shipbuilding, mining, timber and the manufacture of cellulose and paper, which renders pollution a growing concern. Tourism has also been on the rise, owing to the area's natural beauties and to the many mineral springs, attracting visitors for the curative properties of their waters.

6 Summary and Outlook

Most of the basins surrounding the Asian continent can be considered as true marginal seas, contiguous to an outer ocean, enclosed in wide gulfs or partially bounded by peninsulas, island chains or submarine features (e.g. the Seas along the Pacific Ocean western edge). Some are dominated by major wind-driven oceanic currents (e.g. the Arabian Sea or Bay of Bengal, in the northern Indian ocean). Others are disconnected from the outer ocean, with which they have limited exchanges (e.g. the Persian Gulf, the Red Sea), or none at all (e.g. the Caspian Sea), and are essentially controlled by thermohaline processes. In addition, a number of inland freshwater bodies (e.g. Lake Baykal), albeit not marine, should also be included among the major enclosed basins of the Asian continent.

An overview of these basins can only start to sketch out the range of environmental conditions of the Asian continental margins. Understanding the inner workings of the Asian Seas—aiming to preserve their ecological balance, but supporting also the sustainable exploitation of their natural resources—requires appropriate observation systems. Data collection with surface campaigns remains a viable component of much marine research. However, the need to assess large scale marine processes and the prohibitive cost of expeditions at sea—not to mention the recurrence of piracy or political unrest in many coastal countries—call for a growing role of Earth's observations, which offer a wide range of possibilities, complementing in situ data gathering, for the large-scale, long-term assessment of natural settings, water exchanges, basic ecological mechanisms and their main driving forces, as well as anthropogenic impacts on coastal and marine regions.

While the Asian Seas have been probed for a long time, key issues still remain to be studied in detail. The appraisal of surface parameters by means of passive remote sensing techniques—which measure reflected visible and near-infrared sunlight, or surface emissions in the thermal infrared or microwave spectral regions—or of active techniques—which use transmitted impulses of visible or microwave radiation, for a subsequent evaluation of the signal returned by the water surface—promises to help closing the remaining gaps. An in-depth analysis of the specific merits and drawbacks of each technique and spectral region can provide clues to help compose the unique environmental mosaic of the Asian Seas.

Acknowledgements The present review was carried out at the Joint Research Centre (JRC) of the European Commission (EC), as part of research undertaken in the Horizon 2020 Action SEACOAST. The author wishes to express his gratitude to all JRC Colleagues who contributed to this activity. Thanks are also due to M. Gade, Institute of Oceanography, University of Hamburg, for his editorial review of the final paper, and to two anonymous reviewers, whose input allowed the author to improve substantially several sections of the original paper. Images and data were derived, in part, from the Global Marine Information System (GMIS) of the JRC EC.

Bibliography

Introduction, General

Corso W, Joyce PS (1995) Oceanography. Applied science review. Springhouse Corporation, Springhouse, p 182
Emery WJ (2001) Water types and water masses. In: Steele JH, Thorpe SA, Turekian KK (eds) Encyclopedia of ocean sciences, vol 4. Academic Press, San Diego, pp 3179–3187
Grant Gross M (1995) Principles of oceanography, 7th edn. Prentice-Hall Inc., Englewood Cliffs, p 230
International Hydrographic Organization (1953) Limits of oceans and seas. Special publication no. 23, 3rd edn. Imp. Monégasque, Monte Carlo, p 39
Longhurst AR (1998) Ecological geography of the sea. Academic Press, San Diego, p 398
Mann KH, Lazier JRN (2006) Dynamics of marine ecosystems: biological-physical interactions in the oceans. Blackwell Publishing Ltd., Oxford, p 512
Neumann G, Pierson WJ (1966) Principles of physical oceanography. Prentice-Hall, New Jersey, p 1087
Paine L (2015) The sea and civilization: a maritime history of the world. Alfred A. Knopf, Random House LLC, New York; Random House of Canada Ltd., Toronto, p 784
Pedlosky J (1996) Ocean circulation theory. Springer, Berlin, p 466
Pinet PR (1992) Oceanography: an introduction to the planet oceanus. West Publishing Company, St. Paul, p 572
Rahmstorf S (2003) Thermohaline circulation: the current climate. Nature 421(6924):699
Schwartz ML (ed) (2005) Encyclopedia of coastal science. Springer, Dordrecht, p 1213
Sverdrup HU, Johnson MW, Fleming RH (1942) The oceans: their physics, chemistry and general biology. Prentice-Hall Inc., Englewood Cliffs, p 748
Sverdrup KA, Duxbury AC, Duxbury AB (2005) An introduction to the world's oceans, 8th edn. McGraw-Hill Publishers, Boston, p 514
Tomczak M, Godfrey JS (2003a) Regional oceanography: an introduction, 2nd improved edn. Daya Publishing House, New Delhi, p 390

Arctic Ocean Basins

Aagaard K, Carmack EC (1989) The role of sea ice and other fresh water in the arctic circulation. J Geophys Res 94(C10):14485–14498
Alexandrov V, Martin T, Kolatschek J, Eicken H, Kreyscher M, Makshtas AP (2000) Sea ice circulation in the Laptev Sea and ice export to the Arctic Ocean: results from satellite remote sensing and numerical modeling. J Geophys Res 105(C7):17143–17159
Comiso JC, Parkinson CL, Gersten R, Stock L (2008) Accelerated decline in the Arctic sea ice cover. Geophys Res Lett 35(1):L10703. https://doi.org/10.1029/2007GL031972
Grantz A, Johnson L, Sweeney JF (eds) (1990) The arctic ocean region. DNAG series, vol L. Geological Society of America, The Geology of North America, Boulder, p 644
Herman Y (ed) (1989) The arctic seas: climatology, oceanography, geology, and biology. Van Nostrand Reinhold Company, New York, p 888
Jackson HR, Johnson GL (1986) Summary of arctic geophysics. J Geodyn 6(1–4):245–262
Jones EP, Anderson LG, Swift JH (1998) Distribution of atlantic and Pacific waters in the upper Arctic Ocean: implications for circulation. Geophys Res Lett 25(6):765–768

Maykut GA (1982) Large-scale heat exchange and ice production in the central arctic. J Geophys Res 87(C10):7971–7984

Semiletov I, Dudarev O, Luchin V, Charkin A, Shin KH, Tanaka N (2005) The East Siberian Sea as a transition zone between Pacific-derived waters and Arctic shelf waters. Geophys Res Lett 32(10):L10614. https://doi.org/10.1029/2005GL022490

Pacific Ocean Basins

Centurioni LR, Niiler PP, Lee DK (2004) Observations of inflow of Philippine Sea Surface Water into the South China Sea through the Luzon Strait. J Phys Oceanogr 34:113–121

Chang KI, Zhang CI, Park C, Kang DJ, Ju SJ, Lee SH, Wimbush M (eds) (2015) Oceanography of the East Sea (Japan Sea). Springer International Publishing, New York, p 460

Favorite F, Dodimead AJ, Nasu K (1976) Oceanography of the subarctic Pacific region, 1960–71. International North Pacific Fisheries Commission Bulletin, vol 33, p 187

Guan B (1994) Patterns and structures of the currents in Bohai, Huanghai and East China Seas. In: Di Z et al (eds) Oceanology of China Seas, vol 1. Kluwer, Norwell, pp 17–26

Ichikawa H, Beardsley RC (2002) The current system in the Yellow and East China Seas. J Oceanogr 58:77–92

Isobe A (2008) Recent advances in ocean-circulation research on the Yellow Sea and East China Sea shelves. J Oceanogr 64(4):569–584

Kim ST (2012) A review of the Sea of Okhotsk ecosystem response to the climate, with special emphasis on fish populations. ICES J Mar Sci 69(7):1123–1133

Lie HJ, Cho CH (2002) Recent advances in understanding the circulation and hydrography of the East China Sea. Fish Oceanogr 11:318–328

Morton B, Blackmore G (2001) South China Sea. Mar Pollut Bull 42(12):1236–1263

Qu T, Du Y, Meyers G, Ishida A, Wang D (2005) Connecting the tropical Pacific with Indian Ocean through South China Sea. Geophys Res Lett 32(24):L24609, p 4

Reed RK (1990) A year-long observation of water exchange between the North Pacific and the Bering Sea. Limnol Oceanogr 35:1604–1609

Stabeno PJ, Schumacher JD, Ohtani K (1999) The physical oceanography of the Bering Sea. In: Loughlin TR, Ohtani K (eds) Dynamics of the Bering Sea. University of Alaska Sea Grant, Fairbanks, pp 1–28

Su YS, Weng XC (1994) Water masses in the China Seas. In: Di Z et al (eds) Oceanology of China Seas, vol 1, Kluwer, Norwell, pp 3–16

Indian Ocean Basins

Barale V, Gade M (2014) Basic ecosystem dynamics in the Red Sea as seen by sundry remote sensing techniques. In: Barale V, Gade M (eds) Remote sensing of the African Seas. Springer, Heidelberg, pp 337–355

Bower AS, Hunt HD, Price JF (2000) Character and dynamics of the Red Sea and Persian Gulf outflow. J Geophys Res 105(C3):6387–6414

Chatterjee A, Shankar D, Shenoi SSC, Reddy GV, Michael GS, Ravichandran M, Gopalkrishna VV, Rao EPR, Bhaskar TVSU, Sanjeevan VN (2012) A new atlas of temperature and salinity for the North Indian Ocean. J Earth Syst Sci 121:559–593

Chatterjee A, Shankar D, McCreary JP, Vinayachandran PN, Mukherjee A (2017) Dynamics of Andaman Sea circulation and its role in connecting the equatorial Indian Ocean to the Bay of Bengal. J Geophys Res 122(4):3200–3218

Chen G, Han W, Li Y, Wang D (2016) Interannual variability of eastern Indian Ocean upwelling: local versus remote forcing. J Phys Oceanogr 46:789–807
Desai BN (ed) (1992) Oceanography of the Indian Ocean. A.A. Balkema, Rotterdam, p 770
Madhupratap M, Nair KNV, Gopalakrishnan TC, Haridas P, Nair KKC, Venugopal P, Gauns M (2001) Arabian Sea oceanography and fisheries of the west coast of India. Curr Sci 81(4): 355–361
Panikkar NK, Jayaraman R (1966) Biological and oceanographic differences between the Arabian Sea and the Bay of Bengal as Observed from the Indian Region. Proc Ind Acad Sci 64:231–240
Potemra JT, Luther ME, O'Brien JJ (1991) The seasonal circulation of the upper ocean in the Bay of Bengal. J Geophys Res 96:12667–12683
Shankar D, Vinayachandran PN, Unnikrishnan AS (2002) The monsoon currents in the north Indian Ocean. Prog Oceanogr 52:63–120
Shetye SR, Gouveia AD (1998) Coastal circulation in the north Indian Ocean—coastal segment 14, S-W. Sea 11:523–556
Siedler G (1969) General circulation of water masses in the Red Sea. In: Degens ET, Ross DA (eds) Hot brines and recent heavy metal deposits in the Red Sea. Springer, Berlin, pp 131–137
Varkey MJ, Murty VSN, Suryanarayana A (1996) Physical oceanography of the Bay of Bengal and Andaman Sea. Oceanogr Mar Biol Annu Rev 34:1–70

Enclosed Basins

Boomer I, Aladin N, Plotnikov I, Whatley R (2000) The palaeolimnology of the Aral Sea: a review. Quatern Sci Rev 19(13):1259–1278
Hampton SE, Izmesteva LR, Moore MV, Katz SL, Dennis B, Silow EA (2008) Sixty years of environmental change in the world's largest freshwater lake—Lake Baikal, Siberia. Glob Change Biol 14:1947–1958
Ibrayev RA, Ozsoy E, Schrum C, Sur HI (2010) Seasonal variability of the Caspian Sea three-dimensional circulation, sea level and air-sea interaction. Ocean Sci 6:311–329
Kideys AE (2002) Fall and rise of the Black Sea ecosystem. Science 297(5586):1482–1484
Kostianoy AG, Kosarev AN (eds) (2005) The Caspian Sea environment. In: Handbook of environmental chemistry, vol 5P, Springer, Berlin, p 271
Kozhov M (1963) Lake Baikal and its life. Springer Science+Business Media, Dordrecht, p 344
Micklin Ph (2010) The past, present, and future Aral Sea. Lakes Reservoir Res Manage 15(3): 193–213
Ozsoy E, Unluata U (1997) Oceanography of the Black Sea: a review of some recent results. Earth Sci Rev 42(4):231–272
Petr T (1992) Lake Balkhash, Kazakhstan. Int J Salt Lake Res 1(1):21–46
Pinardi N, Masetti E (2000) Variability of the large scale general circulation of the Mediterranean Sea from observations and modelling: a review. Palaeogeogr Palaeoclimatol Palaeoecol 158(3–4):153–173
Propastin PA (2008) Simple model for monitoring Balkhash Lake water levels and Ili River discharges: application of remote sensing. Lakes Reservoir Res Manage 13:77–81
Verkhozina VA, Kozhova OM, Kusner YS (2000) Hydrodynamics as a limiting factor in the Lake Baikal ecosystem. Aquat Ecosyst Health Manage 3(2):203–210

Summary & Outlook

Barale V, Gade M (eds) (2008) Remote sensing of the European Seas. Springer, New York, p 514
Barale V, Gade M (eds) (2014) Remote sensing of the African Seas. Springer, New York, p 428
Bhatt JJ (1978) Oceanography. Exploring the planet ocean. D. Van Nostrand Company, New York, p 322
Peterson RG, Stramma L, Kortum G (1996) Early concepts in charts and circulation. Prog Oceanogr 37:1–115
Tomczak M, Godfrey JS (2003b) Regional oceanography: an introduction, 2nd improved edn. Daya Publishing House, New Delhi, p 390

A Historical Overview of Remote Sensing of the Asian Seas

Leonid M. Mitnik and Martin Gade

Abstract In this chapter we give an overview of national and international satellite missions that have been developed over the past decades. After first earth observation (EO) missions had started in the 1960s, the first remote sensing satellite observing, among others, the Asian Seas were launched in the 1970s. Nowadays, a wealth of earth-orbiting satellites is in place, carrying a variety of sophisticated sensors that can be used for routine surveillance, but also to gain deeper insight into specific processes in the marine environment. While most recent sensors and applications will be described in the following book chapters, here, we focus on historical sensors and missions, and we provide some examples of Remote Sensing of the Asian Seas, in the early days of spaceborne missions.

Keywords History · Satellite mission · Earth Observation · USA Soviet Union · Europe · Japan · India · China · South Korea

1 Introduction

Routine atmospheric and oceanic remote sensing, at first from aircraft and then from satellites, has developed mostly since World War II. In the 1970s and 1980s the USA and USSR designed and launched several new passive and active sensors that started a new era of the retrieval of valuable quantitative information on parameters and processes in the ocean-atmosphere system. Among those sensors were multichannel scanning microwave (MW) radiometers, real aperture radar (RAR) and synthetic aperture radar (SAR), scatterometers, and altimeters (Basharinov et al. 1974; Fu and Holt 1982; Karlin 1999; Mitnik and Victorov 1990; Kramer 2002).

L. M. Mitnik (✉)
V.I. Il'ichev Pacific Oceanological Institute FEB RAS, Vladivostok, Russia
e-mail: mitnik@poi.dvo.ru

M. Gade
Institut für Meereskunde, Universität Hamburg, Hamburg, Germany

V. Barale and M. Gade (eds.), *Remote Sensing of the Asian Seas*,
https://doi.org/10.1007/978-3-319-94067-0_2

From the late 1980s space agencies, organizations, and programs were also created in Europe and in Asian countries such as Japan, India, Republic of China, and South Korea. One of the main tasks of space programs in Asian countries was the development of sensors and remote sensing satellites for the observation not only of the World's Oceans, but also of the coastal zones and open sea areas surrounding these countries. This strongly influenced the selection of sensor characteristics (number of channels, spectral bands, spatial resolution, swath width, etc.) and satellite orbit parameters (low orbiting or geostationary, inclination, time of equator crossing, etc.).

New remote sensing instruments were developed and new sun-synchronous low-orbiting and geostationary satellites were launched by several Asian countries and were supported by sophisticated ground facilities and application developers. They include optical systems with high spatial resolution, some of which with stereoscopic capacity, multispectral meteorological and oceanographic sensor systems, and SAR. Sensors operating in the visible, IR and microwave (MW) range became the main source of spatio-temporal information on ocean colour, sea surface temperature (SST), wind speed, marine pollution, dynamic phenomena in the coastal zones and in the open ocean, on sea ice, as well as on hazardous marine weather systems such as tropical, extratropical and polar cyclones, etc.

The next section briefly summarizes the early satellites built in the USA and Soviet Union from the 1960s, the section thereafter gives an overview of European satellite missions that complemented the American and Soviet missions from the 1980s. Finally, Asian satellite missions devoted to the observation of the Asian Seas are introduced in Sect. 4. For the sake of shortness we won't introduce other national satellite missions, e.g. from Canada (Radarsat-1/2), Argentina (SAC-C/D), or Brazil (CBERS-1-4, jointly with China), nor will we describe the joint American/European Shuttle Missions (SIR-A/B/C) in the 1980s and 1990s.

2 The Early Days: U.S. and Soviet Missions

The early operational satellites carried sensors working in the visual and IR ranges of the electromagnetic spectrum. The first one, the American Television and Infrared Observation Satellite (TIROS)-I, was launched on 1 April 1960. TIROS data were mainly used for meteorological observations, and so were data from the Application Technology Satellites (ATS) that went to replace the TIROS series in 1966 and that acquired cloud distribution images over the next 10 years.

To observe the Earth's surface under cloudy conditions highly sensitive MW radiometers were developed in the Soviet Union and USA. The first two Soviet satellites, each carrying four MW radiometers into space, were Cosmos-243, launched on 23 September 1968, and Cosmos-384, launched on 10 December 1970. The radiometers operated at frequencies between 3.5 GHz and 37 GHz and allowed retrieving SST, near-surface wind speed, total atmospheric water vapour content, total cloud liquid water content, and sea ice parameters (Basharinov et al. 1974; Mitnik 1972). Figure 1 shows the total atmospheric water vapour field over the Asian marginal seas retrieved from Cosmos-243 measurements on 23/24 September 1968 (Gurvich and Demin 1970). Note the typical features of the atmospheric water vapour field, the low values over the Japan and Yellow Seas (10–15 kg/m^2), associated with an inflow of dry air from the Asian continent, the increase in water vapour content from 15 km/m^2 to 50 kg/m^2 over the East China and South China Seas, where the SST increases from 25 to 30 °C, and the enhanced water vapour content around zones with heavy clouds and precipitation. These features were later confirmed by subsequent MW satellite observations from the American Nimbus-7 satellite (Chang et al. 1984).

Probably the most important role in the field of remote sensing of the Asian seas in visible and IR ranges has been playing the American Advanced Very High Resolution Radiometer (AVHRR) flown on satellites of the U.S. National Oceanographic and Atmospheric Administration (NOAA) since October 1978. The first generation, AVHRR/1 flown on TIROS-N and NOAA-6, carried out observations at four spectral bands, while subsequent AVHRR/2 radiometers operated at five spectral channels (from 1981, flown on NOAA-7 to NOAA-14). The third generation, AVHRR/3, is flown since the launch of NOAA-15 in May 1998 and operates at six spectral channels in the visible (red) and (near to far) IR range.

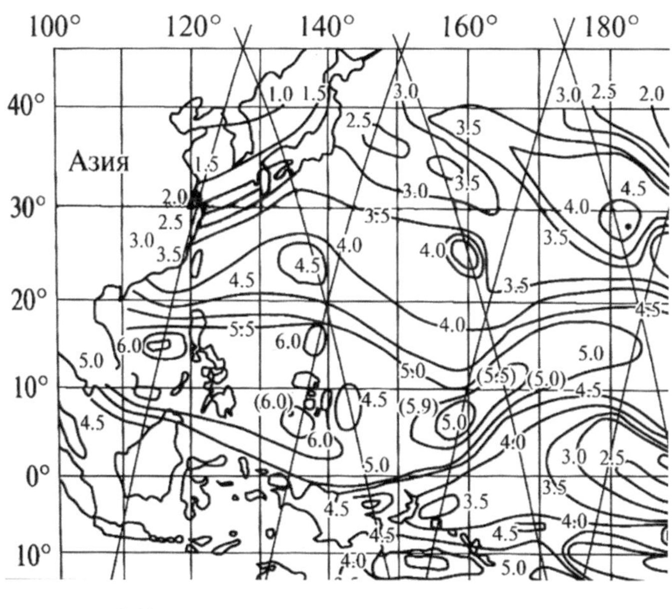

Fig. 1 Water vapour field over the Northwestern Pacific Ocean derived from Cosmos-243 microwave measurements. Slanting lines are the satellite tracks for 23–24 September 1968, numbers near the isolines are precipitable water values in millimetre (1 mm = 10 kg/m^2) (Gurvich and Demin 1970)

AVHRR data have been serving primarily for the retrieval of SST and to study air-sea interactions and the spatial and temporal variability of oceanic dynamic phenomena such as upwellings, fronts, currents, eddies, etc. Figure 2 shows the mean SST in the northern Indian Ocean during the north-east monsoon in January/February (upper row) and south-west monsoon (lower row), for the years 1982 (left column) and 1983 (right column). Clearly visible are regions of coastal upwelling off the Pakistani and Bangladesh coasts in January/February, and along the southern coast of the Arabian Peninsula in July/August.

The growing demand for measurements of chlorophyll concentration, gelbstoff, and particulate matter in the upper water column lead to the development of ocean colour sensors, which are primarily measuring the reflectance in the visual part of the spectrum (which in turn makes them dependent on cloud cover) and whose first sensor in space was the American Coastal Zone Color Scanner (CZCS) aboard NIMBUS-7, launched on 24 October 1978.

CZCS measured in six spectral bands, and examples of first CZCS measurements of chlorophyll concentration in South Asian waters are shown in Fig. 3, in the upper panel for winter (21 Dec–20 Mar) 1978 and in the lower panel for summer (21 Jun–20 Sep) 1979. Main differences can be seen in the Arabian Sea, where the south-west monsoon in summer causes an upwelling of nutrient-rich water, thereby leading to extensive algae blooms. CZCS operated until June 1986. Its follow-on was the Sea-viewing Wide Field-of-View Sensor (SeaWiFS), launched in August 1997, that measured in eight spectral bands and that operated until December 2010. At present observations in 36 visible and infrared bands are carried out by the

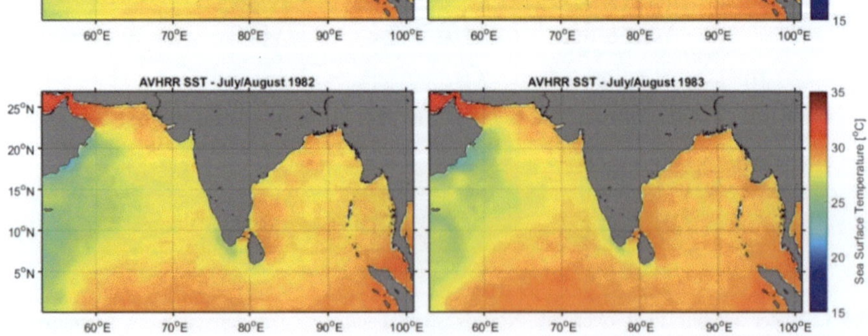

Fig. 2 Mean sea surface temperature derived from AVHRR/2 data. Upper row: January/February; lower row: July/August. Left column: 1982; right column: 1983. *Source* NOAA (2007)

MODerate-resolution Imaging Spectroradiometer (MODIS) on the Terra (Earth Observing System (EOS) AM-1) and on the Aqua (EOS PM-1) satellites that were launched in December 1999 and in May 2002, respectively.

The first earth-orbiting satellite particularly designed for the remote sensing of the Earth's oceans was the American SEAfaring SATellite (Seasat), which was launched on 26 June 1978. Seasat carried a 13.56 GHz radar altimeter (for measurements of sea surface height, SSH), a 14.599 GHz MW scatterometer (ocean winds), MW radiometer (SST), visible and IR radiometer (atmospheric, land and oceanic features), and an L-band SAR operating at 1.275 GHz and horizontal (HH) polarization. Although it ended after only 106 days (on 10 October 1978, likely because of a power failure), the Seasat mission paved the road for a number of later satellites such as TOPEX/Poseidon, QuikSCAT, JASON, Geosat and ERS-1/2.

The most spectacular results Seasat delivered in its short lifetime were probably the L-band SAR images, which, for the first time, showed particular oceanic

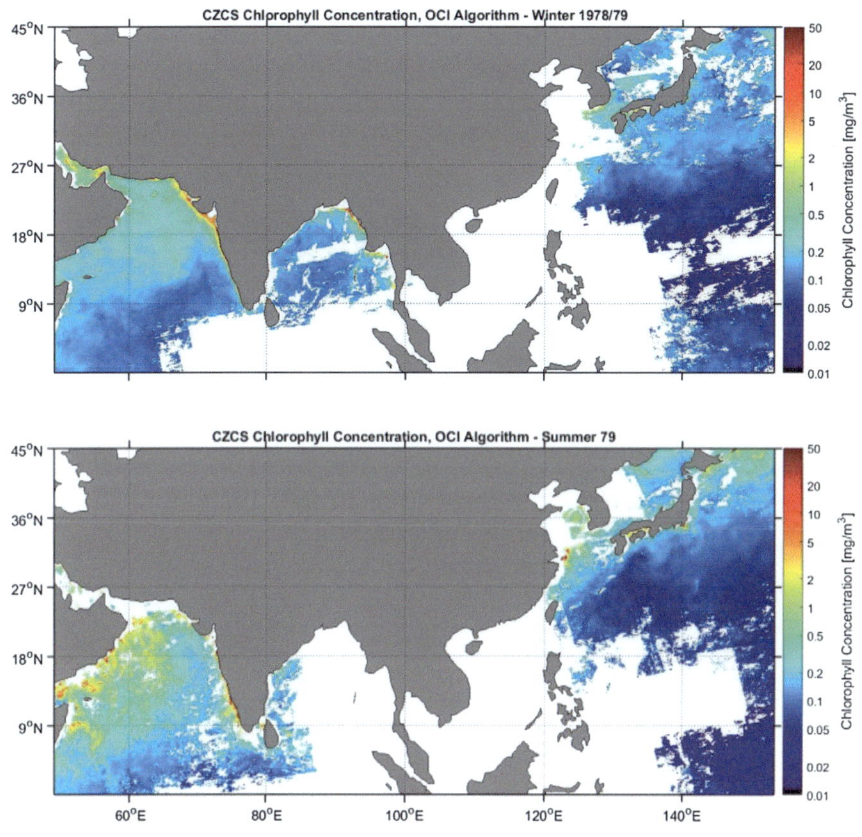

Fig. 3 Comparison of seasonal mean chlorophyll *a* concentrations in South Asian waters measured by CZCS. Upper: winter 1978/79; lower: summer 1979. *Source* NASA (2014)

features such as internal waves, sea ice, marine surface films, underwater bottom topography, oceanic fronts, etc. at high spatial resolution (Fu and Holt 1982). Figure 4 is a composite of three Seasat SAR images (each 100 km × 100 km) of Wrangel Island, between the Chukchi Sea and the East Siberian Sea, acquired on 9 October 1978. Dark areas in the upper image part are grease ice, which is flattening the sea surface seen by the SAR and which is driven by the north-easterly wind. Parallel features in the lower left part are imprints of atmospheric lee waves behind Wrangel Island.

The Soviet OKEAN program was initiated in 1976 and its main objective was the use of active and passive MW sensors for spaceborne ocean remote sensing. The first non-experimental satellite of this series, KOSMOS-1500 (OKEAN-OE-N1), was launched in September 1983 and carried three different sensors that allowed simultaneous measurements, namely an X-band Side-Looking Real Aperture Radar (RAR) (RLSBO), a Side-Looking Scanning MW Radiometer, and the Radio and TV Komplex (RTVK) working at four visible bands (Mitnik and Victorov 1990).

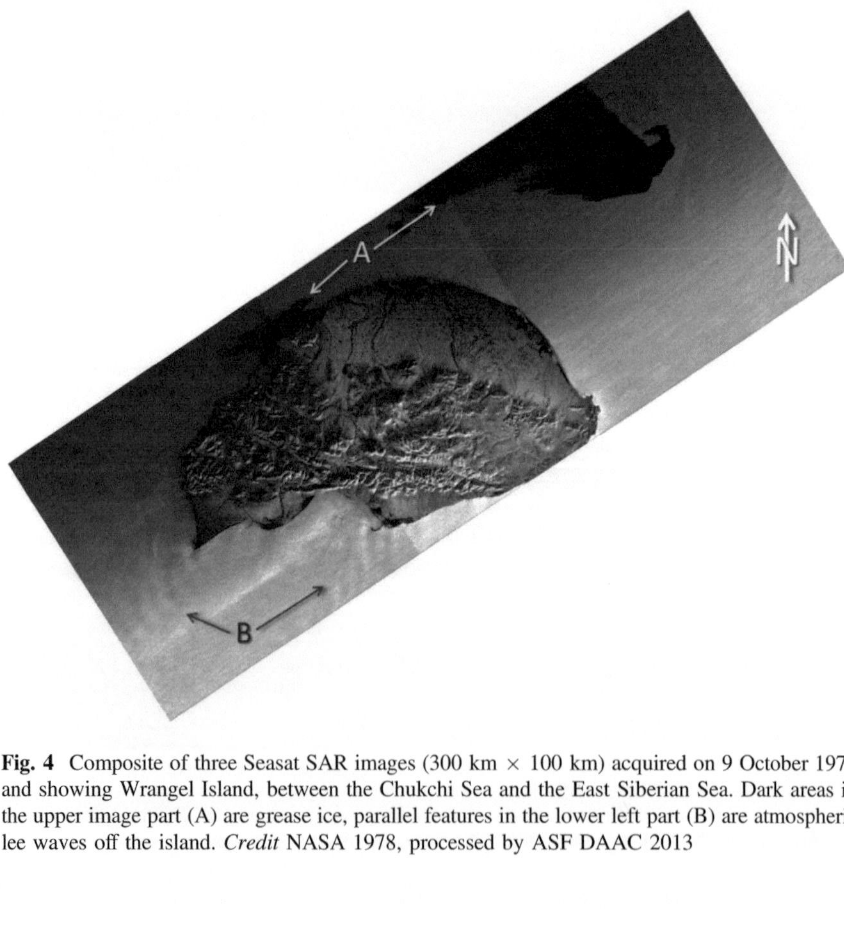

Fig. 4 Composite of three Seasat SAR images (300 km × 100 km) acquired on 9 October 1978 and showing Wrangel Island, between the Chukchi Sea and the East Siberian Sea. Dark areas in the upper image part (A) are grease ice, parallel features in the lower left part (B) are atmospheric lee waves off the island. *Credit* NASA 1978, processed by ASF DAAC 2013

The Russian OKEAN-7 was launched in October 1994 and carried, among others, a Multispectral Scanner of Moderate Resolution (MSU-SK), working at five spectral bands and again the real aperture radar RLSBO. Simultaneous acquisitions of both sensors allowed comparative analyses of atmospheric and oceanic processes, as shown in Fig. 5: an outbreak of cold air in the East and South China Seas manifests in cloud structures visible in the optical imagery, but also in sea surface roughness patterns visible in the RAR image (area 1).

The first Soviet spaceborne SAR systems, both operating at 3.125 GHz (S-band) and horizontal (HH) polarization, were flown on KOSMOS-1870 (launched in July 1987, mission end in July 1989) and ALMAZ-1 (March 1991/October 1992) (Karlin 1999). During the two missions the SAR sensors, Ekor ('Sword') −A and −A1, respectively, were operating from both sides of the spacecraft, and their off-nadir look angle could be changed by rolling the spacecraft about its flight axis (which in turn consumed much fuel, thereby limiting the satellite mission's lifetime). An example of an ALMAZ-1 SAR image is shown in Fig. 6. The image of Tachibana Bay, Japan was acquired on 11 August 1991. Inside the bay (A) fine dark structures are due to surface films, white spots in the upper right part (B) are due to ships, and the bright irregular features in the lower part (C) are due to tidal currents entering the bay through the Hayasakiseto Strait.

3 All-Inclusive: The Early European Satellite Missions

The European Space Agency, ESA, was founded in 1975 (entry into force 1980). After early satellite missions had focussed on space research, ESA's first European Remote Sensing satellite, ERS-1, was launched on 17 July 1991. Its sister satellite, ERS-2, followed on 21 April 1995, and the satellites were operational until March 2000 and July 2011, respectively. The payloads of both satellites were primarily designed to monitor ocean processes at global and meso-scales, as well as the ice coverage in polar regions. The ERS were the first long-term satellite missions involving a large set of active and passive sensors working in the visual and MW spectral ranges.

Both satellites carried the Along-Track Scanning Radiometer (ATSR), which (together with its successors) has provided the most accurate SST measurements so far, thanks to the conical scanning approach and the active mechanical Stirling cooling system. An example of early ATSR data acquired on 1 September 1993 over the East China Sea is shown in Fig. 7: the typhoon *Yancy* was imaged on its way to Japan and can be seen in the center of each strip, representing the ATSR 1.6 μm channel (left), 11 μm channel (middle), and 12 μm channel (right).

ERS-1/2 also carried a 4-channel IR radiometer for measurements of the Earth's surface temperature, a 2-channel nadir-viewing Microwave Radiometer (MWR) for measurements of atmospheric water vapour and cloud liquid water, to improve the accuracy of SST measurements and to provide tropospheric range correction for altimeter measurements, a 13.8 GHz (Ku-band) radar altimeter (RA), and the

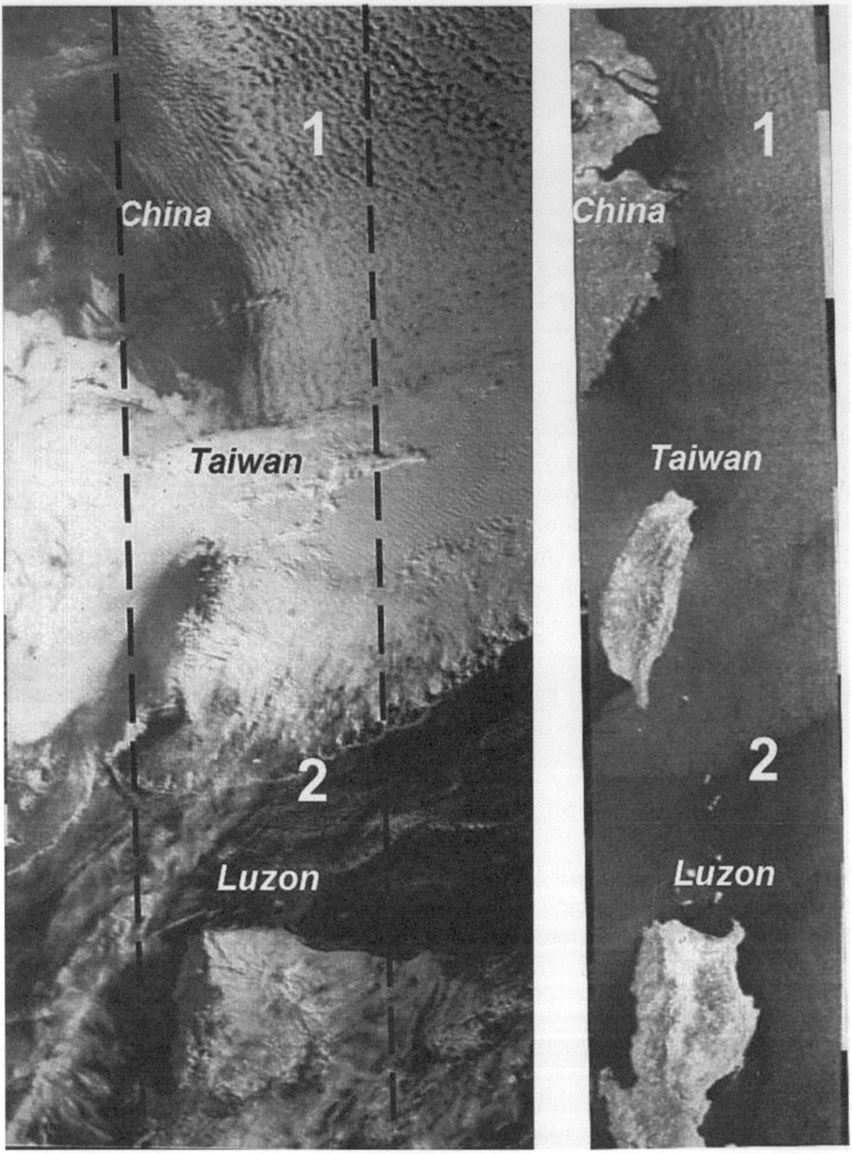

Fig. 5 OKEAN-7 visible (left) and real aperture radar (right) images acquired on 8 January 1996 during an outbreak of cold air over the East and South China Seas associated with mesoscale convective rolls and cells (area 1). The squall line (2) marks the boundary between strong and weak wind areas. The dashed lines in the left panel show the RAR image boundaries

Active Microwave Instrument (AMI), which in turn consisted of a C-band Wind Scatterometer and SAR operating at 5.3 GHz. AMI had three operating modes, the SAR imaging mode, the wind scatterometer mode and the so-called wave mode,

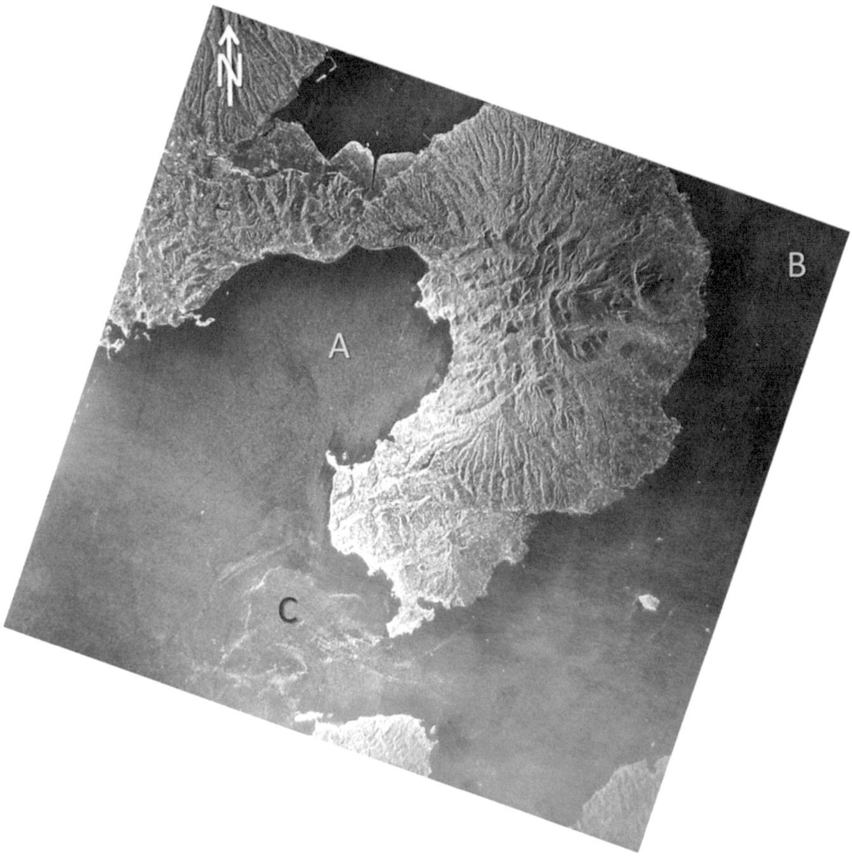

Fig. 6 ALMAZ-1 SAR image acquired on 11 August 1991 showing Tachibana Bay (A), east of Nagasaki, Japan. Bright spots in the upper right corner (B) are due to ships, bright features in the lower part (C) due to tidal currents

in which small (5 km × 5 km) imagettes were recorded every 200 km for ocean wave retrieval. Precise measurements of the marine geoid were made from May 1995 until February 1996, when ERS-1 was put into a (167 days) geodetic orbit.

The European Polar Orbiting Earth Observation Mission (POEM-1) was initiated by ESA in 1991. In 1993, ESA decided to split the mission into the Environmental Satellite, EnviSat, and the Meteorology Operational Programme (MetOp). EnviSat, which was launched on 1 March 2002, followed along the lines of the ERS missions and carried a large set of sensors, including the Advanced ATSR (AATSR), Advanced SAR (ASAR, operating at 5.331 GHz), the 15-band Medium Resolution Imaging Spectrometer (MERIS), Radar Altimeter (RA)-2, a MW Radiometer (MWR), plus a set of sensors for atmospheric studies. Similar to ERS-1/2 EnviSat acquired imagettes in wave mode every 100 km. EnviSat, whose mission ended after a sudden power failure in April 2012, was the largest EO satellite in history.

Fig. 7 ERS-1 ATSR data (500 km × 4200 km) of 1 September 1993, showing the typhoon *Yancy* over the East China Sea. Left: 1.6 μm channel, middle: 11 μm channel, right: 12 μm channel

An example of total suspended matter (TSM) in the Bay of Bengal, as derived from EnviSat MERIS data of 25 April 2003, is shown in Fig. 8. Off the mouths of the Ganges high TSM loads exceeding 40 g/m^3 were measured (red colours). In the open Bay of Bengal two oceanic eddies of approx. 350 km diameter can be seen, with higher concentrations of (advected) TSM at their rims.

ESA's Soil Moisture and Ocean Salinity (SMOS) mission was launched in November 2009 and uses a 2D interferometric 1.4 GHz (L-band) radiometer to provide high-accuracy sea surface salinity (SSS) measurements. Following up on

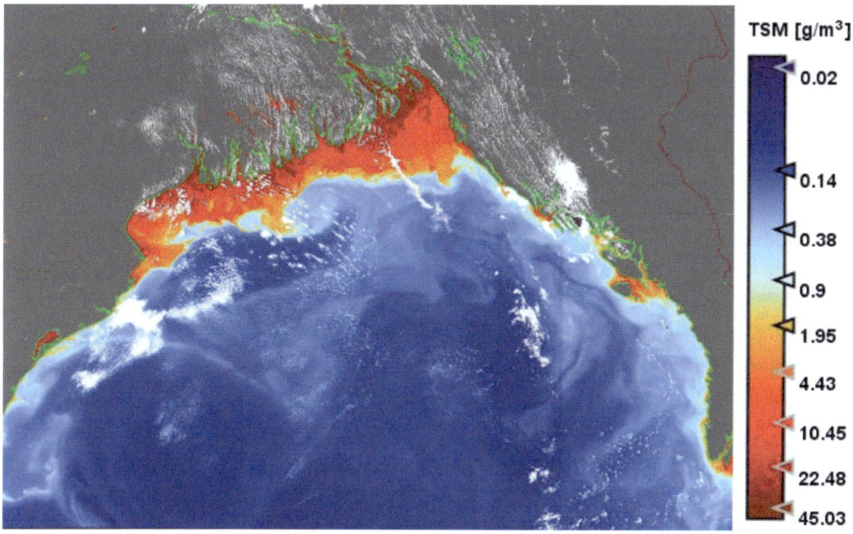

Fig. 8 Total suspended matter (TSM) in the Bay of Bengal, measured by EnviSat MERIS on 25 April 2003. *Data* ESA

EnviSat the next generation of ESA's EO satellites are the Sentinels, launched from April 2014, each of which is carrying specific sensors for land and ocean surveillance (S-1: C-band SAR, S-2: Multispectral Instrument, S-3: Ocean and Land Color Instrument, Sea and Land Surface Temperature Radiometer, Radar Altimeter, MW Radiometer).

4 Recent Asian Satellite Missions

When Asian national space agencies started their satellite missions they had clear ideas of the possibilities of EO at various spectral ranges, of advantages and disadvantages of sensors already tested in flight, of technologies used for satellite data receiving and processing, and of their scientific and operational usage. Four Asian countries have launched civil remote sensing satellites, these being Japan, India, China, and South Korea.

4.1 Japan

The Japanese EO satellite missions began with the Geostationary Meteorological Satellite 1 (GMS-1), *Himawari 1*, launched on 14 July 1977 (the first of a series, whose latest satellite was launched in November 2016). The Japanese Aerospace

Exploration Agency (JAXA), then known as Japanese National Space Development Agency (NASDA), started EO missions using polar-orbiting satellites with the Marine Observation Satellite-1 (MOS-1), *Momo-1*, launched in February 1987 and operating until November 1995. Aboard MOS-1 were a two-channel microwave radiometer, MSR, and the VTIR, a visible/infrared radiometer used for SST retrieval. *Momo-1b* (MOS-1b) was launched in 1990 and conducted global observations, primarily of the oceans, until 1996.

The Japanese Earth Resources Satellite 1, *Fuyo*, (JERS-1) was launched in February 1992 and operated until October 1998. It carried a 1.275 GHz (L-band) HH-polarization SAR and an 8-band visual and IR Optical Sensor (OPS), and it conducted observations used for national land survey, fisheries, environmental preservation and disaster monitoring. The ADvanced Earth Observing Satellite (ADEOS), *Midori*, was launched in August 1996 by JAXA. ADEOS carried two core sensors, the Ocean Color Temperature Scanner (OCTS), which took over the NIMBUS-7 CZCS mission measuring in 12 spectral bands (Kawamura et al. 1998), and the 5-band Advanced Visible and Near-Infrared Radiometer (AVNIR). In addition, six sensors were provided by five organizations in Japan, the US, and France, including the 13.995 GHz (Ku-band) NASA SCATterometer (NSCAT) and the 3-band/15-channel POLarization and Directionality of the Earth's Reflectances (POLDER) sensor, thereby making ADEOS an international and multi-disciplinary mission.

OCTS data provided comprehensive oceanic maps of chlorophyll-a, dissolved matter, and SST, which were used for studies of ocean primary production and carbon cycling, and for environmental monitoring. Due to onboard mechanical problems ADEOS suspended its operation in June 1997. Figure 9 shows an SST map of the East China Sea and Sea of Japan, as measured by OCTS on 1 October 1996. Warm water (red) from the Kuroshio enters the area from the south, while river inflow is a source of cold water (blue) on the Russian coast in the north, close to Vladivostok (upper image centre).

The Tropical Rainfall Measuring Mission (TRMM) was a joint mission of JAXA and NASA launched in November 1997. TRMM's objective was to determine accurate rainfall amounts associated with tropical convective events. For this purpose the satellite carried the first satellite-borne precipitation radar (PR) developed in Japan, the Visible Infrared Scanner (VIRS), and the microwave imager TMI (TRMM Microwave Imager) (Wentz et al. 2000). The combined use of PR and TMI data has greatly improved the estimation of three-dimensional rainfall profiles, e.g. of typhoons over the oceans, which were rarely observed before TRMM. TRMM operated until June 2015.

Figure 10 is an example of a TRMM all-weather SST map of the north-western Pacific derived from TMI data (right panel) merged with data from the Stretched-Visible Infrared Spin Scan Radiometer (S-VISSR) aboard GMS-5 (left panel) (Guan and Kawamura 2004). Daily cloud-free SST products of 0.05° grid size were generated for three regions, Kuroshio, Asia-Pacific, and Pacific, from October 1999 until September 2000.

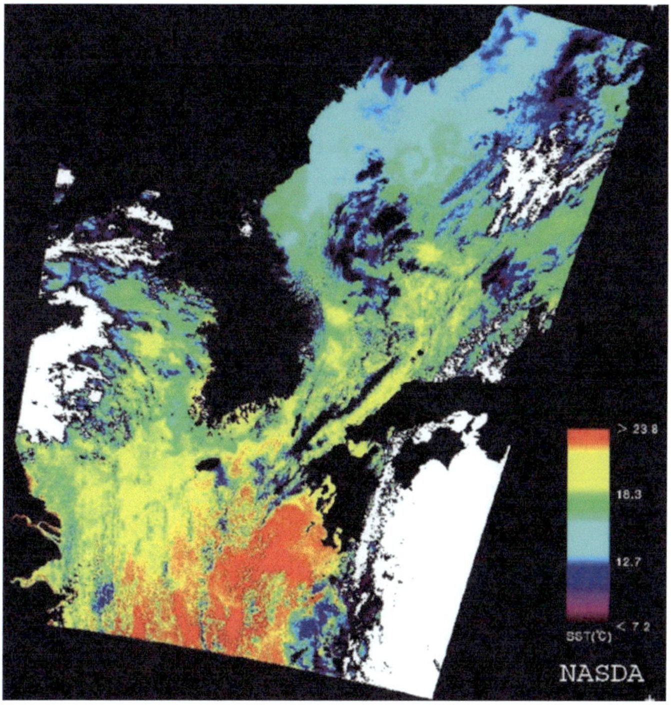

Fig. 9 SST in the Sea of Japan and East China Sea measured by OCTS aboard ADEOS on 1 October 1996. *Credit* JAXA

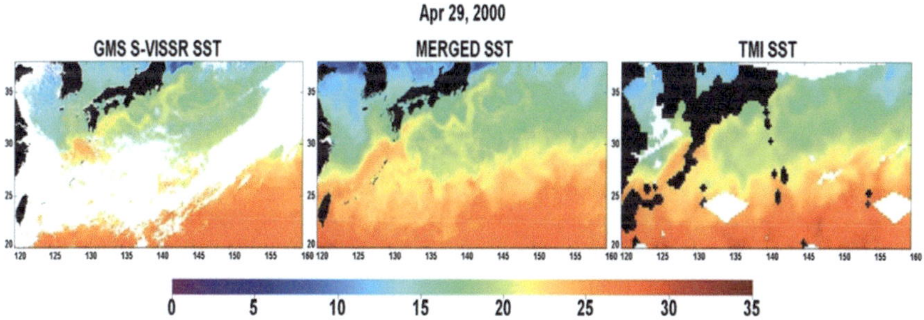

Fig. 10 SST in the north-western Pacific (centre) merged from GMS-5 S-VISSR data (left) and TMI data (right) (Guan and Kawamura 2004)

The legacy of the two-channel microwave radiometer MSR was the Advanced Microwave Scanning Radiometer (AMSR) on ADEOS-II and AMSR-E (AMSR for Earth Observing System) on Aqua (Imaoka et al. 2010). ADEOS-II launched in December 2002 was also equipped with the Global Imager (GLI) of high resolution and multiple wavelength channels based on the OCTS development (Murakami et al. 2006; Sakaida et al. 2006). ADEOS-II's operation ceased in October 2003 due to mechanical problems.

In order to complement optical sensors such as OCTS and GLI Japan launched the Advanced Land Observing Satellite (ALOS-1), *Daichi*, equipped with (among others) the Phased Array type L-band Synthetic Aperture Radar (PALSAR, operating at 1.27 GHz), an upgrade of the SAR aboard JERS-1, and with AVNIR-2, a 4-band high-resolution upgrade of AVNIR. Figure 11 is an AVNIR-2 true-colour composite image showing Mumbai, India, on the mouth of the Ulhas river, which is carrying a high load of sediments into the Arabian Sea. Dark elongated patches in the upper left are due to marine surface films that change the sea surface roughness and, therefore, its sunlight reflectance.

Fig. 11 Mumbai, India, seen by AVNIR-2 aboard ALOS-1 on 23 March 2011. *Credit* JAXA, ESA

Although ALOS was primarily designed for land applications, during the five years of its operation (January 2006 to April 2011) it contributed to the study of sea ice and oceanic dynamic phenomena such as currents, eddies, and internal waves (Mitnik 2008). Its successor, ALOS-2 (*Daichi-2*) launched in May 2014, carries the PALSAR-2, a significant upgrade of the PALSAR, operating at 1.24, 1.26 and 1.28 GHz (L band). In addition to the 10 m resolution survey mode inherited from ALOS-1 PALSAR-2 allows acquisitions at higher resolution (1 m × 3 m pixel) spotlight mode.

The highest spatial resolution among MWR has the AMSR2 onboard the Global Change Observation Mission Water (GCOM-W1) satellite (*Shizuku*) launched in May 2013. 14 microwave channels operating at vertical and horizontal polarizations allow estimating atmospheric water vapor content, cloud liquid water content, rain rate, sea surface wind speed, sea ice parameters etc., thereby having a key value for studies of the water cycle and climate change (Imaoka et al. 2010; Kachi et al. 2013).

4.2 India

In 1962 the Indian National Committee on Space Research (INCOSPAR) was founded and in 1969 the Indian Space Research Organization (ISRO) was established. After the first Indian satellite, *Aryabhatta*, was launched on 19 April 1975 the first Indian EO satellites, *Bhaskara 1* and *Bhaskara 2*, were launched in 1979 and 1981, respectively. Both satellites were equipped with (visible and NIR) cameras and the SAtellite MIcrowave Radiometer, SAMIR. Thereafter, India initiated the indigenous IRS (Indian Remote Sensing Satellite) program to support the national economy in operational fields such as marine fisheries and coastal management.

OceanSat-1, or IRS-P4, was the first Indian satellite built specifically for ocean observations. It was launched on 26 May 1999 and carried the Ocean Colour Monitor (OCM) working at 8 spectral bands, a Multi-frequency Scanning Microwave Radiometer (MSMR) operating at four MW frequencies at vertical and horizontal polarizations. The ISRO OceanSat-2 spacecraft was launched in September 2009 and was equipped with the 13.515 GHz (Ku-band) scatterometer OSCAT to measure ocean winds, with OCM-2 to measure chlorophyll concentrations, phytoplankton blooms, suspended sediments, and atmospheric aerosols. During its life time, until February 2014, OSCAT has demonstrated the importance and significance of the ocean wind vector data in weather modelling and prediction and paved the road for continuations of the global ocean surface wind vector retrieval.

Indian's Radar Imaging SATellite (RISAT) series started with RISAT-2, launched in April 2009 and carrying an X-band SAR operating at 9.59 GHz. RISAT-1 was launched in April 2012 and carries a C-band SAR (5.35 GHz), which is used for civilian EO (Valarmathi et al. 2013). Figure 12 shows a RISAT-1 SAR image of

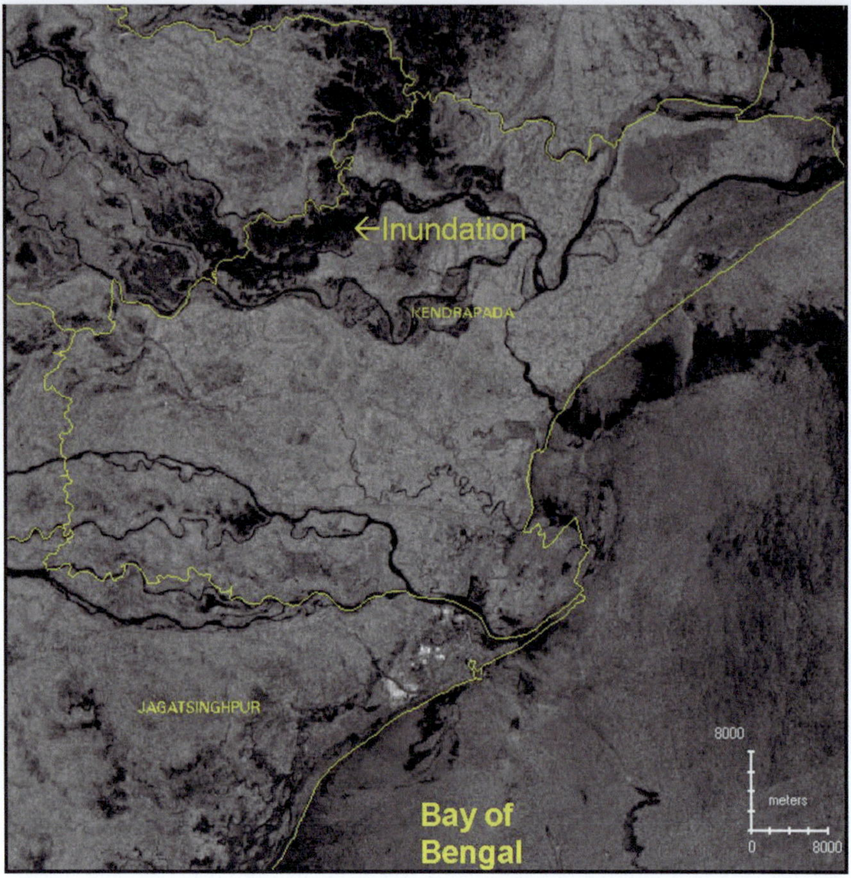

Fig. 12 RISAT-1 SAR image of the Mahanadi river delta on the north-eastern Indian coast, acquired on 16 October 2013, after the landfall of the tropical cyclone Phailin. *Credit* ISRO

the Mahanadi river delta on the north-eastern Indian coast, acquired after the landfall of the tropical cyclone *Phailin*, which was connected with severe rainfall in that region. Dark areas in the right image centre are caused by the river plume in the Bay of Bengal, the irregular dark patch on the bottom is oil pollution from a vessel (bright spot on its upper end).

4.3 China

The Peoples Republic of China launched its first satellite, known as *China 1* or *Mao 1*, on its own "*Long March*" space rocket on 24 April 1970. The 160-kg electronic ball floated around the Earth blaring the patriotic song "*The East Is Red*".

Since 1988, China has launched nine meteorological satellites, five of which are sun-synchronous, and four are geo-stationary.

China's first and second ocean satellites, HY (*Haiyang*)-1A/B, were respectively launched in May 2002 and April 2007 and carried an improved Chinese Ocean Color and Temperature Scanner (COCTS), a 10 channel visible and infrared radiometer, and the Coastal Zone Imager (CZI), a four-channel CCD camera (IOCCG, 2011). HY-1A operated until April 2004.

HY-2 (*Haiyang-2*)/Ocean-2A is a second-generation ocean observation/monitoring satellite series approved by the China National Space Administration (CNSA) in February 2007. HY-2A was launched on 15 August 2011 and represents a follow-up of the HY-1A/B missions, with the overall objective to measure ocean dynamics and environmental parameters with MW sensors (Dong et al. 2004). An example of ocean wind fields derived from HY-2A scatterometer data acquired on 8 August 2012 is shown in Fig. 13. Two cyclones, east of Kamchatka and southeast of Japan, can be delineated through their circular shape and increased wind speed.

Gaofen is a series of Chinese civilian high resolution (*gao fen*) optical and microwave EO satellites, of which six have been successfully launched by the end of 2016. GF-1 and its successor GF-2 (launched in April 2013 and August 2014,

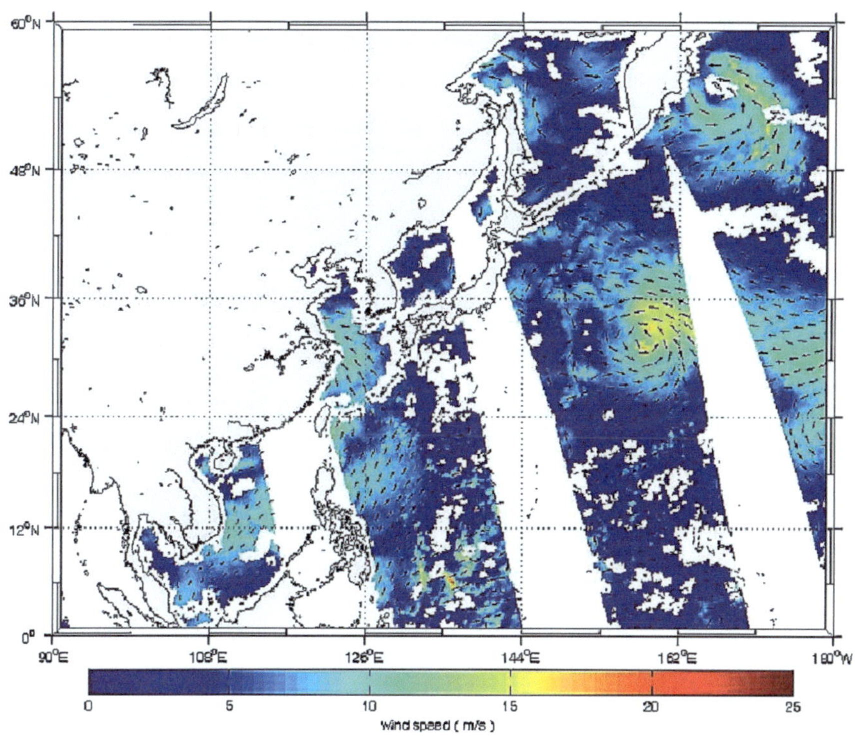

Fig. 13 Sea surface wind retrieval from HY-2A/SCAT (8 August 2012). *Credit* State Key Laboratory of Remote Sensing Science

respectively) carry medium- and high-resolution multispectral cameras and so do GF-9 (September 2014) and GF-8 (June 2015). GF-4 (December 2015) is a geostationary satellite equipped with two staring optical (visible and infrared) sensors. Finally, GF-3 was launched in August 2016 and is the series' first satellite devoted to ocean surveillance, carrying a 5.3 GHz (C-band) SAR. Figure 14 shows a GF-3 Standard Stripmap (SS) SAR image of the South China Sea, south-east of Hainan (upper left corner), acquired on 19 June 2017. Imprints of internal waves can be seen as bright and dark lines in most parts of the image.

4.4 South Korea

South Korea's space agency, the Korea Aerospace Research Institute (KARI) was founded in 1989. The first Korean EO satellite, KOrea Multi-Purpose SATellite-1 (KOMPSAT-1), *Arirang-1*, was launched on 20 December 1999 and carried the

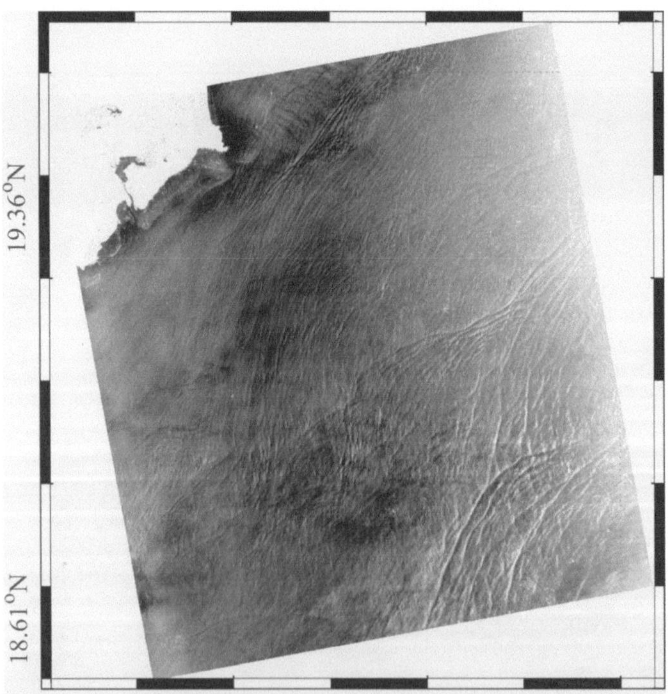

Fig. 14 GF-3 SAR image (130 km × 130 km) of the South China Sea, off Hainan, acquired on 19 June 2017. Imprints of oceanic internal waves can be seen in most parts of the image

panchromatic Electro-Optical Camera (EOC) and the 6 (of 8)-band Ocean Scanning Multispectral Imager (OSMI), which was used for ocean colour measurements, primarily of the seas surrounding the Korean peninsula. The mission of KOMPSAT-1 ended in July 2007. KOMPSAT-2 (*Arirang-2*) launched in July 2006 carries only a panchromatic Multi-Spectral Camera (MSC). Further satellites of the KOMPSAT series, launched from 2012, are carrying high-resolution optical sensors and an X-band SAR.

The first satellite measuring ocean colour from a geostationary orbit, the Communication, Ocean and Meteorological Satellite (COMS or GEO-KOMPSAT-1, *Cheollian-1*) was launched in June 2010. Its primary sensor is the 8-band Geostationary Ocean Color Imager (GOCI), which is observing the Yellow and East China Seas, Sea of Japan, and the north-western Pacific Ocean off Japan (Fig. 15). Using this sensor coastal dynamics can be observed, and tracked, at high spatial resolution.

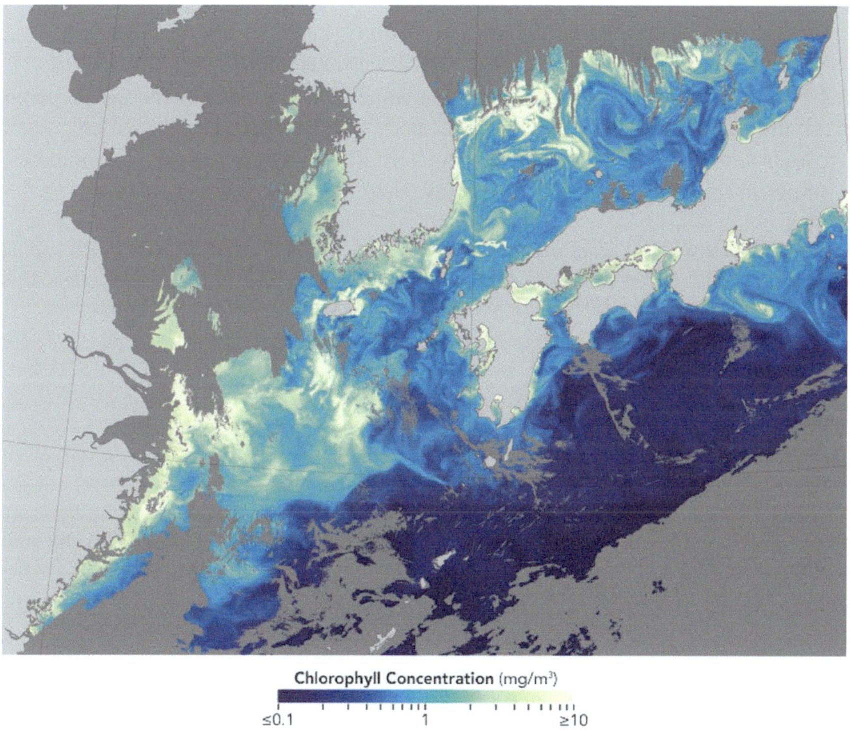

Fig. 15 Chlorophyll concentration in the East China Sea and Sea of Japan, as measured by GOCI on COMS on 12 May 2016. *Image credit* NASA Earth Observatory

5 Summary and Outlook

We have provided a short review of spaceborne remote sensing of Asian Seas, starting from the 1960s, when only the USA and USSR built and operated spaceborne sensors. In the 1970s and 1980s, first European and Japanese satellites were launched, and later on, Indian, Chinese and Korean satellites followed (among others).

Early satellite sensors were passive, working in the visual and IR range of the electromagnetic spectrum and measuring, e.g. atmospheric water vapour content. Passive MW sensors then allowed first measurements of SST and thereby, manifestations of dynamics of the upper ocean layer. The next generation of optical sensors, ocean colour sensors, have been used from the 1970s to study the distribution of chlorophyll-like pigments and particulate matter in the upper layers of the Asian Seas, allowing, among others, studies of harmful algae blooms (HAB).

In addition, active microwave sensors such as scatterometers, altimeters and SAR have been used from the late 1970s to measure global wind fields over the World's oceans (and over the Asian Seas), sea surface height, ocean currents, sea-level rise and the marine geoid, and to study small- to mesoscale oceanic and coastal features, e.g. oceanic internal waves, currents and eddies, and sea ice.

Several future satellite missions are planned, or are already being prepared, involving more accurate and sophisticated sensors, and also involving more countries and national space agencies. As a result, the Asian Seas will be monitored in future in much greater detail than they were during the past decades.

Acknowledgements This work was partly supported by the FEB RAS grants 18-I-010 and the JAXA Project F10. The authors thank ESA for the Envisat data, NOAA's National Climatic Data Center for AVHRR data, NASA's OBPG for CZCS data, and ASF DAAC for Seasat data.

References

Basharinov AE, Gurvich AS, Egorov ST (1974) Microwave emission of the earth as a planet (Radioizluchenie Zemli kak planeti). Nauka Publishing House, Moscow, p 188 (in Russian)

Chang HD, Hwang PH, Wilheit TT, Chang ATC, Staelin DH, Rosenkranz RW (1984) Monthly distribution of precipitable water from the Nimbus-7. J Geophys Res 89(D4):5328–5334. https://doi.org/10.1029/Jd089iD04P053228

Dong X, Xu K, Liu H, Jiang J (2004) The radar altimeter and scatterometer of China's HY-2 satellite. In: Proceedings of IGARSS 2004, USA, 20–24 Sept 2004

Fu L-L, Holt B (1982) Seasat views oceans and sea ice with synthetic aperture radar. JPL Publication, pp 81–120. Jet Propulsion Laboratory, California Institute of Technology, Pasadena, California

Guan L, Kawamura H (2004) Merging satellite infrared and microwave SSTs—methodology and evaluation of the new SST. J Oceanogr 60(5):905–912

Gurvich AS, Demin VV (1970) Total atmospheric humidity measured by the Cosmos-243 satellite. Atmos Oceanic Phys 6:771–779

Imaoka K, Kachi M, Kasahara M, Ito N, Nakagawa K, Oki T (2010) Instrument performance and calibration of AMSR-E and AMSR2. Int Arch Photogram Remote Sens Spec Inf Sci 38(8): 13–18

Kachi M, Imaoka K, Maeda T, Okuyama A, Naoki K, Hori M, Kasahara M, Ito N (2013) Calibration and validation results of the advanced microwave scanning radiometer 2 (AMSR2). In: Proceedings of 19th American meteorological society AMS satellite meteorology, oceanography, and climatology conference, Vienna, Austria, 16–20 Sept 2013

Karlin LN (ed) (1999) Atlas of synthetic aperture radar images of the ocean acquired by ALMAZ-1 satellite. GEOS Publishing House, Moscow

Kawamura H, OCTS Team (1998) OCTS mission overview. J Oceanogr 54:383–399

Kramer HJ (2002) Observation of the Earth and its environment. Survey of missions and sensors, 4th edn. Springer, Berlin, p 1514

Mitnik LM (1972) Measurements of total atmospheric water vapor content over the ocean with the use of microwave radiometers onboard "Kosmos-243" satellite. In: Proceedings of central aerological observatory, no 103, pp 64–72 (in Russian)

Mitnik LM (2008) Advanced land observing satellite PALSAR observations of the oceanic dynamic phenomena in the coastal zone. In: Proceedings of IGARSS'08, USA, vol 2, pp 351–354. https://doi.org/10.1109/igarss.2008.4779000

Mitnik LM, Victorov SV (eds) (1990) Radar Sensing of the Earth's surface from space (Radiolokatsiya poverkhnosti Zemli iz kosmosa). Hydrometeoizdat Publishing House, Leningrad, 200 pp (in Russian)

Murakami H, Sasaoka K, Hosoda K et al (2006) Validation of ADEOS-II GLI ocean color products using in-situ observations. J Oceanogr 62(3):373–393. https://doi.org/10.1007/s10872-006-0062-6

NASA Goddard Space Flight Center, Ocean Biology Processing Group (2014) Coastal zone color scanner (CZCS) data, NASA OB.DAAC, Greenbelt, MD, USA. http://oceandata.sci.gsfc.nasa.gov/CZCS/Mapped/Monthly/4km/chlor_a. Accessed 02 Dec 2016. Maintained by NASA Ocean Biology Distributed Active Archive Center (OB.DAAC), Goddard Space Flight Center, Greenbelt MD

National Climatic Data Center/NESDIS/NOAA/U.S. Department of Commerce (2007) NOAA optimum interpolation 1/4 degree daily sea surface temperature analysis, version 2. Research Data Archive at the National Center for Atmospheric Research, Computational and Information Systems Laboratory. http://rda.ucar.edu/datasets/ds277.7/. Accessed 01 Dec 2016

Sakaida F, Hosoda K, Moriyama M, Murakami H, Mukaida A, Kawamura H (2006) Sea surface temperature observation by global imager (GLI)/ADEOS-II: algorithm and accuracy of the product. J Oceanogr 62(N 3):311–319

Valarmathi N, Tyagi RN, Kamath SM et al (2013) RISAT-1 spacecraft configuration: architecture, technology and performance, radar imaging satellite-1. Curr Sci 104(4):462–471. http://www.currentscience.ac.in/Volumes/104/04/0462.pdf

Wentz FJ, Gentemann C, Smith D, Chelton D (2000) Satellite measurements of sea surface temperature through clouds. Science 288:847–850

An Introduction to Optical Remote Sensing of the Asian Seas: Chinese Dedicated Satellites and Data Processing Techniques

Yan Bai, Lianghong Jiang, Xianqiang He and Vittorio Barale

Abstract The application of optical remote sensing in the Asian Seas is introduced, by reviewing the achievements of ocean colour dedicated missions and data processing techniques in China. Two satellites (HY-1A and HY-1B) have already been launched as part of the Chinese long-term (2015–2025) ocean colour observation program, which foresees the launch of more than 15 orbital platforms. Data processing techniques, including the development of a vector radiative transfer model, as well as calibration, polarization correction, and atmospheric correction algorithms, have all been developed specifically for these missions, with particular attention to the peculiar environmental conditions of the greater China Sea. Progress on the atmospheric correction effectiveness in highly turbid waters in the greater China Sea coastal regions is also examined.

Keywords Ocean colour · Remote sensing · Asian Seas · Data processing technique · Atmospheric correction

1 Introduction

Optical properties of marine waters have been used for centuries to characterize both water quality types and their geographical distribution (Maffione 2001). From the 1600s until the early 1800s, the colour and transparency of seawater were determined essentially by visual inspection, using the human eye as a readily accessible light detector. It was only in 1865 that Fr. Pietro Angelo Secchi—an Italian Jesuit, Director of the Observatory at the Pontifical Gregorian University, then the Roman College—during a scientific cruise of the papal yacht *Immacolata*

Y. Bai (✉) · L. Jiang · X. He
State Key Laboratory of Satellite Ocean Environment Dynamics, Second Institute of Oceanography, State Oceanic Administration, Hangzhou, China
e-mail: baiyan@sio.org.cn

V. Barale
European Commission, Joint Research Centre, Ispra, Italy

V. Barale and M. Gade (eds.), *Remote Sensing of the Asian Seas*,
https://doi.org/10.1007/978-3-319-94067-0_3

Concezione, started to systematically experiment with, and then report about, a very simple device, later identified as the Secchi Disk, for measuring water transparency in water bodies (Wernand 2010). The instrument is a circular disk, 30 cm in diameter, in origin plain white, but later modified—in 1899, by George C. Whipple, an American civil engineer—by dividing it into quadrants painted alternately black and white. The disc is attached to a graded line, to be lowered down in the water column. The depth at which the disk is no longer visible, known as the Secchi depth, is related to water turbidity and has been used extensively as a standard measure of water transparency (Preisendorfer 1986).

While Secchi Disks are still being used the world over, and have generated possibly the longest instrumental datasets quantifying (optically) the world's oceans, in the early 1900s the development of photographic techniques, and later on that of underwater radiometers, caused a quantum leap forward in the study of water optical properties. Scandinavian scientists pioneered theoretical and experimental work on optical processes of natural seawater, and helped form the foundation of modern marine optical research (Hojerslev 1994). In particular, Nils Gunnar Jerlov proposed the concept of an optical classification for oceanic waters, the Jerlov water types, while his books—Optical Oceanography, published in 1968, revised in 1976 and published under the new title Marine Optics—remain widely referenced to this very day (Jerlov 1968, 1976). However, it was with the development of satellite oceanography techniques that optical observations of the sea surface, in particular within or near the visible spectral range, commonly referred to under the generic term of "ocean colour" remote sensing, begun to provide a real novel look at a score of bio-geo-chemical and physical processes of the sea. Early photographs taken from orbit by astronauts equipped with a held-held camera had already shown promising glimpses of possible applications in the area of coastal monitoring. Then, after much development and airborne experiments, the Coastal Zone Color Scanner (CZCS) was launched by the US National Aeronautics and Space Administration (NASA), in October 1978, on board the Nimbus-7 satellite, with the goal to image surface patterns of sea water constituents, and conceivably of measuring planktonic pigments as well (Acker 2015).

The CZCS opened a completely uncharted scientific territory to oceanographic research. Ultimately, it changed the way in which we understand the marine environment, and the role it plays on this planet (Austin 1993). In spite of its success, the CZCS remained one of a kind for more than a decade, after its demise in 1986, before new sensors started being outfitted, to provide a new chance for optical observations of the sea surface from Earth's orbit. Finally, in 1996, two similar versions of the German Modular Optoelectronic Scanner (MOS) were launched. Also in August 1996, the Japanese Ocean Color and Temperature Scanner (OCTS) flew on board the ADEOS platform, until the satellite premature failure in April 1997, and later in the same year the French first Polarization and Directionality of Earth Reflectances (POLDER) sensor took to the sky. And then again, the Sea-viewing Wide Field-Of-View Sensor (SeaWiFS) of NASA was launched in September 1997 on a dedicated small spacecraft, soon to be followed, in 1999, by Moderate Resolution Imaging Spectroradiometer (MODIS), on the Terra orbital platform, and by the

Japanese Ocean Color Imager (OCI), the Indian Ocean Colour Monitor (OCM) and the Korean Ocean Scanning Multispectral Imager (OSMI).

While many Space Agencies the world over were becoming more and more active in the field of ocean optical observations, plans were being made in Europe by the European Space Agency (ESA) for the revolutionary Medium Resolution Imaging Spectrometer (MERIS), which would finally reach Earth orbit in 2002. But ocean colour sensors started to become more and more popular, at the beginning of the new century: in 2002, MERIS was followed by a second MODIS, Aqua version, of NASA, the Japanese Global Imager (GLI), the Chinese Moderate Resolution Imaging Spectroradiometer (CMODIS) and the Chinese Ocean Colour and Temperature Scanner (COCTS) and Coastal Zone Imager (CZI), to be doubled in 2007, and the French POLDER-2. The later was shadowed by POLDER-3, in 2004; by the Indian OCM-2, and the Hyperspectral Imager for the Coastal Ocean (HICO) on the International Space Station, in 2009; by the Korean Geostationary Ocean Color Imager (GOCI), in 2010; by the Visible Infrared Imager Radiometer Suite (VIIRS), by NASA, in 2011; and by the Ocean Land Colour Instrument (OLCI) on board Sentinel-3A, by ESA and EUMETSAT, in 2016.

Today, in 2017, at least a dozen more sensors are ready to be launched in the next decade or so. As the amount of data increases, and the computer power for data processing reaches unprecedented levels, the algorithms required to derive bio-geo-chemical parameters from the remotely sensed signal still continue to refine (Gordon 2010). In fact, the term ocean colour actually refers to the visible light spectrum as observed at the sea surface—i.e. the total spectral radiance $L_t(\lambda)$ received and measured by the remote sensor, where λ identifies the radiation wavelength—which is only related, by the processes of absorption and scattering, to the concentration of various water constituents, immeasurable directly from space orbit. Moreover, the retrieval of environmental parameters from ocean colour depends chiefly on evaluating non-marine (atmospheric) contributions to the data, which can be as high as 90% of the observed signal. First, passive observations of natural sunlight reflected by the Earth's surface are limited to a cloudless, clear atmosphere, owing to the scattering properties of water droplets of clouds or haze (which interfere with the water-leaving radiance $L_w(\lambda)$ propagating upwards from the water surface into the field of view of the orbital sensor). But even in case of a clear atmosphere, accurate atmospheric corrections must be performed, in order to remove $L_{path}(\lambda)$, the path radiance due to Rayleigh scattering and aerosol scattering in the atmosphere, as well as other undesired contributions to the total signal, such as the whitecap reflectance radiance $L_{wc}(\lambda)$, and to retrieve quantitative estimates of the water constituents' concentration. This, in turn, must be based on modeling the relationship between the concentration itself and water optical properties, or related parameters such as light attenuation coefficients.

Surface waters are commonly classified according to the kind of constituents shaping their optical properties. In a first, more general case, usually typical of the pelagic zone (the so-called case 1 waters, first introduced by Morel and Prieur 1977), optical properties are dominated by biological agents such as planktonic pigments (phytoplankton, microscopic unicellular algae, in particular, which contain green chlorophyll-like pigments). In a second, quite special case, typical of coastal areas (the so-called case 2 waters, again according to Morel and Prieur 1977) and recurrent in the Asian Seas, optical properties are often dominated by the impact of high continental runoff, so that the presence of biological agents is accompanied by that of other optically active materials, such as suspended sediments, degrading organic materials and other particulate or dissolved substances. Since the assessment of the atmospheric signal seen by an orbital sensor is based primarily on measurements in the red and near-infrared (or NIR, in brief) spectral region—which are affected the most by the presence of different water constituents, typical of case 2 waters—in general the difficulties of using ocean colour data will increase as the distance from the coast, and from coastal runoff sources, decreases. The depth over which the water-leaving signal—carrying information on the water constituents—is integrated, is essentially the 1st optical depth of the water column (that at which the solar irradiance falls to $1/e$ (0.37) of its value just below the surface). This depth can be on the order of 10's of m, in ideal conditions such as those provided by clear oceanic waters, but can decrease to a few m, or even down to cm, in turbid coastal waters, depending on the nature and concentration of the constituents present in the water column.

Remote sensing of sea surface optical properties finds applications in determining the water constituents presence, nature and concentration, leading to estimates of phytoplankton biomass and productivity, of sediment transport and distribution, and of water attributes as related to runoff plumes and currents. These indicators allow the analysis of phenomena like water quality and plankton blooms, coastal eutrophication, bio-geo-chemical cycles, coastal and fluvial runoff, sediment transport, water dynamics, depth variations in shallow waters, exchanges between coastal zone and open sea, and so on. The exploitation of ocean colour data in studies of bio-geo-chemical cycling is also having an impact on the understanding of climate changes. In fact, the marine component of the carbon cycle, particularly in mesotrophic or eutrophic costal areas, plays an important role in shaping the climatic variability of the Earth—a topical subject, these days, for our environmental-conscious society. Recent developments of such applications include the integration of both in situ and remote sensing data, and their assimilation in ecological models, to provide forecasting capabilities on a score of environmental issues, from local plankton blooming to global climatic change.

In the following, we shall introduce the application of optical remote sensing in the Asian Seas, by reviewing the achievements of ocean colour dedicated missions and data processing techniques in China. First, previous and planned Chinese ocean colour missions will be introduced. Then, data processing techniques developed for these missions will be briefly presented. Finally, progress on the atmospheric correction effectiveness in highly turbid waters in the greater China Sea coastal regions will be examined. The long-term (2015–2025) space program of China foresees the launch of more than 15 ocean satellites, before its completion. The Chinese ocean-related missions comprise three series of satellites, i.e. ocean colour satellites, ocean dynamics satellites, and high spatial or temporal resolution monitoring satellites. Up to now, three experimental ocean orbital platforms have already been launched, including two ocean colour satellites (HY-1A and HY-1B), and one ocean dynamics satellite (HY-2A). Based on these experimental missions, a series of operational ocean satellites will be launched in the next few years, and will contribute to the endeavour of the global ocean remote sensing community. With the launch of these satellites, data processing techniques for the retrieval of ocean colour parameters have also been advanced. For example, a vector radiative transfer model of the coupled ocean-atmosphere system and an operational atmospheric correction algorithm have been developed by He et al. (2007, 2008a, 2010). Generally, the waters along the Chinese coasts are highly turbid, constituting a challenge for optical remote sensing in these regions (He et al. 2012). Conversely, the Chinese coastal regions are the ideal test site for investigating the performance of ocean colour retrieval algorithms in turbid waters. In the past decade, many atmospheric correction algorithms for the Chinese coastal waters have been developed, due to the failure of operational atmospheric correction algorithms in the near-coastal regions of several Asian Seas.

2 Chinese Ocean Colour Satellites

The first Chinese ocean colour satellite, HY-1A, was launched on May 15, 2002, and its mission ended in April 2004 (Pan et al. 2004). Then, HY-1B was launched on April 11, 2007 (He et al. 2008b). Table 1 presents all basic parameters of both satellites. Although the designed life for HY-1B was three years only, the satellite remained in orbit for about nine years, until January 2016. Note that both HY-1A and HY-1B were experimental, and not operational, satellites, so their mission focused mainly on regional rather than global observations.

Table 1 HY-1A and HY-1B basic satellite parameters

Satellite	HY-1A	HY-1B
Launched time	09:50 AM on May 15, 2002	11:27 AM on April 11, 2007
Launch rocket	CZ-4B	CZ-2C
Satellite platform	CAST968	CAST968
Satellite weight (kg)	367	422.5
Orbit type	Sun-synchronous orbit	Sun-synchronous orbit
Orbit height (km)	From 870 to 798	798
Equator crossing time	From 8:30 AM to 10:00 AM	10:30 AM $\pm$ 30 min
Designed life	Two years	Three years
Observing capacity (each day)	Two orbits around China, one orbit for other regions	Three orbits around China, four/five orbits for other regions
Operation period	May 2002 to April 2004	April 2007 to January 2016

Both HY-1A and HY-1B carried two ocean colour sensors—again, the Chinese Ocean Colour and Temperature Scanner (COCTS), and the Coastal Zone Imager (CZI). The full payload of each satellite is described in Tables 2 and 3, respectively. In comparison with the HY-1A COCTS, the swath of the HY-1B COCTS was increased from 1600 to 3000 km, which shortened the revisit time from three days

Table 2 Sensors aboard the HY-1A satellite

Sensor	COCTS	CZI
Field of viewing (°)	90.0	34.78
Spatial resolution (km)	1.1	0.25
Pixels per line	1024	2048
Swath (km)	1600	500
Revisit time (day)	3	7
Digitization (bit/pixel)	10	12
Band	Channel wavelength 1 402–422 nm 2 433–453 nm 3 480–500 nm 4 510–530 nm 5 555–575 nm 6 660–680 nm 7 730–770 nm 8 845–885 nm 9 10.3–11.4 μm 10 11.4–12.5 μm	Channel wavelength 1 420–500 nm 2 520–600 nm 3 610–690 nm 4 760–890 nm

Table 3 Sensors aboard the HY-1B satellite

Sensor	COCTS	CZI
Field of viewing (°)	114	34.78
Spatial resolution (km)	1.1	0.25
Pixels per line	1664	2048
Swath (km)	3000	500
Revisit time (day)	1	7
Digitization (bit/pixel)	10	12
Band	Channel wavelength 1 402–422 nm 2 433–453 nm 3 480–500 nm 4 510–530 nm 5 555–575 nm 6 660–680 nm 7 740–760 nm 8 845–885 nm 9 10.3–11.4 μm 10 11.4–12.5 μm	Channel wavelength 1 433–453 nm 2 555–575 nm 3 655–675 nm 4 675–695 nm

to one day. Moreover, the CZI bandwidth was narrowed from 80 to 20 nm, to meet the requirements for ocean colour retrieval. In addition, as reported in Table 1, the on-board observing capacity was improved for HY-1B.

Being experimental satellites, both HY-1A and HY-1B experienced some difficulties. For example, HY-1A had a problem with the K mirror that stopped rotating, and HY-1B with the cold space signal being contaminated by the Earth reflectance. Moreover, there were no in-orbit calibration instruments, neither on HY-1A nor on HY-1B. Nevertheless, alternative methods were developed to remedy these deficiencies, and both satellites collected copious ocean colour data.

Figure 1 provides an example of the data collected by the HY-1B/COCTS. The overall performance of the HY-1B/COTS in measuring the total spectral radiance $L_t(\lambda)$ is illustrated in Fig. 2, which shows the cross-calibration coefficients obtained, in May 2007, using SeaWiFS as a reference. Based on the results obtained by the experimental HY-1A and HY-1B missions, the operational ocean colour satellites HY-1C and HY-1D, which have global observing capability, are scheduled to be launched in 2018. In addition, the new generation ocean colour satellites of HY-1E

Fig. 1 False colour image of the HY-1B/COCTS data, collected on Mar 1, 2008. The image shows the rich ocean colour patterns on the shelf of the East China Sea

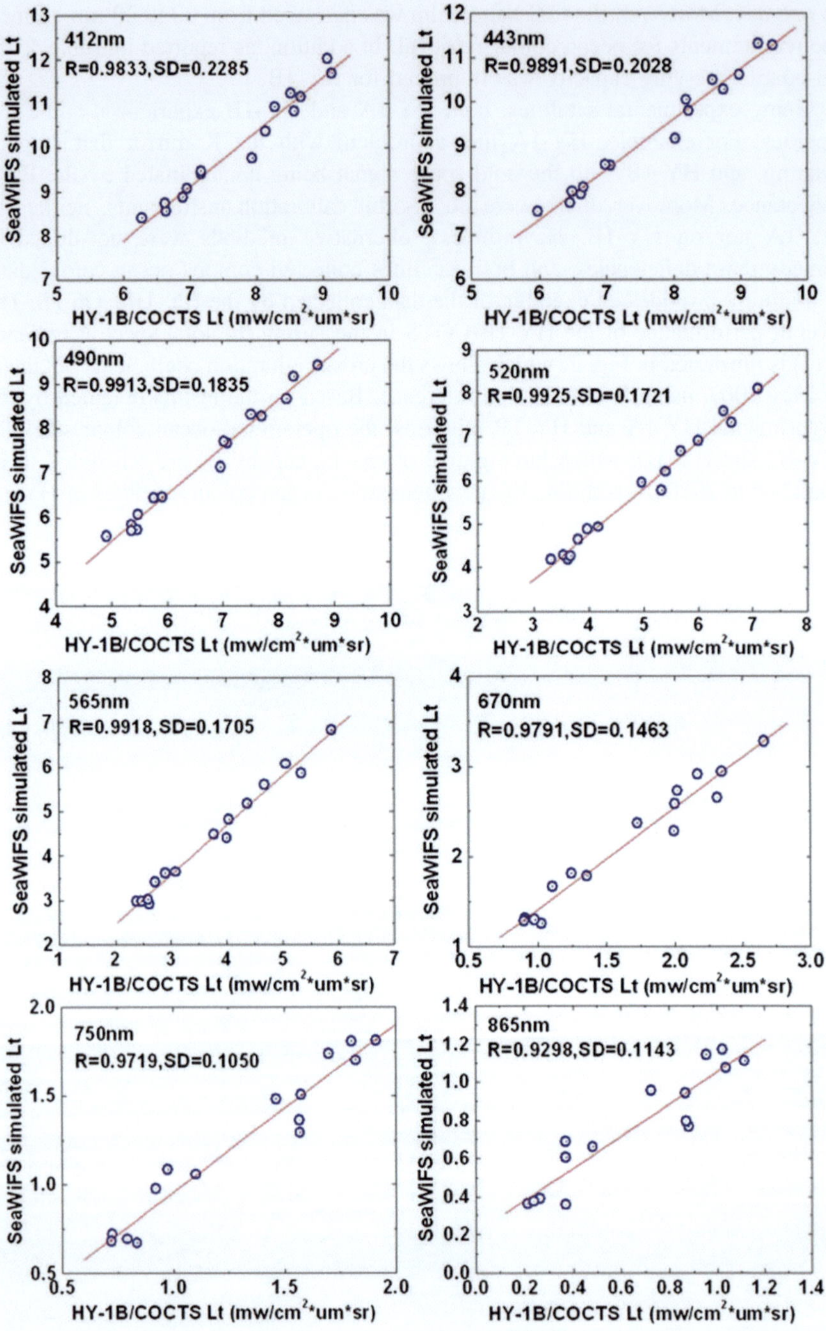

Fig. 2 Cross-calibration coefficients for the HY-1B/COTS in May 2007 using SeaWiFS as a reference. After He et al. (2008a, b)

and HY-1F, which will carry ocean colour sensors similar to the MERIS, on board ENVISAT, and the OLCI, on board Sentinel-3A, with high signal-to-noise ratios, are in the planning stage.

3 Chinese Ocean Colour Satellites: Data Processing

3.1 *Radiative Transfer Model*

The vector radiative transfer model is the foundation of the basic processing technique, the atmospheric correction algorithm, for satellite ocean colour data. Based on the matrix-operator method or the adding-doubling method, a vector radiative transfer model for the coupled ocean-atmosphere system (named PCOART) has been developed by He et al. (2007, 2010). Besides ocean-atmosphere coupling and polarization, the PCOART model also considers sea surface roughness. Validation results show that PCOART can predict the radiance distributions in the atmosphere and the ocean, for both flat and rough sea surfaces, accurately. The atmospheric Rayleigh scattering radiance simulated by PCOART is comparable with the results obtained from the Rayleigh scattering look-up tables of the MODIS, with differences of less than 0.25%. Further, the underwater radiance simulated by PCOART meets the accuracy requirement for the oceanic radiative transfer simulation (He et al. 2010). In addition, PCOART can reproduce the linear polarization radiance measured by the POLDER. A comparison between POLDER observations and PCOART simulations is shown in Fig. 3.

3.2 *In-Orbit Estimation of the Polarization Response Factor*

Polarization response is an important factor, which influences the accuracy of radiance measurements. Since the polarization response effect depends on the observing geometry, a systematic vicarious calibration cannot remove this influence. To meet the uncertainty requirement of retrieved L_w being less than 5%, the polarization sensitivity should be less than 2.5%. However, even for the MODIS, onboard the Terra and Aqua satellites, the polarization responses are as high as 6.0%, which could induce up to 3.8% uncertainty (=6.0% × 70% × 90%, where 70% is the maximum polarization degree, and 90% is the ratio of L_{path} to the L_t) of the measured radiance, and up to 38% uncertainty of the retrieved L_w. Using the pre-launch measured polarization response coefficients, the influence can be corrected partially. However, the polarization response coefficients can change with on-orbit time, and this cannot be monitored by onboard instrument. Moreover, the pre-launch measurement of polarization response coefficients needs specific

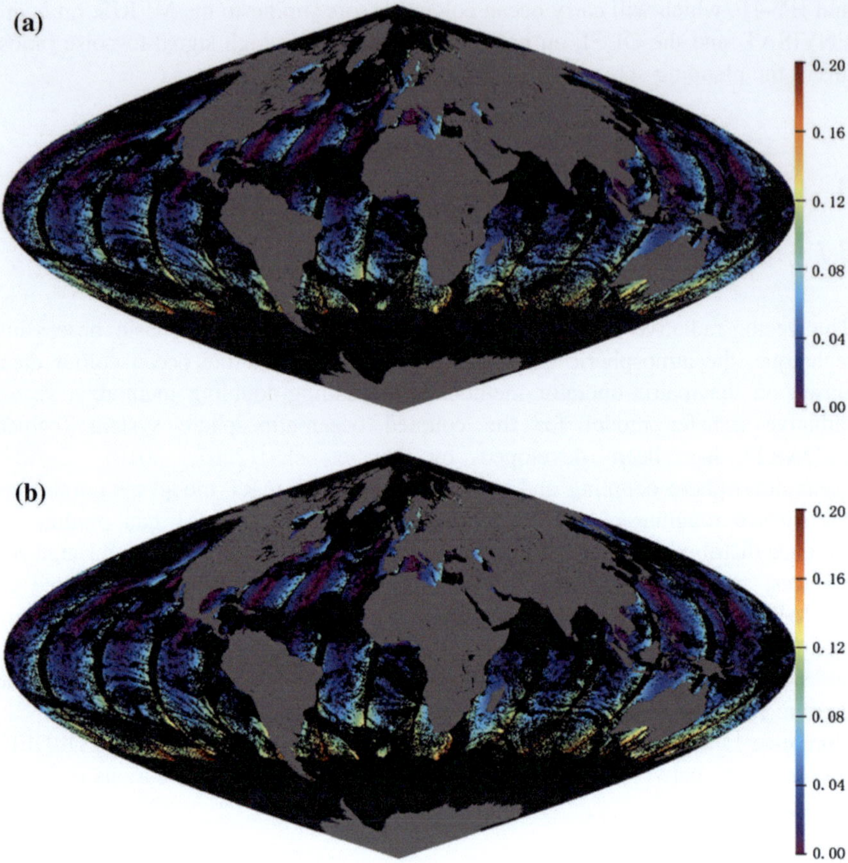

Fig. 3 Comparison of linear polarization reflectance (at 443 nm) between **a** POLDER observations and **b** PCOART simulations. After He et al. (2010)

instruments, and some of the satellite ocean color sensors may not have the pre-launch polarization response coefficients, such as the COCTS. Therefore, on-orbit estimation based on satellite data is necessary for monitoring the polarization response coefficient change of the ocean color sensor with pre-launch measurement, and is essential for the ocean color sensors without pre-launch measurement.

He et al. (2016) proposed a practical method for on-orbit estimation of polarization response, centered on the developed cross-calibration method and polarization simulation by PCOART. The linear polarization components of the Stokes vector entering the sensor are estimated by PCOART by considering the Rayleigh scattering and aerosol scattering. The reference radiance entering the sensor is estimated using the cross-calibration method. Based on the estimated linear polarization components and reference radiance, the polarization response coefficients can be derived by the least squares method.

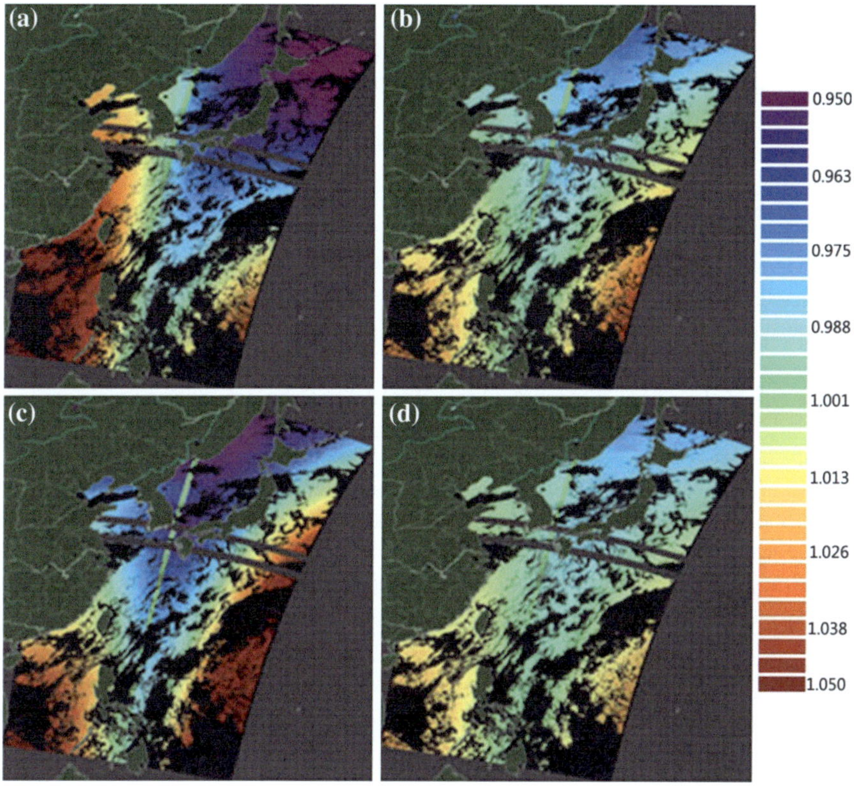

Fig. 4 Distribution of polarization correction amplitude for the HY-1B/COCTS data on Oct. 22, 2007. **a–d** Correspond to the first four bands of the COCTS, respectively. The measured total radiance multiples the correction amplitude gives the corrected radiance

Validations by the Aqua/MODIS data show that the estimated polarization response coefficients are consistent with the pre-launch measurement. Using the developed method, the polarization response sensitivities were derived with the mean values of 17.2, 9.4, 23.2, 7.7, and 4.7% for the first five bands of HY-1B/COCTS, respectively. Figure 4 shows an example of the derived polarization correction amplitude for HY-1B/COCTS using the on-orbit estimated polarization response coefficients. Clearly, the polarization correction amplitude can be up to 5%, with large corrections located at the edges of the scan lines.

3.3 Atmospheric Correction Algorithm

The operational atmospheric correction algorithm for the HY-1B/COCTS is similar to that of the SeaWiFS and MODIS as implemented in the SeaWiFS Data Analysis

System[1] (*SeaDAS*), which uses two NIR wavelengths to estimate aerosol scattering radiances and then extrapolate the results to the visible bands (He et al. 2008a). The total radiance received by COCTS can be expressed by:

$$L_t(\lambda) = L_{path}(\lambda) + t_v(\lambda)L_{wc}(\lambda) + t_v(\lambda)L_w(\lambda) \tag{1}$$

where L_{path} is the atmosphere path radiance due to the Rayleigh scattering and aerosol scattering; L_{wc} is whitecap reflectance radiance; L_w is the desired water-leaving radiance, and t_v is the upward atmosphere diffuse transmittance from the sea surface to the satellite. Note that the sun glint tem is ignored in Eq. (1).

The whitecap radiance can be estimated, using sea surface wind speed, as:

$$\begin{cases} t_v(\lambda)L_{wc}(\lambda) = [c(\lambda) \times R_{wc}(\lambda) \times F_0(\lambda) \times \cos\theta_0 \times t_v(\lambda) \times t_s(\lambda)]/\pi \\ R_{wc}(\lambda) = 0.00000026 \times W^{3.52} \end{cases} \tag{2}$$

where c is the relative whitecap absorption coefficient; F_0 is solar irradiance at the top of the atmosphere; θ_0 is the solar zenith angle; t_s is the downward atmospheric diffuse transmittance from the Sun to the sea surface.

Using Eqs. (1) and (2), we can obtain the whitecap corrected total radiance. Moreover, we need to correct the ozone absorption (and any other atmosphere molecular absorption) as follows:

$$L_t'(\lambda) = [L_t(\lambda) - t_v(\lambda)L_{wc}(\lambda)] \times \exp\left\{\tau_{OZ}(\lambda) \times \left[\frac{1}{\cos\theta_0} + \frac{1}{\cos\theta_v}\right]\right\} \tag{3}$$

Assuming that the water-leaving radiances at the two NIR bands (λ_{nir1} = 750 nm and λ_{nir2} = 865 nm) can be neglected, the ratio between L_{path} and the Rayleigh scattering radiance L_r can be derived as:

$$\begin{cases} R(\lambda_{nir1}) = \frac{L_{path}(\lambda_{nir1})}{L_r(\lambda_{nir1})} = \frac{L_t'(\lambda_{nir1})}{L_r(\lambda_{nir1})} \\ R(\lambda_{nir2}) = \frac{L_{path}(\lambda_{nir2})}{L_r(\lambda_{nir2})} = \frac{L_t'(\lambda_{nir2})}{L_r(\lambda_{nir2})} \end{cases} \tag{4}$$

L_r can be calculated from the pre-generated Rayleigh scattering look-up tables. Based on $R(\lambda_{nir2})$ we can calculate the aerosol optical thickness at λ_{nir2} for each aerosol model as follows:

$$\tau_a(\lambda_{nir2}) = \frac{-b(\lambda_{nir2}) + \sqrt{[b(\lambda_{nir2})]^2 - 4c(\lambda_{nir2})[a(\lambda_{nir2}) - R(\lambda_{nir2})]}}{2c(\lambda_{nir2})} \tag{5}$$

[1]*SeaDAS* is a software package offered by NASA, for the processing, display, analysis, and quality control of ocean color data. See: http://seadas.gsfc.nasa.gov/.

where the quadratic polynomial expanding coefficients, a, b, and c, for each aerosol model can be derived from the pre-generated look-up tables for the R ratio. The atmosphere path radiance can be estimated by the aerosol optical thickness as:

$$R(\lambda) = \frac{L_{path}(\lambda)}{L_r(\lambda)} = a(\lambda) + b(\lambda)\tau_a(\lambda) + c(\lambda)\tau_a(\lambda)^2 \tag{6}$$

while, for each aerosol model, the aerosol optical thickness at λ_{nir1} is:

$$\tau_a(\lambda_{nir1}) = \tau_a(\lambda_{nir2}) \times \frac{c_{ext}(\lambda_{nir1})}{c_{ext}(\lambda_{nir2})} \tag{7}$$

where c_{ext} is the extinct attenuation coefficient on the aerosol model. Then, for each aerosol model, the ratio $R_{model}(\lambda_{nir1})$ between L_{path} and the Rayleigh scattering radiance L_r at λ_{nir1} can be calculated from $\tau_a(\lambda_{nir1})$ using Eq. (6).

Two aerosol models (mod1 and mod2) with $R_{model}(\lambda_{nir1})$ values closest to $R(\lambda_{nir1})$ can then be selected, and the corresponding weight determined by:

$$w = \frac{R_{mod2}(\lambda_{nir1}) - R(\lambda_{nir1})}{R_{mod2}(\lambda_{nir1}) - R_{mod1}(\lambda_{nir1})} \tag{8}$$

Then, the aerosol optical thicknesses at the visible light bands for the two selected aerosol models ($\tau_{a\,mod1}(\lambda)$ and $\tau_{a\,mod2}(\lambda)$) can be calculated using Eq. (7). Also, $R_{mod1}(\lambda)$ and $R_{mod2}(\lambda)$ can be calculated from $\tau_{a\,mod1}(\lambda)$ and $\tau_{a\,mod2}(\lambda)$, respectively, based on Eq. (6). The final aerosol optical thickness $\tau_a(\lambda)$ and $R(\lambda)$ can be estimated as follows:

$$\begin{cases} \tau_a(\lambda) = w \times \tau_{a\,mod1}(\lambda) + (1-w) \times \tau_{a\,mod2}(\lambda) \\ R(\lambda) = w \times R_{mod1}(\lambda) + (1-w) \times R_{mod2}(\lambda) \end{cases} \tag{9}$$

According to Eq. (6), $L_{path}(\lambda)$ can be calculated based on $\tau_a(\lambda)$ and $L_r(\lambda)$. Then, the only unknown in Eq. (1) is the atmosphere diffuse transmittance t_v. For the two selected aerosol models $t_{\upsilon\,mod1}(\lambda)$ and $t_{\upsilon\,mod2}(\lambda)$ can be estimated from the pre-generated look-up tables. Same as Eq. (9), the final $t_\upsilon(\lambda)$ can be derived using the interpolation between $t_{\upsilon\,mod1}(\lambda)$ and $t_{\upsilon\,mod2}(\lambda)$.

Finally, the normalized water-leaving radiance can be derived using:

$$L_{wn}(\lambda) = \frac{L_t'(\lambda) - L_{path}(\lambda)}{t_\upsilon(\lambda)t_s(\lambda)\cos\theta_s}\left(\frac{r}{D}\right)^2 \tag{10}$$

where r and D are instantaneous and mean distances between Sun and Earth.

The key of the above algorithm is the generation of the look-up tables for the Rayleigh scattering, the ratio between path radiance and Rayleigh scattering radiance, and the atmosphere diffuse transmittance. Using the vector radiative transfer model PCOART and the band equivalent Rayleigh scattering optical thickness, the exact look-up tables for the Rayleigh scattering were generated for each band with the relative difference less than 0.25% compared with the results from the Rayleigh-scattering look-up tables in the SeaDAS. For the generation of the look-up tables of the R ratio and the diffuse transmittance, 20 aerosol models were established, including the oceanic aerosol with 98% relative humidity; marine aerosol with relative humidity of 98, 90, 80, 70, and 50%; coastal aerosol with relative humidity of 98, 90, 80, 70, and 50%; troposphere background aerosol with relative humidity of 98, 90, 80, 70, and 50%; and urban aerosol with relative humidity of 98, 90, 70, and 50% (He et al. 2008a).

The validation results using the PCOART simulated top-of-the-atmosphere radiances showed that the atmospheric correction algorithm was accurate, with the retrieval error less than 0.0005 for the water-leaving reflectance $(\pi L_w/F_0 \cos\theta_0)$, meeting the requirement for accurate ocean color atmospheric correction (<0.001). Based on the retrieved normalized water-leaving radiances, the ocean color components such as chlorophyll concentration and suspended particulate matter concentration can be derived with the developed retrieval algorithms. Figures 5 and 6

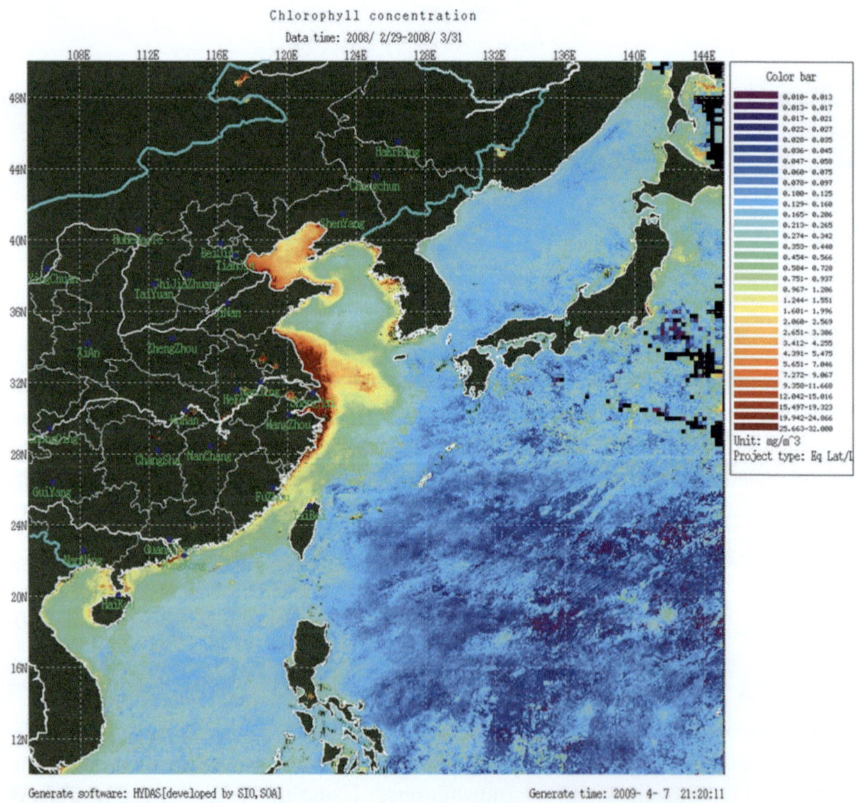

Fig. 5 HY-1B/COCTS monthly mean chlorophyll concentration, March 2008

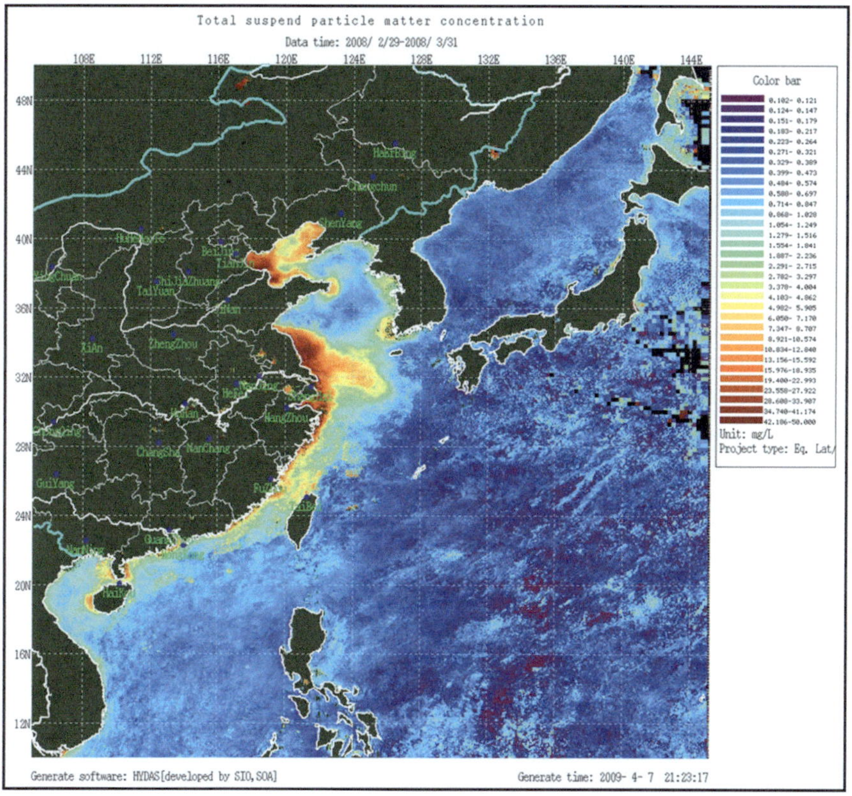

Fig. 6 HY-1B/COCTS monthly mean suspended particulate matter concentration, March 2008

show examples of monthly maps of chlorophyll concentration and suspended particulate matter concentration retrieved from HY-1B/COCTS data. The high chlorophyll and suspended particulate matter concentrations along the coasts and on the continental shelves are well captured by the HY-1B/COCTS.

4 Atmospheric Correction in Turbid Waters Along the Chinese Coastal Regions

Surface waters in the coastal regions off the Chinese mainland are generally highly turbid, especially in the Bohai Sea and near the Chang Jiang river estuary, as shown in Fig. 7. In these highly turbid waters, the operational atmospheric correction based on two NIR bands usually fails, due to a significant contribution of L_w in the NIR bands, so that the assumption of it being negligible at NIR bands is no longer valid. In the past decade, several novel algorithms have been proposed for atmospheric correction in turbid waters along the Chinese coastal regions. Based on the fact that L_w at shortwave-infrared (SWIR) wavelength is always negligible because of strong absorption by pure water, Wang and Shi (2005) used two SWIR bands instead of NIR bands to estimate the aerosol scattering radiances using the same scheme of Gordon and Wang (1994), and observed that L_w matched the in situ data well in the moderately turbid waters of the eastern Chinese seas (Wang et al. 2007). Moreover, for ocean colour sensors lacking the SWIR bands, such as the GOCI, Wang et al. (2012) proposed an iterative scheme to estimate L_w at NIR using a regional relationship between the normalized water-leaving radiance at NIR and diffuse attenuation coefficient at 490 nm, derived from long-term data from the

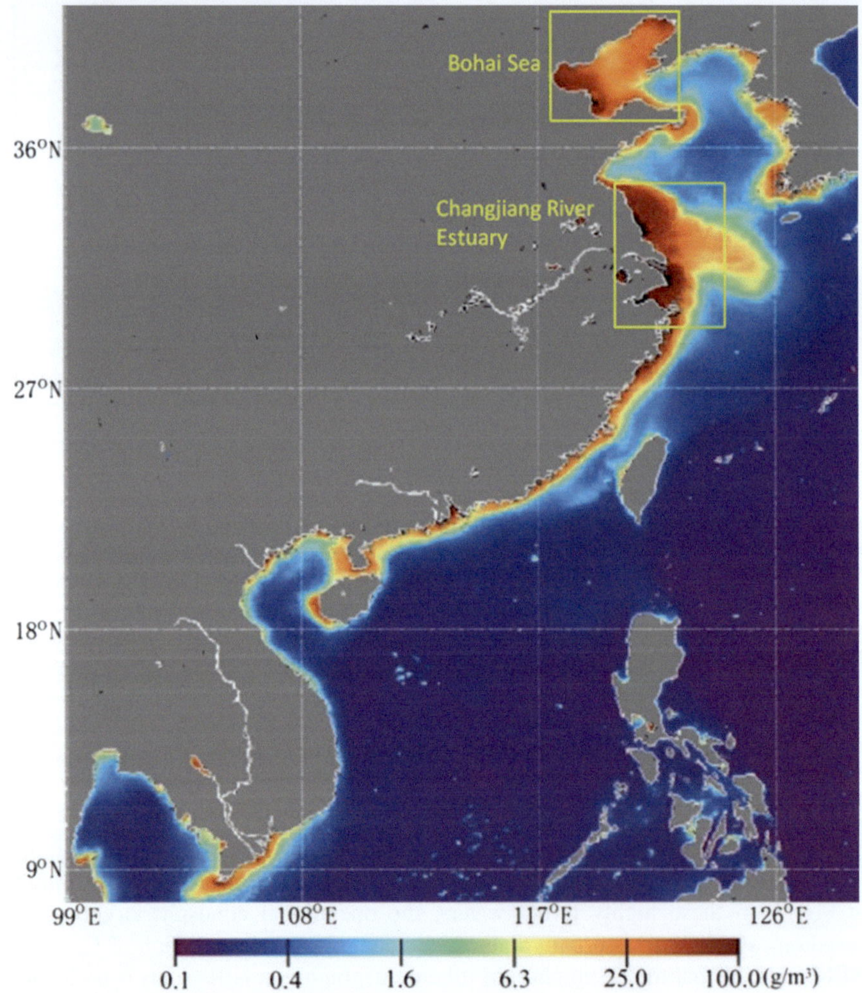

Fig. 7 Mean suspended particulate matter concentration, in the period 2003–2010

MODIS in the western Pacific. In addition, Mao et al. (2014) developed an algorithm to estimate the aerosol scattering reflectance from ocean colour satellite data using the look-up tables of in situ water-leaving reflectance and the assumption of the Angstrom law of the aerosol scattering. Yet, the algorithm only focused on the retrieval of the aerosol scattering reflectance.

Recently, He et al. (2012) proposed a practical atmospheric correction algorithm (called the UV-AC) for highly turbid waters. Instead of conventionally using NIR or SWIR bands as references, the UV-AC algorithm uses the ultraviolet wavelength as the reference band. The principle of the UV-AC algorithm is based on the fact that L_w at ultraviolet wavelengths can be neglected, when compared with that at visible or even NIR wavelengths in highly turbid waters, due to the strong absorption by detritus and Coloured Dissolved Organic Matter (CDOM) (Fig. 8). Unlike most of the atmospheric correction methods for turbid waters, which depend on regional water optical properties, or even aerosol optical properties, the advantage of the UV-AC algorithm is that it does not need any assumption of water optical properties. Validations by both simulated and in situ datasets shows that the UV-AC algorithm is appropriate for atmospheric correction in highly turbid waters (He et al. 2012, 2013).

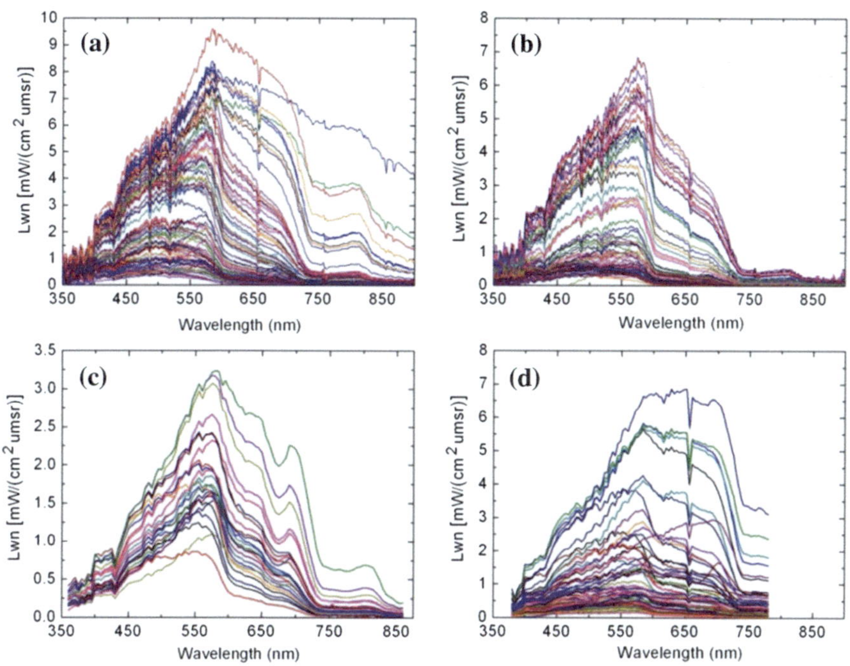

Fig. 8 Normalized in situ water-leaving radiance (L_{wn}) in the Changjiang River Estuary **a** in spring 2003, and **b** in autumn 2003; in the Mississippi River Estuary (**c**); and in the Orinoco River Estuary (**d**). After He et al. (2012)

5 Summary

In the preceding sections, we have reviewed the Chinese ocean colour satellite missions and the data processing techniques that have been developed to tackle the specific conditions of the highly turbid near-coastal waters typical of several Asian Seas. China launched two ocean colour satellites, the HY-1A and the HY-1B, in May 2002 and April 2007, respectively. Two ocean colour sensors, the COCTS and the CZI, were carried aboard both orbital platforms. Both HY-1A and HY-1B were experimental satellites, and focused only on ocean colour observations in the marginal seas off the Chinese mainland and their surrounding areas. Presently, two operational ocean colour satellites, the HY-1C and the HY-1D are being manufactured, and are scheduled to be launched in 2018, with global ocean colour observing capability. Moreover, the next generation Chinese ocean colour satellites, i.e. that of the HY-1E and the HY-1F, is at the planning stage.

Together with the successful launches of these Chinese ocean colour satellites, the corresponding data processing techniques were also developed. A vector radiative transfer model for the coupled ocean-atmosphere system has been developed, to provide a baseline for the data processing to be carried out. Moreover, a cross-calibration method was proposed, in order to remedy the deficiency of any on-orbit calibration instrument on both the HY-1A and the HY-1B, and a practical method for on-orbit estimation of polarization response was also developed, in order to estimate and correct the polarization response effects on the radiance measurement. Based on the radiative transfer model, an operational atmospheric correction algorithm has been developed, specifically for data processing of the Chinese ocean colour satellites, and the core look-up tables have been generated. In addition, several algorithms have been proposed for the atmospheric correction in turbid waters along the Chinese coastal regions, to replace the failing operational atmospheric correction algorithm in these regions.

Acknowledgements We thank Dr. Zoujun You for editing the manuscript. This work was supported by the National Basic Research Programme ("973" Programme) of China (grant #2015CB954002), the National High Technology and Development Program of China (grant 2014AA123301), the National Natural Science Foundation of China (NSFC) (grants #41676170, #41676172, #41476155, and #41621064), and the "Global Change and Air-Sea Interaction" project of China (grant # GASI-03-03-01-01).

References

Acker JG (2015) The color of the atmosphere with the ocean below: a history of NASA's ocean color missions. CreateSpace Independent Publishing Platform, 362pp

Austin RW (1993) Optical remote sensing of the oceans: BC (before CZCS) and AC (after CZCS). In: Barale V, Schlittenhardt PM (eds) Ocean colour: theory and applications in a decade of CZCS experience. Kluwer Academic Publishers, Dordrecht, pp 1–15

Gordon HR (2010) Some reflections on thirtyfive years of ocean color remote sensing. In: Barale V, Gower JFR, Alberotanzam L (eds) Oceanography from space: revisited. Springer, Dordrecht, pp 289–306

Gordon HR, Wang M (1994) Retrieval of water-leaving radiance and aerosol optical thickness over the oceans with SeaWiFS: a preliminary algorithm. Appl Opt 33(3):443–452

He XQ, Pan DL, Bai Y, Zhu QK, Gong F (2007) Vector radiative transfer numerical model of coupled ocean-atmosphere system using matrix-operator method. Sci China Ser D Earth Sci 50(3):442–452

He XQ, Bai Y, Pan DL, Zhu QK (2008a) The atmospheric correction HY-1B/COCTS. Proc SPIE 7147:714717. https://doi.org/10.1117/12.813244

He XQ, Bai Y, Pan DL, Zhu QK, Gong F (2008b) Primary analysis of the ocean color remote sensing data of the HY-1B/COCTS. Proc SPIE 7150:715003

He XQ, Bai Y, Zhu QK, Gong F (2010) A vector radiative transfer model of coupled ocean-atmosphere system using matrix-operator method for rough sea-surface. J Quant Spectrosc Radiat Transfer 111:1426–1448

He XQ, Bai Y, Pan DL, Tang JW, Wang DF (2012) Atmospheric correction of satellite ocean color imagery using the ultraviolet wavelength for highly turbid waters. Opt Express 20(18): 20754–20770

He XQ, Bai Y, Pan DL, Huang NL, Dong X, Chen JS, Chen CTA, Cui QF (2013) Using geostationary satellite ocean color data to map the diurnal dynamics of suspended particulate matter in coastal waters. Remote Sens Environ 133:225–239

He XQ, Pan DL, Bai Y, Mao ZH, Wang TY, Hao ZZ (2016) A practical method for on-orbit estimation of polarization response of satellite ocean color sensor. IEEE Trans Geosci Remote Sens 54(4):1967–1976

Hojerslev NK (1994) A history of early optical oceanographic instrument design in Scandinavia. In: Spinrad RW, Carder KL, Perry MJ (eds) Ocean Optics. Oxford University Press, New York, pp 118–147

Jerlov NG (1968) Optical oceanography. Elsevier oceanography series, vol 5. Elsevier, Amsterdam, 194pp

Jerlov NG (1976) Marine optics. Elsevier oceanography series, vol 14. Elsevier, Amsterdam, 230pp

Maffione RA (2001) Evolution and revolution in measuring ocean optical properties. Oceanogr Ocean Opt 14(3):9–14 (Special Issue)

Mao ZH, Pan DL, Hao ZZ, Chen JY, Tao BY, Zhu QK (2014) A potentially universal algorithm for estimating aerosol scattering reflectance from satellite remote sensing data. Remote Sens Environ 142:131–140

Morel A, Prieur L (1977) Analysis of ocean color. Limnol Oceanogr 22:709–722

Pan DL, He XQ, Zhu QK (2004) In-orbit cross-calibration of HY-1A satellite sensor COCTS. Chin Sci Bull 9(3):521–2526

Preisendorfer RW (1986) Secchi disk science: visual optics of natural waters. Limnol Oceanogr 31:909–926

Wang M, Shi W (2005) Estimation of ocean contribution at the MODIS near-infrared wavelengths along the east coast of the US: two case studies. Geophys Res Lett 32(13):L13606. https://doi.org/10.1029/2005GL022917

Wang M, Tang J, Shi W (2007) MODIS-derived ocean color products along the China east coastal region. Geophys Res Lett 34(6):L06611. https://doi.org/10.1029/2006GL028599

Wang M, Shi W, Jiang LD (2012) Atmospheric correction using near-infrared bands for satellite ocean color data processing in the turbid western Pacific region. Opt Express 20(2):741–753

Wernand MR (2010) On the history of the Secchi disc. J Eur Opt Soc 5. https://doi.org/10.2971/jeos.2010.10013s (Rapid Publications)

An Introduction to Microwave Remote Sensing of the Asian Seas

Martin Gade and Ad Stoffelen

Abstract In this chapter passive and active microwave sensors are introduced, their basic measurement principles are described, and few examples of microwave (MW) remote sensing of the Asian Seas are given. The space-borne sensors under consideration are MW radiometer, altimeter, scatterometer, and synthetic aperture radar (SAR). It is demonstrated that changes in the sea surface roughness can be used to infer ocean surface vector wind fields, but also that this roughness may depend on other parameters such as SST, waves and currents.

Keywords RADAR · Microwaves · Altimeter · Radiometer · Scatterometer SAR · Bragg scattering · Wind vector · Normalized radar cross section

1 Introduction

Microwaves are commonly referred to as that part of the electromagnetic wave spectrum with wavelengths ranging from 1 mm to 1 m, corresponding to frequencies between 300 GHz and 300 MHz (Table 1). In most spectral ranges microwaves penetrate through the Earth's atmosphere with only little loss, which loss is due to components like dry gases and water vapor, but also to attenuation by water droplets, particularly at shorter microwave (MW) wavelengths (higher frequencies). Therefore, sensors working in the MW range of the electromagnetic spectrum have been deployed on satellites from the early days of space-borne earth observation (EO) (Mitnik and Gade, this volume).

Naturally, passive and active MW sensors differ technically, and different (atmospheric, oceanic, and land) parameters can be inferred from their recorded signals. Whereas passive MW sensors (radiometers) measure thermal electromagnetic

M. Gade (✉)
Institut für Meereskunde, Universität Hamburg, Hamburg, Germany
e-mail: Martin.Gade@uni-hamburg.de

A. Stoffelen
Royal Netherlands Meteorological Institute (KNMI), De Bilt, The Netherlands

V. Barale and M. Gade (eds.), *Remote Sensing of the Asian Seas*,
https://doi.org/10.1007/978-3-319-94067-0_4

Table 1 Commonly used microwave frequency bands and corresponding wavelengths

Band	Frequency band (GHz)	Wavelength band (cm)
L	1.0–2.0	15.0–30.0
S	2.0–4.0	7.5–15.0
C	4.0–8.0	3.8–7.5
X	8.0–12.0	2.5–3.8
Ku	12.0–18.5	1.6–2.5
K	18.5–24.0	1.3–1.6
Ka	24.0–40.0	0.8–1.3
V	40.0–75.0	0.4–0.8
W	75.0–110.0	0.3–0.4

radiation emitted by the earth and its atmosphere, active MW sensors such as radar transmit microwaves (usually in pulses) and record the radiation that is scattered back from a target. The transmitted and/or received electromagnetic radiation is often polarized, i.e., its electric field is oriented only in a certain direction: vertically polarized means an orientation in the plane of the antenna's nadir and look directions, and horizontally polarized means an orientation orthogonal to that (Fig. 1). A polarimetric radar can measure amplitude and phase relations between the vertically and horizontally polarized transmitted and received MW signals.

Here, we introduce the basic principles of space-borne MW remote sensing, with emphasis on applications in the Asian Seas, starting with (passive) radiometers and continuing with the three main active sensors, altimeters, scatterometers, and synthetic aperture radar (SAR). Further examples will be provided in the chapters herein.

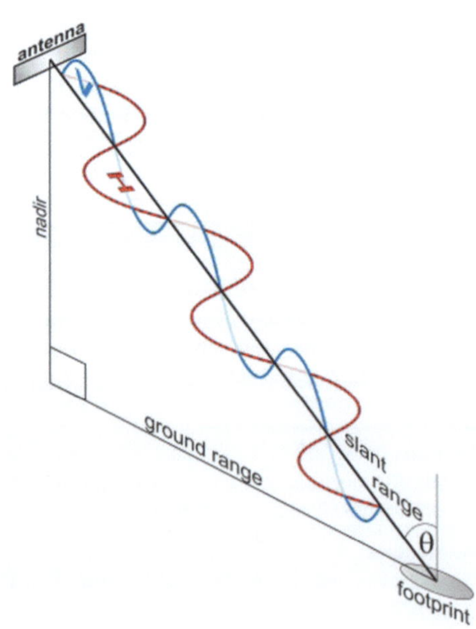

Fig. 1 Schematic demonstrating the polarization of electromagnetic waves. Red: horizontal polarization; blue: vertical polarization. θ is the incidence angle

2 Radiometer

An EO microwave radiometer (MWR) allows deriving meteorological quantities such as vertical temperature and humidity profiles, columnar water vapor content, or columnar liquid water path and precipitation under nearly all weather conditions. MWRs are usually equipped with multiple receiving channels to derive wavelength-dependent emission characteristics of the atmosphere and Earth's surface from the measured brightness temperatures, T_b, whose components are radiations from both the atmosphere and the sea surface (Ulaby et al. 1981):

$$T_b = T_{up} + e\Gamma T_{surf} + (1 - e)\Gamma T_{down} \tag{1}$$

where T_{up}, T_{surf}, and T_{down} are the atmosphere's upwelling radiation, the radiation from the surface, and the atmosphere's downwelling radiation, reflected by the surface, respectively, e is the surface emissivity, and Γ is the transmission in the atmosphere. The horizontal resolution δ of a conventional radiometer is given by its aperture size L, the MW wavelength λ and the distance R from the Earth's surface as (Ulaby et al. 1981):

$$\delta = 1.4\frac{\lambda R}{L} \tag{2}$$

where the factor 1.4 reflects the need for high beam efficiency (Skou 2014).

Whereas in the infrared spectral range the ocean surface T_{surf} mainly depends on the Sea Surface Temperature (SST), at microwaves the wind speed is a driving factor, with increasing emissivity e for increasing wind, due to both increased roughness and foam. However, substantially varying emissions from clouds and precipitation exist, thereby making space-borne measurements of surface parameters more difficult to obtain in cases of precipitation. This is even aggravated by the fact that rain clouds are spatially rather variable on scales of a typical MWR footprint (50 km) and complex to correct for.

Despite those technical constraints polarimetric MWRs have been continuously providing synoptic ocean wind speed patterns with great fidelity since 1987, when the first Special Sensor Microwave Imager (SSM/I) was launched, later on followed by the WindSat Polarimetric Radiometer (launched in 2003). Figure 2 shows a 3-day composite of WindSat wind fields in Southeast Asia measured on 19–21 October 2010.

WindSat measures T_b at different microwave frequencies and polarizations to obtain information on wind direction (Gaiser et al. 2004). Taking advantage of the fact that the horizontally polarized brightness temperature at 19.4 GHz exhibits a strong dependency on wind speed (stronger than the vertically polarized T_b) and that the brightness temperatures at both polarizations depend on wind direction, global wind fields have been measured reaching accuracies better than 2 m/s.

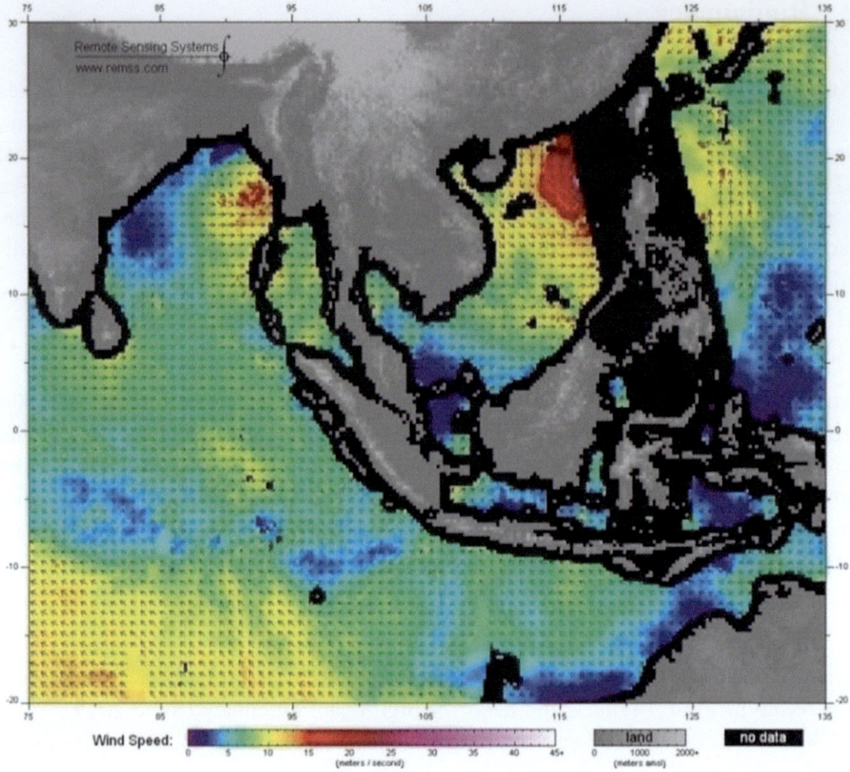

Fig. 2 3-day composite (19–21 Oct 2010) of WindSat wind fields over south-east Asian waters (Wentz et al. 2013). Imprints of the western part of super typhoon *Megi* can be seen in the East China Sea, and of cyclone *Giri* in the eastern Bay of Bengal

In Fig. 2 super-typhoon *Megi* in the East China Sea and the cyclone *Giri* in the Bay of Bengal can be identified through wind speeds exceeding 20 m/s. No wind fields are available within 30–50 km of the coasts, because of T_b contaminations from the land.

At low MW frequencies (L band) T_b also depends on the sea surface salinity (SSS) and on ice cover. Sophisticated MW radiometers such as the Microwave Imaging Radiometer using Aperture Synthesis (MIRAS) of the Soil Moisture and Ocean Salinity (SMOS) mission are taking benefit of this dependence and provide global SSS measurements at accuracies reaching 0.1 PSU (practical salinity units, 1 PSU = 1 g salt per kilogram sea water; typical SSS of the open oceans are around 35 PSU) (Font et al. 2004), along with measurements of sea ice cover and thickness (Tian-Kunze et al. 2013).

3 Altimeter

A radar altimeter is an active microwave instrument that measures the vertical distance to the Earth's surface. Very short electromagnetic pulses are transmitted by the sensor and their echo from the Earth's surface (or, more precisely, from the altimeter footprint) is recorded. The (vertical) distance determines the delay of that echo, and the surface structure the echo's strength and shape. The pulse length, τ, determines the vertical resolution of the altimeter,

$$\Delta h = \frac{c\tau}{2} \tag{3}$$

where c is the speed of light. It follows that extremely short pulses would be needed to realize vertical resolutions on the order of centimeters. At the same time, a higher transmit power would be required to keep the signal-to-noise ratio (SNR) sufficiently high. In practice altimeters use a short pulse to realize a high horizontal resolution, which is much smaller than the width of the antenna beam. The instrument measures the average delay of the half-power signal level at the rising edge of the reflected pulse (see Fig. 5) and by averaging over some 1000 pulse high accuracies of the vertical distance R_{alt} (Fig. 3) are achieved.

The dynamic sea surface height is derived from the measured distance of the altimeter to the sea surface, R_{alt}, the height of the geoid with respect to a reference ellipsoid, h_{geoid}, and corrections for the tide, h_{tide}, and atmospheric pressure, h_{atm}, as follows:

$$h_{dyn} = H_{sat} - R_{alt} - h_{geoid} - h_{tide} - h_{atm} \tag{4}$$

where H_{sat} is the satellite height above the reference ellipsoid. Figure 3 depicts the distances that are relevant for satellite altimeter measurements of ocean surfaces.

Since the beginning of continuous altimeter measurements from space, in the early 1990s, a global sea-level rise of about 70 mm has been measured (approx. 3 mm/year). In addition, precise measurements of the dynamic sea surface height,

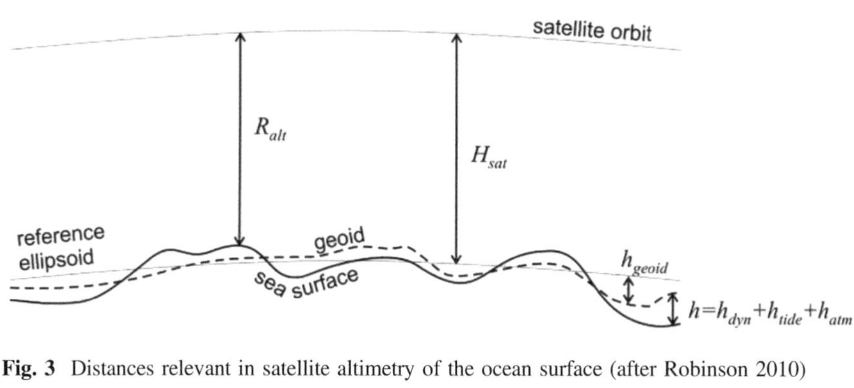

Fig. 3 Distances relevant in satellite altimetry of the ocean surface (after Robinson 2010)

h_{dyn}, allow determining not only regional and global sea levels, but also oceanic circulation (including mesoscale eddies), Rossby and Kelvin waves, and seafloor topography. Figure 4 demonstrates the overall accuracy of a few centimeters, at which modern altimeters perform. The two panels show sea level anomalies (SLA) measured by the radar altimeters aboard the European EnviSat (upper panel) and the French-American Jason-1 (lower panel), each during two consecutive overpasses over the Western Pacific Ocean (green lines in the small maps) on 11 March 2011.

In both panels the blue curves were measured before a 9.0 magnitude earthquake off Japan caused a tsunami, and the red curves show the tsunami waves (whose amplitudes are on the order of decimeters on the open ocean).

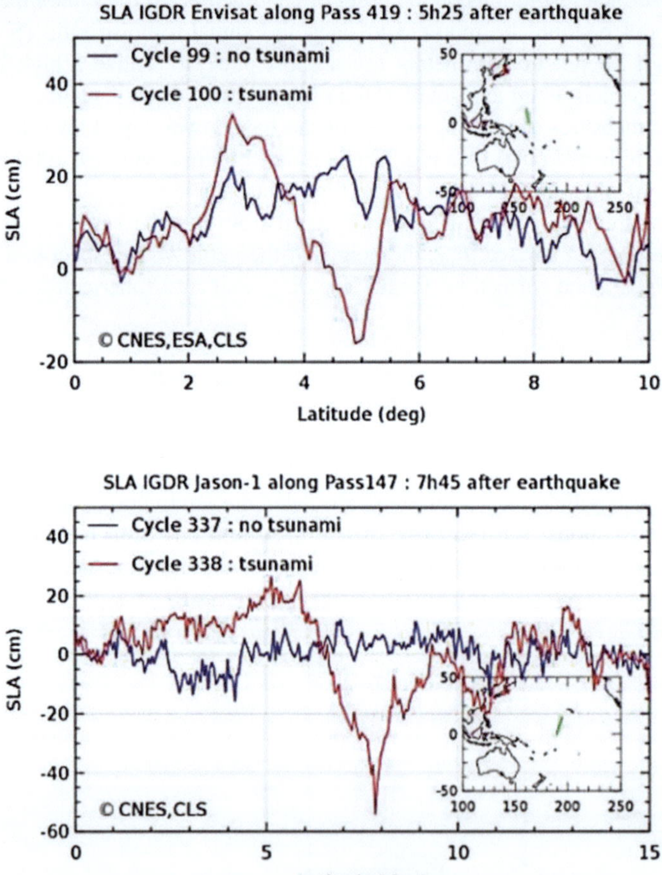

Fig. 4 Sea level anomalies (SLA) for Envisat pass 419 (top; green line in the map) and Jason-1 pass 147 (bottom; green line in the map) for the cycle during the tsunami (in red) and the previous cycle (in blue). *Credits* CNES/ESA/CLS (AVISO 2011)

Measurements of the power and shape of the radar echo are used to determine significant wave height (SWH) and mean square slope (mss) of the ocean surface, but also near-surface winds (although without wind direction). Figure 5 demonstrates the underlying principle (Berger 1972): reflection of the altimeter pulse from a flat surface results in an abrupt (steep) increase in the power of the recorded echo (upper left panel and green curve in the lower panel of Fig. 5); in contrast, rough surfaces lead to a smearing of the pulse's edge, because the altimeter pulse is reflected from the waves' crests earlier than from their troughs (middle and right upper panel, purple and red curves in the lower panel).

The next generation of altimeters such as the Synthetic Aperture Radar (SAR) Interferometer Radar ALtimeter (SIRAL) and the SAR Radar ALtimeter (SRAL) also use the Doppler information of the backscattered signal, thereby allowing wider swaths at finer spatial resolution along the satellite track (down to 300 m). In addition, interferometric techniques deploying two antennas are used to measure slopes perpendicular to the satellite track and to obtain spatial resolutions finer than the swath width.

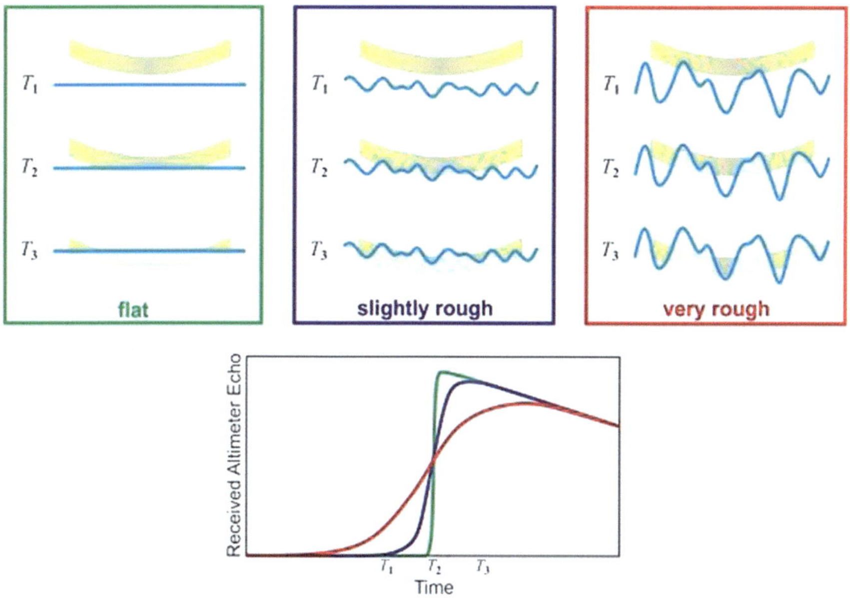

Fig. 5 Upper row: altimeter pulse (in shaded yellow) hitting water surfaces at different sea states (and significant wave height, SWH); lower panel: corresponding shape of the altimeter echo. The rougher the sea surface, the more smeared is the pulse's edge

4 Scatterometer

A radar scatterometer transmits a pulse of microwave energy towards the Earth's surface at oblique incidence angles and measures the backscattered signal power. If the target is wet, or moist, then the penetration depth of the microwaves is rather small, and for saline water it is only a fraction of the microwave's wavelength. Therefore, the microwaves "feel" only the ocean's surface, and the strength of the backscattering mainly depends on the surface roughness (Fig. 6). In case of an ideally flat surface the transmitted microwaves would simply be scattered away from the antenna, whereas with increasing roughness a larger fraction of the transmitted signal is scattered back to the sensor.

It is common to use the normalized radar cross section (NRCS), σ^0, which is the dimensionless radar cross section per unit area and which is computed from the signal power measurement using the distributed target radar equation (Stoffelen 1998),

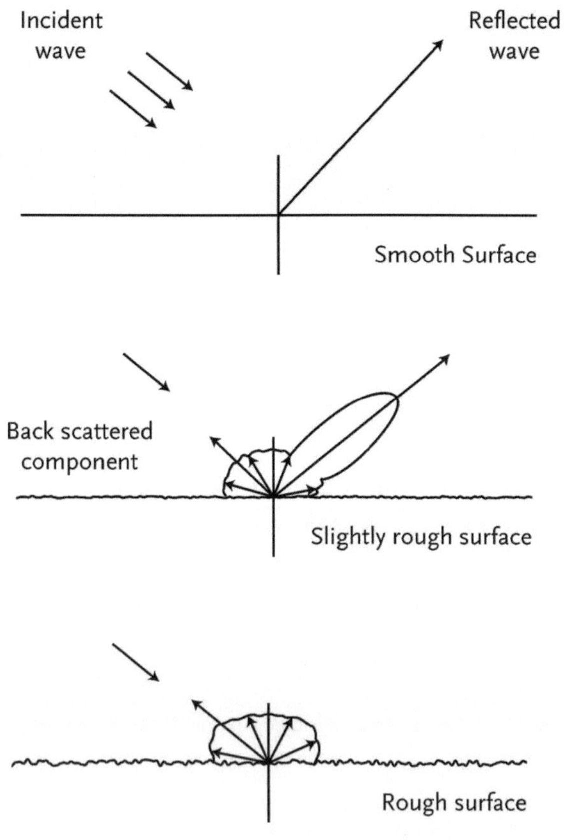

Fig. 6 Schematic representation of microwave reflection and scattering from water surfaces of different roughness (Stoffelen 1998)

$$P_r = \frac{\lambda^2}{(4\pi)^3} \int_F \frac{P_t \sigma^0 G^2}{R^4} dF \tag{5}$$

where P_r and P_t are the received and transmitted power, respectively, F is the area illuminated by the radar, G is the antenna gain and R is the range (distance) of the target. The size of the resolution cell of a scatterometer in ground range (look) direction is determined by the pulse length and incidence angle, and in azimuth (flight) direction by the antenna beam width and the distance to the ground. Scatterometers are shown to be extremely stable, well below 0.1 dB (Belmonte et al. 2017).

The primary application of space-borne scatterometry has been measurements of near-surface winds over the ocean (Stoffelen 1998); hence they are known as wind scatterometers. The radar backscatter over the ocean results from scattering off wind-generated capillary-gravity waves, which are generated by, and generally in equilibrium with, the near-surface wind over the ocean. The scattering mechanism is known as Bragg scattering, which occurs from sea surface waves that are in resonance with the microwaves (Wright 1968), as demonstrated in Fig. 7.

The Bragg resonance condition then reads

$$\lambda_B = \frac{\lambda}{2 \sin \theta} \tag{6}$$

where λ_B is the length of the so-called Bragg waves, and λ is the length of the microwaves hitting the sea surface at incidence angle θ (Fig. 7). The factor 2 in the denominator results from the fact that the MW radiation is incoming and returning, thereby travelling twice the distance. At oblique incidence angles θ the NRCS increases with increasing wind speed (Fig. 8a).

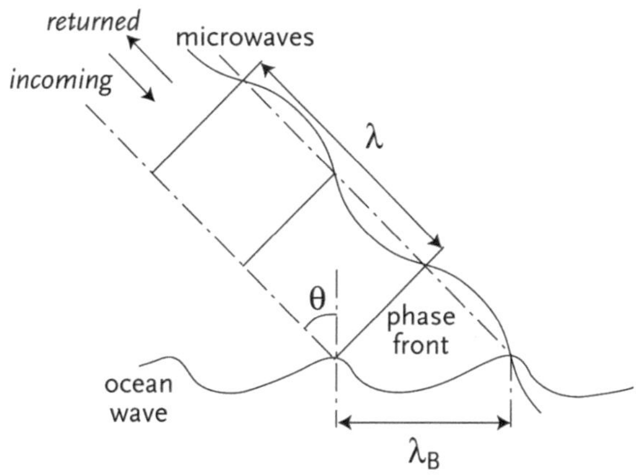

Fig. 7 Schematic showing the Bragg scattering principle: microwaves of length λ hit the water surface at incidence angle θ and are resonantly backscattered by small ocean surface waves of length λ_B (Stoffelen 1998)

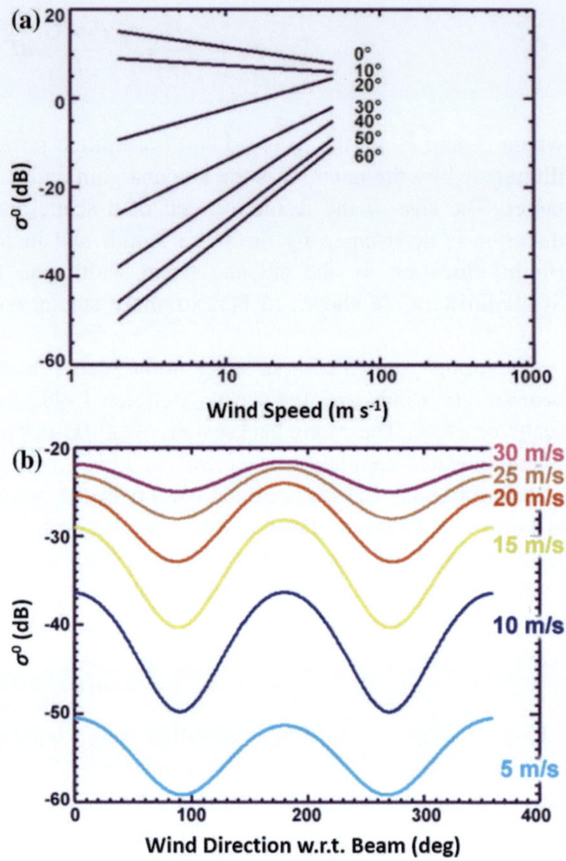

Fig. 8 Dependence of the backscattered radar power on **a** wind speed for different incidence angles as indicated and **b** radar look direction (relative to the wind direction) for different wind speeds as indicated; *credits* Zorana Jelenak, ©UCAR/COMET®

Since these small waves generally grow and propagate in the direction of the wind, the Bragg resonance is strongest in the up- and down-wind scattering directions, while more modest in the cross-wind directions (Fig. 8b). This scattering anisotropy is captured in empirical (Hersbach et al. 2007; Stoffelen et al. 2017) and theoretical (Fois et al. 2015) Geophysical Model Functions (GMF) that describe the relationship between σ^0, wind vector and microwave beam look angles.

A modest limitation in scatterometry exists in the fact that the ocean backscatter anisotropy is dominated by a double harmonic dependency. This is, the ocean return is very similar for viewing directions different by 180° (Fig. 8b). This implies that inverted winds are ambiguous and generally more than one wind vector (2–4) closely fits the measured set of backscatter values. However, processing techniques for wind direction ambiguity removal are well evolved and fail in less than 1% of cases (Portabella and Stoffelen 2004; Lin et al. 2016).

In order to measure the backscattering anisotropy a scatterometer measures several backscatter values over the same target Wind Vector Cell (WVC) on the ocean surface from different azimuth angles. This is realized either by three separate antennae looking into different azimuthal directions (e.g., on ERS1/2 and MetOp-A/B) or by a rotating antenna (e.g., on QuikScat and ScatSat-1). Using the GMF and combining the σ^0 measurements from different azimuth angles the near-surface wind vector over the ocean surface can be obtained by inversion in a wind vector retrieval step. This inversion procedure is non-linear and based on an accurate knowledge of the instrument calibration, noise and GMF.

Figure 9 shows a preliminary, but already quite convincing depiction of a tropical cyclone in the Bay of Bengal, cyclonic storm *Nada* observed on 30 November 2016 by the Indian ScatSat. Despite the preliminary nature of the wind products, the typhoon shows a well-defined eye wall of up to 25 m/s, humidity convergence and land-sea effects.

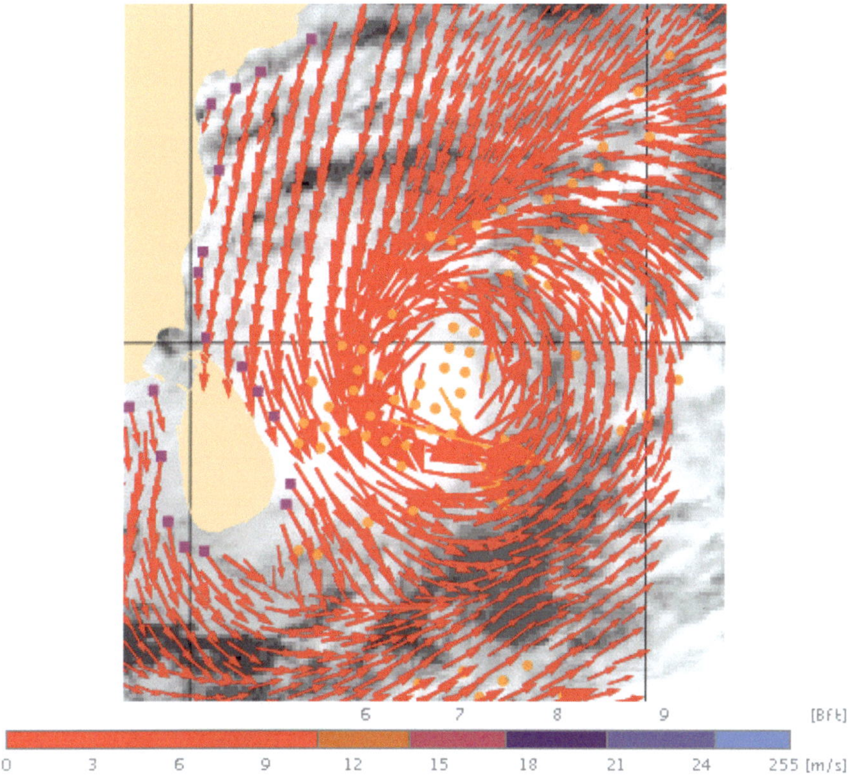

Fig. 9 ScatSat wind vectors showing cyclonic storm *Nada* on 30 November 2016, 3:45 UTC, in the Bay of Bengal (preliminary backscatter processing by ISRO (v1.0), preliminary wind processing by KNMI). The grey background is geostationary infrared imagery. Magenta squares denote land presence, orange arrows a spatially inconsistent WVC and orange dots indicate that the KNMI Quality Control Flag is set. © EUMETSAT/KNMI (Multi-platform viewer: projects.knmi.nl/scatterometer/tile_prod/tile_app.cgi)

5 Synthetic Aperture Radar

The azimuthal resolution of an imaging radar basically depends on the antenna length L (Fig. 10): longer antennas allow imaging at finer resolution, and vice versa. This fact, however, would make it impossible to use a radar aboard a satellite, since the antenna length (i.e., its aperture) would have to be on the order of kilometers to allow 'proper' spatial resolutions of some meters at most. This is why the radar's real aperture is synthetically enhanced by taking advantage of the phase, or Doppler, shift of the backscattered signal (with respect to the transmitted signal): target "A" in Fig. 10 causes a positive Doppler shift, while the antenna is at position "1", zero Doppler shift at position "2", and a negative Doppler shift at position "3". Including this so-called *Doppler history* makes the aperture (synthetically) much longer, and the azimuthal resolution becomes

$$\delta_{SAR} = \frac{L}{2} \tag{7}$$

High spatial resolutions in range direction are realized through pulse compression: a 'longer' pulse is generated together with a broad bandwidth, i.e., the (microwave) frequency is modulated. The echo of such a *chirp* is matched with its original, transmitted signal, thereby resulting in a shortened pulse of higher energy; hence this technique is called *pulse compression*. These techniques, along with further sophisticated additions, allow SAR systems to provide radar images with

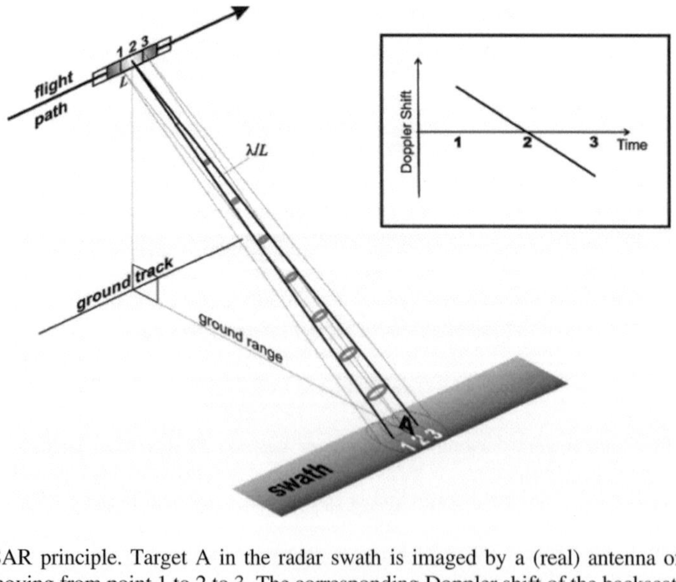

Fig. 10 SAR principle. Target A in the radar swath is imaged by a (real) antenna of length L, which is moving from point 1 to 2 to 3. The corresponding Doppler shift of the backscattered radar pulses is depicted in the small panel on the upper right

pixel sizes reaching well below 1 m^2, even though the data were acquired at orbit heights of several 100 km.

Most space-borne SAR sensors are operating at C or X bands and at incidence angles of 20°–50°, similar to wind scatterometers. Three different acquisition modes are commonly used, Stripmap, ScanSAR, and Spotlight (Fig. 11).

The classical acquisition mode is Stripmap, in which the radar is sensing at constant incidence and azimuth angles (Fig. 11a). In ScanSAR mode multiple parallel swaths are alternatingly scanned by the radar, and the resulting swath is much wider (Fig. 11b; also see the example shown in Fig. 12). In Spotlight mode the azimuthal look direction of the radar is continuously changed in such a way that a single spot at the Earth's surface is imaged for a longer time (Fig. 11c) and at higher spatial resolution, but at the expense of lower spatial coverage.

SAR sensors measure the NRCS, similar to scatterometers. Therefore, GMFs derived from abundant wind scatterometer measurements may be used to infer wind information from SAR (Portabella et al. 2002), but only either wind speed or direction. Since SARs provide a single azimuth view, one cannot infer whether σ^0 variations are due to wind speed (along-view wind vector component) or wind direction (across-view wind vector component) variation. Horstmann et al. (2000) describe that wind direction on the km-scale may be obtained at moderate winds by evaluating wind streaks in SAR images and Mouche et al. (2012) complete the SAR wind vector by Doppler measurements. Table 2 briefly summarizes some qualities of wind measurements carried out by different MW sensors.

Figure 12 demonstrates how wind-speed variations show up on SAR imagery. The Sentinel-1A SAR-C image was acquired on 29 August 2016 over the North-Western Pacific Ocean, off Japan, and shows VV polarization sea surface manifestations of the typhoon *Dindo*. First, note the along-swath and, to a much lesser extent, the across-swath bands, which are calibration artefacts, illustrating the inherent difficulty of SAR calibration, as foot-noted below Table 2. Second, the typhoon's eye can be seen as a dark circular patch, because of the lower wind speed in that area, resulting in a lower NRCS. Curvilinear features in the immediate

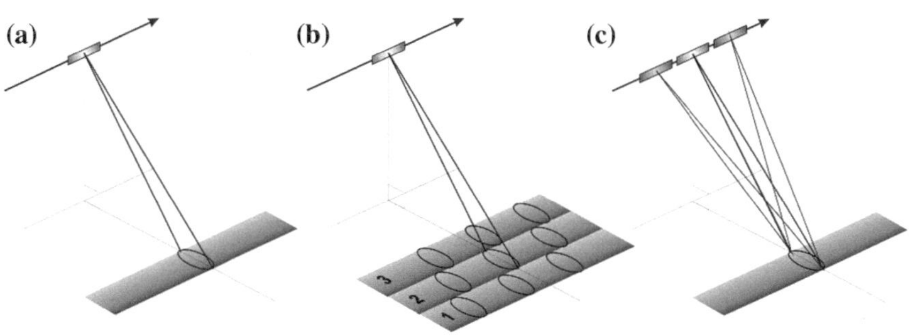

Fig. 11 SAR acquisition modes: **a** Stripmap. **b** ScanSAR. **c** Spotlight

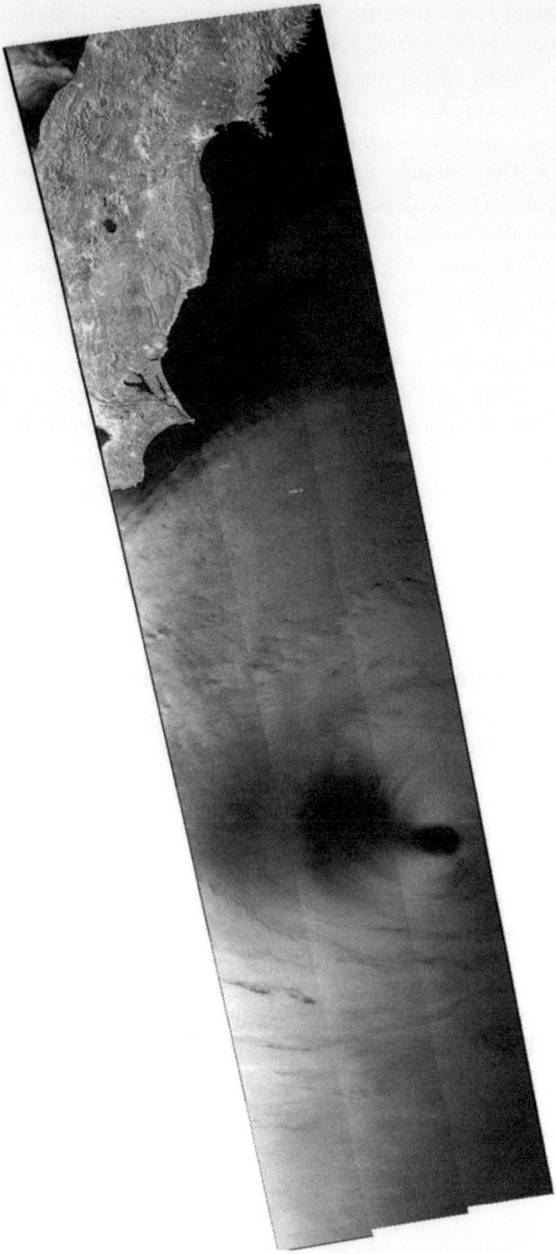

Fig. 12 Sentinel-1A SAR-C image composite (approx. 250 km × 1200 km) of the North-Western Pacific Ocean, off Japan, acquired on 29 August 2016 at 08:33 UTC and showing VV polarization sea surface manifestations of the typhoon *Dindo*. © ESA 2016

Table 2 Characteristics of widely used global ocean satellite wind measurement systems

Quality parameter	Radiometer	Scatterometer	SAR
Wind vector	Only speed[a]	Yes	Partial information[b]
Spatial resolution (km)	50	20	1
Temporal sampling	>Twice daily	>Twice daily	Infrequent
Timeliness[c] (h)	3	½–3	3
Climate data record	Yes, poor in rainy areas	Yes	Fragmented[d]

[a]WindSat provides credible wind direction for speeds above 8 m/s
[b]Only one component is measured, with similar sensitivity to both speed and direction (see text)
[c]Target time, after which data products are available
[d]SAR systems are difficult to calibrate, proven instrument stability over time is relatively low

vicinity of the eye indicate the local wind direction. Vast areas to the west appear also dark, expectedly because here the wind direction was perpendicular to the radar look direction (crosswind).

Irregular patches north of the eye are due to heavy precipitation, which causes additional surface roughness by wind downbursts (Lin et al. 2015) or, at downpour locations, reduced roughness due to induced turbulence in the water by droplet impact, among others (Alpers et al. 2016). In addition, heavy precipitation causes an attenuation of the MW signal in the atmosphere and/or its reflection from hydrometeors (Melsheimer et al. 1998), particularly at the shorter wavelengths. Also note the sharp edge of the typhoon, just off the Japanese coast, and the strong offshore wind in the upper left image corner.

In Fig. 13 an example is shown that demonstrates how surface current variations can manifest in SAR imagery. The image was acquired by the Sentinel-1A SAR-C on 24 August 2017 and shows the northern Bay of Bengal, off the Mouths of the Ganges. Along the entire coast bright features (A) stretch from the coast offshore and are caused by the sea bottom topography in the shallow coastal waters. Variations of the local current over this topography causes surface roughness variations through convergence and divergence (Alpers and Hennings 1984). These roughness variations, i.e., the spatial variation of the Bragg waves' spectral density, can be seen by the SAR.

Further offshore current fronts (B) cause surface roughness variations, and so do oceanic internal waves (C), which can be found all over the lower image part. Also note a narrow band of heavy rain (R) on the coast, which gives rise to a strong surface roughness enhancement (wind downburst), hence to higher image brightness.

The SAR principle builds upon the Doppler shift of the received signal, which is used to achieve higher azimuthal resolutions (see above). This also requires, however, that the signal is not disturbed by any additional Doppler shift, e.g., from a moving target. Such motion causes a vertical shift of the Doppler curve shown in Fig. 10, whose abscissa intersection then deviates from time "2". As a result the target is shifted in the SAR image in azimuthal direction.

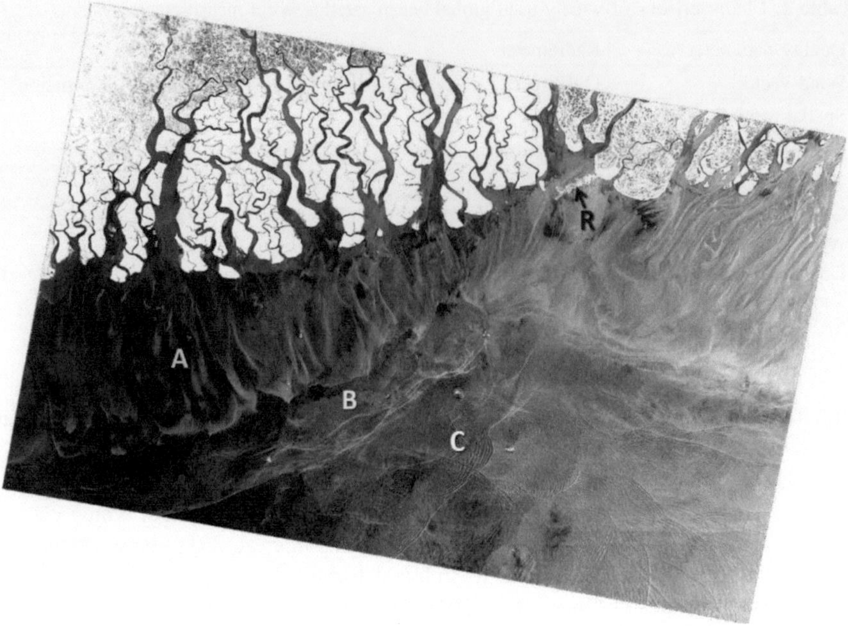

Fig. 13 Sentinel-1A SAR-C image (257 km × 168 km) of the northern Bay of Bengal, off the Mouths of the Ganges, acquired on 24 August 2017 at 23:56 UTC. Variable sea surface currents (see text) modify the sea surface roughness that gives rise to radar backscattering. © ESA 2017

Figure 14 shows examples of this artifact: the underlying Sentinel-1A SAR-C image of the Strait of Hormuz was acquired on 24 December 2016 and shows, among others, various vessels as bright spots. In the fragment on the right some vessels (marked "A") are shifted upwards, off their wakes ("w"), because they were moving away from the SAR [Sentinel-1A was in a descending orbit and was looking from the right (east)]. In contrast, vessel "B" is shifted downwards, because it was moving towards the SAR. This effect, often referred to as *ship-off-the-wake effect*, affects the SAR imaging of all moving targets, and of a moving sea surface in particular.

Figure 14 also shows another artifact: strong scatterers may cause "ghosts" in areas of significantly lower NRCS and at azimuthal distances that depend on the SAR technical setup and on the imaging geometry. All vessels "A" in Fig. 14 have a "ghost" at 4.7 km azimuthal distance (marked by the black lines).

Ocean surface waves are imaged by SAR sensors, and their imprints can be delineated from SAR images; however, depending on their length and height (amplitude) the SAR imaging of those waves may become strongly non-linear (Alpers and Rufenach 1979). If the orbital motion of long ocean waves is too strong (i.e., the waves are too steep), the azimuthal shift of the scatterers in the SAR image can no longer be neglected, thereby making it difficult to infer quantitative information on the wave field from SAR imagery.

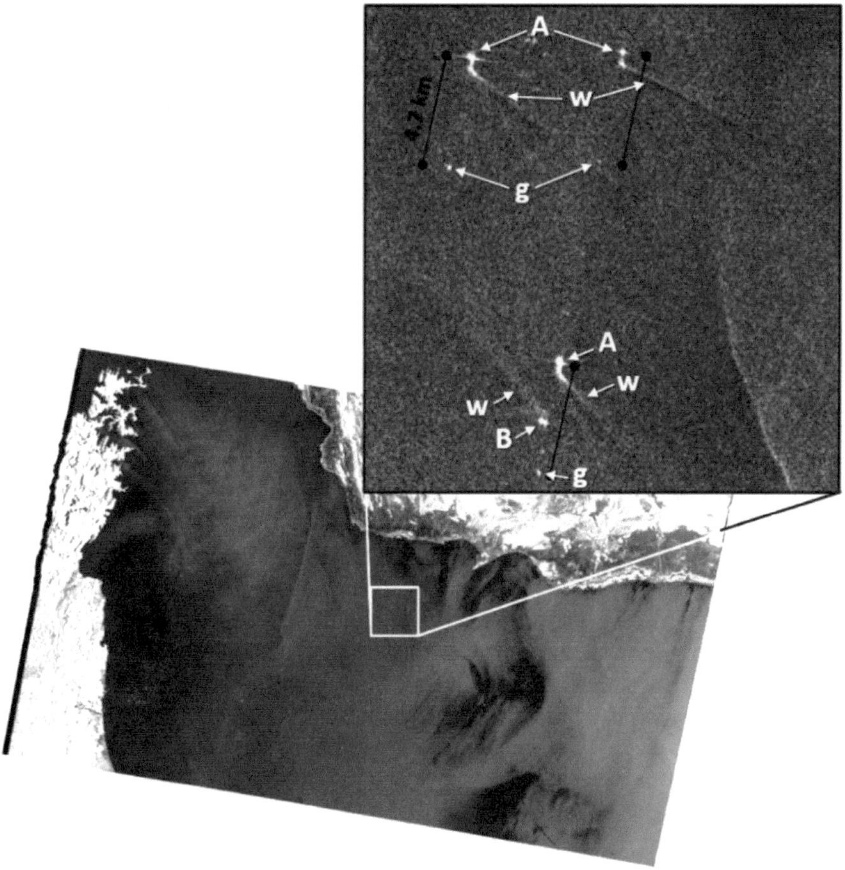

Fig. 14 SAR artifacts in a Sentinel-1A SAR-C image of the Strait of Hormuz acquired on 24 December 2016 at 02:07 UTC (277 km × 170 km). The fragment on the right (20 km × 20 km) shows moving ships ('A' and 'B') that are mapped off their wakes ('w'), as well as ghosts ('g') of the strongly reflecting vessels. SAR data © ESA 2016

6 Summary and Outlook

Passive and active MW sensors have been deployed on earth-orbiting satellites from the early days of satellite oceanography and are capable of measuring a variety of oceanic and atmospheric parameters at sufficiently high spatial resolutions. Whereas radiometer, scatterometer, and altimeter products are global observations used for climate monitoring and as invaluable input for climate models, SAR data are used for small-scale processes and regional monitoring purposes, particularly in coastal waters.

The development of EO wind and stress measurement capabilities is ongoing. This involves expansion of an international virtual constellation (Stoffelen et al., this volume), resolution and coverage enhancement (Vogelzang and Stoffelen 2017), using microwave cross-polarization for higher scatterometer sensitivity to extreme winds (Van Zadelhoff et al. 2014), combination of multiple microwave wavelengths (Wang et al. 2017) and combined wind and wave measurements, among others. New techniques include for example scatterometer Doppler capability to infer mean ocean surface currents (Fois et al. 2016) in addition to wind, and GNSS-R, a method to measure ocean reflections of L-band microwave emissions of opportunity from navigation signals of Global Navigation Satellite Systems (GNSS), see (Zavorotny et al. 2014).

SAR imagery can be used to infer a wealth of oceanic and atmospheric parameters in the marine environment. Marine pollution can be monitored (Gade et al. 1998; Brekke and Solberg 2005; Gade et al., this volume), atmospheric and oceanic dynamics can be quantified on smaller scales (Gade et al. 2013; Li and Ren, this volume), oceanic internal waves can be observed (Alpers and Vlasenko, this volume), and wave and current fields can be calculated (Shao et al., this volume) or can be used to infer underwater bottom topography (Wensink and Campbell 1997; Brusch et al. 2011). Surface currents can be retrieved using data from Along-Track Interferometric SAR systems (Goldstein and Zebker 1987; Romeiser 2014) and by analyses of the Doppler centroid (Johannessen et al. 2008).

However, we also note that most SAR image features can be observed only at certain wind speed ranges: high winds will always be the dominating cause of the surface roughness, thereby hiding the effects that might be of interest. For the sake of shortness, Table 3 provides a list of features that can be seen on SAR imagery, along with their length scales, measurement and imaging parameters, and the wind speed range, at which they are observable (Holt 2004). For a comprehensive scientific background and a detailed description of various SAR applications in the marine environment the reader is referred to Jackson and Apel (2004).

Table 3 Features on SAR imagery, after Holt (2004)

Feature	Scale	Derived measurement	Imaging mechanism	Wind ($m\ s^{-1}$)
Ocean features				
Surface waves	100–600 m wavelength	Wavelength Propagation direction Wave height	Tilt Hydrodynamic Velocity bunching	3–40
Internal waves	0.3–3 km wavelength	Wavelength Direction Amplitude Mixed layer depth	Convergence/divergence Surfactants	2–10
Internal tides	10–20 km	Wavelength Direction	Interaction of centimeter waves/currents/surfactants	3–7

(continued)

Table 3 (continued)

Feature	Scale	Derived measurement	Imaging mechanism	Wind ($m\ s^{-1}$)
Currents and fronts	1–100 km	Location Shear Strain Velocity	Shear/convergence Convergence Wind stress Surfactants	3–10 3–10 3–10 3–7
Eddies	1–200 km diameter	Location and source Diameter Velocity Shear Strain	Shear/convergence Wind stress Surfactants	3–10 3–10 3–7
Shallow water bathymetry	5–50 m depth	Location/change detection Current velocity Depth	Convergence	3–12
Atmospheric features				
Surface winds	>1 km grid	Wind speed Wind direction	Wind stress indirectly via wind rows, models, or sensors	5–25
Roll vortices	1–5 km wavelength	Boundary layer: stratification height Turbulence spectrum Drag coefficient	Wind stress	3–15
Gravity waves	2–10 km wavelength		Wind stress	3–15
Rain cells	2–40 km diameter	Rain rate	Wind stress Rain damping	3–15
Marine surface films				
Biogenic surfactants Mineral oils	>100 m^2 area	Areal extent	Convergence Seeps Ship discharge Run-off	2–8 3–15

Acknowledgements Some of the authors' views expressed in this chapter have been developed through discussions in a wide international forum, among which the International Ocean Vector Winds Science Team (IOVWST) and the International Winds Working Group (IWWG).

References

Alpers W, Rufenach CL (1979) The effect of orbital motions on synthetic aperture radar imagery of ocean waves. IEEE Trans Antennas Propag 27:685–690

Alpers W, Hennings I (1984) A theory of the imaging mechanism of underwater bottom topography by real and synthetic aperture radar. J Geophys Res 89:10529–10546

Alpers W, Zhang B, Mouche A, Zeng K, Wai Chan P (2016) Rain footprints on C-band synthetic aperture radar images of the ocean—revisited. Remote Sens Environ 187:169–185

AVISO (2011) Aviso Users Newsletter. Online available at http://www.aviso.oceanobs.com. Accessed 03 Jan 2017

Belmonte Rivas M, Stoffelen A, Verspeek J, Verhoef A, Neyt X, Anderson C (2017) Cone metrics: a new tool for the intercomparison of scatterometer records. IEEE J Sel Top Appl Earth Obs 10(5):2195–2204

Berger T (1972) Satellite altimetry using ocean backscatter. IEEE Trans Antennas Propag 20(3): 295–309

Brekke C, Solberg AHS (2005) Oil spill detection by satellite remote sensing. Remote Sens Environ 95:1–13

Brusch S, Held P, Lehner S, Rosenthal W, Pleskachevsky A (2011) Underwater bottom topography in coastal areas from TerraSAR-X data. Int J Remote Sens 32:4527–4543

Fois F, Hoogeboom P, Le Chevalier F, Stoffelen A (2015) On the use of cross-polar scattering to observe very high winds. IEEE Trans Geosci Remote Sens. https://doi.org/10.1109/TGRS.2015.2416203

Fois F, Hoogeboom P, Chevalier F, Stoffelen A, Mouche A (2016) DopSCAT: a mission concept for simultaneous measurements of marine winds and surface currents. J Geophys Res. https://doi.org/10.1002/2015JC011011RR

Font J, Lagerloef GSE, Le Vine DM, Camps A, Zanife OZ (2004) The determination of surface salinity with the European SMOS space mission. IEEE Trans Geosci Remote Sens 42:2196–2205

Gade M, Alpers W, Hühnerfuss H, Masuko H, Kobayashi T (1998) The imaging of biogenic and anthropogenic surface films by a multi-frequency multi-polarization synthetic aperture radar measured during the SIR-C/X-SAR missions. J Geophys Res 103:18851–18866

Gade M, Byfield V, Ermakov S, Lavrova O, Mitnik L (2013) Slicks as indicators for marine processes. Oceanography 26(2):138–149

Gaiser PW, St. Germain KM, Twarog EM, Poe GA, Purdy W, Richardson D, Grossman W, Linwood Jones W, Spencer D, Golba G, Cleveland J, Choy L, Bevilacqua RM, Chang PS (2004) The WindSat spaceborne polarimetric microwave radiometer: sensor description and early orbit performance. IEEE Trans Geosci Remote Sens 42(11): 2347

Goldstein RM, Zebker HA (1987) Interferometric radar measurement of ocean surface currents. Nature 328:707–709

Hersbach H, Stoffelen A, de Haan S (2007) An improved C-band scatterometer ocean geophysical model function: CMOD5. J Geophys Res 112(C3). https://doi.org/10.1029/2006jc003743

Holt B (2004) SAR imaging of the ocean surface. In: Jackson CR, Apel JR (eds) Synthetic aperture radar marine user's manual. NOAA NESDIS Office of Research and Applications, Washington DC, pp 25–79

Horstmann J, Koch W, Lehner S, Tonboe R (2000) Wind retrieval over the ocean using synthetic aperture radar with C band HH polarization. IEEE Trans Geosci Remote Sens 38:2122–2131

Johannessen JA, Chapron B, Collard F, Kudryavtsev K, Mouche A, Akimov D, Dagestad KF (2008) Direct ocean surface velocity measurements from space: improved quantitative interpretation of Envisat ASAR observations. Geophys Research Lett 35:L22608

Lin W, Portabella M,. Stoffelen A, Vogelzang J, Verhoef A (2015) ASCAT wind quality under high subcell wind variability conditions. J Geophys Res Oceans 120. https://doi.org/10.1002/2015jc010861

Lin W, Portabella M, Stoffelen A, Vogelzang J, Verhoef A (2016) On mesoscale analysis and ASCAT ambiguity removal. Q Meteorol Soc 142:1745–1756. https://doi.org/10.1002/qj.2770

Melsheimer C, Gade M, Alpers W (1998) Investigation of multifrequency/multipolarization radar signatures of rain cells derived from SIR-C/X-SAR data. J Geophys Res 103:18867–18884

Mouche A, Collard F, Chapron B, Dagestad K-F, Guitton G, Johannessen J, Kerbaol V, Wergeland Hansen M (2012) On the use of doppler shift for sea surface wind retrieval from SAR. IEEE Trans Geosci Remote Sens 50(7):2901–2909

Portabella M, Stoffelen A (2004) A probabilistic approach for SeaWinds data assimilation. Q J R Meteorol Soc 130(596):127–152

Portabella M, Stoffelen A, Johannessen JA (2002) Toward an optimal inversion method for SAR wind retrieval. J Geophys Res 107(C8):1–13

Robinson IS (2010) Discovering the ocean from space. Springer, Heidelberg, p 638

Romeiser R (2014) Ocean applications of interferometric SAR. In: Njoku EG (ed) Encyclopedia of remote sensing. Springer, Heidelberg, pp 426–428

Skou N (2014) Microwave radiometers. In: Njoku EG (ed) Encyclopedia of remote sensing. Springer, Heidelberg, pp 382–385

Stoffelen A (1998) Scatterometry. Ph.D. thesis at the University of Utrecht. ISBN 90-393-1708-9

Stoffelen A, Verspeek J, Vogelzang J, Verhoef A (2017) The CMOD7 geophysical model function for ASCAT and ERS wind retrievals. IEEE J Sel Top Appl Earth Obs 10(5):2123–2134

Tian-Kunze X, Kaleschke L, Maaß N, Mäkynen M, Serra N, Drusch M, Krumpen T (2013) SMOS derived sea ice thickness: algorithm baseline, product specifications and initial verification. Cryosphere Discuss 7:5735–5792

Ulaby FT, Moore RK, Fung AK (1981) Microwave remote sensing—active and passive, volume I: microwave remote sensing fundamentals and radiometry. Addison-Wesley, Reading

Van Zadelhoff GJ, Stoffelen A, Vachon PW, Wolfe J, Horstmann J, Belmonte-Rivas M (2014) Retrieving hurricane wind speeds using cross-polarization C-band measurements. Atmos Meas Tech 7:437–449

Vogelzang J, Stoffelen A (2017) ASCAT ultrahigh-resolution wind products on optimized grids. IEEE J Sel Top Appl Earth Obs 10(5):2332–2339

Wang Z, Stoffelen A, Zhao C, Vogelzang J, Verhoef A, Verspeek J, Lin M, Chen G (2017) An SST-dependent Ku-band geophysical model function for RapidScat. J Geophys Res Oceans 122:3461–3480

Wensink H, Campbell G (1997) Bathymetric map production using the ERS SAR. Backscatter 8:17–22

Wentz FJ, Ricciardulli L, Gentemann C, Meissner T, Hilburn KA, Scott J (2013) Remote sensing systems Coriolis WindSat 3-day environmental suite on 0.25 deg grid, version 7.0.1, South East Asia. Remote sensing systems, Santa Rosa, CA. Available online at www.remss.com/missions/windsat. Accessed 03 Jan 2017

Wright JW (1968) A new model for sea clutter. IEEE Trans Antennas Propag 16:217–223

Zavorotny VU, Gleason S, Cardellach E, Camps A (2014) Tutorial on remote sensing using GNSS bistatic radar of opportunity. IEEE Geosci Remote Sens Mag 2(4):8–45

Part II
Arctic Ocean Marginal Basins

Sea-Ice Parameters from Satellite Remote Sensing

Stefan Kern and Sascha Willmes

Abstract A substantial part of the Asian coast line borders seasonally sea-ice covered waters. This chapter deals with the sea-ice cover along the northern shores facing the Arctic Ocean. Long-term year-round, sustained monitoring of this sea-ice cover is, on the one hand, crucial for shipping, off-shore activities, near-coastal transport, and marine safety. On the other hand, it is crucial to better understand recent and predict future sea-ice cover changes—changes which already have an impact on coastal erosion and regional as well as large-scale weather conditions. This chapter deals with observations of the sea-ice cover in the Siberian sector of the Arctic Ocean and its changes obtained from data of satellite passive microwave sensors. In addition, ways are presented to monitor polynyas and fast ice by means of microwave, optical, and infrared satellite remote sensing. Polynyas form frequently along the fast-ice cover of the Asian Arctic coast during winter. Their role for the Arctic Ocean sea-ice volume and ocean water mass modification is discussed. Furthermore, methods to estimate sea-ice thickness by means of satellite observations are described and illustrated for thin ice.

Keywords Sea ice · Sea-ice concentration · Sea-ice thickness · Sea-ice volume Fast ice · Polynya · Thin ice · Microwave radiometry · Thermal infrared

1 Introduction

Sea ice is frozen sea water floating on polar and sub-polar oceans. During the freezing season it acts as a blanket strongly reducing ocean-atmosphere heat exchange. During the freezing season snow covered sea ice reflects up to 90% of the incident solar radiation; reflection from bare ice is considerably smaller but with

S. Kern (✉)
Integrated Climate Data Center, Universität Hamburg, Hamburg, Germany
e-mail: stefan.kern@uni-hamburg.de

S. Willmes
Umweltmeteorologie, Universität Trier, Trier, Germany

V. Barale and M. Gade (eds.), *Remote Sensing of the Asian Seas*,
https://doi.org/10.1007/978-3-319-94067-0_5

30–60% still 5–10 times more than from open water. During summer it reflects—depending on its coverage with melt water puddles, so called melt ponds—also between 30 and 60% of the incident solar radiation (Perovich 1996). Sea ice substantially reduces the atmosphere-ocean momentum and heat exchange. Drifting sea ice acts as a freshwater and sediment transport medium as it may melt far away from its formation region. Areas subject to sea-ice melt or sea-ice formation are areas of water mass modification by stabilizing or destabilizing the uppermost water layers, respectively. Sea ice is furthermore an important habitat for various plants (algae) and animals and is a serious impediment for shipping and off-shore activities like oil and gas exploration or fishing.

Sea ice exists in both hemispheres. Here we focus on sea ice along the northern shores of Asia, starting with the Kara Sea in the West and ending with the Bering Sea in the East [Barale (this volume)]. Sea ice forming in the Sea of Okhotsk is investigated in Mitnik and Dubina (this volume). For sea ice in the Caspian Sea we refer to Kostianoy et al. (this volume) while sea ice in the Bohai Sea has been studied elsewhere (e.g. Shi and Wang 2012).

Sea ice is an essential component of the Earths' climate system. Its area has declined substantially in the northern Polar Regions during summer and fall (e.g. Meier et al. 2014). It is recognized as an indicator for the amplification of global climate change in the northern Polar Regions. Results obtained from satellite remote sensing have played a major role to achieve this understanding and to better predict the potential future development of the Arctic in terms of climate change and its impact on biology and human activities (Meier et al. 2014). Already today sea-ice retreat potentially influences parts of the Earths' climate system (Francis et al. 2011; Barnhart et al. 2014; Cohen et al. 2014).

This book chapter focuses on satellite remote sensing of sea-ice area (fraction) and thickness for the regions delineated in Fig. 1, and of polynyas and fast ice, the latter two being ubiquitous along the northern Asian shores. A large number of

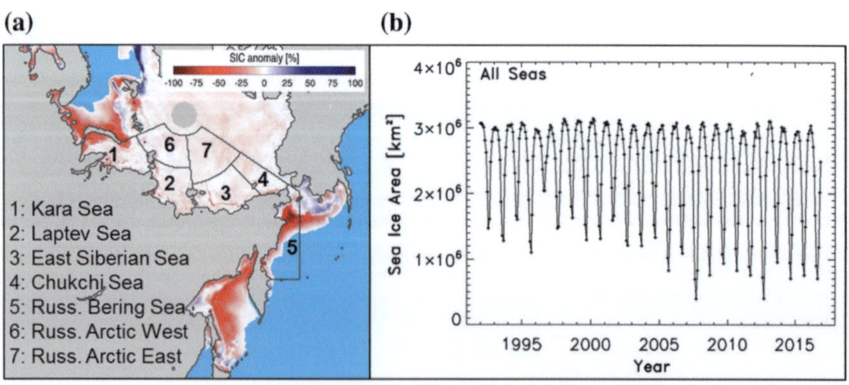

Fig. 1 **a** Location of the regions used superposed on the monthly sea-ice concentration (SIC) anomaly of March 2017 relative to March 1992–2016. **b** Time series of sea-ice area of all seven regions in **a** derived from monthly average SIC[1]

sea-ice properties and parameters cannot be discussed in this chapter. We refer to Dierking and Dall (2007) for discrimination of deformed from level sea ice. We refer to Comiso (2012) and Swan and Long (2012) for discrimination between multiyear and first-year ice. For sea-ice motion, which is an essential parameter to understand sea-ice dynamics in the Arctic Ocean (Spreen et al. 2011; Pfirman et al. 2004) we refer to Girard-Ardhuin and Ezraty (2012), Komarov and Barber (2014), and Sumata et al. (2014). Further, we refer to Brucker and Markus (2013) for snow depth on sea ice, to Markus et al. (2009) for melt season onset and duration, and to Rösel et al. (2012) for melt pond fraction retrieval.

2 Sea-Ice Concentration

Sea-ice concentration (SIC), or sea-ice area fraction, is the percentage coverage of a known area of the sea surface with sea ice. From SIC sea-ice area (SIA) and sea-ice extent (SIE) are computed. The SIE is usually defined as the sum of the area of all grid cells with SIC > 15%. For SIA the actual SIC is used to weigh the contribution of each grid cell area to the total area. SIA equals SIE if every grid cell has SIC = 100%; otherwise SIA < SIE.

In principle SIC can be derived from optical/infrared [Doerffer et al. (this volume)] and active and passive microwave [Mitnik and Gade (this volume)] satellite remote sensing data. To derive SIC with optical data one exploits the difference in surface albedo [Sect. 1 and Smolyanitsky and Bessonov (this volume)]. To derive SIC with thermal infrared (TIR) data one exploits the difference in surface temperature which is substantially lower for (snow covered) sea ice than for open water (see Sect. 3) as long as freezing conditions prevail; (near) isothermal conditions as encountered at the surface during melt prevent SIC retrieval from TIR data. Until recently optical and TIR data were not considered suitable for large-scale SIC retrieval—with few exceptions (e.g. Liu et al. 2016)—because of the polar night (applies to optical data only) and because of clouds; such data have merely been used for small-scale analyses (see Sect. 3) and for validation.

Active microwave observations, particularly synthetic aperture radar [SAR, see Mitnik and Gade (this volume)] allow to discriminate sea ice from open water at spatial resolutions down to a few ten meters. The ambiguities in the SAR signal of both surfaces complicate an automated SIC retrieval using SAR—in addition to the relatively sparse data coverage compared to passive microwave data. SAR data are therefore used for regional ice charting by various national ice services but not yet for large-scale SIC retrieval.

Passive microwave observations allow to discriminate sea ice from open water by means of brightness temperature (TB) differences between both surfaces. TB data are available with complete daily coverage, independent of daylight, and widely independent of clouds. Therefore TB data have been the backbone for SIC retrieval. TB data used for routine SIC retrieval are from the Special Sensor Microwave Imager and Sounder (SSMIS) and Advanced Microwave Scanning

Radiometer (AMSR-E and AMSR2). These sensors measure TB at various frequencies from 6.9 to 91.0 GHz at both horizontal (H) and vertical (V) polarization. TB obtained at the lower frequencies (6.9 and 10.7 GHz) is almost insensitive to atmospheric influence. The spatial resolution achieved with these frequencies is rather coarse: ~50 km. It improves to ~5 km when using AMSR2 89 GHz TB data. At this frequency the atmospheric influence from water vapor and cloud liquid water and wind-induced open water surface roughening can increase substantially. This can lead to positive biases in SIC and patches of spurious SIC in the open ocean. Therefore, deriving SIC is always a trade-off between an as fine as possible spatial (grid) resolution, least sensitivity to atmospheric influence and least sensitivity to the variability in physical and hence radiometric sea ice and snow properties. We refer to Andersen et al. (2007) and Ivanova et al. (2014) for overviews of passive microwave SIC retrieval and limitations.

Here we show results of the Artist Sea Ice (ASI) concentration retrieval algorithm using SSM/I and SSMIS 85 GHz TB polarization difference, i.e. V minus H polarization TB data (Kaleschke et al. 2001). Daily gridded 5-day median filtered ASI SIC, available for the period January 1992 through December 2016[1] are used to compute time series of the monthly total SIA for the Asian regions (Fig. 1a) shown in Fig. 1b. During winter total SIA did not change over the 25-year period shown. During the melt season, however, before 2005 total SIA did not fall below 1.1×10^6 km^2 while after 2005 total SIA was below this value every year with minimum SIA in 2007 and 2012: 400,000 km^2. Time-series of winter (Jan–Mar), spring (April–June), summer (July–Sep) and fall (Oct–Dec) SIA are shown in Fig. 2. In all regions—except the Russian Bering Sea (Fig. 2g)—summer and fall SIA decrease substantially over the period 1992 through 2016. Winter and spring SIA only decrease slightly. Largest decreases of ~50% are observed in the Kara Sea for fall (Fig. 2a) and for summer in the Laptev Sea (Fig. 2b) and East Siberian Sea (Fig. 2c).

3 Polynyas and Fast Ice

Polynyas are extended and recurring areas of open water or thin ice that form in regions where, from a climatological perspective, one would expect a thick and closed sea-ice cover (Barber and Massom 2007). They are persistent features on daily to weekly time scales and thereby represent key areas for Arctic mammals and birds, primary production, the formation of new ice as well for the dynamic and thermodynamic coupling between the ocean and the cold winter atmosphere. Polynyas tend to form adjacent to coast lines or next to the edge of immobile fast ice, which is anchored to the sea floor or to the coast. Heat loss from the ocean to the cold winter atmosphere results in ice formation and salt rejection and the newly formed ice is incorporated into the main drift systems of the Arctic Ocean.

---

[1]http://icdc.cen.uni-hamburg.de/daten/cryosphere/seaiceconcentration-asi-ssmi.html

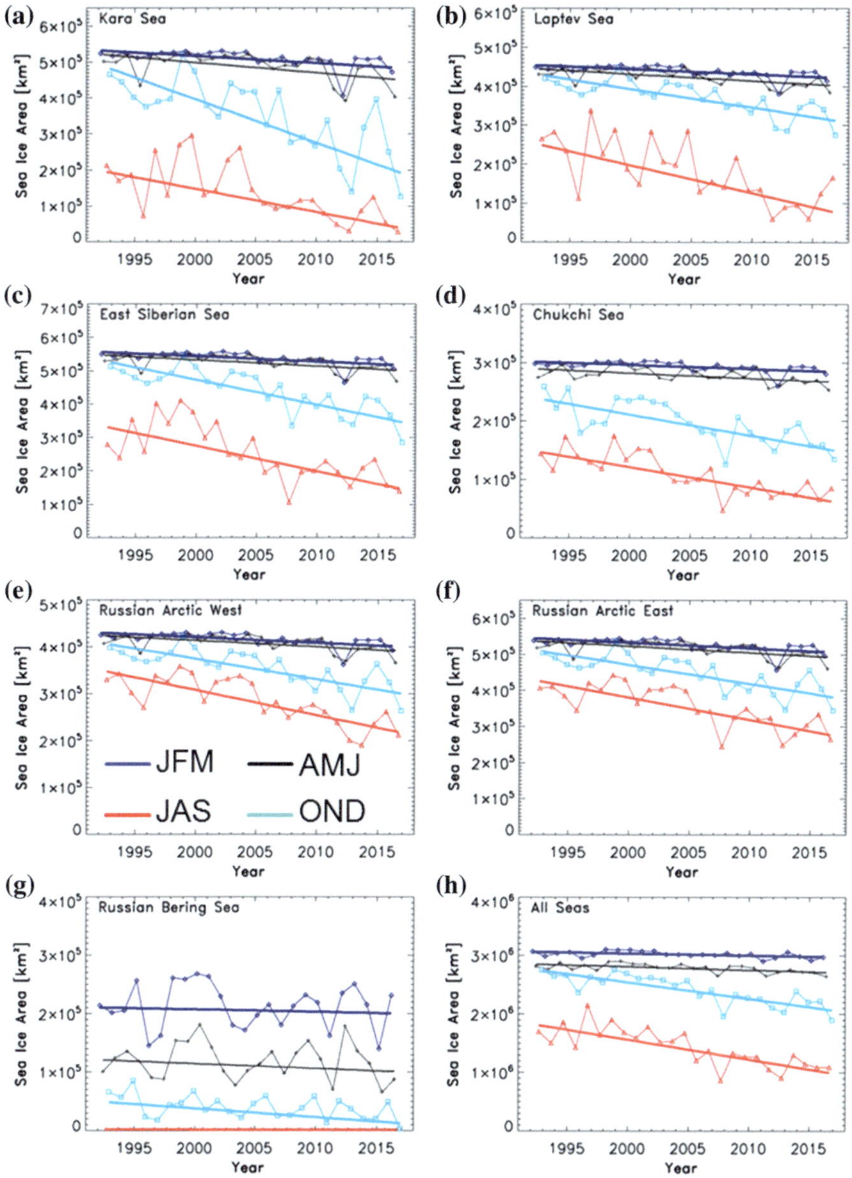

Fig. 2 Time series and trend lines from linear regression analysis of the average seasonal sea-ice area for winter (JFM), spring (AMJ), summer (JAS) and fall (OND) for the seven regions (see Fig. 1a) in images (**a–g**) and for the total sea-ice area of all regions in image (**h**) computed from ASI algorithm SIC data[1]

Several metrics of interest associated with the occurrence of a polynya can be deducted using satellite imagery. First, satellite data can help to identify whether a polynya is present or not. This includes the retrieval of opening and closing dynamics, the polynya area and—if long and homogeneous time series of satellite data are available—a polynya climatology (e.g. Bareiss and Görgen 2005; Kern 2008). Second, since a polynya is per definition covered by open water or thin ice, the thin-ice distribution and the ice-growth within a polynya can be monitored using satellite data.

3.1 Remote Sensing of Arctic Polynyas

Metrics that allow for identification and monitoring of an open polynya from satellite data exist in principle for all wavelength regions from the optical to the microwave range. The choice of data is a trade-of between the required spatial and temporal resolution, the aspired length of the data set, the monitoring season and the availability of clear-sky events.

We first consider optical data. Around a polynya, one can find the most extreme surface albedo contrasts around the globe. On the one hand, open water reflects less than 10% of the incoming shortwave radiation and on the other, snow-covered sea ice with an albedo of around 0.9 exhibits one of the most reflective surfaces on earth. Thereby, optical data from the optical and near-infrared wavelengths can help identifying an open polynya only by means of its low reflectance anomaly with respect to the surrounding sea ice. As clouds, however, exhibit ambiguous brightness values in comparison to sea ice and are not easily captured by automated cloud masks, optical data have been used for a polynya monitoring only in limited case studies, mostly based on manual selection and supervised classification. Furthermore, no reflected sunlight is available during the Arctic winter and hence during most of the season when polynyas in the Siberian Seas can be expected. A case study of a polynya opening in the Laptev Sea is shown in Fig. 3.

The gray values from the near-infrared (NIR) channel of the *Moderate Resolution Imaging Spectroradiometer* (MODIS) indicate the distinct brightness, i.e. reflectance, contrast between the thick ice and the polynya, where a smooth brightness gradient over the snow free thin ice reveals an increase in its thickness from the fast-ice edge towards the mobile pack ice (Fig. 3c).

A better way to obtain thin-ice thickness is offered by thermal infrared (TIR) data. During wintertime, sea ice acts as an insulating layer between the cold atmospheric boundary layer and the relatively warm ocean. This insulation is interrupted where a polynya forms. Thereby, large surface-temperature contrasts are exhibited in the proximity of a polynya, associated with strong fluxes of sensible and latent heat. Surface temperature values can reach from −1.8 °C, i.e. the freezing temperature of sea water, in open water areas down to less than −30 °C over thicker ice, depending on the current atmospheric forcing (Fig. 3d). Satellite TIR imagery and associated surface temperature fields have proven useful for identifying open

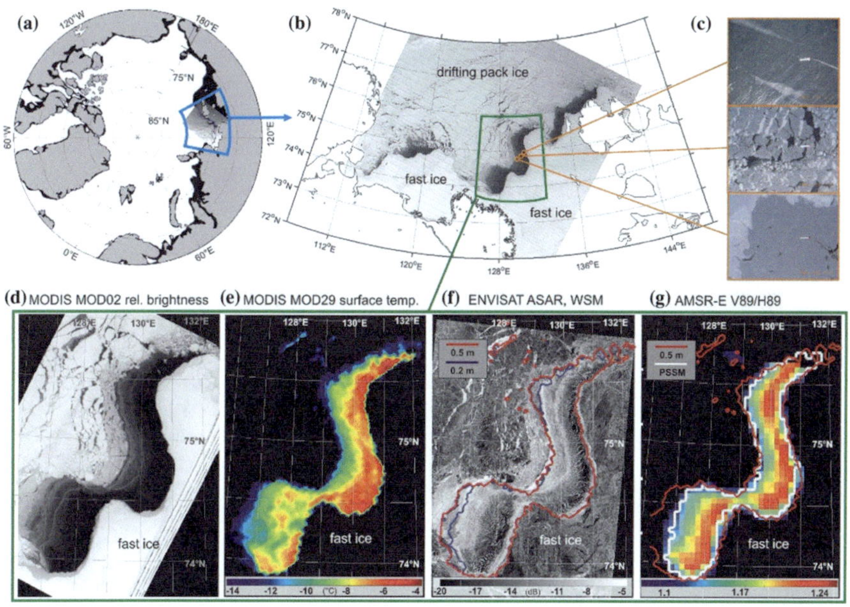

Fig. 3 The Laptev Sea flaw polynya as represented in different types of satellite data and derived products; **a** Arctic overview map and location of the case study, <20 m bathymetry (black), **b** Laptev Sea overview with Moderate Resolution Spectroradiometer (MODIS) near-infrared (NIR) channel image for 29 April, 2008, 1100 UTC, **c** aerial nadir photography from helicopter for three different locations across the polynya (*credits* Thomas Krumpen, AWI), **d** detailed MODIS MOD02 NIR relative brightness of the polynya, **e** associated MODIS MOD29 surface temperature (Riggs and Hall 2006). **f** Envisat Advanced Synthetic Aperture Radar (ASAR) Wide Swath Mode backscatter (© ESA, AO-project AO500, 30 April, 0237 UTC) and iso-lines for thicknesses of 0.05, 0.1, 0.2 and 0.5 m as derived from (**d**), **g** ratio of vertical and horizontal polarized 89 GHz TB from AMSR-E with polynya area as derived from the Polynya Signature Simulation Method (PSSM) (white) and 0.5 m thermal ice thickness (red)

water and thin ice regions in the Arctic. Since the surface temperature is, however, largely dependent on the current atmospheric forcing, local temperature anomalies or temperature gradients are more suitable for a polynya monitoring than the absolute surface temperature data (compare Fig. 4c).

Cloud cover (and daylight) limits polynya monitoring with TIR (and optical) data. Therefore, most of the long-term and large-scale studies on polynya dynamics in the polar seas are based on daily data from passive microwave radiometers (e.g. Willmes et al. 2011; Tamura and Ohshima 2011). In principle the identification of a polynya from passive microwave is based on the pronounced emissivity differences between open water, thin new ice and thick ice (Martin et al. 1992). This difference can be expressed using differences between TBs at different polarizations or frequencies, i.e. the Polarization Ratio (PR = V-polarized TB minus H-polarized TB divided by the sum of the same TBs) and Gradient Ratio (GR = V-polarized higher-frequency TB (e.g. at 37 GHz) minus V-polarized lower-frequency TB

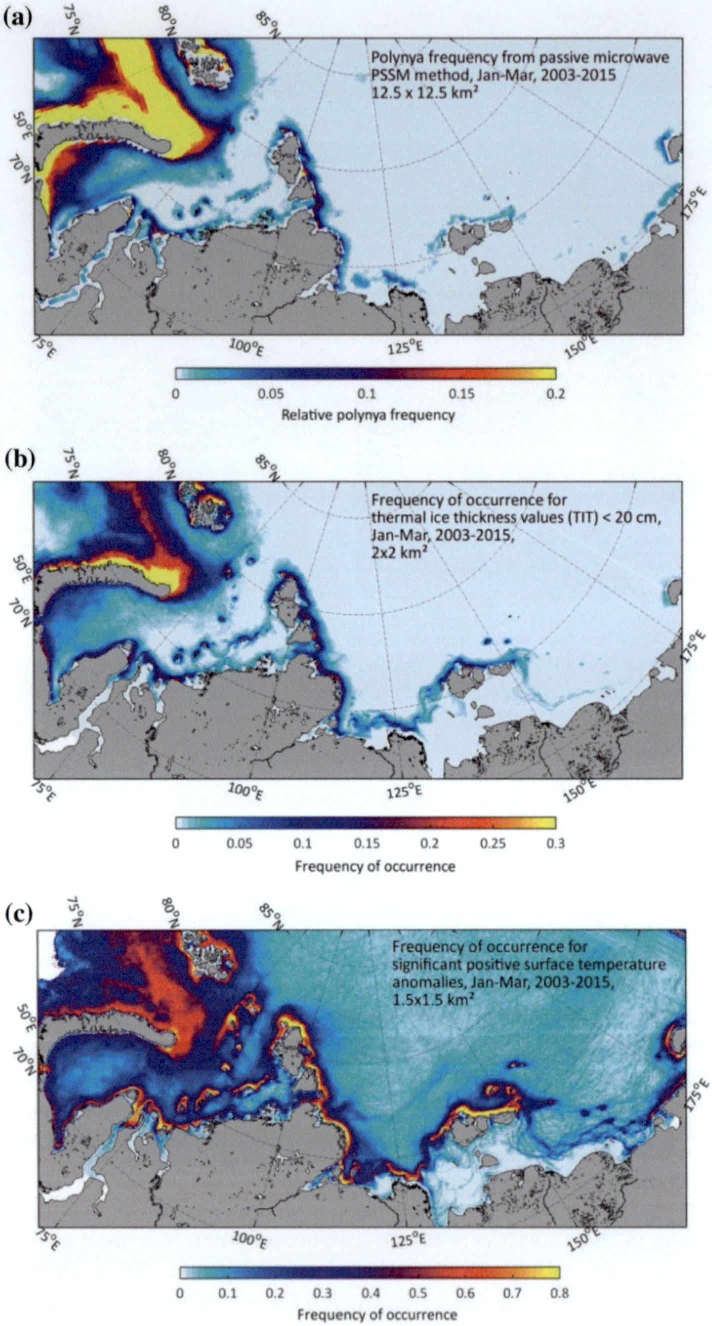
(a)
Polynya frequency from passive microwave
PSSM method, Jan-Mar, 2003-2015
12.5 x 12.5 km²
75°N
80°N
85°N
50°E
70°N
75°E
100°E
125°E
150°E
175°E
0
0.05
0.1
0.15
0.2
Relative polynya frequency
(b)
Frequency of occurrence for
thermal ice thickness values (TIT) < 20 cm,
Jan-Mar, 2003-2015,
2x2 km²
0
0.05
0.1
0.15
0.2
0.25
0.3
Frequency of occurrence
(c)
Frequency of occurrence for
significant positive surface temperature
anomalies, Jan-Mar, 2003-2015,
1.5x1.5 km²
0
0.1
0.2
0.3
0.4
0.5
0.6
0.7
0.8
Frequency of occurrence

◂**Fig. 4** **a** Polynya frequency for passive microwave method PSSM based on the PR at 37 and 85 GHz (Markus and Burns 1995). **b** Frequency of occurrence for thermal ice thickness values below 20 cm (*image credits* Andreas Preusser, University of Trier). **c** Frequency of occurrence for significant positive surface temperature anomalies (Willmes and Heinemann 2016), each for 2003–2015, January–March

(e.g. at 19 GHz) divided by the sum of the same TBs). A simple PR mapping often reveals the emissivity gradients over thin ice with variable thicknesses and can be used to identify a polynya and associated thin-ice thickness distributions (Fig. 3f). Some studies have indicated that SIC < 70% is a reasonable threshold to distinguish between polynya and thick ice areas (e.g. Adams et al. 2011). A more sophisticated method allows an additional separation of open water and thin ice, based on the sensitivity of passive microwave PR to thin ice and open water. It combines the lower atmospheric influence at microwave channels in the 30 to 40 GHz range with the higher spatial resolution near 90 GHz. This is accomplished by applying a threshold to the high-frequency PR and adjusting this threshold iteratively until the resulting classification agrees best with coincident low-resolution PR (Fig. 4a, Polynya Signature Simulation Method (PSSM); Markus and Burns 1995; Kern 2008).

The surface microwave emissivity and the associated microwave TB exhibit large contrasts not only between open water and thick sea ice, but also show a gradual dependency on the thickness of thin ice. The microwave emissivity increases with thickness and is more pronounced at lower frequencies and at horizontal polarization (Naoki et al. 2008; Fig. 3f, g). Thin-ice thickness of up to 0.2 m can be estimated with passive microwave satellite data (Tamura et al. 2011).

The main limitation of using passive microwave data is the coarse spatial resolution. One way to circumvent this is to use active microwave data. The radar backscattering coefficient of thin sea ice as typical for polynyas varies significantly during ice growth (Onstott 1992). Ideally, it decreases when new ice forms in open water due to a weakened surface roughness. Afterwards, it starts to increase as a result of increasing volume scattering and hence thin ice thickness as the young ice grows. Increasing surface roughness due to deformation or frost flowers causes, however, also an increase in the backscattering coefficient. This makes active microwave imagery of polynyas ambiguous and complicates deduction of polynya features (Willmes et al. 2010). Therefore, most SAR-based polynya monitoring studies are limited to case-studies or apply a supervised classification of the polynya area from backscatter. Alternatively, surface information from SAR data can be used to complement passive microwave retrievals or detailed case studies.

3.2 *Remote Sensing of Fast Ice*

The Siberian Arctic is characterized by the most extensive fast-ice areas in the Arctic Ocean. In the Laptev Sea, fast ice can extend up to 500 km off the coast, representing closed, immobile sea-ice areas that follow mainly the 10–20 m bathymetry line while the variability of its extent is driven by variations in surface winds and air temperature (Bareiss and Görgen 2005). Fast ice is a crucial element of the Arctic sea-ice system for several reasons. The fast-ice edge determines the location where polynyas form and the vast fast-ice areas store discharged fresh water from Siberian Rivers during freeze-up and release it during summer melt, contributing thereby to the Arctic freshwater cycle.

Fast ice shows several optical and microwave signatures that can be utilized for a monitoring from satellite data. Fast ice is highly reflective, especially when it is snow covered and—even more important—shows a low variability of the local reflectance since it is immobile and thereby not affected by leads and cracks. During wintertime, the same holds for the surface temperature, such that for temporally averaged (e.g. weekly) surface-temperature data, fast ice is characterized by a lower average temperature in comparison to the mobile pack ice (compare Fig. 4c, where fast ice shows a low frequency of positive surface temperature anomalies).

The radar backscatter signature of fast ice in SAR data is ambiguous and strongly dependent on surface characteristics, e.g. ice surface roughness (Fig. 3f). However, a manual detection of the fast-ice edge from sequences of SAR data is applicable by identifying its mainly immobile edge and can thereby provide time series of the fast-ice evolution and edge dynamics (Selyuzhenok et al. 2015).

In passive microwave data, the GR of AMSR-E TB data from 89 and 36 GHz has been suggested to provide a valuable discriminant between fast ice and thin, mobile pack ice, because the snow on fast ice during winter is hypothesized to be colder and thicker than in the region of mobile ice (Iwamoto et al. 2014).

4 Sea-Ice Thickness

Sea-ice thickness (SIT) integrates over thermodynamic and dynamic processes and hence reflects changes in thermodynamic forcing, e.g. via the air temperature, and in dynamic forcing, e.g. via the wind speed. Therefore SIT carries more information about the past freezing (and melting) season(s) than SIC does. SIT influences heat flux through sea ice particularly for thin ice as is described in Sect. 3. However, also changes in SIT of thicker sea ice can have an impact on the surface heat budget of the Arctic Ocean (e.g. Gerdes 2006).

Since the first SIT map was derived from satellite altimetry of the European Remote Sensing Satellites ERS1/2 (Laxon et al. 2003), satellite radar and laser altimetry has become the backbone for today's SIT retrieval in the Arctic Ocean. A satellite altimeter measures the runtime of a pulse between the sensor and the surface. By knowing the position of the sensor in orbit relative to the Earths' reference ellipsoid with centimeter accuracy the runtime measurement can be converted into the freeboard. The freeboard is the elevation of the ice (or snow) surface relative to the sea surface. Currently, the latter still needs to be approximated from the altimeter data themselves—either via statistical methods (Kwok and Cunningham 2008) or via lead detection methods (Ricker et al. 2014).

The freeboard is subsequently converted into SIT assuming that the sea ice is in isostatic balance. This requires assumptions (based on observations) about water, snow and ice density and snow depth. Uncertainties in these parameters determine SIT uncertainty which ranges between 0.2 and 1.0 m—depending on input parameter uncertainties and SIT (Kwok and Cunningham 2008; Zygmuntowska et al. 2014; Kern et al. 2015). For the Asian Sea regions (see Fig. 1a), except the Russian Bering Sea, ICESat laser altimetry provides average SIT < 1 m in Oct/Nov of years 2003 through 2007 and average SIT between 1 and 2 m in Feb/Mar of the respective next year. For MYI regions, i.e. mainly Russian Arctic West, and coastal areas of the Chukchi Sea, SIT may exceed 3 m (Kwok et al. 2009). The same holds for observations of the radar altimeter CryoSat-2 for winters 2010/11–2013/14 (Kwok and Cunningham 2015).

Radar and laser altimeter measurements become less certain in areas of sea ice thinner than about 0.5 m because of the small freeboard. Therefore thin sea ice needs special treatment as is described in the following paragraphs.

With increasing ice thickness a decrease in surface temperatures can be expected (Yu and Rothrock 1996). However, the ice-surface temperature is also governed by the radiation balance and turbulent heat fluxes, which obscures a simple relationship between surface temperature and ice thickness. The *thermal ice thickness* (Fig. 4b) can, however, be estimated from TIR satellite data using a 1-D thermodynamic energy balance model. Several assumptions have to be made to apply this model. First, the temperature profile through the ice is assumed to be linear, which holds for an ice thickness of up to ~ 0.5 m (Drucker et al. 2003). Second, the water temperature at the ice underside is constant and at its freezing point (usually 271.35 K, in areas with strong river freshwater input the freezing point can expected to be higher). Third, there is no snow on the ice. The ice thickness can then be calculated by TIR satellite data, i.e. surface temperatures, and meteorological data, e.g. from atmospheric reanalysis. To simplify the radiation balance to the longwave terms, thereby avoiding issues with directional shortwave reflectance, the method is applied for nighttime scenes only. The method is based on the condition, that the conductive heat flux through the ice equals the total heat flux to the atmosphere. This implies that all energy loss at the ice surface is compensated by the conductive heat flux through the ice. For thin-ice thicknesses up to 0.2 m derived with this method Adams et al. (2013) give a mean absolute error of ± 4.7 cm. Figure 4 summarizes the general patterns of the polynya frequency which become visible when high-resolution TIR data are used instead of passive microwave data. For daily observations, however, microwave data have the distinct advantage that they are independent of clear-sky conditions.

The second method is based on TB data measured in L-Band at a frequency of 1.4 GHz by the Soil Moisture and Ocean Salinity (SMOS) sensor. For nearly 100% SIC and freezing conditions, TB measured at L-Band varies with thin-ice thickness (Kaleschke et al. 2012) because at these frequencies microwave radiation still penetrates the sea ice. The L-Band radiation penetration through thin ice is mainly a function of its thickness, its temperature and its salinity. The colder and fresher the sea ice the larger is the penetration depth and the larger is the maximum thin sea-ice thickness (TSIT) to be retrieved from SMOS for typical Arctic Ocean winter sea ice: ~0.7 m. SMOS data are independent of cloud cover, but radio frequency interferometry (RFI), which distorts the TB measurement, and a considerably coarser spatial resolution of SMOS TB data compared to, e.g. CryoSat-2, limit the applicability for sea-ice thickness retrieval.

Here we use daily gridded TSIT derived from SMOS TB data with the approach described in Kaleschke et al. (2012) and Tian-Kunze et al. (2014) for the winter seasons, defined as Oct 15 to Apr 14, 2010/11 through 2016/17[2]. This data is available together with an uncertainty estimate based on TB uncertainties and retrieval parameters. We computed monthly average TSIT for the regions delineated in Fig. 1a. A monthly average is computed across months, i.e. from Oct 15 to Nov 14, Nov 15 to Dec 14, and so on, to maximize the number of monthly values. Only those grid cells are used where (i) daily TSIT < 0.7 m, (ii) δ TSIT < TSIT (i.e. we allow 100% uncertainty), and (iii) daily TSIT is based on more than ten valid SMOS TB data. The value of 0.7 m was chosen because it corresponds to the approximate maximum retrievable SMOS TSIT (see above). This value also demarcates the transition between "thin first-year ice" and "medium thick first-year ice" according to WMO nomenclature (WMO 2014).

Figure 5, top, illustrates average monthly TSIT computed for winter 2012/13. It shows the rapid freeze-up of the regions Laptev to Chukchi Sea from Oct/Nov to Nov/Dec. While a considerable region still has TSIT < 0.7 m in Nov/Dec, TSIT is above 0.7 m in Dec/Jan almost everywhere in the mentioned three regions—except towards the Bering Sea and at Wrangel Island. In contrast, large parts of the Kara Sea remain ice free until Dec/Jan. Complete freeze-up of the Kara Sea occurred as late as Mar/Apr (Fig. 5). The Laptev Sea polynya (see Sect. 3) can be identified clearly in the average SMOS TSIT maps of Jan/Feb and Feb/Mar (Fig. 5, top, red circles). The Laptev Sea polynya is the main reason for a considerable fraction of thin sea ice in the Russian Arctic regions near the coast even during mid-winter.

Figure 5a–d show the inter-annual variability of the thin sea-ice volume, i.e. TSIT times thin ice SIA, observed in regions 1–4 (see Fig. 1). The above-mentioned late freeze-up in the Kara Sea in winter 2012/13 is confirmed in Fig. 5a by a maximum thin-ice volume of 110 km^3 occurring in period Jan/Feb. It is associated with the largest thin-ice extent of 320,000 km^2, for the seven years shown, and an average ice thickness in the thin-ice area of 0.35 m. In Jan/Feb 2017 the thin-ice extent was similarly large but the ice thickness was only 0.25 m explaining the smaller thin-ice volume in Jan/Feb 2017. The commonly comparably large Kara Sea thin-ice volume during Jan/Feb through Mar/Apr is partly caused by polynya activity and partly caused by phases of sea-ice retreat even during winter. A common feature observed in the Laptev, East Siberian and Chukchi Seas (Fig. 5b–d) is a peak in thin-ice volume in Nov/Dec or Dec/Jan. This peak is associated with fall freeze-up. In contrast to the Kara Sea, elevated thin-ice volumes

[2]http://icdc.cen.uni-hamburg.de/daten/cryosphere/l3c-smos-sit.html

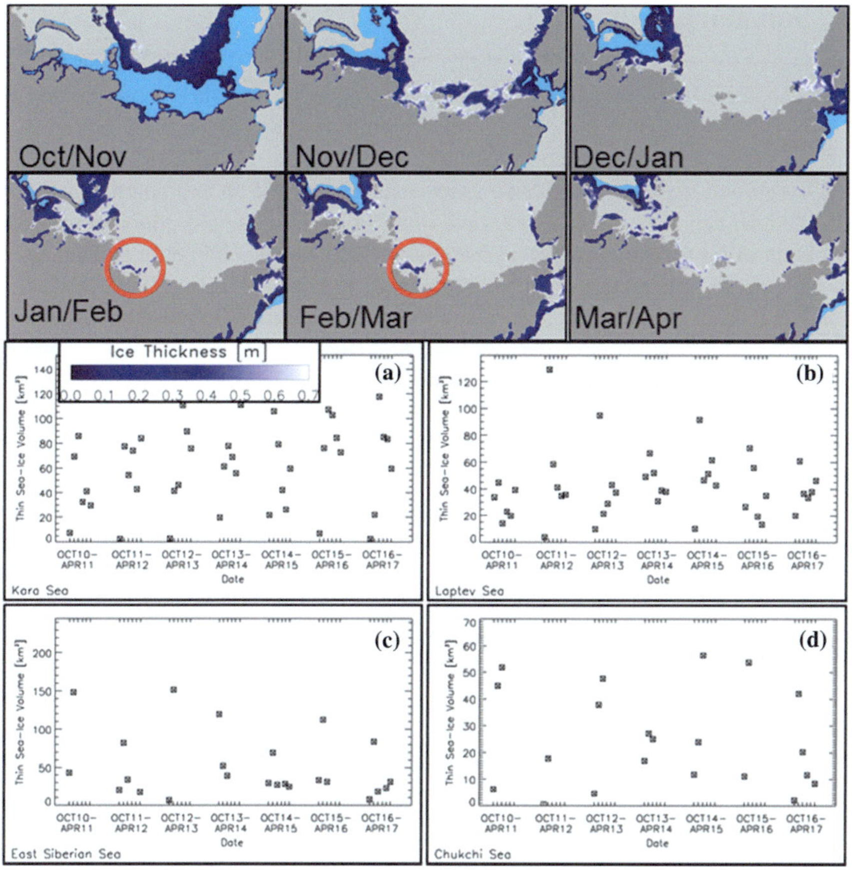

Fig. 5 Top: Average thin ice thickness (<0.7 m) derived from daily gridded SMOS thin ice thickness[2] for monthly periods Oct. 15–Nov. 14, Nov. 15–Dec. 14, etc., for winter 2012/13; dark grey: land, light grey: SMOS thin ice thickness >0.7 m or no SMOS thin ice thickness; cyan: SMOS thin ice thickness < 0.02 m and/or <7 days/period with valid data. Red circles are referred in the text. Bottom: Thin ice volume from SMOS sea-ice thickness and ASI sea-ice concentration (see Sect. 2) averaged over the same monthly periods for winters 2010/11 through 2016/17 for regions 1 to 4 (see Fig. 1)

occurring later in winter are mostly driven by polynyas. The Laptev Sea is the most active of these three regions. For the seven years shown, the Dec/Jan through Mar/Apr average Laptev Sea thin-ice volume is largest for winter 2014/15: $\sim$50 km^3 associated with the maximum average thin-ice extent of 140,000 km^2 and an average ice thickness in the thin-ice area of 0.4 m. For comparison: in winter 2012/13 (see Fig. 5 top) average thin-ice extent and thickness were $\sim$60,000 km^2 and $\sim$0.5 m, respectively.

5 Summary

This chapter introduced and discussed some aspects of satellite remote sensing of sea ice in the Russian Arctic seas. Particular emphasis was given to the retrieval of the long-term daily sea-ice concentration, to monitoring of polynyas and fast ice, and to estimation of sea-ice volume associated with freeze-up and polynya activity. This chapter cannot be taken as a review of past work in this field. Much more work has been done focusing on the regions dealt with in this chapter, e.g. Danielson et al. (2011), Mahoney et al. (2008), and Polykov et al. (2003); see also Smolyanitsky and Bessonov (this volume) and Mitnik and Dubina (this volume).

We presented sea-ice area derived from 85 GHz satellite microwave radiometry. Albeit the larger atmospheric noise at this frequency, reasonable estimates of the sea-ice cover can be retrieved. These are available at a relatively fine grid resolution of 12.5 km for a 25-year long time series, allowing more detailed analyses of sea-ice cover variability closer to the coast and around the many islands along the Russian coast. With the launch of AMSR2 in 2012 sea-ice area fractions can be retrieved now with an even finer grid resolution of 3.125 km (Beitsch et al. 2014). We illustrated various ways to retrieve polynya area and the sea-ice thickness and associated ice production. Polynyas in the Russian Arctic—particularly those in the Laptev Sea—are among the most important "ice factories" of the Arctic Ocean. Being the border between the fast ice, and the thicker seasonal and multiyear drift ice further north, the polynyas along the Russian Arctic coast are important areas to monitor for shipping purposes and navigation strategies. We described how sea-ice thickness can be retrieved from satellite altimetry and from low-frequency satellite microwave radiometry and illustrated how thin-ice area and volume in the Russian Arctic varied over the past few years.

References

Adams S, Willmes S, Heinemann G, Rozman P, Timmermann R, Schröder D (2011) Evaluation of simulated sea-ice concentrations from sea-ice/ocean models using satellite data and polynya classification methods. Polar Res 30. https://doi.org/10.3402/polar.v30i0.7124

Adams S, Willmes S, Schroeder D, Heinemann G, Bauer M, Krumpen T (2013) Improvement and sensitivity analysis of thermal thin-ice retrievals. IEEE Trans Geosci Remote Sens 51:3306–3318. https://doi.org/10.1109/tgrs.2012.2219539

Andersen S, Tonboe R, Kaleschke L, Heygster G, Pedersen LT (2007) Intercomparison of passive microwave sea ice concentration retrievals over the high-concentration Arctic sea ice. J Geophys Res 112:C08004. https://doi.org/10.1029/2006JC003543

Barber DG, Massom RA (2007) The role of sea ice in Arctic and Antarctic polynyas. In: Smith JO, Barber DG (eds.) Polynyas, windows to the world. Elsevier oceanography series, vol 74. Amsterdam

Bareiss J, Görgen K (2005) Spatial and temporal variability of sea ice in the Laptev Sea: analyses and review of satellite passive-microwave data and model results, 1979 to 2002. Glob Planet Change 48:28–54. https://doi.org/10.1016/j.gloplacha.2004.12.004

Barnhart KR, Overeem I, Anderson RS (2014) The effect of changing sea ice on the physical vulnerability of Arctic coasts. Cryosphere 8(5):1777–1799. https://doi.org/10.5194/tc-8-1777-2014

Beitsch A, Kaleschke L, Kern S (2014) Investigating high-resolution AMSR2 sea ice concentrations during the February 2013 fracture event in the Beaufort Sea. Remote Sens 6:3841–3856

Brucker L, Markus T (2013) Arctic-scale assessment of satellite passive microwave-derived snow depth on sea ice using Operation IceBridge airborne data. J Geophys Res Oceans 118:2892–2905. https://doi.org/10.1002/jgrc.20228

Cohen J, Screen JA, Furtado JC, Barlow M, Whittleston D, Coumou D, Francis J, Dethloff K, Entekhabi D, Overland J, Jones J (2014) Recent Arctic amplification and extreme mid-latitude weather. Nat Geosci 7(2234):627–637. https://doi.org/10.1038/NGEO2234

Comiso JC (2012) Large decadal decline of the Arctic multiyear ice cover. J Clim 25:1176–1193. https://doi.org/10.1175/JCLI-D-11-00113.1

Danielson S, Curchitser E, Hedstrom K, Weingartner T, Stabeno P (2011) On ocean and sea ice modes of variability in the Bering Sea. J Geophys Res 116:C12034. https://doi.org/10.1029/2011JC007389

Dierking W, Dall J (2007) Sea-Ice deformation state from synthetic aperture radar imagery—part I: comparison of C- and L-band and different polarization. IEEE Trans Geosci Remote Sens 45 (11):3610–3622

Drucker R, Martin S, Moritz R (2003) Observations of ice thickness and frazil ice in the St. Lawrence island polynya from satellite imagery, upward looking sonar, and salinity/temperature moorings. J Geophys Res 108(C5):3149. https://doi.org/10.1029/2001JC001213

Francis OP, Panteleev GG, Atkinson DE (2011) Ocean wave conditions in the Chukchi Sea from satellite and in situ observations. Geophys Res Lett 38:L24610. https://doi.org/10.1029/2011GL049839

Gerdes R (2006) Atmospheric response to changes in Arctic sea ice thickness. Geophys Res Lett 33:L18709. https://doi.org/10.1029/2006GL027146

Girard-Ardhuin F, Ezraty R (2012) Enhanced Arctic sea ice drift estimation merging radiometer and scatterometer data. IEEE Trans Geosci Remote Sens 50(4):2639–2648

Ivanova N, Johannessen OM, Pedersen LT, Tonboe RT (2014) Retrieval of Arctic sea ice parameters by satellite passive microwave sensors: a comparison of eleven sea ice concentration algorithms. IEEE Trans Geosci Remote Sens 52(11):7233–7246

Iwamoto K, Ohshima KI, Tamura T (2014) Improved mapping of sea ice production in the Arctic Ocean using AMSR-E thin ice thickness algorithm. J Geophys Res Oceans 119:3574–3594. https://doi.org/10.1002/2013JC009749

Kaleschke L, Lüpkes C, Vihma T, Haarpaintner J, Bochert A, Hartmann J, Heygster G (2001) SSM/I sea ice remote sensing for mesoscale ocean-atmosphere interaction analysis. Can J Remote Sens 27(5):526–537

Kaleschke L, Tian-Kunze X, Maaß N, Mäkynen M, Drusch M (2012) Sea ice thickness retrieval from SMOS brightness temperatures during the Arctic freeze-up period. Geophys Res Lett 39: L05501. https://doi.org/10.1029/2012GL050916

Kern S (2008) Polynya area in the Kara Sea, Arctic, obtained with microwave radiometry for 1979–2003. Geosci Remote Sens Lett 5(2):171–175. https://doi.org/10.1109/LGRS.2008.916831

Kern S, Khvorostovsky K, Skourup H, Rinne E, Parsakhoo ZS, Djepa V, Wadhams P, Sandven S (2015) The impact of snow depth, snow density, and ice density on sea ice thickness retrieval from satellite radar altimetry: results from the ESA-CCI sea ice ECV project round robin exercise. Cryosphere 9(1):37–52. https://doi.org/10.5194/tc-9-37-2015

Komarov AS, Barber DG (2014) Sea ice motion tracking from sequential dual-polarization RADARSAT-2 images. IEEE Trans Geosci Remote Sens 52(1):121–136

Kwok R, Cunningham GF (2008) ICESat over Arctic sea ice: estimation of snow depth and ice thickness. J Geophys Res 113:C08010. https://doi.org/10.1029/2008JC004753

Kwok R, Cunningham GF (2015) Variability of Arctic sea ice thickness and volume from CryoSat-2. Phil Trans R Soc A 373:20140157. https://doi.org/10.1098/rsta.2014.0157

Kwok R, Cunningham GF, Wensnahan M, Rigor I, Zwally HJ, Yi D (2009) Thinning and volume loss of the Arctic Ocean sea ice cover: 2003–2008. J Geophys Res 114:C07005. https://doi.org/10.1029/2009JC005312

Laxon S, Peacock N, Smith D (2003) High interannual variability of sea ice thickness in the Arctic region. Nature 425:947–950

Liu Y, Key J, Mahoney R (2016) Sea and freshwater ice concentration from VIIRS on Suomi NPP and the future JPSS satellites. Remote Sens 8(6):523. https://doi.org/10.3390/rs8060523

Mahoney AR, Barry RG, Smolyanitsky V, Fetterer F (2008) Observed sea ice extent in the Russian Arctic, 1933–2006. J Geophys Res 113:C11005. https://doi.org/10.1029/2008JC004830

Markus T, Burns BA (1995) A method to estimate subpixel-scale coastal polynyas with satellite passive microwave data. J Geophys Res 100(C3):4473–4487. https://doi.org/10.1029/94JC02278

Markus T, Stroeve JC, Miller J (2009) Recent changes in Arctic sea ice melt onset, freezeup, and melt season length. J Geophys Res 114:C12024. https://doi.org/10.1029/2009JC005436

Martin S, Steffen K, Comiso JC, Cavalieri DJ, Drinkwater M, Holt B (1992) Microwave remote sensing of polynyas. In: Carsey FD (ed) Microwave remote sensing of sea ice. American Geophysical Union, Washington, D. C. https://doi.org/10.1029/gm068p0291

Meier WN, Hovelsrud GK, van Oort BEH, Key JR, Kovacs KM, Michel C, Haas C, Granskog MA, Gerland S, Perovich DK, Makshtas A, Reist JD (2014) Arctic sea ice in transformation: a review of recent observed changes and impacts on biology and human activity. Rev Geophys 51:185–217. https://doi.org/10.1002/2013RG000431

Naoki K, Ukita J, Nishio F, Nakayama M, Comiso JC, Gasiewski A (2008) Thin sea ice thickness as inferred from passive microwave and in situ observations. J Geophys Res 113: C02S16. https://doi.org/10.1029/2007jc004270

Onstott RG (1992) SAR and Scatterometer signatures of sea ice, in microwave remote sensing of sea ice. In: Carsey FD (ed) Microwave remote sensing of sea ice. American Geophysical Union, Washington, D.C. https://doi.org/10.1029/gm068p0073

Perovich DK (1996) The optical properties of sea ice. U. S. Cold Region Research and Engineering Laboratory. Monograph 96-1, Hanover, N. H., 25pp

Pfirman S, Haxby W, Eicken H, Jeffries M, Bauch D (2004) Drifting Arctic sea ice archives changes in ocean surface conditions. Geophys Res Lett 31:L19401. https://doi.org/10.1029/2004GL020666

Polyakov IV, Alekseev GV, Bekryaev RV, Bhatt US, Colony R, Johnson MA, Karklin VP, Walsh D, Yulin AV (2003) Long-term variability in Arctic marginal seas. J Clim 16(12): 2078–2085

Ricker R, Hendricks S, Helm V, Skourup H, Davidson M (2014) Sensitivity of CryoSat-2 Arctic sea ice freeboard and thickness on radar-waveform interpretation. Cryosphere 8:1607–1622. https://doi.org/10.5194/tc-8-1607-2014

Riggs GA, Hall DK, Salomonson VV (2006) MODIS sea ice products user guide to collection 5. Technical report, National Snow and Ice Data Center, University of Colorado, Boulder, CO, USA

Rösel A, Kaleschke L, Birnbaum G (2012) Melt ponds on Arctic sea ice determined from MODIS satellite data using an artificial neural network. Cryosphere 6:431–446. https://doi.org/10.5194/tc-6-431-2012

Selyuzhenok V, Krumpen T, Mahoney A, Janout M, Gerdes R (2015) Seasonal and interannual variability of fast ice extent in the southeastern Laptev Sea between 1999 and 2013. J Geophys Res Oceans 120:7791–7806. https://doi.org/10.1002/2015JC011135

Shi W, Wang M (2012) Sea ice properties in the Bohai Sea measured by MODIS-Aqua: 2. Study of sea ice seasonal and interannual variability. J Mar Syst 95:41–49

Spreen G, Kwok R, Menemenlis D (2011) Trends in Arctic sea ice drift and role of wind forcing: 1992–2009. Geophys Res Lett 38:L19501. https://doi.org/10.1029/2011GL048970

Sumata H, Lavergne T, Girard-Ardhuin F, Kimura N, Tschudi MA, Kauker F, Karcher M, Gerdes R (2014) An intercomparison of Arctic ice drift products to deduce uncertainty estimates. J Geophys Res Oceans 119:4887–4921. https://doi.org/10.1002/2013JC009724

Swan AM, Long DG (2012) Multiyear Arctic sea ice classification using QuikSCAT. IEEE Trans Geosci Remote Sens 50(9):3317–3326

Tamura T, Ohshima KI (2011) Mapping of sea ice production in the Arctic coastal polynyas. J Geophys Res 116:C07030. https://doi.org/10.1029/2010JC006586

Tian-Kunze X, Kaleschke L, Maaß N, Mäkynen M, Serra N, Drusch M, Krumpen T (2014) SMOS-derived thin sea ice thickness: algorithm baseline, product specifications and initial verifications. Cryosphere 8:997–1018. https://doi.org/10.5194/tc-8-997-2014

Willmes S, Heinemann G (2016) Sea-Ice wintertime lead frequencies and regional characteristics in the Arctic, 2003–2015. Remote Sens 8(1):4. https://doi.org/10.3390/rs8010004

Willmes S, Krumpen T, Adams S, Rabenstein L, Haas C, Hendricks S, Heinemann G, Hölemann, J (2010) Cross-validation of polynya monitoring methods from multi-sensor satellite and airborne data: a case study. Can J Remote Sens 36(Suppl S1): S196–S210

Willmes S, Adams S, Schroeder D, Heinemann G (2011) Spatiotemporal variability of sea-ice coverage, polynya dynamics and ice production in the Laptev Sea between 1979 and 2008. Polar Res 30:5971. https://doi.org/10.3402/polar.v30i0.5971

WMO–JCOMM Expert Team on Sea Ice (2014) WMO Sea ice nomenclature, WMO Publication no. 259, volume I—Terminology and codes, volume II—Illustrated glossary and III—International system of sea-ice symbols, March 2014

Yu Y, Rothrock DA (1996) Thin ice thickness from satellite thermal imagery. J Geophys Res 101:25753–25766. https://doi.org/10.1029/96jc02242

Zygmuntowska M, Rampal P, Ivanova N, Smedsrud LH (2014) Uncertainties in Arctic sea ice thickness and volume: new estimates and implications for trends. Cryosphere 8:705–720. https://doi.org/10.5194/tc-8-705-2014

Ocean Colour Remote Sensing in the Laptev Sea

Birgit Heim, Bennet Juhls, Ekaterina Abramova, Astrid Bracher, Roland Doerffer, Rafael Gonçalves-Araujo, Sebastian Hellman, Alexandra Kraberg, Feodor Martynov and Paul Overduin

Abstract The Laptev and Eastern Siberian shelves are the world's broadest shallow shelf systems. Large Siberian rivers and coastal erosion of up to meters per summer deliver large volumes of terrestrial matter into the Arctic shelf seas. In this chapter we investigate the applicability of Ocean Colour Remote Sensing during the ice-free summer season in the Siberian Laptev Sea region. We show that the early summer river peak discharge may be traced using remote sensing in years characterized by early sea-ice retreat. In the summer time after the peak discharge, the spreading of the main Lena River plume east and north-east of the Lena River Delta into the shelf system becomes hardly traceable using optical remote sensing methods. Measurements of suspended particulate matter (SPM) and coloured dissolved organic matter (cDOM) are of the same magnitude in the coastal waters of Buor Khaya Bay as in the Lena River. Match-up analyses of in situ chlorophyll-a (Chl-a) show that standard *Medium Resolution Imaging Spectrometer* (MERIS) and *Moderate Resolution Imaging Spectroradiometer* (MODIS) satellite-derived Chl-a is not a valid remote sensing product for the coastal waters and the inner shelf region of the Laptev Sea. All MERIS and MODIS-derived Chl-a products are overestimated by at least a factor of ten, probably due to absorption by the

B. Heim (✉) · B. Juhls · A. Bracher · R. Gonçalves-Araujo · S. Hellman
A. Kraberg · P. Overduin
Alfred Wegener Institute Helmholtz Center for Polar and Marine Research, Potsdam and Bremerhaven, Germany
e-mail: birgit.heim@awi.de

B. Juhls
Institute for Space Sciences, Freie Universitaet Berlin, Berlin, Germany

B. Juhls
GEOMAR Helmholtz Centre for Ocean Research, Kiel, Germany

E. Abramova
Lena Tiksi Reserve, Tiksi, Russia

E. Abramova · F. Martynov
Saint Petersburg State University, Saint Petersburg, Russia

R. Doerffer
Helmholtz Centre Geesthacht HZG, Institute for Coastal Research, Geesthacht, Germany

V. Barale and M. Gade (eds.), *Remote Sensing of the Asian Seas*,
https://doi.org/10.1007/978-3-319-94067-0_6

extraordinarily high amount of non-algal particles and cDOM in these coastal and inner-shelf waters. Instead, Ocean Colour remote sensing provides information on wide-spread resuspension over shallows and lateral advection visible in satellite-derived turbidity. Satellite Sea Surface Temperature (SST) data clearly show hydrodynamics and delineate the outflow of the Lena River for hundreds of kilometres out into the shelf seas.

Keywords Ocean colour · Laptev Sea · Lena Delta · Lena River
Arctic coastal waters · cDOM · Chl-a · MERIS · MODIS

1 Introduction

Together with the Eastern Siberian shelf the Laptev Sea region is part of the world's broadest shelf system (Fig. 1). The surface waters are ice-covered for about nine months each year. The Siberian river discharge peaks in early June when sea and river ice are still present. During and after the breakup of the river ice, the sea ice around the deltas and on submarine shallows is still grounded and the sediment transport seems to be forced through pathways beneath the ice (Wegner et al. 2005; Janout et al. 2016; Hölemann et al. 1999) and via flow over the ice (Eicken et al. 1997, 2005).

The Lena River Delta is one of the world's largest. The Lena River discharges through myriads of river branches, mainly through the northern and eastern part of the delta (Charkin et al. 2011; Fedorova et al. 2013). Several geomorphological terraces and large parts of the Siberian coastline, islands and hinterland are characterized by Pleistocene ice-rich and organic-rich sediment sequences called Yedoma (or Ice Complex) (Schirrmeister et al. 2002). These ice-rich sedimentary coastal zones are prone to erosion at rates of up to decametres per summer (Lantuit et al. 2011, 2012; Günther et al. 2013) and deliver large masses of organic matter into coastal waters (Vonk et al. 2012, 2014; Rachold et al. 2004).

Russian-Swedish-US and Russian-German marine programs have studied the Laptev Sea shelf for several years, providing long-term oceanography (e.g. Wegner et al. 2013; Janout et al. 2016) and new insights into organic geochemistry (e.g. Semiletov et al. 2013; Vonk et al. 2012, 2014).

In their overview of organic matter dynamics, Semiletov et al. (2013) show that the source of cDOM in the Eastern Siberian shelf waters is fluvial and that coastal erosion provides elevated concentrations of Particulate Organic Carbon (POC) in coastal waters. Vonk et al. (2012) find that old POC from the Ice Complex quickly settles close to the coast, whereas modern POC originating from the fluvial discharge is distributed widely over the shelf system. Winterfeld et al. (2015a, b) and Gonçalves-Araujo et al. (2015) find considerable degradation of POC and dissolved organic matter on the Laptev Sea Shelf. Russian cruises measuring optical variables (SPM, Chl-a, cDOM and fluorescent Dissolved Organic Matter (fDOM)) are currently being undertaken by the Shirshov Institute of Oceanology (SIO, Russia).

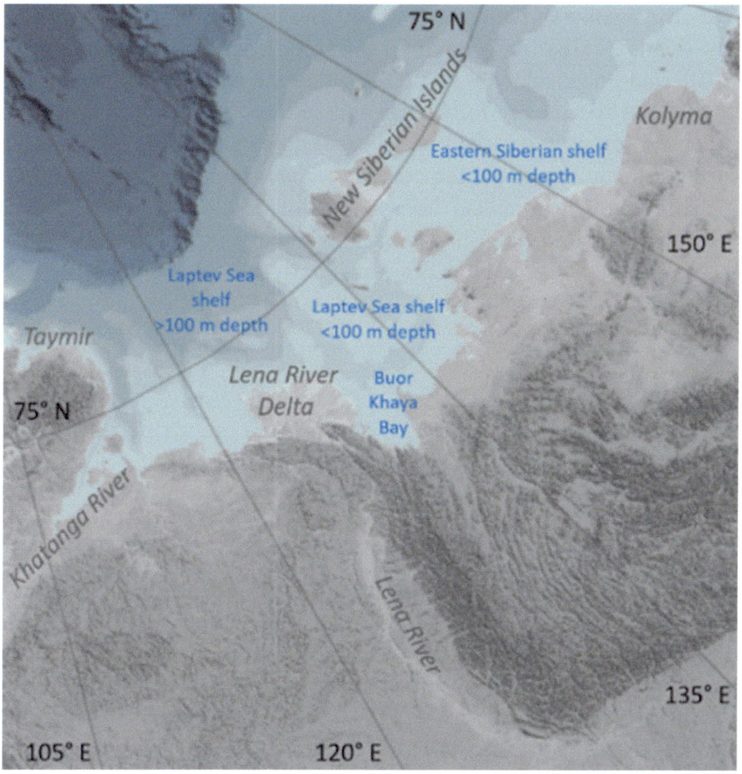

Fig. 1 Overview of central Northern Siberia with the Laptev Sea and Eastern Siberian Sea shelf region. The Laptev Sea is bounded by the Taimyr Peninsula to the west and the New Siberian Islands to the east. The shelf extends 400 km from the coast out to the 100 m isobath in the Laptev Sea, and more than 1000 km in the East Siberian Sea. The Lena River Delta is the largest delta in the Arctic Ocean region

Nonetheless, large regions, specifically of the shallow portions of the Laptev Sea and Siberian shelves, remain under-sampled. The numerous shallows are dangerous and avoided by ocean-going Arctic ship expeditions.

Ocean Colour remote sensing could provide a synoptic view of the spatial distribution of phytoplankton and turbidity but is impeded in polar regions, however, by several limitations: the polar night conditions, low solar elevations, only a few ice-free months and the persistence of clouds and fog during these summer months (International Ocean Colour Coordinating Group 2000). Another limitation lies in the fact that considerable terrestrial input makes the Arctic shelf waters optically complex. A number of groups have addressed or are working on these limitations. For North American arctic coastal and shelf waters, the MALINA project introduced a wide range of new bio-optical insights for the terrestrially influenced waters of the Beaufort Sea region (e.g. Bélanger et al. 2013; Doxaran et al. 2012; Matsuoka et al. 2007, 2013). The SIO in Russia has undertaken a long-term investigation of bio-optics and particles related to satellite remote sensing in the Russian arctic seas (e.g. Burenko et al. 2001a, b, 2004, 2010). And Hessen et al. (2010) and Pozdnyakov et al. (2005) study the southern Kara Sea producing MODIS products specifically for Chl-a and SPM.

Nonetheless, few Ocean Colour remote sensing studies of the Laptev Sea region exist. In this chapter, we show the capabilities and limitations of satellite remote sensing for providing bio-optical information for this region. In the process, we summarize some relevant published work in which remote sensing applications for the surface waters of the Laptev Sea Region were explored (Örek et al. 2013; Wegner et al. 2013; Heim et al. 2014; Vonk et al. 2014; Gonçalves-Araujo et al. 2015; Janout et al. 2016).

2 Materials and Methods

2.1 *In Situ Hydrography and Biogeochemistry*

Russian-German research programs have operated coastal and ocean-going ship expeditions and year-round oceanographic moorings in the Laptev Sea region. The moorings provide a hydrographic record from 2007 on from the central Laptev Sea shelf (Janout et al. 2016). Several ship expeditions (Table 1) sampled data on SPM, cDOM and Chl-a. The 'LENA' expeditions sampled the coastal and shallow inner shelf areas whereas the 'TRANSDRIFT' expeditions used marine vessels in the inner and outer shelves.

We mostly used in situ data for satellite match-up analyses from LENA10 and the TRANSDRIFT expedition in 2010 because several cloud-free MERIS acquisitions were available for the summer of 2010. We also show LENA2013 in situ data that provide fDOM and help to distinguish between terrestrial versus marine autochthonous sources of organic matter. Rare in situ data on bio-optical coefficients come from the LENA11 expedition in the central Lena River delta, which was conducted immediately after the river ice break-up at the end of June in 2011 (Örek et al. 2013).

2.2 *Satellite Data Processing and Analyses*

Since 2010, the CoastColour project (http://www.coastcolourCoastcolour.org/) of the European Space Agency (ESA) has incorporated the Kara Sea and Laptev Sea into its program for evaluating the applicability of Ocean Colour remote sensing in Arctic coastal waters. We processed MERIS Level-1 data using BEAMVISAT4. 10© with the MERIS Case2Regional (C2R) processor for coastal application (Doerffer and Schiller 2008) for the sea-ice-free months (July, August, and

Table 1 Overview of Russian-German coastal and marine ship expeditions in the Laptev Sea (2008–2013) that provided in situ data used in published works exploring remote sensing applications for the Laptev Sea region (Örek et al. 2013; Wegner et al. 2013; Heim et al. 2014; Gonçalves-Araujo et al. 2015)

Expedition code	Date	Case study region (publication)	In situ data
LENA08	8–14 Aug 2008	Buor-Khaya Gulf sampling around Lena River mouth (Heim et al. 2014)	cDOM[a], SPM[b]
LENA09	17–28 Aug 2009	Buor-Khaya Gulf sampling around Lena River mouth (Winterfeld et al. 2015a, b)	cDOM[a], SPM[b]
LENA10	29 July–7 Aug Sept 2010	Transects throughout Buor-Khaya Gulf (Heim et al. 2014; Kraberg et al. 2013)	cDOM[a], CTD[c]
TRANSDRIFT XII 2007 TRANSDRIFT XIV 2008	22 Aug–22 Sept 2007 5–21 Sept 2008	Inner and outer Laptev Sea Shelf (Wegner et al. 2013)	Chl-a[d], SPM[b], CTD[c]
TRANSDRIFT XVII 2010	9–20 Sept 2010	Inner and outer Laptev Sea Shelf (Heim et al. 2014)	Chl-a[d], DOM[a], SPM[b], CTD[c]
LENA2011	26 June–4 July 2011	Interior Lena River Delta (Örek et al. 2013)	Chl-a[d], cDOM[a], ACS[e]
LENA2013	1–7 Sept 2013	Transects throughout Buor-Khaya Gulf (Gonçalves-Araujo et al. 2015; Winterfeld et al. 2015a, b)	Phytoplankton species, fDOM[f], cDOM[a], SPM[b], CTD[c]

[a]cDOM: Specord 200, OSL-RU; [b]SPM filter weighing method, AWI & GEOMAR-DE; [c]Conductivity–Temperature–Depth meter (CTD), AWI-DE; [d]Chl-a: TD-700 fluorimeter, Otto Schmidt Laboratory, OSL-RU; [e]ACS: Wetlabs Hyperspectral Absorption-Attenuation Meter, HZG-DE; [f]fDOM: Aqualog Fuorescence Spectrometer, AWI-DE

September) from 2006 to 2011 (Heim et al. 2014). Since ENVISAT MERIS operations ceased in 2012, MODIS satellite products were processed for analysis of data from the LENA2013 expedition.

The C2R processing modules (in BEAM VISAT and SENTINEL toolboxes), use neural networks to inversely model water-leaving reflectance from calculated top-of-atmosphere reflectance after ozone, water vapour and surface pressure corrections. Inverse modelling then simultaneously derives all specific absorption and scattering coefficients (for water, Chl-a, SPM and cDOM) from the water-leaving reflectance. The bio-optical forward model itself is specifically parameterized to coastal waters with a wide value range for absorption and scattering. C2R-derived Chl-a and SPM are determined from empirically derived algorithms between the specific absorption and scattering coefficients and the concentrations of Chl-a and SPM, respectively. C2R-derived first attenuation depth, is the inverted attenuation coefficient representing the potential layer from which 90% of the signal is received by the satellite sensor.

We also investigated the standard CoastColour remote sensing products for the Laptev Sea and applied (Heim et al. 2014) the global MERIS Chl-a algorithm, OC4Mev6, and the Arctic-specific MERIS Chl-a algorithms OC4L (Wang and Cota 2003) and OC4P (Cota et al. 2004).

3 Results

We give an overview of:

1. In situ data which demonstrates that terrestrial organic matter is widespread in the Laptev Sea. Optical properties from cDOM and fDOM are the key to decipher the marine or terrestrial dominance at the sampling stations.
2. In situ hydrography highlighting that the upper mixed layer extends down several meters, leading to widespread resuspension and therefore turbidity on the wide shallow shelf system.
3. The evaluation of MERIS-derived satellite products using in situ data from ship expeditions in summer 2010.
4. Spatio-temporal dynamics of turbidity and Sea Surface Temperature (SST) during the spring and early summer peak discharge and during the sea-ice-free summer conditions with much lower river discharge.

3.1 Spatio-temporal Dynamics of Surface Biogeochemistry from Optical Measurements

In situ data showed that elevated organic matter was widely distributed northward and eastward of the Lena River Delta for more than 100 km from the land. Higher organic matter concentration and higher Humification IndeX (HIX, Zsolnay et al. 1999) was associated with increased freshwater influence (Gonçalves-Araujo et al. 2015; Heim et al. 2014). The highest HIX was sampled in summer 2013 at the Lena River mouths whereas the lowest HIX but highest Biological Index (BIX, Huguet et al. 2009) values, corresponding to increased marine origin, were sampled in the northern delta region to the west of the main Lena River discharge (Fig. 2) where highest salinities were also observed. For more details, we refer to (Gonçalves-Araujo et al. 2015).

In contrast to cDOM, Chl-a concentrations were rather low in coastal waters, and have maximum concentrations comparable to open ocean moderate blooms. Multi-year TRANSDRIFT expedition data also showed moderate surface Chl-a concentrations in the inner shelf of <1–1.5 μg L^{-1} and lower concentrations around 0.5 μg L^{-1} Chl-a for the outer shelf waters.

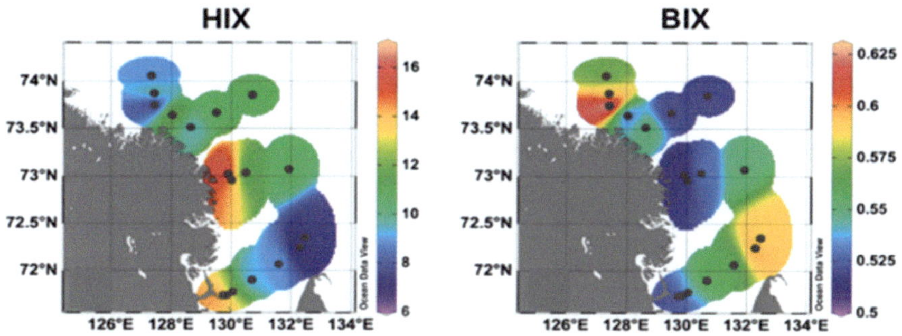

Fig. 2 The spatial distribution of fDOM indices northward and eastward of the Lena River Delta extrapolated from samples on the LENA13 ship expedition in September 2013 (black circles) (Gonçalves-Araujo et al. 2015). fDOM-derived Humification IndeX (HIX) is shown on the left and Biological Index (BIX) on the right. High HIX values indicate high humification of organic matter while high BIX values indicate higher autochthonous contributions

3.2 Laptev Sea Hydrography and Lena River Terrestrial Input

Statistical evaluations of hydrographical data from the Laptev Sea by Dmitrenko et al. (2008) differentiate between normal summers with cyclonic atmospheric circulation and summers with anticyclonic atmospheric circulation. During cyclonic summers, west winds divert the Lena River outflow eastwards and the Eastern Siberian coastal current system develops. Wegner et al. (2013) show how, during the summer of 2007, the surface salinity greatly exceeded the climatic mean and Laptev Sea waters were highly transparent in MERIS Ocean Colour satellite data. During anticyclonic summers, in contrast, the Lena River outflow is directed northward and terrestrial matter is transported directly to the outer shelf. Wegner et al. (2013) describe the summer of 2008 as an example of this less frequent situation, when geochemical signatures of the Lena River were found over much of the Laptev Sea shelf and MERIS Ocean Colour satellite data revealed lower transparency in the shelf waters rather than the high transparency visible in MERIS satellite data for the summer of 2007.

The Laptev Sea is usually stratified during the summer months. The wind-mixed upper layer reaches down to the sea bottom in shallow areas such as the shallow ramp around the Lena Delta and the numerous shoals and shallows. Wind events cause re-suspend sediment, leading to high turbidity in the water above these areas. In 2008 and 2010, we measured Secchi depths of <0.5 m in the Lena River and coastal waters. Measured in situ salinities were lower than brackwater salinity (<5 salinity) to the north and east of the Lena River Delta. The Lena River outflow brought with it a wide range of SPM concentrations additionally due to resuspension. In August 2010 (LENA10), the pycnocline lay at about 8 m depth in most of the Buor Khaya Gulf. After storm events in September 2010 (T-XVII), it sank to below 12 m water depth, a fact visible as increased turbidity over a larger region in MERIS satellite data (Heim et al. 2014).

In summary, due to wide-spread resuspension events, low salinities occurred together with a wide value range of SPM concentrations for Lena River, coastal and inner-shelf waters. SPM is thus not tracer for the Lena river freshwater input during summer months.

3.3 Evaluation of Satellite-Derived Chl-a, SPM and CDOM

Results of match-up analyses from the LENA and the TRANSDRIFT expeditions are described in Heim et al. (2014). The match-up analyses show that MERIS C2R backscatter performed well with in situ SPM. Match-ups of in situ cDOM were not as successful, suggesting that the concentration of particulate matter dominates the water-leaving reflectance over cDOM.

Örek et al. 2013 calculated for Lena River waters a three-fold lower conversion from absorption to Chl-a concentration then the global mean. The phytoplankton seemed to be well adapted to low water transparency. Therefore, Heim et al. (2014) used the regionally specific conversion factor of 7.8 (in Örek et al. 2013) instead of the global factor of 21 to convert the phytoplankton absorption coefficient to concentrations of Chl-a. This method still overestimated Chl-a in coastal and shelf waters by an order of magnitude compared to in situ data (Heim et al. 2014). The over-estimation of Chl-a concentration is common to all standard NASA and ESA processing algorithms and Chl-a CoastColour products. When compared to in situ observations of cDOM, MERIS-derived cDOM (MERIS C2R_a_gelbstoff) was underestimated by an order of magnitude in coastal and shelf waters (Heim et al. 2014).

The high cDOM in surface waters on the shelf creates strong absorption in the same wavelength range used to derive absorption of phytoplankton. This creates problems for all standard NASA and ESA Chl-a algorithms and the coastal processors C2R and CoastColour, which operationally attribute the main contribution of this absorption to Chl-a, creating abnormally high satellite-derived Chl-a concentrations.

3.4 Spatio-temporal Dynamics of Turbidity and SST

For the Laptev Sea region, usable MERIS and MODIS satellite data time series range between daily to monthly temporal resolution, limited by high cloud coverage. Turbidity can be visualized in optical remote sensing products by, for example, high values of satellite-derived SPM, high reflectance, or low values of the satellite-derived first attenuation depth. MERIS and MODIS turbidity reveals meanders with amplitudes of 20–30 km and filaments and eddies with diameters up to 100 km. The temporal resolution of all these spatially well-defined features can be very high: between one and a few days (Heim et al. 2014). Filaments and meanders may change abruptly during this time span.

MODIS SST was used to visualize the Lena River outflow during summer months when the river water was several degrees warmer and spread over the colder shelf waters. The thermal contrast did not decrease until the end of September, when it did so slowly. On 24 August 2013, one week before the LENA13 expedition, warm Lena River water could be traced south-eastward of the Lena River Delta using MODIS SST (Fig. 3). On a relatively cloud-free MERIS SST acquisition on 26 August 2013, colder water masses surfacing east of the Buor-Khaya Gulf became visible (Fig. 3). Two weeks later on 11 September 2013, the SST spatial pattern indicated eastward transport of Lena River water via coastal currents along the eastern Buor Khaya Gulf (Fig. 3).

The turbidity-related spatial pattern in the MODIS red reflectance shows a contrasting situation between SST and turbidity in August 2013 (Fig. 3): we find high turbidity above the shallow delta ramp and coastal shallows to the north and west of the Lena River Delta. At the same time, these most turbid areas are the coldest surface waters detected in MODIS SST. This contradictory spatial pattern is corroborated by in situ geochemistry: The most northern LENA13 transect stations were marine-influenced, with relatively high autochthonous-marine contributions (e.g. low HIX and high BIX, Fig. 2), and occurred together with the coldest SST waters.

The expectation that high SPM clearly delineates the main Lena river plume east of the Lena River Delta does not hold during the summer. Instead, from mid-July until the end of September in all investigated years, turbidity is frequently visible where determined by a combination of stratification and bathymetry, on the wide shallow ramp around the Lena River Delta and above widely distributed shallows and shoals.

Unfortunately, cloud cover obscures the region from optical and thermal remote sensing most frequently during the time of the peak river discharge from June to early July when SPM and organic matter concentrations are highest. Few successful satellite acquisitions exist for this early summer period. A MERIS acquisition in early July 2009 (2009-07-01) showed turbid river water flowing over the ice cover surrounding the delta. In 2011, however, sea ice had already retreated by the end of June. This provided an opportunity in early July (2011-07-03) to get a snap-shot of the Lena River peak river discharge traversing northwards through sea-ice-free waters to the outer shelf (Fig. 4).

Another rare opportunity to trace the peak river discharge into the Laptev Sea shelf system occurred in 2012, when a low-pressure system opened a large polynya along the land-fast-ice edge. Janout et al. (2016) show clear evidence of the Lena River outflow to the central shelf (Fig. 6) from MODIS SST and MODIS red band reflectance from 14 June 2012. They corroborate this observation of Lena River water with mooring data that shows elevated Acoustic Doppler Current Profilers ADCP backscatter values at 5 m depth, indicating higher concentrations of particles in early June compared to winter base values (Fig. 5).

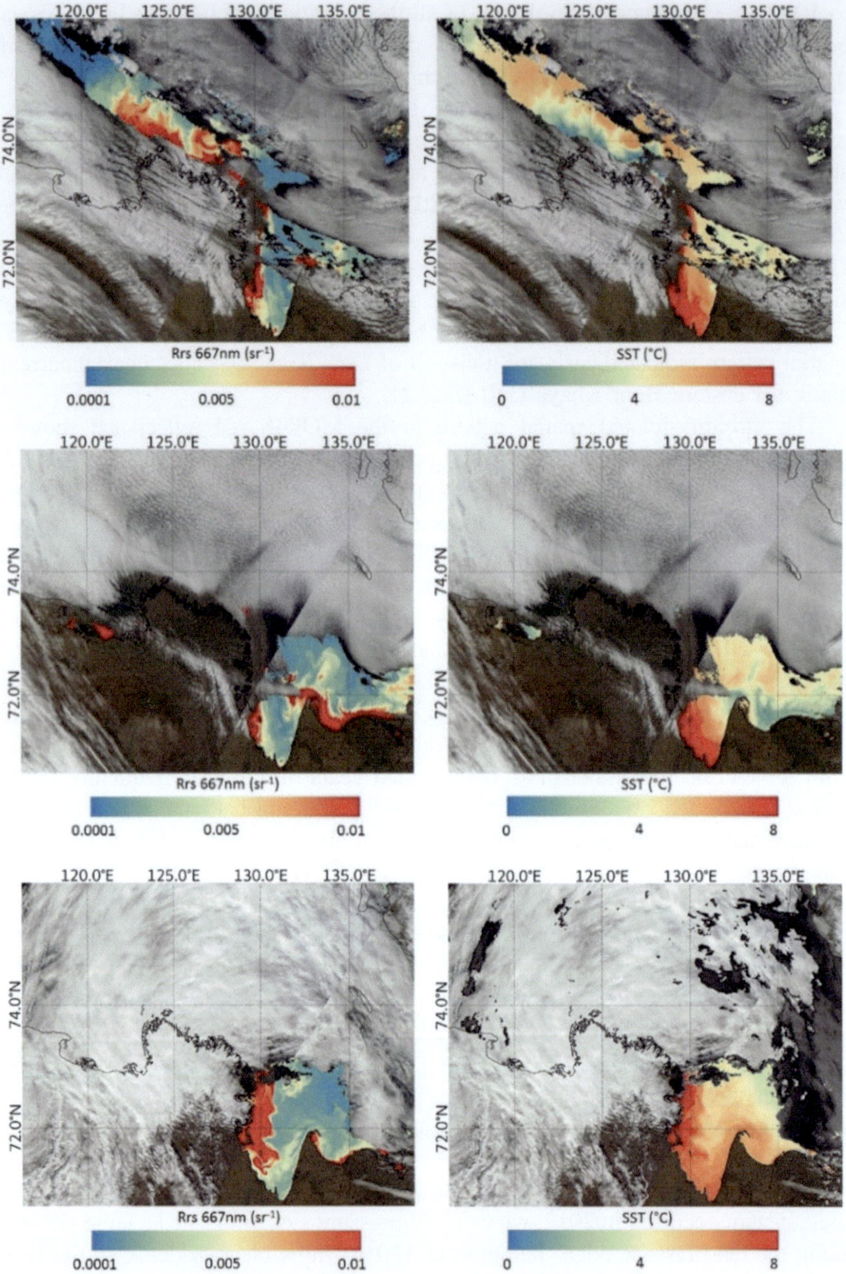

Fig. 3 MODIS L2 colour-coded red Remote Sensing Reflectance (Rrs667 nm) (left) and MODIS L2 SST (right) images. Acquisition dates are 24 and 26 August 2013 for the upper and middle images, and 11 September 2013 for the bottom images

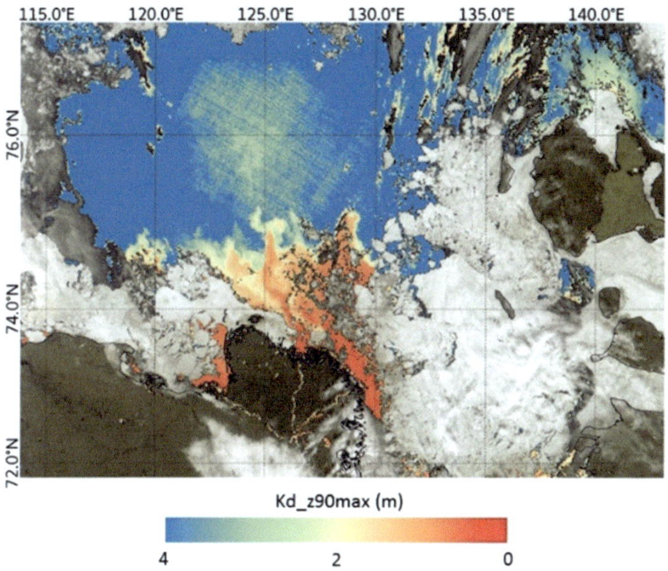

Fig. 4 MERIS CoastColour product of the first attenuation depth [m], on 3 July 2011. The Lena River peak discharge is outlined well by C2R 'attenuation depth'. In 2011 sea-ice retreated exceptionally early and the southern Laptev Sea had open water in early July

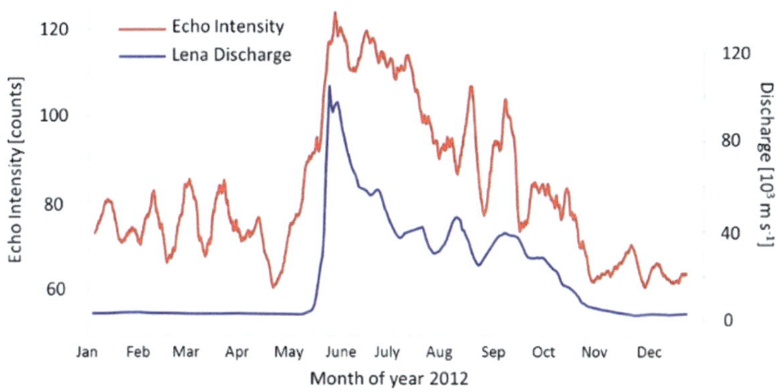

Fig. 5 The time series of Acoustic Doppler Current Profilers (ADCP) backscatter (counts) data from 2012 confirm the breakthrough of the Lena River outflow (blue line, Arctic Great Rivers Observatory, http://www.arcticgreatrivers.org/) via elevated values in echo intensity (red line)

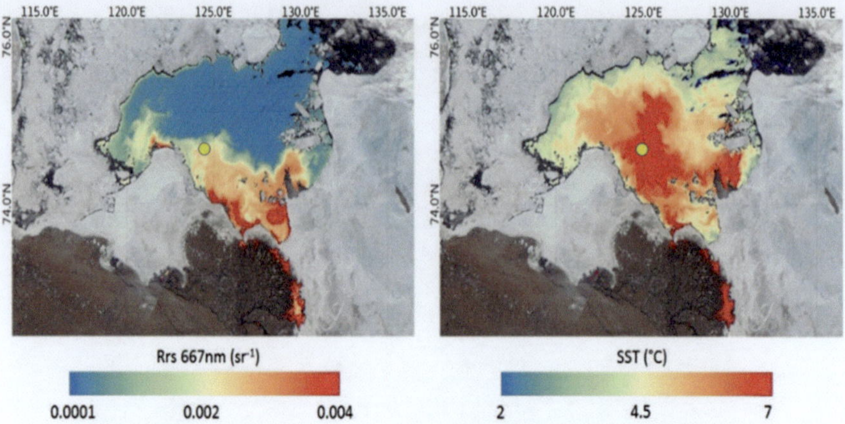

Fig. 6 Colour coded MODIS reflectance at 667 nm (left image) and MODIS SST (right image) on 14 June 2012. The Lena River outflow is visible as high reflectance and high SST values. The yellow circle indicates the location of the mooring

4 Discussion and Outlook

To assess the validity and applicability of satellite-derived optical parameters for the Laptev Sea region, we reviewed Russian-German ship expedition and mooring data. The evaluation experiments undertaken by Heim et al. (2014) show that the highly dynamic hydrography in the southern Laptev Sea region encumbers match-up analyses. However, results in Heim et al. (2014) confirmed that optical satellite-derived parameters related to turbidity can reasonably be used to trace the surface hydrodynamics of the Laptev Sea region.

In contrast, all operational Chl-a satellite products seem to perform poorly in arctic coastal waters of the Russian seas. Vetrov et al. (2008) discussed a Chl-a overestimation factor of around 5 for the Laptev Sea but it is unclear how many samples of which type were used in this estimate. Hessen et al. (2010) describe chlorophyll concentrations nearly an order of magnitude too high for specifically processed MODIS products covering the southern Kara Sea with the Ob Estuary from the summer of 2003 (Pozdnyakov et al. 2005). Örek et al. (2013) showed the difficulty of using bio-optical measurements to determine phytoplankton from water-leaving reflectance close to the mouth of the Lena River because phytoplankton absorption contributed less than 10% to the overall absorption. Their measurements in the organic-rich and turbid Lena River waters revealed that the specific phytoplankton absorption coefficient in the Lena is higher and more effective than the global mean coefficient. We incorporated the enhanced phytoplankton absorption capacity in the MERIS C2R processing, but still found an over ten-fold C2R Chl-a overestimation that we attribute to the elevated cDOM of the inner-shelf (Heim et al. 2014).

The standard assumption in all operational Ocean Colour global algorithms attributes the main share of absorption to phytoplankton absorption. We assume that, due to absorption by cDOM, the high number of terrestrial non-algal particles and the very effective phytoplankton absorption, all global ESA and NASA Chl-a products show Chl-a concentrations that are overestimated by at least a factor of 10 for the Siberian inner shelf regions. This region-specific overestimation is the highest reported to date for arctic coastal waters. In situ cDOM in the South-Eastern Laptev Sea is also the highest reported for arctic coastal waters.

During the ice-free season, the main Lena River discharge flows out northward and eastward of the Lena River Delta into coastal waters with similar turbidity and organic matter concentration. Multi-year expedition data show that the southern Laptev Sea is characterised by a wide-spread freshwater signal and high cDOM of the same order of magnitude as in river water. This finding is also valid for POC, nutrients, pCO_2 and oxygen saturation (Semiletov et al. 2013). For SPM, this is mainly due to resuspension events above the shallow bank around the Lena River Delta and the numerous submarine shallows of the Laptev Sea. Gavrilov et al. (2003) cite observations made on Russian seismic expeditions during the summer months that show that high local turbidity can be observed above these submarine banks (Lisitsin et al. 2000).

The focus of Ocean Colour remote sensing in the Laptev Sea and Eastern Siberian Sea should therefore not be on absolute quantification of phytoplankton or carbon stocks and fluxes but can exploit the qualitative applications of Ocean Colour satellite data. Meanders and filaments that were made visible via turbidity and SST suggests that a high frontal instability is regularly generated in the southern Laptev Sea. Weingartner et al. (1998) also visualize meanders, filaments and eddies for the Eastern Siberian Shelf. We suggest that the visual analysis of hydrographic features made visible by turbidity and by SST is of value for geo-biochemical and geophysical applications. For example, Vonk et al. (2014) use the MERIS information from 2006-2011 to extract the zones of continuous turbidity. Kraberg et al. (2013) discuss high spatial heterogeneity in phytoplankton composition in August 2010 in the Buor Khaya Gulf. Meanders, filaments and eddies may carry chemical and biological fields and could lead to patchy distribution of plankton in the Laptev See region.

Acknowledgements This work is supported by the German Science Foundation (DFG 4575), the Helmholtz Climate Initiative REKLIM, and the Helmholtz Impulse Fond (HGF Young Investigators Group Phytooptics). RG is funded by a Ph.D. fellowship from CAPES–Brazil in collaboration with the German Academic Exchange Service (DAAD), and a visiting fellowship from the Helmholtz Graduate School for Polar and Marine Research (POLMAR–AWI). The Russian–German marine expeditions are supported and organized by AWI and GEOMAR (DE) and the Arctic and Antarctic Research Institute (AARI, RU). The expeditions are an integral part of joint Russian–German projects funded by the German Federal Ministry for Education and Research and the Russian Ministry of Education and Science. The ESA ENVISAT project (MERIS-ID 5504) provided MERIS data. BEAM-VISAT and SNAP are open-source software initiated by ESA and developed by Brockmann Consult (DE). The C2R processor was developed as a joint effort between the HZG Research Centre, Institute for Coastal Research and Brockmann Consult under ESA contract.

References

Bélanger S, Cizmeli SA, Ehn J, Matsuoka A, Doxaran D, Hooker S, Babin M (2013) Light absorption and partitioning in Arctic Ocean surface waters: impact of multiyear ice melting. Biogeosciences 10:6433–6452

Burenkov VI, Ershova SV, Kopelevich OV, Sheberstov SV, Shevchenko VP (2001a) An estimate of the distribution of suspended matter in the Barents Sea waters on the basis of the SeaWiFS satellite ocean color scanner. Oceanology 41(5):622–628

Burenkov VI, Vedernikov VI, Ershova SV, Kopelevich OV, Sheberstov SV (2001b) Use of data from satellite ocean color scanner SeaWiFS for assessment of bio-optical characteristics of the Barents Sea. Oceanology 41(4):461–468

Burenkov VI, Vazyulya SV, Kopelevich OV, Sheberstov SV (2004) Spatio-temporal variability of the suspended matter distribution in the surface layer of the White Sea from the data of the SeaWiFS satellite color scanner. Oceanology 44(4):461–468

Burenkov VI, Goldin YuA, Kravchishina MD (2010) The distribution of the suspended matter concentration in the Kara Sea in September 2007 based on ship and satellite data. Oceanology 50(5):799–806

Charkin AN, Dudarev OV, Semiletov IP, Kruhmalev AV, Vonk JE, Sánchez-García L, Karlsson E, Gustafsson O (2011) Seasonal and interannual variability of sedimentation and organic matter distribution in the Buor-Khaya Gulf: the primary recipient of input from Lena River and coastal erosion in the southeast Laptev Sea. Biogeosciences 8:2581–2594

Cota GF, Wang J, Comiso JC (2004) Transformation of global satellite chlorophyll retrievals with a regionally tuned algorithm. Remote Sens Environ 90:373–377

Dmitrenko IA, Kirillov SA, Tremblay LB (2008) The long-term and interannual variability of summer freshwater storage over the Eastern Siberian shelf: implication for climate change. J Geophys Res 113:C03007

Doerffer R, Schiller H (2008) MERIS Regional coastal and lake case 2 water project atmospheric correction. GKSS Research Center, Geesthacht, p 42

Doxaran D, Ehn J, Bélanger S, Matsuoka A, Hooker S, Babin M (2012) Optical characterisation of suspended particles in the Mackenzie River plume (Canadian Arctic Ocean) and implications for ocean colour remote sensing. Biogeosciences 9:3213–3229

Eicken H, Reimnitz E, Alexandrov V, Martin T, Kassens H, Viehoff T (1997) Sea-ice processes in the Laptev Sea and their importance for sediment export. Cont Shelf Res 17(2):205–233

Eicken H, Dmitrenko I, Tyshko K, Darovskikh A, Dierking W, Blahak U, Groves J, Kassens H (2005) Zonation of the Laptev Sea landfast ice cover and its importance in a frozen estuary. Glob Plan Change 48:55–83

Fedororova I, Chetverova A, Bolshiyanov D, Makarov A, Boike J, Heim B, Morgenstern A, Overduin P, Wegner C, Kashina V, Eulenburg A, Dobrotina E, Sidorina I (2013) Lena Delta hydrology and geochemistry. Biogeosciences 10:20179–20237

Gavrilov AV, Romanovskii NN, Romanovsky VE, Hubberten HW, Tumskoy VE (2003) Reconstruction of ice complex remnants on the Eastern Siberian Arctic shelf. Permafrost Perigl 14:187–198

Gonçalves-Araujo R, Stedmon CA, Heim B, Dubinenkov I, Kraberg A, Moiseev D, Bracher A (2015) From fresh to marine waters: characterization and fate of dissolved organic matter in the Lena River Delta region, Siberia. Front Marine Sci 2:108

Günther F, Overduin PP, Sandakov AV, Grosse G, Grigoriev MN (2013) Short-and long-term thermo-erosion of ice-rich permafrost coasts in the Laptev Sea region. Biogeosciences 10:4297–4318

Heim B, Abramova E, Doerffer R, Günther F, Hölemann J, Kraberg A, Lantuit H, Loginova A, Martynov F, Overduin PP, Wegner C (2014) Ocean colour remote sensing in the Southern Laptev Sea: evaluation and applications. Biogeosciences 11(15):4191–4210

Hessen DO, Carroll JL, Kjeldstad B, Korosov AA, Pettersson LH, Pozdnyakov D, Sørensen K (2010) Input of organic carbon as determinant of nutrient fluxes, light climate and productivity in the Ob and Yenisey estuaries. Estuar Coast Shelf Sci 88:53–66

Hölemann J, Schirmacher M, Prange A, Kassens H (1999) Geochemistry of surficial and ice-rafted sediments from the Laptev Sea (Siberia). Estuar Coast Shelf Sci 49:45–59

Huguet A, Vacher L, Relexans S, Saubusse S, Froidefond JM, Parlanti E (2009) Properties of fluorescent dissolved organic matter in the Gironde Estuary. Org Geochem 40:706–719

Sathyendranath S (ed) (2000) Remote sensing of ocean colour in coastal, and other optically-complex waters. Reports of the International Ocean-Colour Coordinating Group (IOCG), Dartmouth, Canada 3:140

Janout M, Hölemann J, Juhls B, Krumpen T, Rabe B, Bauch D, Wegner C, Kassens H, Timokhov L (2016) Episodic warming of near-bottom waters under the Arctic Sea ice on the central Laptev Sea shelf. Geophys Res Let 43(1):264–270

Kraberg AC, Druzhkova E, Heim B, Loeder MJG, Wiltshire KH (2013) Phytoplankton community structure in the Lena Delta (Siberia, Russia) in relation to hydrography. Biogeosciences 10:7263–72777277. https://doi.org/10.5194/bg-10-7263-2013

Lantuit H, Atkinson D, Overduin PP, Grigoriev M, Rachold V, Grosse G, Hubberten H-W (2011) Coastal erosion dynamics on the permafrost-dominated Bykovsky Peninsula, North Siberia, 1951–2006. Polar Res 30:7341

Lantuit H, Overduin PP, Couture N, Wetterich S, Are F, Atkinson D, Brown J, Cherkashov G, Drozdov D, Forbes D, Graves-Gaylord A, Grigoriev M, Hubberten HW, Jordan J, Jorgenson T, Ødegård RS, Ogorodov S, Pollard W, Rachold V, Sedenko S, Solomon S, Steenhuisen F, Streletskaya I, Vasiliev A (2012) The Arctic coastal dynamics database: a new classification scheme and statistics on Arctic permafrost coastlines. Estuar Coast 35(2):383–400

Lisitsin AP, Shevchenko VP, Burenkov VI (2000) Hydrooptics and suspended matter of Arctic Seas. Atmos Ocean Opt 13:61–71

Matsuoka A, Huot Y, Shimada K, Saitoh S, Babin M (2007) Bio-optical characteristics of the Western Arctic Ocean: implications for ocean color algorithms. Can J Remote Sens 33:503–518

Matsuoka A, Hooker SB, Bricaud A, Gentili B, Babin M (2013) Estimating absorption coefficients of colored dissolved organic matter (cDOM) using a semi-analytical algorithm for southern Beaufort Sea waters: application to deriving concentrations of dissolved organic carbon from space. Biogeosciences 10:917–927

Örek H, Doerffer R, Röttgers R, Boersma M, Wiltshire KH (2013) Contribution to a bio-optical model for remote sensing of Lena River water. Biogeosciences 10:7081–7094

Pozdnyakov DV, Korosov AA, Pettersson LH, Johannessen OM (2005) MODIS evidences the river run-off impact on the Kara Sea trophy. Int J Remote Sensing 26:364–368

Rachold V, Eicken H, Gordeev VV, Grigoriev MN, Hubberten H-W, Lisitzin AP, Shevchenko VP, Schirrmeister L (2004) Modern terrigenous organic carbon input to the Arctic Ocean. In: Stein R, Macdonald RW (eds) Organic carbon cycle in the Arctic Ocean: present and past. Springer, Berlin, pp 33–55

Schirrmeister L, Siegert Ch, Kunitzky VV, Grootes P, Erlenkeuser H (2002) Late quaternary ice rich permafrost sequences as a palaeoenvironmental archive for the Laptev Sea region in Northern Siberia. Int J Earth S 91:154–167

Semiletov IP, Shakhova NE, Pipko II, Pugach SP, Charkin AN, Dudarev OV, Kosmach DA, Nishino S (2013) Space-time dynamics of carbon and environmental parameters related to carbon dioxide emissions in the Buor-Khaya Bay and adjacent part of the Laptev Sea. Biogeosciences 10:5977–5996

Vetrov EA, Romankevich EA, Belyaev NA (2008) Chlorophyll, primary production, fluxes, and balance of organic carbon in the Laptev Sea. Geochem Int 46:1055–1063

Vonk JE, Sánchez-García L, van Dongen BE, Alling V, Kosmach D, Charkin A, Semiletov IP, Dudarev OV, Shakhova N, Roos P, Eglinton T, Andersson A, Gustafsson O (2012) Activation of old carbon by erosion of coastal and subsea permafrost in Arctic Siberia. Nature 489:137–140

Vonk JE, Semiletov IP, Dudarev OV, Eglinton TI, Andersson A, Shakhova N, Charkin A, Heim B, Gustafsson O (2014) Preferential burial of permafrost-derived organic carbon in Siberian-Arctic shelf waters. J Geophys Res Oceans 119(12):8410–8421

Wang J, Cota GF (2003) Remote sensing reflectance in the Beaufort and Chukchi Seas: observations and models. Appl Opt 42:2754–2765

Wegner C, Hölemann JA, Dmitrenko I, Kirillov SA, Kassens H (2005) Seasonal variations in sediment dynamics on the Laptev Sea shelf (Siberian Arctic). Glob Planet Change 48:126–140

Wegner C, Bauch D, Hölemann JA, Janout MA, Heim B, Novikin A, Kassens H, Timokhov L (2013) Interannual variability of surface and bottom sediment transport on the Laptev Sea shelf during summer. Biogeosciences 10:1117–1129

Weingartner TJ, Cavalieri DJ, Aagaard K, Sasaki Y (1998) Circulation, dense water formation, and outflow on the Northeast Chukchi Shelf. J Geophys Res 103:7647–7661

Winterfeld M, Goñi M, Just J, Hefter J, Mollenhauer G (2015a) Characterization of particulate organic matter in the Lena River delta and adjacent nearshore zone, NE Siberia; Part 2: Lignin-derived phenol compositions. Biogeosciences 12:2261–2283

Winterfeld M, Laepple T, Mollenhauer G (2015b) Characterization of particulate organic matter in the Lena River delta and adjacent nearshore zone, NE Siberia—Part I: Radiocarbon inventories. Biogeosciences 12(12):3769–3788

Zsolnay Á, Baigar E, Jimenez E, Steinweg B, Saccomandi F (1999) Differentiating with fluorescence spectroscopy the source of dissolved organic matter in soils subjected to drying. Chemosphere 38:45–50

Ship Traffic in the Asian Arctic Seas

T. Eriksen, H. Greidanus, M. Vespe and C. Santamaria

Abstract This chapter quantifies shipping intensities, and maps out shipping patterns and their changes over seasons and over years, in the Kara, Laptev and East Siberian Seas. Use is made of two main data sources: first, position reports from the AIS automatic ship reporting system, in particular using the Norwegian AIS satellites; and second, ship detections derived from satellite radar images, in particular from the EU's Sentinel-1. It is seen that the ship traffic in the three seas is seasonally dominated by the winter ice cover, which practically halts all shipping in the Laptev and East Siberian Seas for a large part of the year; in the Kara Sea, the busiest of the three, some routes remain used also in winter. Over the last years, the number of ships observed by satellite AIS has grown, especially in the Kara Sea. Concerning shipping type, practically no fishing activity is seen (in contrast to the Barents Sea); most traffic seems to be transport and exploration.

Keywords Maritime surveillance · Ship traffic · AIS · Synthetic aperture radar Arctic shipping

1 Introduction

Human activity in the Arctic is increasing, attracted by resources such as hydrocarbons, ores and fish, by the promise of shortened transport routes and by technology and a more clement climate which makes the Arctic more accessible. Monitoring these activities can enhance the safety of those living and working in the harsh environment and help determine how the environment is affected. Much of the human activity is reflected in shipping patterns and intensities. This paper discusses some approaches to monitoring them in the Asian Arctic waters.

T. Eriksen
Norwegian Defence Research Establishment (FFI), Kjeller, Norway

H. Greidanus (✉) · M. Vespe · C. Santamaria
European Commission—Joint Research Centre (JRC), Ispra, Italy
e-mail: harm.greidanus@ec.europa.eu

V. Barale and M. Gade (eds.), *Remote Sensing of the Asian Seas*,
https://doi.org/10.1007/978-3-319-94067-0_7

Today, the most efficient tool for ship traffic monitoring is the Automatic Identification System (AIS) (International Maritime Organization 2000; International Telecommunication Union 2010). A ship using AIS broadcasts short messages containing the ship's identification and geographic location, plus further data about its movements, properties and voyage, in the VHF radio band. The UN's International Maritime Organisation (IMO) globally mandates the use of AIS on many types of vessels[1] (International Maritime Organization 2000), including all polar class ships (International Maritime Organization 2002). The AIS position reports are transmitted at intervals between 2 and 10 s when ships are sailing, allowing the ships to be tracked by any receiver within VHF line of sight. AIS receivers can be mounted on satellites, allowing global coverage. From a satellite in low Earth orbit the swath width is approximately 5000 km, but the satellite can only collect messages broadcasted by a ship for approximately 10 min as it passes over.

The second relevant tool for ship traffic monitoring is satellite Synthetic Aperture Radar (SAR). A SAR makes images of the Earth surface in the microwave spectrum. Over the seas, this enables the detection of ships, independent (unlike optical imaging satellites) of clouds or day/night conditions. This detection is regardless of whether the ship uses AIS, but the detection does not give ship identification, only a coarse classification. Small ships will however remain undetected, with a cut-off size depending on a series of conditions including radar resolution, ship characteristics, ice conditions and sea state. Under favourable conditions, ships down to half the radar resolution may be detected, i.e. down to 10–25 m considering typical resolutions used for maritime surveillance. The sensitivity is less for higher wind speed and sea state and for non-metallic ships. Even though satellite SAR can only make images of a particular place infrequently and satellite AIS can only detect reporting ships, the combination of satellite AIS and satellite SAR provides a means to assess and monitor the ship traffic in remote seas.

In this chapter the two technologies will be applied to the Kara, Laptev and East Siberian Seas. The Arctic waters further to the east—the Chukchi and Bering Seas which straddle Asia and America—are only discussed for comparison in the analysis of AIS data. The Barents Sea, which is a European sea, is not treated (ship presence in the west of the Barents Sea was discussed by Santamaria et al. (2016)). The location of the seas is shown in Fig. 1. To derive some geospatial statistics, polygons that limit the three seas have been used as obtained from Wikipedia (2016) which follows International Hydrographic Organization (1953).

As the main limiting factor for Arctic shipping, sea ice is briefly discussed to start with.

[1]All vessels with a gross tonnage of 300 and upwards in international traffic, cargo ships of 500 gross tonnage and upwards not engaged on international voyages, and all passenger ships.

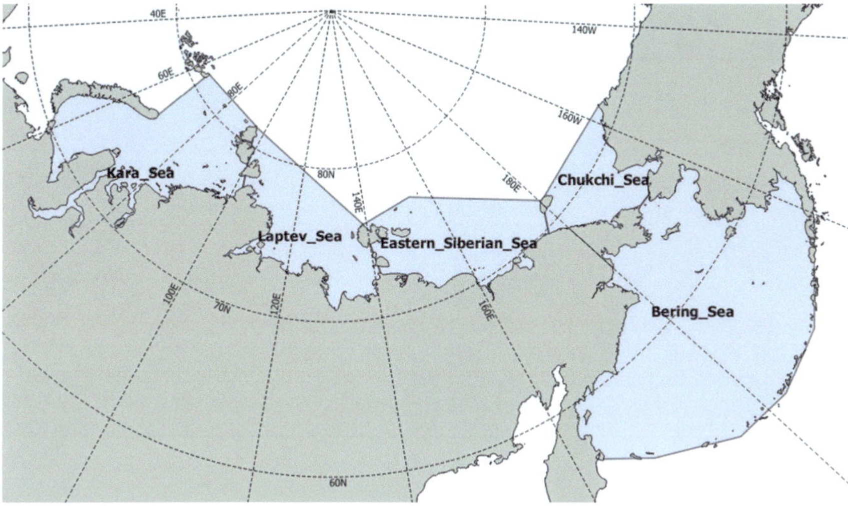

Fig. 1 Polygons that define the limits of the Kara Sea, Laptev Sea and East Siberian Sea—the left three areas. Also drawn in are the Chukchi Sea and the Bering Sea (the right two areas)

2 Sea Ice Cover

To visualise the spatial and temporal extent of the sea ice cover, maps were made to show (a) the number of days in the year that the sea is ice covered, and (b) the day in the year that is in the middle of the ice-free period. This was done for the three seas for the year 2015. As input data, daily sea ice cover maps taken from the Multisensor Analyzed Sea Ice Extent—Northern Hemisphere (MASIE-NH) products were used (National Ice Center 2017). The daily images at 4 km resolution were stacked in time, and for each geographical pixel on sea and each day, 0 was assigned if the pixel was ice covered and 1 if the pixel was open sea. The sum for each pixel gives the number of ice-free days, which is converted to the number of ice-covered days by subtraction from 365. Using day number from 1 to 365, the average day number of the ice-free days gives the middle of the ice-free period for each pixel. This procedure gives meaningful results as long as the year starts (1 January) and ends (31 December) with ice cover in each pixel; this was indeed the case.

The results are shown in Fig. 2. These figures reveal well-defined spatial patterns for the two quantities. The number of ice-covered days (left panel) varies from 0 (blue—never ice) to 364 (red) and 365 (white—always ice, the circumpolar area at centre-left). The day in the year that is the middle of the ice-free period (right panel) has a colour scale that starts with blue for day number (1 + 365)/2 (which is 2 July) where there is never ice cover, and goes up to red for day 280 (7 October); the permanently ice-covered circumpolar area is coloured white as in the left panel.

Towards the north of the Kara Sea, shown at the bottom one-third of the map, only a few days were ice free, while in the southwest as well as off the Northern tip

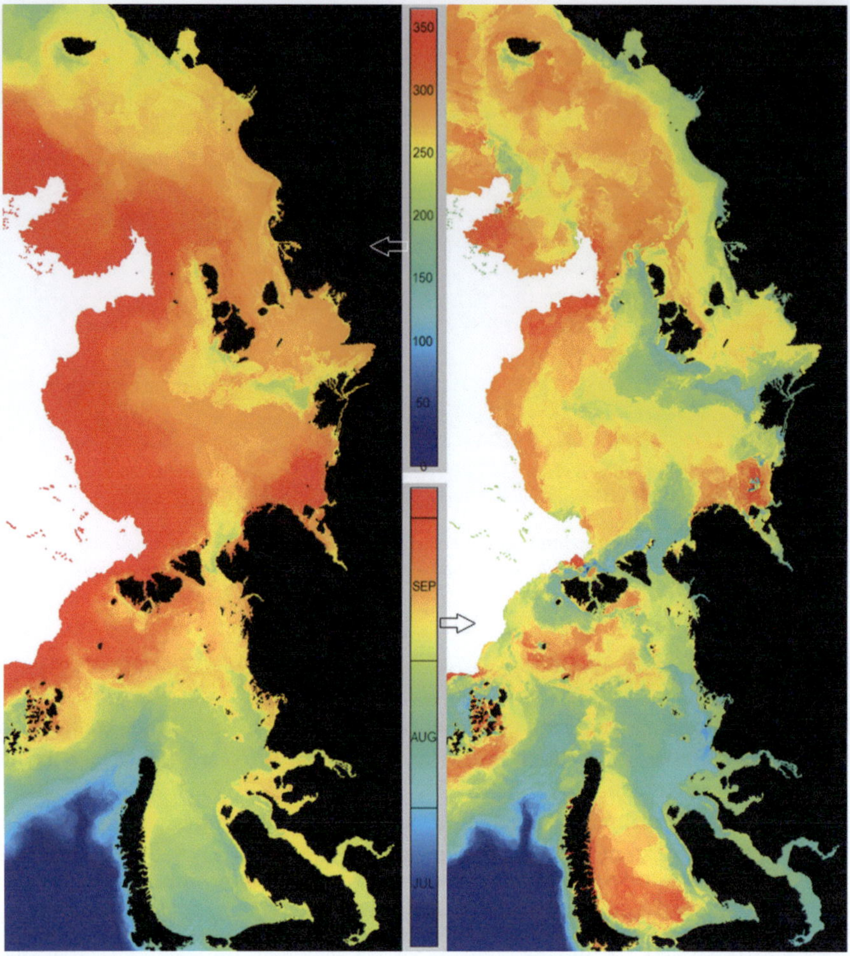

Fig. 2 Left: The number of ice-covered days in (from bottom to top) the Kara, Laptev and East Siberian Seas in the year 2015. Land is black, white is all year ice covered. Right column: The day of the middle of the ice-free period. See main text for further explanation. Derived from the NSIDC/NIC MASIE product

of Novaya Zemlya, more than half of the year (over 200 days) was ice free. The ice-free period was earlier (early August) in the south-central part, but later (end September) in the northern and southwestern parts.

For the Laptev Sea (middle one-third of the figure), the north-western part as well as a part in the south west remained ice covered throughout the year. Other parts were open up to only less than half the year. The ice-free period was centred earliest, around 10 August, in the west and southeast of centre, and latest, around end September, in the southwest.

The East Siberian Sea, the top one-third, had (always referring to 2015) the least open water. Most of the north remained frozen and most areas in the south had only up to 130 days of open water. The northern areas that opened up did so around the end of September, but most of the rest 10 days earlier, and in the south, closer to the coast, earlier again, around 1 September.

3 Shipping Statistics from AIS

3.1 The AIS Data

Globally, about 150,000 ships carry AIS equipment now. Shipping in the Arctic Seas is of course quite limited, so only a very small fraction of those will be found there. A number of 2272 different ships seen during the month of August 2014 (the busiest of the year) in the entire Arctic (>67°) is reported by Eriksen and Olsen (2018). Data from the AISSat-1 and AISSat-2 satellites were used in that study. From approximately 630 km altitude, these satellites have a swath width for AIS message reception of about 5000 km across. They are in polar orbit, meaning that the swath of each passes over the Arctic about 15 times per day.

Although AIS is a powerful tool, the ship traffic picture it yields is not complete or fully correct. In the first place, as mentioned in the introduction, only a part of the ships are subject to AIS carriage. Secondly, only a fraction of the AIS messages transmitted by ships are successfully picked up by a satellite. Messages can be missed and errors can be introduced due to low signal level or interference. Interference can be from external sources or from too many simultaneous messages in the receiver's range, though the latter is not likely to occur in the Arctic. Furthermore, reception is limited by temporal coverage. A ship remains in view of an overpassing satellite for at most 15 min, and shorter if it is toward the edge of the satellite footprint. A point at 75° latitude will be covered about 10% of the time by one satellite. Ships transmit dynamic information with an interval of between 3 min (at anchor) and 2–10 s when sailing, depending on their speed. Finally, ships may deliberately switch off their AIS. For these reasons, the ship traffic seen on AIS is only an approximation of what is really present—even for the sub-set of ships that are subject to AIS carriage requirement (International Maritime Organisation 2000).

3.2 Ship Traffic Spatial Distribution

Figures 3 and 4 give an impression of the geographic distribution of the shipping activity in summer and in winter. In each figure, the left plot shows each received ship position during a month with colours indicating type of ship and activity: Slow moving fishing vessels are coloured red and are interpreted to be engaged in fishing. Slow moving vessels of other types than fishing are coloured green, and are interpreted to be exploring. Fast moving vessels are coloured blue; these are merchants, cargo ships, tankers or transiting fishing ships. Finally, the tracks of

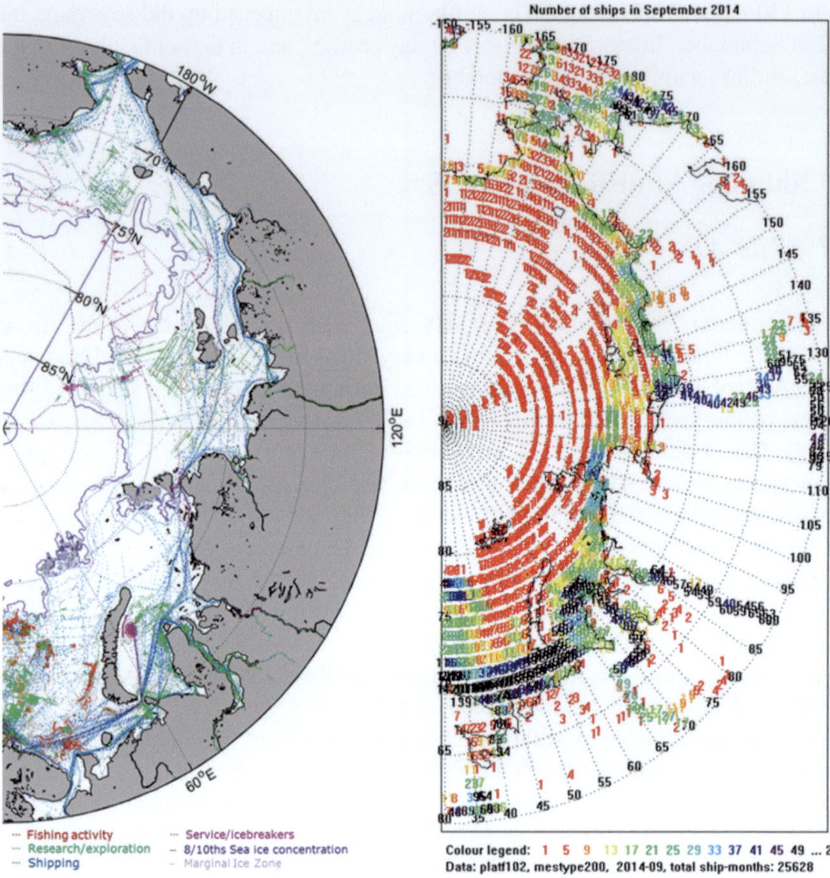

Fig. 3 AIS positions received during the month of September 2014 (left) and the corresponding number of different ships seen per square degree (right). In the left plot, the ice cover extent is drawn in white bounded by a purple line; see main text for colour coding of the tracks

icebreakers and similar service ships are coloured pink. The right plot shows the count of ships (unique MMSI numbers[2]) detected per month on a $1° \times 1°$ grid, plotted as numbers in the centre of the grid cells using a rainbow colour scale; from red for low numbers, to violet for 48 different ships per cell during the month, and black for 49 and higher.

While the busiest month in 2014 for the whole Arctic was August, it was September[3] in the quadrant of the Kara, Laptev and East Siberian Seas (Eriksen and Olsen 2018), coincident with their maximum open water as seen in the previous

[2]Maritime Mobile Service Identity—A unique identifier for ships broadcasting on AIS.

[3]A similar analysis for shipping in Arctic waters that included data for September 2015 was shown by Eriksen et al. (2016).

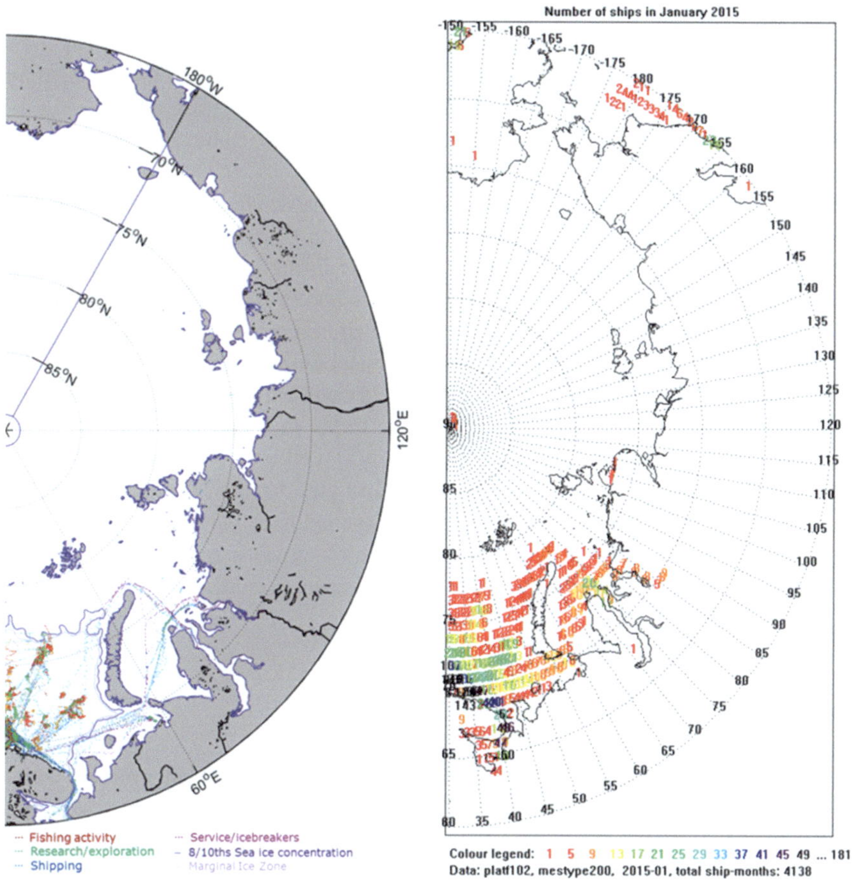

Fig. 4 AIS positions (left) and number of different ships (right) in January 2015

section. The ship positions in the left plot of Fig. 3 show that in all three seas, there are clearly distinguished traffic routes that cross them; transit routes show up in blue, and the green tracks show rectangular patterns indicative of surveying. Traffic that follows the coast is also seen in all of the seas. A few pink tracks show icebreakers going into the ice, even up to the North Pole. In the right plot it can be seen that more than 30 ships are crossing the Kara Sea, and that the number at Bely Island (73°N, 71°E) is higher than 50. The number of ships travelling across the Laptev Sea is less than 20, but the choke point (73°N, 142°E) of the Laptev Strait shows a number of 30. Along the coast in the East Siberian and Chukchi Seas the number of ships is 10–20. Note that this method cannot be expected to detect each ship traveling at high speed in all the grid cells; at 60°N, 1° in longitude is 30 nautical miles, at 70°N it is only 20 nautical miles. Hence the highest numbers observed can be considered the minimum number of ships that are actually in the area. The activity associated with survey ships and ice breakers in the left plot is seen as low numbers with red colour in the right one; two ships have been at the North Pole, and no more than two ships seem to have been in any of the sea areas except in the ship traffic lanes. Also the inland waterways show up in the map, extending far south into the continent. The Yenisei River at 90°E has the highest ship counts, more than 50, the Lena river further East (130°E) has numbers near 40, and the Ob River to the west (75°E) near 20.

The map of January 2015 (Fig. 4), on the other hand, shows the period that the Kara, Laptev and East Siberian Seas as well as the Chukchi Sea and the inland waterways are completely ice covered. AIS traffic in the Laptev and East Siberian Seas has disappeared, but in the Kara Sea, still some of the main routes remain in use, even if they are in the ice. In the plot to the right it can be seen that the number of ships at Bely Island is around 20, in the Yenisei Gulf it is 5–10, and 5–10 ships pass both north and south of Novaya Zemlya. In contrast, the Barents Sea (bottom) remains mostly ice free and with considerable traffic in both seasons.

The sum of the ship counts of all the grid cells, referred to as the total ship-months in the plot, is used as an aggregated figure for the ship activity. For the area plotted in Figs. 3 and 4, the number of ship-months is almost six times higher in September 2014 than in January 2015, which gives a quantification of the difference in activity between the summer and winter seasons.

3.3 Ship Traffic Per Flag State

It is possible to group the ships per flag state, using the MMSI number which identifies the ship in the AIS message; its first three digits are the Maritime Identification Digits that determine the ship's flag state (International Telecommunication Union 2017). This was done separately for each sea, using the polygons of Fig. 1. The resulting number of ships pertaining to the month of August 2015 that was the busiest for the three seas in 2015 is plotted in Fig. 5. The data labels representing the flag state use the two-letter country codes defined in ISO 3166—country codes, from the International Organization for Standardization (2015).

As already shown in Figs. 3 and 4, the colour coding in Fig. 5 indicates that most ships are in the Kara Sea (red). The East Siberian Sea has the lowest number of ships. Furthermore, Russian-flagged ships are the most abundant. The second category, indicated as "noFl", is the count of MMSI numbers from ships for which the first three digits do not correspond to a valid country code, most likely due to errors in the MMSI numbers set in the AIS equipment. Then follow a few flag states with some significant presence in the Kara Sea—Netherlands, Belgium, Bahamas and Cyprus, tapering off into a list of more flag states with ever lower numbers. The sum of the ship counts for the three sea areas for all flag states gives a total of 483 ship-months, of which 390 ship-months are for Russian ships in the three seas. Hence, Russian-flagged ships contribute 81%, and if we add 30 ship-months of noFl (assuming the erroneous MID codes are local traffic) together these amounts to 87%, for the month of August 2015.

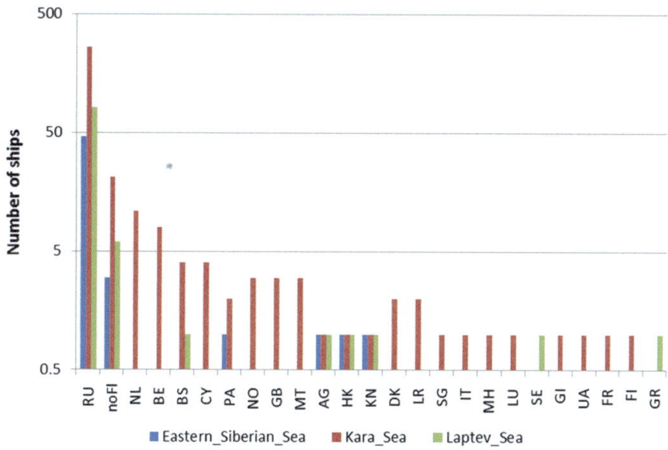

Fig. 5 Number of different ships seen on AIS during the month of August 2015, per flag state (horizontal axis) and per sea (colour). Note the logarithmic scale

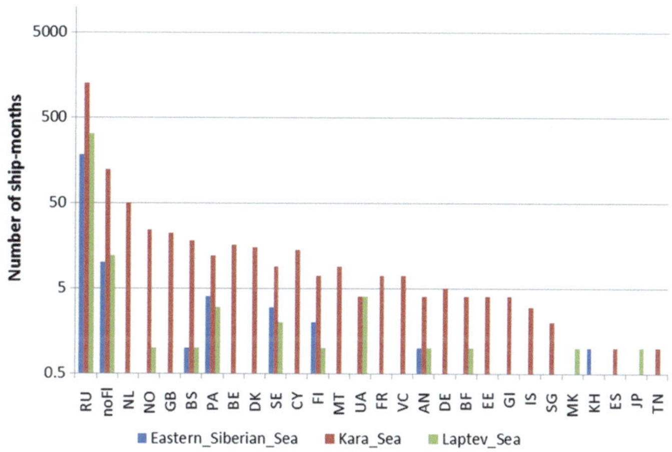

Fig. 6 Number of ship-months seen on AIS during the year 2014, per flag state (horizontal axis) and per sea (colour)

The study of ship traffic per flag state can be extended from one month to the full year. The result for 2014 is shown in Fig. 6. Adding the number of ships from each flag state and in each sea for 12 months gives the total number of ship-months. We do not differentiate between an observation of the same or a new ship for each month that it is seen. The result is a sum of the 12 monthly histograms of 2014, of which Fig. 5 was one example (although from 2015).

The two figures, for the busiest month and for the whole year, show much the same pattern. A notable difference is that for the full-year histogram, some Nordic flag states have moved up on the list of number of ship-months in the Kara Sea: Norway, Denmark and Sweden, as well as Great Britain.

The total for the three seas and all flag states is 2181 ship-months, of which a 1767 are by Russian ships and 144 are noFl, giving percentages similar to that of August 2015; of 81% for Russian flagged ships, 88% if also noFl is assumed Russian.

3.4 Ship Traffic Evolution

Satellite AIS data collected with AISSat-1 (launched 12 July 2010) represent more than five years of maritime traffic data from the Arctic. The monthly number of ships seen by AISSat-1 since its start is plotted in Fig. 7. This is done for each of the seas separately, following the polygons of Fig. 1, in a similar manner as that presented in sectors of 45° longitude in (Eriksen and Olsen 2018).

The temporal curves for the Kara, Laptev and East Siberian Seas are dominated by a strong seasonal pattern. Ships are for the most part only present in the summer and autumn months, between June and November, peaking in August or September. Only the Kara Sea shows some ship traffic in the rest of the year, consistent with having less ice cover than the other two. Overall, the Kara Sea shows more ship traffic than the Laptev and East Siberian Seas. For comparison, the figure also includes the graphs for the Chukchi and Bering Seas. The Chukchi Sea ship traffic behaves much like that of the Laptev and East Siberian Seas. The Bering

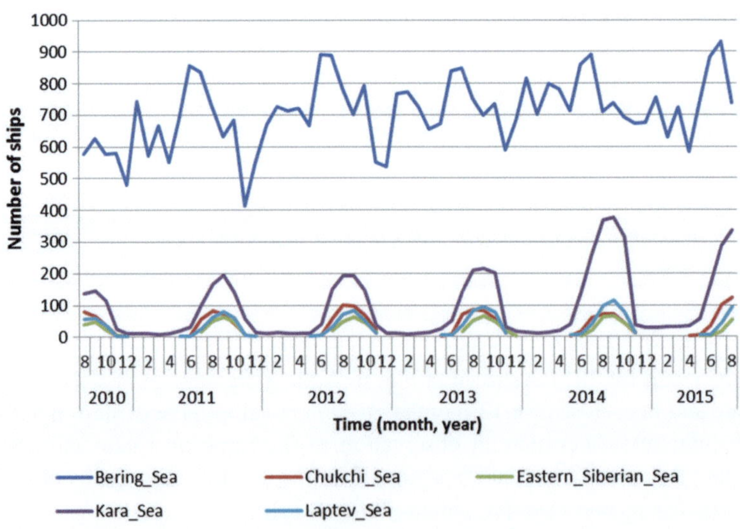

Fig. 7 The number of ships seen on AIS for each month between August 2010 and August 2015. The counts are separated by sea, colour coded as indicated

Sea on the other hand, extending much further to the south, does not show as a whole the typical Arctic behaviour—its ship traffic is much higher and much less dependent on season.

The monthly number of different ships seen in the Kara Sea had a yearly maximum of around 200 for 2011–2013. From 2013 to 2014 this maximum jumped from 216 to 376, and in 2015 the number seems to reach a similar level. The growth from September 2013 to September 2014 is 376 − 216 = 160 ships, or 74% (160/216). To check whether the growth in numbers can be attributed to an actual increase in the shipping in the Kara Sea or rather to an increase in the fraction of ships carrying AIS, the global AIS data base was queried for the 376 ships (of September 2014) for three months from August to October 2013: 252 ships were found and 124 ships were not found. Hence, as a minimum, the real increase in ships visiting the Kara Sea from 2013 to 2014 is 252 − 216 = 36 ships, or 17% (36/216). Considering the 124 ships that were not appearing in the global AIS data from 2013 and therefore contribute to a global increase in AIS-visible traffic in 2014, any of these could or could not have been in the Kara Sea in 2013, so this increase of 124 ships (57%) cannot be unambiguously attributed to a local increase in shipping. Further uncertainties may be present due to ships changing ownership and therefore MMSI number. Grouping the 124 newly seen ships by flag state, 103 or 83% (103/124) are likely to be local traffic (95 ships from Russia and 8 noFl, assuming the erroneous MID codes are local traffic), and 21 ships are from other countries.

4 Ship Presence from Satellite Imaging

4.1 Ship Detection in Arctic Waters

Several satellite SAR systems are available today. From their polar orbits, as for the AIS satellites, the polar zones can be observed more frequently than the lower latitudes. Sentinel-1 is the satellite SAR of the Copernicus Earth Observation program of the EU, operated by ESA (European Space Agency). It is the only satellite SAR that makes its data available for free. The areas imaged by Sentinel-1 and the imaging modes are determined by a predefined observation scenario. The two image products that are routinely acquired over the Arctic are Interferometric Wide Ground Range Detected High resolution (IW-GRDH, 250 km image size at 20 m resolution) and Extra Wide Medium resolution (EW-GRDM, 400 km at 90 m) (European Space Agency 2013). The higher resolutions enable the detection of smaller ships, but over smaller areas. Images from the Sentinel-1 archive have been used to assess the ship traffic in the three seas.

Ship detection in SAR images is based on the fact that a ship produces a stronger radar reflection (from its metal structures) than the surrounding sea (from the rough sea surface). An introduction to the topic can e.g. be found in Pichel et al. (2004); extensive reviews have been given by Crisp (2004) and Arnesen and Olsen (2004); and recent references can be found in Greidanus et al. (2017). Sea ice and icebergs,

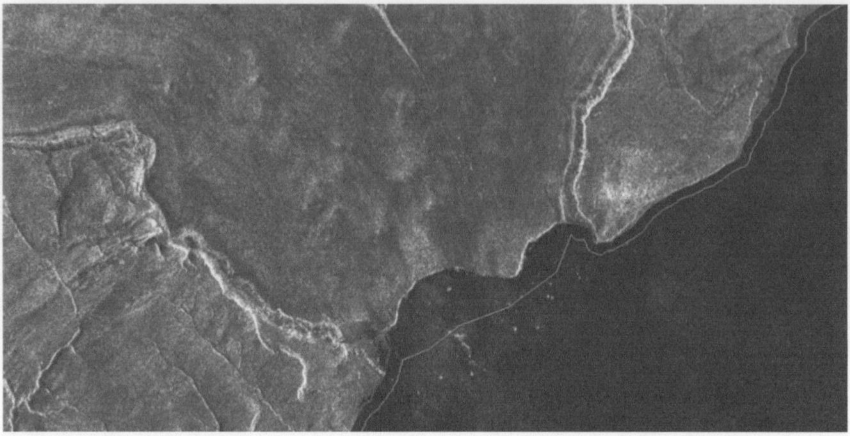

Fig. 8 Part of the NE coast of Novaya Zemlya around lat 75.73°, lon 64.31°. The bright dots on the (dark) sea are interpreted to be icebergs that calve off the glacier which flows into the sea here. Sentinel-1 IW image, VV polarisation. The white line shows the buffered land mask

however, also produce significant radar echoes, which are very difficult to distinguish from those of a ship. Reliable ship detection with radar in the presence of sea ice is therefore still a serious challenge. An example is shown in Fig. 8. The bright dots on the dark sea background, here thought to be caused by floating pieces of ice as they are just in front of a calving glacier, look exactly like ship radar signatures. Ship—iceberg discrimination is e.g. discussed in Power et al. (2013) and Denbina et al. (2015). With high resolution (Bentes et al. 2016) more is possible, but that is not always available. Brekke and Anfinsen (2011) discuss ship detection within the sea ice.

The land in the images must be masked to prevent spurious detections on land. Publicly available global coastline maps are used for that. However, these are not always accurate enough. Therefore, the land masks are extended by a buffer zone. This means that ships very near the coast will not be found.

As is the case with AIS, the ship traffic picture from SAR is incomplete and contains errors. False detections may come e.g. from icebergs and uncharted rocks or reefs but also from meteorological and oceanographic phenomena. Small ships will not be detected against the sea clutter background (as specified in Sect. 1), and even bigger ones may be masked out by a rough sea (which increases the clutter).

4.2 Correlation Between Reporting Ships and SAR-Detected Ships

A few Sentinel-1 images in ESA's archive have been analysed for the presence of ships. This was done with the SUMO ship detection software (Kourti et al. 2005;

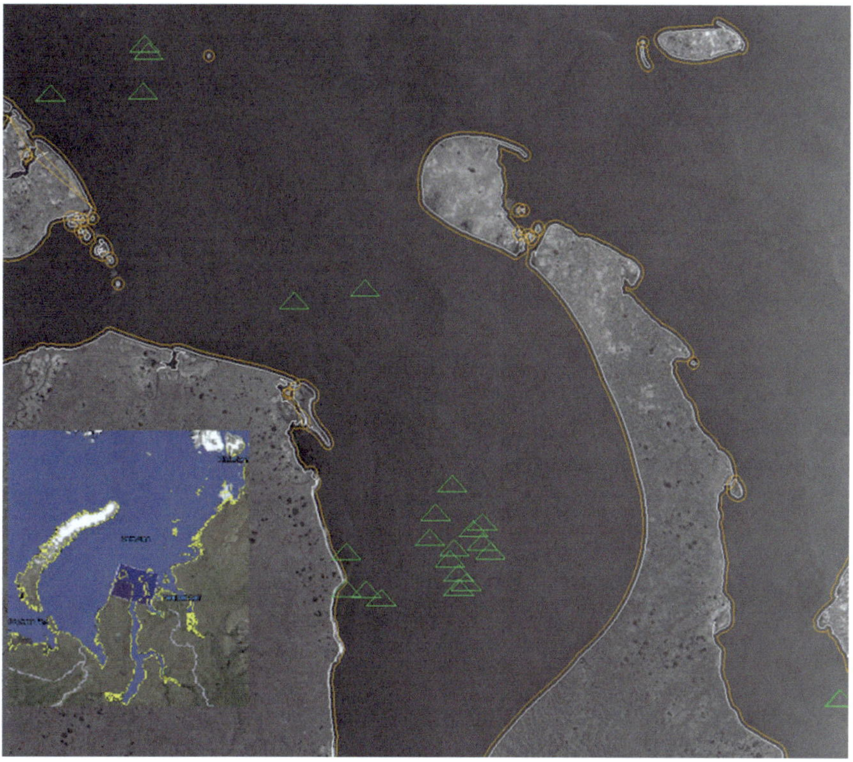

Fig. 9 Part of a Sentinel-1 IW image over the Kara Sea. Detected ships are indicated by green triangles. The orange line is a land mask, buffered outward by 1 km. The inset (on Google Earth) shows the location of the image

Greidanus et al. 2017). SUMO is a Constant False Alarm Rate (CFAR) detector. It finds bright pixels that, based on the probability density function (PDF) of the radar sea clutter background, have only a very low probability of being a false alarm that is constant across the image. The K distribution is used as the model PDF of the background, with a nominal false alarm rate of 10^{-7}. Neighbouring detected pixels are clustered into one target. The images were analysed automatically but the results visually verified and corrected where needed. Only IW-GRDH images outside of sea ice cover, away from the ice edge, were selected. The presence of ice was derived from visual inspection of the image. An example of ships detected in such an image is shown in Fig. 9.

The positions of the ships detected in the SAR images can be compared with the positions of the ships that are reporting on AIS. To that end, the ships' AIS position reports—which are only received at the times of AIS satellite overpasses—must be interpolated to the times of the SAR images. When the AIS-interpolated reporting ship position falls close enough to a SAR-detected ship position, the two are said to be correlated and referring to the same ship. An uncorrelated SAR ship detection indicates the presence of a ship that is not reporting on AIS. An uncorrelated interpolated AIS position indicates a ship that is not detected by the satellite SAR. However, it should be noted that these attributions come with some uncertainty, because of the inaccuracies and incompleteness of the AIS and SAR ship traffic pictures as discussed, and because the interpolation of the reported ship positions introduces uncertainties.

Sixteen Sentinel-1 IW images were analysed this way: 10 from the Kara Sea, 2 from the Laptev Sea, and 4 from the East Siberian Sea. A total of 128 ships were detected by SAR and 86 ships were detected with AIS (interpolated to be inside the SAR image frames at the time of imaging). Correlating the datasets, 69 ships are seen in both datasets, 59 are seen on SAR only, and 17 on AIS only. This would indicate an overall fraction of 41% (59/145) of ships in those seas that are not reporting on AIS. On the other hand, there are 69% more (145/86-1) ships seen on SAR and AIS together than on AIS alone. However, the uncertainties discussed above have to be kept in mind.

4.3 Ship Movements Through Sea Ice

Although the individual detection of ships in sea ice in satellite SAR images still presents a challenge, alternative analyses are possible. From the Sentinel-1 archive, 67 images in EW mode and HH polarisation over the same location in the Yenisei Gulf (on the Kara Sea) were selected, acquired during the late summer to winter season between 9 October 2014 and 20 April 2015. The images were mixed ascending and descending passes, and mixed GRDM and GRDH products. From each image, the same geographic sub-area was cut out, which was resampled to North up and smoothed to a common resolution. The resulting time series shows how the water in the Gulf freezes up into a distinct pattern of high and low radar backscatter areas. Between 10 October and 6 December, sea ice appears, which first moves around in patches, but then starts to remain fixed in place, first linked to the coast but finally everywhere in the Gulf. The sea ice pattern that is formed by 6 December remains fixed during all of the winter up to the last image in the series of 20 April. From 6 December onward, a bright line appears crossing lengthwise through the Gulf, and from 13 January 2015, another line branches off, which in time become brighter than the first—see Fig. 10.

The bright lines can be associated with ship routes through the sea ice. Figure 11 plots the positions of 3.5 months' worth of AIS messages from the same period. The AIS tracks coincide exactly with the bright lines in the SAR images; the less-travelled track is the earlier one.

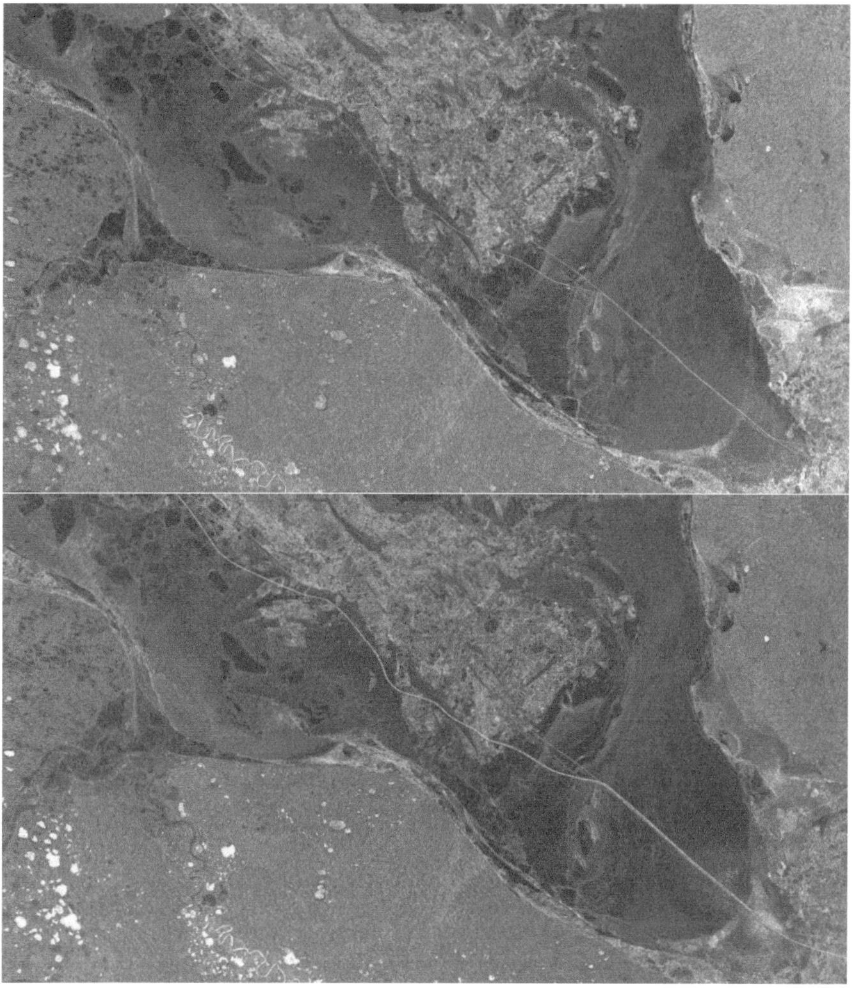

Fig. 10 Yenisei Gulf. Top: 13 January 2015. Bottom: 20 April 2015

The phenomenon that the sea ice and ship lanes remain fixed does not occur every year, however. Sentinel-1 images from the winter of 2015–2016 show that the ice cover breaks up from time to time, forcing the ship lanes through the ice to change.

Fig. 11 AIS messages received between 1 January and 22 April 2015 (3.5 months) plotted in the same area as the Sentinel-1 images from Fig. 10. The total number of messages is 5088

5 Conclusions

Much of the Asian Arctic seas—Kara, Laptev and East Siberian—are ice covered for more than 9 months of the year, especially in the north and east. The Kara Sea is more open, with an ice-free period in 2015 of up to 7–8 months in places. The Kara Sea also shows the highest ship traffic of the three, with up to almost 400 different ships seen on AIS in the month of September 2014, which is the busiest month of the year for each of the three seas. In the summer months, shipping moves mostly along the coast but also along routes further from the coast, and for surveying purposes even further north. In contrast to the Barents Sea, no fishing activity is seen on AIS. In the winter months, AIS shipping stops in the Laptev and East Siberian Seas, and only a few routes remain in use in the Kara Sea. The total count of ship-months, adding one count for each calendar month that a certain MMSI number is seen on AIS, adds up to 2181 over all three seas and all of 2014. Of that, the Russian-flagged part is estimated to 81%; the fraction is somewhat lower in the Kara Sea and higher in the Laptev and East Siberian Seas.

Considering the evolution in the amount of ships in the different seas, the monthly number seen in the Kara Sea shows a marked increase from September 2013 to September 2014, whereas increases are much smaller or absent in all other years and in the other seas. The increase is from 216 to 376 ships, i.e. by 160 ships or 74%. As 252 of the ships were already observed sailing in the summer of 2013, $252 - 216 = 36$ ships, or 17%, is a minimum estimate for the growth in the Kara sea. The other part of the increase, 124 ships (or 57%), can either be ships that are new to the area or ships that were already there but started to use AIS.

A small-sample trial using 16 Sentinel-1 images in ice-free conditions indicates that there are about 70% more ships in the area that can be detected by satellite SAR than what is seen on AIS. Sentinel-1 images also show how routes through the winter sea ice can remain fixed in place in the frozen pack ice for the whole winter season.

The quoted numbers carry uncertainties of the order of maybe tens of percent due to the nature of satellite AIS and SAR measurements. Nonetheless, these two tools provide valuable insights into human activities and their developments over time in remote places such as the Asian Arctic.

Acknowledgements Sentinel-1 images are © Copernicus 2014–2015. Satellite AIS data from AISSat-1 and AISSat-2 were obtained courtesy of the Norwegian Coastal Administration. A coastline of OpenStreetMap, © OpenStreetMap contributors, was used.

References

Arnesen TN, Olsen RB (2004) Literature review on vessel detection. Norwegian Defence Research Establishment (FFI), Kjeller

Bentes C, Frost A, Velotto D, Tings B (2016) Ship-iceberg discrimination with convolutional neural networks in high resolution SAR images. In: Proceedings of the EUSAR 2016, pp 491–494

Brekke C, Anfinsen SN (2011) Ship detection in ice-infested waters based on dual-polarization SAR imagery. IEEE Geosci Remote Sens Lett 8(3):391–395

Crisp DJ (2004) The state-of-the-art in ship detection in synthetic aperture radar imagery. Defence Science and Technology Organisation, Information Sciences Laboratory, Edinburgh

Denbina M, Collins M, Atteia G (2015) On the detection and discrimination of ships and icebergs using simulated dual-polarized RADARSAT constellation data. Can J Remote Sens 41(5):363–379. https://doi.org/10.1080/07038992.2015.1104630

Eriksen T, Olsen Ø (2018) Vessel tracking using automatic identification system data in the arctic. In: Hildebrand LP, Brigham LW, Johansson TM (eds) Sustainable shipping in a changing arctic, Springer International Publishing. https://doi.org/10.1007/978-3-319-78425-0

Eriksen T, Ommundsen A, Gade B, Aronsen M (2016) Maritime situational awareness in the high north. In: Proceedings of the maritime knowledge discovery and anomaly detection workshop. European Commission—Joint Research Centre. European Union, 2016, pp 87–90. ISBN 978-92-79-61301-2 (PDF), https://doi.org/10.2788/025881

European Space Agency (2013) Sentinel-1 user handbook. GMES-S1OP-EOPG-TN-13-0001, 1 Sept 2013

Greidanus H, Alvarez M, Santamaria C, Thoorens F-X, Kourti N, Argentieri P (2017) The SUMO ship detector algorithm for satellite radar images. Remote Sens 9(3):246. https://doi.org/10.3390/rs9030246

International Hydrographic Organization (1953) Limits of oceans and seas. Special publication no. 23 (S-23), 3rd edn. Monaco

International Maritime Organisation (2000) International convention for the safety of life at sea "SOLAS" (1974) Chapter V "Safety of navigation". AIS was only introduced by amendment in December 2000

International Maritime Organisation (2002) Guidelines for ships operating in arctic ice-covered waters. MSC/Circ. 1056, MEPC/Circ. 399, 23 Dec 2002

International Organization for Standardization (2015) Country Codes—ISO 3166. http://www.iso.org/iso/country_codes. Accessed 29 July 2015

International Telecommunication Union (2010) Technical characteristics for a universal shipborne automatic identification system using time division multiple access in the VHF maritime mobile band. ITU-R M.1371-4

International Telecommunication Union (2017) Table of maritime identification digits. http://www.itu.int/en/ITU-R/terrestrial/fmd/Pages/mid.aspx. © ITU 2017. Accessed 21 Mar 2017

Kourti N, Shepherd I, Greidanus H, Alvarez M, Aresu E, Bauna T, Chesworth J, Lemoine G, Schwartz G (2005) Integrating remote sensing in fisheries control. Fish Manage Ecol 12:295–307

National Ice Center and National Snow and Ice Data Center (2017) Compiled by F. Fetterer, M. Savoie, S. Helfrich, and P. Clemente-Colón. 2010, updated daily. Multisensor Analyzed Sea Ice Extent—Northern Hemisphere (MASIE-NH), Version 1 [geotiff/4km/all surface/region 0]. NSIDC: National Snow and Ice Data Center, Boulder. http://dx.doi.org/10.7265/N5GT5K3K. Accessed 23 Mar 2017

Pichel WG, Clemente-Colon P, Wackerman CC, Friedman KS (2004) Ship and wake detection. In: Jackson CR, Apel JR (eds) Synthetic aperture radar marine user's manual. Department of Commerce, National Oceanic and Atmospheric Administration, Washington, DC, pp 277–303

Power D, Yue B, Youden J (2013) Summary of previous research in iceberg and ship detection and discrimination in SAR. Report R-13-060-1098, Rev. 4.0, DRDC-RDDC-2014-C79

Santamaria C, Greidanus H, Fournier M, Eriksen T, Vespe M, Alvarez M, Fernandez Arguedas V, Delaney C, Argentieri P (2016) Sentinel-1 contribution to monitoring maritime activity in the arctic. In Proceedings of the ESA living planet symposium. Prague, Czech Republic, ESA SP-72, 9–13 May 2016

Wikipedia page "List of seas". https://en.wikipedia.org/wiki/List_of_seas. Accessed 2 May 2016

Part III
Pacific Ocean Marginal Basins

The Sea of Okhotsk: Scientific Applications of Remote Sensing

Leonid M. Mitnik and Vyacheslav A. Dubina

Abstract The characteristics of the Sea of Okhotsk have not been adequately explored due to the severe environmental conditions and satellite measurements are particular appealing for their study. Results of remote sensing of the various oceanic and atmospheric dynamic phenomena in the Sea of Okhotsk are reviewed. The multisensor data analysis is considered as the most effective approach for the phenomena detection, interpretation and operational applications. Particular emphasis has been placed on the use of passive and active microwave techniques. Microwave remote sensing provide estimates of the sea surface temperature, surface wind speed, sea ice parameters, atmospheric water vapor content, cloud liquid water content, as well as allows mapping of current and eddy boundaries, the surface manifestations of the internal waves, bottom topography signatures, the organized wind speed variations in the marine boundary layer of the atmosphere.

Keywords Sea of Okhotsk · Currents · Productivity · Internal waves
Sea ice · Marginal ice zone · Atmospheric cyclones · AMSR-E
AMSR2 · ALOS PALSAR · Envisat · Sentinel-1A · MODIS · VIIRS
ASCAT

1 Introduction

Although the Sea of Okhotsk is located in temperate latitudes, its many hydrometeorological characteristics are close to those of the polar ocean: severe winter with low air temperature and enhanced cyclonic activity, extended periods of ice cover. Research work, fishery and transport operations, oil recovery and new oil platform construction are conducted here throughout the year. Research and economic

L. M. Mitnik (✉) · V. A. Dubina
V.I. Il'ichev Pacific Oceanological Institute FEB RAS, Vladivostok, Russia
e-mail: mitnik@poi.dvo.ru

V. Barale and M. Gade (eds.), *Remote Sensing of the Asian Seas*,
https://doi.org/10.1007/978-3-319-94067-0_8

activities are rather complicated in winter period. Strong winds accompanying deep mid-latitude and polar cyclones significantly influence the sea surface roughness and air-sea interaction, as well as formation, development, and destruction of ice cover, position and structure of the ice edge. From this follows the need of gaining spatial information about wind and wave characteristics, sea surface temperature, water dynamic and color and ice cover.

The common practice for estimating oceanic and atmospheric conditions is the joint use of satellite data obtained in the visible, infrared and microwave ranges. Only space measurements can provide a pattern of geophysical parameter distribution over the whole sea practically at a single moment. Cloudiness and low solar altitude in winter limit the possibilities of surveying in the visible range. There are fewer limitations in the infrared (IR) range: satellite measurements are also made in the dark, however, clouds hinder the sea surface sensing. Solving cloudiness problem can be made by using the microwave (MW) radiometric (passive) and radar (active) data. The weak influence of weather conditions, the independence of solar illumination, the possibility of estimating sea surface and ice characteristics, and a number of meteorological parameters all point to high priority of multichannel MW data in spite of the low resolution of satellite passive MW radiometer and large revisit time of satellite Synthetic Aperture Radars. In the paper, this approach is demonstrated while considering the oceanic and atmospheric phenomena and processes in the Sea of Okhotsk. The oceanic phenomena include water circulation, currents, eddies of various scales, internal waves, upwellings and sea ice. Among the atmospheric phenomena the main attention is given to extratropical cyclones, polar lows and mesoscale convective cells and rolls.

The broad arsenal of satellite sensors is applied to characterize quantitatively the phenomena under study. The visible and IR images from Terra and Aqua MODIS (Moderate Resolution Imaging Spectroradiometer) and Landsat sensors provide spatial distribution of cloudiness, and (under clear sky) maps of sea surface temperature (SST) and chlorophyll-*a* concentration Chl-*a* (ocean color). Scatterometers onboard QuikSCAT, Oceansat-2, MetOp-A and MetOp-B satellites serve as a source of data on sea surface wind speed (SWS) and direction. Oceanic dynamic phenomena and surface wind speed W signatures were revealed on the Synthetic Aperture Radar (SAR) images acquired by Envisat, ALOS-1/2 (Advanced Land Observing Satellite) and Sentinel-1A satellites. The quantitative information on SST, SWS, total water vapor content V and total cloud liquid water content Q were obtained by processing of passive MW measurements from DMSP (Defense Meteorological Satellite Program), Aqua, TRMM (Tropical Rain Measuring Mission), GCOM-W1 (Global Change Observation Mission-Water) and GPM (Global Precipitation Measurements) satellites (Katsaros et al. 2014). All sensors were used to study sea ice concentration and characteristics. Sea ice served also as an indicator of surface circulation of various scales.

2 Circulation and Productivity

The Sea of Okhotsk is one of the most highly productive areas of the World Ocean. The increased biological productivity of the sea is largely due to its water circulation. The straits connecting the Sea of Okhotsk to the Japan Sea and Pacific Ocean allow water exchange between the basins, which manifests itself in fields of sea surface temperature, sea surface roughness and ocean color. The surface circulation in the sea is complex and characterized by the presence of several cyclonic gyres. The main gyre is formed by the East Sakhalin, North Okhotsk, and West Kamchatka Currents (Fig. 1). The second gyre covers a southwestern part of the sea where intense vortices are observed in warm and cold seasons (SHH 1998; Dubina et al. 2012; Ohshima et al. 2002). One of the most sustained element of the circulation is the Soya Warm Current, a branch of the Tsushima Current. The currents can show the variable contrasts against the surrounding waters (Fig. 2). In winter the contrast of currents, eddies and other dynamic features may be distinctly larger due to the pack ice (Shevchenko et al. 2004; Ebuchi et al. 2006).

The increased productivity is observed in the coastal zone and in the Terpeniya Bay where Chl-*a* concentration reaches very high values—up to 10–30 mg/l (probably due to algorithm error in the coastal productive waters) (Fig. 2b). The Chl-*a* values are much lower in the central parts of the main and the second gyres: typically <0.3–0.4 mg/l. There is no clear boundary of the East Sakhalin Current (ESC). The thermal contrasts (Fig. 2a) are detected at several places only likely due to the data averaging. This conclusion follows from Fig. 3 showing the visible and IR images taken on 16 October 2007. The greenish waters of the ESC are traced from Cape Elizabeth **1** to Cape Terpeniya **2** (Fig. 3a). The undulating boundary parts the ESC from the blue offshore waters. The width of the current flow is 25–65 km. The second ESC branch formed over the continental slope is seen as a band of the colder water **3** (Fig. 3b). The branch does not appear in the color image (Fig. 3a). Positions of maximum ocean color and SST gradients do not align (Fig. 3).

The Amur River, one of the greatest rivers of the world, flows into the Amur Liman, a narrow and shallow strait connecting the Sea of Okhotsk with the Japan Sea. The river runoff contains a large quantity of suspension. Most turbid low salinity waters fill the northern Amur Liman **4** and enter the Sakhalin Bay **5** where they interact with the clearer sea waters (Fig. 3a).

The velocity of tidal currents in the Sea of Okhotsk can reach 1 m/s and more (Dubina et al. 2012; Ohshima et al. 2002; Shevchenko et al. 2004). Tidal currents in the vicinity of Shantarsky Islands, Northeastern Sakhalin, Iona Island, Kashevarov Bank, Shelikhov Bay and Kuril Straits (Fig. 1) lead to intense water mixing and internal wave generation (Kowalik and Polyakov 1999; Shevchenko et al. 2004; Mitnik and Dubina 2007).

The West Kamchatka, North Okhotsk and East Sakhalin Cold Currents (Fig. 1) are usually not delineated on SAR images, which can be explained by the weak gradients of current velocity and/or by the low SST gradients near their boundaries.

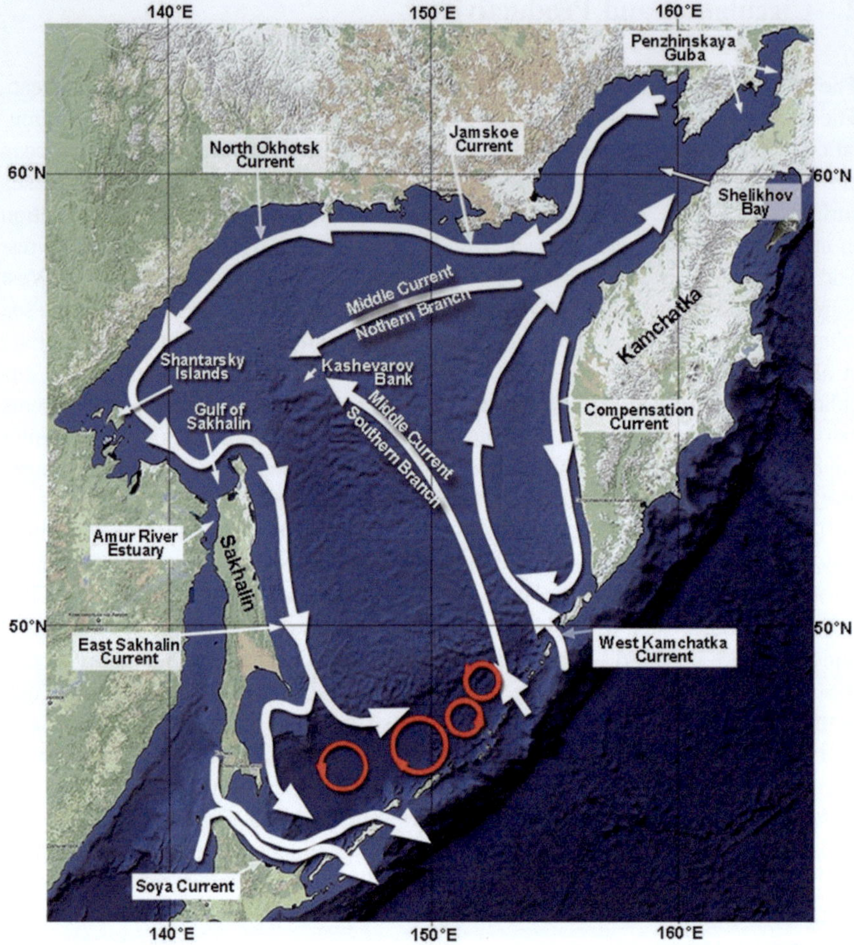

Fig. 1 Main currents of the Okhotsk Sea (Mitnik and Dubina 2010a)

The ship observations support this conclusion. Sometimes the current boundaries manifest themselves in SAR images as small contrast lines especially when the directions of currents and tidal currents coincide (Mitnik et al. 2007).

Unlike the above mentioned currents, the Soya Warm Current is characterized by the presence of sharp gradients of current velocity and SST (Ebuchi et al. 2006, 2009; Mitnik et al. 2005, 2007). SST gradients can reach 1 K/km and more as follows from processing of NOAA AVHRR and Landsat ETM + thermal images. The form and location of SAR signatures and thermal fronts (Fig. 4) are in a good agreement that follows from joint analysis of Envisat and ALOS SAR with MODIS, NOAA AVHRR and Landsat ETM + IR images acquired in 1993–2010 (Mitnik et al. 2005, 2007; Mitnik and Dubina 2010a, b).

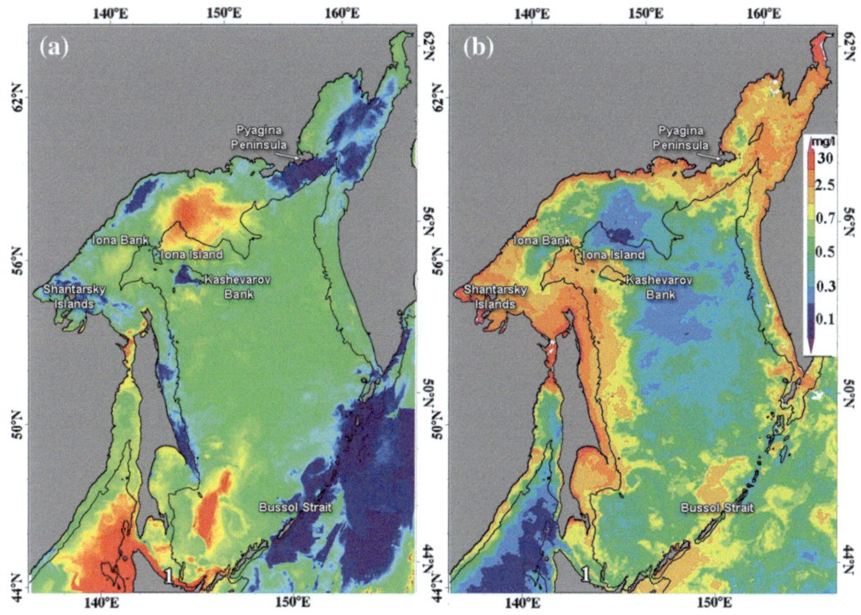

Fig. 2 The equalized false color infrared image composed of the Aqua and Terra MODIS data acquired on 1–3 September 2009 (**a**) and monthly averaged Chl-*a* concentration in September 2009 (**b**). The dark lines are 200-m isobath, **1** is the Soya Warm Current

There are tidal currents near the Eastern Sakhalin coast, in the Kashevarov Bank area and other regions. The current velocity of two-dimensional currents near the Northeast Sakhalin can reach 1 m/s. The time series of satellite optical and SAR and ground radar images of drift ice allow to distinguish the components of ice movement caused by tidal current, wind and surface current. Their relative contribution to the observable drift changes along the Sakhalin coast (Shevchenko et al. 2004; Ohshima et al. 2002).

3 Internal Waves

Numerous packets of internal waves (IWs) were revealed on the satellite SAR images. The Terra and Aqua MODIS visible images were also screened to improve temporal and spatial resolution. Parameters of IWs such as number of waves in a packet, their wavelengths, a crest length, and propagation velocity were estimated from the images. Propagation velocity of IWs in the open sea was computed from spacing between packets generated by the successive tides as well as from displacement of the internal wave signatures on Envisat and ERS-2 SAR images taken 30 min apart and on Terra and Aqua MODIS images taken in about 100 min

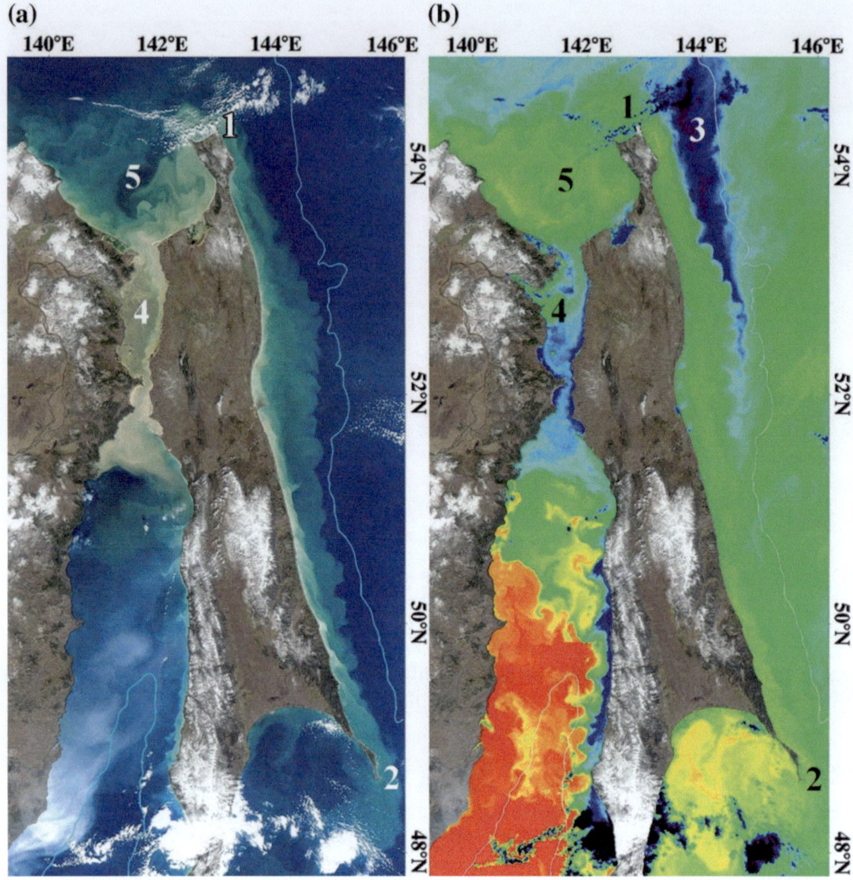

Fig. 3 The visible true color (**a**) and infrared false color (**b**) images of the Sakhalin shelf taken by Aqua MODIS on 16 October 2007 at 01:45 UTC; **1** is Cape Elizabeth, **2** is Cape Terpeniya, **3** is the second branch of the East Sakhalin Current, **4** is the Amur Liman, **5** is the Sakhalin Bay

apart. The variable currents associated with IWs can significantly modify the short wind waves with the wavelengths <0.5 m. As a result, the IW signatures are observed both by radars and visually from ships and satellites under moderate winds.

IWs were observed on SAR images from April till December. They were detected in the coastal zone and in the open sea and characterized by the broad range of parameters such as wavelength, width of a packet, crest length, etc.

The complexity of the signatures of IWs revealed on satellite SAR images of the Sea of Okhotsk indicates that there are numerous sources of IW generation. Internal wave packets propagating on the shore form sometimes the common elongated fronts length of which can reach ≈500 km (Mitnik et al. 2005). The distance between the fronts varies from ≈30 to ≈40 km.

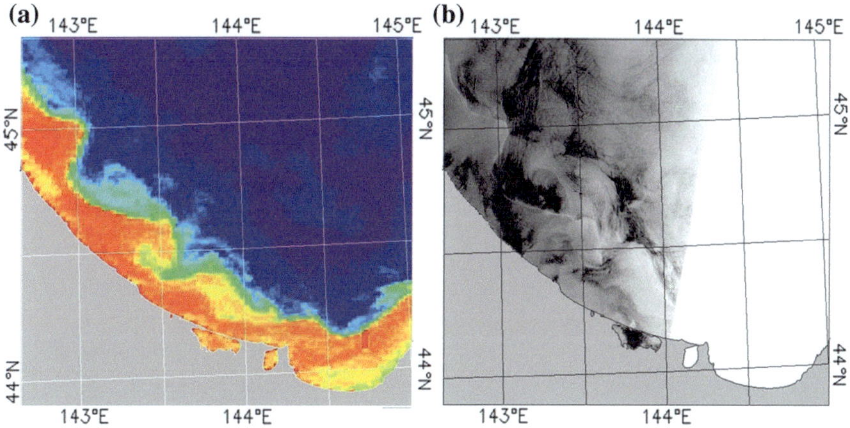

Fig. 4 The Soya Warm Current eddies on 15 September 2004: **a** Terra MODIS thermal infrared image at 00:45 UTC and **b** Envisat ASAR image at 12:05 UTC

Velocity of tidal currents in the Sea of Okhotsk can reach 1 m/s and more (Mitnik and Dubina 2006, 2010a; Mitnik et al. 2007; Ohshima et al. 2002). Interaction of tidal flows with the bottom features such as underwater mountains and canyons and a continental shelf in the vicinity of Shantarsky Islands, Northeast Sakhalin, Iona Island, Kashevarov Bank, Shelikhov Bay and Kuril Straits results in internal wave generation in waters with stable stratification. They can be particularly pronounced where pycnocline is well developed.

The complexity of signatures is typical for internal waves in the Aniva Bay (Fig. 5). Aniva Bay is characterized by the increased biological productivity and intensive fishery. Analysis of satellite SAR images have shown that the IWs are rather common phenomenon in a warm period when there are strong-stratified waters in the Bay to the north of the Soya Warm Current (Mitnik and Dubina 2007, 2010a, b; Mitnik et al. 2005, 2007). Combination of stratification, currents and variable bottom topography is favorable for internal wave generation. Several packets of the IWs propagating mainly westward and eastward are visible on Envisat ASAR (Fig. 5a) and ALOS PALSAR (Phased Array type L-band Synthetic Aperture Radar) (Fig. 5b) images. A width of the wave packets is 1–10 km. The length of several first crests exceeds 40 km. Within the packets, the wavelengths Λ appear to be monotonically decreasing, from the front to the rear. So for packet A, Λ decreases from 1.8 to 0.9 km, for packet B from 2.5 to 2.0 km, and for packet C from 0.9 to 0.5 km (Fig. 5c). Weak low-contrast IWs were also detected within the Soya Warm Current.

A fixed source of internal wave generation in a Northern part of the Sea is the Kashevarov Bank with depths of about 200–250 m (Figs. 1 and 2) due to intense diurnal tides currents (Kowalik and Polyakov 1999). IW packets propagate in different directions and interact with packets formed by other nonuniformities of bottom topography and Sakhalin shelf slope (Dubina and Mitnik 2007).

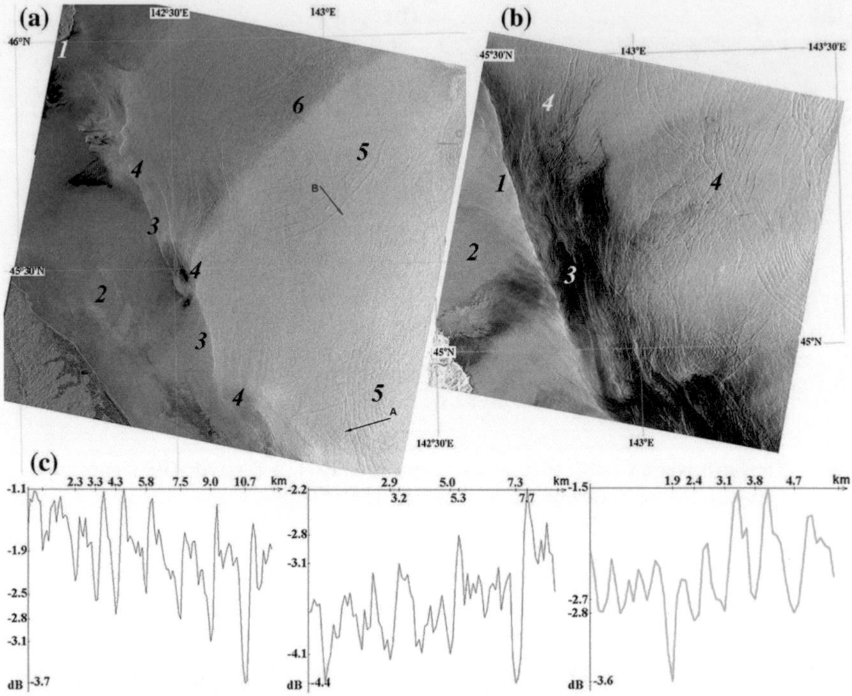

Fig. 5 The Soya Warm Current (SWC) and internal waves on the images obtained by Envisat ASAR on 15 August 2005 at 00:49 UTC (**a**) and ALOS-1 PALSAR on 17 September 2006 at 01:15 UTC (**b**). The numbers on (**a**): **1** is Cape Krilion, **2** is the Soya Warm Current, **3** is current front, **4** are the eddy centers in the SWC boundary zone, **5** are packets of IWs in the Aniva Bay, color arrows **A**, **B**, and **C** show the direction of their propagation; the profiles of the radar backscatter along the transects A, B, and C are given in (**c**), **6** is the atmospheric front. Numbers on (**b**): **1** is the current front, **2** is the warm current water, **3** is the cold water, **4** are IWs

Increased current speed near the headlands and straits combined with bottom risings is favorable for IW generation. IWs are observed in the areas of Cape Krilion, Cape Terpeniya, Kuril Straits etc. (Mitnik 2009; Mitnik and Dubina 2012). The diurnal tidal flow is considerably intensified around the Kuril Straits, with a maximum flow speed of over 2 m/s. In numerical simulations have shown that energetic internal waves are formed with amplitudes exceeding 100 m, resulting in wave breaking. Ship observations of internal waves in the Kuril Straits are, however, very limited (Nakamura et al. 2010). At the same time numerous trains of nonlinear IWs were found on ASAR and PALSAR images of the Kuril Strait area both in the Sea of Okhotsk and in the Pacific Ocean. They were detected around the southern, central and northern Kuril Islands.

The database of IWs in the Sea of Okhotsk was built up. It includes ERS-1/2, Envisat and ALOS-1 SAR images and Terra and Aqua visible images (Mitnik 2008; Mitnik and Dubina 2010a). The parameters of internal waves in the Sea of Okhotsk can be summarized as follows: The packet length is 20 km, the number of waves in a packet is ≥ 25, the maximum L is 5.8 km, the minimum $L < 0.1$ km, the leading crest length is 4–105 km, the maximum light band width is 500 m, the maximum dark band width is 800 m, the propagation speed is 0.6 m/s. All the analyzed satellite images were summarized to produce the most complete and detailed maps showing the spatial distribution of the internal waves (Mitnik and Dubina 2007).

4 Sea Ice

The information about sea ice is used in activities associated with transportation, fishing, disaster mitigation and commercial development in the Sea of Okhotsk. The first ice appears at the end of October–beginning of November in the western and north-eastern regions of the sea. Cold air from the East Siberia produces sea ice and pushes away the new ice from a coast. Such conditions maintain the coastal polynyas and continuous formation of thin ice which are the characteristic features in the northern part of the sea, along the Sakhalin, in the Terpeniya and Aniva bays. After two months the western and north-eastern regions connect and ice drifts southward filling the central region of the sea. The maximum ice extent occurs in mid-March (on 20 February in 2015) and thereafter the ice extent retreats, and all ice disappears by the beginning of June. The Sea of Okhotsk ice occurs as fast ice and pack ice. Fast ice lies in the coastal area. The estimated mean maximum annual ice thickness distributes geographically from 0.4 to 1.1 m that is a real threat to transport operations and oil production. The pack ice is of dynamic nature, driven by winds and currents. The drift speed of ice is as much as 35 km/day on average in coastal regions off Sakhalin and Hokkaido islands (SHH 1998; Shevchenko et al. 2004).

Different physical parameters are responsible for formation of signal variations in the various spectral ranges. The brightness of visible images is defined by ice types and snow cover. The albedo of ice increases from the values close to water albedo (grease ice, dark nilas) up to snow albedo (Fig. 6b). Dependence of albedo on ice thickness is observed till the ice thickness reaching 30–40 cm (gray and gray-white ice). The brightness temperature measured by infrared sensor T_{BIR} is a function of the underlying surface temperature T_s which decreases from temperature of open water near the ice edge and in polynyas (around 0 °C) to surface temperature of thick snow-covered gray-white and white ice which, in turn, depends on air temperature (Adams et al. 2013; Lebedev and Paramonov 2001) (Fig. 6c). Temperature of young ice (especially snowless) is defined not only by ice thickness and air temperature but also by wind speed. Passive MW measurements provide the low resolution information with good spatial and temporal coverage. Brightness temperatures of the sea ice at MW frequencies depend on frequency, polarization, incidence angle and type of ice that is used in the retrieval algorithms for ice concentration and characteristics (Comiso 2010; Heygster et al. 2012). Time series of this information is the basis for the global and regional (Sea of Okhotsk) climate studies.

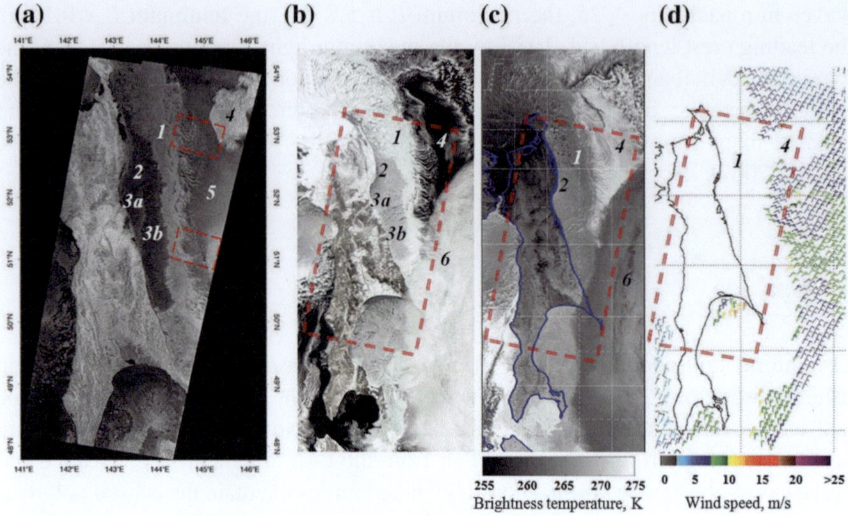

Fig. 6 Ice cover around Sakhalin on 5 February 2015 obtained by Sentinel-1A SAR at 20:38 UTC (**a**), Aqua MODIS visible image at 03:00 Гр (**b**), Suomi NPP VIIRS infrared image at 03:30 UTC (**c**) and MetOp-B scatterometer derived wind field at 10:22 UTC (**d**). Dashed red rectangles on (**b–d**) mark SAR image boundaries. 1—the band of white and grey-white snow-covered ice; 2, 3a, 3b—thin ice with small differences in the surface roughness and snow cover, 4—new (pancake) ice, 5—open sea, 6—clouds

The brightness of the SAR images is proportional to the backscatter (Normalized Radar Cross Section—NRCS) variations which are defined mainly by the underlying surface roughness, salinity and dielectric permittivity and depend on the polarization and wavelength of sensing signals and incidence angle. The contribution of the volume scattering to the NRCS is small for the first year ice but can be increased in ridged/deformed ice. The variety of ice types, conditions of ice formation and evolution, snow cover thickness and properties, surface wind speed and other factors tends to ambiguous interpretation of brightness variations of SAR images, discriminating ice types and open water within ice cover and in the marginal ice zone (Fig. 6a). Brightness variability on SAR imagery of the sea surface is described by geophysical model function (GMF) for particular combination of SAR and sensing geometry characteristics and also on wind speed and direction. Increase in the sea surface roughness is accompanied by the increase of the NRCS (brightness of radar image). Brighter areas indicate the strips of ridges and ridged ice. The areas of the decreased roughness appear on SAR images as dark spots, narrow and wide bands. These features can be caused by fracture zones at low wind speed and also by grease ice that damps small-scale wind waves. Grease ice is distinguished near the ice edge and in the open sea against the rough water surface background where it marks the areas of new ice formation.

In 1983–2000, the sea ice in the Sea of Okhotsk was studied with Side-Looking Real Aperture Radar (RAR) from Kosmos-1500, -1766 and Okean series satellites.

RAR operates at wavelength of 3.15 cm at vertical polarization. A swath width is 460 km at a ground resolution of 1–3 km (Mitnik and Kalmykov 1992; Mitnik and Victorov 1990). The first Envisat ASAR images of the Sea of Okhotsk were taken in December 2002–March 2003. Current satellite SARs operate at X-, C- and L-bands at vertical, horizontal and cross polarizations. A swath width changed from 20–70 to 350–500 km at a ground resolution of 3–10 and 100–150 m, correspondingly. SAR is a powerful tool for detailed ice mapping through its high resolution providing information on the ice type, concentration, roughness, floe size and velocity that illustrate Figs. 6 and 7.

Radar contrasts enhance in the marginal ice zone (MIZ) due to the great difference in the scattering properties of various ice types and the sea surface. Irregular ocean circulation features and wind action aligns the loose and broken ice in the MIZ into identifiable types of ice-ocean patterns including ice bands, jets edge waves, eddies, and mushroom like structures with dimensions ranging between 2–3 and 100–120 km. Coastal polynyas are frequently observed (Mitnik 2008; Mitnik and Dubina 2007; 2010a, b).

Grease ice formation and its changing to pancake ice illustrates Envisat-1 ASAR image of the Northern Sakhalin and the surrounding waters (Fig. 7). Sakhalin Bay **1** is covered by thick white ice. The boundary of the consolidated ice **2** marks an narrow dark-gray band **3** consisting of grease ice mainly. Grease ice is observed further north-northeast (upper part of the image) where it forms in the open sea **4** under action of the cold dry northern winds. Elongated low-contrast bands show wind direction in the open sea area **4** from northeast to southwest. Dark-gray area **5** is an indicator of intense grease ice formation. Grease ice has the increased viscosity and damps the small-scale sea surface roughness. As a result, area **5** has negative radar contrast against the open sea **4**. Downstream **5** there is area **6** with alternating narrow dark (grease ice) and gray-white (pancake ice) bands. They are the surface manifestations of Langmuir circulation. Area **7** is characterized by light tone indicating that the almost all sea surface is covered by pancake ice which is characterized by the high backscatter values.

Sea ice is an excellent tracer of the complicated water movement (Wakatsuchi and Oshima 1990; Mitnik and Dubina 2010a, b). Figure 8 shows the ALOS-1 PALSAR image taken on 10 February 2009 when the SST contrasts in the Soya Warm Current region near Hokkaido coast are weak if any. The image was taken in fine beam mode with the incidence angle of 34.1°, swath width of 70 km and spatial resolution of 10 m. Several sea ice eddies and a mushroom-like structure are seen in the image. The most remarkable features are an cyclonic eddy **1** and a mushroom-like structure (vortex dipole) consisting of anticyclonic eddy **2a,** cyclonic eddy **2c**, stem **3**, and cap **4** covering both eddies. Circular eddy **1** of approximately 17 km in size consists of two spirals converging to its center. The cyclonic eddy **2c** is approximately 6 × 9 km in size and has well defined spiral structure as opposite to anticyclonic one. Another small cyclonic eddy **5** is located near the ice edge to the east of **2c**. All eddies have high contrast against the background. They are very likely young structures formed mainly by pancake ice.

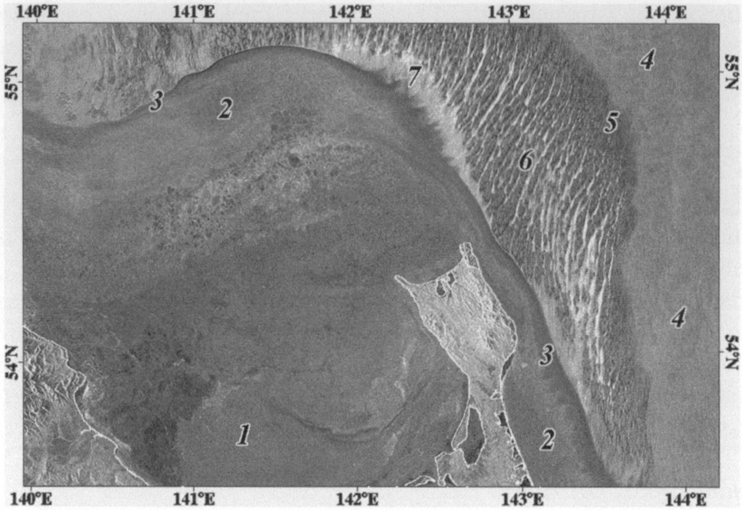

Fig. 7 Envisat ASAR image of the Northwest Okhotsk Sea and Sakhalin acquired on 18 January 2006 at 12:07 UTC: **1** Sakhalin Bay with white ice, **2** consolidated ice, **3** band of grease ice, **4** open sea, **5** grease ice, **6** alternating bands of pancake ice (bright) and grease ice (dark), **7** pancake ice

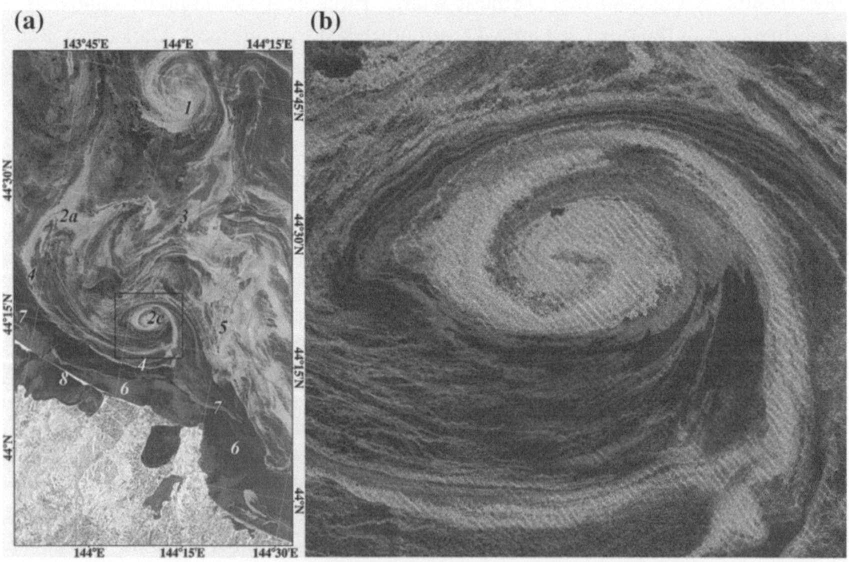

Fig. 8 ALOS-1 PALSAR image with HH-polarization acquired on 10 February 2009 at 12:46 UTC showing cyclonic eddies and mushroom-like structures **1–5** in sea ice, open water **6**, ice band **7** near Hokkaido coast and Saroma Lake **8** (**a**). Dashed rectangle marks the boundaries of enlarged fragment (**b**) with a spiral cyclonic eddy **2c** and swell propagating southwestward

The low brightness of the ice-free water **6** between Hokkaido coast and the nearest ice band **7** evidences that the wind speed was low during PALSAR sensing. At the same time, southwestward swell manifest itself as the inclined lines in the areas covered by pancake ice in particular in the center of eddy **2c** and in spirals (Fig. 8b). The brightness variations result from modulation of the incidence angle by swell. The swell wavelength is 220 m. Swell is also detected in the open sea as the low contrast lines (areas **6**). The variations of the sea ice roughness are seen in Saroma Lake **8** (Mitnik and Dubina 2010a).

5 Weather Systems with Strong Wind

During winter, severe weather events over the Okhotsk Sea are caused by synoptic-scale extratropical cyclones, mesoscale cyclones and cold air outbreaks. Their passage can be accompanied by rains, snowfalls and damaging winds. These weather systems occur as a result of interaction of cold and dry Siberian air masses with the relatively warm sea surface. Winter storms can also travel from the south deepening over the warm Kuroshio waters.

The large size of the extratropical cyclones (ETCs) and their relative frequency create the potential for widespread damage while wind speeds in these storms seldom reach those of typhoon. The winds inside ETC, as the largest source of momentum for the ocean surface, affect the full range of ocean movement, from individual surface waves to complete current systems. Winds along the ocean surface regulate interaction between the atmosphere, ocean and sea ice via modulation of the air-sea-ice exchanges of heat, moisture, gases, and particulates. Fields of the sea surface wind speed W and temperature T, total atmospheric water vapour content V and total cloud liquid water content Q characterize quantitatively the evolution of ETC structure and the air sea interaction during its life circle.

The intense mesoscale cyclones with a comma or spiral cloud system are often observed over the Okhotsk Sea in a cold season (Gurvich et al. 2008; Mitnik 2009; Mitnik et al. 2007; Rasmussen and Turner 2003). The Aqua and Terra MODIS and NOAA AVHRR visible and infrared images and V fields retrieved from passive microwave measurements have been used for mesoscale cyclone detection and satellite scatterometer data—for wind field retrieval. Sea surface wind distribution with the significantly better spatial resolution can be inferred from the backscatter variations caused by the wind speed and direction changes and measured by satellite RAR (Mitnik and Victorov 1990) and SAR (Mitnik 2009). Fields of the surface wind speed W, water vapor V and cloud liquid water Q in the cyclones can be derived from the satellite passive microwave measurements (Bobylev et al. 2010; Gurvich et al. 2008; Mitnik and Mitnik 2003, 2010; Mitnik et al. 2007; Zabolotskikh et al. 2013, 2014, 2015).

When cold and stably stratified continental and polar air masses move over the relatively warm sea surface, the strong atmospheric convection is developed. It manifests itself in organized large eddies (roll clouds) and small scale fronts close to

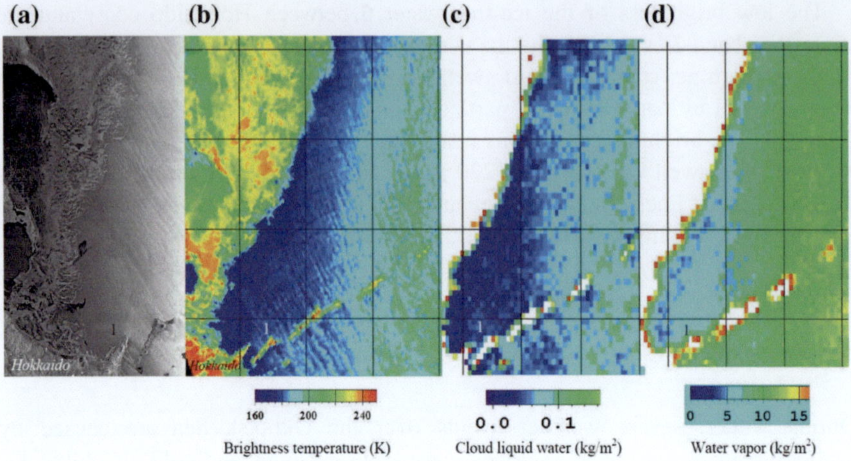

Fig. 9 Sea ice and fields of atmospheric parameters in the Okhotsk Sea on 26 January 2008: ALOS-1 PALSAR image acquired at 00:50 UTC (**a**), brightness temperature at 89.0 GHz with horizontal polarization (**b**), cloud liquid water content (**c**) and atmospheric water vapor content (**d**) derived from Aqua AMSR-E measurements at 02:50 UTC. 1—Kunashir Island

the coast or ice edge. Further downstream, deeper convection is observed and structure of clouds changes at first to transition form and then to open and closed convective cells (Etling and Brown 1993; Brűmmer 1999; Liu et al. 2006; Mitnik and Mitnik 2009).

Figure 9 indicates the relationships between ice and wind conditions and the changing integrated atmospheric parameters during cold air outbreak in the Okhotsk Sea. Weather conditions on 26 January 2008 were determined by rear area of a vast deep cyclone in the Pacific Ocean. Wind speed in the Okhotsk See exceeded 20 m/s. The difference of water and air temperatures reached 20–24 °C near the ice edge and 10–11 °C near the Southern Kuril Islands. The *V* values varied from ≈1.5 to ≈3.0 kg/m^2 near Sakhalin coast. MODIS, SAR and AMSR-E images were carried out with 2 h difference and less. Fields of cloudiness, surface wind and brightness temperature correspond in structure to each other (Mitnik and Mitnik 2007, 2009; Mitnik et al. 2007; Mourad and Walter 1996).

6 Conclusions

The severe environmental conditions hamper the scientific and operational data acquisition in the Okhotsk Sea. The usage of multi-satellite/multisensor sensing is the most comprehensive approach for the oceanic and atmospheric phenomena and processes study. The passive and active microwave techniques are applicable for the quantitative data retrieval independently on sun illumination and cloudiness.

The possibilities of MW sensors were demonstrated as applied to signature detection of the sea surface circulation, currents and eddy boundaries, internal waves, various types of sea ice, etc. Eddies, eddy streets, internal waves and other dynamic structures are sensitive indicators of the sea surface circulation. The interaction of nonperiodic and periodic currents with the bottom elevations, sills, shelf edge and headlines is accompanied with the generation of the diversified mesoscale and submesoscale oceanic phenomena which are reflected in the fields of the sea surface roughness, temperature, and Chl-*a* registered by various satellite sensors that emphasizes the advantage of multisensory measurements for their study. MW signatures revealed by different satellite sensors are in a good agreement. Availability of several satellite microwave sensors allows tracing the evolution of both the oceanic and atmospheric phenomena with better temporal resolution. It is especially important in the Okhotsk Sea for monitoring of sea ice, deep extratropical cyclones and for the short-living polar lows which have a pronounced effect on activities associated with transportation, fishing, disaster mitigation, oil deposit and other commercial development.

Acknowledgements This work was partly supported by the FEB RAS grant 18-I-010 and the JAXA Project F10. Authors thank the JAXA for the Aqua AMSR-E and GCOM-W1 AMSR2 and ALOS PALSAR data, the ESA for the Envisat and Sentinel-1A SAR images, the NASA for the MODIS and VIIRS data and NOAA for ASCAT wind data.

References

Adams S, Willmes S, Schröder D, Heinemann G, Bauer M, Krumpen T (2013) Improvement and sensitivity analysis of thermal thin-ice thickness retrievals. IEEE Trans Geosci Remote Sens 53:3306–3318

Bobylev L, Zabolotskikh E, Mitnik LM, Mitnik ML (2010) Atmospheric water vapor and cloud liquid water retrieval over the Arctic Ocean using satellite passive microwave sensing. IEEE Trans Geosci Remote Sens 48:283–294

Brűmmer B (1999) Roll and cell convection in wintertime Arctic cold air outbreaks. J Atmos Sci 56:2613–2636

Comiso J (2010) Polar Ocean from space. Springer, Heidelberg, p 430

Dubina VA, Fayman PA, Zhabin IA, Ponomarev VI, Kuzlyakina YA (2012) The Okhotsk Sea currents: satellite imagery and numerical simulation. Curr Probl Remote Sens Earth Space 9(1): 206–212 (in Russian)

Ebuchi N, Fukamachi Y, Ohshima KI, Shirasawa K, Ishikawa M, Takatsuka T, Daibo T, Wakatsuchi M (2006) Observation of the Soya Warm Current using HF radar. J Oceanogr 62:47–61

Ebuchi N, Fukamachi Y, Ohshima KI, Wakatsuchi M (2009) Subinertial and seasonal variations in the Soya Warm current revealed by HF ocean radars, coastal tide gauges and a bottom-mounted ADCP. J Oceanogr 65:31–43

Etling D, Brown RA (1993) Roll vortices in the planetary boundary layer: a review. Bound Layer Meteorol 65:215–248

Gurvich IA, Mitnik LM, Mitnik ML (2008) Mesoscale cyclogenesis over the Far Eastern Seas: study based on satellite microwave radiometric and radar measurements. Investig Earth Space 5:58–73. (Issledovanie Zemli iz Kosmosa, in Russian)

Heygster G, Alexandrov V, Dybkjær G et al (2012) Remote sensing of sea ice: advances during the DAMOCLES project. Cryosphere 6(6):1411–1434

Katsaros KB, Mitnik LM, Black PG (2014) Microwave instruments for observing tropical cyclones. In: Tang D, Sui G (eds) Typhoon impacts and crisis management. Springer, Heidelberg, pp 5–61

Kowalik Z, Polyakov I (1999) Diurnal tides over Kashevarov Bank, Sea of Okhotsk. J Geophys Res 104:5361–5380

Lebedev GA, Paramonov AI (2001) Determination of the physical characteristics of sea ice from satellite infrared sensing. Meteorol Hydrol 2:72–80 (in Russian)

Liu AQ, Moore GWK, Tsuboki K, Renfrew IA (2006) The effect of the sea-ice zone on the development of boundary-layer roll clouds during cold air outbreaks. Bound Layer Meteorol 118:557–581

Mitnik LM (2008) Advanced land observing satellite PALSAR observations of the oceanic dynamic phenomena in the coastal zone. Proc IGARSS 2:351–354

Mitnik LM (2009) Mesoscale atmospheric vortices in the Okhotsk and Bering Seas: results of satellite multisensor study. In: Nihoul CJ, Kostianoy AG (eds) Influence of climate change on the changing Arctic and Sub-Arctic conditions. Springer, Dordrecht, The Netherlands, pp 37–56

Mitnik LM, Dubina VA (2006) Surface currents in the Japan and Okhotsk Seas: study with satellite SAR. In: Proceedings of IGARSS 2006, Denver, Colorado, pp 2394–2395

Mitnik L, Dubina V (2007) Spatial-temporal distribution and characteristics of internal waves in the Okhotsk and Japan Seas studied by ERS-1/2 SAR and Envisat ASAR. In: Proceedings of ENVISAT symposium 2007, 23–27 Apr 2007, ESA SP-636, Montreux, Switzerland

Mitnik LM, Dubina VA (2010a) Satellite SAR sensing of the Okhotsk Sea: geographical and seasonal features, ESA Publication SP-686, Bergen, Norway, 28 June–2 July 2010

Mitnik LM, Dubina VA (2010b) Interpretation of SAR signatures of the sea surface: multisensory approach. In: Barale V, Gower JFR, Alberotanza L (eds) Oceanography from space, revisited. Springer, Dordrecht, pp 113–130

Mitnik L, Dubina V (2012) Satellite SAR sensing of oceanic dynamics in the Kuril straits area. In: Proceedings of IGARSS 2012, Munich, Germany, 22–27 July 2012, pp 7632–7635

Mitnik LM, Dubina VA, Shevchenko GV (2005) ERS SAR and Envisat ASAR observations of oceanic dynamic phenomena in the southwestern Okhotsk Sea. In: Envisat & ERS ESA Symposium, ESA SP-572, Salzburg, Austria, 6–10 Sept 2004

Mitnik LM, Kalmykov AI (1992) Structure and dynamics of the Sea of Okhotsk marginal ice zone from "Ocean" satellite radar sensing data. J Geophys Res 97(C5):7429–7445

Mitnik LM, Mitnik ML (2003) Retrieval of atmospheric and ocean surface parameters from ADEOS-II AMSR data: comparison of errors of global and regional algorithms. Radio Sci 38 (4):8065. https://doi.org/10.1029/2002RS002659

Mitnik LM, Mitnik ML (2007) Passive and active microwave sensing of cold air outbreaks over the Northwest Pacific Ocean. In: Proceedings of IGARSS'07, Barcelona, 23–27 July

Mitnik LM, Mitnik ML (2009) Modeling of microwave characteristics of organized mesoscale convection over the ocean. Currt Probl Remote Sens Earth Space 6(2):147–154 (in Russian)

Mitnik LM, Mitnik ML (2010) AMSR-E advanced wind speed retrieval algorithm and its application to marine weather systems. In: Proceedings of IGARSS 2010, Hawaii, pp 3224–3227

Mitnik LM, Mitnik ML, Dubina VA (2007) Remote radiophysical sensing of the ocean—atmosphere system. In: Akulichev VA (ed) Far Eastern Seas of Russia. Book 4. Physical methods of research. Nauka Publishing House, Moscow, pp 449–537 (in Russian)

Mitnik LM, Victorov SV (eds) (1990) Radar sensing of the earth surface from space. Hydrometeoizdat Publishing House, Leningrad, 200pp

Mourad PD, Walter BA (1996) Viewing a cold air outbreak using satellite-based synthetic aperture radar and advanced very high resolution radiometer imagery. J Geophys Res 101:16391–16400

Nakamura T, Kawasaki Y, Kono T, Awaj T (2010) Large-amplitude internal waves observed in the Kruzenshtern strait of the Kuril Island Chain and possible water transport and mixing. Cont Shelf Res 30:598–607

Ohshima KI, Wakatsuchi M, Fukamachi Y, Mizuta G (2002) Near-surface circulation and tidal currents of the Okhotsk Sea observed with satellite-tracked drifters. J Geophys Res 107:3195. https://doi.org/10.1029/2001JC001005

Rasmussen EA, Turner J (2003) Polar lows. Cambridge University Press, Cambridge, UK, 612pp

Sea Hydrometeorology and Hydrochemistry (SHH) (1998) The Okhotsk Sea. Issue 1 Hydrometeorological conditions. In: Glukhovskiy BK, Goptarev IP, Terziev FC (eds) Gidrometeoizdat, Sankt Petersburg, 343pp (in Russian)

Shevchenko GV, Rabinovich AB, Thomson RE (2004) Sea-ice drift on the Northeastern shelf of Sakhalin Island. J Phys Oceanogr 34:2470–2491

Wakatsuchi M, Ohshima KI (1990) Observations of ice-ocean eddy street off the Hokkaido coast in the Sea of Okhotsk through radar images. J Phys Oceanogr 20(4):585–594

Zabolotskikh EV, Mitnik LM, Chapron B (2013) New approach for severe marine weather study using satellite passive microwave sensing. Geophys Res Lett 40(13):3347–3350

Zabolotskikh EV, Mitnik LM, Chapron B (2014) GCOM-W1 AMSR2 and MetOp-A. ASCAT wind speeds for the extratropical cyclones over the North Atlantic. Remote Sens Environ 147:89–98. https://doi.org/10.1016/j.rse.2014.02.016

Zabolotskikh EV, Mitnik LM, Reul N, Chapron B (2015) New possibilities for geophysical parameter retrievals opened by GCOM-W1 AMSR2. IEEE J Sel Top Appl Earth Obs Remote Sens 8(9):1–14. https://doi.org/10.1109/JSTARS.2015.2416514

Phytoplankton and Primary Production in the Japan Sea

Joji Ishizaka and Keiko Yamada

Abstract The Japan Sea is a marginal sea surrounded by the Japanese archipelago, Korea Peninsula, Sakhalin Island, mainland China and Russian landmasses which isolate it from the East China Sea, Pacific Ocean, and Okhotsk Sea. In this chapter, studies of phytoplankton and primary production in the Japan Sea using ocean colour satellite data have been summarized. Satellite chlorophyll-a (Chl-a) concentrations were generally well correlated with in situ Chl-a, except in low chlorophyll-a regions within the central region of the Japan Sea, where satellite estimates were higher. A marked seasonality in Chl-a was observed in association with spring and fall blooms over large areas of Japan Sea, with the spring bloom commencing around the subpolar front in the south moving northwards. Overall, interannual variations in Chl-a, appeared to be related to large scale global climate variations. Despite observations of increasing trends in Chl-a, reports of decreases in primary production require that we continue to better understand the role of physical forcing factors in particular, how mesoscale eddies and typhoons that are unique to this region, regulate primary production and phytoplankton biomass distribution. This review also draws attention to the lack of ocean colour algorithms for discriminating phytoplankton functional groups from space. The availability of satellite derived information on phytoplankton functional groups is essential for ocean biogeochemical and fisheries research. This study also emphasizes the need for more bio-optical measurements and algorithm development efforts to fill this gap.

Keywords Japan Sea · Ocean colour · Chlorophyll-a · Primary production Phytoplankton

---

J. Ishizaka (✉)
Institute for Space-Earth Environmental Research,
Nagoya University, Nagoya, Japan
e-mail: jishizaka@nagoya-u.jp

K. Yamada
Keimyung University, Daegu, Korea

V. Barale and M. Gade (eds.), *Remote Sensing of the Asian Seas*,
https://doi.org/10.1007/978-3-319-94067-0_9

1 Introduction

The Japan Sea is an enclosed sea surrounded by the Eurasian continent on the west and the islands of Japan to the east (Fig. 1). Its total area is about 1,000,000 km^2. The average depth is 1750 m and its maximum depth is about 3742 m, which is fairly deep for an enclosed sea. It is connected to the East China Sea via the Tsushima Strait, to the Pacific Ocean via the Tsugaru Strait and to the Okhotsk Sea via the Soya and Tatar Straits.

This ocean is often called miniaturized ocean because it is known that there is deep water, Japan Proper Water, formation. The Tsushima Warm Current is the relatively strong warm current which flows from south through the Tsushima Strait into the Japan Sea. This current displays characteristics of a Western Boundary Current, although it has three branches, with two flowing along the Japanese and Korean coasts, and the other branch in between those branches. The Liman Current is a cold current flowing from north to south, and it interacts with the northward flowing Tsushima current. A subpolar front separates the southern warm water and northern cold water, and it is well known that the interaction of these two current systems results in the formation of eddies in this area.

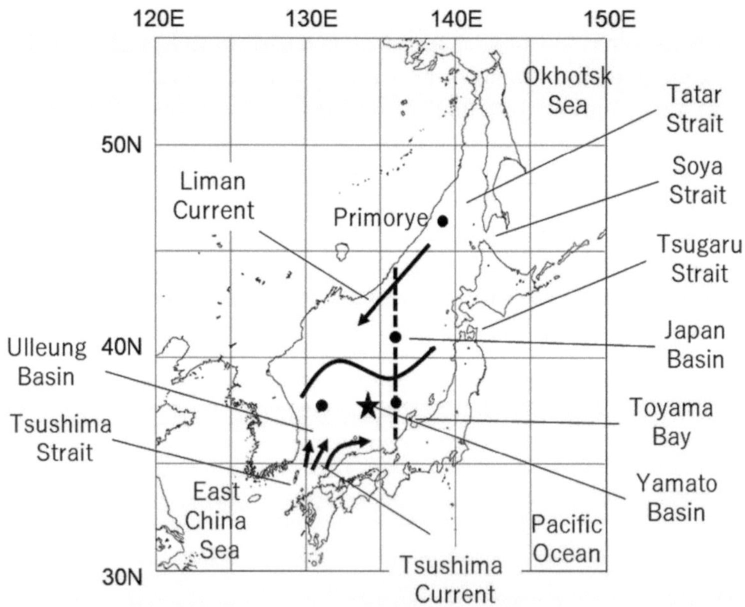

Fig. 1 Map of the Japan Sea. The thick line indicates the approximate position of the subpolar front. The arrows from north and south indicates Liman and Tsushima Currents, respectively. The dashed line indicates the section used for time series by Yamada et al. (2004) and utilized in Fig. 3. The star indicates the position of the long-term observation of the Japan Metrological Agency used by Chiba and Saino (2002) and Yamada and Ishizaka (2006). The four black circles indicate the position of primary production estimated by Yamada et al. (2005)

Because the Japan Sea is bordered by several countries, undertaking shipboard cruises in these politically complex waters continues to present a significant impediment to further progress in understanding biological oceanographic processes and biochemical cycling in this region. For this reason, remote sensing is very important as it is the only available, and perhaps the most ideal tool to study this area as a complete system.

This review is a synthesis of phytoplankton and primary production in the Japan Sea using ocean colour remote sensing. It encompasses the following topics: (1) seasonal variation of chlorophyll-a (Chl-a) and spring bloom, (2) primary production, (3) long-term trends of Chl-a and primary production, (4) influence of eddies, and (5) influence of typhoons.

2 Seasonal Variation of Chl-a and Spring Bloom

Ocean colour remote sensing is important tool for studying the ocean biogeo-chemical processes because it can detect phytoplankton in seawater, and it is the only means of measuring basin-wide biological parameters that are directly related to ocean biogeochemical processes and ecosystem dynamics. Estimation of in-water parameters, such as Chl-a, requires atmospheric correction and in-water algorithms. Atmospheric correction procedures are particularly problematic because of the presence of yellow Asian dust or absorptive aerosol over this region, which if not accounted for accurately can lead to errors in Chl-a estimates (Li et al. 2003). Despite this challenge, satellite derived estimates of Chl-a have been shown to be fairly accurate, but problematic in areas of low Chl-a ($\sim$0.1 mg m^{-3}) in the central region of the Japan Sea where the satellite values are almost two fold higher than those measured in situ (Yamada et al. 2004). Another driver of error in satellite-derived Chl-a values is the presence of rivers around the Japan Sea. Some of these rivers may discharge sediment and coloured dissolved organic matter (CDOM) rich, and their inflow into the Japan Sea can also influence the accuracy of satellite derived Chl-a concentrations. One example of the impact of riverine influence is within Peter the Great Bay (Salyuk et al. 2010). These observations suggest that verification and development of more accurate atmospheric and in-water algorithms for the Japan Sea will need to continue, especially in areas that come under the influence of rivers, to further improve estimations of Chl-a concentrations from space.

Seasonal variation of phytoplankton pigment (Chl-a + phaeopigments) was first described in detail by Kim et al. (2000) using Coastal Zone Color Scanner (CZCS) data. Typical spring and fall blooms were clearly seen in this area, and the spring bloom was observed to start from the south and to move towards the north with the increase of temperature. They also compared the development of the spring bloom with the evolution of the mixed layer and concluded that critical depth theory (Sverdrup 1953) is applicable to this area.

Yamada et al. (2004) described seasonal and inter-annual variation of Chl-a in the Japan Sea using data from Ocean Color and Temperature Scanner (OCTS) and Sea-viewing Wide Field-of-view Sensor (SeaWiFS) data from 1997 to 2002. As shown in Figs. 2 and 3 from even longer data set with Moderate Resolution Imaging Spectroradiometer (MODIS) on Aqua (EOS PM) satellite, the high quality and long time series data, clearly shows the regional differences in the timing of the spring bloom. The spring bloom was observed to start from the subpolar front and in the south in March, north of the front, along the Primorye and Hokkaido coast in April, and in the middle of the Japan Basin in May.

Recently, Jo et al. (2014) used wavelet analysis to clarify the regional and temporal variations of SeaWiFS Chl-a data during 1998–2007. They found that variation of the Chl-a with 6-month cycle corresponding to spring and fall blooms and that this seasonal trend was persistent almost throughout the Japan Sea. Jo et al. (2014) also reported that 6 months variance was highest with strong spring and fall blooms in Japan basin and Ulleung Warm Eddy. Elsewhere, the subpolar front displayed both 6 and 12 months variance, with two weak blooms, and the East Korea Bay and Yamato Basin displayed 12-month variance associated with a strong spring bloom. These results indicate the existence of variability in the seasonality of satellite Chl-a in different areas within the Japan Sea.

Yamada et al. (2004) were able to show large interannual variations in the timing of the spring bloom and were able to relate it to wind forcing using data obtained from Special Sensor Microwave/Imager (SSM/I). Strong and weak winter-time wind years during La Niña and El Niño years were seen to be responsible for late and early start timings of the spring bloom, respectively. The difference could be about 1 month (Fig. 3). Kim et al. (2007) also analyzed spring and fall blooms of the subpolar front area using SeaWiFS and MODIS Chl-a and Active Microwave Instrument-wind (AMI-wind), NASA Scatterometer (NSCAT) and Quick Scatterometer (QuickSCAT) derived wind data from 1997 to 2003. They found that the spring bloom happened 6–15 days after wind stress weakened, and fall bloom started 3–9 days after wind strengthened. They also pointed out that sporadic wind events may also influence the timing of those blooms.

It is well known that the entire north Pacific Ocean experienced a large-scale climate regime shift from 1980s to 1990s. Satellite derived Chl-a would have been an important variable to understand how the ecosystem changed as a result of this regime shift. Information on the timing of phytoplankton bloom, often called phenology (cf. Edwards and Richardson 2004), is important to understand the variation of fish production (Platt et al. 2003). In the Japan Sea, Chiba and Saino (2002) used long-term phytoplankton and zooplankton community data collected four times a year, west of the Yamato Basin by the Japan Metrological Agency that the warm (cold) regime in 1980s (1990s) resulted in an early (later) spring bloom (Fig. 1).

To circumvent the issue of lack of satellite Chl-a data before 1996, Yamada and Ishizaka (2006) made an attempt to estimate the timing of the spring bloom using in situ data of the west of the Yamato Basin and pairing these with SeaWiFS derived spring bloom parameters. They used Gaussian fitting to estimate the

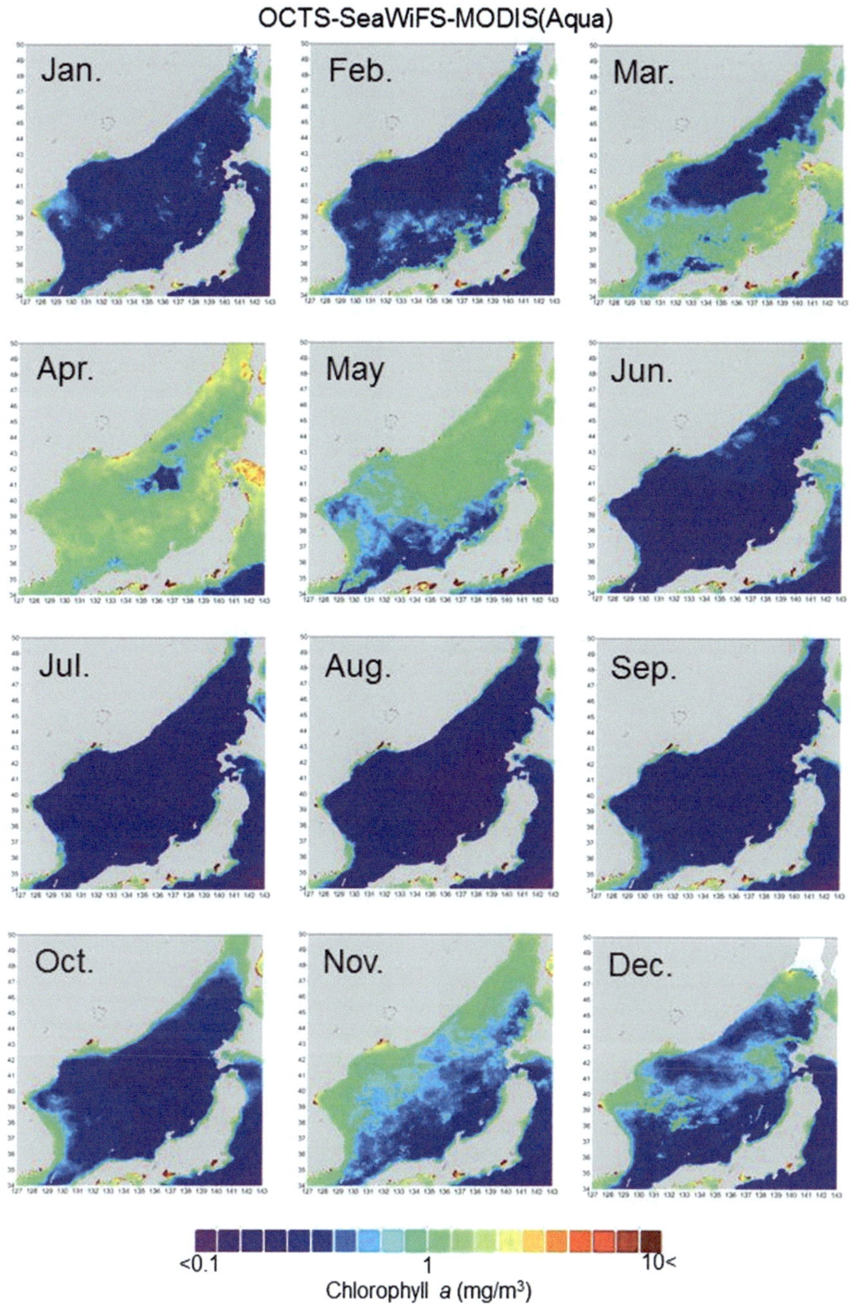

Fig. 2 Monthly Chl-a composite of OCT-SeaWiFS-MODIS (Aqua) from 1997 to 2016

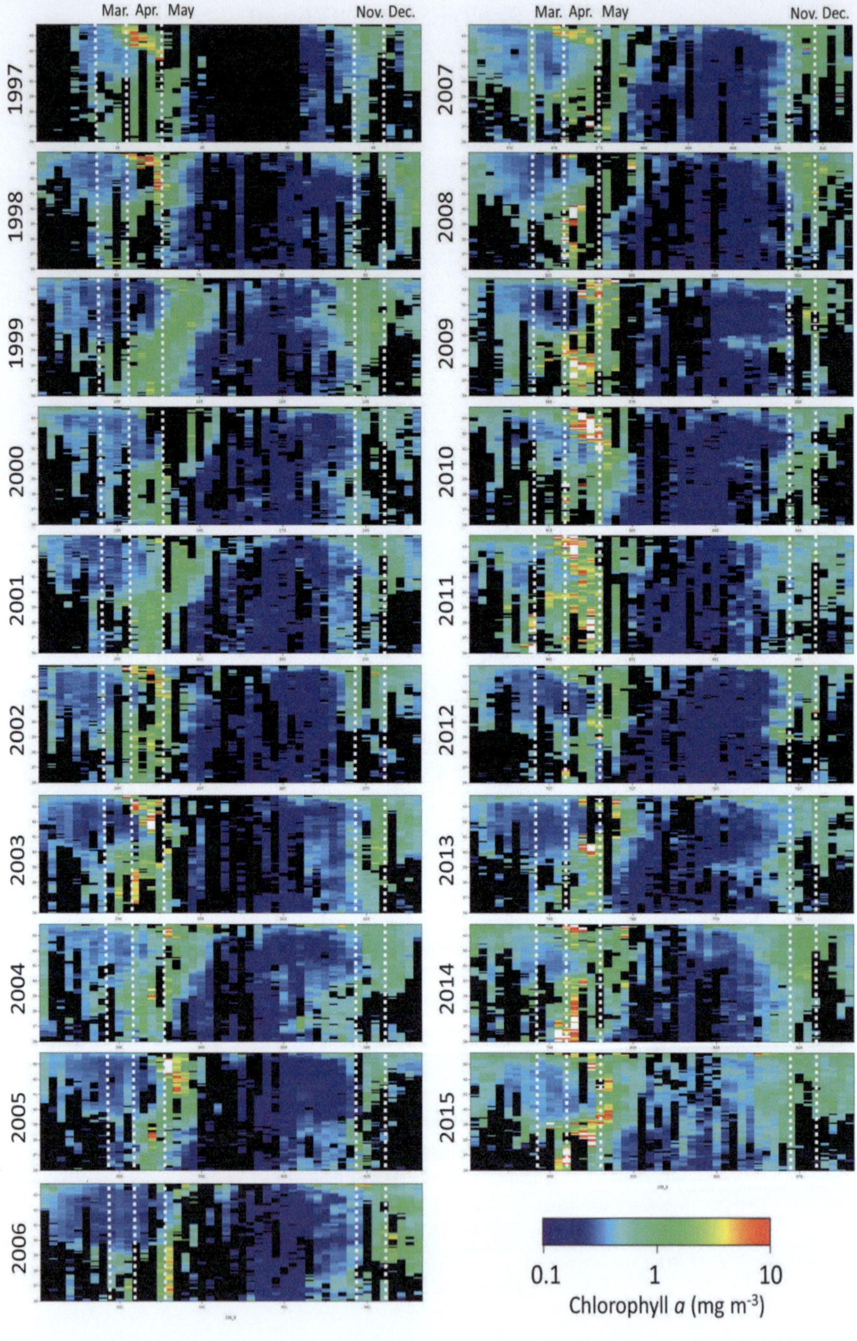

Fig. 3 A Hovmoeller plot of Chl-a along 135°E (averaged value between 134.5 and 135.5°E) using OCTS-SeaWiFS-MODIS (Aqua) time series from 1997 to 2016. The white dashed lines indicate 1st of February, March, April, November, and December

parameters of spring bloom development at a fixed station in the Yamato Basin. The spring bloom start and end timing was significantly correlated with the winter wind and ship-observed stratification data. They used the correlations to reproduce the past spring bloom and found it was early and shorter in mid-1980s. This early spring bloom predicted from the environments was consistent with the spring bloom timing expected from the phytoplankton and zooplankton community by Chiba and Saino (2002). On the other hand, Zuenko and Pershina (2011) observed late spring blooms during the 1980s along the coast of Japan in the Japanese sardine spawning ground, by fitting SeaWiFS Chl-a data with sea surface temperature (SST) and precipitation.

Yoo and Kim (2004) compared spring blooms of the southwestern Japan Sea in 1997 observed by OCTS and in 1999 observed by SeaWiFS and found significant differences between those years. The spring bloom was intensive, long and productive in the subpolar front, but the one in Ulleung Basin was suppressed in 1997. In 1999, the spring bloom in both areas was shorter, and the productivity was much lower in 1999 than in 1997. They further discussed the relation to volume transport of Tsushima Warm Current and possible relation to ENSO.

Jo et al. (2007) also analyzed the spring bloom of the northern Japan Sea during 1992–2002 with Total Ozone Mapping Spectrometer (TOMS) data for transport of Asian dust. They reported that the onset of the early spring bloom in the northern Japan Sea coincided with wet deposition of Asian dust and suggested the possibility of iron influence on the early spring bloom.

Yamada et al. (2004) pointed out that the spring bloom timing along the Pri-morye coast may be also influenced by melting ice in Tatar Strait. More recently, Park et al. (2014) focused on understanding the relationship between ice melt and the onset of the spring bloom and found that spring-time SeaWiFS Chl-a along the Prymnorye coast from 1998 to 2007 correlated significantly with the amount of SSM/I detected sea ice during winter of the previous year in Tatar Strait. They suggested that this was either due to the possible influence of nutrient inputs from the melting ice and/or the development of stratification caused by freshwater derived from ice-melt.

3 Primary Production

Yamada et al. (2005) estimated primary production in the Japan Sea using a two-phytoplankton-community algorithm of Kameda and Ishizaka (2005), which is a modified version of Vertically Generalized Production Model (VGPM; Behrenfeld and Falkowski 1997). Kameda and Ishizaka (2005) modified the parameter of optimum primary productivity per Chl-a (P^B_{opt}) to a function of not only temperature but also Chl-a assuming that phytoplankton is composed by two different size community; large (small) dominated in high (low) Chl-a water with low (high) primary production per Chl-a. The Kameda and Ishizaka (2005) model was compared with several other primary production models used in various regions of global ocean and was reportedly one of the best models for the relatively shallow coastal region (Saba et al. 2011). Yamada et al. (2005) verified the model with in situ primary production data in the Japan Sea and found that the algorithm performed fairly well, although their verification data set was very small. Accuracy of the model was also reported by Yoon et al. (2012) and Joo et al. (2014) with a much larger dataset from the water off Korea.

Yamada et al. (2005) estimated primary production in the Japan Sea from 1998 to 2002 with SeaWiFS Chl-a, Photosynthetically Active radiation (PAR) and AVHRR SST. Their satellite derived estimates of annual primary production were 170, 161, 191 and 222 gC m^{-2} $year^{-1}$ along the Russian coast, in the middle of the Japan Basin, the south-eastern and the south-western regions (Fig. 1), respectively. Primary production in the south-western region along the Korean coast was highest. They also reported that the contribution of spring bloom to annual primary production was as high as 42%. Interannual variation of primary production was also high during the spring bloom, due to the winter convective mixing and the influx of high nutrient waters into the euphotic zone.

4 Long-Term Trends of Chl-a and Primary Production

Lee et al. (2014) compared 1998–2001 SeaWiFS (re-processing 2010.0) and 2008–2011 MODIS-Aqua (reprocessing 2012.0) Chl-a, and found MODIS-Aqua Chl-a during the 2008–2011 was significantly higher than the SeaWiFS Chl-a for the 1998–2001 period in many regions of the Japan Sea, especially along the coastal areas and the sub-polar front during the spring bloom. Although Lee et al. (2014) did not find any significant changes in the environment; they suggested that the increase in anticyclonic eddies, the increase in winter SST, and changes in the light regime as possible cause of increases in primary productivity. On the other hand, long term trends of primary production studied by Joo et al. (2016) using MODIS-Aqua data and the model of Kameda and Ishizaka (2005) estimated primary production at 246.8 gC m^{-2} $year^{-1}$ for the whole basin, with a significant decreasing trend of 13% per 10 years during the period between 2003 and 2013. Joo et al. (2016) suggested that the shallowing trend of the mixed layer could be the cause of the decrease in primary production, although it still remains unclear whether this was part of the Pacific Decadal Oscillation or part of a longer trend related to global warming. However, the authors did not provide any clues as to how the decrease in primary production was related to the increase of Chl-a reported by Lee et al. (2014).

Increasing trends of Chl-a can also be noticed in the northern Japan Sea in the more recent version of Hovmoller diagram updated until 2016 (Fig. 3). Unfortunately, global analysis of the combined data of satellite, in situ and model (Gregg and Rousseaux 2014) data do not cover this marginal sea, and presently we do not have enough information to conclude whether this trend is accurate.

Thus, further analysis is required for assessing the long-term tread of Chl-a and primary production in this area.

Terauchi et al. (2014) also analyzed the trend, as well as the magnitude of change of Chl-a over a period from 1998 to 2009 in Toyama Bay, one of several small bays of Japan in the Japan Sea, using a locally tuned SeaWiFS 1998–2009 Chl-a product. The objective of their analysis was to use the information obtained as preliminary assessment of eutrophication within the bay. This analysis helps delineate the bay into 6 areas; High-Increasing, High-No trend, High-Decreasing, Low-Increasing, Low-No trend and Low-Increasing. Their analysis indicated that the most of the outer part of the bay was Low-No trend, except for the head of the bay where it displayed High-No trend or High-Increase. There are several rivers in this area and observations of nutrient loadings into the Japan Sea suggests that total nitrogen from Jinzu River increased during 1986–2008 although it was stable after 1998. The High-No trend and High-Increase at the head of the bay may be related to the eutrophication by the river. They suggested a similar approach for preliminary assessment of eutrophication in the large area for Northwest Pacific Action Plan (NOWPAP) of United Nations Environmental Program (UNEP).

5 Influence of Mesoscale Eddies

It is known that the Japan Sea is a region with very active mesoscale eddies. Hyun et al. (2009) studied the biomass and production of phytoplankton as well as bacteria in the Ulleung Basin with ship observations in early April 2006. Through the use of MODIS SST and Chl-a, they found that biomass and production of phytoplankton was higher along the rim than at the center of the anti-cyclonic Ulleung Warm Eddy within the Ulleung Basin. They also reported that bacterial production was higher along the rim of the eddy than at the center; although the biomass of bacteria was constant throughout the eddy, which they attributed to active grazing by microzooplankton. Hyun et al. (2009) were also able to show that a large fraction of the biomass and production of phytoplankton in the upwelling region along the coast of east Korea was eventually transported to the Ulleung Basin by the Ulleung Warm Eddy thereby enhancing bacterial and benthic production within the basin.

Yoo and Park (2009) used SeaWiFS Chl-a and AVHRR SST data from 1998 to 2006 and QuickSCAT wind data to analyze various effects of changes in physical drivers, including coastal upwelling, Tsushima Current, eddy and river discharge, on biological production in the south-western area of the Japan Sea. They found that the high production in the southwestern Japan Sea was mainly supported by the wind-driven upwelling along the Korean coast except in winter. They also found that the transport of the upwelled water via the Ulleung Western Eddy to Ulleung Basin was variable but important.

The idea that anticyclonic Ulleung Western Eddies enhance productivity of this area is continues to be a matter of debate. Studies by Yoo and Park (2009) revealed results that were in contrast to those of Ahn et al. (2005) who suggested enhancement of Chl-a from the SeaWiFS images. On the other hand, Lim et al. (2012) were suggested to observe a clear enhancement of primary production by the eddy during early spring using hourly images of Geo-stationary Ocean Color Imager (GOCI), the world first ocean colour sensor on a geostationary satellite (Ryu and Ishizaka 2012). They suggested the stratification at the ring of the eddy may enhance the high Chl-a.

Spring-time images of MODIS Chl-a overlaid on sea level anomaly maps obtained from Archiving, Validation and Interpretation of Satellite Oceanographic data (AVISO) of 2003 (Fig. 4) showed clearly that anticyclonic eddies in Ulleung and Yamato Basins can delay the onset of the spring bloom to coincide with late blooms in the Japan Basin (Maúre et al. 2017). However, it is still unclear how much primary production can be enhanced by the presence of the eddies.

Using shipboard data alone without satellite images, Kim et al. (2012) were suggested to observe significant enhancement of subsurface production by the Ulleung Western Eddy during the summers of 2005 and 2007. Although the euphotic column was strongly stratified and surface Chl-a was low, Kim et al. (2012) were able to observe high sub-surface Chl-a and nutrients in 2005 without the transport of highly productive, high biomass coastal upwelling water into the Ulleung Basin. They suggested that nutrient supply by an intra-thermocline eddy structure, characterized by a subsurface lens of relatively homogeneous water, supplied nutrients to enhance Chl-a at the base of the euphotic zone (McGillicuddy et al. 1999) at least in 2005.

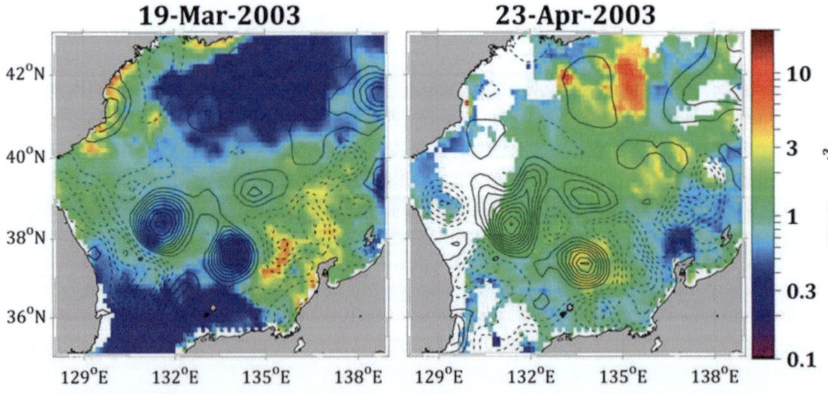

Fig. 4 Weekly MODIS Chl-a averaged from 19 March and 23 April, 2003 overlaid with AVISO Sea Level Anomaly

6 Influence of Typhoon

The influence of typhoons (or hurricanes) on planktonic ecosystems has gained attention in the global ocean as well as in the East Asian Marginal Seas (Subrahmanyam et al. 2002; Siswanto et al. 2007). Recent warming trends and climate changes appear to be causing an increase in the magnitude and the number of typhoons in the western Pacific Ocean. Since satellite ocean colour can serve as a useful tool for analysis of the impacts of typhoons on planktonic ecosystems, understanding how climate-driven changes in typhoon magnitude and the number can impact phytoplankton distribution is a promising area for future research in the Japan Sea. In general, typhoons tend to become weaker as they move into the Japan Sea. However, despite this weakening, ship-board and satellite observations suggest significant enhancement of production following passage of a typhoon in the Japan Sea (Son et al. 2006). The observations of Son et al. (2006) were made during the passage of a category 1 typhoon Megi, which traversed over the Japan Sea in 2004. Using MODIS Chl-a and SST as well as satellite-tracked drifters equipped with temperature and conductivity sensors, they were able to observe significant cooling of the surface waters associated with increases in Chl-a. Following the passage of the typhoon, a 3 °C decrease of SST was observed along its track which lasted for almost 2 weeks. Surface nutrient concentrations estimated by the linear regression with SST, showed an increase during this period. Chl-a increase was about 70% on average along the pass, and this was supported by increases of 90, 40, and 35% for NO_3, PO_4, and SiO_2 estimated by the decrease of SST.

7 Conclusions

This short review summarizes studies that have been undertaken in the Japan Sea using ocean colour remote sensing. These studies include: (1) seasonal variation of chlorophyll-a (Chl-a) and spring bloom, (2) primary production, (3) long-term trends of Chl-a and primary production, (4) influence of eddies, and (5) influence of typhoon. A promising area of ocean colour remote sensing is identification of phytoplankton functional groups (IOCCG 2014). At present, specific research that uses remotely sensed ocean colour data to identify phytoplankton functional group in the Japan Sea is non-existent, except for global information available from the Chl-a based model that discriminates phytoplankton based on cell size such as microplankton which are dominated by diatoms, and nanoplankton which are dominated by Prymnesiophytes (Hirata et al. 2011). Further studies of phytoplankton groups and their impact on satellite ocean color are required to take this research forward. As we have described, ocean colour remote sensing is presently the most promising tool to investigate biological oceanographic and biogeochemical processes in geopolitically complex marginal seas. The development of common robust algorithms and international verification activities, are necessary to improve the accuracy of satellite ocean color information for monitoring long-term changes within such ecosystems.

References

Ahn YH, Shanmugam P, Chang KI, Moon JE, Ryu JH (2005) Spatial and temporal aspects of phytoplankton blooms in complex ecosystems off the Korean Coast from satellite ocean color observations. Ocean Sci J 40(2):67–78

Behrenfeld MJ, Falkowski PG (1997) Photosynthetic rates derived from satellite-based chlorophyll concentration. Limnol Oceanogr 42:1–20

Chiba S, Saino T (2002) Interdecadal change in the upper water column environment and spring diatom community structure in the Japan Sea: an early summer hypothesis. Mar Ecol Prog Ser 231:23–35

Edwards M, Richardson AJ (2004) Impact of climate change on marine pelagic phenology and trophic mismatch. Nature 430:881–884

Gregg WW, Rousseaux CS (2014) Decadal trends in global pelagic ocean chlorophyll: a new assessment integrating multiple satellites, in situ data, and models. J Geophys Res Oceans 119(9):5921–5933

Hirata T, Hardman-Mountford NJ, Brewin RJW, Aiken J, Barlow R, Suzuki K, Isada T, Howell E, Hashioka T, Noguchi-Aita M, Yamanaka Y (2011) Synoptic relationships between surface chlorophyll-a and diagnostic pigments specific to phytoplankton functional types. Biogeosciences 8:311–327

Hyun JH, Kim D, Shin CW, Noh JH, Yang EJ, Mok JS, Kim SH, Kim HC, Yoo S (2009) Enhanced phytoplankton and bacterioplankton production coupled to coastal upwelling and an anticyclonic eddy in the Ulleung basin, East Sea. Aqua Micro Ecol 54(1):45–54

IOCCG (2014) Phytoplankton functional types from space. In: Sathendranath S (ed) Reports of the International Ocean-Colour Coordinating Group, No. 15., IOCCG, Dartmouth, Canada, pp. 156

Jo CO, Lee JY, Park KA, Kim YH, Kim KR (2007) Asian dust initiated early spring bloom in the northern East/Japan Sea, Geophys Res Lett 34(5)

Jo CO et al (2014) Spatial distribution of seasonality of SeaWiFS chlorophyll-a concentrations in the East/Japan Sea. J Mar Sys 139:288–298

Joo H, Park JW, Son S, Noh JH, Jeong JY, Kwak JH, Saux-Picart S, Choi JH, Kang CK, Lee SH (2014) Long-term annual primary production in the Ulleung Basin as a biological hot spot in the East/Japan Sea. J Geophys Res Oceans 119(5):3002–3011

Joo H, Son S, Park JW, Kang JJ, Jeong JY, Lee CIl, Kang CK, Lee SH (2016) Long-term pattern of primary productivity in the East/Japan Sea based on ocean color data derived from MODIS-Aqua. Remote Sens 8(1)

Kameda T, Ishizaka J (2005) Size-fractionated primary production estimated by a two-phytoplankton community model applicable to ocean color remote sensing. J Oceanogr 61(4):663–672

Kim SW, Saitoh SI, Ishizaka J, Isoda Y, Kishino M (2000) Temporal and spatial variability of phytoplankton pigment concentrations in the Japan sea derived from CZCS images. J Oceanogr 56(5):527–538

Kim HC, Yoo SJ, Oh IS (2007) Relationship between phytoplankton bloom and wind stress in the sub-polar frontal area of the Japan/East Sea. J Mar Sys 67(3–4):205–216

Kim D, Yang EJ, Kim KH, Shin CW, Park J, Yoo S, Hyun JH (2012) Impact of an anticyclonic eddy on the summer nutrient and chlorophyll a distributions in the Ulleung basin, East Sea (Japan Sea). ICES J Mar Sci 69(1):23–29

Lee SH, Son S, Dahms HU, Park JW, Lim JH, Noh JH, Kwon JI, Joo HT, Jeong JY, Kang CK (2014) Decadal changes of phytoplankton chlorophyll-a in the East Sea/Sea of Japan. Oceanology 54(6):771–779

Li LP, Fukushima H, Frouin R, Mitchell BG, He MX, Uno I, Takamura T, Ohta S (2003) Influence of submicron absorptive aerosol on sea-viewing wide field-of-view sensor (SeaWiFS)-derived marine reflectance during Aerosol Characterization Experiment (ACE)-Asia. J Geophyic Res Atmos 108(D15)

Lim JH, Son S, Park JW, Kwak JH, Kang CK, Son YB, Kwon JN, Lee SH (2012) Enhanced biological activity by an anticyclonic warm eddy during early spring in the East Sea (Japan Sea) detected by the geostationary ocean color satellite. Ocean Sci J 47(3):377–385

Maúre ER, Ishizaka J, Sukigara C, Mino Y, Aiki H, Matsuno T, Tomita H, Goes JI, Gomes HR (2017) Mesoscale eddies control the timing of spring phytoplankton blooms: A case study in the Japan Sea. Geophys Res Lett 44(21):11115–11124

McGillicuddy DJ, Johnson R, Siegel DA, Michaels AF, Bates NR, Knap AH (1999) Mesoscale variations of biogeochemical properties in the Sargasso Sea. J Geophys Res Oceans 104(C6): 13381–13394

Park KA, Kang CK, Kim KR, Park JE (2014) Role of sea ice on satellite-observed chlorophyll-a concentration variations during spring bloom in the East/Japan sea. Deep-Sea Res I 83:34–44

Platt T, Fuentes-Yaco C, Frank KT (2003) Spring algal bloom and larval fish survival. Nature 423:398–399

Ryu JH, Ishizaka J (2012) GOCI data processing and ocean applications. Ocean Sci J 47(3):221

Saba VS et al (2011) An evaluation of ocean color model estimates of marine primary productivity in coastal and pelagic regions across the globe. Biogeosciences 8(2):489–503

Salyuk P, Bukin O, Alexanin A, Pavlov A, Mayor A, Shmirko K, Akmaykin D, Krikun V (2010) Optical properties of Peter the Great Bay waters compared with satellite ocean colour data. Int J Remote Sens 31(17–18):4651–4664

Siswanto E, Ishizaka J, Yokouchi K, Tanaka K, Tan CK (2007) Estimation of interannual and interdecadal variations of typhoon-induced primary production: a case study for the outer shelf of the East China Sea. Geophys Res Lett 34(3). https://doi.org/10.1029/2006gl028368

Son SH, Platt T, Bouman H, Lee DK, Sathyendranath S (2006) Satellite observation of chlorophyll and nutrients increase induced by Typhoon Megi in the Japan/East Sea. Geophys Res Lett 33(5). https://doi.org/10.1029/2005gl025065

Subrahmanyam B, Rao KH, Rao NS, Murty VSN, Sharp RJ (2002) Influence of a tropical cyclone on Chlorophyll-a Concentration in the Arabian Sea. Geophys Res Lett 29(22). https://doi.org/10.1029/2002gl015892

Sverdrup HU (1953) On conditions for the vernal blooming of phytoplankton. J Cons Explor Mer 18:287–295

Terauchi G, Tsujimoto R, Ishizaka J, Nakata H (2014) Preliminary assessment of eutrophication by remotely sensed chlorophyll-a in Toyama Bay, the Sea of Japan. J Oceanogr 70(2):175–184

Yamada K, Ishizaka J (2006) Estimation of inter-decadal change of spring bloom timing, in the case of the Japan Sea. Geophys Res Lett 33(2). https://doi.org/10.1029/2005gl024792

Yamada K, Ishizaka J, Yoo S, Kim HC, Chiba S (2004) Seasonal and interannual variability of sea surface chlorophyll a concentration in the Japan/East Sea (JES). Prog Ocean 61(2–4):193–211

Yamada K, Ishizaka J, Nagata H (2005) Spatial and temporal variability of satellite primary production in the Japan Sea from 1998 to 2002. J Oceanogr 61(5):857–869

Yoo S, Kim H (2004) Suppression and enhancement of the spring bloom in the southwestern East Sea/Japan Sea. Deep Sea Res Part II: Top Stud Oceanogr 51(10–11):1093–1111

Yoo S, Park J (2009) Why is the southwest the most productive region of the East Sea/Sea of Japan? J Mar Sys 78(2):301–315

Yoon JE, Park J, Yoo S (2012) Comparison of primary productivity algorithms for Korean Waters. Ocean Sci J 47(4):473–487

Zuenko YI, Pershina EA (2011) Variations of sea surface temperature and dates of spring bloom at spawning grounds of Japanese sardine in the Japan Sea. Izvestia TINRO (Newsletters of Pacific Fisheries Research Center) 165:251–264 (in Russian)

Complementary Remote Sensing Observations of the Tsushima Warm Current Patterns

Kaoru Ichikawa, Yutaka Yoshikawa, Akihiko Morimoto, Ken-ichi Fukudom and Jong-Hwan Yoon

Abstract When monitoring exchanges between adjacent marginal seas, measurements of the volume transport through connecting straits are essential. In such cases, acoustic Doppler current profilers (ADCPs) mounted at the bottom of ferries that regularly cross such straits can provide useful platforms for directly monitoring strait currents. However, since their observations would naturally be confined to each ferry's set route, it would be impossible for them to observe all relevant current patterns. In addition, a ship-mounted ADCP is very expensive to install, maintain, and (when necessary) remove. In this chapter, we will describe alternative indirect remote sensing techniques that can be used to observe surface velocities in the Korea Strait between Japan and Korea. One such technique involves a high-frequency (HF) ocean radar system that can provide synoptic views of the surface velocity field with high resolutions in both space and time, even though additional processes are required to separate the geostrophic ocean currents from the ageostrophic tidal and wind-driven currents. Alternatively, slopes of the sea surface dynamic height (SSDH) can provide a geostrophic component of the surface velocity. Unfortunately, the satellite altimeters that are most commonly used to measure the SSDH field in open oceans are unsuitable for use in narrow straits.

K. Ichikawa (✉) · J.-H. Yoon
Research Institute for Applied Mechanics, Kyushu University, Kasuga, Japan
e-mail: ichikawa@riam.kyushu-u.ac.jp

J.-H. Yoon
e-mail: yoon@riam.kyushu-u.ac.jp

Y. Yoshikawa
Divison of Earth and Planetary Sciences, Graduate School of Science, Kyoto University, Kyoto, Japan
e-mail: yosikawa@kugi.kyoto-u.ac.jp

A. Morimoto
Center for Marine Environmental Studies, Ehime University, Matsuyama, Japan
e-mail: amorimoto@ehime-u.ac.jp

K. Fukudom
National Institute of Technology, Toyama College, Toyama, Japan
e-mail: dome@nc-toyama.ac.jp

V. Barale and M. Gade (eds.), *Remote Sensing of the Asian Seas*,
https://doi.org/10.1007/978-3-319-94067-0_10

Instead, the use of global navigation satellite systems (GNSS) is showing promise as a way to obtain the coastal SSDH, even though proper spatial smoothing processes are still required.

Keywords Tsushima Warm Current · HF ocean radar · Coastal altimetry GNSS · Sea surface dynamic height · ADCP

1 Introduction

Compared to deep open oceans such as the Pacific, the volume content of marginal seas is generally so small that water exchange as inflow/outflow in a strait becomes significantly important. Indeed, most Asian marginal seas are connected with each other through straits, such as the Taiwan Strait north of Taiwan that connects the South China Sea and the East China Sea, and the Korea Strait south of Korea that connects the East China Sea and the Japan Sea. When considering the ongoing high demands for information on ocean currents from fisheries, such as fish egg and larvae transport estimations, as well as the need to control the engineering and environmental impacts from vigorous human activities (including ship routing and marine debris), the accurate monitoring of the currents in these straits is an essential concern for oceanographers, marine engineers, and environmental scientists.

An acoustic Doppler current profiler (ADCP) mounted at the bottom of a ferry that regularly crosses a strait provided the most accurate and convenient method to monitor currents in the strait. The ADCP transmits sound pulses downward in several directions and measures the Doppler frequency shifts of their echoes reflected from small floating particles at various depths. Assuming that these particles move along with the surrounding water, we can obtain the vertical profiles of vector velocities relative to the ferry movements, except for the layers near the sea surface and close to the seafloor where echoes from materials other than floating particles are present. By adding the ground speed of the ferry determined by Global Navigation Satellite Systems (GNSS) positioning, these relative velocities are converted to the absolute velocities relative to the ground. In the Korea Strait between Hakata, Japan, and Pusan, Korea, the ferry "*New Camellia*" (19,961 Gross Tons) directly measures all of the daily vertical profiles of vector velocities along the ship route shown in Fig. 1 (Takikawa et al. 2005; Fukudome et al. 2010). Although an extended series of observations is necessary to accurately separate tidal constituents from the hour-fixed velocity measurements (Fukudome et al. 2010), inter-annual, seasonal, and intra-seasonal variations of the northeastward volume transport across the Straits, commonly referred to as the Tsushima Warm Current (TWC), are fully described using these data. Although the origin of the TWC is not yet fully understood (Isobe 1999), warm water either from the Kuroshio Current or from the Taiwan Warm Current is carried into the Japan Sea by the TWC along the northern coast of the Japanese main island (Fig. 1), thereby providing significant poleward heat transport from the East China Sea (Hirose and Fukudome 2006).

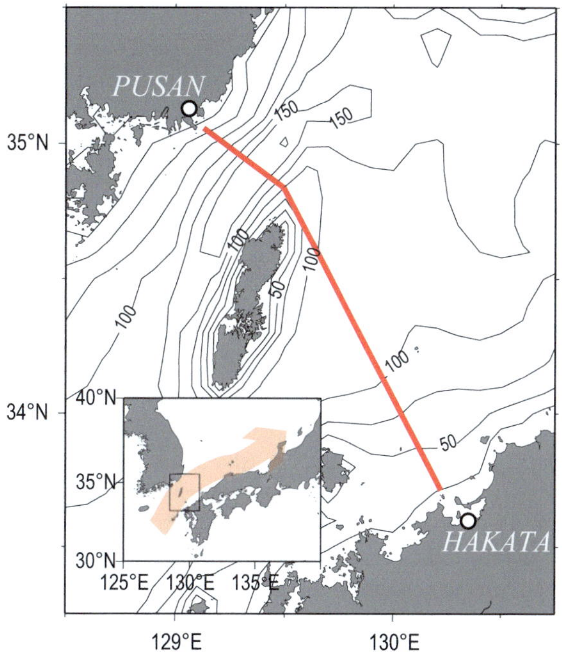

Fig. 1 Location of the Korea Strait (small map at the left corner) and a standard ship route of "*New Camellia*" between Pusan and Hakata. Bottom topographic contours are also shown in the figure (after Takikawa et al. 2005). The broad arrow in the small map shows the general path of the TWC)

Recently, a similar system has been introduced in the Taiwan Strait in order to measure the Taiwan Warm Current (Chen et al. 2016).

Nevertheless, practical applications involving ship-mounted ADCP measurements suffer from at least two specific difficulties. The first is the lack of the horizontal distribution of the current structure. Since the measurements are limited to those taken on fixed ship routes, the spatial distribution of the velocity normal to the routes is not available. This is important because the flows passing through narrow channels in straits are energetic, which means small-scale variations can be expected. In such conditions, the lack of horizontal views may cause misinterpretations of variations at a given position. However, it should also be noted that cross-strait volume transport calculations will not significantly influenced by neglecting such small-scale variations because they will be eventually averaged out by horizontal integration. The other problem is a more practical issue; installing ADCPs on ferry boats is a very expensive endeavor that has to be negotiated with the ship owners and operators. In addition, routine maintenance expenses, and cost of their eventual removal are considerable.

Remote-sensing techniques, either from satellites or from land/ships, can be expected to provide solutions to those problems. In this chapter, we describe practical examples of two indirect surface velocity measurement schemes selected from the dense observation system developed for the TWC: high-frequency (HF) ocean radars (Sect. 2) and sea surface dynamic height (SSDH) measurements (Sect. 3).

HF ocean radars measure the Doppler frequency shift of backscattered radar echoes, which is proportional to the radial component of the speed of backscattering materials on the sea surface. In most cases, the backscattering materials are ocean wind waves with several-meter wavelengths, whose phase propagation speeds are known. After subtracting the phase speed of the wind waves, the measured Doppler velocity provides the drift speed of the body mass on which the wind waves are present, i.e., the surface layer velocity. This velocity contains not only the geostrophic ocean currents such as the TWC, but also ageostrophic tidal and wind-driven components, the separation of which will be described in Sect. 2.

In contrast, the SSDH relates solely to the geostrophic velocity at the sea surface, in a manner that the speed of a surface current in geostrophic balance is proportional to the orthogonal SSDH gradient. SSDH values obtained by satellite altimeters are used to describe various oceanographic variations in open oceans, and are already commonly used to monitor surface velocity variations. However, the use of altimeter data in coastal areas is somewhat complicated because those instruments were originally designed for use in open ocean areas (Cipollini et al. 2013). Near land, altimeter data are often lost due to waveform (echo intensity data) contamination (Gommenginger et al. 2011). Since the traditional algorithms used by altimeters assume uniform backscattering at the sea surface within a footprint, unexpected reflections caused by other objects (such as ships, land, or even calm water surfaces in semi-closed bays) can cause algorithm malfunctions. Various improvements have been proposed, such as the development of new algorithms (e.g. Passaro et al. 2014; Wang and Ichikawa 2017) and the fabrication of new altimeters that are properly designed for coastal usage, e.g. AltiKa (Vincent et al. 2006) and CryoSat-2 (Passaro et al. 2016). Nevertheless, a certain amount of SSDH data acquired near land will always be unsuitable for analysis, which is a serious problem when analyzing narrow straits.

An alternative method to obtain the coastal SSDH distribution is to use precise GNSS positioning. Slant ranges from GNSS satellites, whose positions are precisely known, are determined from the measured travel time of the transmitted microwaves to a GNSS receiver. By referring more than four GNSS satellites, the exact position of the receiver can be obtained theoretically. However, in practice, the estimated position may include errors of several meters because these slant ranges are not precisely determined due to various factors: i.e., errors of the on-board clocks of each satellite, and travel time delay in the ionosphere and the atmosphere. Precise positioning is made possible by the real-time kinematic (RTK) analysis of the GNSS data, which accurately determines the relative position of the receiver to adjacent fixed reference sites by interferometric positioning (e.g. Pullen and Rife 2009). Although the height estimation (with errors of several centimeters for individual measurements) is generally less accurate than the horizontal positioning (with errors of a few centimeters), the precise RTK GNSS could be used to measure the coastal SSDH. In Sect. 3, coastal SSDH measurements by satellite altimeters are first examined in Sect. 3.1. Then, an alternative SSDH measurement method using RTK GNSS data is examined in Sect. 3.2.

2 Surface Velocity Field Observed by HF Ocean Radars

Seven HF ocean radar sites have been deployed and operated by the Research Institute for Applied Mechanics (RIAM), Kyushu University for more than 15 years (Yoshikawa et al. 2006, 2010). Five crossed-loop antenna type radars (SeaSonde, CODAR Ocean Sensors) and two digital-beam-forming array antenna type radars (Nagano Japan Radio Co. Ltd.; NJRC) measure each radial velocity component along the radial lines shown in Fig. 2. These radial velocities are used to produce the surface velocity field every hour at approximately 6 km grids.

The surface velocity obtained by the HF ocean radar consists of three major components: geostrophic velocity, tidal current, and wind-driven current (Tokeshi et al. 2007; Yoshikawa et al. 2007). The tidal current is generally the strongest velocity component in shallow and narrow straits, but accurate separation of this component is possible since the sampling intervals of HF ocean radars are short enough with respect to the tidal periods, unlike daily measurements by a ferry. The wind-driven current is often estimated linearly from independent wind data by using the deflection angle from the wind direction and the speed factor, or the speed ratio of the wind-driven current to the wind (Tokeshi et al. 2007; Yoshikawa et al. 2007). As the nature of boundary layer currents, both the speed and direction of

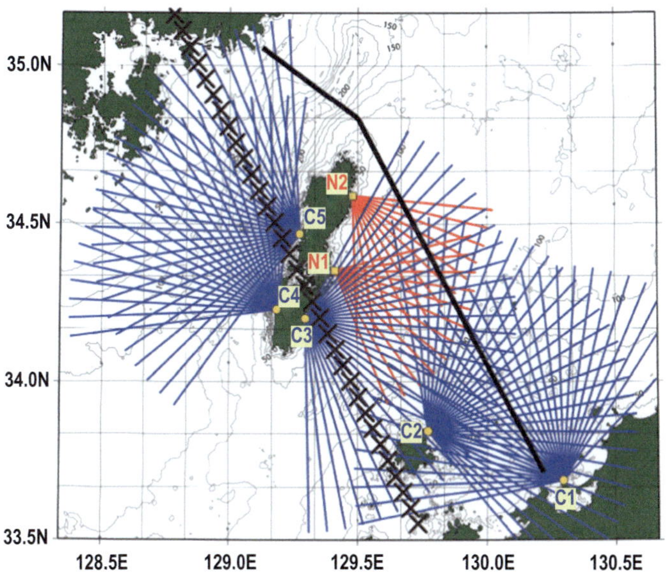

Fig. 2 Locations of the HF ocean radar sites (CODAR; C1–C5 and NJRC; N1 and N2; after Yoshikawa et al. 2010). Radial directions from the sites are indicated by blue (for C1–C5) and red lines (for N1 and N2) and the vector surface velocity is determined at the crossover points of these radial lines. The bold line and crosses indicate the ship route of "*New Camellia*" shown in Fig. 1 and the sub-satellite track of the Jason-1 altimeter, respectively

wind-driven current depend strongly on depth, so their relevant factors will depend on the corresponding depth of the currents measured by the HF ocean radars. Since HF ocean radars measure surface currents by the drift speed of wind waves with particular wavelengths that satisfy the Bragg backscatter condition, which is determined by the radar wavelength, the corresponding depths of the HF ocean radar measurements (approximately a few meters) depend on the HF ocean radar frequency (Yoshikawa et al. 2007). Since these parameters are also known to depend on seasonal viscous conditions in the surface Ekman layer (Yoshikawa and Masuda 2009), estimations of those parameters should be conducted for each location and each season.

Figure 3 shows the five-year averaged monthly mean surface velocity fields in November, with and without the wind-driven currents (Yoshikawa et al. 2010). While tidal currents should be averaged out in such long-term averages, the stable northerly winds that blow in this area in winter produce long-term mean southward and/or southwestward wind-driven currents. The southeastward currents are found east of Tsushima Island (around 129° 35′E and 34° 25′N) and north of Hakata (around 130° 10′E and 33° 50′N) in Fig. 3a, but they cross the topographic contours. The southeastward currents so revealed are confined solely to the surface Ekman layer since the geostrophic current component in Fig. 3b turns to follow along the topographic contours of these areas. Note that changes between Fig. 3a, b are not significant in the western channel and at the center of the eastern channel, where the long-term mean geostrophic TWC are significantly strong with respect to the long-term mean wind-driven current component. This also suggests that the dominance of the geostrophic, tidal, and wind-driven current components would depend on both the location and the time-scales of interest.

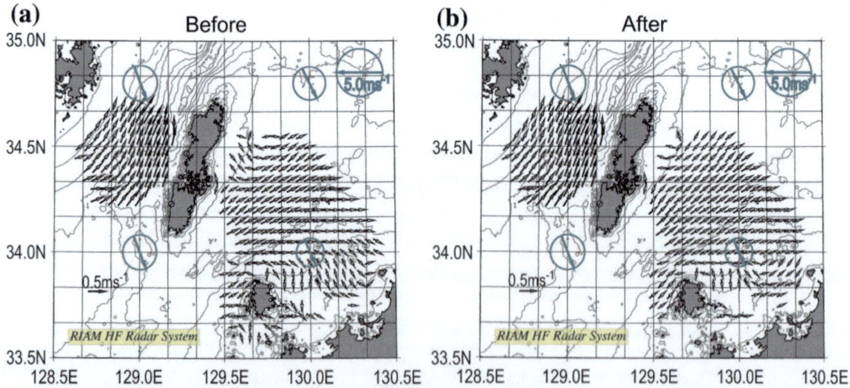

Fig. 3 Five-year averaged monthly mean velocity field in November before (**a**) and after (**b**) the removal of the wind-driven current component (after Yoshikawa et al. 2010). Arrows in circles indicate the 5-year averaged monthly mean wind, and contours with thin lines indicate bottom topography

The annual mean geostrophic velocity field (Fig. 4a) clearly shows the TWC both in the eastern and western channels, together with a regional counter-clockwise circulation north of Hakata (south of 34°N). The temporal anomaly component from the mean was further analyzed by the Empirical Orthogonal Function (EOF) method (Yoshikawa et al. 2010). We found that the first mode (which explains 79% of the total variance) shows seasonal variations, with the spatial current pattern shown in Fig. 4b in summer (and the reverse direction in winter). In the western channel, the current structure of the temporal anomaly component in Fig. 4b is as the same as the mean in Fig. 4a, thereby suggesting the stronger (or weaker) TWC in summer (or winter). Meanwhile, in the eastern channel, evident cyclonic and anticyclonic eddy-like structures are present at 34° 20′N and 34°N, respectively. This result shows that the strong TWC tends to flow in the middle of the channel around 34° 10′N in summer, with recirculating counter currents east of Tsushima Island and north of Hakata. In winter, when the EOF score is less-significantly negative, the TWC tends to be weaker and broader. These horizontal structure changes of the TWC cannot be fully recognized by the along-route ship-mounted ADCP data observations discussed in Sect. 1.

3 Sea Surface Dynamic Height Observations

Although HF ocean radars provide valuable data, such systems are expensive and intensive work is necessary to maintain them properly, thereby suggesting that the second practical difficulty inherent with the ship-mounted ADCP setup described in Sect. 1 has not been fully resolved. Therefore, other feasible methods, such as monitoring currents by the SSDH, are also required.

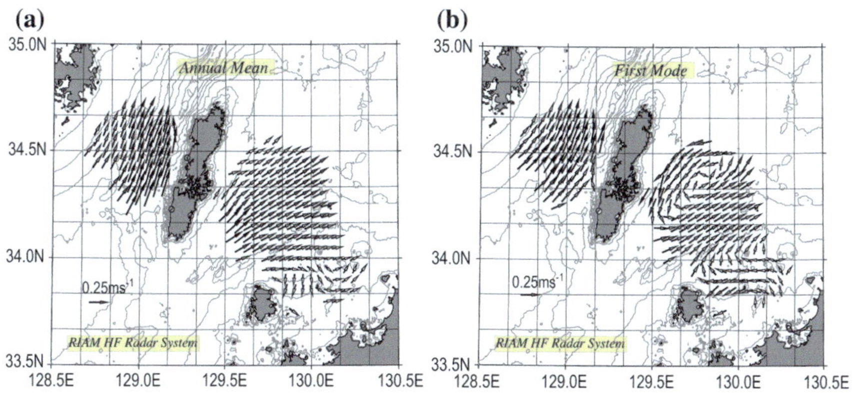

Fig. 4 Five-year averaged annual mean surface geostrophic velocity field (**a**) and the dominant temporal anomaly component (**b**) (after Yoshikawa et al. 2010). The arrows and contours are the same as those used in Fig. 3

The easiest way to obtain SSDH variations at a particular point is to use long-term tide gauge records with removal of tidal components imposed either by tide-killer filters or the harmonic analysis. The SSDH difference between two points should be related to the horizontally integrated surface geostrophic current between those two points. Actually, temporal changes of the SSDH differences determined by detided tide gauge records on both sides of the Korea or Taiwan Straits have been found to be correlated with the horizontally integrated volume transports obtained by the ship-mounted ADCPs (Takikawa and Yoon 2005; Chen et al. 2016). Note that the conversion factor of the SSDH difference to the ADCP-determined TWC volume transport varies seasonally, as the vertical current profile of the TWC changes. More specifically, the TWC is surface intensified in summer, while it is more barotropic in winter when the density stratification is weaker. Meanwhile, this factor remains almost constant for the shallow and barotropic Taiwan Warm Current (Chen et al. 2016).

Furthermore, if the spatial SSDH distribution at any location is obtained, instead of the SSDH difference between fixed tide gauge stations, the surface geostrophic velocity at a given location can be estimated. Therefore, along with HF ocean radars, SSDH observations within straits provide useful tools for monitoring velocity variations in straits.

3.1 SSDH by Satellite Altimeters

This subsection will begin with an examination of the use of the satellite altimetry in the Korea Strait. The along-track regional Jason-1 altimeter data product that was developed, validated, and distributed by Center for Topographic studies of the Ocean and Hydrosphere (CTOH)/Laboratorie d'Etudes en Géophysique et Océanographie Spatiales (LEGOS), France, is used in this study, along with the application of an additional tidal correction produced by the harmonic analysis of the nine major (M_2, S_2, N_2, K_2, K_1, O_1, P_1, Q_1, Mf) constituents (Tokeshi et al. 2007). Due to the sparse sub-satellite tracks of Jason altimeters, only one track (number 36) is found crossing the HF ocean radar coverage in the Strait (Fig. 2). Furthermore, since numerous data points near lands were masked due to contaminations, only 550 samples were available during the 5.5-year period (from December 2002 to April 2008) that overlapped with HF ocean radar data.

The SSDH temporal anomaly (SSDHA) from the 5.5-year mean is determined, and the geostrophic velocity anomaly component normal to the sub-satellite track is calculated from the along-track SSDHA data. The corresponding velocity anomaly component at the same time and location is determined from the HF ocean radar data corrected with the tidal and wind-driven currents described in Sect. 2, and those two velocity components are compared (Fig. 5). Since the correlation coefficient (0.09) is low, it is obvious that the two velocity data sets behave differently.

In order to evaluate the geostrophic velocity noises or contaminations caused by the ageostrophic components in each data set, the altimetry data along-track spatial

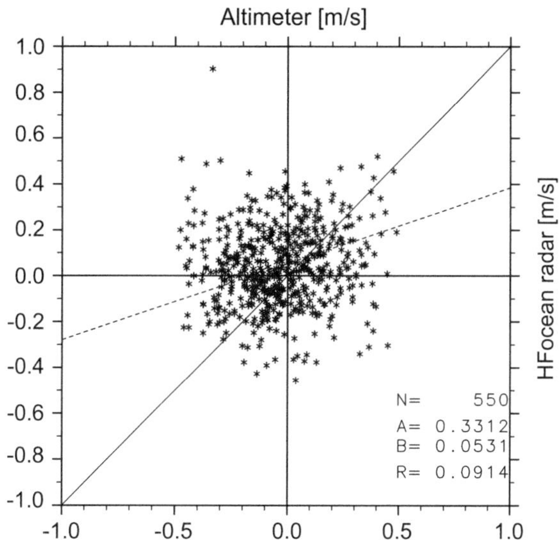

Fig. 5 Scatter plot for the geostrophic velocity anomaly component normal to the sub-satellite track, determined from the altimeter data and extracted from the HF ocean radar data. The number of samples (N), the slope (A), and the intersect (B) of the regression line (broken line), and correlation coefficient (R) are shown in the figure

smoothing and HF data temporal averaging are examined as described in Tokeshi et al. (2007). The correlation coefficient for each combination of spatial and temporal smoothing parameters (Fig. 6a) indicates that altimeter data spatial smoothing provides a more effective way to improve the correlation, thereby suggesting contamination by ageostrophic components in the altimeter data (Kashima et al. 2003). Although the scatter plot for the smoothed data (Fig. 6b) is improved from the original unsmoothed data in Fig. 5, it should be noted that the number of valid comparisons is reduced to just 62. This is because the channels are so narrow that continuous sections of along-track data longer than 28 km can only be obtained in the middle of the channels.

3.2 SSDH by Global Navigation Satellite System

In the previous subsection, we explained why SSDH observations by satellites altimeters are inadequate in narrow straits. In this subsection, an alternative method that uses GNSS to measure the SSDH is examined.

In the present study, we have deployed a RTK GNSS receiver system (NetR5, Trimble) since September 2011 on the "*New Camellia*" ferry in order to obtain the SSDH values along the ship tracks (Ichikawa et al. 2013). The height of the antenna H is determined every 30 s by a RTK software (RTNet, Hitachi Zosen), analyzing the GNSS data received on the ferry and at three fixed Japanese GNSS Earth Observation Network (GEONET) reference sites within the Strait, which are maintained by the Geospatial Information Authority of Japan, together with the International GNSS Service final orbits (Dow et al. 2009).

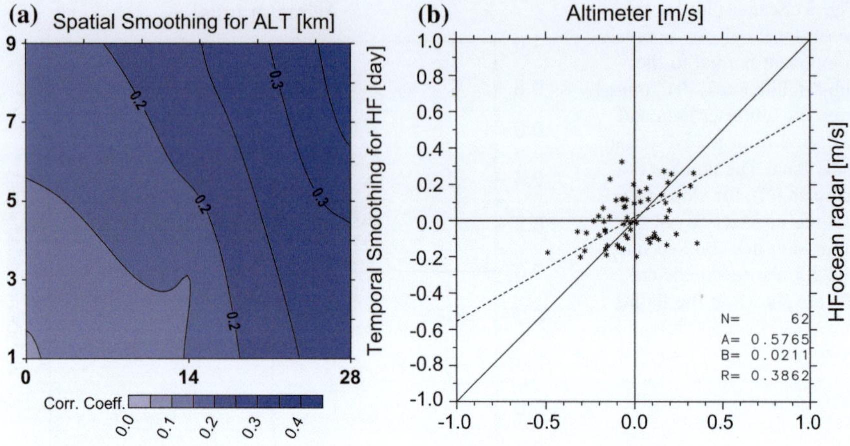

Fig. 6 Correlation coefficient for various combinations of temporal and spatial smoothing parameters (**a**). The scatter plot is the same as that used in Fig. 5, but with 28-km smoothed altimeter data and 9-day averaged HF ocean radar data (**b**)

The SSDH ζ $(t, \boldsymbol{r})$ at a time t and location $\boldsymbol{r}$ is estimated from the relation

$$\zeta(t,\boldsymbol{r}) + \xi(t,\boldsymbol{r}) = H(t,\boldsymbol{r}) - \{Z_G(\boldsymbol{r}) + Z_T(t,\boldsymbol{r}) + H_S(t)\}, \tag{1}$$

where Z_G is the EGM2008 geoid height (Pavlis et al. 2012), Z_T is the tidal height calculated by a local tide model based on altimeters (Morimoto 2009), H_S is the antenna height variation caused by the ship motion monitored by the draft and tilt gauges installed on the ferry, and ξ is the unknown ageostrophic height component that is not balanced with the geostrophic velocity. The observed height $\zeta + \xi$ is then interpolated at the standard points on the ship route by the Gaussian smoothing with an 8-km decorrelation scale.

For comparison, the SSDH ζ_{ADCP} is calculated from the detided velocity at the top layer (18-m depth) of the ship-mounted ADCP data described in Sect. 1 (Fukudome et al. 2010) while assuming a geostrophic balance. Note that the wind-driven current component is not removed from the ADCP data, but it should be less significant at this depth. The 5-month mean values from September 2011 to January 2012 (Fig. 7) show good agreements between the SSDH determined by the ship-mounted ADCP $\bar{\zeta}_{ADCP}$ and the height $\bar{\zeta} + \bar{\xi}$ obtained from the ship-mounted GNSS receiver; the difference is within just a few centimeters. This agreement reveals that both the ageostrophic component ξ and the short-term variations $Z_T + H_S$ become negligibly small in the long-term means, and that the GNSS observations H and the geoid height Z_G are sufficiently accurate.

On the other hand, unlike the mean values, the temporal anomaly components of the ADCP-determined SSDH $\zeta'_{ADCP}(t,\boldsymbol{r}) = \zeta_{ADCP}(t,\boldsymbol{r}) - \bar{\zeta}_{ADCP}(\boldsymbol{r})$, and the GNSS-obtained height $\zeta'(t,\boldsymbol{r}) + \xi'(t,\boldsymbol{r})$ behave differently. In an example from 4

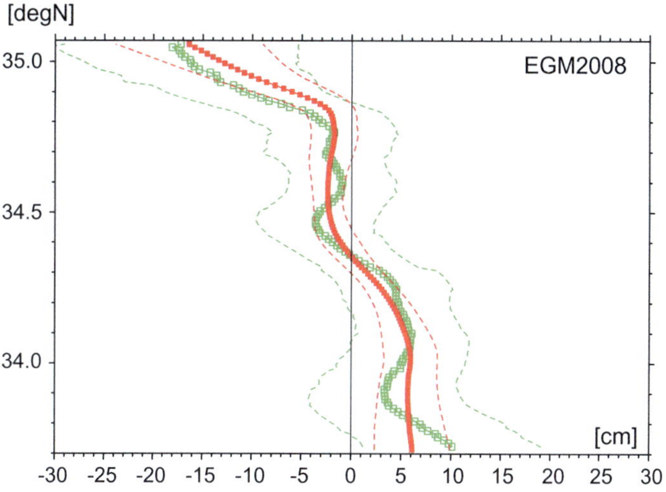

Fig. 7 Five-month mean SSDH $\bar{\zeta}_{ADCP}(\boldsymbol{r})$ from the ship-mounted ADCP (red) and the mean height $\bar{\zeta}(\boldsymbol{r}) + \bar{\xi}(\boldsymbol{r})$ from the ship-mounted GNSS receiver (green) along the ship track shown in Fig. 1. Standard deviation of the mean is shown by broken lines

December 2011 (Fig. 8), we found that the GNSS-obtained height $\zeta' + \xi'$ contains small-scale variations with 30–40 km wavelengthes, while the two anomalies ζ'_{ADCP} and $\zeta' + \xi'$ are almost equal when an along-track 30-km smoothing is applied to $\zeta' + \xi'$, as indicated by the broken lines. These results suggest the significant presence of the smaller-scale ageostrophic component $\xi(t, \boldsymbol{r})$. It should also be noted that the spatially smoothed altimetry data result in a better comparison with the geostrophic current estimated from the HF ocean radar in Sect. 3.1.

4 Summary and Discussion

Volume transports across straits, which are important for estimation of water exchanges and the resident times in marginal seas, are often monitored by ADCPs mounted at the bottom of ferries regularly crossing those straits. However, these observations cannot provide the spatial distribution of the currents in the direction normal to the ship route. In contrast, since HF ocean radar systems can provide surface velocity fields with high resolutions in both space and time, they are useful for monitoring spatial current variations in narrow straits. In the Korea Strait, width, location, and strength variations of the TWC and its recirculation are clearly described by the HF ocean radar system deployed by RIAM, Kyushu University.

Unfortunately, the SSDH satellite altimetry observations in the Korea Strait are not as useful as they are in open oceans due to the sparse locations of sub-satellite tracks and the loss of usable data near land. As a result, there is very little such data

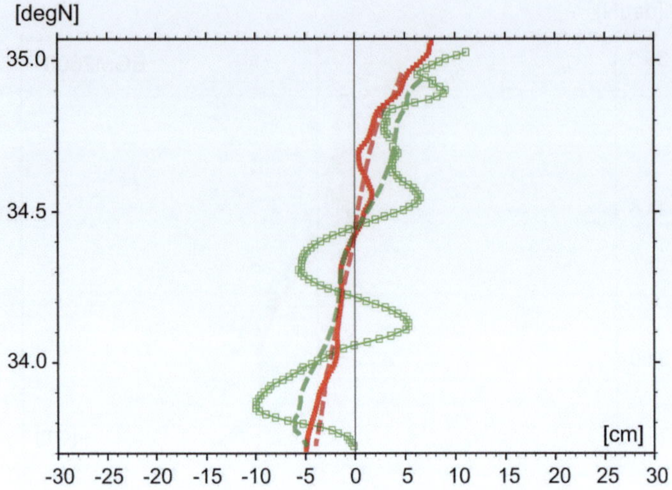

Fig. 8 Temporal anomaly ζ'_{ADCP} (red) and $\zeta' + \xi'$ (green) along the ship track on 4 December, 2011. The 30-km smoothed values are also plotted by bold broken lines

available. This problem is common for any narrow strait. However, as long as proper spatial smoothing is applied, the SSDH estimation produced by the GNSS receiver mounted at the top of the ferry "*New Camellia*" was found to result in good agreement with the ship-mounted ADCP data. Furthermore, since the costs of installing and maintaining a ship-mounted GNSS receiver are much lower than those of the ship-mounted ADCP, SSDH observations by GNSS could provide a feasible alternative method for monitoring volume transports in narrow straits.

In Sect. 3.2, the small-scale variations present in the GNSS observations (Fig. 8) are considered mainly due to the real ageostrophic anomaly component $\xi'(t, \boldsymbol{r})$, but they could simply be temporally variable tide errors Z_T or height variations produced by ship motions H_S. If these errors have zero-mean random features, they will be diminished in the mean in Fig. 7, but present in the anomaly component in Fig. 8. These results indicate that additional SSDH observations, take independently from the ferry, will be necessary in order to study this matter in greater detail.

At the end, an example of the three different data sets are shown in Fig. 9 for comparison. More specifically, the ship-mounted ADCP (red line in Fig. 9b), the HF ocean radar (Fig. 9a), and the ship-mounted GNSS receiver (green line in Fig. 9b). All data sets were acquired on 7 November 2011. A significant cyclonic eddy is clearly seen in the HF ocean radar data, centered at 34° 20′N just west of the ship route (Fig. 9a). The presence of this eddy is also recognized both in the ship-mounted ADCP and GNSS receiver data (Fig. 9b) as a negative SSDHA around 34.4°N. The agreement of these three different data sets, all showing the eddy east of Tsushima Island, is a very promising result. However, the other significant small-scale variations at 33.8°N and 34.7°N seen in $\zeta' + \xi'$ could not be identified in the ADCP-derived SSDHA ζ'_{ADCP}, even though small undulations are present in the

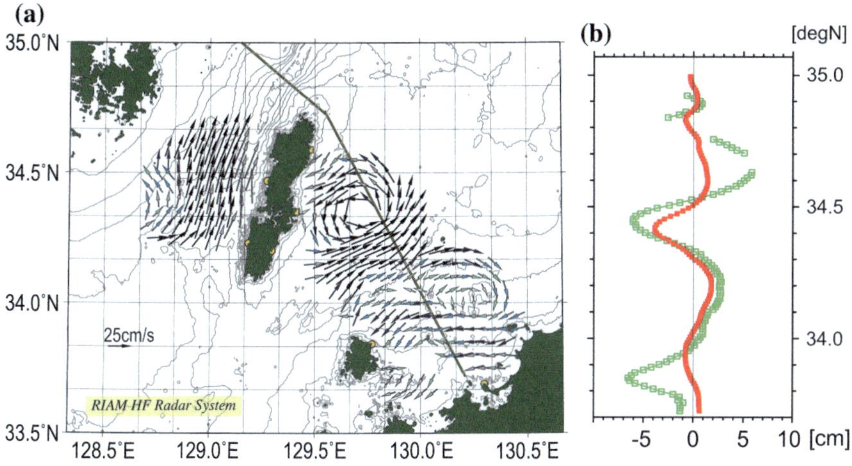

Fig. 9 The surface geostrophic current field obtained by the daily-averaged HF ocean radar on 7 November 2011 (**a**), and the same figure as Fig. 8 but 7 November 2011 (**b**). The ship route is plotted in the left panel (a) for reference. Vectors in panel (**a**) are the same as Fig. 3, and the colors in the panel (**b**) are the same as Fig. 8

adjacent. Moreover, these undulations are outside the HF ocean radar coverage. Further studies will be conducted to compare these different types of observations.

Acknowledgements This work is partially supported by the Japan Society for the Promotion of Science (JSPS) KAKENHI Grant Number JP25247076.

References

Chen HW, Liu CT, Matsuno T, Ichikawa K, Fukudome K, Yang Y, Doong DJ, Tsai WL (2016) Temporal variations of volume transport through the Taiwan Strait, as identified by three-year measurements. Cont Shelf Res 114:41–53

Cipollini P, Benveniste J, Miller L, Picot N, Scharroo R, Sturb T, Vandemark D, Vignudelli S, Zoffoli S, Andersen O, Bao L, Birol F, Coelho E, Deng X, Emery W, Fenoglio L, Fernandes J, Gomez-Enri J, Griffin D, Han G, Hausman J, Ichikawa K, Kostianoy A, Kourafalou V, Labroue S, Ray R, Saraceno M, Smith W, Thibaut P, Wilkin J, Yenamandra S (2013) Conquering the coastal zone: a new frontier for satellite altimetry. In: Ouwehand L (ed) Proceedings of 20 years of progress in radar altimetry, Venice Italy. ESA SP-710. ISBN 978-92-9221-274-2. https://doi.org/10.5270/esa.sp-710.altimetry2012

Dow JM, Nelian RE, Rizos C (2009) The international GNSS service in a changing landscape of Global Navigation Satellite Systems. J Geod 83:191–198. https://doi.org/10.1007/s00190-008-0300-3

Fukudome K, Yoon JH, Ostrovskii A, Takikawa T, Han IS (2010) Seasonal volume transport variation in the Tsushima Warm Current through the Tsushima Straits from 10 Years of ADCP Observations. J Oceanogr 66:539–551

Gommenginger C, Thibaut P, Fenoglio-Marc L, Quartly G, Deng X, Gomez-Enri J, Chanellor P, Gao Y (2011) Retracking altimeter waveforms near the coasts. In: Vingudeli S et al (eds) Coastal altimetry. Springer, Berlin, pp 61–100. https://doi.org/10.1007/978-3-642-12796-0_4. ISBN 978-3-642-12795-3

Hirose N, Fukudome K (2006) Monitoring the Tsushima Warm Current improves seasonal prediction of the regional snowfall. Sci Online Lett Atmos 2:61–63

Ichikawa K, Morimoto A, Fukudome K, Yoon JH Yoshikawa Y (2013) Coastal sea surface dynamic height monitoring by GPS mounted on a ferry boat. In: Ouwehand L (ed) Proceedings of 20 years of progress in radar altimetry, Venice, Italy. ESA SP-710. https://doi.org/10.5270/esa.sp-710.altimetry2012. ISBN 978-92-9221-274-2

Isobe A (1999) On the origin of the Tsushima Warm Current and its seasonality. Cont Shelf Res 19:117–133

Kashima M, Imawaki S, Umatani S, Uchida H, Hashibe Y, Ichikawa H, Fukasawa M (2003) Geostrophy in the intermediate and deep layers of the Kuroshio and its recirculation regions south of Japan. J Oceanogr 59:291–301

Morimoto A (2009) Evaluation of tidal error in altimetry data in the Asian Marginal Seas. J Oceanogr 65:477–485

Passaro M, Cipollini P, Vignudelli S, Quartly GD, Snaith HM (2014) ALES: a multi-mission adaptive subwaveform retracker for coastal and open ocean altimetry. Rem Sens Env 145: 173–189

Passaro M, Salvatore D, Graham QD, Snaith HM, Beneviste J, Cipollini P, Lucas B (2016) Cross-calibrating ALES Envisat and CryoSat-2 Delay-Doppler: a coastal altimetry study in the Indonesian Seas. Adv Space Res 58:289–303. https://doi.org/10.1016/j.asr.2016.04.011

Pavlis NK, Holmes SA, Kenyon SC, Factor JK (2012) The development and evaluation of the Earth Gravitational Model 2008 (EGM2008). J Geophys Res 117:B04406. https://doi.org/10.1029/2011JB008916

Pullen S, Rife J (2009) Differential GNSS: accuracy and integrity. In: Gleason S, Gebre-Egziabher D (eds) GNSS applications and methods. Artech House, Boston, 508 p. ISBN 978-1-59693-329-3

Takikawa T, Yoon JH (2005) Volume transport through the Tsushima Straits estimated from sea level difference. J Oceanogr 61:699–708

Takikawa T, Yoon JH, Cho KD (2005) The Tsushima Warm Current through Tsushima Straits estimated from Ferryboat ADCP Data. J Phys Oceanogr 35:1154–1168. https://doi.org/10.1175/JPO2742.1

Tokeshi R, Ichikawa K, Fujii S, Sato K, Kojima S (2007) Estimating the geostrophic velocity obtained by HF radar observations in the upstream area of the Kuroshio. J Oceanogr 63: 720–771

Vincent P, Steunou N, Caubet E, Phalippou L, Rey L, Thouvenot E, Verron J (2006) AltiKa: a Ka-band Altimetry Payload and system for operational altimetry during the GMES period. Sensors (Basel) 6:208–234

Wang XF, Ichikawa K (2017) Coastal waveform retracking for Jason-2 Altimeter data based on along-track echograms around the Tsushima Islands in Japan. Rem Sens 9:762. https://doi.org/10.3390/rs9070762

Yoshikawa Y, Masuda A (2009) Seasonal variations in the speed factor and deflection angle of the wind-driven surface flow in the Tsushima Strait. J Geophys Res 114:C12022. https://doi.org/10.1029/2009jc005632

Yoshikawa Y, Masuda A, Marubayashi K, Ishibashi M, Okuno A (2006) On the accuracy of HF radar measurement in the Tsushima Strait. J Geophys Res 111:C04009. https://doi.org/10.1029/2005jc003232

Yoshikawa Y, Matsuno T, Marubayashi K, Fukudome K (2007) A surface velocity spiral observed with ADCP and HF radar in the Tsushima Strait. J Geophys Res 112:C06022. https://doi.org/10.1029/2006jc003625

Yoshikawa Y, Masuda A, Marubayashi K, Ishibashi M (2010) Seasonal variations of the surface currents in the Tsushima Strait. J Oceanogr 66:223–232. https://doi.org/10.1007/s10872-010-0020-1

Remote Sensing of the Kuroshio Current System

Kaoru Ichikawa

Abstract Variations of the Kuroshio Current System, which is the western boundary current of the North Pacific, influence both large-scale climate changes and regional waters. Satellite altimeters have been widely used to monitor various aspects of the Kuroshio, such as its position and volume transport. Although both meanders of the Kuroshio axis and isolated mesoscale eddies result in similar mesoscale temporal anomaly, they can be distinguished when the mean sea surface dynamic height is retrieved, which cannot be obtained by altimetry data alone. This enables us to describe the shedding and merging of offshore mesoscale eddies and the development of the stable large meander south of Japan. Spatial and temporal resolutions of satellite altimetry data are, however, not sufficient to describe fast-moving small-scale variations of the Kuroshio itself. The high-resolution surface velocity field around the Kuroshio can be directly obtained by high-frequency (HF) ocean radars, which can provide thorough descriptions of the Kuroshio in terms of variations of the maximum speed of the Kuroshio and the displacement of its axis position. In addition, short-term variations of the Kuroshio, such as responses to extraordinary winds of a typhoon, can be studied quantitatively.

Keywords Kuroshio current system · Satellite altimeter · HF ocean radar
Sea surface dynamic height · Surface drifters · Kuroshio extension
Mesoscale eddies · Kuroshio axis · Volume transport · Meander

1 Introduction

The Kuroshio Current System consists of the upstream Kuroshio east of Taiwan, the Kuroshio along the shelf break of the East China Sea, the Kuroshio south of Japan between the Tokara Strait and the Izu-Ogasawara Ridge, and the Kuroshio Extension (KE) east of Japan (Fig. 1). As the western boundary current of the

K. Ichikawa (✉)
Research Institute for Applied Mechanics, Kyushu University, Kasuga, Japan
e-mail: ichikawa@riam.kyushu-u.ac.jp

V. Barale and M. Gade (eds.), *Remote Sensing of the Asian Seas*,
https://doi.org/10.1007/978-3-319-94067-0_11

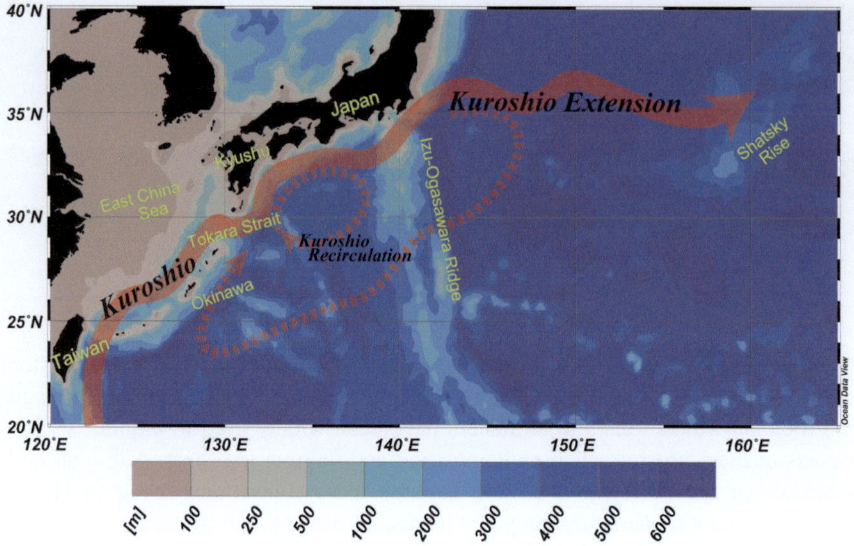

Fig. 1 Schematic map of the Kuroshio Current System with bathymetry contours

North Pacific, the Kuroshio Current System is one of the most energetic currents in the world and carries huge volumes of warm water from lower to higher latitudes, and thus is important not only to adjacent marginal seas (Guo et al. 2006) but also to the global climate system. Therefore, variations of the Kuroshio Current System are routinely monitored by several regular cruises by research institutes, but these cruises cannot cover the entire length of the Kuroshio Current System.

Since the water carried by the Kuroshio is significantly warmer than the surrounding water, the high contrast of the sea surface temperature (SST), or the Kuroshio SST front, provides remotely sensible estimates of the position of the Kuroshio, although the SST front does not exactly follow the axis of the Kuroshio due to complex modifications by air-sea interaction and mixing processes. In order to estimate the volume and heat transports of the Kuroshio Current System, however, its speed and width, which are not directly observable based on the SST field, must be known. Meanwhile, satellite altimeters observe temporal variations of the sea surface dynamic height (SSDH), which is directly related to the geostrophic velocity at the sea surface. Therefore, satellite altimeters would be one of the best tools for monitoring variations of the Kuroshio transport and positions.

Nevertheless, there are several limitations in the use of satellite altimetry, especially in monitoring the Kuroshio. First, although the temporal anomaly component of the SSDH is well observed by satellite altimeters, the temporal mean SSDH is not available due to contamination of the uncertainty of the present geoid models. The geoid models have been drastically improved by satellite geodetic missions, such as the Gravity Recovery and Climate Experiment (GRACE) and Gravity Field and Steady-State Ocean Circulation Explorer (GOCE), as well as other geodetic missions using satellite altimeters, so that a large-scale geoid is well established. However, the accuracy of geoid models remains insufficient for spatial scales smaller than a few hundred kilometers. Since the width of the Kuroshio is as narrow as these scales, the missing mean SSDH of the Kuroshio still needs to be estimated separately.

Another limitation of the satellite altimetry is low resolution in both time and space. The SSDH is observed only at a point beneath a satellite, so that the same location will not be observed until the satellite returns after completing one cycle. Moreover, no observations are available between two adjacent sub-satellite tracks. Typical values are 10 days and 330 km or 35 days and 75 km. These sparse and intermittent samplings can be improved by the use of multiple satellite altimeters and are not a serious problem for typical offshore mesoscale eddies that slowly propagate westward because their temporal and spatial scales are much larger. However, some meanders of the Kuroshio path that are small and quickly advect downstream could be missed, even if several altimeters are used simultaneously. As described in Sect. 2, several processing schemes have been proposed to overcome these limitations.

A more direct solution for overcoming these limitations is to use high-frequency (HF) ocean radars. Although the spatial coverage of HF radar is limited compared to satellite observations, HF radar systems can provide surface velocity fields with high spatial and temporal resolutions. The use of HF radar to monitor the Kuroshio is described in Sect. 3, followed by a discussion and summary in Sect. 4.

2 Observation Using Satellite Altimeters

In this section, methods are introduced to relax the above-described limitations; namely, the missing temporal mean SSDH (Sect. 2.1) and sparse and intermittent samplings (Sect. 2.2). Since these are fundamental problems in SSDH observations by nadir-looking satellite altimeters, it is difficult to solve these problems using altimeters alone. Since these limitations are inherent only in satellite altimetry, observations by other than satellite altimeters may not be subject to these limitations. Therefore, the combination of altimetry observations with other observations is of key interest in this section.

2.1 Retrieval of the Missing Mean SSDH

Note first that the retrieval of the mean SSDH is area dependent. In the KE, where the movement of the axis often exceeds its width, the magnitudes of the SSDH anomaly (SSDHA) and the mean SSDH become comparable. Actually, the SSDHA is so large in this area that the mean SSDH can be estimated from the SSDHA alone, assuming the horizontal Gaussian structure of the current around the axis and large displacement of the axis (Qiu 1995). In the Kuroshio south of Japan and the along-the-shelf break of the East China Sea, the axis movement is restricted by the northern topographic boundary so that the anomaly results in an SSDH that is smaller than the mean SSDH. Therefore, in order to obtain the total SSDH, namely, the anomaly plus mean SSDH, a more precise mean SSDH is necessary in the Kuroshio, rather than in the KE.

A naïve but effective method is to use the climatological mean SSDH based on long-term hydrographic observations as the mean SSDH (Ichikawa and Imawaki 1994; Qiu and Chen 2005). Figure 2 shows time series of the total SSDH field estimated from a Geosat altimeter using this method. In the figure, a cyclonic ring at 33°N pinched off in June 1987 from a southward meander of the KE at 145°E and then moved westward to coalesce in September 1987 at 141°E with the large meander of the Kuroshio south of Japan, and eventually advected downstream. The figure also shows the formation of the stable large meander of the Kuroshio south of Japan in December 1986, although the mean SSDH there may be less reliable than in the KE.

Note that both meanders and rings result in similar isolated closed contours in the SSDHA field. As such, the total SSDH is especially useful for distinguishing these features. Rings detached from jets and mesoscale eddies in open ocean, often identifiable by isolated closed contours or by Okubo-Weiss parameters (Itoh and Yasuda 2010), propagate westward due to the nature of the Rossby waves (Chelton et al. 2007).

Longer time series of the KE path (Fig. 3) determined by a similar method clearly exhibit the highly variable nature of the KE, as described earlier. Moreover, inter-annual variations of the location of the KE axis and its stability are evident in the figure. The location of the KE axis tends to move southward during active meandering periods (e.g., 1995–1997) as compared to that during inactive periods (e.g., 2002–2004) (Qiu and Chen 2005; Bessières et al. 2013). During the inactive periods, especially in the upstream region west of 153°E (i.e., west of the Shatsky Rise), the KE axis takes almost the same path, which is approximately identified by the stable path length value. Moreover, the eddy kinetic energy calculated from the SSDHA field is large (small) in the active (inactive) periods, whereas the KE strength, which is determined by the SSDH difference across the KE axis, tends to rather decrease in the active periods (Qiu and Chen 2005). These results are strongly subject to the retrieval of the missing mean SSDH.

More precise estimation of the mean SSDH can be accomplished using contemporary data observed during the same period as the altimetry data, instead of the climatological long-term mean. When the total SSDH or geostrophic velocity is observed, the temporal mean component can be extracted by subtracting the anomaly component observed by the altimeter. As the number of such estimations of the mean component increases, the statistically averaged temporal mean SSDH becomes more reliable. Therefore, hydrographic observations (Imawaki et al. 2001; Andres et al. 2008) or surface velocity observations by current meters (Ichikawa et al. 2004) along repeated cruise lines are often used in this method, although their spatial coverage becomes very limited. The precise retrieval of the mean SSDH

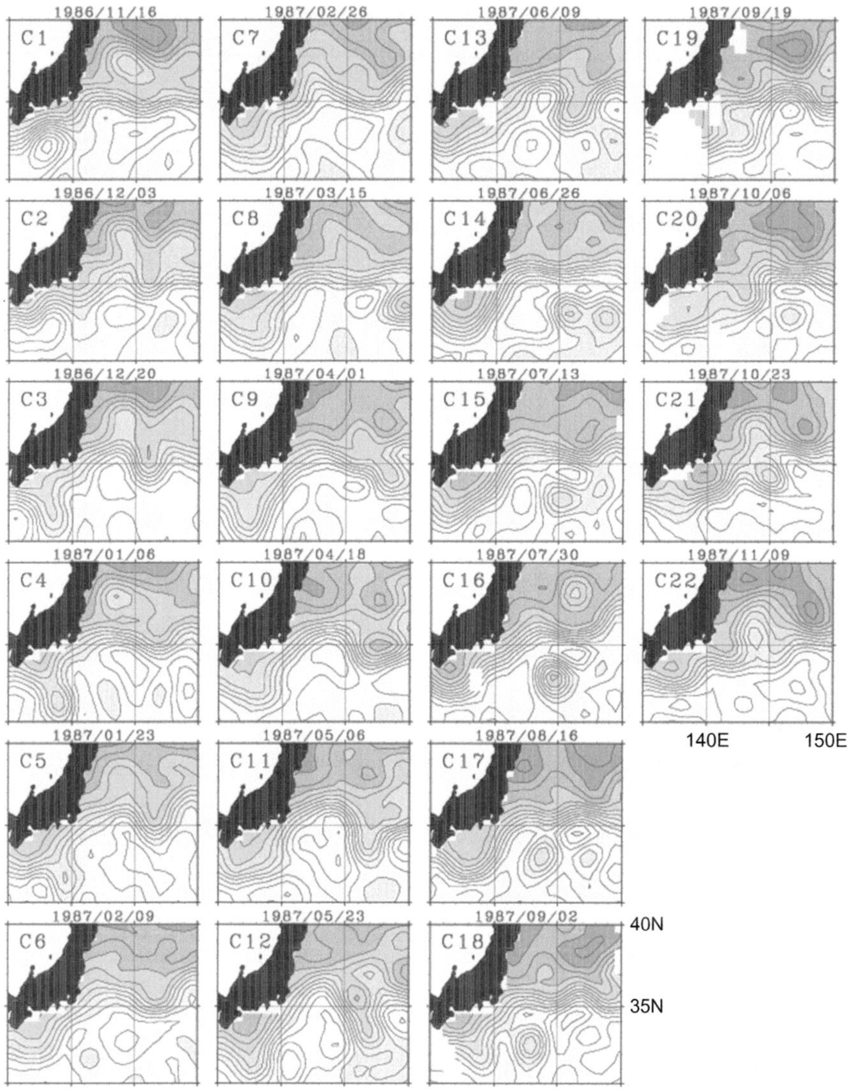

Fig. 2 The 17-day averaged total SSDH field from November 1986 to November 1987 determined using the Geosat SSDHA and the climatological mean SSDH (after Ichikawa and Imawaki 1994). The contour intervals are 0.1 m, and the lower SSDH is shaded heavily. Formation of the stable large meander south of Japan (C3–C5), a cyclonic ring that pinched off from a southward meander of the KE (C13–C14), and the coalescence of the westward-propagating ring toward the Kuroshio large meander (C18–C20) are observed

enables us not only to estimate the quantitative Kuroshio volume transport but also to distinguish the Kuroshio from offshore mesoscale eddies and the regional recirculation (Imawaki et al. 2001).

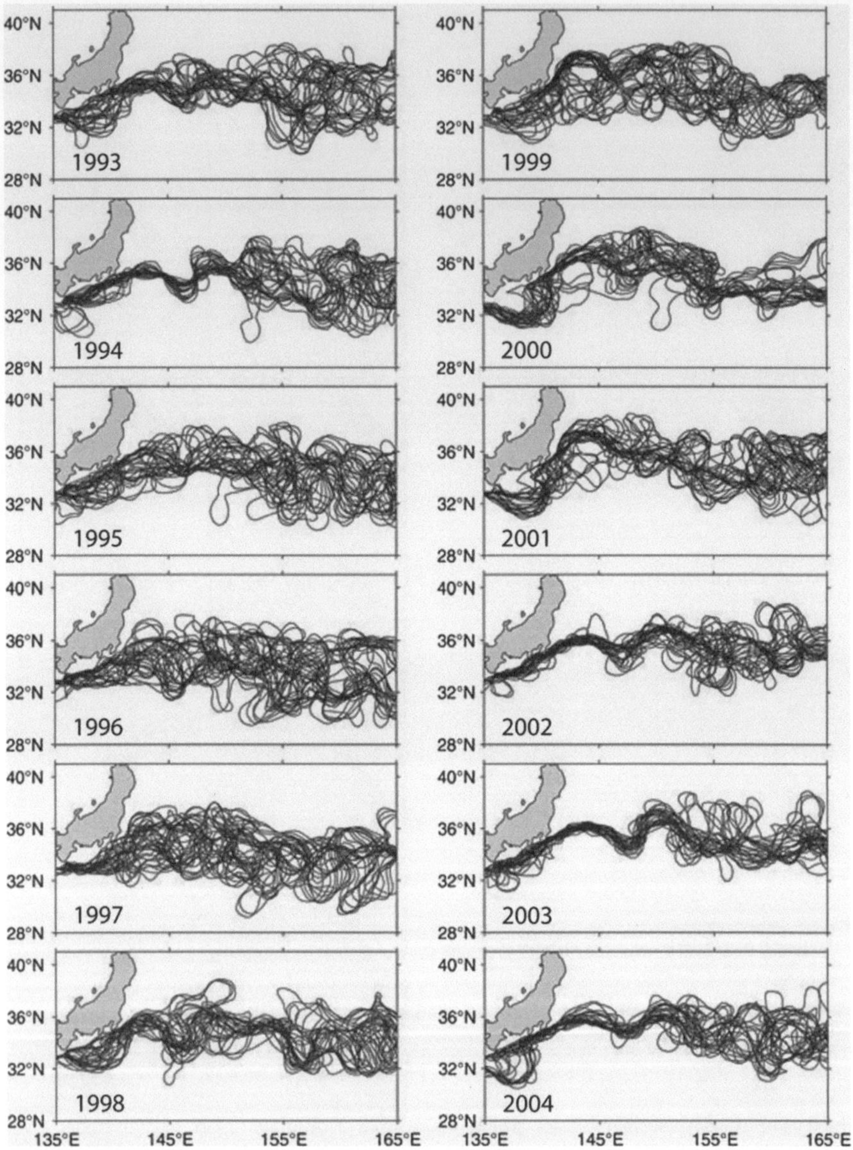

Fig. 3 Yearly paths of the Kuroshio Current System plotted every 14 days (after Qiu and Chen 2005)

The trajectories of a large number of surface drifters (Uchida and Imawaki 2003; Maximenko et al. 2009; Ambe et al. 2004) are also often used in this method, which can satisfy both the requirements of the spatial coverage and the number of observations. Even in the area south of Japan, where the accurate mean SSDH is required, the Kuroshio can be so well defined by this method that the position, speed, and width of the Kuroshio can be statistically discussed (Fig. 4). As explained at the beginning of Sect. 2.1, the position of the Kuroshio axis is quite variable east of the Izu-Ogasawara Ridge (140°E), whereas it is quite stable west of the Izu-Ogasawara Ridge (Fig. 4a). Along 140°E (line G in Fig. 4a), three peaks are

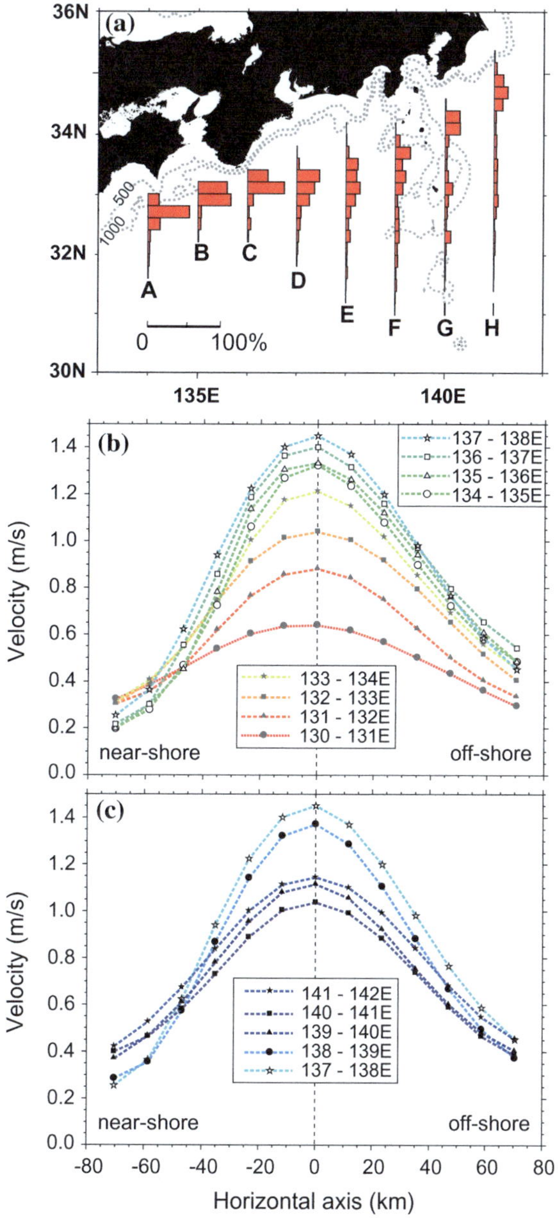

Fig. 4 **a** Histogram of the Kuroshio axis position with bathymetry contours. **b** Horizontal distributions of current speed near the Kuroshio axis west of 138°E, and **c** these east of 137°E. The statistics are based on combined data from satellite altimeters and surface drifting buoys from October 1992 to December 2000 (after Ambe et al. 2004). Bathymetry contours are shown in the upper panel

found: north of 34°N, at 33°N, and north of 32°N. These peaks suggest that the Kuroshio selectively flows only at these locations, i.e., gaps of the Izu-Ogasawara Ridge. In addition, the Kuroshio west of the Izu-Ogasawara Ridge has a significant asymmetrical current structure around the Kuroshio axis (Fig. 4b), suggesting the frictional influence of the topographic boundary. Note also that the accurate description of the Kuroshio provides a better reference for numerical models (e.g., Endoh et al. 2011).

2.2 Independent Kuroshio Indices with High Resolutions

Note that the variations of the Kuroshio Current System described in the previous subsection are based on the SSDHA data obtained by satellite altimeters. Therefore, the spatial and temporal resolutions are limited by sparse and intermittent sampling configurations. Moreover, since proper observations of satellite altimeters are not available on land or in shallow seas, the SSDH close to these boundary areas tends to be unreliable when spatially interpolated. Variations of the Kuroshio Current System itself are, therefore, less reliable in the altimetry observations, especially for the Kuroshio west of the Izu-Ogasawara Ridge, as compared with the observations of offshore variations.

On the other hand, if we use some other data independent of the intermittent observations of the satellite altimeters, we could introduce quantitative and more reliable indices of the Kuroshio variations themselves. Then, the variations of the Kuroshio defined by these indices could be interpreted in relation to some remote variations, such as mesoscale eddies off the Kuroshio axis or far offshore areas, which can be precisely observed by altimeters, even with sparse and intermittent observations.

Tide gauge data are often used for these indices, which are well distributed along the Kuroshio path in the Japanese southern coast and have sufficient temporal resolution (Ichikawa 2001; Zhang and Ichikawa 2005). At the Tokara Strait between Kyushu and Okinawa islands around 130°E, tide gauge stations are available on both the northern and southern sides of the Kuroshio, so that the Kuroshio volume transport and the position of the Kuroshio have been estimated using these tide gauge data (Kawabe 1995). The temporally lagged cross correlation between the Kuroshio volume transport determined by the tide gauge data in the Tokara Strait and the surrounding SSDHA observed by satellite altimeters (Fig. 5) clearly exhibits sequential interactions between offshore mesoscale eddies and the Kuroshio. Anticyclonic (cyclonic) eddies represented by positive (negative) SSDHAs are known to propagate westward (Chelton et al. 2007). Anticyclonic (cyclonic) eddies at 23°N eventually merged into the offshore side of the Kuroshio south of the Okinawa islands around 127°E and advected downstream along the Kuroshio west of the Okinawa islands to reach the southern end of the Tokara Strait within 20 days, resulting in an increase (decrease) of the volume transport in the Tokara Strait. These sequential interactions can be traced in Fig. 5 following the

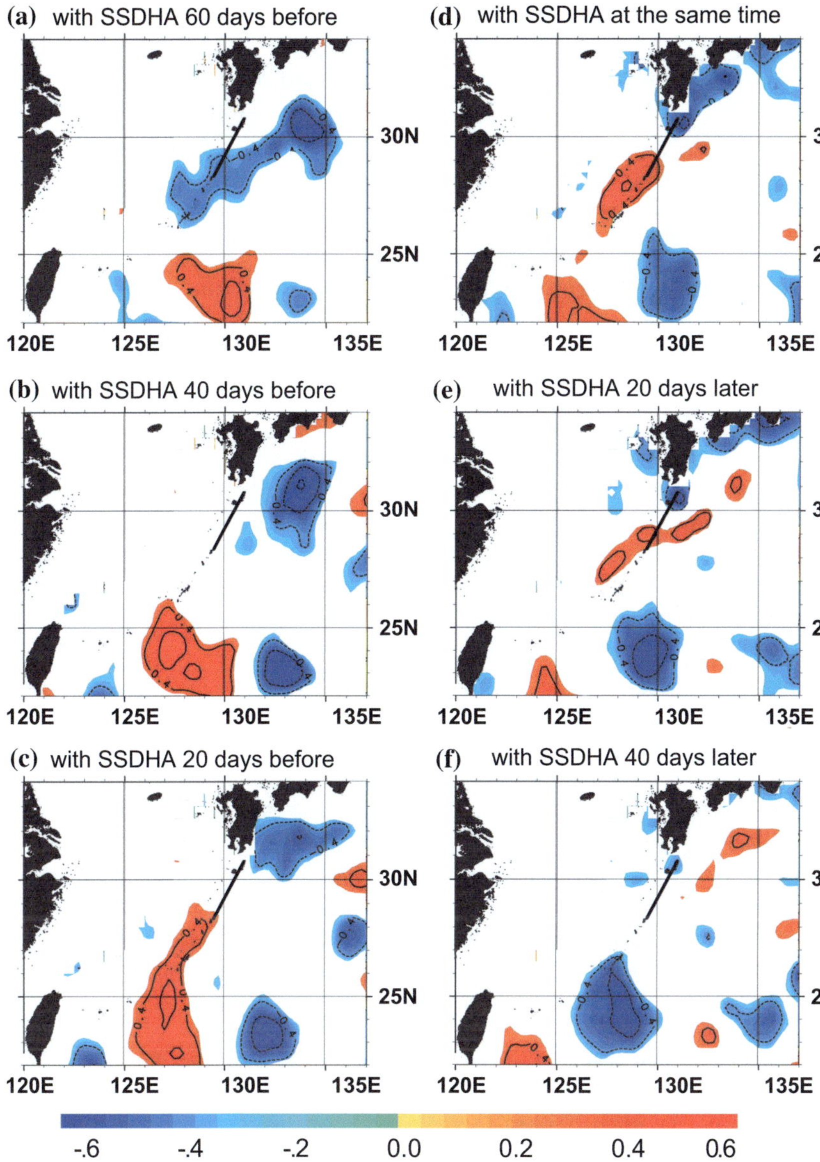

Fig. 5 Spatial distribution of cross correlation coefficients between the Kuroshio volume transport estimated by tide gauges in the Tokara Strait (bold lines around 130°E) and the SSDHA **a** 60 days before, **b** 40 days before, **c** 20 days before, **d** on the same day, **e** 20 days after, and **f** 40 days after (following Ichikawa 2001). The contour intervals are 0.2, and positive (negative) values are shown in red (blue). Statistically insignificant values at the 95% confidence level are masked

temporal lags, as the westward shift of a positively correlated area around 23°N (Fig. 5a, b), its elongated shape west of the Okinawa islands (Fig. 5c, d), and its stretched shape along the offshore side of the Kuroshio downstream of the Tokara Strait (Fig. 5e, f). Recall that the positive correlation area indicates that positive (negative) SSDHA tends to be found there when the volume transport in the Tokara Strait is increased (decreased).

3 Observations Using HF Ocean Radar Systems

Although the areal coverage is limited to a circular area of approximately 200 km in radius, an HF ocean radar system alone can provide the total surface velocity field with quite high spatial and temporal resolutions. This means that indices of the Kuroshio variations as in Sect. 2.2 can be provided by an HF ocean radar system (Ichikawa et al. 2008). Examples of the daily-averaged surface velocity field in the upstream Kuroshio area north of Yonaguni and Ishigaki Islands, which are located northeast of Taiwan (Fig. 6a, b), exhibit clear variations in speed and position of the Kuroshio axis. The Kuroshio axis moved northward in May 2003 (Fig. 6a) with the lower maximum speed and moved southward with increased maximum speed in September 2003 (Fig. 6b).

3.1 Combination with Satellite Altimeters

Positive and negative SSDHAs to the north and south of the Kuroshio axis around 123°E (Fig. 6c, d) are consistent with the decrease/increase of the Kuroshio speed, assuming geostrophic balance. The maximum sea surface speed along 123.5°E, where the largest displacement of the Kuroshio axis is observed in Fig. 6a, b, is taken, and the mesoscale temporal variations of these Kuroshio speed indices are then extracted by a band-pass filter for 10–240 days for statistical comparison with the surrounding SSDHA.

The cross correlation between the mesoscale variability of the Kuroshio speed and the surrounding SSDHA is shown in Fig. 7. The figure shows the downstream advection of the area with significant positive correlation, including its elongation east of Taiwan. This suggests that the positive (negative) SSDHA, i.e., anticyclonic (cyclonic) eddies, at approximately 21°N eventually produces increasing (decreasing) Kuroshio speed at 123.5°E after approximately 40 days. In addition, a small area with negative correlation is found advected downstream on the coastal side of the Kuroshio axis (i.e., on the continental shelf of the East China Sea), although the estimation of the SSDHA is less reliable near the boundary. Unfortunately, the Kuroshio variations on the continental shelf were not well described by the HF ocean radar, because the surface velocity field over the shelf was not fully covered by the radar data used in the present study. Remember that

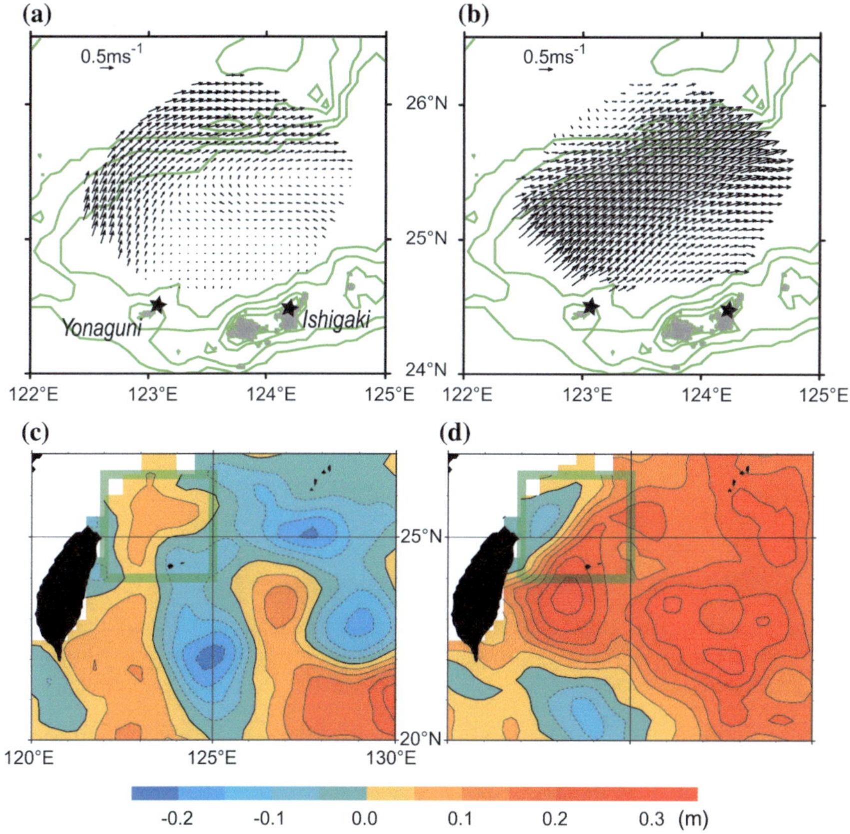

Fig. 6 Daily-averaged surface velocity field on **a** 1 June 2003 and **b** 13 September 2003, and the SSDHA field **c** on 30 May 2003 and **d** 7 September 2003 (following Ichikawa et al. 2008). The locations of antennas (star symbols) and bathymetry contours are shown in the upper panels. The green boxes in the lower panels represent the areas shown in the upper panels. The contour intervals of the SSDHA are 0.05 m, and positive (negative) SSDHA is shown in red (blue)

areal coverage is an inherent limitation of the HF ocean radar. Nevertheless, variations of the upstream Kuroshio have been found related to variations of the downstream Kuroshio path (Kashima et al. 2009; Usui et al. 2013).

3.2 *Surface Velocity Field with Higher Resolutions*

In Fig. 7, we focus on only mesoscale variations of the Kuroshio speed index in order to match the offshore SSDHA obtained by satellite altimeters. However, the HF ocean radar can observe much shorter-period variations (Morimoto et al. 2009; Takahashi et al. 2009). For example, a large displacement of a band of strong

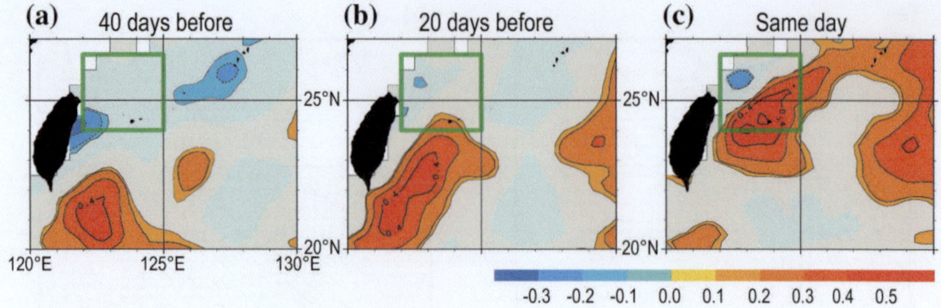

Fig. 7 Spatial distribution of cross correlation coefficients between the mesoscale (10–240 days) variations of the Kuroshio speed at 123.5°E determined using the HF radar and the SSDHA **a** 40 days before, **b** 20 days before, and **c** on the same day (following Ichikawa et al. 2008). The analysis period is from July 2001 to January 2005. The contour intervals are 0.1, and statistically insignificant values at the 90% confidence level are masked. The green boxes are the same as in Fig. 6

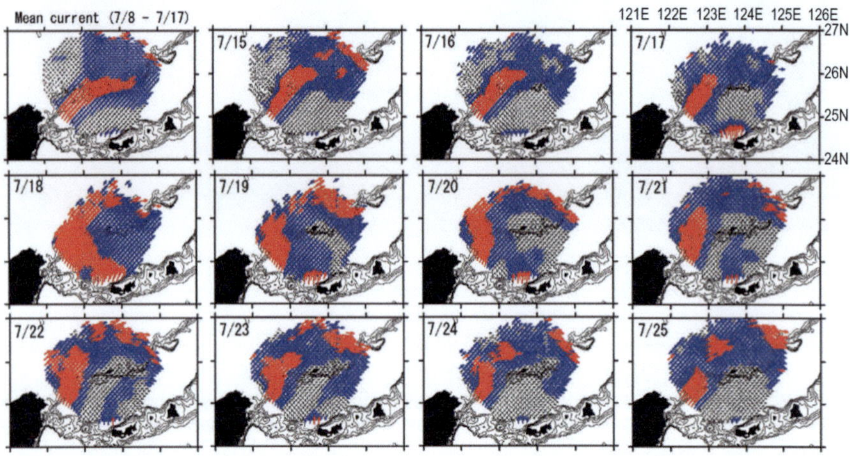

Fig. 8 Daily averaged surface velocity field observed by HF ocean radars before and after the passage of a typhoon *Haitang* at the southwest corner of the panel on 18 July 2005 (following Morimoto et al. 2009). Vectors having speeds that exceed 1 (0.5) ms^{-1} are shown in red (blue). The ten-day averaged velocity field before passage is also shown in the panel in the top-left corner

surface currents onto the continental shelf was recorded during the passage of a typhoon in 2005 in the upstream Kuroshio area (Fig. 8), which is poorly described by intermittent satellite altimetry observations.

Some part of the surface current is directly driven by wind forcing (Tokeshi et al. 2007), which flows independently from the Kuroshio beneath, but the strong current shown in Fig. 8 would correspond to the Kuroshio in the surface layer because it took several days to recover the original position, even though no strong wind remained. The density structure on the continental shelf was altered by strong vertical mixing induced by typhoon winds, so that the Kuroshio was able to intrude into the shelf, and it took some time to recover the original density state by re-stratification (Morimoto et al. 2009).

4 Summary and Discussion

Satellite altimetry can provide wide coverage of the surface geostrophic velocity field and so should be useful for monitoring variations, such as path variations and volume transport, in the Kuroshio Current System. Unfortunately, the temporal mean component is not available due to contamination of the present geoid model with spatial scales smaller than a few hundred kilometers, but several methods have been proposed to retrieve the missing mean using data other than satellite altimeter data. The mean component is important in order to distinguish between an offshore mesoscale eddy and a meander of the Kuroshio path, or between increasing/decreasing of the Kuroshio speed and displacement of the axis, all of which result in similar SSDHAs.

Sparse and intermittent samplings of satellite altimeters are also limitations of monitoring variations of the Kuroshio, where small and fast-moving meanders are often found. By incorporating indices of the Kuroshio variations defined by other observations with more dense samplings, satellite altimetry can provide a wide view of remotely related variations outside of the Kuroshio. Actually, interaction with offshore mesoscale eddies and downstream advection along the offshore Kuroshio edge are commonly observed both in the upstream Kuroshio and the Kuroshio south of Japan. In the near future, sampling of the SSDH may be drastically improved by the use of wide-swath altimeters, which can observe a map of the sea surface height field at off-nadir angles by the interferometry technique. The Surface Water and Ocean Topography (SWOT) mission by the National Aeronautics and Space Administration (NASA) of the United States and Le Centre national d'Études Spatiales (CNES) of France has been approved (Fu et al. 2012) to be launched in 2020, and the Coastal and Ocean Measurement mission with Precise and Innovative Radar Altimeter (COMPIRA) mission by the Japanese Aerospace eXploration Agency (JAXA) is being planned (Nakamura et al. 2012).

Using high-resolution observations by the HF ocean radar, the Kuroshio variations are well described, without these inherent limitations of satellite altimeters. Not only the mesoscale displacements of the Kuroshio axis but also small and fast-moving meanders of the Kuroshio path can be described by the HF ocean radar system, together with its maximum speed variations (Takahashi et al. 2009). Moreover, even much shorter changes caused by the passage of a typhoon can be traced as a complete time series of the surface velocity field (Morimoto et al. 2009).

However, the area coverage is limited to an approximately 200-km radius from the antenna due to reflection by the ionosphere, so that the Kuroshio may not be fully covered. The best way to overcome this limitation is to place radar sites on both the coastal and offshore sides of the Kuroshio. A collaborative study between

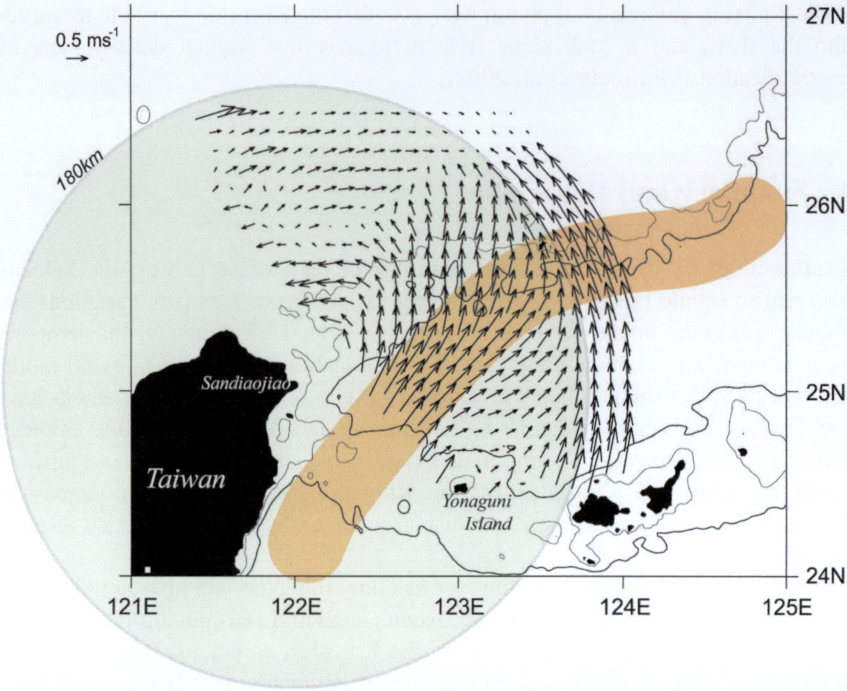

Fig. 9 Surface velocity averaged over 15–17 June 2012 determined from two radars at Yonaguni and Sandiaojiao (after Ichikawa et al. 2013). The schematic position of the Kuroshio and the bottom topography are illustrated. A circle having a radius of 180 km centered at the Sandiaojiao station is also shown in the panel

Japan and Taiwan has been started in order to fully cover the Kuroshio from HF radar sites on the coastal and offshore sides (Ichikawa et al. 2013). As shown by preliminary three-day observations conducted in 2012 (Fig. 9), the upstream Kuroshio near the continental shelf break, including the small area with significant negative correlation shown in Fig. 7, is fully covered by the radar sites. In addition to these promising results, however, Fig. 9 also includes unrealistic or discontinuous velocity fields, such as the strong northward current around 124°E crossing the shelf break. As some of these unrealistic currents are located just outside of the 180-km-radius circle centered at Sandiaojiao station, Taiwan, further collaborative studies are needed, including the investigation of quality control of the data.

Acknowledgements The present study was supported in part by Japan Society for the Promotion of Science (JSPS) KAKENHI Grant Number JPH05821.

References

Ambe D, Imawaki S, Uchida S, Ichikawa K (2004) Estimating the Kuroshio Axis South of japan using combination of satellite altimetry and drifting buoys. J Oceanogr 60:375–382. https://doi.org/10.1023/B:JOCE.0000038343.3148.fe

Andres M, Park JH, Wimbush M, Zhu XH, Chang KI, Ichikawa H (2008) Study of the Kuroshio/Ryukyu current system based on satellite-Altimeter and in situ measurements. J Oceanogr 64:937–950

Bessières L, Rio MH, Dufau C, Boone C, Pujol MI (2013) Ocean state indicators from MyOcean altimeter products. Ocean Sci 9:545–560. https://doi.org/10.5194/os-9-545-2013

Chelton DB, Schlax MG, Samelson RM, de Szoeke RA (2007) Global observations of large oceanic eddies. Geophys Res Lett 34:LL15606. https://doi.org/10.1029/2007gl030812

Endoh T, Tsujino H, Hibiya T (2011) The effect of Koshu Seamount on the formation of the Kuroshio Large Meander South of Japan. J Phys Oceanogr 41:1624–1629. https://doi.org/10.1175/JPO-D-11-074.1

Fu LL, Alsdorf D, Morrow R, Rodriguez E, Mognard N (2012) SWOT: the surface water and ocean topography mission—wide-swath altimetric measurement of water elevation on earth. JPL Publication 12-05, 228pp

Guo XY, Miyazawa Y, Yamagata T (2006) The Kuroshio onshore intrusion along the Shelf Break of the East China Sea: the origin of the Tsushima Warm Current. J Phys Oceanogr 36:2205–2231

Ichikawa K (2001) Variation of the Kuroshio in the Tokara Strait induced by meso-scale eddies. J Oceanogr 57:55–68. https://doi.org/10.1023/A:1011174720390

Ichikawa K, Imawaki S (1994) Life history of a cyclonic ring detached from the Kuroshio Extension as seen by the geosat altimeter. J Geophys Res 99C8:15953–15966

Ichikawa K, Gohda N, Arai M, Kaneko A (2004) Monigoring surface velocity from repeated ADCP observations and satellite altimetry. J Oceanogr 60:365–374

Ichikawa K, Tokeshi R, Kashima M, Sato K, Matsuoka T, Kojima S, Fujii S (2008) Kuroshio variations in the upstream region as seen by HF radar and satellite altimetry data. Int J Remote Sens 29:6417–6426. https://doi.org/10.1080/01431160802175454

Ichikawa K, Yang WC, Morimoto A, Yoshikawa Y, Sugitani S, Chiang WS, Lai JW, Liang EY, Liu CT, Lee CW, Yufu K, Kyushima M, Fujii S, Senju T, Ide Y (2013) Preliminary results of the Japan-Taiwan joint survey on combining ocean radar data in the Kurosho Upstream Region. Ocean Sci J 48:141–148. https://doi.org/10.1007/s12601-013-0011-4

Imawaki S, Uchida H, Ichikawa H, Fukasawa M, Umatani S, ASUKA Group (2001) Satellite altimeter monitoring the Kuroshio transport south of Japan. Geophys Res Lett 28:17–20

Ito S, Yasuda I (2010) Characteristics of mesoscale eddies in the Kuroshio-Oyashio Extension Region detected from the distribution of the sea surface height anomaly. J Phys Oceanogr 40:1018–1034

Kashima M, Ito S, Ichikawa K, Imawaki S, Umatani S, Uchida H, Setou T (2009) Quasiperiodic small meanders of the Kuroshio off Cape Ashizuri and their inter-annual modulation caused by quasiperiodic arrivals of mesoscale eddies. J Oceanogr 69:73–80. https://doi.org/10.1007/s10872-09-007-7

Kawabe M (1995) Variations of current path, velocity and volume transport of the Kuroshio in relation with the large meander. J Phys Oceanogr 25:3103–3117

Maximenko N, Niiler P, Rio MH, Melnichenko O, Centurion L, Chambers D, Zlotnicki V, Galpern B (2009) Mean dynamic topography of the ocean derived from satellite and drifting Buoy Data using three different techniques. J Atmosph Ocean Tech 26:1910–1919. https://doi.org/10.1175/2009JTECHO672.1

Morimoto A, Kojima S, Jan S, Takahashi D (2009) Movement of the Kuroshio axis to the northeast shelf of Taiwan during typhoon events. Estuarine Coastal Shelf Sci 82:547–552

Nakamura R, Uematsu A, Nakajima Y Yajima Y, Sato R, COMPIRA Team (2012) X-band interferometric SAR sensor for the Japanese altimetry mission COMPIRA. In: Space aeronautical and navigational electonics (SANE2012-85), pp 163–168

Qiu B (1995) Variability and engergetics of the Kuroshio Extension and its recirculation gyre from the first two-year TOPEX data. J Phys Oceanogr 25:1827–1842

Qiu B, Chen S (2005) Variability of the Kuroshio Extension Jet, recirculation gyre, and mesoscale eddies on decadal time scales. J Phys Oceanogr 35:2090–2103. https://doi.org/10.1175/JPO2807.1

Takahashi D, Guo XY, Morimoto A, Kojima S (2009) Biweekly periodic variation of the Kuroshio axis northeast of Taiwan as revealed by ocean high-frequency radar. Cont Shlf Res 29:1896–1907

Tokeshi R, Ichikawa K, Fujii S, Sato K, Kojima S (2007) Estimating the geostrophic velocity obtained by HF radar observations in the upstream area of the Kuroshio. J Oceanogr 63:720–771

Uchida H, Imawaki S (2003) Eulerian mean surface velocity field derived by combining drifter and satellite altimeter data. Geophys Res Lett 30. https://doi.org/10.1029/2002gl016445

Usui N, Tsujino H, Nakano H, Matsumoto S (2013) Long-term variability of the Kuroshio path south of Japan. J Oceanogr 69:637–670. https://doi.org/10.1007/s10872-013-0197-1

Zhang ZZ, Ichikawa K (2005) Influence of the Kuroshio fluctuations on sea level variations along the south coast of Japan. J Oceanogr 61:979–985

Remote Sensing of Korean Tidal Flats

Joo-Hyung Ryu and Yoon-Kyung Lee

Abstract In Korea, coastal reclamation was constructed as a project to create vast rice fields while simultaneously serving a symbolic role demonstrating the country's capacity for reconstruction that began in 1953. Korean tidal flats are increasingly being changed by various construction projects. This chapter reviews the remote sensing techniques used to monitor Korean tidal flats and suggests appropriate techniques for meeting monitoring targets for the effective management of tidal flats. Three different monitoring targets were examined: topography, sedimentary facies, and bio facies. Waterline method and SAR interferometry have been used for generating the intertidal digital elevation model (DEM). Sedimentary facies of the tidal flats can be classified into the three categories mud, mixed, and sand, at a spatial resolution of 4 m. A potential map for macro benthos was generated with high accuracy based on the spatial variables such as exposure time and sedimentary facies map. The details of those data can be further enhanced by the use of an unmanned aerial vehicle (UAV) and new satellite system. Thematic maps based on remote sensing can help improve policy decisions from a management perspective. In this study, contents were constructed and summarized according to the research of Ryu et al. (Ocean Coast Manage 102:458–470, 2014).

Keywords Remote sensing · Tidal Flat · Topography · Sedimentary facies
Biofacies estimation

J.-H. Ryu (✉) · Y.-K. Lee
Korea Ocean Satellite Center, Korea Institute of Ocean Science & Technology, 385 Haeyang-ro, Yeongdo-gu, Busan Metropolitan City 49111, Republic of Korea
e-mail: jhryu@kiost.ac.kr

V. Barale and M. Gade (eds.), *Remote Sensing of the Asian Seas*,
https://doi.org/10.1007/978-3-319-94067-0_12

1 Introduction

Tidal flats, which are a mixture of seawater and fresh water environments, have economic, social, ecological and cultural value as habitats for a diverse range of living organisms, as coastal protection against storms, as buffer zones for seawater from land-based pollutants, as opportunities for formal and informal education and landscape diversity for watching birds, marine mammals etc. (Brooks et al. 2006; Kirwan and Murray 2007; Wilson et al. 2005). Furthermore, tidal flats are emerging area of interest due to the sequestration of carbon dioxide from the atmosphere and the potential for moderating climate change (Sondak and Chung 2015). However, tidal flats have been significantly restructured and damaged by human activities such as urbanized on a large scale and are threatened by a rise in sea level (Kirwan et al. 2010; Ryu et al. 2004).

Field surveys for monitoring aspects of tidal flats (e.g., topography, sedimentary facies and biofacies distribution) are restricted due to the short exposure time between tides, physical difficulties with the work (Ryu et al. 2002; Tyler et al. 1996). Moreover, field surveys are constrained because of limited points and lines introducing errors during interpolation (Tyler et al. 1996). It may not adequately represent changes in large tidal flats (Boak and Turner 2005). To overcome these limitations, remote sensing data combined with field surveys have been used as complementary data since the late 1980s (Cracknell 1999). The advantages of using remote sensing include the potential to perform time series analysis, the estimation of quantitative budget of sediments and the use of archived images for the monitoring of tidal flats where survey data are not available. Remote sensing techniques have been used to monitor tidal flats and attempt to identify the most appropriate methods for facilitating effective tidal flat management in an efficient and cost-effective manner. Thus, remote sensing techniques such as optical image, SAR (Synthetic Aperture Radar), and LiDAR (Light Detection And Ranging) have been used for accurate mapping of tidal flats (Adam et al. 2010; Heygster et al. 2010; van der Wal and Herman 2007; Zhao et al. 2008).

In Korea, the use of remote sensing techniques to monitor tidal flats began in 1986 (Yoo 1986). Ryu et al. (2002) used remote sensing to study the Gomso tidal flat. The study introduced waterline extraction, the concept of effective exposed area, and a consideration of the remnant surface water effect with respect to spectral reflectance measurements. Despite the limited number of researchers using remote sensing to study Korean tidal flats, a diverse range of remote sensing techniques have been used to study tidal flats, and the results have been published in peer reviewed journals. We focused on the papers about remote sensing techniques related to topography, sedimentology, and ecology because environments in a tidal flat include diverse variables with complex interactions among tidal topography, sedimentary facies continuous exposure duration, tidal wetness and tidal biomass.

2 Korean Tidal Flats

In South Korea, tidal flats cover a total area of about 2500 km^2. About 2100 km^2 of tidal flats are located on the west coast, while only remaining 400 km^2 of tidal flats are situated on the south coast. The tidal flat area could be calculated from the satellite images, based on the traditional Korean conception that Getbol (viz. tidal flats) are sedimentary shores with the seaward boundaries representing the mean lowest low level (MLLL). An alternative calculation is based on the nautical chart, rather than satellite imagery, using MLLL as the seaward boundary for tidal flats. The total mudflat area calculated by this method is estimated to be ca. 2500 km^2 at present (Koh and Khim 2014).

Vast tidal flats have formed along coastal areas of Korea as shown in Fig. 1. The distribution of the Korean tidal flats, from north to south was described below by the name and location. The tidal range in Incheon, located in the central coastal area, is ~7.3 m, whereas the tidal range in Mokpo, located in the south, is ~3 m (Ryu et al. 2014). There are various types of tidal flats, including closed tidal flats in Hamhae, Haenam, and Muan, semi-closed tidal flats in Gomso, and open tidal flats in Baeksu. Moreover, Korean tidal flats have two distinctive features: (a) they are composed of fine grains, and (b) the sand barriers are poorly developed (Ryu et al. 2008). Table 1 summarizes information about the Korean tidal flat areas in terms of administrative districts, which consist of 5 provinces and over 20 counties adjacent to the existing tidal flats, with additional information for the reclamation projects in the given area (MLTM 2010). At this time, the detailed description was only given to the tidal flats in the West Sea of Korea as those accounts for the majority (ca. 84% to the total areas) of the tidal flats in Korea. Overall, the most extensive tidal flat areas adjoining the West Sea are located in Gyeonggi Bay, Saemangeum, and Jeollanam-do, with respective areas of 873, 400, and 1037 km^2 (Koh and Khim 2014). However, the total area of tidal flats along the Korean coast has decreased due to reclamation projects that have occurred since the Japanese colonial period. Most of the West Sea in Korea was embanked during the past several decades in order to promote land earning for new agricultural areas and new industrial areas by the continuing reclamation projects of the tidal flats (Table 1), which would have resulted in great loss of the natural tidal flat areas. Especially, it should be noted that the Saemangeum tidal flat was barraged in 2006, by the construction of a 33.9 km world longest dike. The Saemangeum reclamation project involved the conversion of 28,300 ha of tidal flats and 11,800 of shallow estuaries, equivalent to 60% of the size of Singapore or two thirds the area of Seoul, into rice-fields and new land for industry and tourism (Cho et al. 2008; Hong et al. 2010).

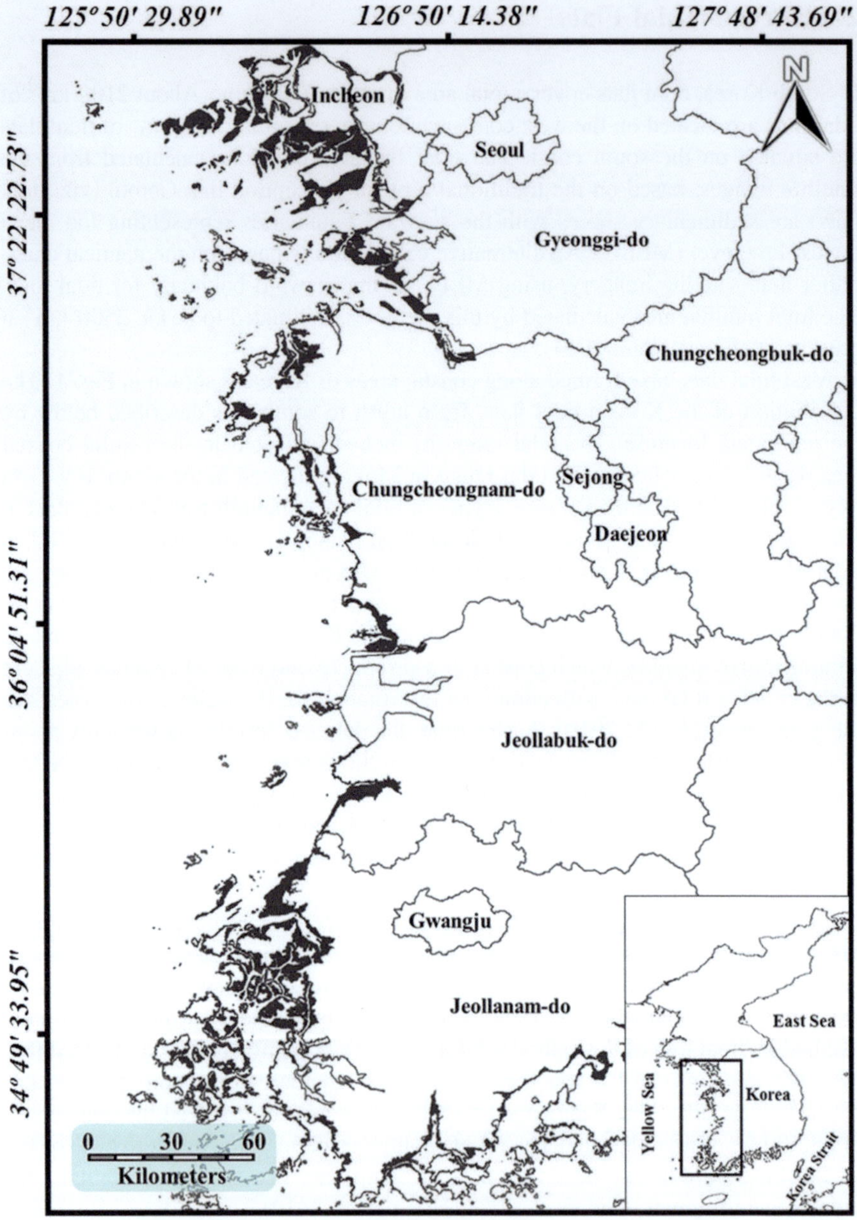

Fig. 1 Map showing the tidal flats belong to the province in the West Sea

Table 1 Areas of the Korean tidal flats belong to the province in the West Sea (MLTM 2010)

Province Country	Tidal flat area		Major reclamation projects in the given areas
	km^2	(%)	
West Sea	2080	(84)	
Gyeonggi-do	873	(35)	Gimpo, Yeongjong, Shihwa
Chungheongnam-do	359	(14)	Taean, Hongseong, Gunjang
Jeollanam-do	118	(5)	Saemangeum
Jeollabuk-do	731	(29)	Yeongsan R.
South Sea	409	(16)	
Total	2489		

3 Intertidal DEM Generation

The topography of the Gomso Bay tidal flats was studied using a remote sensing technique in early 2000. The accuracy of the intertidal DEM produced using the waterline method depends on how many images with various tidal heights are used for generating the DEM, how the waterline is extracted from the image, and how the exact altitude is assigned to the extracted waterline. Waterlines have been extracted from the NIR, SWIR, and TIR bands to identify the most effective band for precisely determining the waterline (Ryu et al. 2002). Ryu et al. (2008) identified structural and decadal changes in the Gomso tidal flats by subtracting the comparable intertidal DEM constructed using the waterline method of Ryu et al. (2002). The local sediment trends were analyzed by a systematic comparison of the horizontal differences between the actual waterlines from optical imagery and modeled waterlines from the reference DEM (Lee et al. 2011b). In the analysis of full-polarization synthetic-aperture radar (SAR) images, Bragg wave resonance was able to discern discrepancies in the waterline detected on C-, L-, and P-band SAR imagery, revealing that a film of remnant surface water had caused the mislocation of the waterline (Kim et al. 2007). The bar morphology of the Han Estuary was investigated using multi-temporal optical imagery and C-band SAR imagery (Yang and Ouchi 2012). In conventional repeat-pass SAR systems, it is difficult to obtain coherent interferometic SAR pairs over tidal flats due to the different exposure time according to the acquisition date, dielectric constant affected by variations in moisture content and amount of remnant water, and correlation length depended by dielectric constant (Ryu et al. 2014; Won et al. 2003). Intertidal DEMs with 30 cm vertical resolution were successfully generated by radar interferometry using cross-tandem data from the European Remote sensing Satellite-Environmental Satellite (ERS-ENVISAT), which can overcome the limitations of atmospheric delay and changes in soil moisture content (Park et al. 2012). The TanDEM-X formation and system parameters were adjusted in accordance with the science goal during the TanDEM-X Science Phase that was planned for total 15 months, from October 2014 to December 2015. Large spatial baseline and high resolution imaging mode gave us a great opportunity to generate the unprecedented intertidal

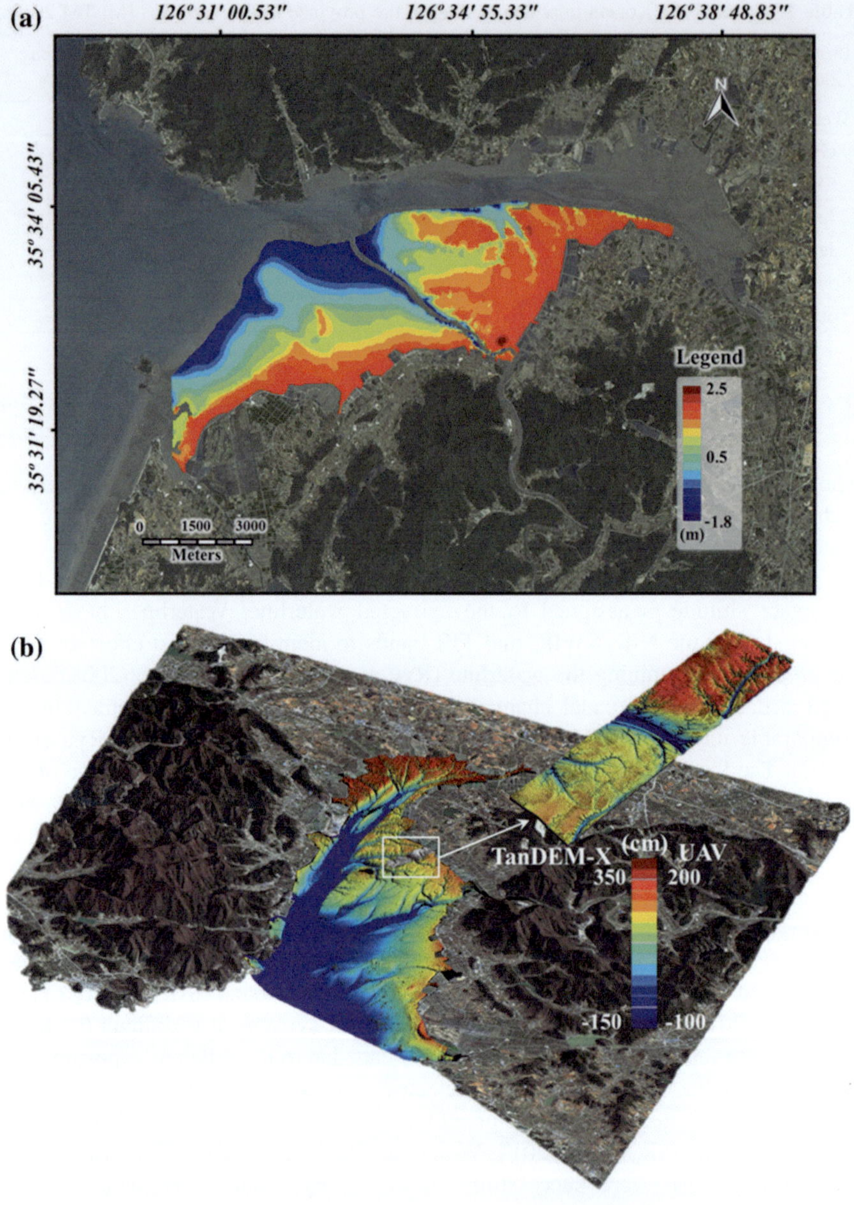

Fig. 2 The examples of a digital elevation model constructed using **a** the waterline method based on the Landsat ETM+, **b** interferometric SAR based on the TanDEM-X (grey scale) and triangulation method based on the UAVs images (color scale). Intertidal DEMs generated by waterline method using 7 waterlines from 2009 to 2012 has 30 m spatial resolution with an accuracy of 11.9 cm r.m.s.e (Lee et al. 2015). The TanDEM-X over Gomso bay were acquired using bistatic and stripmap (SM) modes on the July 12, 2015. Interferometric Intertidal DEMs generated by TanDEM-X has 5 m spatial resolution with an accuracy of 15 cm r.m.s.e. UAVs image was acquired on the September 1, 2015. Intertidal DEMs generated by UAVs image has 17 cm spatial resolution with an accuracy of 10 cm r.m.s.e. (DLR blog 2016)

DEM with significantly improved vertical accuracy. As a results, highly accurate intertidal DEMs with 5-m resolution and an accuracy of 15 cm r.m.s.e were generated based on the TanDEM-X interferometric SAR technique because TanDEM-X allows the acquisition of the coherent interferometric SAR pairs with no time lag (bistatic mode) or approximately 10-second temporal baseline (monostatic mode) between master and slave SAR image (Lee et al. 2016). Recently, Unmanned Aerial Vehicles (UAVs) have been used to obtain images on the coastal area with high spatial (dozen of cm) and temporal resolution (with desired acquisition time). Highly accurate intertidal DEMs with 17 cm resolution and an accuracy of 10 cm r.m.s.e. were generated from aerial triangulation method using rotary-wing UAV as shown in Fig. 2. Ground control points still need to be set up in tidal flat to correct the UAV position errors although interior and exterior orientation was corrected using camera calibration (Kim et al. 2016).

4 Tidal Flats Surface Mapping

Surface sediment classification has been studied using optical images from Landsat TM/ETM+, IKONOS, Kompsat-2/3, and UAV. The case study at the Gomso tidal flat led to the recommendation of 0.25 mm as the critical grain size to classify surface sediment distribution using Landsat ETM+ band 4, while Landsat ETM+ band 5 was recommended for detecting the surface water cover (Ryu et al. 2004). The spatial distribution of surface sediments was quantitatively estimated based on spectral reflectance in high-resolution images of the Hwang-do tidal flat because this area is good for testing the classification of surface sedimentary facies due to the clear separation between sedimentary facies (Choi et al. 2010b). Object-based classification using high-spatial-resolution imaging (IKONOS) was applied for the surface sedimentary facies map, as shown in Fig. 3b. The relationship between the resultant surface sediment map and the spectral reflectance and intertidal topography measurements was determined based on the frequency ratio model (Choi et al. 2010b). The integration of spatial data (e.g., tidal channel density) and data from IKONOS using the near-infrared (NIR) band provided an accurate map of the surface sedimentary facies in the tidal flats (Choi et al. 2011d). Local topography and local structure, such as tidal channels and creeks, emerged as factors affecting surface sedimentary facies (Ryu et al. 2004). Fractal analysis of the tidal channel distribution effectively enhanced the classification of surface sedimentary facies (Eom et al. 2012). The geostatistical mapping procedure was applied to IKONOS imagery to estimate grain-size distribution (Park et al. 2009a). The surface roughness (which is related to fishery activities), texture, and surface grain size of airborne polarimetric SAR image were estimated using the Integral Equation Method (IEM) (Park et al. 2009b). Variations in backscattering coefficients due to changes in the moisture content of mud flats were observed in an indoor experiment using SAR imagery (Lee et al. 2011a). Recently, the spatial distribution about 30-m of surface sediments was quantitatively estimated based on UAV images of the

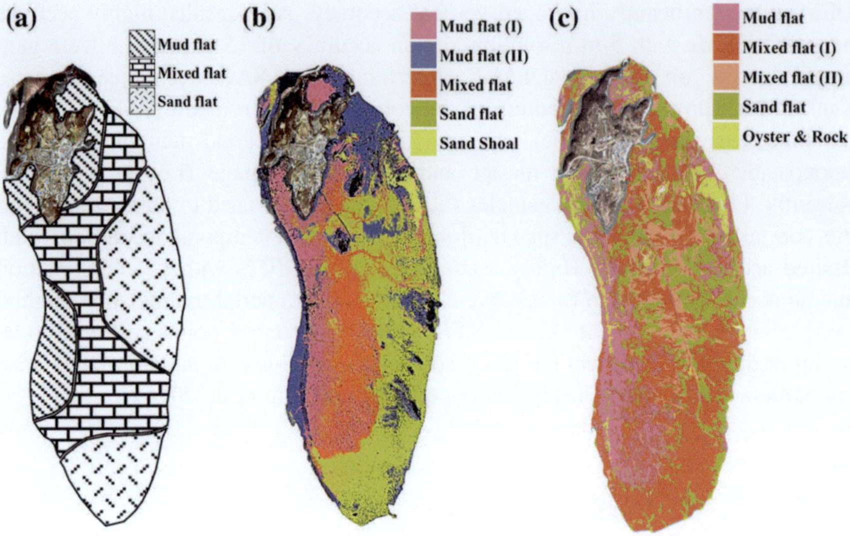

Fig. 3 The examples of Hwangdo tidal flats surface mapping results applied by traditional in situ interpolation (**a**), object-based classification derived from IKONOS image with 90.7% overall accuracy (Choi et al. 2010b) (**b**), object-based classification derived from Kompsat-3 image with 83% overall accuracy (**c**)

Hwang-do tidal flat to test the fine classification of surface sedimentary facies. Basin could be a parameter to determine the sedimentary facies base on the comparison of grain composition, remnant water coverage derived from 1 by 1 m quadrat, elevation and channel characteristics.

5 Bio Facies Mapping

Tidal bio facies such as microphytobenthos, macrozoobenthos, and halophytes have been analyzed using remotely sensed data as shown in Fig. 4. Seasonal changes in the microphytobenthos in the Geunso Bay tidal flats were detected by a series of Landsat images (Choi et al. 2010a). Distribution maps of macrozoobenthos have been generated based on the integration of thematic maps and information regarding the status of the benthic environment. Spatial databases were constructed from factors related to the distribution of macrozoobenthos, such as the distribution of sedimentary facies, topography, local slope, channel distribution, and reflectance. Correlations between the distribution of macrozoobenthos and environmental factors were quantitatively calculated using a frequency ratio model based on a geographic information system (GIS), logistic regression, and an artificial neural network. Potential habitat mapping of macrozoobenthos was performed based on the weighted value for each factor as shown in Fig. 5 (Choi et al. 2011a, b, c; Lee et al. 2013). Herbaceous halophytes were mapped based on the backscattering

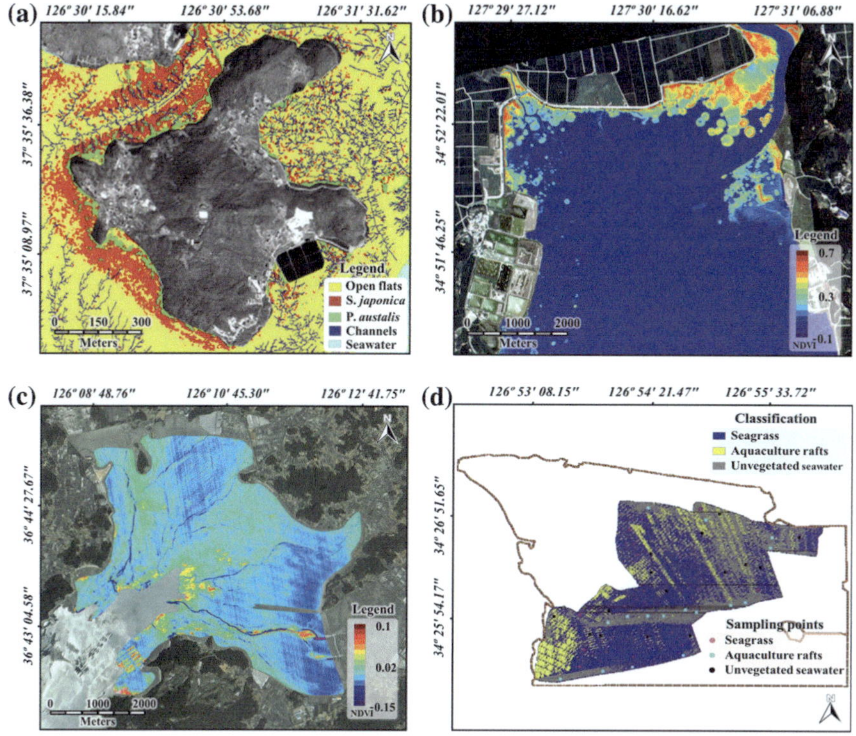

Fig. 4 **a** Classification map generated according to decision tree using TerraSAR-X images and a tidal channel vector for halophyte mapping in Ganghwa tidal flats (Lee et al. 2012), **b** NDVI image induced by Kompsat-2 image for detecting *Phragmites Communis* (Common Reed) in Suncheon Bay (Hong et al. 2015), **c** NDVI image induced by Kompsat-2 image for detecting diatom and *Zostera japonica* in Geunso Bay (Eom et al. 2011), **d** Classification map generated based on the Mahalanobid Distance Classification (MDC) classifier on Kompsat-2 image (Kim et al. 2015)

coefficients in TerraSAR-X data (Lee et al. 2012). It has been determined that winter is the best season to distinguish between *Phragmites australis* and *Suaeda japonica*, while *S. japonica* exhibits the greatest backscattering in summer at HH polarization. The patch of *S. japonica* was easily detected in UAV images because of its high spatial resolution. High spatial resolution from UAV image can distinguish the boundary between S. japonica and non-vegetated tidal flat despite of the lack of and NIR band for vegetation index. The location of submarine ground discharge, which affects various species of fauna and flora inhabiting the tidal flats, was observed in TerraSAR-X and RADARSAT-2 data using the IEM and Oh model (Kim et al. 2011).

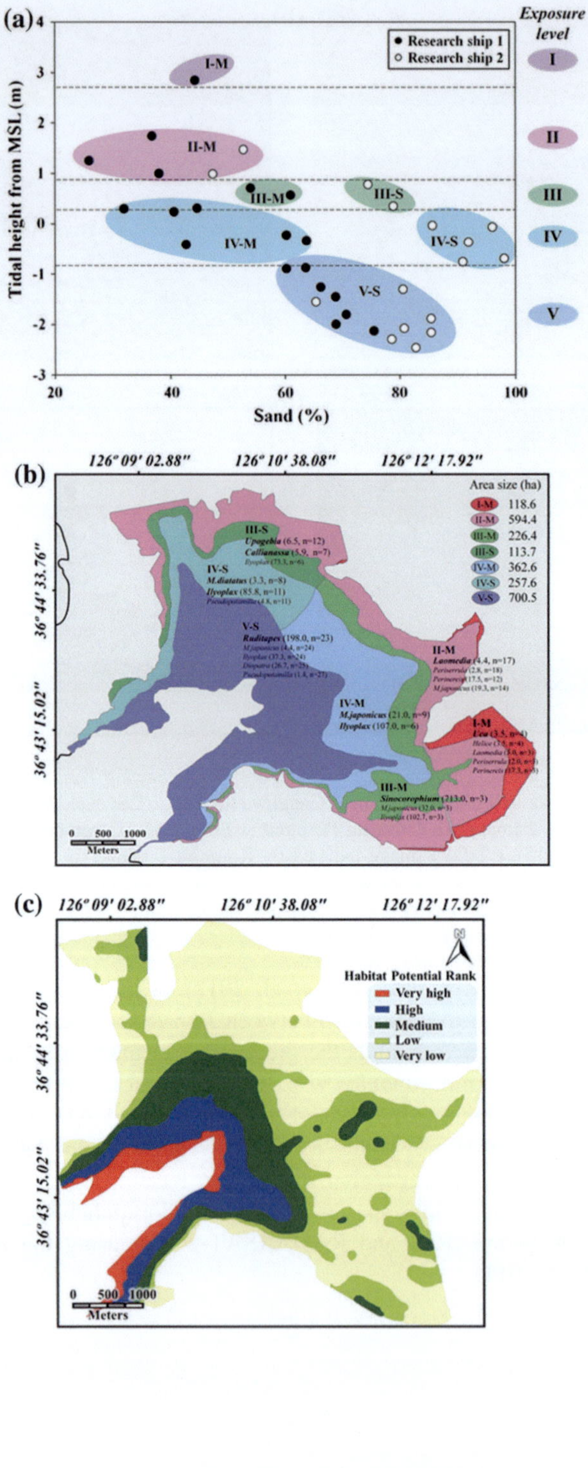
(a)
Exposure level
I
II
III
IV
V
Research ship 1
Research ship 2
I-M
II-M
III-M
III-S
IV-M
IV-S
V-S
Tidal height from MSL (m)
Sand (%)
4
3
2
1
0
-1
-2
-3
20
40
60
80
100
(b)
126° 09' 02.88"
126° 10' 38.08"
126° 12' 17.92"
36° 44' 33.76"
36° 43' 15.02"
Area size (ha)
I-M 118.6
II-M 594.4
III-M 226.4
III-S 113.7
IV-M 362.6
IV-S 257.6
V-S 700.5
III-S
Upogebia (6.5, n=12)
Callianassa (5.9, n=7)
IV-S
M.diatatus (3.3, n=8)
Ilyoplax (85.8, n=11)
V-S
Ruditapes (198.0, n=23)
II-M
Laomedia (4.4, n=17)
IV-M
M.japonicus (21.0, n=9)
Ilyoplax (107.0, n=6)
I-M
III-M
Sinocorophium (713.0, n=3)
0 500 1000
Meters
(c)
126° 09' 02.88"
126° 10' 38.08"
126° 12' 17.92"
36° 44' 33.76"
36° 43' 15.02"
Habitat Potential Rank
Very high
High
Medium
Low
Very low
0 500 1000
Meters

◂**Fig. 5** **a** Distribution of each station with respect to sand composition and tidal height from MSL. Macrofaunal station groups determined from the cluster analysis are superimposed. **b** Classification of station groups and their dominant species in the Geunso tidal flats. The number represents the mean density (m^{-2}) of each species (KORDI 2010). **c** Map of potential habitat of Ruditapes philippinarum generated from artificial neural network model (Lee et al. 2014)

6 Conclusions

This review has shown that remote sensing techniques can contribute to the monitoring of the topography, sedimentary facies, and bio facies of tidal flats in Korea. The case studies revealed that the levels of thematic mapping varied according to the type of remote sensing imagery used (i.e., optic or microwave). Not only optical images with high spatial and middle spatial resolution but also SAR images with C- and X-bands are used for intertidal DEM generation. Although the accuracy of the topographic maps is enhanced by using high-spatial resolution images such as IKONOS, Quickbirds, and Kompsat, Landsat TM/ETM+ is still the most widely used due to the large amount of archived data, convenience of time-series analysis, and data that can be obtained free of charge. Recently, UAVs have been widely used to obtain the information of tidal flats with high spatial resolution. UAVs applications help to understand not only seasonal morphologic change but also evaluate morphologic change by sporadic events such as floods and typhoons. Reflectance determined from Landsat TM/ETM+ and ASTER is often used for analyzing halophytes and macrobenthos, but they cannot be used to determine precise distributions. Optical images with a spatial resolution of 4 m (high spatial resolution) effectively map the distribution of macrobenthos. X-band SAR images distinguish the boundary of perennial and annual halophytes. Surface grain size can be classified by various sensors; however, high-spatial-resolution optical images generate more detailed sedimentary facies maps. C-band and X-band SAR can also be used to generate surface sedimentary maps using the IEM model. A combination of different sensors improves the intertidal surface mapping because additional information enhance distinguishing different intertidal surface cover types based on its spectral, spatial and textural characteristics. Sentienl-2 with NIR and SWIR will be useful to understand the remnant water contents on intertidal surface. Few national thematic maps such as intertidal DEMs and channel distribution have been derived based on the airborne LiDAR data only for the specific interested area. Intertidal DEMs can be generated by the waterline method from optical, SAR and LiDAR imagery. In particular, various waterlines with different tidal heights within a short period are needed for accurate intertidal DEM generation. In the case of using Landsat TM/ETM+ with 16-day revisits, still widely used for intertidal DEM generation, data should be archived at least within a year and a half for intertidal DEM generation because it is difficult to obtain data without cloud and seasonal variation due to the monsoons in the winter and typhoons in the summer. Therefore, morphological changes due to seasonal effects are hard to detect from intertidal DEMs generated by the waterline method. Precise intertidal

DEMs can be generated by interferometry in TerraSAR-X tandem data. Several intertidal DEMs for various seasons can be easily generated because the interferometric coherence of paired images over tidal flats is high not only mud dominance but also sand dominance due to the no time lag between master and slave SAR image. Precise intertidal DEMs can also be generated using airborne LiDAR data. Bathymetric LiDAR systems employ a blue-green laser with a typical wavelength of 530 nm to maximize water penetration depth. However, high turbidity is the most limiting factor in discriminating water and tidal flats using LiDAR. Precise intertidal DEMs can also be generated using airborne UAVs data. However, coverage of intertidal DEM is very limited due to the short operation time of UAVs.

Using Landsat TM/ETM+, coarse sediment (grain size >0.025 mm) is the practical grain size for classifying surface sediment distribution. Classification into three types of sediment (mud, mixed, and sand) has been developed for images with a spatial resolution of 30 m, such as those provided by Landsat TM/ETM+. However, a grain size >0.0625 m is the practical grain size when using high spatial resolution data. If the sedimentary facies are dominated by a specific type, classification is limited based on reflectance. Tidal channels, another topographic feature, should be considered in image-based sediment classification to enhance the distribution results. A GIS can be effectively applied as a basic analysis tool in the assessment of spatial relationships between the occurrence of macrobenthos species and related environmental factors. Based on these relationships, potential maps of macrobenthos have been successfully generated. The accuracy of the potential maps is enhanced by including other factors that affect the distribution of benthos, such as temperature. Seasonal variations and abrupt changes in the benthos associated with environmental changes are difficult to analyze because spatial variables such as intertidal DEMs, tidal channel density, and sedimentary facies cannot be generated on a seasonal basis. Halophytes with thick density (perennial grass such as *P. australis*) can be distinguishable, and halophytes having low density (e.g., annual plants such as *S. japonica*) can be differentiated due to the texture difference from the barren tidal flats in high spatial-resolution UAVs image. Species with similar reflectance characteristics cannot be distinguished in optical images due to low spatial and spectral resolution. In X-band SAR, low-density halophytes can be discriminated from perennial grass and barren tidal flats. Seasonal variations in the halophyte density should be evident in TanDEM-X data. To enhance the accuracy of thematic maps and to develop thematic maps for other factors in tidal flats, it is necessary not only to use the various sensors but also to develop a new classification scheme. Temperature is an important influence on species distribution in tidal flats. The accuracy of the extracted waterline is higher using the thermal band. However, studies of tidal flats using the thermal band should be considered in the initial stages of a study. Hyperspectral images of various durations that provide data for different seasons are effective in generating accurate maps for benthos and halophytes. High-resolution image derived from UAVs can be used to generate detailed DEMs and channel distributions. UAVs are a good solution for overcoming the limitations presented by tidal flats. Multispectral, hyperspectral, thermal, SAR, LiDAR sensors can be loaded onto manned and unmanned aerial vehicles to obtain data for the desired areas and periods with the desired sensor. Geophysical information extracted from the optical and microwave wavelengths has proven to be useful for characterization of the chemical composition and morphology of surfaces. Therefore, radar sensors have the potential to provide additional information that may be combined with optical sensor data for improving land classification and reducing confusion among land covers (Henderson et al. 2002; van der Wal and Herman 2007). In case of fusion between low-resolution multispectral image and higher resolution SAR image, the resultant data products have advantages of high spatial resolution, structural information (from SAR image), and spectral resolution (from optical and infrared band) (Pal et al. 2007). In case of the fusion airborne LiDAR system combined with multispectral or hyperspectral image, researchers can not only generate accurate classification maps but also estimate biochemical and biophysical parameters of wetland vegetation such as water content, biomass, and leaf area index (Klemas 2013). An integrated analysis of sedimentary facies in tidal flats and the distribution of suspended sediment in coastal water are necessary for effective monitoring of the sedimentary environment in coastal and tidal flats. For this integrated analysis, data from marine observation satellites with low spatial resolution and high temporal resolution are required.

References

Adam E, Mutanga O, Rugege D (2010) Multispectral and hyperspectral remote sensing for identification and mapping of wetland vegetation: a review. Wetl Ecol Manage 18:281–296

Boak EH, Turner IL (2005) Shoreline definition and detection: a review. J Coast Res 21:688–703

Brooks TM, Mittermeier RA, de Fonseca GAB, Gerlach J, Hoffmann N, Lamoreux JF, Mittermeier CG, Pilgrim JD, Rodrigues ASL (2006) Global biodiversity conservation priorities. Science 313:58–61

Cho JY, Son JG, Song CH, Hwang SA, Lee YM, Jeong SY, Chung BY (2008) Integrated nutrient management for environmental-friendly rice production in salt-affected rice paddy fields of Saemangeum reclaimed land of South Korea. Paddy Water Environ 6:263–273

Choi JK, Ryu JH, Eom JA, Rho SM, Nho JH (2010a) Analysis on the sedimentary environment and microphytobenthos distribution in the Geunso bay tidal flat using remotely sensed data. J Wetl Res 12(3):67–78

Choi JK, Ryu JH, Lee YK, Yoo HR, Woo HJ, Kim CH (2010b) Quantitative estimation of intertidal sediment characteristics using remote sensing and GIS. Estuar Coast Shelf Sci 88:125–134

Choi JK, Oh HJ, Koo BJ, Ryu JH, Lee SR (2011a) Spatial Polychaeta habitat potential mapping using probabilistic models. Estuar Coast Shelf Sci 93:98–105

Choi JK, Oh HJ, Koo BJ, Ryu JH, Lee SR (2011b) Crustacean habitat potential mapping in a tidal flat using remote sensing and GIS. Ecol Model 222:1522–1533

Choi JK, Oh HJ, Koo BJ, Lee SR, Ryu JH (2011c) Macrobenthos habitat mapping in a tidal flat using remotely sensed data and a GIS-based probabilistic model. Mar Pollut Bull 62:564–572

Choi JK, Ryu JH, Eom J (2011d) Integration of spatial variables derived from remotely sensed data for the mapping of the tidal surface sediment distribution. J Coast Res SI64:1653–1657

Cracknell AP (1999) Remote sensing techniques in estuaries and coastal zones. Int J Remote Sens 19:485–496

DLR blog (2016) http://www.dlr.de/blogs/en/home/tandem-x/How-researchers-use-the-latest-Earth-observation-data.aspx. Accessed 20 March 2017

Eom J, Choi JK, Ryu JH, Won JS, Noh JH (2011) Spectral characteristics of microphytobenthos on the surface sediments in Geunso-bay tidal flat. In: International Symposium on Remote Sensing 2011

Eom J, Choi JK, Ryu JH, Woo HJ, Won JS, Jang S (2012) Tidal channel distribution in relation to surface sedimentary facies based on remotely sensed data. Geosci J 16:127–137

Henderson FM, Chasanm R, Portolese J, Hart T (2002) Evaluation of SAR-optical imagery synthesis techniques in a complex coastal ecosystem. Photogramm Eng Remote Sens 68:839–846

Heygster G, Dannenberg J, Notholt J (2010) Topographic mapping of the German tidal flats analyzing SAR images with the waterline method. IEEE Trans Geosci Remote 48:1019–1030

Hong SK, Koh CH, Harris RR, Kim JE, Lee JS, Ihm BS (2010) Land use in Korean tidal wetlands: impacts and management strategies. Environ Manage 45:1014–1026

Hong SH, Chun SS, Eom JA (2015) The 2009-based detailed distribution pattern and area of Phragmites communis-dominant and Suaeda japonica-dominant communities on the Suncheon-bay and Beolgyo estuarine wetlands. J Wetl Res 17(1):26–37

Kim DJ, Moon WM, Park SE, Kim JE, Lee HS (2007) Dependence of waterline mapping on radar frequency used for SAR images in intertidal areas. IEEE Geosci Remote Sens 4:269–273

Kim DJ, Moon WM, Kim G, Park SE, Lee H (2011) Submarine groundwater discharge in tidal flats revealed by space-borne synthetic aperture radar. Remote Sens Environ 115:793–800

Kim K, Choi JK, Ryu JH, Jeong HJ, Lee K, Park MG, Kim KY (2015) Observation of typhoon-induced seagrass die-off using remote sensing. Estuar Coast Shelf Sci 154:111–121

Kim BJ, Lee YK, Ryu JH, Lee SK, Kim KL (2016) DEM generation of intertidal zone in Korea using unmanned aerial vehicle. In: Proceedings IEEE international geoscience and remote sensing symposium. https://doi.org/10.1109/igarss.2016.7730749

Kirwan ML, Murray AB (2007) A coupled geomorphic and ecological model of tidal marsh evolution. Proc Natl Acad Sci USA 105(15):6118–6122

Kirwan ML, Guntenspergen GR, D'Alpaos A, Morris JT, Mudd SM, Temmerman S (2010) Limits on the adaptability of coastal marshes to rising sea level. Geophys Res Lett 37:L23401. https://doi.org/10.1029/2010GL045489

Klemas V (2013) Airborne remote sensing of coastal features and processes: an overview. J Coast Res 29:239–255

Koh CH, Khim JS (2014) The Korean tidal flat of the Yellow Sea: physical setting, ecosystem and management. Ocean Coast Manage 102:398–414

KORDI (Korea Ocean Research Development Institute) (2010) Studies on sediments, waters and biota to understand major environmental factors in rehabilitation of degraded tidal flats

Lee H, Chae H, Cho SJ (2011a) Radar backscattering of intertidal mudflats observed by Radarsat-1 SAR images and ground-based scatterometer experiments. IEEE Trans Geosci Remote 49:1701–1711

Lee YK, Ryu JH, Choi JK, Soh JG, Eom J, Won JS (2011b) A study of decadal sedimentation trend changes by waterline comparisons within the Ganghwa tidal flats initiated by human activities. J Coast Res 27:857–869

Lee YK, Park JW, Choi JK, Oh Y, Won JS (2012) Potential uses of TerraSAR-X for mapping herbaceous halophytes over salt marsh and tidal flats. Estuar Coast Shelf Sci 115:366–376

Lee S, Park I, Koo BJ, Ryu JH, Choi JK, Woo HJ (2013) Macrobenthos habitat potential mapping using GIS-based artificial neural network models. Mar Pollut Bull 67:177–186

Lee S, Choi JK, Park IH, Koo BJ, Ryu JH, Lee YK (2014) Application of geospatial models to map potential Ruditapes philippinarum habitat using remote sensing and GIS. Int J Remote Sens 35(1):3875–3891

Lee YK, Ryu JH, Choi JK, Lee S, Woo HJ (2015) Satellite-based observations of unexpected coastal changes due to the Saemangeum Dyke construction, Korea. Mar Pollut Bull 97:150–159

Lee SK, Ryu JH, Lee YK, Kim KL (2016) Tidal flat digital elevation model (DEM) construction by means of Tandem-X. In: Proceedings 2016 IEEE international geoscience and remote sensing symposium https://doi.org/10.1109/igarss.2016.7729081

MLTM (Ministry of Land, Transport, and Maritime Affairs) (2010) The Basic Survey of Coastal Wetland in 2008

Pal SK, Majumdar TJ, Bhattacharya AK (2007) ERS-2 SAR and IRS-1C LISS III data fusion: a PCA approach to improve remote sensing based geological interpretation. ISPRS J Photogramm Remote Sens 61:281–297

Park NW, Jang DH, Chi KH (2009a) Integration of IKONOS imagery for geostatistical mapping of sediment grain size at Baramarae beach, Korea. Int J Remote Sens 30:5703–5724

Park SE, Moon WM, Kim DJ (2009b) Estimation of surface roughness parameter in intertidal mudflat using airborne polarimetric SAR data. IEEE Trans Geosci Remote 47:1022–1031

Park JW, Choi JH, Lee YK, Won JS (2012) Intertidal DEM generation using satellite radar interferometry. Kor J Remote Sens 28(1):121–128

Ryu JH, Won JS, Min KD (2002) Waterline extraction from Landsat TM data in a tidal flat e a case study in Gomso Bay, Korea. Remote Sens Environ 83:442–456

Ryu JH, Na YH, Won JS, Doerffer R (2004) A critical grain size for Landsat ETM+ investigations into intertidal sediments: a case study of the Gomso tidal flats, Korea. Estuar Coast Shelf Sci 60:491–502

Ryu JH, Kim CH, Lee YK, Won JS, Chun SS, Lee S (2008) Detecting the intertidal morphologic change using satellite data. Estuar Coast Shelf Sci 78:623–632

Ryu JH, Choi JK, Lee YK (2014) Potential of remote sensing in management of tidal flats: a case study of thematic mapping in the Korean tidal flats. Ocean Coast Manage 102:458–470

Sondak CFA, Chung IK (2015) Potential blue carbon from coastal ecosystems in the Republic of Korea. Ocean Sci J 50(1):1–8

Tyler AN, Sanderson DCW, Scott EM, Allyson JD (1996) Accounting for spatial variability and fields of view in environmental gamma ray spectrometry. J Environ Rad 33:213–235

van der Wal D, Herman PMJ (2007) Regression-based synergy of optical, shortwave infrared and microwave remote sensing for monitoring the grain-size of intertidal sediments. Remote Sens Environ 111:89–106

Wilson MA, Costanza R, Boumans R, Liu S (2005) Integrated assessment and valuation of ecosystem goods and services provided by coastal systems. In: Wilson, JG (ed) The intertidal ecosystem: the value of Ireland's Shores. Royal Irish Academy, Dublin, pp 1–24

Won JS, Na YH, Kim SW (2003) Tidal flat DEM generation by satellite remote sensing. Proc 2003 IEEE Int Geosci Remote Sens Symp 3, 2116–2118

Yang CS, Ouchi K (2012) Analysis of bar morphology using multi-temporal and multi-sensor satellite images: example from the Han Estuary, Korea. Mar Geol 311:17–31

Yoo HR (1986) Three dimensional study of the tidal flat geomorphology in the Kum River estuary by multidate image analysis. J. Korean Soc Remote Sens 2(2):95–105

Zhao B, Guo H, Yan Y, Wang Q, Li B (2008) A simple waterline approach for tidelands using multi-temporal satellite images: a case study in the Yangtze Delta. Estuar Coast Shelf Sci 77:134–142

SAR Remote Sensing of Intertidal Flats in Korea

Byung-Hun Choe and Duk-jin Kim

Abstract Extensive intertidal flats along the west coast of the Korean Peninsula form very unique coastal ecosystems providing habitat for a variety of benthic species. Spatial variation of physical surface properties and structures in intertidal flats can be associated to different polarimetric Synthetic Aperture Radar (SAR) signatures, which have a great potential for intertidal flat mapping. We find that complex and rough surface structures of densely populated wild oyster reefs can cause very strong diffused-multiple scattering and depolarization at C- and X-bands, while not at a relatively longer wavelength, L-band. Along the outer margin of intertidal flats adjacent to the land, very dark radar backscattering signatures are observed, which result from specular reflection on shallow water puddles formed by Submarine Groundwater Discharge (SGD). This chapter demonstrates distinct radar scattering signatures of wild oyster habitat and SGD in intertidal flats based on polarimetric SAR observations and theoretical scattering models, and suggests how they can be quantitatively mapped using multi-frequency polarimetric SAR. Finally, the correlation between the distribution of wild oyster habitat and the occurrence of SGD is discussed.

Keywords Polarimetric SAR · Intertidal flats · Oyster · Submarine Groundwater Discharge (SGD) · Polarimetric target decomposition · Radar backscattering model

1 Introduction

The large tidal range in the West Sea (Yellow Sea) of the Korean Peninsula has given rise to development of extensive intertidal flats along the western and southern coasts, which are up to approximately 3% (2800 km^2) of the South Korean

B.-H. Choe
Department of Earth Sciences, University of Western Ontario, London, Canada

D. Kim (✉)
School of Earth and Environmental Sciences, Seoul National University, Seoul, South Korea
e-mail: djkim@snu.ac.kr

V. Barale and M. Gade (eds.), *Remote Sensing of the Asian Seas*,
https://doi.org/10.1007/978-3-319-94067-0_13

territory (Koh 2001). The intertidal zones form very unique costal ecosystems between the land and the sea where diverse benthic flora and fauna inhabit, and abundant organic/inorganic matter and nutrients from both freshwater and seawater make them more productive than any other part on land. In terms of coastal environment monitoring and fishery resource management, there has been a great need for efficient mapping techniques for intertidal flats to supplement extremely hard field works in mudflats and spatial constraints. It has spurred a variety of application of remote sensors and there has been a great advance in intertidal flat mapping using optical and SAR sensors. Optical remote sensing has significantly contributed to waterline extraction (e.g., Ryu et al. 2002), sediment classification (e.g., Ryu et al. 2004; Brockmann and Stelzer 2008; Choi et al. 2010), and benthic habitat mapping (e.g. salt marshes (Belluco et al. 2006), mussel/oyster beds (Choi et al. 2011; Le Bris et al. 2016)). However, it is not easy to characterize distinct spectral signatures depending on specific targets on intertidal flats as applied for vegetation or geological mapping. Spectral reflectance is susceptible to variable water content on intertidal flat surfaces (Ryu et al. 2004). Also, frequent cloud cover in the coastal areas and very limited acquisition timeframes at low tide during the daytime make it difficult to obtain good quality imagery. In this respect, SAR provides a great advantage with its capability to acquire independent of sunlight and cloud coverages. Radar backscattering depends on physical surface parameters such as surface roughness and dielectric properties. In particular, intertidal flats are ideal for investigating radar scattering signatures caused by surface roughness as most of intertidal flat surfaces are highly saturated with moisture content over 40% and radar backscattering is mainly affected by surface roughness (Choe et al. 2012).

Early studies have tried to classify sediment types based on variation in radar backscattering (van der Wal et al. 2005) and to quantitatively estimate surface roughness parameters relating to sediment types by applying radar scattering model inversion methods (Gade et al. 2008; Park et al. 2009; Choe and Kim 2011). Recently, several studies have observed distinctly strong radar backscattering signatures from halophytes and mussel beds, which can be differentiated enough to map their habitats from mudflats based on a single polarization data (Lee et al. 2012; Gade et al. 2014). Those findings are based on a single polarization data, but it is suggested that different surface roughness and physical structures of benthic habitats in intertidal flats can be greatly observed from SAR sensors particularly with a fully polarimetric capability. Dehouck et al. (2011) and Nieuwhof et al. (2015) confirmed that oyster beds and salt marshes are differentiated from mudflats by using dual polarimetric SAR including a cross-polarization channel (i.e., HV; transmitting through a horizontally polarized channel and receiving through a vertically polarized channel or VH; transmitting through a vertically polarized channel and receiving through a horizontally polarized channel). Polarimetric decomposition parameters from quad polarimetric SAR data were applied to classify salt marshes by combining with optical sensor-derived parameters (van Beijma et al. 2014).

This study introduces polarimetric SAR investigations from oyster reefs and SGD in intertidal flats with a case study of Jebu Island in the west coast of the Korean Peninsula. Oysters inhabit on intertidal flat surfaces and form very rough reef structures combined with their sharp and jagged shells, while most shellfishes are obscured beneath the surfaces. Exposed oyster reefs on intertidal flats have apparently distinct surface roughness (Choe et al. 2012). On the other hand, water puddles are present along the outer margin of the intertidal flats adjacent to the land and have much more reflective surfaces than sea surfaces with capillary waves by wind, which are likely to cause specular reflection on the water surface. They are highly related to groundwater discharge of this region (Kim et al. 2011). Here, we report what polarimetric SAR signatures are observed from oyster reefs and groundwater discharge, and explain how their surface properties are related to the polarimetric SAR observations based on theoretical radar scattering models and field measurements, and suggest how they can be quantitatively mapped using multi-frequency (L-, C-, X-band) SAR. Finally, we discuss why oysters are densely populated in the intertidal flats where extensive groundwater discharge is present. Since SGD transports rich inorganic nutrients dissolved in fresh groundwater originated from land, it is highly associated with abundance and productivity of benthic species in the coastal zone (Burnett et al. 2003; Hwang et al. 2010; Hosono et al. 2012).

2 Geological Setting of Jebu Island

Jebu Island is an inverted triangle shaped island of approximately 1 km^2 located on the mid-western coast of the Korean peninsula (37° 10′N, 126° 37′E) (Fig. 1). The intertidal flats are widely developed along the inner bays between Jebu Island and the mainland and representative of the west coast intertidal flats. Due to the maximum tidal level at a range of 9 m or higher, the road connecting the island to the

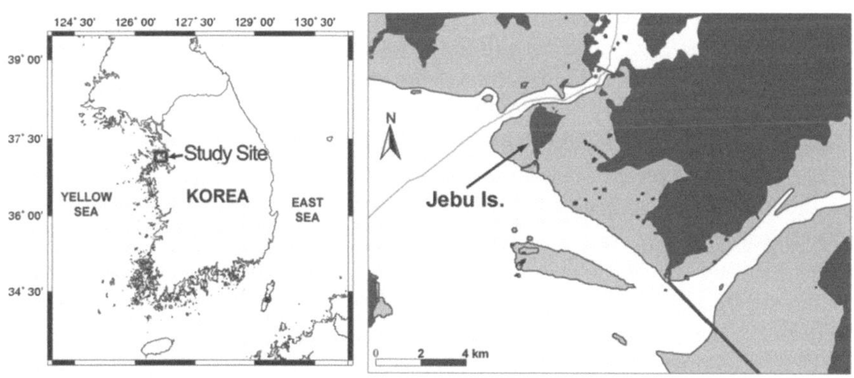

Fig. 1 Location (left) and intertidal flat areas (right) of Jebu Island. The black, white, and gray colors in the right figure represent land, sea, and exposed intertidal flats, respectively. The black dots along a transect on the right figure indicate the location of measuring the surface parameters for groundwater discharge. Figure from Kim et al. (2011)

mainland appears and disappears between the low and high tides, which is also known for a phenomenon of "Moses Miracle". The island is separated from the mainland with the road immerged at high tide, and intertidal zones of over 20 km^2 are exposed at low tide. The sediments are loam or sandy loam soils mainly composed of sand, silt, and clay. Tidal channels are formed around the central intertidal flats with a little slope toward the south and tidewater ebbs and flows along the channels between the island and the mainland carrying abundant nutrients and plankton, where oyster beds are densely populated. Shellfish (e.g., oyster, Manila clam) and small octopuses are main fishery resources of this region.

In addition, several investigations have shown that the groundwater system of this area is influenced by seawater intrusion (Jeen et al. 2001; Kim et al. 2003). The regional bedrocks are Precambrian granitic gneiss and schist complex and overlain by Quaternary alluvial deposits of mud, clay, and fine sand at a thickness between 9 and 28 m (Jeen et al. 2001). The annual precipitation from January, 2007 to December, 2008 was about 1330 mm, and most precipitation was concentrated from July to September resulting in higher groundwater levels during the rainy season. The lowest and highest groundwater levels measured at two monitoring wells in Jebu Island during the period were −1 and 3 m based on the water table, respectively (source from Korea Meteorological Administration).

3 Polarimetric SAR Signatures of Oyster Reefs

3.1 Multiple-Diffused Scattering and Depolarization

Relatively very strong radar backscattering in the cross-polarization channels (i.e., HV and VH) of C-band (5.405 GHz) RADARSAT-2 quad polarimetric data (at an incidence angle of 36°) was observed around main tidal channels in the central intertidal flats of Jebu Island (Fig. 2a), unlike the co-polarization channels (i.e., HH and VV) showing little variation within the intertidal flats (Fig. 2b). Even though it is generally not expected from intertidal flat surfaces of very smooth muddy sediments, the amount of HV and VH backscattering coefficients was over −18 dB.

Based on the cross-polarized backscattering signature, we further investigated by applying the Freeman-Durden (FD) three component decomposition, which can describe target surfaces with three scattering components modelled by a very smooth Bragg surface (single-bounce scattering), a dihedral corner (double-bounce scattering), and a forest canopy of randomly oriented thin cylinder-like scatters (multiple-diffused or volume scattering) (Freeman and Durden 1998). Most areas of the intertidal flats showed dominant single-bounce scattering as expected, but it was noted that multiple-diffused scattering is also observed dispersedly in the isolated central parts of the intertidal flats at C-band (Fig. 2c). The normalized multiple-diffused scattering ratio to total three scattering components (i.e., $M/(S+D+M)$ where S, D, and M are single-bounce scattering, double-bounce scattering, and multiple-diffused scattering, respectively) was greater than 0.7, which is

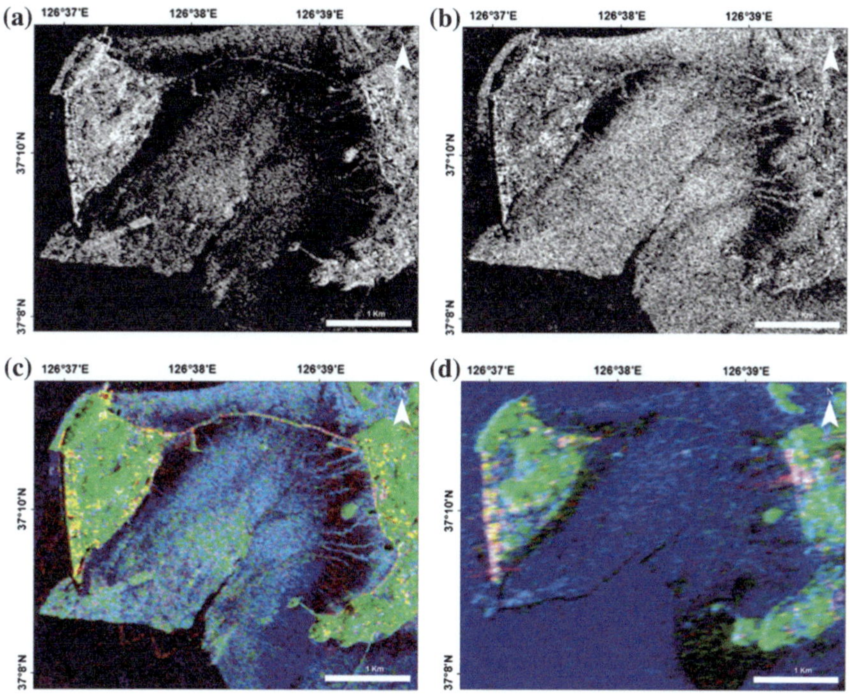

Fig. 2 RADARSAT-2 backscattering coefficient images (**a**; VH polarization, **b**; VV polarization) and Freeman-Durden decomposition results from RADARSAT-2 (**c**, November 20, 2008) and ALOS PALSAR (**d**, April 13, 2007). The FD RGB composites represent double-bounce scattering (red), multiple-diffused scattering (green), and single-bounce scattering (blue), respectively. Figure modified from Choe et al. (2012)

comparable to the values observed from forest and vegetated areas surrounding the intertidal flats. Our field observation and GPS tracking have confirmed these areas are coincident with where wild oyster reefs are densely populated. However, the FD result from L-band (1.27 GHz) ALOS PALSAR showed only dominant single-bounce scattering from both oyster reefs and mudflats, except for two small forested islets on the eastern margin of the intertidal flats (Fig. 2d).

For a quantitative investigation on the change of polarization status (i.e., target depolarization) observed from oyster reefs, we applied additional polarimetric parameters such as cross-polarized ratio (i.e., ratio of cross-polarized backscattering to co-polarized backscattering (HV/HH or VH/VV)), co-polarized phase difference (i.e., phase difference between HH and VV co-polarized channels ranging from $-\pi$ to π degrees), and co-polarized correlation coefficient (i.e., correlation coefficient between HH and VV co-polarized channels ranging from 0 to 1) (Choe et al. 2012). For example, very rough surfaces or complex forest canopy relative to radar wavelength transmitted can cause multiple-diffused or volume scattering resulting in polarimetric change of transmitted radar signals, which relatively increases

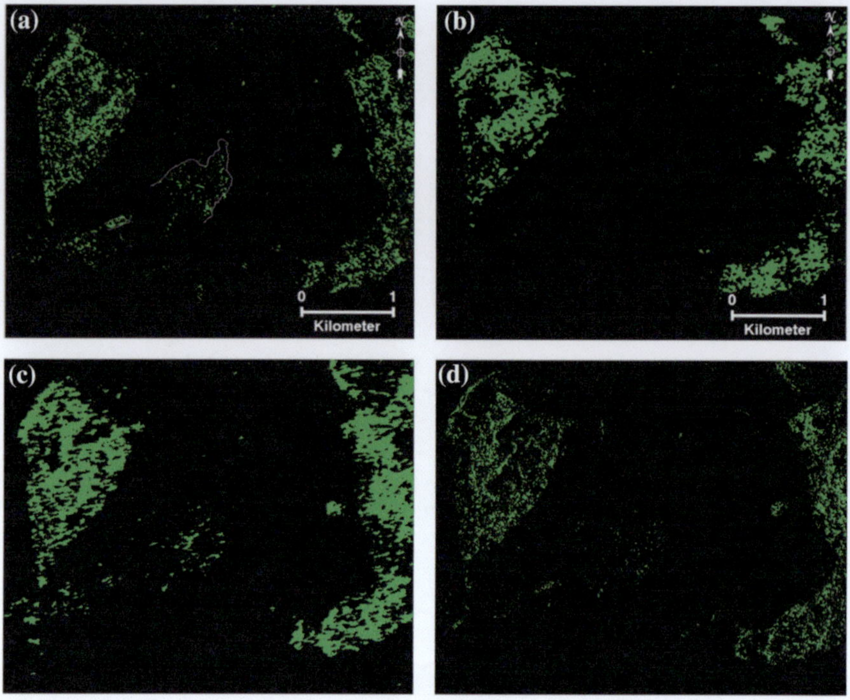

Fig. 3 Cross-polarized ratio images from **a** RADARSAT-2 (November 20, 2008), **b** ALOS PALSAR (April 13, 2007), **c** Envisat ASAR (June 10, 2005), and **d** TerraSAR-X (November 20, 2008). The green dots represent the areas greater than −8 dB (**a**), −8 dB (**b**), −11 dB (**c**), and −8 dB (**d**), respectively. The white lines in figure (**a**) represent GPS tracking lines obtained along the boundaries of oyster reefs in the field. Figure modified after Choe et al. (2012)

backscattering in the cross-polarized channels and decreases the correlation between the co-polarized channels.

The cross-polarized ratios of oyster reefs were over −8 dB, which is much greater than those of mudflats below −15 dB (Fig. 3a). The co-polarized phase difference was derived randomly over 30° (π/6) or below −30° (−π/6), whereas mudflats showed values nearly approaching to 0° (refer to Fig. 7 in Choe et al. 2012). Similarly, the co-polarized correlation coefficients of oyster reefs ranged below 0.4, which are much smaller than those of mudflats showing values over 0.7 (refer to Fig. 7 in Choe et al. 2012). The results showed that very strong depolarization is observed from oyster reefs at C-band, resulting in a significant increase in the cross-polarized HV and VH backscattering and very variable phase difference and low correlation between the co-polarized HH and VV channels. It suggests that very rough and complex oyster reef surfaces generate multiple-diffused scattering and depolarization at C-band.

In addition to quad-polarimetric RADARSAT-2 data, dual polarimetric Envisat ASAR (C-band, 5.331 GHz) and TerraSAR-X (X-band, 9.65 GHz) data at difference incidence angles were acquired over the same target area and they also confirmed oyster reef signatures using the cross-polarized backscattering coefficient and cross-polarized ratio. The C-band Envisat ASAR data observed the cross-polarized backscattering coefficients greater than −18 dB and the cross-polarized ratios greater than −11 dB from oyster reefs (Fig. 3c), even though they were acquired at a lower incidence angle of 20° with its low spatial resolution of 30 m per pixel. The cross-polarized backscattering coefficients and cross-polarized ratios derived from the TerraSAR-X data at the incidence angle of 23° were greater than −16 dB (−14 dB in very densely populated oyster reefs) and −8 dB, respectively (Fig. 3d).

In particular, the cross-polarized backscattering coefficients at X-band were about 2 dB higher than at C-band. Given that TerraSAR-X observations at even higher incidence angles over 40° also showed the cross-polarized backscattering coefficients greater than −18 dB, this indicates that the depolarization effect occurred in oyster reefs increases at higher frequency domain even considering the effect of incidence angles. L-band ALOS PALSAR quad polarimetric data (at an incidence angle of 23°), on the other hand, does not show distinct differences within the intertidal flats. Any significant polarimetric signatures of oyster reefs were not observed from FD and depolarization parameters (Figs. 2d and 3b). Oyster reefs have a similar range with mudflats in the cross-polarized ratio (less than −15 dB), the co-polarized phase difference (approaching to 0°), and the co-polarized correlation coefficient (over 0.7), and consequently were not discriminated from mudflats at L-band. The result shows different scattering features of oyster reefs at multi-frequency regimes.

3.2 *Radar Backscattering Model Simulation and Ground-Based Scatterometer Measurements*

Field campaigns were conducted in the intertidal flats of Jebu Island to measure surface profiles and volumetric soil moisture (see Choe et al. (2012) for the field measurements in detail). Root Mean Square (RMS) heights (i.e., vertical variation represented by root mean squared z values of a surface profile) and correlation length (i.e., horizontal variation of a surface profile determined by the surface auto correlation function) were calculated from the surface profiles (Fig. 4). The average RMS height and correlation length were 1.6 and 3.3 cm for oyster reefs, and were 0.6 and 11.3 cm for mudflats. The measurements confirmed oyster reefs have very rough surfaces with about 2.7 times higher RMS heights and about 3.5 times shorter correlation length compared to mudflats. The volumetric soil moisture was measured at over 50% in both of oyster reefs and mudflats. The dielectric constant as an input parameter for radar backscattering model simulation was converted from the average volumetric soil moisture using an empirical model by Hallikainen et al. (1985), and the dielectric constant of oyster shells composed of calcium carbonate was additionally considered for oyster reefs (see Choe et al. 2012). Based on the

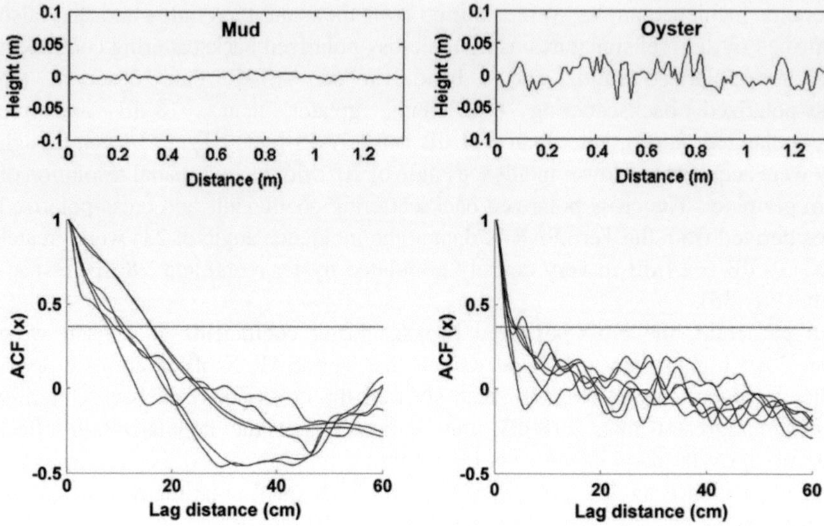

Fig. 4 Surface profile examples (upper) and auto-correlation function (ACF) plots (lower) of a total of six surface profiles collected (left; mudflats, right; oyster reefs). The dashed lines in the right figures represent y = 1/e, and the lag distance (x) intersected with y = 1/e is determined as correlation length. Figure modified from Choe et al. (2012)

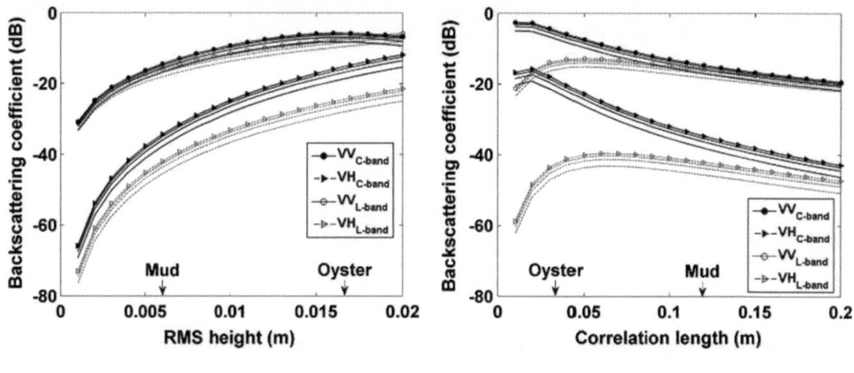

Fig. 5 IEM model simulations for C-band and L-band frequencies. VH and VV backscattering coefficients were generated as functions of RMS height (left, with a constant correlation length of 11.7 cm) and correlation length (right, with a constant RMS height of 0.6 cm). The dielectric constants of oyster reefs were calculated by assuming that the proportions of oysters relative to sediments are 0 (symbols), 25, 50, and 75%. Figure from Choe et al. (2012)

in situ measurements, polarimetric backscattering signatures of oyster reefs and mudflats were simulated by applying the Integral Equation Method (IEM) model (Fung et al. 1992) (Fig. 5). The IEM is theoretically modelled by approximate integral equations assuming a randomly rough dielectric surface and calculates radar backscattering coefficients depending surface parameters (i.e., RMS height, correlation length, dielectric constant) and radar sensor parameters (i.e., polarization, radar frequency, incidence angle) (Fung et al. 1992).

The simulated result showed that the cross-polarized backscattering coefficient (σ^0_{vh}) increases far more significantly than the co-polarized backscattering coefficient (σ^0_{vv}), as the RMS height increases and the correlation length decreases. In other words, the ratio of the cross-polarized backscattering to the co-polarized backscattering increases as surface roughness increases from the mudflat range to the oyster reef range. In particular, the difference between mudflats and oyster reefs was notable at the relatively short C-band wavelength. The dielectric constant, on the other hand, affected very little to the radar backscattering depending on the variation of volumetric soil moisture at over 50% or the proportion of oysters to sediments. Although the dielectric constant of oyster shells is assumed to be relatively very low, the effective dielectric constant of oyster reefs was still high due to the surrounding sediments saturated with high moisture contents. The IEM model simulation confirmed the radar backscattering in intertidal flats is mainly controlled by surface roughness and it increases dominantly in the cross-polarization at C-band according to the increase of surface roughness.

However, it appears that the difference between oyster reefs and mudflats is not relatively distinct at L-band (approximately 24 cm in wavelength) which is about four times longer than C-band (5.7 cm). It can be simply explained according to the Rayleigh criterion for rough surfaces ($h > \lambda/8\cos\theta$, where h is the standard deviation of surface heights, λ is the wavelength, and θ is the incidence angle) (Ulaby et al. 1986). It was indicated that oyster reefs are rough surfaces exceeding the criterion (e.g. h >0.8 cm, at an incidence angle of 36°) and mudflats are smooth surfaces at C-band, while they are both classified as smooth surfaces at L-band. It demonstrates that the difference in surface roughness of oyster reefs and mudflats can be detected differently depending on the radar wavelength. As a result, the radar backscattering sensitivity to the surface roughness of oyster reefs is much greater at C-band, which can cause multiple-diffused scattering and depolarization effects.

In addition, quad-polarimetric backscattering coefficients were measured from oyster reefs and mudflats using a ground-based scatterometer system with C-/X-band antennas (Fig. 6). Oyster reefs showed relatively strong backscattering signatures in all polarization channels. The radar backscattering from oyster reefs was not affected by incidence angles, whereas it rapidly decreased in mudflats as the incidence angle increased. Also, the difference between oyster reefs and mudflats became more obvious as the incidence angle increased. It suggests that incidence angles over 30° are more effective for differentiating oyster reefs and mudflats. In particular, a significant increase in the cross polarizations (HV and VH) was observed from oyster reefs, and it was a little more noticeable at X-band than at C-band. The average cross-polarized backscattering coefficient and the cross-polarized ratio at incidence angles between 20° and 50° were −19 dB and −8.4 dB at C-band, and were −13.5 dB and −6.3 dB at X-band, respectively. In the co-polarized channels of HH and VV, oyster reefs represented different backscattering characteristics. Mudflats showed higher backscattering in the VV polarization

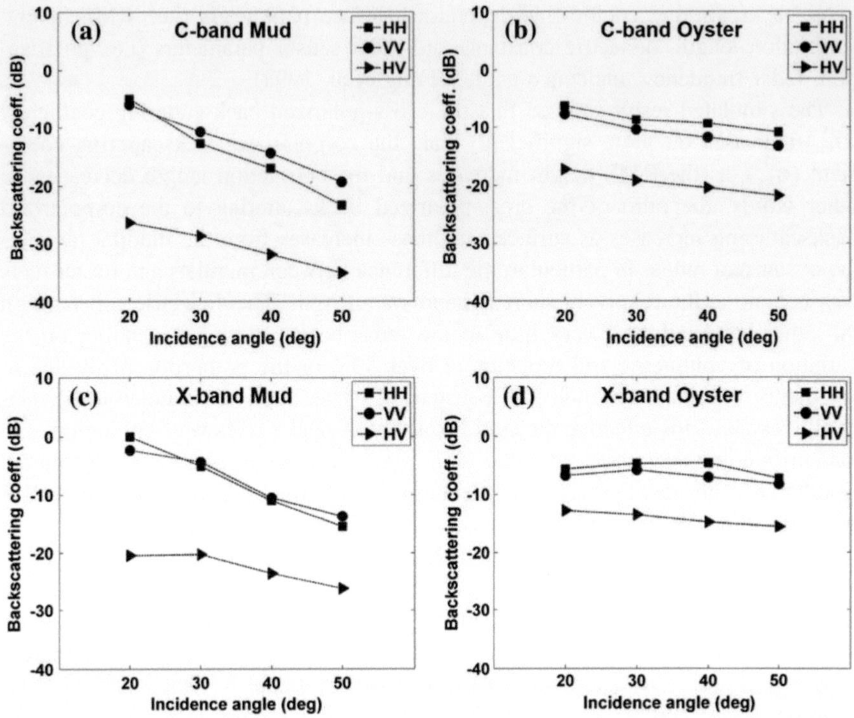

Fig. 6 Ground-based scatterometer measurements. C-band measurements in mudflats (**a**), C-band measurements in oyster reefs (**b**), X-band measurements in mudflats (**c**), and X-band measurements in oyster reefs (**d**). A total of twenty measurements were collected from each of

than in the HH polarization. It is a general backscattering feature of very smooth bare soils like mudflats, and the difference between HH and VV polarizations decreases as the surface roughness increases (Oh 2004). Oyster reefs, on the other hand, generated slightly higher backscattering in the HH polarization, which can be observed from forest or vegetated areas. On average, the HH-polarized backscattering coefficients are about 2 dB higher than the VV-polarized backscattering coefficients at C-band, and 1.5 dB higher at X-band.

4 SAR Signatures of Groundwater Discharge in Intertidal Flats

In the SAR image of Jebu intertidal flats where oyster signatures were detected, belt shaped and relatively darker areas along the upper part of the intertidal flats were also observed (Fig. 7). In order to investigate what made such darker features in the

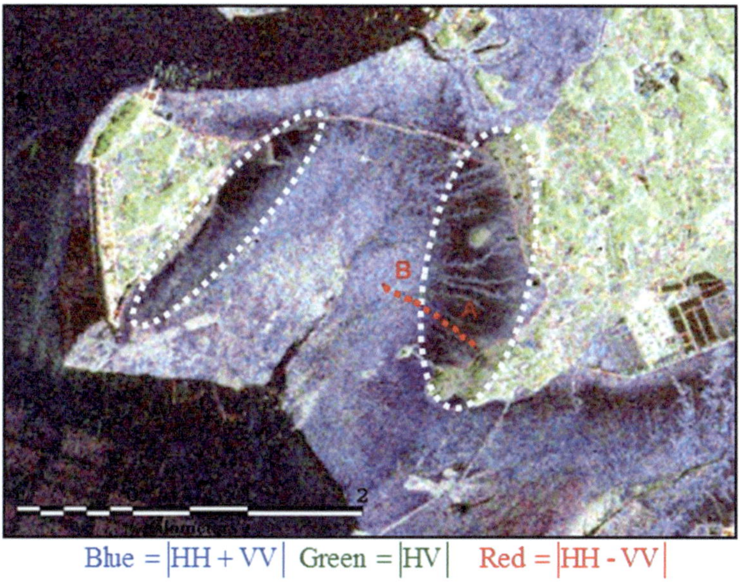

Fig. 7 RADARSAT-2 Pauli composite image of Jebu Island. Relatively dark features (white dashed areas) are observed along the upper intertidal flats. The red dots indicate the spots measuring the surface parameters for groundwater discharge. Figure from Kim et al. (2011)

SAR image, in situ measurement was conducted. Surface roughness and soil moistures were measured along a transect line. Backscattering coefficient values were calculated from IEM and Oh models along the transect line by entering the surface parameters obtained from field measurements (Fung et al. 1992; Oh et al. 2002), and they were compared to the extracted backscattering coefficient values from SAR data (Fig. 8). The Oh model is a semi-empirical model based on lab measurements of multi-frequency and multi-angular radar backscattering responses over bare soil surfaces with different roughness and moisture contents, which provides empirical formulas between surface parameters and polarimetric radar scattering (Oh et al. 2002). The model-simulated backscattering coefficient values have not always agreed with the SAR-derived values. Specifically, lower SAR backscattering was observed in area A, while similar radar backscattering values were estimated in area B.

Here we have focused more on the geological setting of the intertidal flats to find what can cause this low SAR backscattering. Our in situ investigations confirmed that area A was partially covered with remnant water in puddles, while there was almost no water in area B. Because the surface roughness parameters were measured only from exposed sediments, excluding the water puddle regions, the backscattering coefficients calculated from IEM and Oh models do not account for the contribution of these water puddles. However, the SAR-derived backscattering coefficients should include the influence of these water puddles, which lowers the

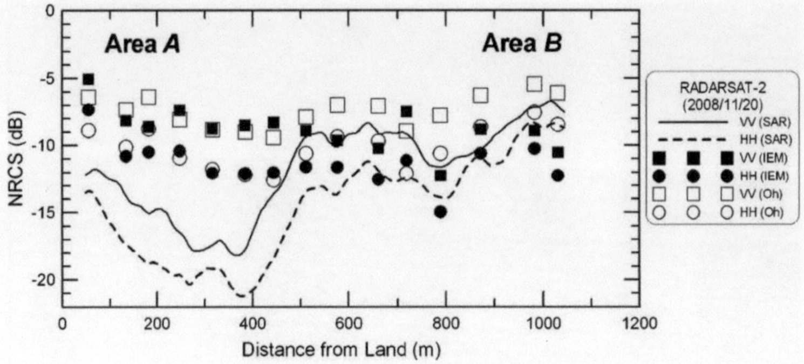

Fig. 8 Backscattering coefficient values extracted from SAR data and numerical radar scattering models (IEM and Oh models) along the transect between Area A and Area B in Fig. 7. The solid and dashed lines represent the SAR extracted backscattering coefficient values for VV and HH polarizations, respectively. The dark squares and circles represent the calculated values using IEM model for VV and HH polarizations, and the white squares and circles represent the calculated values using Oh model, respectively. Figure from Kim et al. (2011)

backscattering coefficient values due to the specular reflection on water puddle surfaces. We concluded that these water puddles were generated from groundwater discharge, because the Area A is closer to the land than the sea. If the water in the puddles is remnant sea water during ebb tide, the water puddles is more likely to be located closer to the sea, rather than close to the land. However, this is clearly not the case. So, the numerous water puddles near the land are related to the geological setting of these areas, which eventually led to conclude that these water puddles are attributed to groundwater discharge. A high occurrence of groundwater discharge was observed in Area A during our in situ measurement.

According to many studies, groundwater discharge at coastal regions plays an important role in transporting land nutrients to coastal oceans (Burnett et al. 2003; Moore 2006; Slomp and Van Cappellen 2004). We found that the strontium isotope ratio (Sr^{87}/Sr^{86}) of the oyster shells in Jebu intertidal flat showed a higher value than that of seawater. In general, the ratio of strontium 87–86 is greater in groundwater than in seawater (Paytan et al. 1993; Négrel et al. 2004). Thus, the groundwater discharge seemed to play an important role for wild oysters densely populated in intertidal flats.

5 Concluding Remarks

Multi-frequency polarimetric SAR showed a great potential for mapping oyster habitats and groundwater discharge signatures in intertidal flats, based on their distinct surface roughness and specular reflections. Very rough surfaces of oyster reefs even beyond the range of the surrounding mudflats can generate very distinct polarimetric signatures at the frequencies higher than C-band, particularly increasing in cross-polarized backscattering (i.e., HV, VH) relative to co-polarized backscattering (i.e., HH, VV). These signatures can be quantitatively mapped depending on frequency based on polarimetric target decomposition and depolarization parameters (e.g., cross-polarized ratio, co-polarized phase difference, co-polarized correlation coefficient) by using currently orbiting RADARSAT-2 (C-band, quad-pol), Sentinel-1 (C-band, dual-pol), TerraSAR-X (X-band, quad-pol), and COSMO-SkyMed (X-band, dual-pol) SAR sensors. Low SAR backscattering signatures observed in the upper part of the intertidal flats can indicate the presence of water, which can possibly be associated to groundwater discharge. Further in situ investigation and strontium isotope analysis may be needed to confirm the correlation with groundwater discharge.

References

Belluco E, Camuffo M, Ferrari S, Modenese L, Silvestri S, Marani A, Marani M (2006) Mapping salt-marsh vegetation by multispectral and hyperspectral remote sensing. Remote Sens Environ 105:54–67

Brockmann C, Stelzer K (2008) Optical remote sensing of intertidal flats. In: Barale V, Gade M (eds) Remote sensing of the European seas. Springer, Heidelberg, pp 117–128

Burnett W, Bokuniewicz H, Huettel M, Moore W, Taniguchi M (2003) Groundwater and pore water inputs to the coastal zone. Biogeochemistry 66:3–33

Choe BH, Kim D (2011) Retrieval of surface parameters in tidal flats using radar backscattering model and multi-frequency SAR data. Kor J Remote Sens 27:225–234

Choe BH, Kim D, Hwang JH, Oh Y, Moon WM (2012) Detection of oyster habitat in tidal flats using multi-frequency polarimetric SAR data. Estuar Coast Shelf Sci 97:28–37

Choi JK, Ryu JH, Lee YK, Yoo HR, Woo HJ, Kim CH (2010) Quantitative estimation of intertidal sediment characteristics using remote sensing and GIS. Estuar Coast Shelf Sci 88:125–134

Choi JK, Oh HJ, Koo BJ, Lee S, Ryu JH (2011) Macrobenthos habitat mapping in a tidal flat using remotely sensed data and a GIS-based probabilistic model. Mar Pollut Bull 62:564–572

Dehouck A, Lafon V, Baghdadi N, Roubache A, Rabaute T (2011) Potential of TerraSAR-X imagery for mapping intertidal coastal wetlands. In: Proceedings of 4th TerraSAR-X science team meeting, Oberpfaffenhofen, Germany

Freeman A, Durden SL (1998) A three-component scattering model for polarimetric SAR data. IEEE Trans Geosci Remote Sens 36:969–973

Fung A, Li Z, Chen K (1992) Backscattering from a randomly rough dielectric surface. IEEE Trans Geosci Remote Sens 30:356–369

Gade M, Alpers W, Melsheimer C, Tanck G (2008) Classification of sediments on exposed tidal flats in the German Bight using multi-frequency radar data. Remote Sens Environ 112:1603–1613

Gade M, Melchionna S, Stelzer K, Kohlus J (2014) Multi-frequency SAR data help improving the monitoring of intertidal flats on the German North Sea Coast. Estuar Coast Shelf Sci 140:32–42

Hallikainen MT, Ulaby FT, Dobson MC, El-rayes MA, Wu LK (1985) Microwave dielectric behavior of wet soil-part 1: empirical models and experimental observations. IEEE Trans Geosci Remote Sens 23:25–34

Hosono T, Ono M, Burnett WC, Tokunaga T, Taniguchi M, Akimichi T (2012) Sptial distribution of submarine groundwater discharge and associated nutrients within a local coastal area. Environ Sci Technol 46:5319–5326

Hwang DW, Kim G, Lee WC, Oh HT (2010) The role of submarine groundwater discharge (SGD) in nutrient budgets of Gamak Bay, a shellfish farming bay, in Korea. J Sea Res 64: 224–230

Jeen SW, Kim JM, Ko KS, Yum B, Chang HW (2001) Hydrogeochemical characteristics of groundwater in a mid-western coastal aquifer system, Korea. Geosci J 5:339–348

Kim Y, Lee K, Jung H (2003) Hydrogeological properties of coastal aquifer in Hwaseong area: implication for tidal effects. J Geol Soc Korea 39:385–390 (in Korean with English abstract)

Kim D, Moon WM, Kim G, Park SE, Lee H (2011) Submarine groundwater discharge in tidal flats revealed by space-borne synthetic aperture radar. Remote Sens Environ 115:793–800

Koh CH (2001) The Korean tidal flat: environment, biology and human. Seoul National University Press, Seoul

Le Bris A, Rosa P, Lerouxel A, Cognie B, Gernez P, Launeau P, Robin M, Barillé L (2016) Hyperspectral remote sensing of wild oyster reefs. Estuar Coast Shelf Sci 172:1–12

Lee YK, Park JW, Choi JK, Oh Y, Won JS (2012) Potential uses of TerraSAR-X for mapping herbaceous halophytes over salt marsh and tidal flats. Estuar Coast Shelf Sci 115:366–376

Moore WS (2006) The role of submarine groundwater discharge in coastal biogeochemistry. J Geochem Explor 88:389–393

Négrel P, Petelet-Giraud E, Widory D (2004) Strontium isotope geochemistry of alluvial groundwater: a tracer for groundwater resources characterization. Hydrol Earth Syst Sci 8:959–972

Nieuwhof S, Herman PM, Dankers N, Troost K, van der Wal D (2015) Remote sensing of epibenthic shellfish using synthetic aperture radar satellite imagery. Remote Sens 7:3710–3734

Oh Y (2004) Quantitative retrieval of soil moisture content and surface roughness from multipolarized radar observations of bare soil surfaces. IEEE Trans Geosci Remote Sens 42:596–601

Oh Y, Sarabandi K, Ulaby F (2002) Semi-empirical model of the ensemble-averaged differential Mueller matrix for microwave backscattering from bare soil surfaces. IEEE Trans Geosci Remote Sens 40:1348–1355

Park SE, Moon WM, Kim D (2009) Estimation of surface roughness parameter in intertidal mudflat using airborne polarimetric SAR data. IEEE Trans Geosci Remote Sens 47:1022–1031

Paytan A, Kastner M, Martin EE, Macdougall JD, Herbert T (1993) Marine barite as a monitor of seawater strontium isotope composition. Nature 366:445–449

Ryu JH, Won JS, Min KD (2002) Waterline extraction from Landsat TM data in a tidal flat: a case study in Gomso Bay, Korea. Remote Sens Environ 83:442–456

Ryu JH, Na YH, Won JS, Doerffer R (2004) A critical grain size for Landsat ETM+ investigations into intertidal sediments: a case study of the Gomso tidal flats, Korea. Estuar Coast Shelf Sci 60:491–502

Slomp CP, Van Cappellen P (2004) Nutrient inputs to the coastal ocean through submarine groundwater discharge: controls and potential impact. J Hydrol 295:64–86

Ulaby FT, Moore RK, Fung AK (1986) Microwave remote sensing: active and passive-Vol II: radar remote sensing and surface scattering and emission theory. Artech House, Norwood MA

van Beijma S, Comber A, Lamb A (2014) Random forest classification of salt marsh vegetation habitats using quad-polarimetric airborne SAR, elevation and optical RS data. Remote Sens Environ 149:118–129

van der Wal D, Herman PMJ, Wielemaker-van den Dool A (2005) Characterisation of surface roughness and sediment texture of intertidal flats using ERS SAR imagery. Remote Sens Environ 98:96–109

Remote Sensing for Improved Forecast of Typhoons

Yuei-An Liou, Ji-Chyun Liu, Fabrice Chane-Ming, Jing-Shan Hong, Ching-Yuang Huang, Po-Kuan Chiang and Samuel Jolivet

Abstract This chapter presents the advantages of remote sensing in various aspects of tropical cycles or typhoons for the purpose of improved forecasts. A variety of variables influencing the typhoons become our concerns and sequentially investigated. First of all, to effectively predict the rainfall associated with a landfalling typhoon, the ground-based Global Positioning System (GPS) zenith total delay is combined with Doppler radar data through data assimilation algorithms. Subsequently, discussions on a natural phenomenon of interactions among two typhoons with and without tropical depressions (TDs) are elaborated. Remote sensing imagery and image processing techniques are applied to analyze relevant interactions and physical responses, including TDs' appearance, development, interaction and how they merge. Then, the application of remote sensing observational data in numerical modeling for the study of atmospheric gravity waves, especially during the occurrence of asymmetric tropical cyclones is presented.

Y.-A. Liou (✉) · P.-K. Chiang
Center for Space and Remote Sensing Research,
National Central University, Taoyuan, Taiwan
e-mail: yueian@csrsr.ncu.edu.tw

Y.-A. Liou · C.-Y. Huang · P.-K. Chiang
Taiwan Group on Earth Observations, Zhubei, Taiwan

J.-C. Liu
Electrical Engineering Department, Chien Hsin University
of Science and Technology, Taoyuan, Taiwan

F. Chane-Ming
Laboratoire de l'Atmosphère et des Cyclones, Université de la Réunion,
UMR8105, CNRS-Météo France-Université, La Réunion, France

J.-S. Hong
Central Weather Bureau, Taipei, Taiwan

C.-Y. Huang
Department of Atmospheric Sciences, National Central University,
Taoyuan, Taiwan

S. Jolivet
Meteobooking, Bossonnens, Switzerland

V. Barale and M. Gade (eds.), *Remote Sensing of the Asian Seas*,
https://doi.org/10.1007/978-3-319-94067-0_14

Finally, a brief introduction is given to the oceanic surface wind measurement from different satellites with already demonstrated or potential impacts on typhoon simulations and predictions. Note that atmospheric and oceanic parameters derived from observations of Global Navigation Satellite System (GNSS) receivers onboard low Earth orbit (LEO) satellites, i.e., FORMOSAT-3/COSMIC and to-be-launched FORMOSAT-7/COSMIC-2, are also discussed within the selected topics. The importance of implementing remote sensing technology in the investigation and forecast of typhoons is the conclusion.

Keywords Remote sensing · Typhoon · Forecast · Atmospheric gravity wave GPS · GNSS-R

1 Introduction

This chapter aims to demonstrate the effectiveness of remote sensing technology in the investigation and improved forecasts of typhoons in the western Pacific Ocean where typhoons are considered as one of the most critical hazards and causes for disasters due to their resulting torrential rains, high winds, and associated storm surges.

Torrential rains from decaying typhoons can produce extensive urban and river flooding, and trigger landslides and debris flows in mountainous regions. For example, 2965 mm of rainfall was measured over a four-day period during Typhoon Morakot (2009), triggering massive mudslides and severe flooding and resulting in 673 deaths and 26 missing persons (Wu 2013) in Taiwan. Accurate prediction of typhoon track, intensity, wind fields, and reliable quantitative precipitation forecast (QPF) is crucial for taking precautions to protect human life, agriculture, and industry.

However, many challenges remain to achieve highly accurate predictions of interest. Among them, we chose to discuss four issues incorporating remote sensing technology in the following sections of this chapter: (1) How much can space-borne sensor-observed water vapor improve the QPF (Sect. 2); (2) What would the tracks and intensity of the typhoons be modified under the circumstance when there exist two typhoons with or without additional tropical depressions (Sect. 3); (3) How would atmospheric gravity waves (GWs) evolve during the occurrence of asymmetric tropical cyclones (TCs) (Sect. 4); and (4) How large improvements can be achieved when oceanic surface wind measurements are assimilated into numerical weather prediction (NWP) models (Sect. 5)?

2 Predicting Rainfall of a Typhoon by Assimilating GPS-Derived Zenith Total Delay Combined with Doppler Radar

Water vapor, as a source of precipitation, plays an important role in maintaining the structure and evolution of typhoons. GPS sensing has been demonstrated as a reliable and accurate method to measure the total column atmospheric water vapor, also called precipitable water vapor (PWV) (Liou et al. 2001). Its application in meteorology is commonly called GPS Meteorology, such as calibrations of radiosondes (Ciesielski et al. 2010) and satellite observations (Birkenheuer and Gutman 2005), investigation of typhoon landfall processes (Liou and Huang 2000), evolution of monsoonal systems (Kursinski et al. 2008), etc. Taiwan Central Weather Bureau set up a network of 169 GPS stations to observe PWV with relatively high temporal and spatial resolution that is not achieved by any other observing system.

Assimilation of GPS-derived PWV into NWP systems has been demonstrated to effectively improve moisture and rainfall forecasts (Gutman and Benjamin 2001; Bennitt and Jupp 2012; Mahfouf et al. 2015), and tropical cyclone intensity forecasts (Shoji et al. 2011). Prediction of short-duration extreme rainfalls (e.g., >100 mm/3 h) is challenging with limited predictability due to lack of observations, vigorous precipitation processes, and strong system evolution. The key to enhance predictability of short-duration extreme rainfall events depends on high resolution and advanced numerical model forecasts. A rapidly updated numerical system to assimilate radar, GPS PWV, and surface observations is critical to enhance predictability for short-duration extreme rainfall.

Assimilation of radar observations improves model QPF (Xiao et al. 2005). Radar observations contribute radial wind and rain/snow water information to the model initial condition. However, due to lack of water vapor information, radar observation limits its impact within the first few hours (Wang et al. 2013). In contrast, assimilation of GPS PWV with radar observations is helpful not only to extend the impact of radar data assimilation but also to improve the accuracy of model QPF.

In this section, near real-time continuous island-wide water vapor observations are shown to contribute to long-term climate monitoring and short-term weather forecasting positively. The impact of GPS PWV on model QPF in Typhoon Dujuan (2015) is demonstrated. Typhoon Dujuan moved westward and made landfall at approximately 1040 UTC (Coordinated Universal Time) on 28 September 2015, exiting Taiwan at 1700 UTC on the same day (Fig. 1a). Before landfall, Dujuan reached a maximum wind speed of approximately 51 m/s. The composited column maximum radar reflectivity at 0700 UTC on 28 September 2015 shows the clear eye, active eyewall, and an enhanced rainband due to terrain lifting of typhoon circulation in northeastern Taiwan (Fig. 1b), west of the typhoon eye. It is the most critical issue for QPF to predict the rainfall associated with such a rainband system.

Two data assimilation experiments were designed to highlight the impact of GPS PWV on model QPF. The control run is the experiment with assimilating radar observations only, including radial wind and radar reflectivity. The assimilation of radial wind and reflectivity is expected to correct model wind and rain water. The GPS PWV run assimilates radar observation and GPS PWV, as well as is expected to correct the model moisture field additionally. Figure 2 shows the difference between two experiments in initial water vapor mixing ratio. The figure illustrates that assimilation of GPS PWV makes the initial moisture field drier than that in the control run. The impact of GPS PWV in the model initial condition

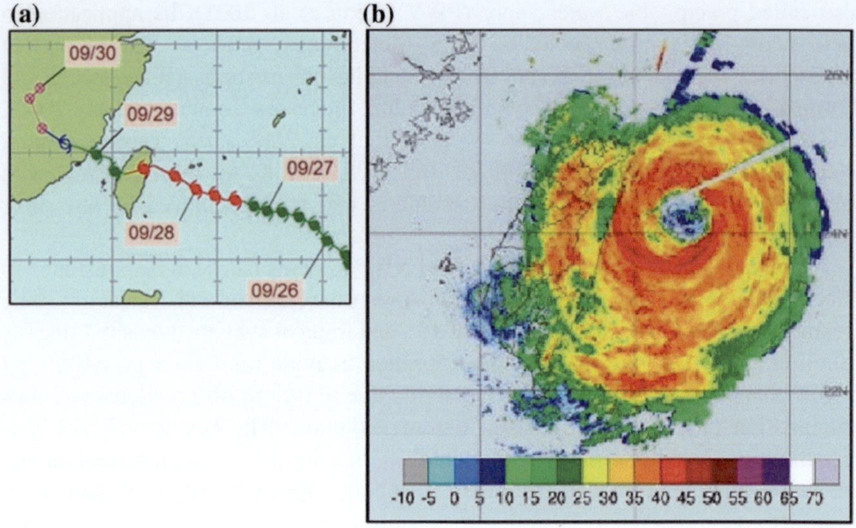

Fig. 1 **a** The best track of Typhoon Dujuan (2015); and **b** The composited column maximum radar reflectivity at 0700 UTC on 28 September 2015. Reproduced from (Typhoon Database 2015, http://rdc28.cwb.gov.tw/)

occurs not only over Taiwan Island but also extends to roughly 200 km offshore although GPS observations are only available over Taiwan Island.

Figure 3a–c show the observations, model 0–6 h QPF for the control and GPS PWV runs, respectively. They show that the GPS PWV run produces less rainfall over the Southwestern Taiwan. Rainfall near typhoon center over the Western Taiwan in the GPS PWV run is less than that in the control run. Apparently, results in the GPS PWV run are more consistent with the observation as shown in Fig. 3a. It is also shown that assimilation of GPS PWV can improve short-range rainfall forecast during a typhoon landfall period. In addition to GPS PWV, the use of more remote sensing observations, especially from the GPS radio occultation (RO) and Global Navigation Satellite System (GNSS) reflectometry observations of the future FORMOSAT-7/COSMIC-2 mission, are expected to benefit the improvement of the short-duration extreme rainfall system. This is of most critical importance for disaster reduction and prevention.

3 Interactions Among Two Typhoons—A New Aspect of Fujiwhara Effect

A natural phenomenon of interactions among two typhoons (or TCs) with or without TD is discussed in this section by using remote sensing imagery and image processing techniques. Cases with and without physical influences of TDs on two

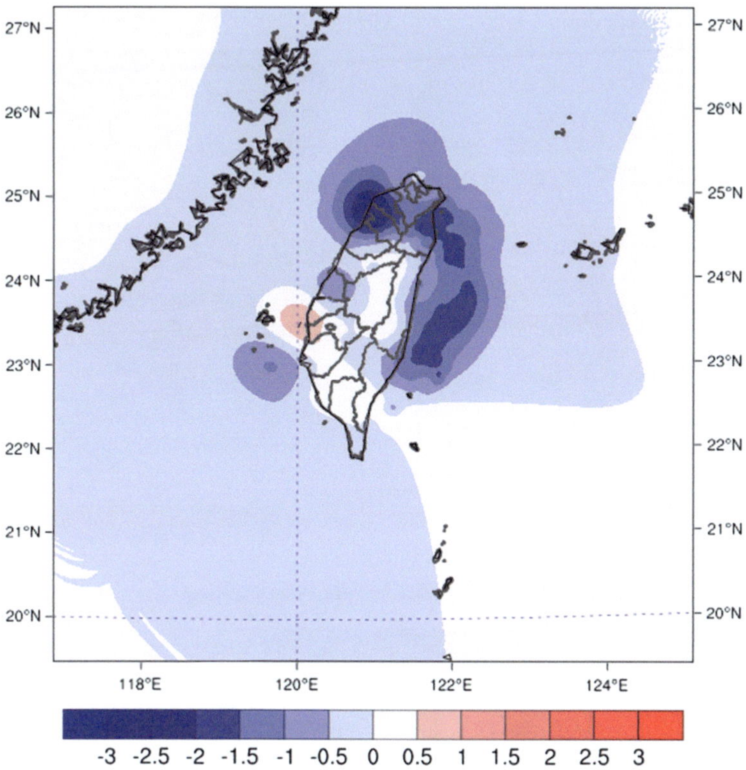

Fig. 2 The difference for the initial water vapor mixing ratio (shading, g/kg) at model level 10 between the control run and GPS PWV run. The initial time is at 1200 UTC on 28 September 2015

typhoons are examined through analyses of four parameters (distance, height difference, rotation, and size ratio). Relevant responses include TD's appearance, development, interaction, and mergence.

3.1 Fujiwhara Effect

When two tropical cyclones are located with a distance of about 1400 km, the related phenomenon is called cyclone-cyclone interaction or "Fujiwhara effect" (Fujiwhara 1923). In general, stronger TCs have larger sizes, bigger height difference, faster rotation speed, etc. When two stronger TCs come close, they start to interact with each other at distance of approximately 1400 km. In contrast, two weaker TCs must be closer within a distance of around 1100 km before interaction. If one TC is stronger than the other, the interaction would happen at distances between 1200 and 1300 km. However, two questions remain: (1) whether Fujiwara

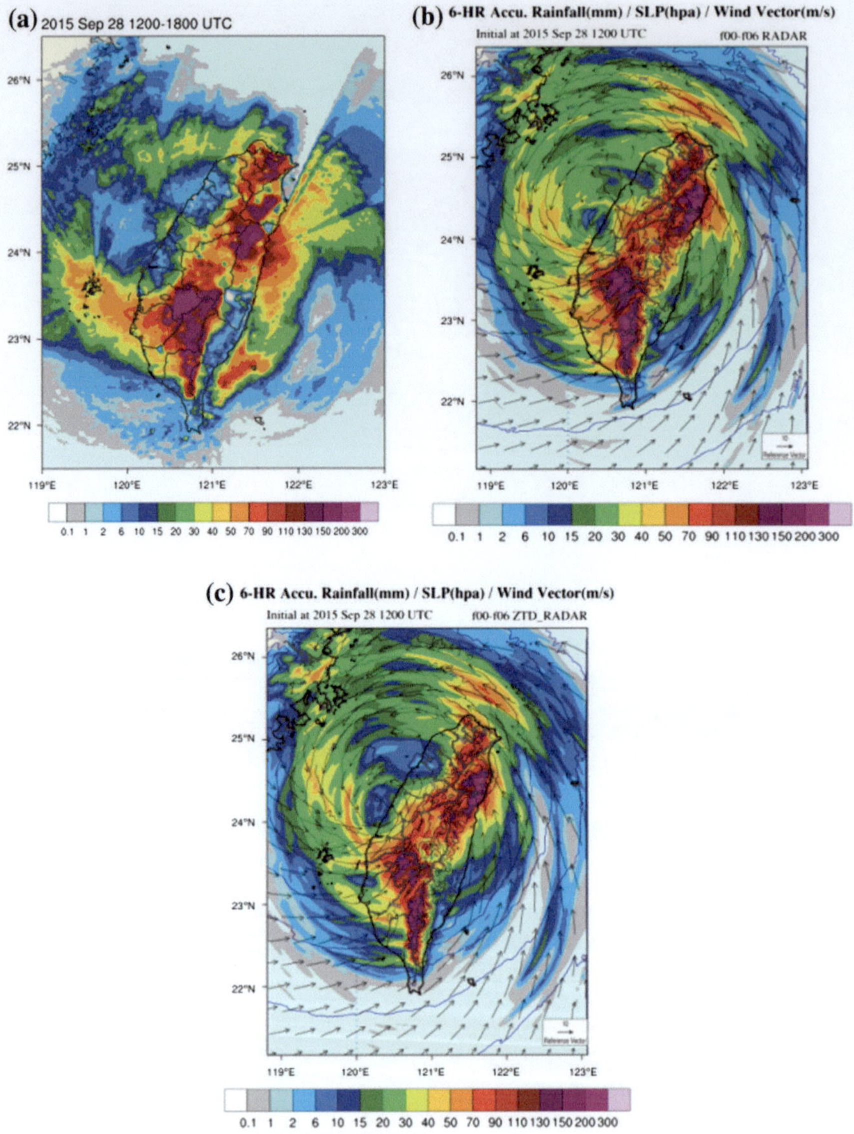

Fig. 3 The 6-h accumulated rainfall (mm) from 1200 UTC through 1800 UTC on 28 September 2015: **a** observation; and model 0–6 h QPF for **b** the control run, and **c** the GPS PWV run

effect still exists, if distance between two typhoons is more than 1500 km; and (2) what interactions (i.e. depression-cyclone interactions) will occur when lower/ smaller TDs are located between two higher/larger cyclones with closer distances between them and the major TC's than between the two TC/s themselves?

3.2 Satellite Images and Methodology

To observe the cloud structure in detail, the short infrared channel (Platnick et al. 2003; King et al. 2003; Huang et al. 2004; Hong et al. 2007; Minnis et al. 2011) with 3.7 μm (Platnick et al. 2003) was used by satellite instruments. The image reconstruction technique (IRT) was then applied to examine the satellite images for further information. For describing depression-cyclone and related interactions, four parameters (i.e. distance, height difference, rotation, and sizes of typhoons and TDs), cloud images of typhoons, tracks of typhoons, and time series of interactions were employed. Subsequently, to investigate the behaviors of typical cyclone-cyclone interactions and TD-inlaid cyclone-cyclone interactions, two cases were utilized (1) Typhoons Fitow and Danas (2013) as a case of typical cyclone-cyclone interactions; and (2) Typhoons Tembin and Bolaven (2012) as an example of TD-inlaid cyclone-cyclone interactions.

3.3 Interactions Among Two Typhoons Without Tropical Depression

Typhoons Fitow and Danas (2013) were observed from satellite cloud images. Initially, they were separated by a distance of 1600 km (i.e. more than 1400 km of typical cyclone-cyclone interaction). The typical interaction could not occur between the two typhoons. Then, Typhoon Danas moved quickly, and Typhoon Fitow became weaker after touching the mainland. Although the distance between the two typhoons was decreased to 1200 km, there was no mid TD presence between them. It leads to the conclusion that the cyclone-cyclone interaction did not happen.

3.4 Interactions Between Two Typhoons with Two Tropical Depressions

Figure 4a–c present three representative cloud images (t_1, t_3, and t_5) of Typhoons Tembin and Bolaven (2012), taken at six times (t_1–t_6, 1st time interval = 5 h and the rest of the intervals = 4 h since t_1 = 12h32, August 21) between August 21 and 22. In Fig. 4a, TD_1 appears between two typhoons. The developed TD_1 and TD_2 demonstrate two sets of depression-cyclone interactions via individual interactions "Tembin-and-TD_1" and "Bolaven-and-TD_2" presented in Fig. 4b. These interactions could be observed by inspecting the convections between them. Firstly, the mid TDs, Tembin-and-TD_1 located approximately 910–980 km apart with a height difference of 2 km at t_2–t_6. Then, Bolaven-and-TD_2 came close to around 780–920 km apart with a height difference of 3.5 km at t_2–t_4. Two sets of

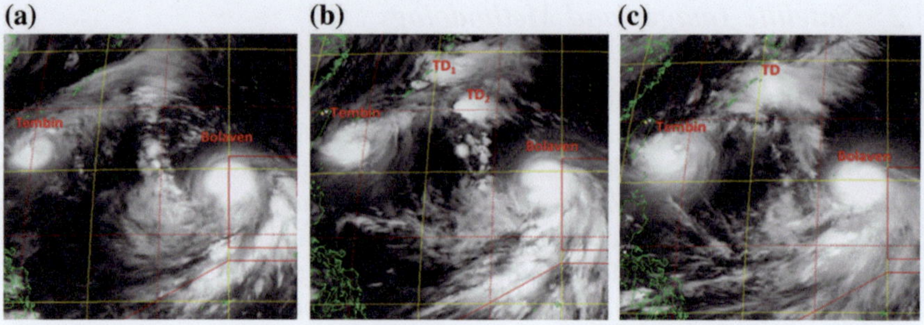

Fig. 4 Three representative time cloud images of Typhoons Tembin and Bolaven and TDs (2012): **a** at t_1 = 12h32, August 21, **b** at t_3 = 21h32, August 21, and **c** at t_5 = 05h32, August 22. Reproduced from (Liu et al. 2015; Liou et al. 2016)

depression-cyclone interactions occurred accordingly with the consequence to be assumed as a combined depression-cyclone interaction. In the end, TD_1 and TD_2 were merged and formed a new TD as presented in Fig. 4c.

Figure 4a–c reveal a natural evolution starting from a TD's appearance (at t_1), two TDs' interactions (at t_3), to their mergence into a single TD (at t_5). Tracks of two typhoons and two TDs are plotted in Fig. 5. To elaborate further, Typhoon Tembin moved towards west, while Typhoon Bolaven moved towards northwest within t_1–t_6. The direction of TD_1's movement changed counterclockwise to the south after interacting with Typhoon Tembin within t_2–t_6. On the other hand, the direction of TD_2's movement was changed counter-clockwise and directed to Typhoon Bolaven within t_2–t_4.

For the TD-inlaid cyclone-cyclone case, the observed distance between two typhoons was 1600 km, indicating no typical cyclone-to-cyclone interactions. However, when typhoon (TC) Tembin-and-TD_1 (Bolaven-and-TD_2) came close to 910 (/780) km apart with a height difference factor of 2 (/3.5) km, the combined

Fig. 5 Tracks of Typhoons Tembin and Bolaven and two TDs (2012) in the period of August 21–22. Reproduced from (Liu et al. 2015; Liou et al. 2016)

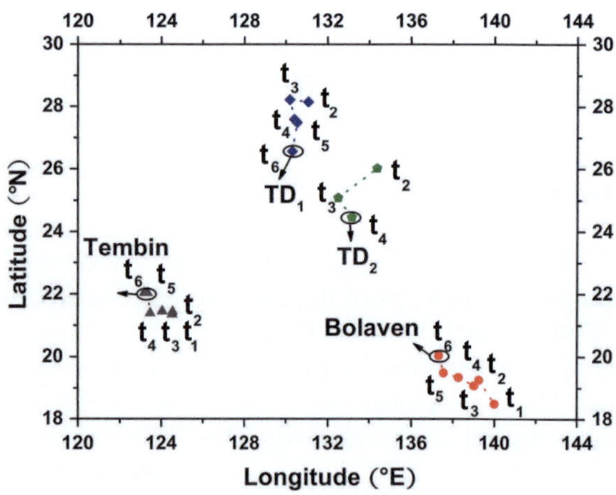

depression-cyclone interactions were formed. TD_1 and TD_2 merged and then formed a new TD. This newly formed TD provided upward convections to make typhoon(s) to start rotating and generate strengthened power, with the effect of the TC to TC interaction (Liou et al. 2016).

According to the Fujiwhara effect, typical cyclone-cyclone interaction does not happen in the case presented here. However, the long-distance typhoon-typhoon interaction may occur, owing to two sets of depression-cyclone interactions and subsequent mergence of two tropical depressions (i.e. TD_1 and TD_2). Hence, one typhoon (i.e. Tembin) will indirectly interact with the other typhoon (i.e. Bolaven). In addition to distance, the rest of the three parameters of concern (i.e. height difference, rotation, and size) can be viewed as appropriate indicators to analyze and discuss, for the purpose of further elaborating the depression-cyclone interactions. Through discussions of the four parameters, the effect of the two TDs on the two typhoons can be considered as an alternative approach to depict interactions of two typhoons with a distance (1500 km) apart of more than 1400 km with 100 km deviation. Note that the threshold distance to define the cyclone-cyclone interactions can be characterized by the generalized empirical formulas presented by Liou et al. (Liou et al. 2016). The defined interactions can be helpful to be an indicator for improved forecasts of typhoons. Readers of this chapter are referred to Liou et al. (2016) for further details.

4 Atmospheric Gravity Waves During the Occurrence of Asymmetric Tropical Cyclones

Atmospheric GWs affect the dynamics of the middle atmosphere when they propagate upward from troposphere to higher altitudes (Fritts and Alexander 2003). Besides transport of chemical constituents, deposition of GW momentum and induced mixing are key processes in atmospheric chemistry. At larger scales, both observations and modeling support that GWs must be adequately simulated in the global circulation and climate models (Kafando et al. 2015). Several studies reveal that the parameterization of GWs is crucial for representation of quasi-biennial oscillation (QBO) in the tropical stratosphere. From 1995, the international effort supported by SPARC (Stratospheric Processes and their Role in Climate) focuses on climatology of GW parameters, wave activity and sources from observations to improve parameterization in global and climate models. However, this issue remains unresolved today despite much progress in modeling and GW parameterization achievements during the last decade. Indeed, convective sources are complex and poorly represented by models unlike other primary GW sources such as orography and jet imbalance. Thus, from 2012, the SPARC community decided to focus on sources and new descriptions of the variability of GWs for global and climate models (Cámara et al. 2014). TCs are among the most destructive convective phenomena, which form over warm tropical ocean basins. With an average of 80 TCs in the Tropics every year, they are significant threats for the island and coastal regions. Possible effects of climate variability and change on TC dynamics and related damages are one of the most important topics in climate research in the Tropics (Kuleshov et al. 2009). Therefore, several proxies for TC activity are developed for long-term studies. Analyzing radiosonde data, Chane-Ming et al. (2002) observed the production of a significant amount of GW energy during TC Hudah (2000) on Madagascar and in Mozambique. Chane-Ming et al. (2010) analyzed the characteristics of convective GWs observed in the upper troposphere (UT) and lower stratosphere (LS) for intense TC Dina (January 2002) and Faxai (December 2001) in the South West basin of the Indian Ocean and Northwest Pacific, respectively. For the two TCs, GW energy activity increased in the LS during TC intensification and was shown to be strongly correlated with the intensity of the TC. SSMI (Special Sensor Microwave Imager) and GMS-5 (Geostationary Meteorological Satellite-5, i.e. Himawari 5) satellite images reveal an intense asymmetric TC Faxai on 22 December 2001 at 1800 UTC near its peak intensity (Fig. 6a).

Recent increase in spatial and temporal resolution of numerical weather prediction models as well as improvement in data assimilation, especially assimilation of GPS radio occultation bending angles from late December 2006, led to improved stratospheric temperature distribution over tropical oceans as well as the description of TCs (internal structure and trajectory) in comparison with observations. The analyses of ECMWF (European Centre for Medium-Range Weather Forecasts) and Meso-NH (Mesoscale Non-Hydrostatic Model) are in agreement with radiosonde and GPS radio occultation data; those indicate the evidence of a dominant TC-related quasi-inertia mesoscale GW in the LS, more intense during TC intensification (Chane-Ming et al. 2014). A wavenumber-1, responsible of the asymmetry of the wind fields, can be regarded as a source of leading inertia GW with horizontal wavelengths of 400–800 km during intensification and landfall. In particular, FORMOSAT-3/COSMIC radio occultation data are used to derive GW activity and characteristics of vertical wavelengths over the ocean and validate modeling of TC environment. Horizontal GW parameters can also be derived from RO data (Chane-Ming et al. 2016). Figure 6b and c visualize asymmetric structures in TC Soudelor (2015) during its peak of intensity on August 4 near Saipan and when approaching Taiwan on 7 August 2015.

In comparison with Fig. 6a, better horizontal resolution of satellite images are now available for the description of TC morphology with time. Figure 6d displays the spatial distribution of 22 GPS RO profiles from 6 to 9 August during TC Soudelor. GPS RO profiles on 7 August enable us to probe the TC inner core as well as its environment. More profiles will be available in the next two years when the FORMOSAT-7/COSMIC-2 program will produce 12,000 sounding profiles per day, compared to the approximate 2,000 soundings per day currently generated by FORMOSAT-3/COSMIC. The asymmetric structure of TC Soudelor is well observed on the distribution of horizontal wind intensity derived from operational ECMWF analyses with $0.125° \times 0.125°$ horizontal resolution on 7 August at 1200 UTC and 4-km heights during the beginning of the landfall of TC Soudelor on

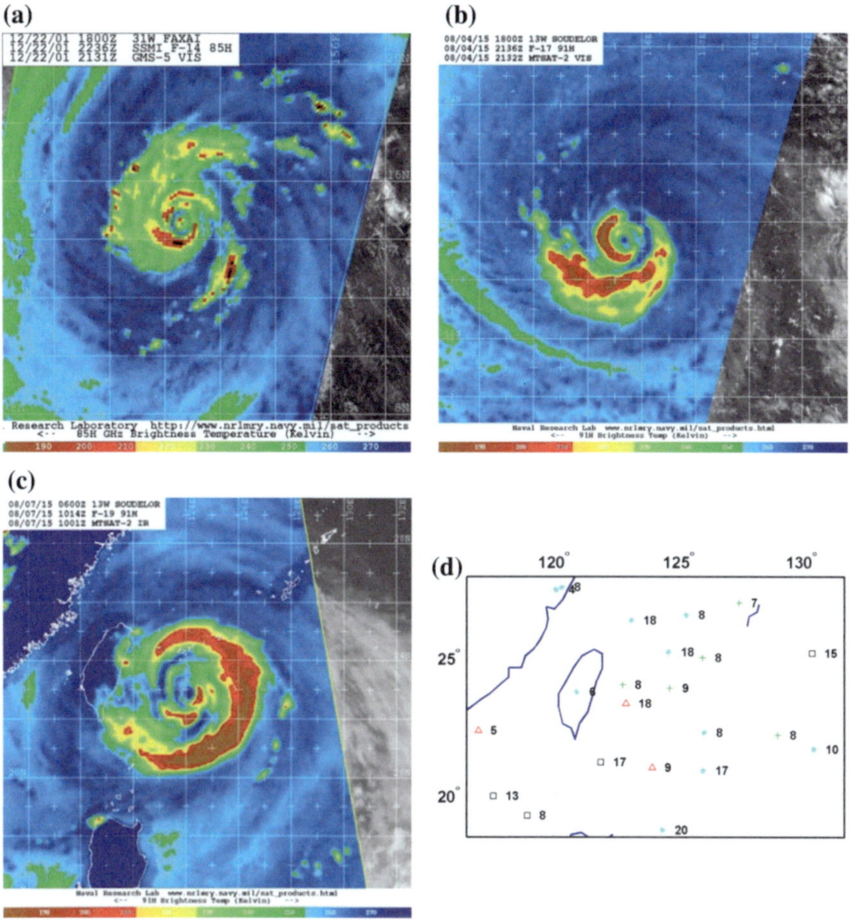

Fig. 6 Satellite images for a category-5 typhoon: **a** TC Faxai on 22 December 2001 at 1800 UTC near its peak intensity, **b** TC Soudelor near Saipan (4 August 2015), **c** TC Soudelor near Taiwan (7 August 2015), and **d** Distribution of FormoSat-3/COSMIC GPS RO data during TC Soudelor on 6, 7, 8 and 9 August 2015 marked with multiplication sign, plus sign, triangle, square symbols, respectively. Numbers indicate hours (UTC) for corresponding dates (symbols)

Taiwan (Fig. 7a). The simulated structure is more contracted probably because of the difference in time between ECMWF and satellite images and increased effects of Taiwan orography on TC Soudelor during the landfall. Figure 7b and c show evidence of concentric mesoscale GW structures with horizontal wavelengths of 100–200 km in the UT and propagating upward in the LS.

In conclusion, recent remote sensing imagery and observations have proved to be important for future studies on TC such as TC asymmetry and induced-

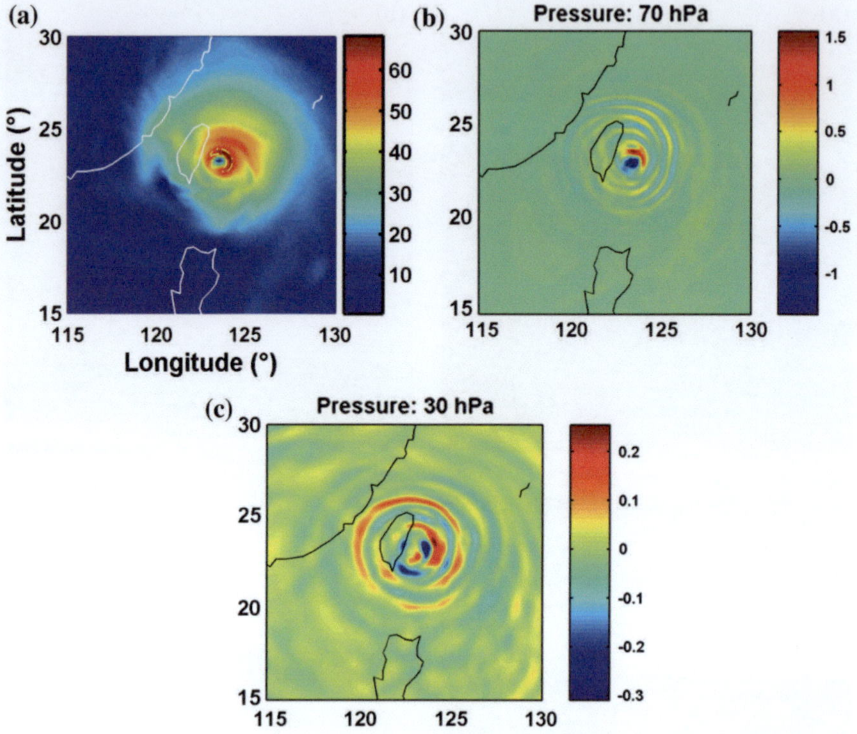

Fig. 7 **a** Distribution of horizontal wind intensity (ms-1) derived from operational ECMWF analyses with 0.125° × 0.125° horizontal resolution on 7 August at 1200 UTC and 4-km heights. The yellow dash-dot line indicates wind values >60 ms^{-1}. Corresponding vertical velocity in the pressure coordinate (Pa s^{-1}), **b** in the upper troposphere at 70 hPa, and **c** in the lower stratosphere at 50 hPa

stratospheric GWs. At present, a preliminary climatological study based on satellite images of asymmetric TCs, RO data and signatures of GWs in ECMWF is performed for some TC basins.

5 Oceanic Surface Wind Measurement for Typhoon Simulations and Predictions

The impacts of oceanic surface observations, mostly the surface wind, on typhoon simulations and predictions have been demonstrated by previous research (Huang et al. 2005). Here, a brief introduction is given to the oceanic surface wind measurement from different satellites with already demonstrated or potential impacts on typhoon simulations and predictions.

5.1 Scatterometer

The NASA QuikSCAT (Quick Scatterometer) is an Earth observation satellite carrying the SeaWinds scatterometer (an active radar scatterometer), aiming at measuring the surface wind speed and direction over the ice-free global oceans. QuikSCAT measured winds in measurement swaths 1800 km wide centered on the satellite ground track with no nadir gap, thus capable of collecting at least one vector wind measurement over 93% of the global oceans each day. QuikSCAT provided measurements of the wind speed and direction, which is referenced to 10 m above the sea surface at a spatial resolution of 12.5–25 km.

Multiple on-orbit scatterometer platforms have been built recently by several countries, including the operational European Organisation for the Exploitation of Meteorological Satellites (EUMETSAT) Advanced Scatterometer (ASCAT) on MetOp-A and MetOp-B, India's Oceansat-2 scatterometer and China's HaiYang-2A (HY-2A) scatterometer, as well as current QuikSCAT and future NASA scatterometer missions in development. These scatterometers have replenished the measurements of oceanic surface wind since the termination of QuikSCAT measurement.

These scatterometer observations from QuikSCAT have a wide array of applications and have contributed to climatological studies, weather forecasting, as well as hurricane/typhoon analysis. QuikSCAT oceanic wind has been used in operational tropical cyclone analysis and forecasting at the National Hurricane Center to identify and locate the center of tropical cyclones, estimate its intensity, and wind radii analysis (Von Ahn et al. 2006; Chang and Jelenak 2006). In research, these data of surface wind information also appear helpful to improve forecast on typhoon track and intensity. Figure 8 shows an example of the impact of QuikSCAT winds in the Pacific Typhoon Nari (2001), which exhibits severe undulation over the metropolitan Taipei and southwest coast of Taiwan (Huang et al. 2005). As the GPS RO data are assimilated into a regional model, the offshore track off the northwestern coast of Taiwan in the sensitivity experiment without the RO data is moved inland to be more consistent with the best track. As the QuikSCAT wind data are assimilated together, the track ends up closer the observed position. Care about the QuikSCAT underestimate on wind intensity over 30 m s^{-1} needs to be taken into account. Through data assimilation, the impact of such underestimate on forecast can be alleviated by providing larger observation errors and/or limited application to regions always from the intense vortex core.

5.2 Global Navigation Satellite System-Reflectometry (GNSS-R)

Global Navigation Satellite System (GNSS) (including GPS, GLONASS (Russia), GALILEO (Europe), etc.) reflectometry leverages signals of opportunity to remotely

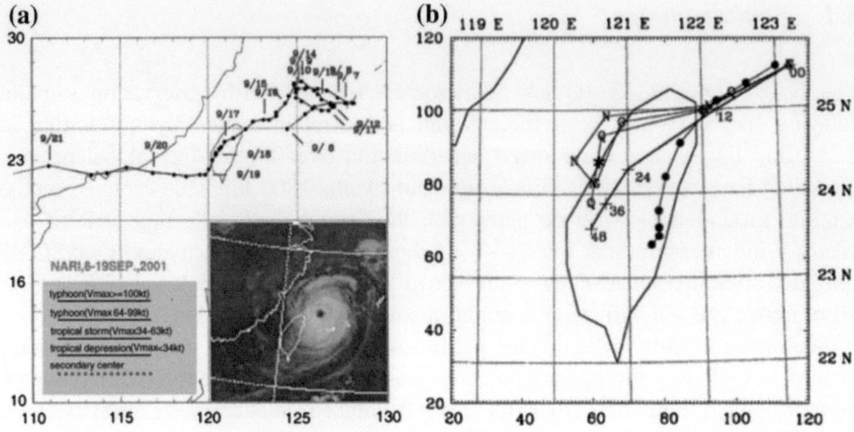

Fig. 8 **a** The best tracks from CWB for Typhon Nari (Sep. 2001) and the satellite images at 0032 UTC 16 Sep 2001, and **b** the simulated tracks for Nari every 12 h marked by N (no GPS data), G (with GPS data) and Q (with both GPS and uikSCAT wind data), and the best track from CWB (plus signs) and from JTWC (solid circles) at an interval of 6 h. Reproduced from (Huang et al. 2005)

sense the Earth's surface for a variety of science investigations. However, ionospheric refraction also affects GNSS reflections detected at low Earth orbit (LEO). While multi-frequency GNSS enables the elimination of most of the ionospheric error, single-frequency missions are still susceptible to this ranging delay.

The assimilation of real GNSS-R observations (when available) in neural atmospheric models and ionospheric models, will be under way to improve the model forecast. However, the simulated inverted wind speed by GNSS-R must be evaluated first and then assimilated into typhoon/hurricane models. Another potential source of GNSS-R will be from the FORMOSAT-7/COSMIC-2 mission, which is a US-Taiwan bilateral satellite program for continuing the FORMOSAT-3/COSMIC mission and aiming to provide global high-resolution RO data for use in operational weather models as well as meteorological, ionospheric and geodetic research. National Space Program Organization (NSPO) in Taiwan has contributed to provide such great opportunities for both research and operation. Onboard the six LEOs, the RO data measured from FORMOSAT-3/COSMIC launched in April 2006 take the advantages at high vertical resolution (several meters in the troposphere) and global even coverage (but at a relative coarser resolution of about 300 km). These RO data have proven valuable for a variety of climate and weather analyses (Ho et al. 2014) and in improving typhoon track prediction (Chen et al. 2015). FORMOSAT-7/COSMIC-2 is constituted for two phases, and each phase will launch six LEOs to take global RO measurements with cross links to GNSS. In the second phase starting in late 2018, a conceptual option of the second payload from the Taiwan community will be hopefully targeted at GNSS-R for a mission life of 5 years.

GNSS RO data from both phases of FORMOSAT-7/COSMIC-2 have been shown to cause a remarkable improvement on a forecast of severe regional weather through observing system simulation experiments (OSSEs) (reports available from NSPO). The distributions of simulated RO data from both phases, each providing around 500 points within 48 h in the regional model domain, are about three to four times FORMOSAT-3/COSMIC measurements. For measurements at lower latitudes in the western North Pacific where typhoons originate and develop, the benefits of the second phase on the measurement of oceanic surface wind from GNSS-R need to be considered particularly. The inverted oceanic properties (including wind speed) from the simulated GNSS-R observations of FORMOSAT-7/COSMIC-2 can be evaluated by OSSE as well to reveal their potential impacts on typhoon/hurricane prediction.

6 Conclusions

A variety of variables influencing the typhoons, including (1) zenith total delay/precipitable water vapor of the atmosphere, radial wind and radar reflectivity (by radar), and radio occultation reflectometry observations; (2) distance, height difference, rotation and size ratio of the typhoon; (3) atmospheric gravity waves; and (4) oceanic surface observations (especially ocean winds), are sequentially investigated in this chapter. Firstly, a rapidly updated data numerical system to assimilate the Doppler radar, GPS-derived PWV, and surface meteorological observations is presented to provide the extreme rainfall of a typhoon event. Then, two sets of cyclone-cyclone interactions with and without TDs are examined. Satellite imageries were used as an alternative approach to depict interactions of two typhoons with the distance between them of more than 1400 km. Subsequently, FORMOSAT-3/COSMIC radio occultation data were used to derive gravity wave information and to examine the characteristics of vertical wavelengths over the ocean and validate modeling of the tropical cyclone environment. Finally, the observational data from different satellites are used to investigate the potential impacts on numerical modeling related to oceanic surface wind measurement for improved typhoon simulations and predictions. In summary, remote sensing technology is broadly applied and is to be considered as an effective tool in the field of typhoon modeling and forecasts for disaster prevention and reduction.

References

Bennitt GV, Jupp A (2012) Operational assimilation of GPS zenith total delay observations into the Met Office numerical weather prediction models. Mon Weather Rev 140(8):2706–2719

Birkenheuer D, Gutman S (2005) A comparison of GOES moisture-derived product and GPS-IPW data during IHOP-2002. J Atmos Ocean Technol 22(11):1838–1845

Cámara A, Lott F, Hertzog A (2014) Intermittency in a stochastic parameterization of nonorographic gravity waves. J Geophys Res Atmos 119(21):11905–11919

Chane-Ming F, Roff G, Robert L, Leveau J (2002) Gravity wave characteristics over Tromelin Island during the passage of cyclone Hudah. Geophy Res Lett 29(6)

Chane-Ming F, Chen Z, Roux F (2010) Analysis of gravity-waves produced by intense tropical cyclones. Ann Geophys 2:531–547 (Copernicus GmbH)

Chane-Ming F, Ibrahim C, Barthe C, Jolivet S, Keckhut P, Liou Y-A, Kuleshov Y (2014) Observation and a numerical study of gravity waves during tropical cyclone Ivan (2008). Atmos Chem Phys 14(2):641–658

Chane-Ming F, Vignelles D, Jegou F, Berthet G, Renard JB, Gheusi F, Kuleshov Y (2016) Gravity-wave effects on tracer gases and stratospheric aerosol concentrations during the 2013 ChArMEx campaign. Atmos Chem Phys 2016:1–45. https://doi.org/10.5194/acp-2015-889

Chang P, Jelenak Z (2006) NOAA operational satellite ocean surface vector winds requirements workshop report. NOAA/NESDIS, Miami, Florida

Chen Y-C, Hsieh M-E, Hsiao L-F, Kuo Y-H, Yang M-J, Huang C-Y, Lee C-S (2015) Systematic evaluation of the impacts of GPSRO data on the prediction of typhoons over the northwestern Pacific in 2008–2010. Atmos Meas Tech 8(6):2531–2542

Ciesielski PE, Chang W-M, Huang S-C, Johnson RH, Jong-Dao Jou B, Lee W-C, Lin P-H, Liu C-H, Wang J (2010) Quality-controlled upper-air sounding dataset for TiMREX/SoWMEX: development and corrections. J Atmos Ocean Technol 27(11):1802–1821

Fritts DC, Alexander MJ (2003) Gravity wave dynamics and effects in the middle atmosphere. Rev Geophys 41:1003

Fujiwhara S (1923) On the growth and decay of vortical systems. Q J R Meteorol Soc 49(206):75–104

Gutman SI, Benjamin SG (2001) The role of ground-based GPS meteorological observations in numerical weather prediction. GPS Solut 4(4):16–24

Ho S-P, Yue X, Zeng Z, Ao CO, Huang C-Y, Kursinski ER, Kuo Y-H (2014) Applications of COSMIC radio occultation data from the troposphere to ionosphere and potential impacts of COSMIC-2 data. Bull Am Meteorol Soc 95(1):ES18–ES22

Hong G, Yang P, Huang H-L, Baum BA, Hu Y, Platnick S (2007) The sensitivity of ice cloud optical and microphysical passive satellite retrievals to cloud geometrical thickness. IEEE Trans Geosci Remote Sens 45(5):1315–1323

Huang H-L, Yang P, Wei H, Baum BA, Hu Y, Antonelli P, Ackerman SA (2004) Inference of ice cloud properties from high spectral resolution infrared observations. IEEE Trans Geosci Remote Sens 42(4):842–853

Huang C-Y, Kuo Y-H, Chen S-H, Vandenberghe F (2005) Improvements in typhoon forecasts with assimilated GPS occultation refractivity. Weather Forecast 20(6):931–953

Kafando P, Chane-Ming F, Petitdidier M (2015) Stratospheric variability of wave activity and parameters in equatorial coastal and tropical sites during the West African monsoon. Clim Dyn 1–24

King MD, Menzel WP, Kaufman YJ, Tanré D, Gao B-C, Platnick S, Ackerman SA, Remer LA, Pincus R, Hubanks PA (2003) Cloud and aerosol properties, precipitable water, and profiles of temperature and water vapor from MODIS. IEEE Trans Geosci Remote Sens 41(2):442–458

Kuleshov Y, Chane-Ming F, Qi L, Chouaibou I, Hoareau C, Roux F (2009) Tropical cyclone genesis in the Southern Hemisphere and its relationship with the ENSO. Ann Geophys Atmos Hydrosp Space Sci 6:2523–2538

Kursinski E, Bennett R, Gochis D, Gutman S, Holub K, Mastaler R, Minjarez Sosa C, Minjarez Sosa I, van Hove T (2008) Water vapor and surface observations in northwestern Mexico during the 2004 NAME Enhanced Observing Period. Geophys Res Lett 35(3)

Liou Y-A, Huang C-Y (2000) GPS observations of PW during the passage of a typhoon. Earth Planets Space 52(10):709–712

Liou Y-A, Teng Y-T, Van Hove T, Liljegren JC (2001) Comparison of precipitable water observations in the near tropics by GPS, microwave radiometer, and radiosondes. J Appl Meteorol 40(1):5–15

Liou Y-A, Liu J-C, Wu M-X, Lee Y-J, Cheng C-H, Kuei C-P, Hong R-M (2016) Generalized empirical formulas of threshold distance to characterize cyclone-cyclone interactions. IEEE Trans Geosci Remote Sens (99):1–11. https://doi.org/10.1109/tgrs.2016.2519538

Liu J-C, Liou Y-A, Wu M-X, Lee Y-J, Cheng C-H, Kuei C-P, Hong R-M (2015) Analysis of interactions among two tropical depressions and typhoons Tembin and Bolaven (2012) in pacific ocean by using satellite cloud images. IEEE Trans Geosci Remote Sens 53(3):1394–1402. https://doi.org/10.1109/tgrs.2014.2339220

Mahfouf J-F, Ahmed F, Moll P, Teferle FN (2015) Assimilation of zenith total delays in the AROME France convective scale model: a recent assessment. Tellus A 67

Minnis P, Sun-Mack S, Young DF, Heck PW, Garber DP, Chen Y, Spangenberg DA, Arduini RF, Trepte QZ, Smith WL Jr (2011) CERES edition-2 cloud property retrievals using TRMM VIRS and Terra and Aqua MODIS data—part I: algorithms. IEEE Trans Geosci Remote Sens 49(11):4374–4400

Platnick S, King MD, Ackerman SA, Menzel WP, Baum BA, Riédi JC, Frey RA (2003) The MODIS cloud products: algorithms and examples from Terra. IEEE Trans Geosci Remote Sens 41(2):459–473

Shoji Y, Kunii M, Saito K (2011) Mesoscale data assimilation of Myanmar cyclone Nargis Part II: Assimilation of GPS-derived precipitable water vapor. J Meteorol Soc Jpn 89(1):67–88

Typhoon Database (2015) Central Weather Bureau, Taiwan. http://rdc28.cwb.gov.tw/

Von Ahn JM, Sienkiewicz JM, Chang PS (2006) Operational impact of QuikSCAT winds at the NOAA Ocean Prediction Center. Weather Forecast 21(4):523–539

Wang H, Sun J, Fan S, Huang X-Y (2013) Indirect assimilation of radar reflectivity with WRF 3D-Var and its impact on prediction of four summertime convective events. J Appl Meteorol Climatol 52:889–902

Wu C-C (2013) Typhoon Morakot: key findings from the journal TAO for improving prediction of extreme rains at landfall. Bull Am Meteorol Soc 94(2):155–160

Xiao Q, Kuo Y-H, Sun J, Lee W-C, Lim E, Guo Y-R, Barker DM (2005) Assimilation of Doppler radar observations with a regional 3DVAR system: impact of Doppler velocities on forecasts of a heavy rainfall case. J Appl Meteorol 44(6):768–788

Mapping of Sea Surface Wind and Current Fields in the China Seas Using X-Band Spaceborne SAR

Xiao-Ming Li and Yong Zheng Ren

Abstract This chapter presents some examples of mapping sea surface wind and currents field in the China Sea using the new generation of spaceborne Synthetic Aperture Radar (SAR), TerraSAR-X and TanDEM-X with high spatial resolution. With respect to the sea surface wind field, we choose two cases to show contrasting sea surface wind patterns downstream of an archipelago in the Bohai Sea, the bright wind jet and the dark wind wake. This comparison highlights the capability of mapping the sea surface wind field at high spatial resolution using X-band SAR and the dedicated Geophysical Model Function. The tidal current in the HangZhou Bay, East China Sea, is a typical semi-diurnal current with a maximum speed up to 3 m/s in the spring tide. We demonstrate a novel application for deriving tidal current in the HangZhou bay based on the satellite constellation data from TerraSAR-X and TanDEM-X operating in the pursuit monostatic mode. The short temporal interval, on the scale of a few seconds, for the two satellite acquisitions over the same area is a unique advantage for mapping tidal current fields with great spatial and temporal variations.

Keywords Synthetic Aperture Radar · X-band SAR · Sea surface wind
Tidal current

1 Introduction

Sea surface wind, waves and currents are the three most important parameters at the air-sea interface. Since the launch of the first civil spaceborne Synthetic Aperture Radar (SAR) aboard the SEASAT mission in 1978 and particularly since the

---

X.-M. Li (✉) · Y. Z. Ren
Institute of Remote Sensing and Digital Earth, Chinese Academy of Sciences, Beijing, China
e-mail: lixm@radi.ac.cn

X.-M. Li
Hainan Key Laboratory of Earth Observation, Sanya, China

V. Barale and M. Gade (eds.), *Remote Sensing of the Asian Seas*,
https://doi.org/10.1007/978-3-319-94067-0_15

ERS-1/SAR mission in 1991, scientists have developed algorithms and methods for deriving sea surface wind, wave and current fields at high spatial resolution, on the scale of a few kilometers. The capability of mapping sea surface wind, waves and currents using spaceborne SAR is of particular importance for coastal monitoring, where the sea state often exhibits great spatial and temporal variations.

Among the three parameters, the derivation of sea surface wind field from spaceborne SAR data is at the most mature development stage. The general methodology to derive the sea surface wind field from SAR is to apply an empirical Geophysical Model Function (GMF) to the calibrated SAR image using wind direction from external sources or wind streaks visible in SAR imagery. The GMF is related to the Normalized Radar Cross Section (NRCS) through the sea surface wind speed, direction and radar incidence angle. In the past decades, the C-band (5.3 GHz) GMF families, e.g., the widely used CMOD4 (Stoffelen and Anderson 1997), CMOD-IFR (Quilfen et al. 1998) and CMOD5 (Hersbach et al. 2007), were developed for the C-band Scatterometers in Vertical-Vertical (VV) polarization to derive the sea surface wind field over a large swath with a spatial resolution of 25 or 12.5 km. Since ERS/SAR and ENVISAT/ASAR are also operated in the C-band, the CMOD functions are adapted for the C-band SAR data to map the sea surface wind field at high spatial resolution for different applications, e.g., meso-scale wind (Lehner et al. 1998), katabatic wind (Li et al. 2007), Bora events (Signell et al. 2010), coastal upwelling (Li et al. 2009), and offshore wind farming (Christiansen and Hasager 2005). Combined with the Polarization Ratio (PR) models (Thompson et al. 1998; Mouche et al. 2005; Zhang et al. 2011), which convert the NRCS in Horizontal-Horizontal (HH) polarization to that in VV polarization, the CMOD functions are also applied to derive the sea surface wind field from the HH polarization SAR data.

A major obstacle in deriving sea surface wind fields from SAR data using GMFs is that wind directions and speeds cannot be retrieved simultaneously because SAR sensors have only a single antenna, unlike scatterometers, which have multiple beams. Therefore, sea surface wind directions have to be acquired first to retrieve sea surface wind fields from spaceborne SAR data. Visible wind streaks in SAR images are indications of sea surface wind direction (Gerling 1986). It is generally accepted that wind streaks in SAR images are induced by roll vortices in the marine atmospheric boundary layer (Mourad et al. 2000; Zhao et al. 2016), and it is assumed that the wind streaks are aligned with the sea surface wind directions. There are generally two methods available to derive sea surface wind directions from SAR imagery depending on these visible wind streaks. One method is the Fast Fourier Transform (FFT) method (Lehner et al. 1998) applied in the frequency domain because the sea surface wind direction is perpendicular to the FFT spectral peaks of the wind streaks. The second method, the Local Gradient (LG) (Koch 2004; Wang and Li 2016) method, is applied in the spatial domain. However, for operational purposes (Monaldo 2000), wind directions from Numerical Weather Prediction (NWP) models are widely used for sea surface wind retrieval from spaceborne SAR data.

The launch of the X-band (9.6 GHz) spaceborne SAR satellites TerraSAR-X (TSX) and Cosmo-SkyMed (CSK) in 2007, which can yield data up to 1 m spatial resolution, has drawn considerable attention. Different from the C-band SAR that can adapt the CMOD functions to retrieve sea surface wind field, there are no X-band GMFs available. Therefore, efforts have been made to develop independent X-band SAR GMF to derive a high spatial resolution wind field from TSX and CSK data. Linear (Ren et al. 2012) (called XMOD1) and non-linear (Li and Lehner 2014) (XMOD2) X-band GMFs were developed, which are dedicated to TSX data for retrieving the sea surface wind field. In comparison with in situ buoy measurements, the retrieved sea surface wind speed using XMOD2 has a bias of −0.29 m/s and RMSE of 1.46 m/s (Li and Lehner 2014). A brief description of XMOD2 is given in Sect. 2, prior to presenting the cases of mapping coastal wind field using TSX data.

While sea surface wind retrieval from spaceborne SAR is at a mature stage and some operational services are also available, studies of deriving sea surface current field from spaceborne SAR data are ongoing. The along-track interferometry (ATI) technique has demonstrated the potential for deriving the sea surface current field (Goldstein and Zebker 1987; Goldstein et al. 1989) and was applied to the Shuttle Radar Topography Mission (SRTM) (Romeiser et al. 2005) and spaceborne TSX and TDX SAR data (Romeiser et al. 2010; 2014). In addition to the ATI technique, the SAR Doppler centroid method (Chapron et al. 2005; Johannessen et al. 2008) was also developed to derive the sea surface current. Both the ATI and Doppler centroid methods yield line-of-sight components of sea surface vectors. Different from the two methods described above using the SAR complex signal, the Maximum Cross Correlation (MCC) method for deriving sea surface velocities is based on pattern matching of sea surface features appearing in sequential satellite remote sensing images. The method was first demonstrated by Emery et al. (1986), using data from the Advanced Very High Resolution Radiometer (AVHRR). In these sequential AVHRR images, the sea surface temperature gradient patterns induced by advection were spatially matched to derive sea surface velocities. The method was extended later to SAR data (or both SAR and optical images) by tracking the appearance of slicks or macro-algal blooms (e.g., demonstrated in Lyzenga and Marmorino 1998; Liu and Hsu 2009; Gade et al. 2012; Qazi et al. 2014). These sea surface slicks, or algal blooms, are presented in SAR images as dark or bright patches due to the reduced or enhanced sea surface radar backscatter, which have been illustrated using both theory and experiments (e.g., by Gade et al. 1998; Ciappa et al. 2010).

In Sect. 3 of this chapter, we present case studies for deriving coastal tidal current fields with significant spatial and temporal variations using the MCC method, and based on SAR data acquired from a unique configuration of TSX and TDX, the pursuit monostatic mode.

2 Derivation of the Sea Surface Wind Field Using TSX and TDX Data

Prior to presenting the case studies of coastal sea surface wind fields, a brief description of the applied X-band SAR GMF is presented.

2.1 *The X-Band GMF XMOD2*

Different from the spaceborne scatterometer, which automatically obtains global sea surface roughness measurements, SAR-obtained data depends primarily on the commands from the ground segmentation. Therefore, it is very difficult to obtain a large quantity of collocated TSX data and in situ measurements for determining the functions relating σ_0 with radar image geometry (i.e., incidence angle θ), sea surface wind speed (v) and direction (ψ). Interestingly, it has been found that the σ_0 values derived from TSX and TDX data are statistically close to those simulated for C-band radar using in situ buoy measurements, as shown in Fig. 1.

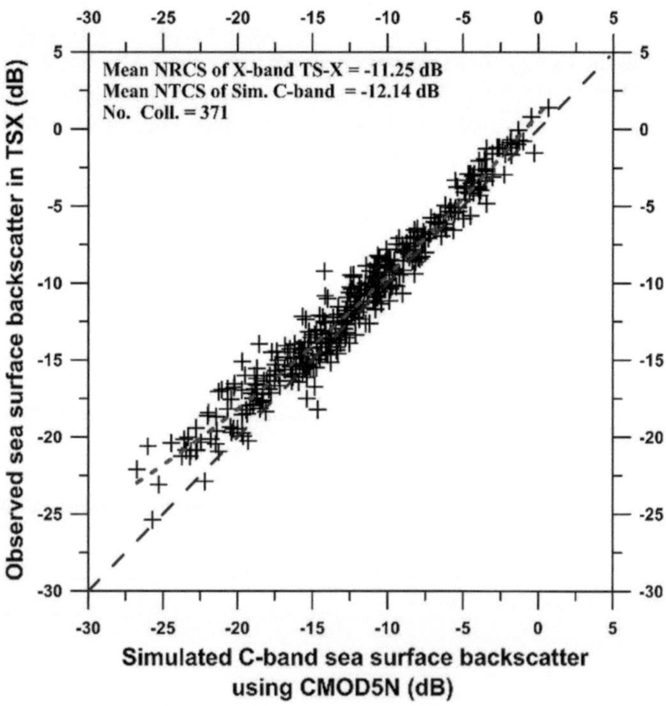

Fig. 1 Comparisons of σ_0 measured by TSX/TDX with those simulated for C-band radar using CMOD5N. The input sea surface wind speed and directions are obtained from buoys, which are temporally and spatially collocated with the TSX/TDX data

The similarity of sea surface radar backscatter between C- and X-band SAR suggests that we may use the well-developed CMOD functions as a prototype for developing the X-band SAR GMF for sea surface wind field retrieval. We use the general expression proposed in CMOD5 for XMOD2, as follows:

$$z(v, \phi, \theta) = B_0^p(v, \theta)(1 + B_1(v, \theta)\cos\phi + B_2(v, \theta)\cos 2\phi) \quad (1)$$

where the isotropic term B_0, upwind/downwind amplitude B_1 and upwind/crosswind amplitude B_2 are functions of incidence angle and sea surface wind speed v at 10 m height. The relative direction $\phi = \psi - \alpha$ is the angle between wind direction ψ and radar look direction α. The coefficients of the functions in XMOD2 were determined using the collocated in situ buoy measurements and NWP model results. A detailed description of XMOD2 development is presented in (Li and Lehner 2014).

Figure 2a, b show the simulated sea surface radar backscatter using XMOD2 for incidence angles of 20°, 30° and 40° in the upwind and cross wind, respectively. In the plots, the same simulations for C-band radar using CMOD5 are also presented for comparison. For the upwind, the transition in the difference between X- and C-band radar backscatter shows a dependence on the sea surface wind speed. For the three incidence angles, the transition appears at approximately 8, 6, and 4 m/s, respectively. However, in the crosswind, the simulated X-band radar backscatter is generally larger than that of C-band, without dependence on the sea surface wind speed. For the validation of XMOD2, an independent in situ buoy dataset was used. Figure 3 shows the scatter diagram, which yields a bias of −0.29 m/s and an RMSE of 1.46 m/s.

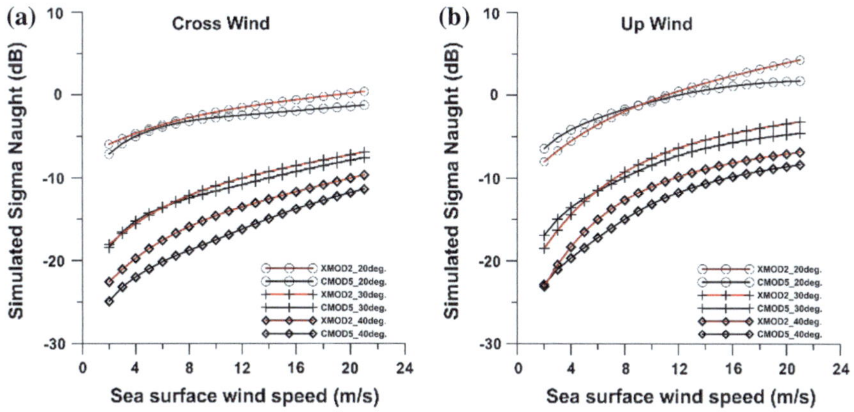

Fig. 2 Simulated sea surface radar backscatter using XMOD2 for incidence angles of 20°, 30° and 40° against sea surface wind speed in the **a** upwind and **b** cross wind. The red lines present the same simulation but using CMOD5 for C-band radar data

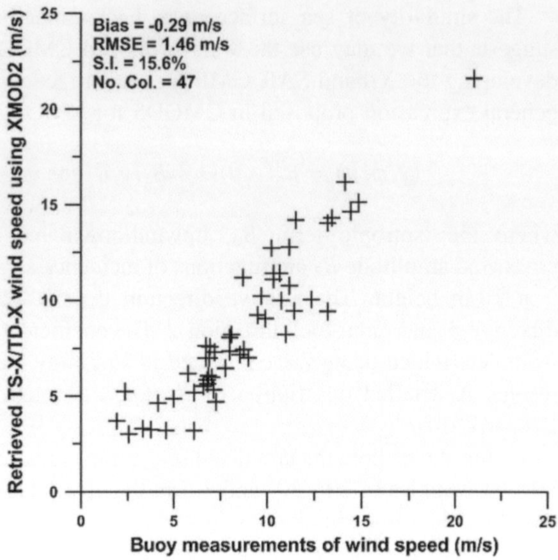

Fig. 3 Comparison of the retrieved TSX/TDX sea surface wind speed using XMOD2 with in situ buoy measurements obtained from July 2012 to January 2013

2.2 *Examples of Mapping Coastal Sea Surface Wind in the Bohai Sea*

Two interesting cases of mapping sea surface wind in the Bohai Sea using high resolution TSX data in ScanSAR mode are presented in Fig. 4. The TSX ScanSAR mode image has a pixel size of 8.25 m and a swath width of 100 km. The two images were both acquired over the Miao Dao archipelago in the Bohai Sea, as indicated by the scattered gray parts in the middle of the images. The Miao Dao archipelago is located between the Shandong (the large gray part in the maps) and Liaodong Peninsula in Northern China. The islands have a variable orography, with hills generally lower than 150 m.

The sea surface wind speed for the two cases were retrieved based on XMOD2. The input sea surface wind directions are derived according to the visible wind streaks in the two cases. The retrieved sea surface wind speed has a grid size of 500 m. As shown, the sea surface winds exhibit different patterns downstream of the islands. For the case on March 29, 2011 (Fig. 4a), clearly visible wind jets with sea surface windspeeds greater than 12 m/s were formed in the lee of the northern three islands (Beihuangcheng Island, Daqin Island, and Tuoji Island), whereas the wake was observed in the lee of the two southern large islands (Daheishan Island and Nanchangshan Island). In the other case on April 28, 2012 (Fig. 4b), in contrast, wakes were formed in the lee of the two northern islands. For the middle islands, however, we observe that both the wakes and jets are present.

The formation of a wake or jet is closely related to the atmospheric conditions and obstacle dimensions. We found a number of TSX cases in the same area showing different wake and jet patterns in the lee of the islands. In this chapter, we

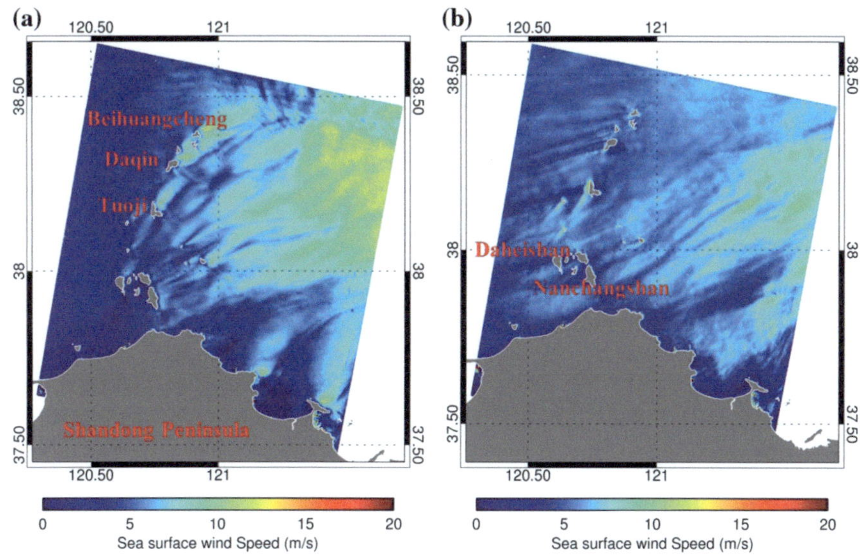

Fig. 4 Derived sea surface wind speed maps for the cases in the Bohai Sea: **a** acquired on March 29, 2011 at 22:09 UTC and **b** acquired on April 28, 2012 at 22:09 UTC. The sea surface wind directions are southwesterly for case (**a**) and northeasterly for (**b**), respectively

do not investigate how wakes and jets are formed in detail, while only highlight the capability of high-resolution spaceborne SAR for mapping coastal surface wind using these two interesting cases.

3 Derivation of Sea Surface Current Field Using TSX and TDX Data

TSX and TDX can operate in the pursuit monostatic mode, i.e., the two satellites image the same area over a short interval temporal on the scale of a few seconds, as shown in Fig. 5. The imaged sea surface features from the two continuous SARs with such a short temporal interval show some correlation. This correlation between the TSX and TDX pursuit monostatic mode images provides an opportunity to derive the sea surface current field based on the Maximum Cross Correlation (MCC) method. This method is particularly effective for the tidal current field that has significant temporal variations, while the short temporal interval of TSX and TDX pursuit monostatic images can minimize de-correlation. In the following subsection, the methodology is described briefly.

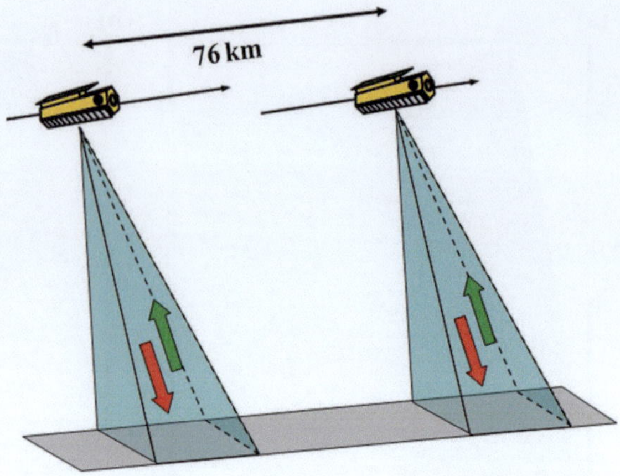

Fig. 5 Illumination of TSX and TDX operation in the pursuit monostatic mode. Courtesy of Thomas Busch at the German Aerospace Center (DLR)

3.1 Methodology

The MCC method uses a time series of satellite images and adopts the automatic feature-tracking method for pattern recognition to extract sea surface velocities. The data pair consists of two images. A region in the first SAR image is selected as template T and a larger region in the second one is selected as search region S, which has the same central position as template T, as shown in Fig. 6. Template T is allowed to move vertically and horizontally by overlapping within search region S. The search region under the coverage of the template is called sub-region S'. The cross correlation coefficient γ is used to measure the similarity between T and S':

$$\gamma(p,q) = \frac{\sum\sum\left[S'^{(x+p,y+q)} - \overline{S'^{(x+p,y+q)}}\right]\left[T(x,y) - \overline{T(x,y)}\right]}{\left(\sum\sum\left[S'^{(x+p,y+q)} - \overline{S'^{(x+p,y+q)}}\right]^2\sum\sum\left[T(x,y) - \overline{T(x,y)}\right]^2\right)^{\frac{1}{2}}} \quad (2)$$

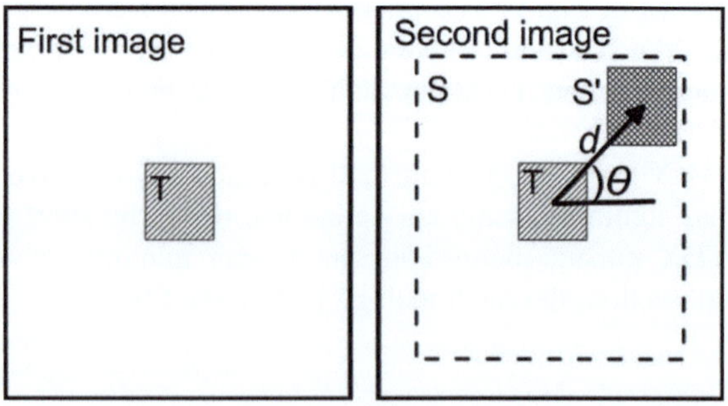

Fig. 6 Demonstration of the MCC method used to derive the sea surface velocity using analyzing time series satellite images

$$C = \frac{d}{\Delta t} = \frac{\sqrt{(p_{max}\Delta x)^2 + (q_{max}\Delta y)^2}}{\Delta t} \tag{3}$$

$$\theta = \tan^{-1}\left(\frac{q_{max}\Delta y}{p_{max}\Delta x}\right) \tag{4}$$

where *(m, n)* is the center of template T and (i, j) is the location of sub-region S′. Location *(i′, j′)* in the second SAR image that produces the maximum cross correlation with respect to location (m, n) in the first SAR image indicates the most likely displacement due to sea surface advection. Using the maximum correlation coefficient, we can calculate the corresponding space displacement d and then derive the tidal current speed according to the space displacement and time interval between the first and second SAR images (Eq. 3). The direction of motion is given by θ (Eq. 4).

The template size and searching length in the MCC method have impacts on the calculated results. If the template and search length are too small, the method cannot extract effective features. On the other hand, a too-large template size and search length will result in a loss of fine structure in the tidal current fields, as well as a large computational cost. Therefore, an appropriate template size and search length needs to be determined according to some a priori knowledge of tidal current fields in the area of interest (AOI). In the following subsection, determining appropriate parameters for the MCC method is presented in the two examples from the Hangzhou Bay in the East China Sea.

3.2 Examples of Mapping Coastal Tidal Current Field Using SAR Constellation Data

In this section, two cases of mapping coastal tidal current field in the Hangzhou Bay, East China Sea are presented. The irregular semi-diurnal tide, where the M_2 constituent is the dominant tidal component, is the main driving force for water flow in the bay. The tidal amplitude is 3–4 m at the bay summit and 4–6 m further upstream (Xie et al. 2009). The specific topography often makes the tidal velocities higher than 3 m/s in the spring tides. Moreover, the tidal current presents an alternating feature, namely, it flows from east to west during flood and reverses during ebb. Our previous study showed that the tidal current here has very strong interactions with offshore wind farm turbine piles in the bay, which even form visible wakes with lengths beyond 500 m in high resolution SAR images (Li et al. 2014). Therefore, we chose the Hangzhou Bay as an appropriate test site for the proposed method to derive the tidal current field.

Case 1: Mapping the tidal current field using TSX and TDX pursuit monostatic mode data

The TSX (left panel) and TSX (right) in Fig. 7 were acquired in the pursuit monostatic mode in the Hangzhou bay on December 3, 2014 at 09:44:44 UTC and 09:44:54 UTC, respectively, at a 10 s temporal interval. Both are in the Stripmap mode with a pixel size of 3 m. The bright lines appearing in the middle of the images are Donghai Bridge, which connects the City of Shanghai with the Yangshan harbor. The two images were co-registered prior to applying the MCC method for deriving the tidal current field.

As mentioned previously, the template size and search length of the MCC method have a considerable effect on the calculated sea surface velocities. Therefore, some empirical knowledge of the tidal current fields in the AOI is necessary to determine the appropriate template size and search length for the MCC method. According to the rough estimation based on the tidal chart, the maximum tidal current speed when the SAR data were acquired for the area was approximately 1.4 m/s, corresponding to a sea surface displacement of approximately 14 m in 10 s (i.e., the temporal interval between the TSX and TDX acquisitions). Therefore, the matching template size was set to 5 $\times$ 5 pixels, which corresponds to a spatial resolution of 15 $\times$ 15 m for the derived tidal current fields, and is used to extract the tidal currents in the bay. To tolerate the model prediction bias and ensure that all features of the tidal current fields in the bay can be tracked, the maximum searching length was set to 8 pixels (i.e., 24 m in space), which corresponds to a maximum tidal current speed of 2.4 m/s over a temporal interval of 10 s.

Using the parameter configurations described above, the calculated MCC coefficient for the TSX/TDX data pair is shown in Fig. 8a. The tidal current field is subsequently derived using a threshold MCC coefficient greater than 0.8, as shown in Fig. 8b. The black arrows in the figure indicate the sea surface flows, which were derived based on the maximum cross correlation coefficients via Eq. 4. The arrows indicate the re-sampled results of the tidal current vectors to provide clarification in the figure. The westward surface flows suggest a tidal flood phase at the moment of SAR acquisitions.

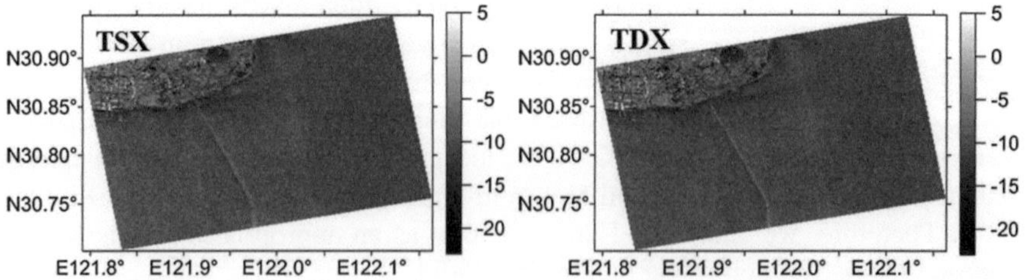

Fig. 7 Radiometrically calibrated TSX (left) and TDX (right) data acquired over the Donghai Bridge in Hangzhou Bay in the East China Sea at 9:44:44 (TSX) and 9:44:54 UTC (TDX) on December 3, 2014

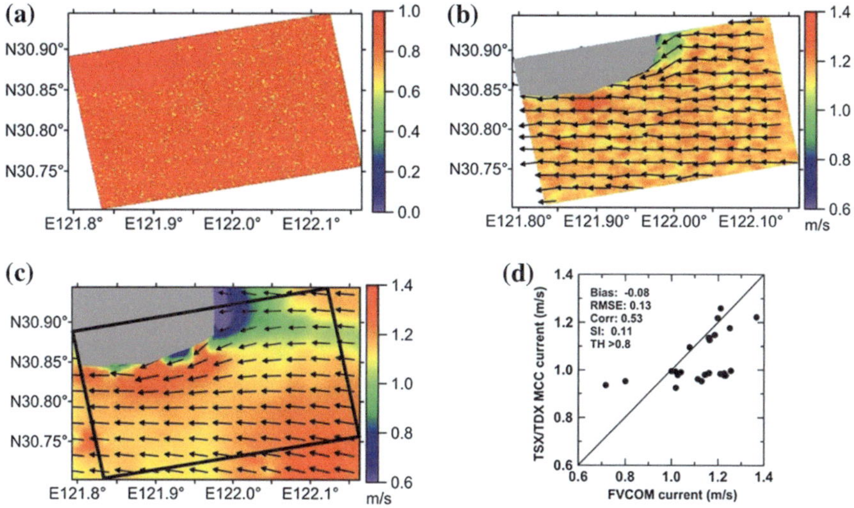

Fig. 8 **a** Calculated maximum cross-correlation coefficients from the TSX/TDX data pairs shown in Fig. 7; **b** derived tidal current field using the MCC coefficients greater than 0.8; **c** predicted tidal current field from the FVCOM at 9:50 UTC on December 3, 2014, where the black rectangle in the image represents the spatial coverage of the TSX/TDX image; **d** comparison of the SAR derived tidal current speed with the FVCOM simulation results. Note that the SAR retrieval was re-sampled to the same grid size as the model simulation

To verify the SAR derived tidal current field, we conducted a model simulation based on the FVCOM, which is available in a regular grid size of 3 km. Figure 9c shows the FVCOM tidal current field at 09:50 UTC on December 3, 2014, close to the TSX and TDX data acquisition time at 09:44 UTC. Figure 9d shows the comparison of the SAR derived tidal current speed with the model simulation, with a bias of −0.08 m/s and an RMSE of 0.13 m/s, suggesting good agreement with the model simulation. In this case, MCC coefficients greater than 0.8 were used to derive tidal current speed. We also verified the results using different threshold of MCC coefficients, which are described in detail in our recent paper by Ren et al. (2017).

In this case, the TSX and TDX operating as a satellite constellation were exploited to derive the tidal current field based on the MCC method. As there are more spaceborne SAR sensors in orbit, there are also possibilities that SAR sensors from different agencies can also form a virtual satellite constellation data pair. These pairs could satisfy the requirement of short temporal interval to derive tidal current field using the MCC method. In the following subsection, we present another case at the same site using the virtual pursuit monostatic mode data formed by TDX and the X-band SAR sensor Cosmo-SkyMed (CSK).

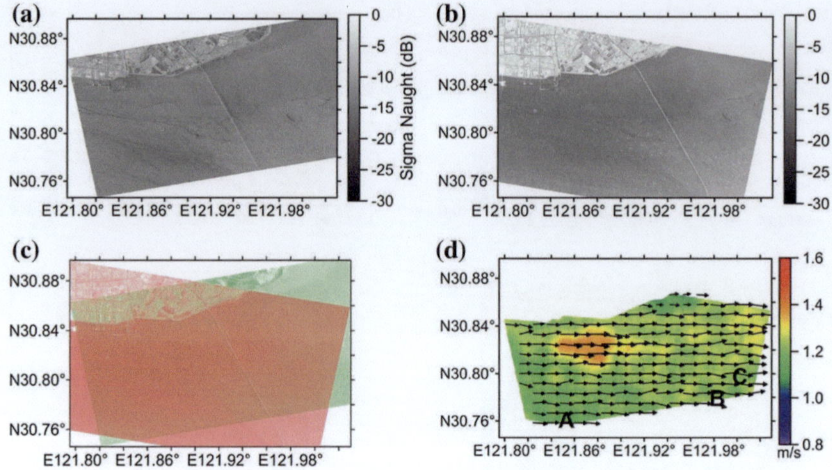

Fig. 9 Radiometrically calibrated TSX (**a**) and CSK (**b**) images acquired in Hangzhou Bay on July 22, 2015, at 9:44:45 and 9:44:47 UTC; **c** the two overlapping SAR scenes acquired in the Bay; **d** the derived tidal current field using MCC method in the overlapping area of the two SAR images

Case 2: Mapping the tidal current field using a TSX and CSK data pair

Figure 9a, b show TDX and CSK SAR images over the north of Hangzhou Bay acquired at 09:44:45 UTC and 09:44:47 UTC (scene initiation time) on July 22, 2015, respectively. It should be noted that the temporal interval of the overlapping area (Fig. 9c) of the two SAR images varies from 2 to 6 s because of the different orbit directions of the two satellites.

Similar to the first case study, the template size was also set to 5 × 5 pixels (15 × 15 m spatially), but the search length was set to 5 pixels (15 m) due to the shorter temporal interval between the two images in the MCC method to derive the tidal current fields. The retrieved tidal current fields with a MCC coefficient greater than 0.8 are shown in Fig. 9d. The eastward flow suggests an ebb at the moment of the SAR acquisitions. A higher tidal current speed of approximately 1.4 m/s was observed in the area of the Luchaogang dock, which is also observed in the same area in the previous case. This higher tidal current speed region was very likely induced by the artificially altered coastal zones from the dock construction. In the figure showing the retrieved tidal current field, three sites where in situ ADCP measurements are available are marked by A, B and C. The A (Luchaogang station) and C (Nanhuizui station) sites are the two fixed stations where the ADCPs were moored, and the B site is where in situ measurements were obtained during the cruise conducted in the bay.

Table 1 shows the comparison of derived tidal current velocity with the in situ measurements at the three sites. For comparison at station A, the differences between the SAR retrievals and ADCP measurements are relatively large at

Table 1 Comparison of retrieved tidal current velocities with in situ ADCP measurements

Stations	ADCP measurements	SAR retrieval results
A (N30.75°, E121.85°)	0.88 m/s, 89.87°	1.17 m/s, 91.34°
B (N30.77°, E121.98°)	1.12 m/s, 111.56°	1.11 m/s, 105.06°
C (N30.80°, E122.0°)	1.31 m/s, 95.00°	1.22 m/s, 94.64°

approximately 0.3 m/s (33%). However, the differences at the station B and C are generally less than 0.1 m/s (<7%). The temporal interval between the two SAR images in the overlapping area (Fig. 9c) for station A was approximately 2.4 s, which was shorter than those in station B by 0.6 s and C by 1.5 s, respectively, because of the different orbit directions of the two satellites. Therefore, the excessively short temporal interval between two SAR images for site A may have induced bias in the retrieval because of the actual displacement, which was smaller than one pixel. This result should partially explain why the difference between the SAR retrieval and ADCP measurements in station A was larger than those in the other two sites. If the ADCP measurements from the tidal current fields are treated as the ground truth, the comparisons suggest that the derived results from the SAR constellation data using the MCC method can yield accurate estimations of regional tidal current fields.

Both cases presented in the two previous subsections show that high-resolution SAR constellation data pairs in short temporal interval can be used to derive variable tidal current field based on the traditional MCC method. This can yield full current vectors, which has not been achieved previously, even using the existing ATI technique. Although the TSX/TDX pursuit monostatic mode is a unique configuration that has not been implemented for other SAR sensors, the second case study using the TDX/CSK data pair indicates that the proposed method has wide applicability. This is particularly true because of the increasing availability of spaceborne SAR constellation data, including the Sentinel-1A&B and the planned RADARSAT Constellation Mission.

4 Summary

While the C-band SARs have been widely exploited over the last two or more decades, as indicated by the many achievements and applications presented in this book, the applications of X-band SAR to ocean remote sensing and coastal monitoring have also started to draw considerable attention. In this chapter, we presented a few case studies that derived the basic marine-meteo parameters for sea surface wind and current fields from the X-band spaceborne SAR sensors TSX, TDX and CSK.

Similar to the widely used C-band Scatterometer GMF, we developed an X-band SAR GMF to derive the sea surface wind field at high spatial resolution from TSX/

TDX data, which is a traditional and effective method for prospective operational service as well as for ocean and coastal studies. The two contrasting cases in the Bohai Sea suggest the potential for high spatial resolution spaceborne SAR to map the fine structure of complicated coastal surface wind and their downstream wake and jet patterns induced by the interaction of different atmospheric conditions and local orography.

The derivation of surface current vectors at high spatial resolution from spaceborne SAR data has been established in previous work. However, both the recently achieved spaceborne ATI technique from TSX and TDX and the Doppler Centroid method applied to the ENVISAT/ASAR and Sentinel-1 SAR data, can yield only the current components in the line of sight instead of full vectors. Using a conventional tracking method, i.e., the MCC method, we found it is possible to derive coastal current field using SAR data pairs with short temporal intervals. In particular, as more and more spaceborne SAR are in orbit and operate in constellation, based on this simple but effective method, there are opportunities to map the coastal current field with great spatial and temporal variations.

Since the launch of SEASAT in 1978, spaceborne SAR has demonstrated its unique capability and played an important role in marine and coastal remote sensing. Until now, spaceborne SAR has not been widely exploited, unlike other spaceborne microwave remote sensing techniques, such as the scatterometer, radar altimeter and radiometer, for many reasons. However, it is believed that as the number of SAR sensors from different agencies in orbit increase and as new techniques are developed, many previous shortcomings will be overcome, and these new sensors will provide even more capabilities in marine and coastal remote sensing.

Acknowledgements The copyright of the presented TSX and TDX data belongs to the German Aerospace Center (DLR). These data were provided by DLR to the authors via several AO proposals. The present study was partially supported by grants from the National Natural Science Foundation of China (No. 41406198), the "Pioneered Hundred Talents Program, Chinese Academy of Sciences" and the Hainan Key S&T Programme (No. ZDKJ2016015).

References

Chapron B, Collard F, Ardhuin F (2005) Direct measurements of ocean surface velocity from space: interpretation and validation. J Geophys Res Oceans. https://doi.org/10.1029/2004jc002809

Christiansen MB, Hasager CB (2005) Wake effects of large offshore wind farms identified from satellite SAR. Remote Sens Environ 98:251–268

Ciappa A, Pietranera L, Coletta A, Jiang XW (2010) Surface transport detected by pairs of COSMO-SkyMed ScanSAR images in the Qingdao region (Yellow Sea) during a macro-algal bloom in July 2008. J Mar Syst 80:135–142

Emery WJ, Thomas AC, Collins MJ, Crawford WR, Mackas DL (1986) An objective method for computing advective surface velocities from sequential infrared satellite images. J Geophys Res Oceans 91(C11):12865–12878

Gade M, Alpers W, Huhnerfuss H, Wismann VR, Lange PA (1998) On the reduction of the radar backscatter by oceanic surface films: scatterometer measurements and their theoretical interpretation. Remote Sens Environ 66:52–70

Gade M, Seppke B, Dreschler-Fischer L (2012) Mesoscale surface current fields in the Baltic Sea derived from multi-sensor satellite data. Int J Remote Sens 33(10):3122–3146

Gerling TW (1986) Structure of the surface wind field from the Seasat SAR. J Geophys Res Oceans 91(C2):2308–2320

Goldstein RM, Zebker HA (1987) Interferometric radar measurement of ocean surface currents. Nature 328:707–709

Goldstein RM, Zebker HA, Barnett TP (1989) Remote sensing of ocean currents. Science 246 (4935):1282–1285

Hersbach H, Stoffelen A, De Haan S (2007) An improved C-band scatterometer ocean geophysical model function: CMOD5. J Geophys Res Oceans. https://doi.org/10.1029/2006jc003743

Johannessen JA, Chapron B, Collard F, Kudryavtsev V, Mouche A, Akimov D, Dagestad KF (2008) Direct ocean surface velocity measurements from space: improved quantitative interpretation of Envisat ASAR observations. Res Lett, Geophys. https://doi.org/10.1029/2008GL035709

Koch W (2004) Directional analysis of SAR images aiming at wind direction. IEEE Trans Geosci Remote Sens 42(4):702–710

Lehner S, Horstmann J, Koch W, Rosenthal W (1998) Mesoscale wind measurements using recalibrated ERS SAR images. J Geophys Res Oceans 103(C4):7847–7856

Li XM, Lehner S (2014) Algorithm for sea surface wind retrieval from TerraSAR-X and TanDEM-X data. IEEE Trans Geosci Remote Sens 52(5):2928–2939

Li X, Zheng W, Pichel WG, Zou CZ, Clemente-Coln P (2007) Coastal katabatic winds imaged by SAR. Geophys Res Lett. https://doi.org/10.1029/2006gl028055

Li XM, Li X, He M (2009) Coastal upwelling observed by multi-satellite sensors. Sci China Ser D Earth Sci 52(7):1030–1038

Li XM, Chi L, Chen X, Ren Y, Lehner S (2014) SAR observation and numerical modeling of tidal current wakes at the East China Sea offshore wind farm. J Geophys Res Oceans. https://doi.org/10.1002/2014jc009822

Liu AK, Hsu MK (2009) Deriving ocean surface drift using multiple SAR sensors. Rem Sens 1 (3):266–277

Lyzenga DR, Marmorino GO (1998) Measurement of surface currents using sequential synthetic aperture radar images of slick patterns near the edge of the Gulf Stream. J Geophys Res Oceans 103:18769–18777

Monaldo F (2000) The Alaska SAR demonstration and near-real-time synthetic aperture radar winds. J Hopkins APL Tech D 21(1):75–79

Mouche AA, Hauser D, Daloze JF, Gurin C (2005) Dual-polarization measurements at C band over the ocean: results from airborne radar observations and comparison with ENVISAT ASAR data. IEEE Trans Geosci Remote Sens 43(4):753–769

Mourad PD, Thompson DR, Vandemark DC (2000) Extracting fine-scale wind fields from synthetic aperture radar images of the ocean surface. J Hopkins APL Tech D 21:108–115

Qazi WA, Emery WJ, Fox-Kemper B (2014) Computing ocean surface currents over the coastal California current system using 30-Min-Lag sequential SAR images. IEEE Trans Geosci Remote Sens 52(12):7559–7580

Quilfen Y, Chapron B, Elfouhaily T, Katsaros K, Tournadre J (1998) Observation of tropical cyclones by high resolution scatterometry. J Geophys Res 103:77677786

Ren Y, Lehner S, Brusch S, Li X, He M (2012) An algorithm for the retrieval of sea surface wind fields using X-band TerraSAR-X data. Int J Remote Sens 3(23):7310–7336

Ren Y, Li XM, Gao GP, Busche TE (2017) Derivation of sea surface tidal current from spaceborne SAR constellation data. IEEE Trans Geosci Remote Sens 55(6). https://doi.org/10.1109/tgrs.2017.2666086

Romeiser R et al (2005) Current measurements by SAR along-track interferometry from a space shuttle. IEEE Trans Geosci Remote Sens 43(10):2315–2324

Romeiser R, Suchandt S, Runge H, Steinbrecher U, Grunler S (2010) First analysis of TerraSAR-X along-track InSAR-derived current fields. IEEE Trans Geosci Remote Sens 2 (48):280–829

Romeiser R, Runge H, Suchandt S, Kahle R, Rossi C, Bell PS (2014) Quality assessment of surface current fields from TerraSAR-X and TanDEM-X along-track interferometry and Doppler centroid analysis. IEEE Trans Geosci Remote Sens 52(5):2759–2772

Signell RP, Chiggiato J, Horstmann J, Doyle JD, Pullen J, Askari F (2010) High-resolution mapping of Bora winds in the northern Adriatic Sea using synthetic aperture radar. J Geophys Res Oceans. https://doi.org/10.1029/2009jc005524

Stoffelen A, Anderson D (1997) Scatterometer data interpretation: estimation and validation of the transfer function CMOD4. J Geophys Res 102:5767–5780

Thompson DR, Elfouhaily TM, Chapron B (1998) Polarization ratio for microwave backscattering from the ocean surface at low to moderate incidence angles. Proc Geosci Remote Sens Symp 3:1671–1673

Wang YR, Li XM (2016) Derivation of sea surface wind directions from TerraSAR-X data using the local gradient method. Remote Sens. https://doi.org/10.3390/rs8010053

Xie D, Wang Z, Gao S, De Vriend HJ (2009) Modeling the tidal channel morphodynamics in a macro-tidal embayment, Hangzhou Bay, China. Cont Shelf Res 29(15):1757–1767

Zhang B, Perrie W, He Y (2011) Wind speed retrieval from RADARSAT-2 quad polarization images using a new polarization ratio model. J Geophys Res Oceans. https://doi.org/10.1029/2010jc006522

Zhao Y, Li XM, Sha J (2016) Sea surface wind streaks in spaceborne synthetic aperture radar imagery. J Geophys Res: Oceans 121(9):6731–6741

Retrieval of Winds and Waves from Synthetic Aperture Radar Imagery

Weizeng Shao, Xiaofeng Li and Xiaofeng Yang

Abstract Recent developments of winds and waves retrieval from synthetic aperture radar (SAR) are introduced. winds were retrieved using the geophysical model function (GMF) together with polarization ratio (PR) model. An ocean surface wave retrieval algorithm, named Parameterized First-guess Spectrum Method (PFSM), is used to extract wave parameters from X-band TerraSAR-X/TanDEM-X (TS-X/TD-X) and C-band Sentinel-1 (S-1) SAR images over whole ocean, in particular, there are several cases at China seas. This theoretic-based algorithm relies on the first-guess wave spectrum produced by SAR-derived wind speed. More recent, several studies have made an attempt to retrieve significant wave height (SWH) and mean wave period (MWP) based on the cutoff wavelength in azimuth (or satellite flight) direction. A new semi-empirical algorithm for wave parameters retrieval at C-band, which depends on the radar incidence angle, wave propagation angle relative to range direction and azimuthal cutoff wavelength, has been implemented for S-1 SAR images and then validated against moored buoys. Although existing waves retrieval algorithms work for C- and X-band SAR with about 0.6 m error of SWH, the accuracy of retrieval waves is expected to be improved in the future.

Keywords Wind · Wave · Algorithms · SAR

W. Shao (✉)
Zhejiang Ocean University, Zhejiang, China
e-mail: shaoweizeng@zjou.edu.cn

X. Li
National Oceanic and Atmospheric Administration (NOAA),
College Park, MD, USA

X. Yang
Institute of Remote Sensing and Digital Earth,
Chinese Academy of Sciences, Beijing, China

V. Barale and M. Gade (eds.), *Remote Sensing of the Asian Seas*,
https://doi.org/10.1007/978-3-319-94067-0_16

1 Introduction

With high spatial resolution, relatively large swath and all-weather imaging capability, Synthetic Aperture Radar (SAR) is an efficient instrument providing data to derive sea surface wind and wave parameters. SAR sensors usually operate in C-band [ERS-1/2, ENVISAT-ASAR, Radarsat-1/2, Sentinel-1 (S-1), Gaofen-3 (GF-3)], L-band (SEASAT, ALOS-1/2), and X-band (such as TerraSAR-X (TS-X) with its twins TanDEM-X (TD-X), and Cosmo-SkyMed). The TS-X and TD-X were launched in 2007 and 2010 respectively. Their orbital height and period are 514 km and 100 min. Both satellites carry an X-band (9.8 GHz) SAR sensor with high spatial resolution up to 0.25 m for 'staring spotlight' mode. S-1 SAR satellite was launched by European Space Agency (ESA) in April 2014. It operates in C-band with different polarizations and beam modes. Wind and wave monitoring are the two main marine applications of SAR in all-weather conditions (Chapron et al. 2001), especially in tropical cyclones (Li et al. 2002, 2013).

Geophysical model function (GMF) is popular used for winds retrieval. GMF describes an empirical relationship between normalized radar cross-section (NRCS) in vertical-vertical (VV) polarization and a wind vector. There are several C-band GMFs e.g., CMOD4 (Stoffelen and Anderson 1997), CMOD-IFR developed at Institut Francais de Recherche pour Exploitation de la MER (IFREMER) (Quilfen et al. 1998), CMOD5 (Hersbach et al. 2007), CMOD5N for neutral winds (Hersbach 2010) and the new C-SARMOD (Mouche and Charpon 2015). X-band GMFs include XMOD1 (Ren et al. 2012) and XMOD2 (Li and Lehner 2014). These GMFs provide a convenient application for wind retrieval from SAR and have been successfully implemented for various C-band SAR data over the last few decades, e.g., ERS-1/2 (Lehner and Horstmann 1998), Envisat-ASAR (Yang et al. 2011), Radarsat-1/2 (Shao et al. 2014) and Sentinel-1A/1B (Monaldo et al. 2016; Lin et al. 2017) within about 2 m/s error of wind speed. As GMFs are applied for SAR in horizontal-horizontal (HH) polarization, polarization ratio (PR) models at C- (Zhang et al. 2011) and X-band (Shao et al. 2016a, b), which are described as the ratio between NRCS in VV and HH polarization, are usually used together with GMFs.

Traditionally, the methodology of wave retrieval from SAR is based on the SAR mapping mechanism (Hasselmann and Hasselmann 1991). The major modulations of waves on SAR have been well studied over many decades, e.g., tilt modulation, hydrodynamic modulation and velocity bunching (Alpers et al. 1981; Alpers and Bruning 1986). Both tilt and hydrodynamic modulations are linear mapping mechanisms. However, velocity bunching is caused by the relative motion between satellite platform and sea surface, resulting in a Doppler frequency shift in the azimuth direction (the direction of satellite flight is defined as the azimuth direction and radar look direction is the range direction). This velocity bunching causes wavelength smaller than a specific value (or cutoff wavelength) in the azimuth direction cannot be detectable and peak of the SAR spectrum rotates toward the range direction (Kerbaol et al. 1998). These two effects create the difficulty of directly inverting SAR intensity spectra to wave spectra.

The basic scattering physics that is commonly used in wave retrieval algorithms, i.e., MPI, SPRA and PFSM, are not radar frequency dependent. Historically, the first wave algorithm named the 'Max-Planck Institute Algorithm' (MPI) (Hasselmann et al. 1996; Schulz-Stellenfleth et al. 2005) was based on C-band ERS SAR data. Later on, the 'Semi Parametric Retrieval Algorithm Scheme' (SPRA) (Mastenbroek and de Valk 2000) was developed on the basis of the same modulation mechanism. These algorithms use empirical functions, such as Joint North Sea Wave Project (JONSWAP) (Hasselmann et al. 1973) to invert the wind-sea wave spectrum, and satellite scatterometer to get wind data. The SPRA algorithm then does a forward simulation until the simulated SAR spectrum is closest to the actual SAR image spectrum. The difference between the simulated SAR spectrum of the best-inverted wind-sea wave spectrum and the observed SAR intensity spectrum is supposed to be determined by the swell. The SPRA scheme is easier to implement than the MPI. A validation study using moored buoy data showed that the SPRA is better than the MPI (Voorrips et al. 1997). Theoretically, the error of the SPRA scheme is embedded in the swell SAR spectrum retrieval. To overcome the model-induced swell spectrum error, the Parameterized First-guess Spectrum Method (PFSM) was developed (Sun and Guan 2006; Sun and Kawamura 2009; Shao et al. 2015; Lin et al. 2017). Firstly, the SAR intensity spectrum is divided into two parts: the SAR portion of wind-sea and the linear-mapping swell. As for the wind-sea portion, the best first-guess spectrum is obtained by searching for the most suitable parameters in an empirical wave model such as JONSWAP, Pierson-Moskowitz (PM) etc., and then the MPI scheme is employed for the wave spectrum retrieval. The linear-mapping swell spectrum is acquired by directly inverting the given portion of the SAR spectrum generated by swell. Thus, PFSM, which is based on a SAR imaging mechanism (Alpers et al. 1981) like MPI, allows for the quick estimation of wave parameters similar to SPRA.

Empirical wave retrieval algorithms, without calculating complex MTF, were developed by the SAR group at the German Aerospace Center (DLR), and include CWAVE_ERS for ERS-2 SAR (Schulz-Stellenfleth et al. 2007), CWAVE_ENV for Envisat-ASAR (Li et al. 2011) and XWAVE for X-band TerraSAR-X SAR (Pleskachevsky et al. 2016; Shao et al. 2017). CWAVE allows direct retrieval of significant wave height (SWH) from SAR wave mode data at a fixed incidence angle of about 23°. However, these are not conductive to the operational application of wave retrieval from various SAR data. Interestingly, cutoff wavelength is related to not only wind speed but also SWH (Vachon et al. 1994; Stopa et al. 2016). Recently, some work has been devoted for developing the significant wave height retrieval algorithms by using only SAR-derived cutoff wavelength (Wang et al. 2012; Ren et al. 2015; Shao et al. 2016a, b; Grieco et al. 2016; Stopa et al. 2017). Collectively, it was found that SWH was linearly related to cutoff wavelength. We have developed an empirical algorithm for wave parameters retrieval from C-band SAR, called CSAR_WAVE (Shao et al. 2016a, b, 2017), including the terms of cutoff wavelength in azimuth direction, radar incidence angle and wave propagation direction.

In this chapter, winds were retrieved from VV-polarization S-1 and TS-X SAR images by using the mature winds retrieval methodology, that is new GMF at C-band and X-band together with PR models. Then, the algorithm PFSM, which has been successfully applied for wave retrieval from C- and X-band SAR data. Retrieval results of SWH and MWP were compared to European Centre for Medium-Range Weather Forecasts (ECMWF) reanalysis grids data and National Data Buoy Center (NDBC) buoy measurements of the National Oceanic and Atmospheric Administration (NOAA). In addition, the validation of CSAR_WAVE is also presented.

2 Data

In our work, two types of SAR image, including X-band TS-X/TD-X and C-band S-1, are used for winds and waves retrieval. These images are collocated with ECMWF reanalysis data or NDBC buoys.

2.1 *Description of X-Band TS-X/TD-X Image*

The wind speed is needed for wave retrieval of algorithm PFSM. ECMWF produces numerical forecasts for users worldwide, and has continuously provided global atmospheric-marine reanalysis data with a high spatial resolution daily at intervals of six hours since 1979. In our work, ECMWF reanalysis wind vectors are available at 0.25° grids, while our imagery has a higher spatial resolution ranging from 2 to 6 m in data collections. We directly employed ECMWF reanalysis wind directions for winds retrieval.

As an example, TD-X image acquired in ScanSAR mode overlaid on the wind vectors derived from the VV-polarization TD-X image by using X-band GMF XMOD2, which was taken on 21st October 2010 at 22:05 UTC in South China sea in the period of typhoon Megi, is shown in Fig. 1a. It is shown that contaminations produced by rain exist in the image, where the retrieved winds are significant distorted.

In Fig. 1b, the ambient wind field is from ECMWF at 24:00 UTC, in which the black rectangle represents the coverage of TS-X shown in Fig. 1a. It is obviously found that the SAR image provides a lot more detailed wind information than the model output and the SAR-derived wind speeds are greater than ECMWF reanalysis winds.

A total of 16 HH-polarization TS-X/TD-X images acquired in StripMap mode, which cover the NOAA buoys of NDBC, were collected for validating algorithm PFSM. It is necessary to figure out that X-band PR has to been used together with XMOD2 for winds retrieval from HH-polarization TS-X/TD-X image.

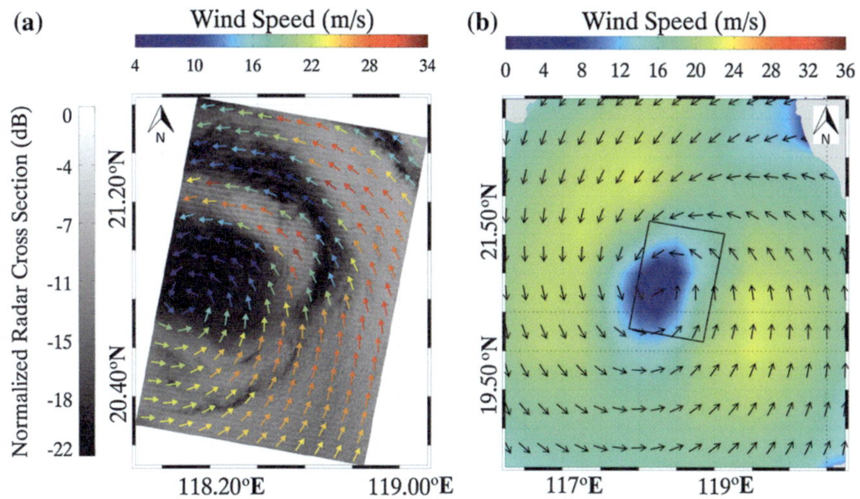

Fig. 1 **a** The quick-look example of VV-polarization TD-X image taken on 21th October 2010 at 22:05 UTC in South China sea after calibration and wind field retrieval in the period of typhoon Megi. **b** ECMWF wind field collected at 24:00 UTC, noting that the black rectangle represents the geographic location of TD-X

2.2 Description of C-Band S-1 Image

We have collected a total of 57 StripMap mode S-1 SAR images acquired in VV-polarization in U.S. coastal waters. The radar incidence angle of these images ranges from 20° to 47°. The coverage of each image contains at least one NDBC buoy. As an example, a SAR image acquired on 31st December 2014 at 02:06 UTC in western U.S. coastal waters is shown in Fig. 2.

The white circles represent the locations of corresponding buoys (IDs: 46114 and 46239). S-1 SAR has a high spatial resolution ranged from 10 to 20 m in our data collection. We extracted two 1024 × 1024 pixel sub-scenes centered at the two buoy locations and shown in Fig. 3a, b. It is necessary to figure out that good-quality power spectra of NRCS of Fig. 3a, b can be obtained by using the two-dimensional Fast Fourier Transform (FFT) method, as shown in Fig. 4a, b. In our work, the sub-scenes extracted from whole SAR, which have good-quality power spectra of SAR, are chosen for wave retrieval.

Additional 50 VV-polarization S-1 SAR images acquired in Interferometric Wide mode have been collected in China Seas, in which the time difference between SAR imaging and ECMWF reanalysis wave data is within three hours. As an example, the S-1 SAR image acquired on 19th September 2015 at 10:16 UTC is

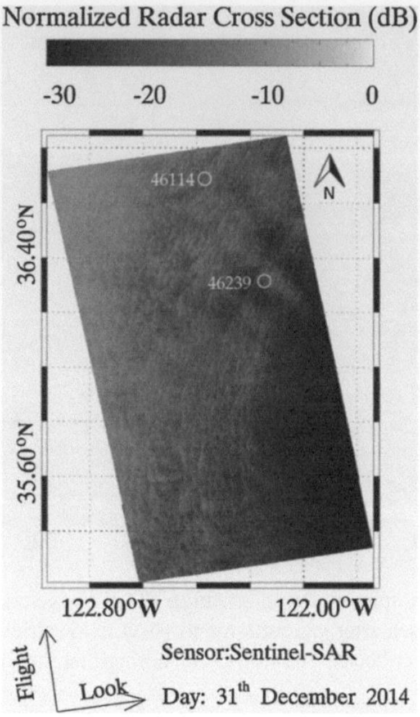

Fig. 2 An example of collected C-band VV-polarization Sentinel SAR image taken on 31st December 2014 at 02:06 UTC in west U.S. coastal area. The observations from buoys were measured at 02:10 UTC

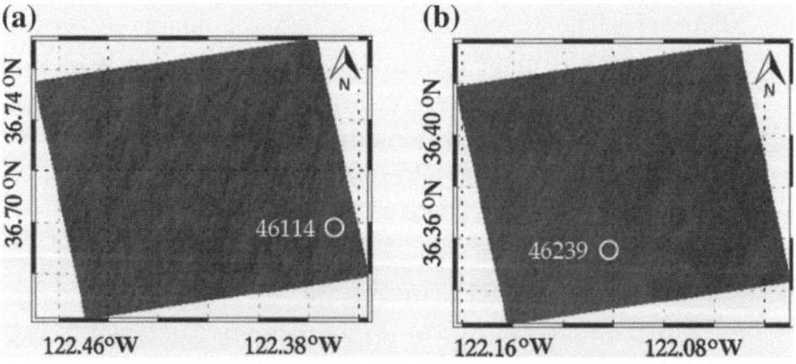

Fig. 3 The quick-look image of sub-scene which covers the location of the buoys **a** ID: 46114; **b** ID: 46239

shown in Fig. 5, in which colored arrows represent the retrieval winds by using the new C-band GMF C-SARMOD after using the ECMWF reanalysis wind directions at a $0.25° \times 0.25°$ grids.

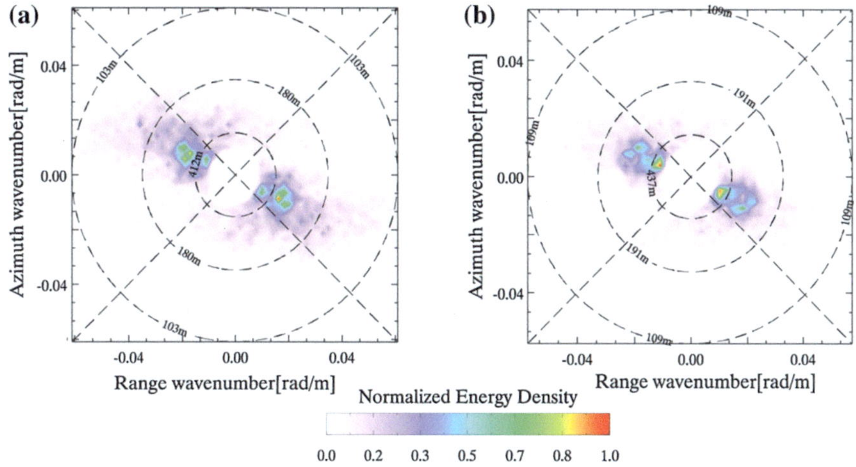

Fig. 4 Two-dimensional SAR spectra by using FFT method **a** corresponding to sub-scene of Fig. 3a; **b** corresponding to sub-scene of Fig. 3b

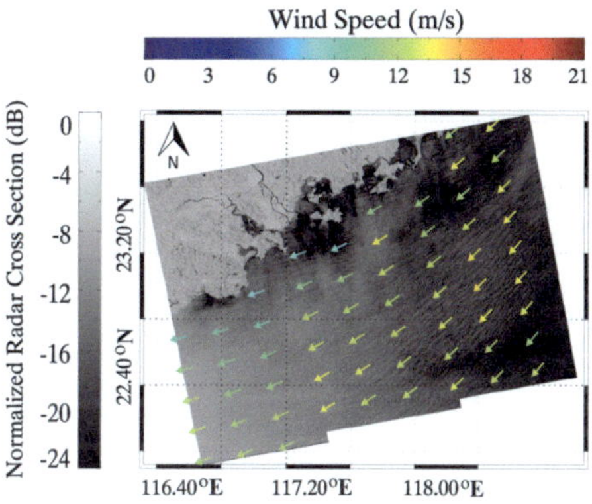

Fig. 5 A case of VV-polarization S-1 image, acquired 19th September 2015 at 10:16 UTC in China Seas, showing the quick-look image overlaid retrieval wind vectors by using GMF C-SARMOD after employing the ECMWF wind directions at a 0.5° × 0.5° grids

3 Methodology

In this section, GMF at C- and X-band SAR for winds retrieval is introduced. Besides, the two algorithms for wave retrieval, including theoretic-based PFSM and a new empirical algorithm, are presented.

3.1 Scheme of Wave Retrieval Algorithm PFSM

Until now, the GMF has successfully been applied for wind retrieval from various C- and X-band SAR data in VV-polarization. The GMF takes the general formulation:

$$\sigma^0 = B_0(1 + B_1 \cos \varphi + B_2 \cos 2\varphi)^p \tag{1}$$

where σ^0 is the linear SAR-measured NRCS in VV-polarization, ϕ is the wind direction related to range direction, p is a parametric constant, B_0 to B_2 are the functions of wind speed at 10 m above sea surface U_{10} and radar incidence angle is θ. CMOD5N is the latest version and the accuracy of wind speeds retrieved from a number of Sentinel-1 SAR images using CMOD5N have been investigated, showing a 1.35 m/s stander deviation (STD) of wind speed validated against scatterometer measurements on aboard Metop-A/B. However, C-SARMOD has not yet been applied for wind field retrieval in the study proposed by Mouche and Chapron (2015). The comparison between SAR-derived winds and NOAA buoys shows a 1.44 m/s RMSE of wind speed by using XMOD2.

PR models, which are described as the ratio between NRCS in VV and HH polarization, are usually used together with GMF as wind retrieval for C-band HH-polarization SAR. Recent research has shown that C-band SAR PR has a linear relation with wind speed (Zhang et al. 2011), in addition to the incidence angle. It is not surprising that the PR model, which includes dependence of sea surface wind speed, performs better than other PR models involving only the dependence of incidence angle, which is stated as follows:

$$PR = \frac{\sigma^0_{VV}}{\sigma^0_{HH}} = P(\theta) U_{10}^{Q(\theta)} \tag{2}$$

where

$$P(\theta) = P_1 \theta^2 + P_2 \theta + P_3 \tag{3}$$

and

$$Q(\theta) = Q_1 \theta + Q_2 \tag{4}$$

where σ^0_{VV} and σ^0_{HH} are the SAR-measured linear NRCS in VV- and HH-polarization respectively, coefficients P_1 to P_3 and Q_1 to Q_2 are tuned by an amount of quad-polarization Radarsat-2 images and the collocated buoy measurements. The dependence of wind speed on X-band PR has been investigated through dual-polarization TS-X/TD-X images and ECMWF reanalysis winds (Shao et al. 2016a, b) following which the PR model proposed in Zhang et al. (2011) was adopted for X-band SAR.

3.2 Scheme of Wave Retrieval Algorithm PFSM

Algorithm PFSM is based on the expression for the forward mapping of a wave spectrum into the SAR image spectrum. Similar to using MPI and SPRA, we need to choose the empirical wave spectrum as the first-guess wave spectrum. JONSWAP wave spectrum is used to produce the first-guess wave spectrum. The progress in PFSM is that the linear-mapping spectrum portion can be separated from the SAR intensity spectrum by calculating the wavenumber threshold. The threshold is determined by Sun and Guan (2006) calculating the threshold of wave number:

$$|k_{sep}| \approx \left(2.87g\frac{v^2}{R^2}\frac{1}{U_{10}^4\cos^2\phi\left(\sin^2\theta\sin^2\phi + \cos^2\phi\right)}\right)^{1/3} \tag{5}$$

where v is the satellite platform velocity, R is satellite slant range (definition is the distance between radar and sea surface), g is gravity acceleration and ϕ is the angle between wave propagation direction and radar look direction. The portion of the linear-mapping SAR spectrum is where the wave numbers are smaller than the separation wave number k_{sep}, and the left portion corresponds to the non-linear wind-sea state.

The reduction of the non-linear mapping to a linear mapping spectrum for wave numbers less than k_{sep} means that velocity bunching can be ignored for this regime (Alpers et al. 1979). Under this condition, the wave spectrum can be inverted directly from the linear part in the SAR spectrum, while for the non-linear part of the SAR spectrum, it is important to search for the most suitable parameters of sea state, including peak phase velocity and wave propagation direction, to produce the first-guess wave spectrum by JONSWAP empirical function. After that, the MPI is applied to invert the wind-sea wave spectrum using the prior best first-guess wave spectrum. Finally, SWH and MWP can be derived from a one-dimension wave number spectrum, combining the wind-sea and swell wave spectra inverted from SAR. The spectrum of a SAR intensity image has two peaks. Wave retrieval based on such an image spectrum has an inherent 180° directional ambiguity. We cannot remove the ambiguity without additional external information; however, SWH H_s and MWP T_{mw} can be directly obtained by using Eqs. (6) and (7) from the inverted two-dimensional wave number spectrum S_k.

$$H_s = 4\sqrt{\int S_k dk} \tag{6}$$

$$T_{mw} = \frac{\int_0^\infty S_k dk}{\int_0^\infty k^2 S_k dk} \tag{7}$$

3.3 Empirical Algorithm for Wave Retrieval at C-Band SAR

It is widely recognized that the SAR image is distorted by the azimuthal Doppler shift in the image location of a moving object when the targets in the SAR scene are moving. This strong non-linear effect is well known as the velocity bunching mechanism. As a result, waves in the azimuth direction with a wave number larger than the critical cutoff value will not be imaged by SAR. Theoretically, SWH and cutoff wavelength are proportional to a zeroth and second moment of the wave spectrum (Hasselmann and Hasselmann 1991; Vachon et al. 1994). respectively. On the other hand, SWH is more sensitive to the degree of nonlinearity than other parameters, e.g., wind speed, wind direction and wave period. Therefore, it is not surprising that SWH can be related to cutoff wavelength. More recently, several studies have made an attempt to retrieve SWH through cutoff wavelength (Wang et al. 2012; Ren et al. 2015; Shao et al. 2016a, b; Stopa et al. 2017). Among these algorithms, the new empirical algorithm, herein called CSAR_WAVE, was developed by Shao et al. (2016a, b), which describes the relation between SWH and cutoff wavelength, radar incidence angle and wave propagation direction.

The CSAR_WAVE exhibited in Shao et al. (2016a, b) is designed as follows,

$$H_s = \left(\frac{\lambda_c}{\beta}\right)(A_1 + A_2 \sin\theta + A_3 \cos 2\phi) + A_4 \tag{8}$$

where

$$\beta = \frac{R(\theta)}{v}$$

where H_s is the SWH, λ_c is the cutoff wavelength, which is estimated from the two-dimensional SAR spectra with Gaussian fit function integrated in the range direction, v is satellite flight velocity, θ is the radar incidence angle, $R(\theta)$ is the slant range at specific incidence angle, ϕ is the peak direction relative to look direction in a SAR spectrum ranged from 0° to 90° here, and matrix A_s are determined by collocating 93 of VV-polarization S-1 SAR images and SWH from NOAA buoy measurements, as shown in Table 1.

The empirical model for MWP retrieval is stated as

$$T_{mw} = H_s\left(\frac{\beta}{\lambda_c}\right)C_1 + C_2 \tag{9}$$

where the coefficient matrix C is determined from the S-1 matchup data by using the least-square-fit method, as shown in Table 2.

Table 1 Values of coefficients A

A_1	A_2	A_3	A_4
0.48	0.26	0.27	0.22

Table 2 Values of coefficients C

C_1	C_2
1.65	5.60

CSAR_WAVE allows retrieving SWH and MWP without any prior information. Moreover, the advantage is that it is not restricted for application unlike to CWAVE_ERS and CWAVE_ENVI, due to CWAVEs were originally designed for ERS-2 and ENVISAT-ASAR SAR wave mode data.

The block diagram of the scheme of semi-empirical algorithm for wave parameters retrieval based on cutoff wavelength estimation is shown in Fig. 6. First, the two-dimensional SAR image spectra is calculated from NRCS by using two-dimensional FFT method. Since the SAR spectra has 180° ambiguity with two peaks, we use the one between 0° and 90°, instead of finding the true wave propagation direction. Secondly, one-dimensional SAR spectra is obtained by integrating the two-dimensional SAR spectra in the range direction in order to estimate the cutoff wavelength. And then SWH is calculated from SAR-derived cutoff wavelength, radar incidence angle and peak wave direction relative to range direction by using Eq. (8). At last, MWP is calculated by using Eq. (9).

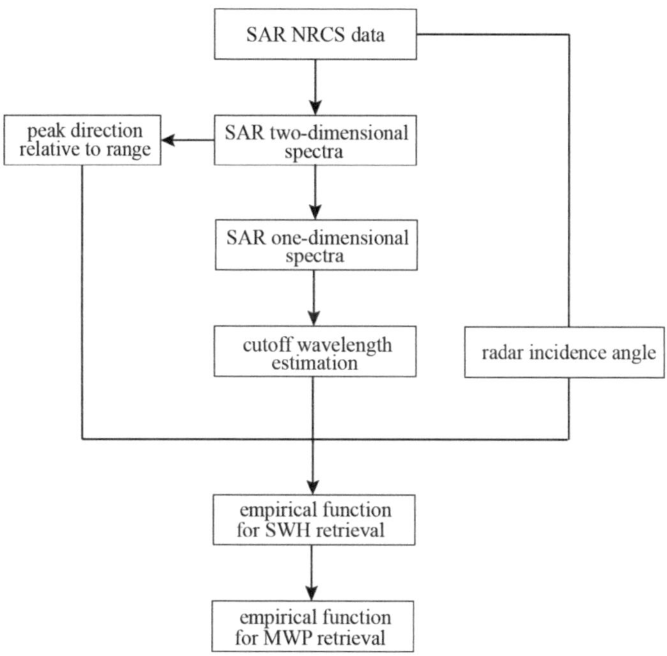

Fig. 6 Block diagram of the scheme of semi-empirical wave parameters retrieval, including SWH and MWP

4 Results

4.1 Validation of the New GMF C-SARMOD

Among 57 matchups with NOAA buoys, there are 52 measurements with winds. Subsequently, the wind speeds using the new C-band GMF C-SARMOD were compared to measurements from NOAA buoys. Figure 7 shows that wind speeds from all buoy measurements versus SAR-derived wind speeds by using C-SARMOD at 2.5 m/s bins ranged from 0 to 20 m/s. The error bar represents the STD of bias of each bin. The comparison shows a 1.63 m/s root mean square error (RMSE) of wind speed and a 1.6 m/s of STD.

As presented in the study proposed by Monaldo et al. (2016), the STD of wind speed when using CMOD5N is less than 1.44 m/s validated against the scatterometer ASCAT wind speeds. The slightly larger STD of wind speed herein is presumed to be due to different wind sources used for comparison. However, results indicate that C-SARMOD works well for wind retrieval from Sentinel-1 SAR data as well as CMOD5N and the wind speed is reliable enough to be used for wave retrieval.

4.2 Validation of Algorithm PFSM

After wind speeds are retrieved using the mature winds retrieval technology. Following that, we can retrieve SWH by using the PFSM algorithm.

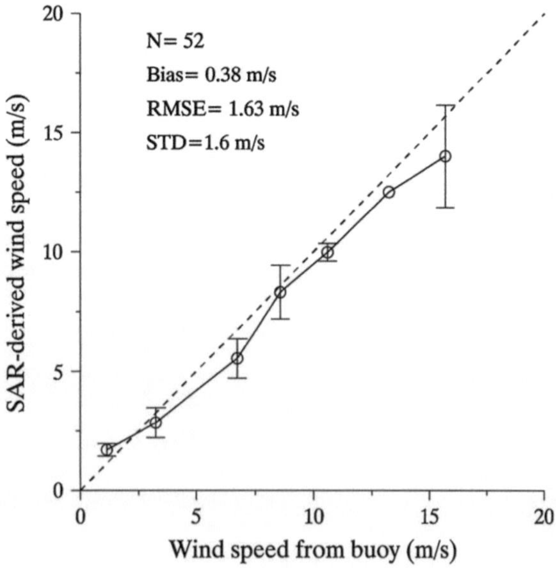

Fig. 7 SAR-derived winds speed using C-SARMOD versus wind speeds from all in situ buoys at average, for 2.5 m/s of wind speed bins between 0 and 20 m/s. Error bar represents one STD after binning the comparison

As shown in Fig. 8a, wind streaks are obviously observed in the HH-polarization TS-X image acquired at 16:19 UTC on December 13th 2010. The SAR-derived wind fields are obtained by using XMOD2 and PR mode at X-band. The two-dimensional TS-X SAR intensity spectrum from area A is exhibited in Fig. 8b. The retrieved two-dimensional wave spectrum with corresponding one-dimensional wave spectrum by using algorithm PFSM, is shown in Fig. 8c, d, respectively. In this case, the ambiguity in the SAR-derived wave direction can be resolved because we can obtain the real wave propagation direction by reference to the wind vector because the wind-sea system in area A is an open sea system far from the land.

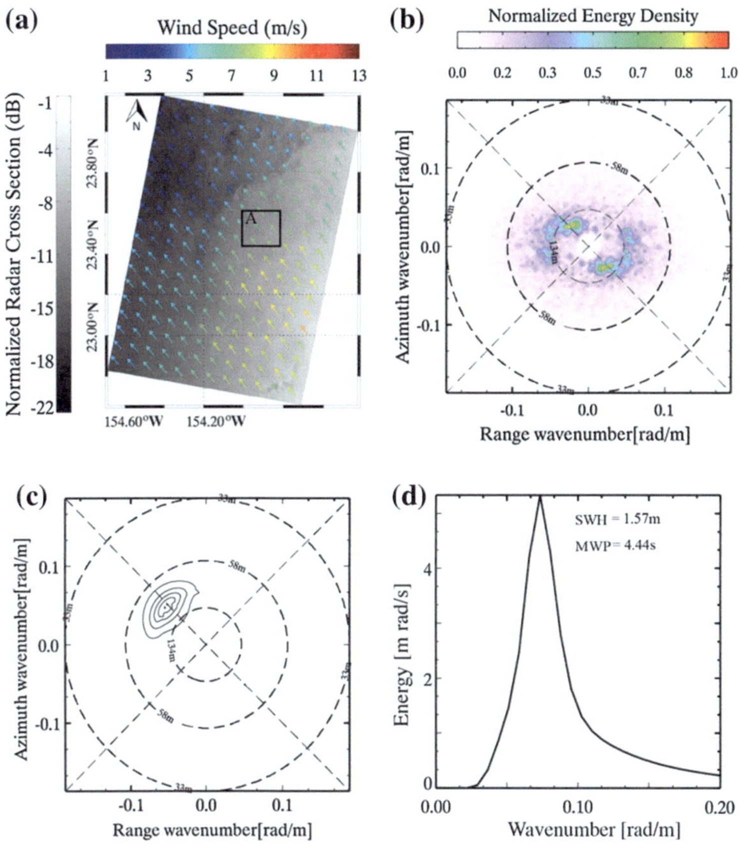

Fig. 8 A case of HH-polarization TS-X image, taken at 16:19 UTC on December 13th, 2010, showing the wind field and wave retrieval **a** wind field retrieval **b** two-dimensional TS-X intensity spectrum **c** two-dimensional wave spectrum retrieval inverted from TS-X **d** one-dimensional wave spectrum retrieval with SWH and MWP

However, as for most cases in our study, it is impossible to remove the ambiguity in the wind-sea direction because the wave propagation changes a lot due to bathymetry, leading to a mismatch between the SAR-derived wind direction and the 180° ambiguity wave direction from the two-dimensional TS-X intensity spectrum. Although the swell direction has an ambiguity similar to wind-sea, both these have little influence on the inversions of SWH and MWP.

Furthermore, 9 HH-polarization TS-X/TD-X images covering in situ buoys are processed using the X-band wind GMF and PFSM wave scheme. The SAR and buoy comparison shows that the RMSE of SWH is 0.26 m with a Scatter Index (SI) of 19.8% and the RMSE of MWP is 0.45 s with an SI of 26.0%, as shown in Fig. 9. In addition, we applied the PFSM algorithm on a total of 16 available TS-X/TD-X images and compared the results with those from the WaveWatch-III model results. The RMSE of SWH is 0.43 m with an SI of 32.8% and the RMSE of MWP is 0.47 s with an SI of 34.7%.

The wave retrieval has a similar accuracy to that of the C-band SAR, which has an SI of about 20% for SWH when validated against measurements from buoys or altimeters (Schulz-Stellenfleth et al. 2005; Li et al. 2011; Wang et al. 2012) and an SI of 38% when comparing the wave retrievals with the WAM model predictions (Monaldo and Beal 1998).

We also applied the algorithm PFSM for additional 50 VV-polarization S-1 SAR images in Interferometric Wide mode at China Seas. The sub-scenes in the images covering the ECMWF locations at a 0.5° × 0.5° grids were used. Retrieval results were compared to ECMWF reanalysis wave data, showing a 0.67 m STD of SWH at 0.5 m bins with a 2.21 s STD of MWP at 1.5 s bins, as exhibited in Fig. 10.

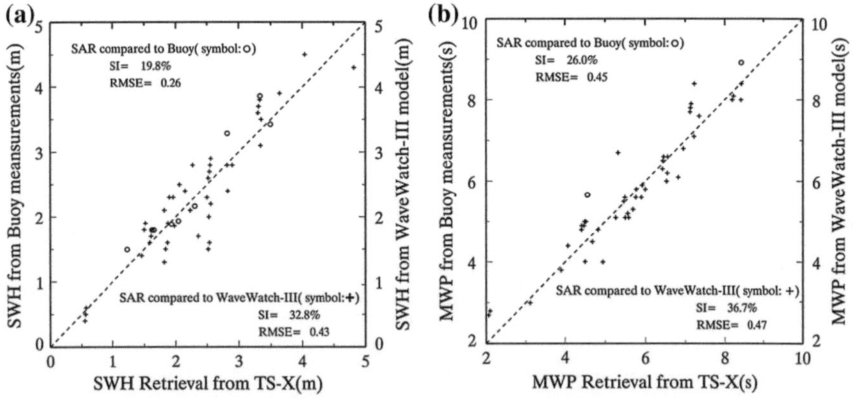

Fig. 9 Wave parameters retrieval results from HH-polarization TS-X/TD-X images are compared to buoy measurements and all TS-X/TD-X images are compared to computation from the WaveWatch-III model **a** SWH comparison **b** MWP comparison

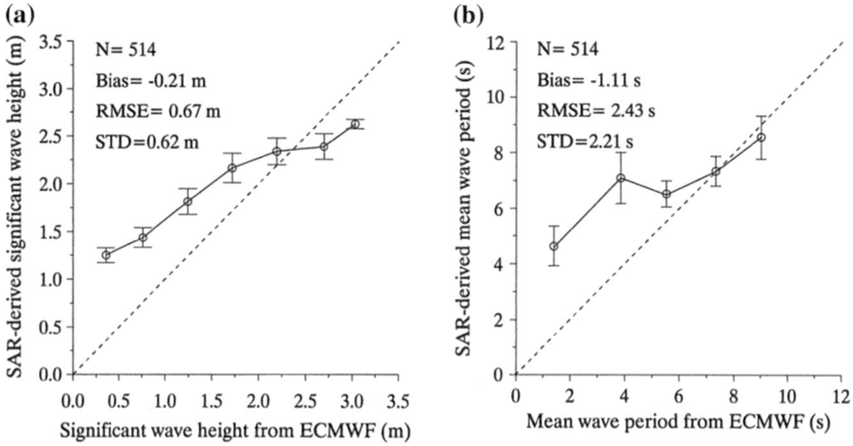

Fig. 10 SAR-derived ocean wave parameter by using algorithm PFSM versus ECMWF reanalysis wave data at average. **a** SWH comparison at 0.5 m bins between 0 and 3.5 m. **b** MWP comparison at 1.5 s bins between 1 and 10 s. Error bar represents the STD of bias of each bin

4.3 Validation of the Empirical Algorithm CSAR_WAVE

The validation of a total of 57 matchups is exhibited in Fig. 11 by using the new empirical algorithm CSAR_WAVE, showing the RMSE of MWP is 0.69 m with an SI of 18.3% and RMSE of MWP is 1.86 s with a 24.8% of SI. Referred to achievements of several researches for wave retrieval from C-band SAR data, STD

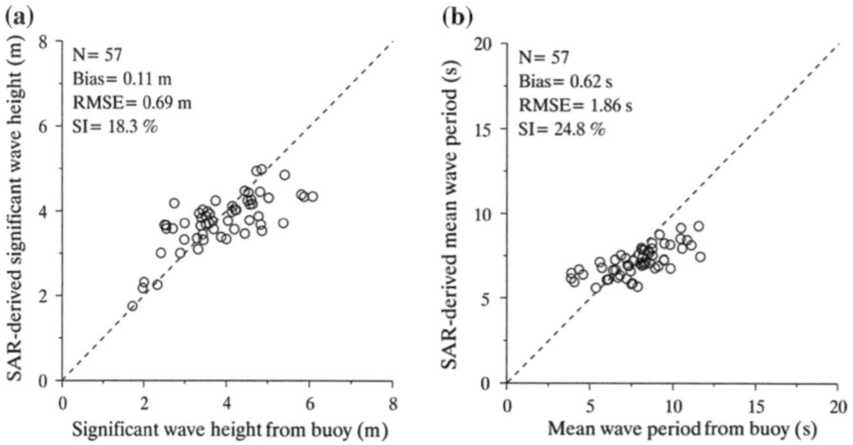

Fig. 11 Wave parameters retrieval results from 57 matchups are compared to buoy measurements. **a** SWH comparison; **b** MWP comparison

of H_s is ranged from 0.4 to 0.7 m against NOAA buoy measurements or numeric wave model results (Mastenbroek and de Valk 2000; Shao et al. 2015; Schulz-Stellenfleth et al. 2005; Li et al. 2010; He et al. 2006). When compared with buoy measurements, the accuracy of retrieved SWH by using CSAR_WAVE reaches 80%. The semi-empirical algorithm herein is expected to work well for various types of C-band SAR data. It is more applicable than existed CWAVE models, because CWAVEs were specifically tuned to derive wave information from ERS-2 and ENVISAT-ASAR wave mode data. The CWAVE model cannot be directly used for wave retrieval from SAR image mode data. In addition, statistical results (an RMSE of 0.69 m of SWH) of our semi-empirical algorithm outperformed the traditional empirical function proposed in Ren et al. (2015) (an RMSE of 0.87 m of SWH).

5 Conclusions

Wind speed and SWH are two of most the important parameters in oceanography research, and SAR has the capability to observe wind and wave in wide spatial coverage. In our work, the accuracy of wind speed and SWH retrieval results from SAR have been simultaneously investigated.

A tuned GMF is often used for wind retrieval from SAR data. The CMOD family functions were developed from NRCS from VV-polarization ERS-1/2, and collocated winds from scatterometer and ECMWF reanalysis wind fields. The new GMF C-SARMOD was derived from ENVISAT-ASAR data in VV-polarization and it has not been validated yet in Mouche and Charpon (2015). Here, 52 matchups in VV-polarization with the NOAA buoys are used for validating the new GMF C-SARMOD. The results show that the STD of wind speed is 1.6 m/s. The accuracy is close to the analysis results as using CMOD5N validated against the scatterometer ASCAT wind speeds (Monaldo et al. 2016).

The algorithm PFSM was employed to retrieve SWH and MWP. A total of 16 TS-X/TD-X images in HH-polarization were collected for validation against results from the WaveWatch-III model, showing that the RMSE of SWH between SAR-derived SWH and the WaveWatch-III model computation is 0.43 m with an SI of 32.8% and the RMSE of the MWP is 0.47 s with an SI of 34.7%. Additional VV-polarization 50 S-1 SAR images taken in the China Seas were implemented and the retrieval results were compared to the ECMWF reanalysis wave data, showing a 0.67 m STD of SWH and a 2.21 s STD of MWP. 57 matchups with NOAA buoys are also used to validate CSAR_WAVE. The validation results show a 0.69 m RMSE of SWH with a 18.3% of SI and 1.86 s RMSE of MWP with a 24.8% of SI. The comparisons indicate that the accuracy of retrieval results by using CSAR_WAVE is acceptable, due to the SI of SWH by using the existing wave retrieval algorithms against to other sources are around 20%.

In summary, the proposed methods are operationally applicable for wind and wave retrieval from C- and X-band SAR images. Moreover, the independent extraction of wind and wave from SAR images due to SAR-derived wind speed is no longer required in the process of wave retrieval when using the empirical CSAR_WAVE model. However, the accuracy of retrieval waves is expected to be improved in the future.

Acknowledgements The X-band TS-X/TD-X images are kindly provided by German Aerospace Center (DLR) through several AOs (No. OCE1656 and OCE2256). We appreciated European Space Agency (ESA) provides C-band S-1 SAR images for free via https://scihub.copernicus.eu. NOAA buoy measurements were accessed via http://www.ndbc.noaa.gov/. ECMWF reanalysis data were openly accessed via http://www.ecmwf.int. Wave data from the WaveWatch-III model were kindly provided by the professional Institut Francais de Recherche pour Exploitation de la MER (IFREMER) group and the data were collected through the server: ftp.ifremer.fr.

References

Alpers W, Bruning C (1986) On the relative importance of motion-related contributions to SAR imaging mechanism of ocean surface waves. IEEE Trans Geosci Remote Sens 86:10529–10546

Alpers W, Rufenach C (1979) The effect of orbital motions on synthetic aperture radar imagery of ocean waves. IEEE Trans Antennas Propag 27:685–690

Alpers W, Ross DB, Rufenach CL (1981) On the detectability of ocean surface waves by real and synthetic radar. J Geophys Res 86:10529–10546

Chapron B, Johnsen H, Garello R (2001) Wave and wind retrieval from SAR images of the ocean. Ann Telecommun 56:682–699

Grieco G, Lin W, Migliaccio M, Nirchio F, Portabella M (2016) Dependency of the Sentinel-1 azimuth wavelength cut-off on significant wave height and wind speed. Int J Remote Sens 37:5086–5104

Hasselmann K, Hasselmann S (1991) On the nonlinear mapping of an ocean wave spectrum into a synthetic aperture radar image spectrum. J Geophys Res 96:10713–10729

Hasselmann K, Barnett TP, Bouws E, Carlson H, Cartwright DE, Enke K, Ewing JA, Gienapp H, Hasselmann DE, Kruseman P (1973) Measurements of wind-wave growth and swell decay during the joint north sea wave project (JONSWAP). Deutches Hydrographisches Institut, Hamburg, Germany

Hasselmann S, Bruning C, Hasselmann K (1996) An improved algorithm for the retrieval of ocean wave spectra from synthetic aperture radar image spectra. J Geophys Res 101:6615–6629

He YJ, Shen H, Perrie W (2006) Remote sensing of ocean waves by polarimetric SAR. J Atmos Oceanic Technol 23:1768

Hersbach H (2010) Comparison of C-Band scatterometer CMOD5.N equivalent neutral winds with ECMWF. J Atmos Ocean Tech 27:721–736

Hersbach H, Stoffelen A, Haan SD (2007) An improved C-band scatterometer ocean geophysical model function: CMOD5. J Geophys Res 112:C03006

Kerbaol V, Chapron B, Vachon PW (1998) Analysis of ERS-1/2 synthetic aperture radar wave mode imagettes. J Geophys Res 115:7833–7846

Lehner S, Horstmann J (1998) Mesoscale wind measurements using recalibrated ERS SAR images. J Geophys Res 103:7847–7856

Li XM, Lehner S (2014) Algorithm for sea surface wind retrieval from TerraSAR-X and TanDEM-X Data. IEEE Trans Geosci Remote Sens 52:2928–2939

Li XF, Pichel WG, He MX, Wu S, Friedman K, Clemente-Colon P, Zhao CF (2002) Observation of hurricane-generated ocean swell refraction at the gulf stream north wall with the RADARSAT-1 synthetic aperture radar. IEEE Trans Geosci Remote Sens 40:2131–2142

Li XM, Lehner S, Rosenthal W (2010) Investigation of ocean surface wave refraction using TerraSAR-X data. IEEE Trans Geosci Remote Sens 48:830–840

Li XM, Lehner S, Bruns T (2011) Ocean wave integral parameter measurements using Envisat ASAR wave mode data. IEEE Trans Geosci Remote Sens 49:155–174

Li XF, Zhang JA, Yang XF, Pichel WG, DeMaria M, Long D, Li ZW (2013) Tropical cyclone morphology from spaceborne synthetic aperture radar. Bull Am Meteorol Soc 94:215

Lin B, Shao WZ, Li XF, Li H, Du XQ, Ji QY, Cai LN (2017) Development and validation of an ocean wave retrieval algorithm for VV-polarization Sentinel-1 SAR data. Acta Oceanol 36:95–101

Mastenbroek C, de Valk CF (2000) A semi-parametric algorithm to retrieve ocean wave spectra from synthetic aperture radar. J Geophys Res 105:3497–3516

Monaldo FM, Beal RC (1998) Comparison of SIR-C SAR wavenumber spectra with WAM model predictions. J Geophys Res 103:18815–18822

Monaldo FM, Jackson C, Li XF, Pichel WG (2016) Preliminary evaluation of Sentinel-1A wind speed retrievals. IEEE J Sel Topics Appl Earth Observ Remote Sens 9:2638–2642

Mouche AA, Chapron B (2015) Global C-Band Envisat, RADARSAT-2 and Sentinel-1 SAR measurements in co-polarization and cross-polarization. J Geophys Res 88:3–9

Pleskachevsky AL, Rosenthal W, Lehner S (2016) Meteo-marine parameters for highly variable environment in coastal regions from satellite radar images, ISPRS. J Photogram 119:464–484

Quilfen Y, Bentamy A, Elfouhaily TM, Katsaros K, Tournadre J (1998) Observation of tropical cyclones by high-resolution scatterometry. J Geophys Res 103:7767–7786

Ren YZ, He MX, Lehner S (2012) An algorithm for the retrieval of sea surface wind fields using X-band TerraSAR-X data. Int J Remote Sens 33:7310–7336

Ren L, Yang JS, Zheng G, Wang J (2015) Significant wave height estimation using azimuth cutoff of C-band RADARSAT-2 single-polarization SAR images. Acta Oceanol Sin 12:1–9

Schulz-Stellenfleth J, Lehner S, Hoja D (2005) A parametric scheme for the retrieval of two-dimensional ocean wave spectra from synthetic aperture radar look cross spectra. J Geophys Res 110:297–314

Schulz-Stellenfleth J, Konig T, Lehner S (2007) An empirical approach for the retrieval of integral ocean wave parameters from synthetic aperture radar data. J Geophys Res 112:1–14

Shao WZ, Sun J, Guan CL, Sun ZF (2014) A method for sea surface wind field retrieval from SAR image mode data. J Ocean Univ China 13:198–204

Shao WZ, Li XF, Sun J (2015) Ocean wave parameters retrieval from TerraSAR-X images validated against buoy measurements and model results. Remote Sens 7:12815–12828

Shao WZ, Zhang Z, Li XM, Wang WL (2016a) Sea surface wind speed retrieval from TerraSAR-X HH-polarization data using an improved polarization ratio model. IEEE J Sel Topics Appl Earth Observ Remote Sens 9:4991–4997

Shao WZ, Zhang Z, Li XF, Li H (2016b) Ocean wave parameters retrieval from Sentinel-1 SAR imagery. Remote Sens 8:707–721

Shao WZ, Wang J, Li XF, Sun J (2017a) An empirical algorithm for wave retrieval from co-polarization X-Band SAR imagery. Remote Sens 9:711–724

Shao WZ, Shen YX, Sun J (2017b) Preliminary assessment of wind and wave retrieval from Chinese Gaofen-3 SAR imagery. Sensors 17:1705–1718

Stoffelen A, Anderson D (1997) Scatterometer data interpretation: derivation of the transfer function, CMOD4. J Geophys Res 102:5767–5780

Stopa JE, Mouche AA (2017) Significant wave heights from Sentinel-1 SAR: validation and applications. J Geophys Res 122:1827–1848

Stopa JE, Ardhuin F, Collard F, Chapron B (2016) Estimating wave orbital velocities through the azimuth cut-off from space borne satellites. J Geophys Res 120:7616–7634

Sun J, Guan CL (2006) Parameterized first-guess spectrum method for retrieving directional spectrum of swell-dominated waves and huge waves from SAR images. Chin J Oceanol Limnol 24:12–20

Sun J, Kawamura H (2009) Retrieval of surface wave parameters from SAR images and their validation in the coastal seas around Japan. J Oceanogr 65:567–577

Vachon PW, Krogstad HE, Paterson JS (1994) Airborne and spaceborne synthetic aperture radar observations of ocean waves. Atmos Ocean 32:83–112

Voorrips C, Makin V, Hasselmann S (1997) Assimilation of wave spectra from pitch-and-roll buoys in a North Sea wave model. J Geophys Res 102:5829–5849

Wang H, Zhu J, Yang JS (2012) A semi-empirical algorithm for SAR wave height retrieval and its validation using Envisat ASAR wave mode data. Acta Oceanol Sin 31:59–66

Yang XF, Li XF, Pichel W, Li ZW (2011) Comparison of ocean surface winds from ENVISAT ASAR, Metop ASCAT scatterometer, buoy Measurements, and NOGAPS model. IEEE Trans Geosci Remote Sens 49:4743–4750

Zhang B, Perrie W, He YJ (2011) Wind speed retrieval from RADARSAT-2 quad-polarization images using a new polarization ratio model. J Geophys Res 116:1318–1323

Ocean Colour Remote Sensing in Chinese Marginal Seas

Xianqiang He and Xiaoyan Chen

Abstract The Chinese marginal seas are some of the largest marginal seas in the world, which include the South China Sea (SCS), East China Sea (ECS), Yellow Sea (YS), and Bohai Sea (BS). The SCS is the largest tropical marginal sea and the ECS is the 11th largest marginal sea in the world. Both are within the world's strongest monsoon area–the East Asian monsoon, with significant seasonal and interannual variations and even long-term changes. Moreover, the Chinese marginal seas receive a tremendous amount of fresh water and terrestrial materials from several world's largest rivers, such as the Changjiang (largest river in the Eurasian continent), Yellow River, Pearl River, and Mekong River, which make the seas more vulnerable to anthropogenic influences. Considering field investigations being limited in space and temporal coverage, satellite ocean colour remote technique has been used widely in the Chinese marginal seas to investigate environmental dynamics at different time scales from diurnal, daily, seasonal to interannual variation. In this chapter, we highlight some of the new techniques and applications of satellite ocean colour remote sensing in Chinese marginal seas in the past few years, including detecting diurnal dynamics of coastal waters, retrieval of marine carbon parameters, monitoring of marine ecologic environment, and satellite oceanographic research.

Keywords Ocean colour remote sensing · Chinese marginal seas
Diurnal dynamic · River plume · Typhoon · Marine carbon cycle

---

X. He (✉) · X. Chen
State Key Laboratory of Satellite Ocean Environment Dynamics,
Second Institute of Oceanography, State Oceanic Administration,
Hangzhou, China
e-mail: hexianqiang@sio.org.cn

V. Barale and M. Gade (eds.), *Remote Sensing of the Asian Seas*,
https://doi.org/10.1007/978-3-319-94067-0_17

1 Introduction

The Chinese marginal seas are some of the largest marginal seas in the world (Fig. 1), which include the South China Sea (SCS), East China Sea (ECS), Yellow Sea (YS) and Bohai Sea (BS). The SCS is the world's largest tropical marginal sea with a total area around 3,500,000 km^2. The averaged depth of the SCS is more than 2000 m, with a maximum depth larger than 5000 m in the central basin. The continental shelves (<200 m) are generally located along the north and southwest coasts in the SCS, and they are narrow along the east coast. In the northern SCS, the deep Luzon Strait with depth range of 2000–5000 m connects the SCS with the

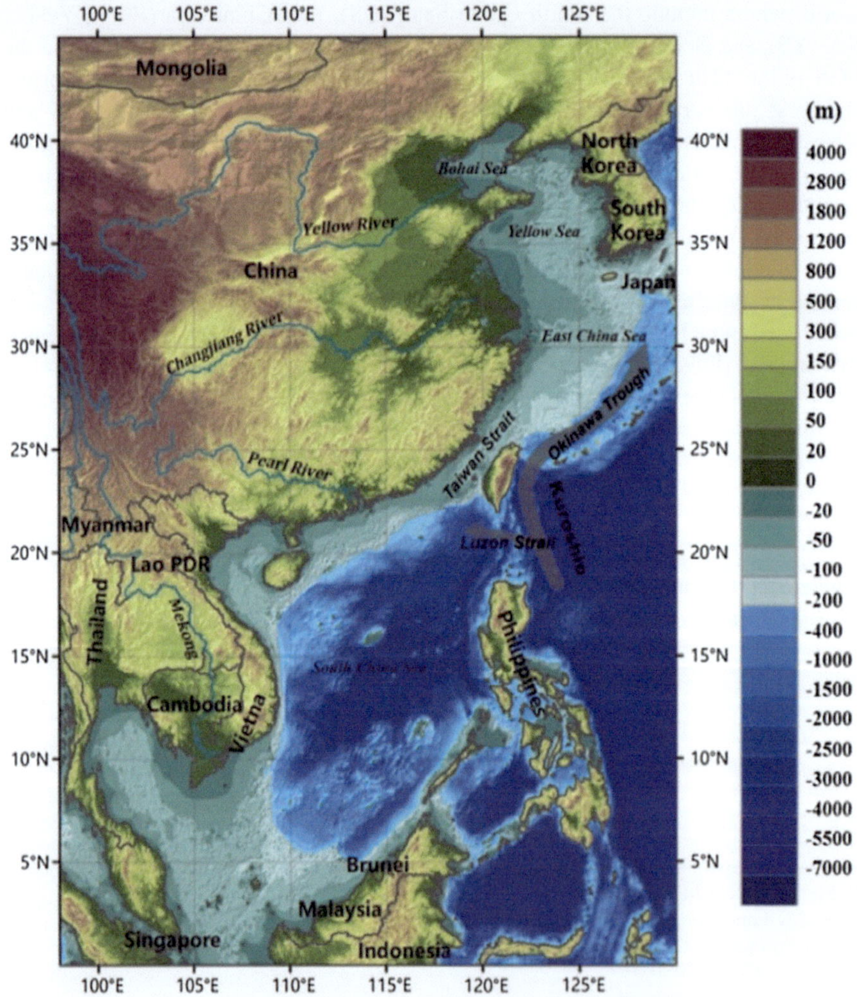

Fig. 1 Location and bathymetry of Chinese marginal seas

Pacific Ocean; and the oligotrophic Kuroshio waters intrude into the SCS via the strait. In addition, the shallow Taiwan Strait with a mean depth about 50 m connects the SCS with the ECS; there is a net SCS watermass transport flowing into the ECS via the strait. Two world's large rivers enter the SCS, one is the Pearl River to the north, and the other is the Mekong River to the southwest. The Pearl River is the 13th largest river in the world in terms of discharge, which delivers about 3.5×10^{11} m^3 of fresh water and 8.5×10^7 t of sediment per year, with 80% of the total discharge occurring in the wet season (April–September). The Mekong River is the eighth largest river in the world with an annual discharge of about 4.7×10^{11} m^3, and delivers about 1.7×10^8 t of sediment per year (Wolanski et al. 1996).

The eastern Chinese seas include the Bohai Sea, Yellow Sea and East China Sea, with areas of about 77,000, 417,000 and 751,000 km^2, and mean depths of 18, 40 and 72 m, respectively. The eastern Chinese seas have one of the widest shelves in the world. Except for the deep Okinawa Trough at the southeast, the water depths are all less than 200 m in the eastern Chinese seas. The eastern Chinese seas are surrounded by lands on the western and northern sides, and bounded by the Kuroshio on the southeastern side. The Kuroshio flows northeastward along the Okinawa Trough, and exchanges waters and materials with the eastern Chinese seas. In addition, the ECS receives waters and materials from the SCS through the Taiwan Strait, and also exports terrestrial materials into the northern SCS in the winter season. The eastern Chinese seas receive a tremendous amount of fresh water and suspended sediments from rivers, including two of the largest rivers in the world, the Changjiang (Yangtze River) and Huanghe (Yellow River). The Changjiang River, the fifth largest river in the world and largest river in the Eurasian continent, flows into the ECS and YS, delivering about 9.25×10^{11} m^3 of fresh water and 4.8×10^8 t of sediment per year (He et al. 2013a).

Both the SCS and eastern Chinese seas are influenced by the world's strongest monsoon—the East Asian monsoon. Their current circulations and biogeochemical processes are complex with significant seasonal variation. Moreover, the monsoon is sensitive to nature climate variability such as the El Niño-Southern Oscillation (ENSO) events, and further influences the interannual variation and long-term change of the marine environment in the SCS and eastern Chinese seas. In addition, with the rapid economic development in China and surrounding countries, lots of terrestrial materials have entered the surrounding coastal waters through rivers, atmospheric dust deposition and ground waters. Thus, the SCS and eastern Chinese seas are vulnerable marginal seas influenced by both nature climate changes and anthropogenic activities. How the marine environment and ecosystems will respond to climate variability is one of the key concerns for oceanographic research concerning the Chinese marginal seas.

Together with the limited spatial-temporal coverage by traditional in situ measurements, satellite ocean colour remote sensing with daily or hourly observations and wide coverage has been widely used to investigate the dynamics of the Chinese marginal seas at time scales from diurnal, daily, seasonal to interannual variation. In this review, we highlight some of the new findings, techniques and applications of satellite ocean colour remote sensing related to the Chinese marginal seas in the past few years, including diurnal dynamic observations in coastal waters, retrieval of marine carbon parameters, monitoring of marine ecologic environment, and satellite oceanographic research. Finally, summary and future perspectives for ocean colour remote sensing in Chinese marginal seas are provided.

2 Diurnal Dynamic Observations by Geostationary Ocean Colour Satellite

The coastal waters of the Chinese marginal seas are highly dynamic, especially in the eastern Chinese seas due to shallow water depth and strong tidal disturbance, combined with huge river inputs. With the successful launch of the world's first geostationary satellite ocean colour sensor GOCI in June 2010, the whole eastern Chinese seas can be observed hourly from 8:30 AM to 3:30 PM Beijing time. He et al. (2013a) demonstrated the unique capacity of GOCI to observe the diurnal variation of the suspended particulate matter (TSM) in the Hangzhou Bay (Fig. 2). They found that the diurnal variation of the TSM in the Hangzhou Bay is significant, and the magnitude of the diurnal variation is of high spatial inhomogeneity, with the maximum up to 5000 mg/l in the central Hangzhou Bay. Moreover, the diurnal variation magnitude of the TSM is larger during the spring tide than during the neap tide. In addition, they suggested that caution should be taken when mapping TSM in the coastal waters using cruise sampling and conventional polar-orbiting satellite data, as the temporal resolution of such data is insufficient for capturing diurnal variation. Lou and Hu (2014) revealed the capacity of GOCI in catching diurnal dynamics of algal blooms along the coast of the ECS. They found

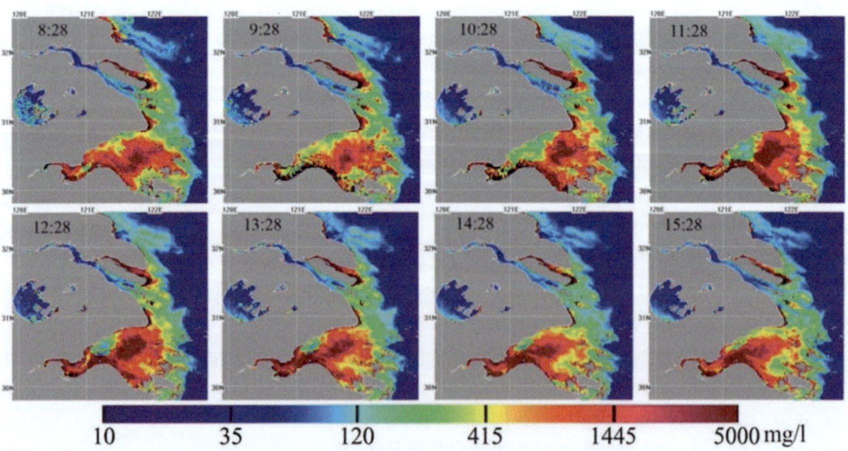

Fig. 2 GOCI-derived hourly TSM in the Hangzhou Bay on 5 April 2011. The marked times on the plots refer to Beijing time. After He et al. (2013a)

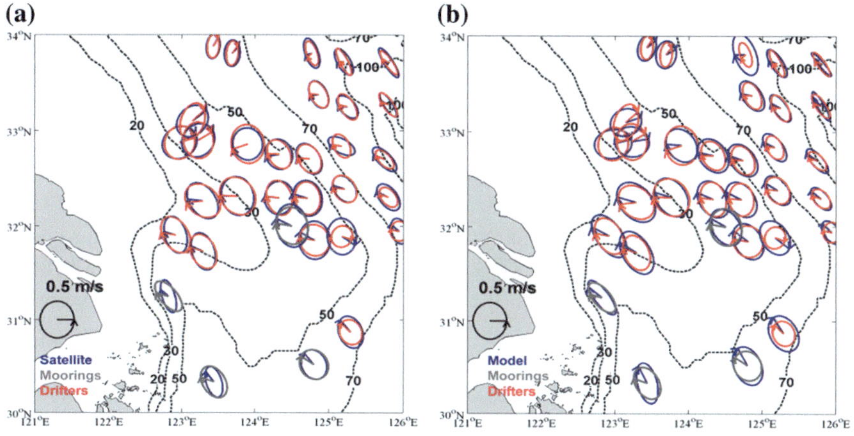

Fig. 3 Comparison of M2 semidiurnal tidal ellipses retrieved by satellite and tidal model (blue) with the in situ observations from four moorings (gray) and 28 drifter locations (red) (a) is for the satellite data, and (b) for the tidal model output. Arrows denote tidal phases relative to GMT. After Hu et al. (2016a)

bloom areas increasing significantly from morning to afternoon. Cheng et al. (2016) investigated the dynamic variation of the TSM in the macro-tidal Yalu River Estuary (YRE) of China using the GOCI data; and they found that the diurnal variation of the TSM was mainly affected by a re-suspension process.

The hourly sequential images of GOCI can also be used to derive sea surface currents with high spatial resolution (Hu et al. 2016a, 2017). Based on the GOCI-derived sea surface currents that include all current portions, Hu et al. (2016a) proposed a novel method to separate the semidiurnal M2 tidal currents and residual currents from the total currents in the ECS. The GOCI-derived M2 tidal currents were consistent with the in situ observations from drifters and moorings, and with the tidal numerical model results with the relative error less than 10% (Fig. 3). Moreover, based on the GOCI-derived hourly TSM imagery, Hu et al. (2016b) found that the diurnal displacement of turbidity front in the Changjiang River Estuary could be more than 10 km in response to the onshore-offshore movement of tidal currents.

3 Satellite Observations of River Plumes

River plumes are critical zones of land-ocean interaction where transformation takes place for the exports of sediments, nutrients and organic materials from land to the sea. Approximately 40% of the total amounts of fresh water and particulate materials transported into the seas comes from the world's 10 largest rivers. Thus, understanding the spatial-temporal variation of large river's plume is crucial to study material transport and biogeochemical processes in marginal seas. Yet, the plume extension is highly dynamic, which is influenced by river discharge, tide, wind, surrounding currents, etc. With daily or hourly observations and wide coverage, satellite ocean colour remote sensing is an ideal tool for monitoring and understanding the plume transportation.

Salinity is a good indicator for river plume extending. Although microwave ra-diometer at L-band, such as SMOS (the Soil Moisture and Ocean Salinity) and Aquarius/SAC-D satellite, can monitor the global sea surface salinity, the coarse spatial resolution (typically 30–300 km) and long revisiting time (3 days or more) makes them difficulty to observe the high spatial-temporal dynamic river plume. With the high spatial (~ 1 km) and temporal (~ 1 day) resolution, salinity retrieved from ocean colour satellite data is practical way to observe the river plume. Based on the in situ data from nine cruises covering all seasons and full salinity ranges in the ECS, Bai et al. (2013) found that the colour dissolved organic matter (CDOM) was generally conservative for all seasons. They established a seasonally uniform algorithm for retrieval of surface water salinity from satellite-derived CDOM. Validation results showed that the absolute errors of satellite-derived salinity were generally less than 1 psu. Using the developed algorithm and time series of SeaWiFS data, Bai et al. (2014) investigated interannual variation of the Changjiang River plume in summertime during 1998–2010 (Fig. 4). They identified three types of Changjiang River plume extension shape in summertime: one is the commonly known northeastward transportation, the second is a case in which most of the plume water crossed the Cheju Strait into the Tsushima-Korea Strait with only a small fraction staying on the shelf of the ECS, and the third is a rare case in which the plume front moved southeastward.

Moreover, they found no significant long-term trend in summertime plume area during 1998–2010, and concluded that the interannual variation was probably regulated by natural variation rather than anthropogenic effects (such as the construction of the Three Gorges Dam).

In the northern SCS, using satellite-derived water transparency and chlorophyll a concentration (Chla) images, Bai et al. (2015b) found that the Pearl River plume could intrude into the Penghu Channel in the southeast part of the Taiwan Strait, which is also the main channel connecting the southern ECS and northern SCS. By analysing the time series of satellite data, they revealed that the Pearl River plume intruding into the Penghu Channel occurred in the summertime of most years. Recently, from daily satellite Chla images in June and July 2015, He et al. (2016) found that a cyclonic eddy entrained the Pearl River plume into the oligotrophic basin of the northern SCS, and induced a rare large phytoplankton bloom in the oligotrophic basin, which was about 500 km in length, 100 km in width and lasting more than 19 days (Fig. 5).

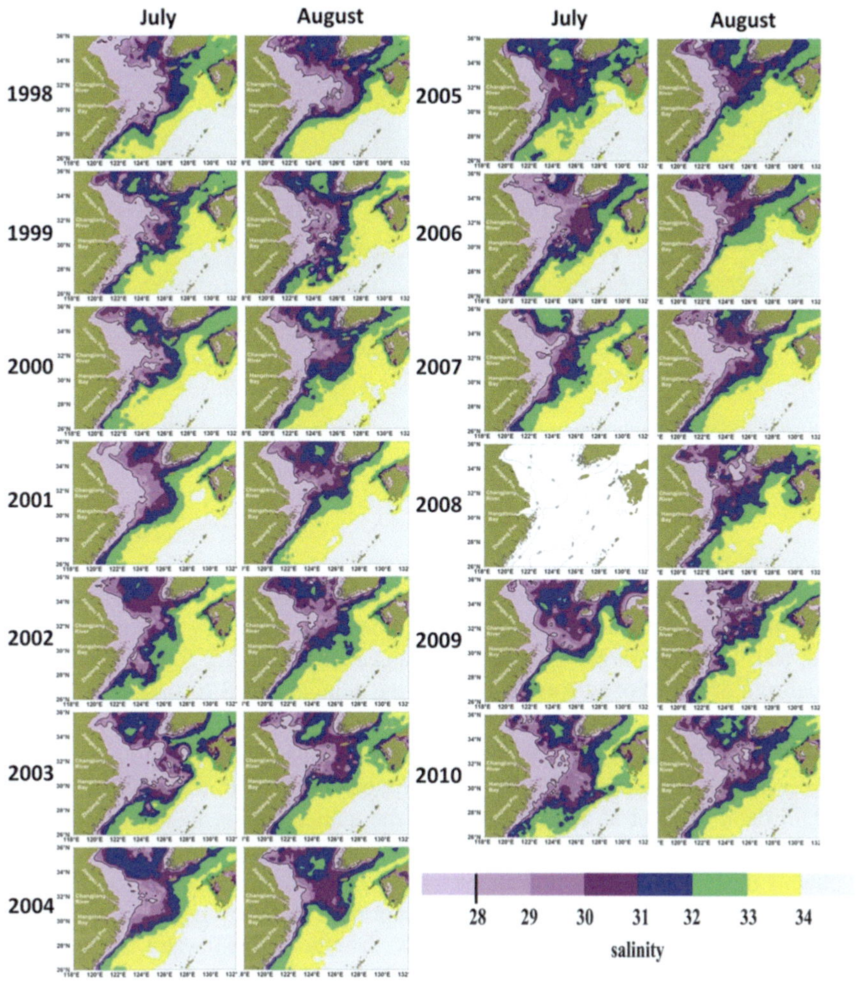

Fig. 4 Satellite-derived monthly distribution of the Changjiang River plume in July and August during 1998–2010. Data were unavailable for July 2008 owing to sensor failure. After Bai et al. (2014)

4 Satellite Monitoring of Typhoon Influences on Marine Environment

The Chinese marginal seas are one of the regions mostly influenced by typhoons. Taking advantage of satellite ocean colour remote sensing, lots of studies have been carried out to reveal the enhanced phytoplankton biomass and primary production by typhoons in recent years (Chen et al. 2012; Zhao et al. 2013; Liu et al. 2013). In addition, Ye et al. (2013) found an interesting case that typhoon not only induced

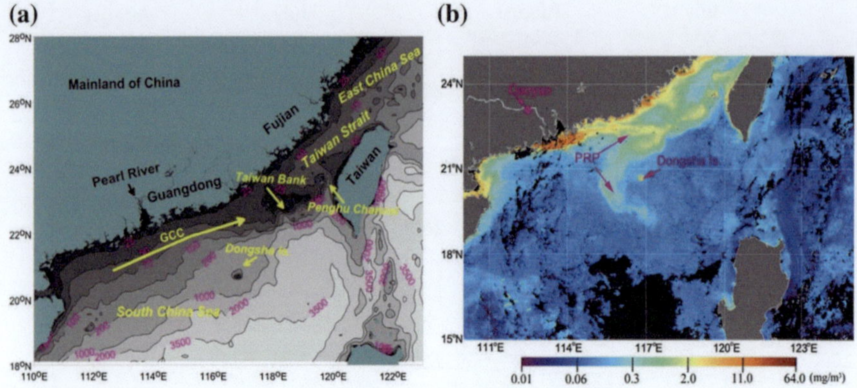

Fig. 5 **a** Location and bathymetry (isobaths in meters) in the northern South China Sea. GCC: the Guangdong Coastal Current. **b** Daily satellite Chla image on 17 June 2015. Black areas were covered by clouds. After He et al. (2016)

surface water blooms but also induced interior water blooms just above the thermocline in the northern SCS after the passage of Typhoon Nuri in August 2008. Chung et al. (2012) reported that diatom abundance increased by approximately 50 times, and the major target fish species changed from skipjack tuna to squid after the passage of Morakot in August 2009 in the southern ECS (Chang et al. 2014). Similarly, Yu et al. (2013) found that fish species number increased by approximately 15% in the northern SCS after the two typhoons of GONI and Koppu in the summer of 2009, respectively. Recently, based on 10-year (2003–2012) Aqua/MODIS data, Chen et al. (2015) estimated the annual carbon fixation induced by typhoons in the SCS, and suggested that the annual-mean carbon fixation due to typhoons in the whole SCS was about 5–15% of the new primary production.

Besides the biological effects, another important effect by typhoons is the terrestrial material transportation. Using daily satellite-derived water transparency images, He et al. (2014) found that the terrestrial materials transported into the southern Okinawa Trough triggered by Typhoon Morakot in August 2009. Moreover, they found that the terrestrial materials from the coast of mainland China could cross the Taiwan Strait thanks to typhoon, and could be further transported to the region northeast of Taiwan (Fig. 6). They suggested that the episodic cyclone-triggered terrestrial material transport could be another source of mud in the southern Okinawa Trough.

Similarly, Li et al. (2015) found that the TSM in the regions near the Minjiang River Estuary dramatically increased during the passage of Typhoon Soulik in July 2013, and the seawater with relatively high sediment concentration was transported to the middle of the Taiwan Strait after being carried along by wind-induced flows.

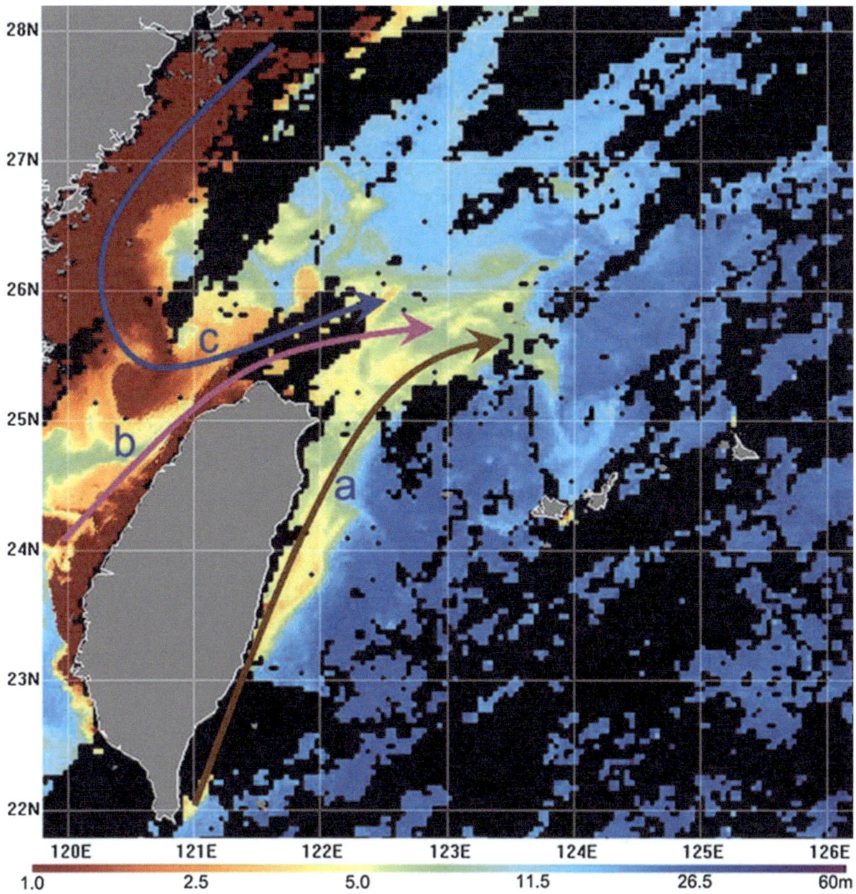

Fig. 6 Schematic map of the three sources of turbid waters transported to the region off northeastern Taiwan triggered by Morakot. Shading indicates satellite-derived water transparency on 14 August 2009. The source (a) was the east coast of Taiwan, (b) was the west coast of Taiwan, and (c) was the coast of mainland China. After He et al. (2014)

5 Satellite Retrieval of Marine Carbon Parameters

The ocean contains the largest amount of actively cycled carbon in the world, and plays a key role in global carbon cycle. In recent years, satellite retrievals of marine carbon parameters in the Chinese marginal seas have been significantly advanced, especially the retrievals of dissolved organic carbon (DOC), particulate organic carbon (POC) and surface water partial pressure of carbon dioxide (pCO_2).

Liu et al. (2014) proposed a novel model to estimate the DOC inventories in both euphotic layer and whole water column from satellite data in the ECS. In their model, surface DOC concentration was retrieved by satellite-derived CDOM and

Chla directly. Then, the vertical profile types of DOC concentration (uniform or stratified) were determined by the water masses classified by satellite data. Finally, the inventory was estimated by integrating the vertical profile of DOC based on the satellite-derived surface DOC and its vertical profile model. In the northern SCS, Pan and Wong (2015) established an empirical algorithm for satellite retrieval of surface DOC concentration from satellite-derived CDOM and sea surface temperature (SST).

As for the satellite retrieval of POC, Hu et al. (2015) developed a regional empirical algorithm in the shelf and basin of the SCS, by establishing a power function relationship between POC and the blue-to-green band ratio of satellite-retrieved remote sensing reflectance. Similarly, using the in situ datasets from four seasonal cruises, Liu et al. (2015a) established an empirical algorithm in the Pearl River Estuary (PRE), which is a linear function of the two reflectance ratios from four bands.

The air-sea flux of carbon dioxide (CO_2) is a more direct parameter in relation to global climate change, as CO_2 is the major greenhouse gas inducing global warming. Field measurements are insufficient to estimate air-sea CO_2 flux due to limited spatial-temporal resolution, especially in highly dynamic marginal seas. In the past decade, satellite remote sensing was shown to be a useful tool contributing to assessments of pCO_2 and subsequent air-sea CO_2 flux. For example, Zhu et al. (2009) developed a regression algorithm for satellite retrieval of pCO_2 in summertime in the northern SCS by establishing a multiple regression relationship between pCO_2 and SST or Chla. Similarly, Tao et al. (2012) established a regression algorithm to estimate pCO_2 from satellite-derived SST and Chla in the YS and BS. Based on a more complex regression approach, Jo et al. (2012) developed a neural network model to estimate pCO_2 using the satellite-derived SST and Chla in the northern SCS.

For complex continental shelves, such as the ECS influenced by the Changjiang River plume, empirical pCO_2 algorithms may not be applicable. Bai et al. (2015a) proposed a "mechanistic semi-analytic algorithm" (MeSAA) to retrieve pCO_2 in the ECS. In the MeSAA algorithm, the total variation of pCO_2 is considered as the sum of individual pCO_2 variations induced by different processes, such as mixing, thermodynamics and biological activities. Theoretically, all factors inducing pCO_2 variation can be added in the algorithm, and the MeSAA can be applied to different regions. Recently, Song et al. (2016) applied the MeSAA algorithm to estimate the pCO_2 in the Bering Sea using different controlling factors from the ECS, and their results illustrated the applicability of the MeSAA in different marginal seas (Fig 7).

Based on time series of satellite ocean colour data, the spatial-temporal distribution and interannual variation of biological characteristics (e.g., Chla, primary production) in the Chinese marginal seas have been investigated widely in recent years (Palacz et al. 2011; Shang et al. 2011; Shi and Wang 2012; Yamaguchi et al. 2012; Liu and Chen 2014; Liu and Wang 2013; Liu et al. 2014). In addition, the Chinese marginal seas are some of the highest primary productivity seas in the world, with phytoplankton blooms occurring frequently. Using 14-year (1998–2011) satellite ocean colour data from SeaWiFS and Aqua/MODIS, He et al. (2013b) investigated

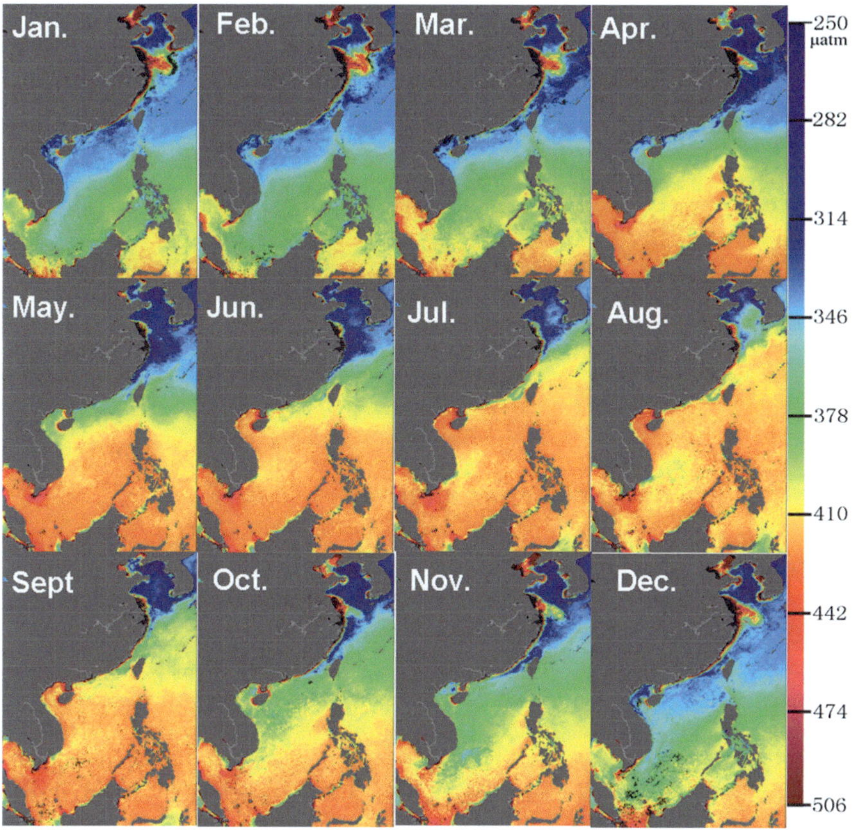

Fig. 7 Satellite-derived monthly sea surface partial pressure of carbon dioxide (pCO_2) in the Chinese marginal seas retrieved by the MeSSA algorithm

the seasonal and interannual variability and long-term change of phytoplankton blooms in the eastern Chinese seas. They found that the blooms showed significant seasonal variation, with most occurrences in spring (April–June) and summer (July–September). Moreover, they found a doubling in bloom intensity in the YS and BS during 1998–2011, which might be suggested to be caused by the increase of nutrients in these regions. In contrast, there was no significant long-term increase or decrease of bloom intensity in the Changjiang River Estuary, which was consistent with the fact that there was no significant change in Changjiang River discharge.

Shang et al. (2014) developed an approach to discriminate dinoflagellate from diatom blooms in the ECS from space. Tao et al. (2015) developed a novel method for discriminating Prorocentrum donghaiense from diatom blooms in the ECS using the MODIS data, based on the distinct spectral differences in the green and red spectral ranges of phytoplankton absorption between the two algal groups (Fig. 8).

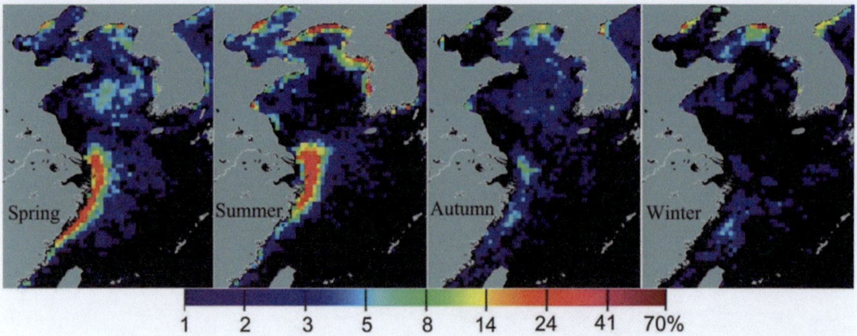

Fig. 8 Satellite-derived bloom occurring frequencies in different seasons in the eastern Chinese seas. After He et al. (2013a)

6 Summary and Future Perspectives

In this chapter, we highlight some of the progresses on the applications of satellite ocean colour remote sensing in the Chinese marginal seas in the past few years. Overall, significant progresses were made. Taking advantage of hourly observations by geostationary ocean colour satellite data from the GOCI, the high diurnal dynamics of the TSM, blooms and turbidity fronts in the eastern Chinese seas were revealed, and the retrieval methods of sea surface currents, tidal currents and residual currents were developed. Moreover, the long-term and high spatial resolution salinity field of the Changjiang River plume was successfully retrieved from the time series of satellite ocean colour data and the extension of the Pearl River plume was captured by the daily satellite ocean colour observations. In addition, the effects of typhoon on marine biological environment were widely investigated, and some new findings could not be observed by the traditional in situ measurements. More encouraging progresses are the achievements on satellite retrievals of marine carbon parameters in the Chinese marginal seas, especially the proposed novel algorithms for the estimations of DOC and POC inventories, and pCO_2.

Although satellite ocean colour remote sensing in both technique development and scientific applications have been significantly advanced in the Chinese marginal seas in the past few years, challenges and unknowns still exist and need to be solved in the near future. First, the quality of satellite ocean colour data is still poor for the turbid waters in the Chinese marginal seas, which is overlooked in the past and current research and applications, especially the long-term change study. Second, separating anthropogenic effects from nature changes is still a challenge, and high quality, long-term consistent ocean colour dataset needs to be established. Third, operational applications of satellite ocean colour remote sensing are preliminary in the Chinese marginal seas, and need to be enhanced.

Acknowledgements The authors would like to thank Dr. Zoujun Yu for editing the manuscript. This work was supported by the National Basic Research Programme ("973" Programme) of China (grant #2015CB954002), the National High Technology and Development Program of China (grant 2014AA123301), the National Natural Science Foundation of China (NSFC) (grants #41676170, #41676172, #41476155, and #41621064), and the "Global Change and Air-Sea Interaction" project of China (grant # GASI-03-03-01-01).

References

Bai Y, Pan DL, Cai WJ, He XQ, Wang DF, Tao BY, Zhu QK (2013) Remote sensing of salinity from satellite-derived CDOM in the Changjiang River dominated East China Sea. J Geophys Res [Oceans] 118(1):227–243. https://doi.org/10.1029/2012jc008467

Bai Y, He XQ, Pan DL, Chen CTA, Kang Y, Chen XY, Cai WJ (2014) Summertime Changjiang River plume variation during 1998–2010. J Geophys Res [Oceans] 119(9):6238–6257. https://doi.org/10.1002/2014jc009866

Bai Y, Cai WJ, He XQ, Zhai WD, Pan DL, Dai MH, Yu PS (2015a) A mechanistic semi-analytical method for remotely sensing sea surface pCO_2 in river-dominated coastal oceans: a case study from the East China Sea. J Geophys Res [Oceans] 120(3):2331–2349. https://doi.org/10.1002/2014jc010632

Bai Y, Huang TH, He XQ, Wang SL, Hsin YC, Wu CR, Zhai WD, Lui HK, Chen CTA (2015b) Intrusion of the Pearl River plume into the main channel of the Taiwan Strait in summer. J Sea Res 95:1–15. https://doi.org/10.1016/j.seares.2014.10.003

Chang Y, Chan JW, Huang YCA, Lin WQ, Lee MA, Lee KT, Liao CH, Wang KY, Kuo YC (2014) Typhoon-enhanced upwelling and its influence on fishing activities in the Southern East China Sea. Int J Remote Sens 35(17):6561–6572. https://doi.org/10.1080/01431161.2014.958248

Chen XY, Pan DL, He XQ, Bai Y, Wang DF (2012) Upper ocean responses to category 5 typhoon Megi in the Western North Pacific. Acta Oceanol Sin 31(1):51–58. https://doi.org/10.1007/s13131-012-0175-2

Chen XY, Pan DL, Bai Y, He XQ, Chen CTA, Kang Y, Tao BY (2015) Estimation of typhoon-enhanced primary production in the South China Sea: a comparison with the Western North Pacific. Cont Shelf Res 111:286–293. https://doi.org/10.1016/j.csr.2015.10.003

Cheng Z, Wang XH, Paull D, Gao J (2016) Application of the geostationary ocean color imager to mapping the diurnal and seasonal variability of surface suspended matter in a Macro-tidal Estuary. Remote Sensing 8(3):244

Chung CC, Gong GC, Hung CC (2012) Effect of Typhoon Morakot on microphytoplankton population dynamics in the subtropical Northwest Pacific. Mar Ecol Prog Ser 448:39–49. https://doi.org/10.3354/meps09490

He X, Bai Y, Pan D, Chen CTA, Cheng Q, Wang D, Gong F (2013a) Satellite views of the seasonal and interannual variability of phytoplankton blooms in the eastern China seas over the past 14 yr (1998–2011). Biogeosciences 10(7):4721–4739. https://doi.org/10.5194/bg-10-4721-2013

He XQ, Bai Y, Pan DL, Huang NL, Dong X, Chen JS, Chen CTA, Cui QF (2013b) Using geostationary satellite ocean color data to map the diurnal dynamics of suspended particulate matter in coastal waters. Remote Sens Environ 133:225–239. https://doi.org/10.1016/j.rse.2013.01.023

He XQ, Bai Y, Chen CTA, Hsin YC, Wu CR, Zhai WD, Liu ZL, Gong F (2014) Satellite views of the episodic terrestrial material transport to the Southern Okinawa Trough driven by typhoon. J Geophys Res [Oceans] 119(7):4490–4504. https://doi.org/10.1002/2014jc009872

He X, Xu D, Bai Y, Pan D, Chen C-TA, Chen X, Gong F (2016) Eddy-entrained Pearl River plume into the oligotrophic basin of the South China Sea. Cont Shelf Res 124:117–124. https://doi.org/10.1016/j.csr.2016.06.003

Hu S, Cao W, Wang G, Xu Z, Zhao W, Lin J, Zhou W, Yao L (2015) Empirical ocean color algorithm for estimating particulate organic carbon in the South China Sea. Chin J Oceanol Limnol 33:764–778

Hu ZF, Pan DL, He XQ, Bai Y (2016a) Diurnal Variability of Turbidity Fronts Observed by Geostationary Satellite Ocean Color Remote Sensing. Remote Sens 8(2):147. https://doi.org/10.3390/rs8020147

Hu ZF, Wang DP, Pan DL, He XQ, Miyazawa Y, Bai Y, Wang DF, Gong F (2016b) Mapping surface tidal currents and Changjiang plume in the East China sea from geostationary ocean color imager. J Geophys Res [Oceans] 121(3):1563–1572. https://doi.org/10.1002/2015jc011469

Hu ZF, Pan DL, He XQ, Song DH, Huang NL, Bai Y, Xu Y, Wang XH, Zhang L, Gong F (2017) Assessment of the MCC method to estimate sea surface currents in highly turbid coastal waters from GOCI. Int J Remote Sens 38(2):572–597

Jo YH, Dai MH, Zhai WD, Yan XH, Shang SL (2012) On the variations of sea surface pCO_2 in the Northern South China Sea: a remote sensing based neural network approach. J Geophys Res [Oceans] 117(c8):1–2

Li YH, Xu XH, Yin XJ, Fang JY, Hu WY, Chen J (2015) Remote-sensing observations of Typhoon Soulik (2013) forced upwelling and sediment transport enhancement in the Northern Taiwan Strait. Int J Remote Sens 36(8):2201–2218. https://doi.org/10.1080/01431161.2015.1035407

Liu FF, Chen CQ (2014) Seasonal variation of chlorophyll a in the South China Sea from 1997–2010. Aquat Ecosyst Health Manage 17(3):212–220. https://doi.org/10.1080/14634988.2014.942211

Liu DY, Wang YQ (2013) Trends of satellite derived chlorophyll-a (1997–2011) in the Bohai and Yellow Seas, China: effects of bathymetry on seasonal and inter-annual patterns. Prog Oceanogr 116:154–166. https://doi.org/10.1016/j.pocean.2013.07.003

Liu HX, Hu ZF, Huang LM, Huang HH, Chen ZZ, Song XY, Ke ZX, Zhou LB (2013) Biological response to typhoon in Northern South China Sea: a case study of Koppu. Cont Shelf Res 68:123–132. https://doi.org/10.1016/j.csr.2013.08.009

Liu ML, Liu XN, Ma AH, Li T, Du ZH (2014) Spatio-temporal stability and abnormality of chlorophyll-a in the Northern South China Sea during 2002–2012 from MODIS images using wavelet analysis. Cont Shelf Res 75:15–27. https://doi.org/10.1016/j.csr.2013.12.010

Liu D, Pan DL, Bai Y, He XQ, Wang DF, Wei JA, Zhang L (2015) Remote Sensing Observation of Particulate Organic Carbon in the Pearl River Estuary. Remote Sens 7(7):8683–8704. https://doi.org/10.3390/rs70708683

Lou XL, Hu CM (2014) Diurnal changes of a harmful algal bloom in the East China Sea: observations from GOCI. Remote Sens Environ 140:562–572. https://doi.org/10.1016/j.rse.2013.09.031

Palacz AP, Xue HJ, Armbrecht C, Zhang CY, Chai F (2011) Seasonal and inter-annual changes in the surface chlorophyll of the South China Sea. J Geophys Res [Oceans] 116(c9):1–2

Pan XJ, Wong GTF (2015) An improved algorithm for remotely sensing marine dissolved organic carbon: climatology in the Northern South China Sea shelf-sea and adjacent waters. Deep Sea Res Part II 117:131–142. https://doi.org/10.1016/j.dsr2.2015.02.025

Pan XJ, Wong GTF, Tai JH, Ho TY (2015) Climatology of physical hydrographic and biological characteristics of the Northern South China Sea shelf-sea (NoSoCS) and adjacent waters: observations from satellite remote sensing. Deep Sea Res Part II 117:10–22. https://doi.org/10.1016/j.dsr2.2015.02.022

Shang S, Dong Q, Lee Z, Li Y, Xie Y, Behrenfeld M (2011) MODIS observed phytoplankton dynamics in the Taiwan strait: an absorption-based analysis. Biogeosciences 8(4):841–850. https://doi.org/10.5194/bg-8-841-2011

Shang SL, Wu JY, Huang BQ, Lin G, Lee Z, Liu J, Shang SP (2014) A new approach to discriminate dinoflagellate from diatom blooms from space in the East China Sea. J Geophys Res [Oceans] 119(7):4653–4668. https://doi.org/10.1002/2014jc009876

Shi W, Wang MH (2012) Satellite views of the Bohai Sea Yellow Sea, and East China Sea. Progr Oceanogr 104:30–45. https://doi.org/10.1016/j.pocean.2012.05.001

Song X, Bai Y, Cai W-J, Chen C-TA, Pan D, He X, Zhu Q (2016) Remote Sensing of Sea Surface pCO_2 in the Bering Sea in Summer Based on a Mechanistic Semi-Analytical Algorithm (MeSAA). Remote Sens 8(7):558

Tao Z, Qin BY, Li ZW, Yang XF (2012) Satellite observations of the partial pressure of carbon dioxide in the surface water of the Huanghai Sea and the Bohai Sea. Acta Oceanol Sin 31 (3):67–73. https://doi.org/10.1007/s13131-012-0207-y

Tao BY, Mao ZH, Lei H, Pan DL, Shen YZ, Bai Y, Zhu QK, Li Z (2015) Novel method for discriminating Prorocentrum donghaiense from diatom blooms in the East China Sea using MODIS measurements. Remote Sens Environ 158:267–280. https://doi.org/10.1016/j.rse.2014.11.004

Wolanski E, Ngoc Huan N, Trong Dao L, Huu Nhan N, Ngoc Thuy N (1996) Fine-sediment dynamics in the Mekong River Estuary, Vietnam. Estuar Coast Shelf Sci 43(5):565–582. https://doi.org/10.1006/ecss.1996.0088

Yamaguchi H, Kim HC, Son YB, Kim SW, Okamura K, Kiyomoto Y, Ishizaka J (2012) Seasonal and summer interannual variations of SeaWiFS chlorophyll a in the Yellow Sea and East China Sea. Progr Oceanogr 105:22–29

Ye HJ, Sui Y, Tang DL, Afanasyev YD (2013) A subsurface chlorophyll a bloom induced by typhoon in the South China Sea. J Mar Syst 128:138–145. https://doi.org/10.1016/j.jmarsys.2013.04.010

Yu J, Tang DL, Li YZ, Huang ZR, Chen GB (2013) Increase in fish abundance during two typhoons in the South China Sea. Adv Space Res 51(9):1734–1749. https://doi.org/10.1016/j.asr.2012.11.019

Zhao H, Han GQ, Zhang SW, Wang DX (2013) Two phytoplankton blooms near Luzon Strait generated by lingering Typhoon Parma. J Geophys Res Biogeosciences 118(2):412–421. https://doi.org/10.1002/jgrg.20041

Zhu Y, Shang SL, Zhai WD, Dai MH (2009) Satellite-derived surface water pCO_2 and air-sea CO_2 fluxes in the northern South China Sea in summer. Progr Nat Sci Mater Int 19(6):775–779. https://doi.org/10.1016/j.pnsc.2008.09.004

Observing Sea Levels in the China Seas from Satellite Altimetry

Xiangbo Feng and Yongcun Cheng

Abstract Due to the threat of global warming, extensive studies of the natural and anthropogenic causes of sea level change have been performed. The use of satellite altimetry contributes enormously to such studies, especially where in situ observations are rare. This chapter highlights the authors' recent investigations of sea level measurements in the China seas made by satellite altimetry. Different sea level components are investigated. Progress is being made towards a better estimation of the ocean tides in the China seas using a comprehensive combination of satellite altimetry products. The seasonal sea level cycle, another crucial component of sea level in the China seas, is also systematically studied by using different analysis approaches. We finally explore the long-term trends and variability of mean sea level by analyzing the latest (1993–2016) satellite altimetry. The relationships between mean sea level and large-scale ocean circulation and climate variability are also examined.

Keywords Satellite altimetry · China seas · Ocean tides · Seasonal variability Inter-annual variability · Mean sea level rise · Climate impact

1 Introduction

The global mean sea level (MSL), one of the most important indicators of climate change, has been widely accepted as increasing because our planet is warming up. The global MSL is estimated to rise by 3.2 ± 0.4 mm/year from the satellite altimetry and 2.8 ± 0.8 mm/year from the tide gauges for the period 1993–2009, and 1.7 ± 0.2 mm/year for 1900–2009 (Church and White 2011). There is concern

X. Feng (✉)
Department of Meteorology, University of Reading, Reading, UK
e-mail: xiangbo.feng@reading.ac.uk

Y. Cheng
Beijing Piesat Information Technology Co. Ltd., Beijing, China

Y. Cheng
Center for Coastal Physical Oceanography, Old Dominion University, Norfolk, USA

V. Barale and M. Gade (eds.), *Remote Sensing of the Asian Seas*,
https://doi.org/10.1007/978-3-319-94067-0_18

that the rate of MSL rise is increasing, but this remains under debate because of the appearance of multi-decadal variability in the existing sea level records (Haigh et al. 2014). Importantly, there are remarkable regional variations of MSL rise, which, using the satellite altimetry data as an example, varies between −10 mm/year to 10 mm/year in last 20 years. As mentioned by the Intergovernmental Panel on Climate Change Fifth Assessment Report (IPCC AR5), the regional MSL changes that differ substantially from the global MSL changes remain largely unexplored, suggesting the necessity of understanding sea level changes in each individual region. The regional distribution of MSL changes is more associated with natural variability and local anthropogenic forcing, and is less related to the factors causing the global MSL rise. Regional sea level is in principle associated with the dynamical redistribution of the ocean heat and water mass through large-scale ocean circulations, which, however, are changing under global warming. Thus, identifying the roles of natural variability in a context of global warming is key when studying regional sea level change.

The China seas connect the Asian continent with the Pacific Ocean, and are taken to be the areas where both large-scale oceanic and atmospheric modes have strong impacts on sea level. Another important feature of the China seas is the shallow continental shelf, which tends to eliminate the impact of large-scale variability from the ocean interior, such as the Pacific Decadal Oscillation (PDO) and Kuroshio Current. In addition, the coastal geomorphology of the China seas has been greatly changed by intensive anthropogenic activities, and this can potentially alter the occurrence of extreme sea level events through modulating the tide regimes (Feng et al. 2015a). All of these aspects, along with MSL rise and tropical cyclone changes, add to the uncertainties in understanding the threat from sea level elevation in these regions (Feng and Tsimplis 2014). The Chinese coasts are heavily populated and have been classified as some of the most vulnerable areas under climate change because of coastal flooding (Hallegatte et al. 2013).

The 1970s saw the start of 'era of satellite altimetry' (Kaula 1970). Since then sea surface height (SSH) measurements from space have made great contributions to the precise monitoring of global and regional patterns of sea level change. A satellite orbiting the Earth and carrying a radar altimeter emits a narrow beam of radar pulses onto an area of the ocean surface. The high-frequency pulses reach the ocean surface and are then reflected back to the satellite. The distance from the satellite to the surface is known as 'altimeter range'. After subtracting the height of the satellite above the reference ellipsoid, the ellipsoidal SSH is obtained. The height of the satellite is determined by using satellite tracking systems such as Global Positioning System (GPS). After removing ocean tides and the inverse barometer (IB) effect, the MSL anomaly is finally derived. With environmental corrections applied, altimetry has accuracy of order cm.

The purpose of this chapter is to highlight our recent investigations of sea level observations in the China seas using the satellite altimetry. The marginal seas of China are specified as the Bohai, Yellow, East China and South China Seas. Different contributions to the sea level for each region are studied, including the tides, the seasonal sea level cycle and MSL.

This chapter is structured as follows. In Sect. 2, we will briefly summarize the conventional observations (and theories) that have been used to study sea levels in the China seas, and highlight the necessity of implementing satellite altimetry in this region. In Sect. 3, the observations of the ocean tides from satellite altimetry are discussed. In Sect. 4, the spatial and temporal variability of the seasonal sea level cycle is further described by employing different analysis methods. In this section, discrepancies of satellite altimetry are also featured. In Sect. 5, using the latest satellite altimetry data (1993–2016), the MSL is investigated from the perspectives of linear trends and inter-annual variability. Relationships with the internal climate variability are also discussed. Finally, the summary and discussions are given in Sect. 6.

2 Why Is Satellite Altimetry Important for Tackling Sea Level Changes in China?

Tides are the most dominant signal in the sea level variations, especially in the marginal seas. They are astronomically forced by gravitational attraction, and have been theoretically resolved by the Equilibrium tide theory since the 19th century (Darwin and Adams 1883). However, the dynamic response of the ocean to the tidal forces causes the Equilibrium tides to depart considerably from the observations. In the marginal seas adjacent to China, shallow-water effects can alter the behavior of tides from that expected by the theory. Good examples are tidal wave resonances, which happen in the Taiwan Strait for semi-diurnal tides, and in the Gulf of Tokin for the diurnal tides. These create the largest amplitudes of M_2 (~ 1.8 m) and O_1 (~ 1 m) respectively, and neither of them can be well predicted by the Equilibrium tide theory. Additionally, other aspects, such as atmospheric forcing, ocean circulations and human activity, also contribute to altering the tidal signals in the real world. Prior to the altimetry era, the harmonic features of tides were estimated empirically based on in situ measurements (e.g. Fang 1986). However, the conventional observations in the China seas are still too sparse to resolve these features, such as the amphidromes of some semi-diurnal tides.

The seasonal cycle, and specifically its annual and semi-annual components, is the most energetic signal amongst the non-tidal variability of sea levels. The gravitational forcing contributes very little to the observed seasonal sea level cycle. Seasonality in meteorological, oceanographic, and hydrological processes mainly constrains the seasonal sea level cycle, but the contribution of each factor varies spatially and temporally (Feng et al. 2015b). Sea level seasonality is usually excluded in the analysis of MSL trends or variability. However, this does not necessarily imply that the seasonal sea level cycle is unimportant. For example, in the China seas, the seasonal sea level cycle can apparently contribute to the occurrence of sea level extremes (Feng and Tsimplis 2014), the seasonal variations of tides, and the seawater-freshwater balance around the river estuaries. Recently, due to concern about climate change, it has become essential to assess the instability of seasonal sea level cycle. Investigating the spatio-temporal variability of the seasonal sea level cycle in the China seas will fit into the suite of sea level studies in the oceans worldwide.

The 1980s saw the start of studying MSL change at the Chinese coasts based on the limited tide gauge records (Emery and You 1981; Emery and Aubrey 1986). These studies warned of the heavy influence of land vertical movement on the estimation of eustatic MSL rise when using the conventional observations. Ren (1994) analyzed 32 long-term tide gauge records, and concluded that the rate of eustatic MSL rise in China in the last 80 years was small (1–2 mm/year). Recently, even longer tide gauge records have been analyzed (e.g. Feng and Tsimplis 2014; Marcos et al. 2012), but it still remains difficult to know the absolute MSL changes due to the lack of geophysical and datum corrections in the early years. Again, tide gauges are too sparse to know the complete patterns and causes for MSL change.

3 Ocean Tides

Seven months of TOPEX/Poseidon (T/P) altimeter data were used by (Mazzega and Berge 1994) to first derive the ocean tides in the China seas. With more T/P data available, the tidal regimes can be better captured with more tidal constituents resolved (e.g. Fang et al. 2004). Now, 24+ years of satellite altimeter observations are available in the primary ground track through the T/P and Jason-1/2/3 missions, which make it possible to resolve most of the tidal constituents in the harmonic analysis (Cheng and Andersen 2011).

Progress has been made towards deriving more tidal constituents in (Cheng et al. 2016a). Four years of the T/P-Jason-1 tandem mission data were combined with 18 years (1993–2010) of primary joint T/P and Jason 1/2 mission time series. In the north China seas, compared with 116 tide-gauge observations, the amplitudes of the four main tides (M_2, S_2, K_1 and O_1) are improved by 31, 38, 7, and 7% respectively when replacing the T/P-Jason primary mission data with the new combined mission data. Figure 1 shows the estimated cotidal charts of the four main tides from the combined mission data. The patterns of the M_2 and S_2 tides are similar in the area because their frequencies are comparable. For the M_2 tide, there are four amphidromes, which are located in the Liaodong Bay, the Yellow River mouth, the North Yellow Sea, and the South Yellow Sea, respectively. Unlike the amphidromes in the Yellow River mouth and the North Yellow Sea, the amphidromes in the Liaodong Bay and the South Yellow Sea are more robust. The location of the amphidromic point for M_2 near the Yellow River mouth is still arguable, which might be related to the change of the Yellow River delta coastline. The amphidromes of the diurnal tides are better represented, with one located in the Bohai Sea and the other in the south Yellow Sea.

However, discrepancies still remain between the above estimations and in situ measurements. Figure 2 shows the differences between tidal amplitudes estimated from tide gauges and altimetry data. M_2 is underestimated by altimetry around the

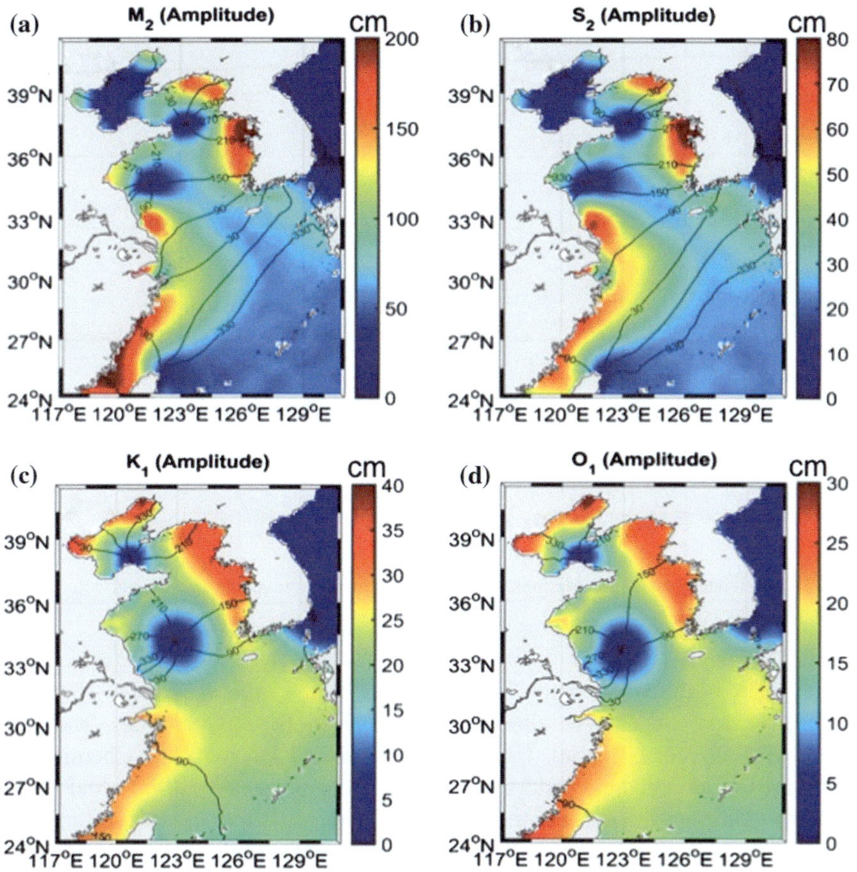

Fig. 1 Cotidal charts of M_2 (**a**), S_2 (**b**), K_1 (**c**) and O_1 (**d**) tides estimated from combined T/P-Jason-1/2 primary and tandem mission data in the Bohai, Yellow and East China Seas. The colors denote the amplitude (cm) and the solid curves denote the Greenwich phase lags

Korean peninsula and Yangtze River (Fig. 2a), presumably due to the biases of altimetry on shallow waters. In the Bohai Sea, M_2 is overestimated by altimetry, as a result of too few altimetry measurements being used because of the data flagging which is applied where the water depth is too shallow (<20 m). The Yellow Sea and Gulf of Thailand are other areas with significant discrepancies, for which river discharge is thought to be responsible. Diurnal tides, for example the K_1 tide (Fig. 2b), are normally underestimated by altimetry, with larger biases where the tidal amplitude is also greater. Both M_2 and K_1 are biased at three offshore stations in the East China Sea, which might be related to the poor representability of altimetry at local scales.

It is worth noting that in the above estimations the 18.6-year nodal cycle of lunar tides are assumed to follow the equilibrium tidal theory, which is about 3.7% of amplitude for semidiurnal tides and over 10% for diurnal tides. However, in the

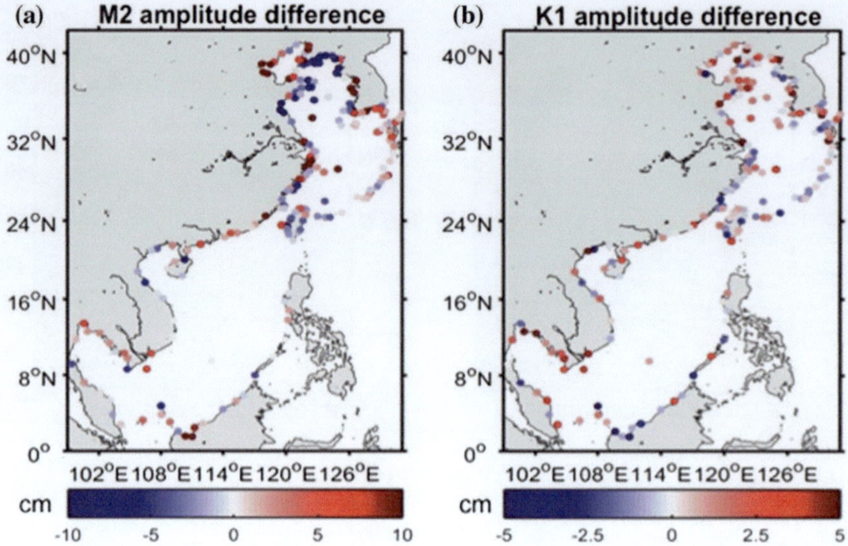

Fig. 2 Differences (cm) of tidal amplitude estimated from tide gauge and satellite altimetry for M_2 (**a**) and K_1 (**b**)

China seas, as the water depth is shallow (<100 m in the north), the effects of bottom friction are able to alter the nodal behaviors of lunar tides from that expected (Feng et al. 2015a). Using 16-year T/P and Jason-1 data, Cherniawsky et al. (2010) found that the nodal cycle of main lunar tides exceeds the nodal values by the equilibrium theory over this region. This suggests that the tidal estimations by altimetry, as mentioned above, contain these biases and need to be re-evaluated in the future.

4 Seasonal Sea Level Variability

4.1 Spatial Variability

Because the seasonal sea level variability is a low-frequency signal, its characteristic spatial patterns occur at large scales. Gridded satellite radar altimeter data that cover the northwest Pacific (0–65°N, 100–170°E) for 1993–2013 were used to investigate the seasonal sea level variability in the China seas. The data were produced by SSALTO/DUACS and distributed by the Archiving, Validation and Interpretation of Satellite Oceanographic data program (AVISO). The data consist of monthly averaged maps of sea level anomalies, corresponding to multi-mission gridded sea surface height anomaly (including Saral, Cryosat-2, Jason-1&2, T/P, Envisat, GFO, ERS-1&2 and Geosat) with respect to a 21-year mean sea level.

The spatial resolution of the gridded altimeter data is 1/4° × 1/4°. Oceanic and atmospheric dynamics are routinely corrected in the mission track data, including the ocean tide, the pole tide and the dynamic atmospheric correction (DAC) (Carrère and Lyard 2003). Monthly sea level data from tide gauges, which are provided by the Permanent Service for Mean Sea Level (Holgate et al. 2013), were also used. The harmonic parameters of the annual and semi-annual cycles were estimated through least squares fitting to the whole monthly records (Feng et al. 2015b).

Figure 3a shows the variance of monthly sea level records. The monthly mean sea level is varying dramatically at regional scales, with the largest variance (300 cm^2) seen in the western boundary currents (i.e. the Kuroshio and Oyashio Currents) and in the Bohai Sea. The seasonal cycle, including annual and semi-annual components, accounts for 60–90% of the sea level variance in the vast majority of areas of the China marginal seas (Fig. 3b). The seasonal cycle becomes less dominant in the ocean interior (explaining 40–50% of variance), with the lowest impacts along the Kuroshio and Oyashio currents whereas the variance is at the largest, indicating the weak seasonality in the large-scale ocean currents. The semi-annual cycle of sea level is only statistically detectable in the shallow-water regions, with amplitudes usually <5 cm (not shown).

The annual cycle has the prodigious influence in the Bohai and East China Seas with amplitudes >20 cm (Fig. 3c). The sea level peaks in December–January in the equatorial area, and peaks in August-November towards the north (Fig. 3d), implying the presence of a baroclinic component that is maintained by the large-scale currents. The phase of the annual cycle is uniform in the northern part of the China Seas, indicating the same forcing is experienced by that region. We confirmed that the surface air pressure and steric height are the two dominant contributors over this region, with each of them having an amplitude of 10–12 cm.

The seasonal cycle in the South China Sea is complicated, with varying spatial variability in the amplitude and phase (Fig. 3c, d), showing that different mechanisms influence the sea level. In the east part of the South China Sea, the sea level is highest when the open ocean is warmest. This is because the steric height, mainly due to the thermal expansion of the upper ocean, is penetrating the South China Sea through the Luzon Straight following the ocean currents. In the west part of the South China Sea where the water depth is shallow, the barotropic response to the local wind becomes more prevailing. For example, the Asian monsoon is changing seasonally in this region, with strong northeast/southwest winds in winter/summer. This forces cyclonic/anticyclonic mesoscale eddies accordingly in winter/summer. The sea level is then dynamically responding to the geostrophic currents, with high/low levels in winter/summer.

Figure 4 shows the comparison of the amplitude of annual sea level cycle between AVISO and tide gauges. At the points closest to the tide gauges, AVISO significantly underestimates the annual cycle at 59 of the 120 stations by 2–9 cm, and overestimates it at two stations. Large underestimates of 5–8 cm are found in the west of the South China Sea, the East China Sea and the Sea of Japan. AVISO is

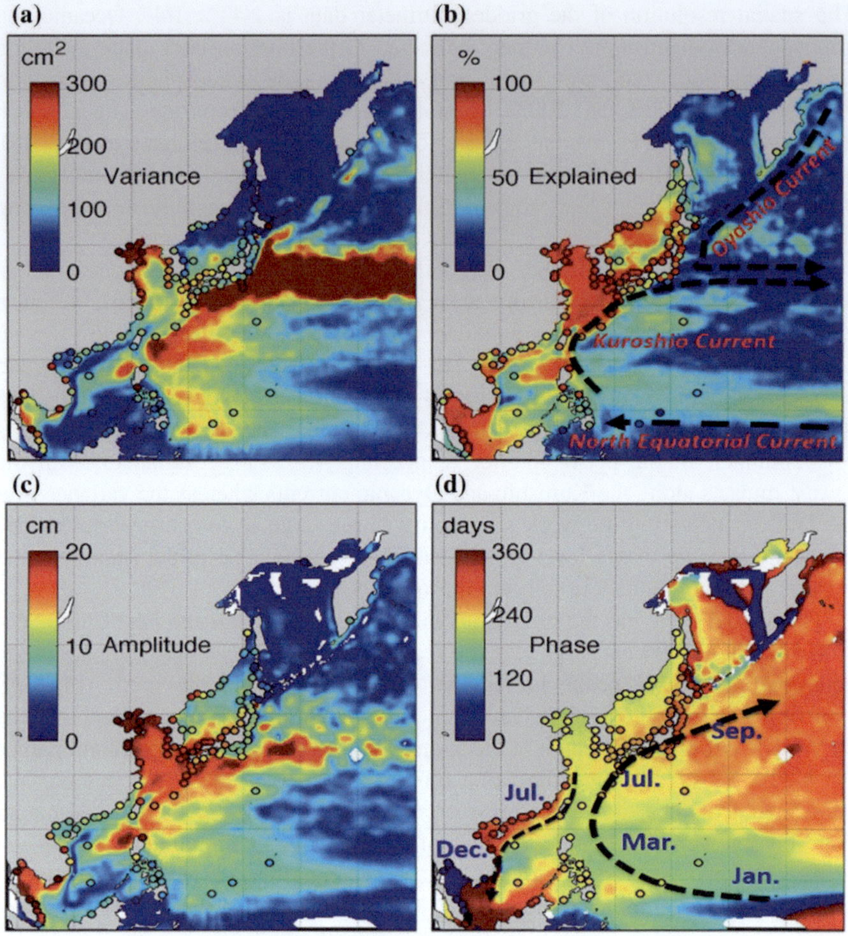

Fig. 3 **a** Variance of monthly sea level anomaly observed over 1993–2013 from tide gauges and AVISO, **b** percentage of the variance explained by the harmonic seasonal cycle (annual and semi-annual cycles), and **c** amplitude and **d** phase of annual sea level cycle, with blank areas showing the estimations not passing the significance test at 95% confidence. In **c** the black dashed lines indicate schematic routes of the Oyashio, the Kuroshio and the North Equatorial Currents, based on the geostrophic velocity data distributed by AVISO, while in **d** the black dashed lines show the changing patterns of annual cycle phase. Please note the isostatic IB effect (DAC) is added back to the AVISO records

also seen to underestimate the semi-annual cycle amplitude (Feng et al. 2015b). The phase of the seasonal cycle is also different between the AVISO and tide gauge estimations, but they are statistically insignificant. These discrepancies are not related to the period applied in the analysis, as they persist when different periods are considered. Thus, we confirm that these features are real biases in satellite altimetry, implying further improvements can be sought in the future.

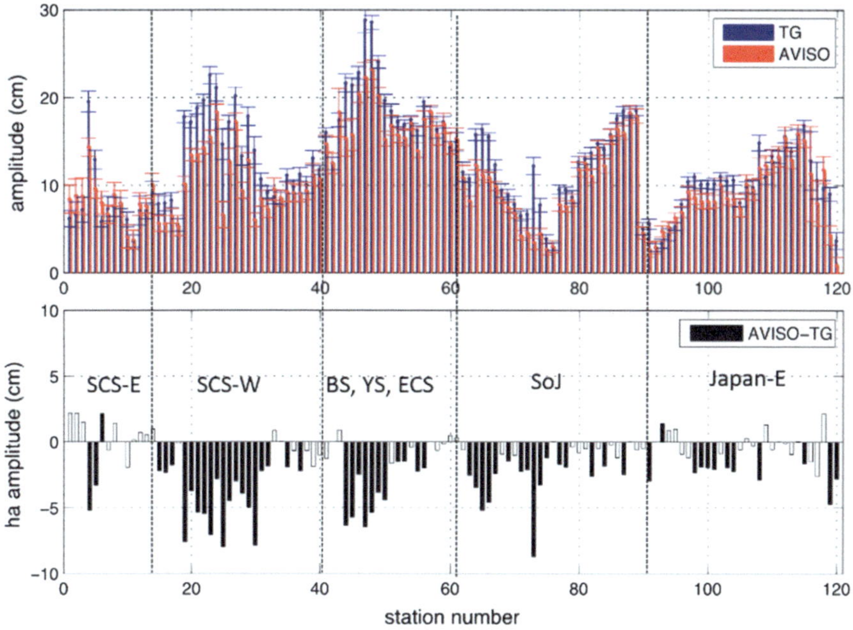

Fig. 4 (top) Annual cycle amplitudes determined by tide gauges and AVISO at the closest points to tide gauges in the marginal seas of Asia, and (bottom) differences of annual cycle amplitudes (AVISO—tide gauges). The tide gauges are divided into different sub-regions as shown in Fig. 3: the east of South China Sea (SCS-E), the west of South China Sea (SCS-W), the East China, Yellow and Bohai Seas (ECS, YS, BS), the Sea of Japan (SoJ), and the east coasts of Japan (Japan-E). The black bars in the bottom panel indicate the differences that pass the significance test, while the grey bars indicate insignificant differences

4.2 *Temporal Variability*

The instability of seasonal sea level cycle with time has also been studied using the Cyclostationary Empirical Orthogonal Function (CSEOF) method. In short, the CSEOF method is based on the theory of harmonizable cyclostationary functions, which can decompose 3D data into a series of spatial modes (Loading Vectors) and a temporal Principle Component Time Series (PCTS). This method is similar to the conventional Empirical Orthogonal Function (EOF) method, but it additionally requires the periods of the signals of interest (e.g. 12-months for the annual cycle) to be imposed a priori. Using this method, the climatology of the spatial patterns of sea level is obtained, along with a time series, which describes the temporal variability of the climatology (e.g. annual cycle). Details of this technique can be found in (Cheng et al. 2016b). The isostatic IB effect (i.e. DAC) is excluded here as its annual cycle has no clear temporal variability.

The PCTS of the first CSEOF mode, which accounts for nearly 64% of the monthly sea level variance in the north China seas (the Bohai, Yellow and East

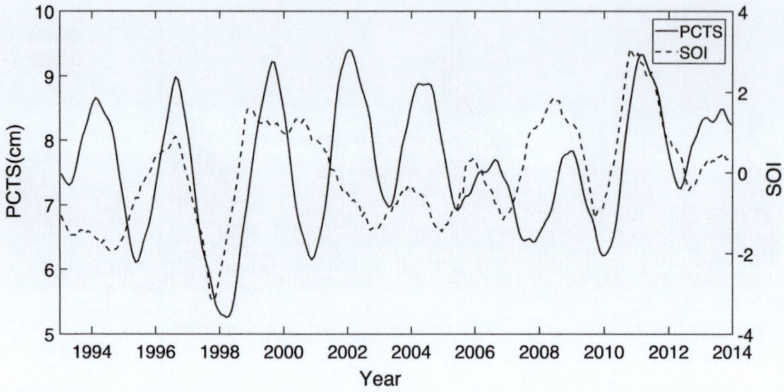

Fig. 5 PCTS (cm) of the first CSEOF mode for the north China seas derived from AVISO for 1993–2013 and 13-month moving average of SOI index

China Seas), display a clear inter-annual variability (Fig. 5). The annual cycle is changing within a range of 5–9 cm. The temporal variability of the annual sea level cycle over this region is closely correlated with the El Niño–Southern Oscillation (ENSO) (corresponding to the opposite Southern Oscillation Index, SOI), but delayed by a few months. This confirms the contribution from steric height, which is moving northward along the western boundary currents (as discussed above). The disagreements between the PCTS and ENSO (e.g. 2000–2002) indicate that other factors, such as the local wind forcing, are also considerably important for the annual sea level cycle in the north China seas.

The temporal variability of the annual sea level cycle in the South China Sea is shown in Fig. 6. The PCTSs of the two largest modes account for 66 and 8% of the monthly sea level variations. The PCTS of the first mode, mostly representing the variability in the north and west of the South China Sea (shallow-water areas) (Cheng et al. 2016b), temporally changes within ±2 cm in strong El-Niño years.

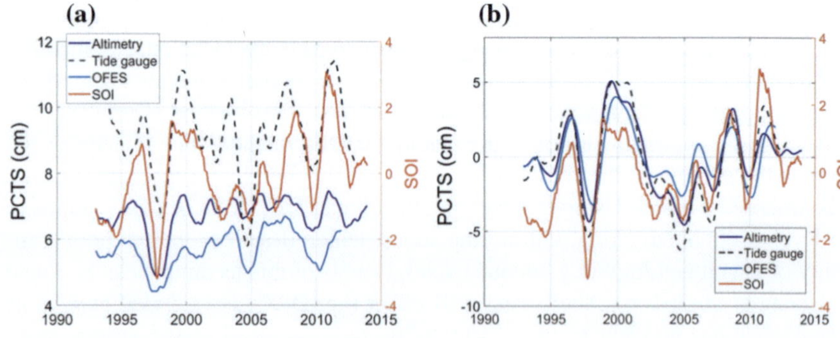

Fig. 6 The PCTS (cm) of the first **a** and second **b** CSEOF modes for the South China Sea derived from AVISO, tide gauges and the OFES model for 1993–2013 and a 13-month moving average of the SOI index

These changes are significantly correlated with the variability of the SOI ($r = 0.4$, above 95% confidence level). This means that the barotropic forcing of sea level variability in these shallow water regions is ENSO-related. The PCTS of the second mode, mostly representing sea level variability in the deep central and eastern regions, changes with larger magnitudes (±10 cm), and it is closely associated with ENSO ($r = 0.6$, above 95% confidence level), confirming the strong impact of steric height in deep waters.

The PCTSs of the two largest modes derived from tide gauges are more variable than those from AVISO (Fig. 6), confirming the discrepancies between altimetry and tide gauge measurements on observing the seasonal sea level cycle that have been indicated in Fig. 4. We have also addressed the question of whether ocean general circulation models (OGCMs) have the capability to reproduce the temporal changes in seasonal sea level cycle over this region. As a demonstration, the OGCMs for the Earth Simulator (OFES), which is forced by NCEP/NCAR atmosphere reanalysis, is included in Fig. 6. Generally speaking, OFES can reasonably capture the observed sea level variability. However, discrepancies still remain in the variability and mean, such as in the 2000s for the first mode. These discrepancies might be due to coastal dynamics that are poorly resolved by the ocean model over this region, such as coastal ocean upwelling which is highly dependent on the quality of atmospheric forcing.

5 Mean Sea Level

5.1 Trends

MSL trends and variations in the China seas are estimated by analyzing the 24-year (1993–2016) gridded altimetry product AVISO. The linear trend is estimated by least squares fitting to the time series after removing the climatological cycle. The IB effect and ocean tides are also subtracted. We note that the altimetry data contains the variations of the geoid. Strictly speaking, the temporal variations of the geoid must be corrected before estimating the long-term MSL trend. However, it still remains difficult to remove this signal over the entire altimetry era due to the lack of persistent observations, such as the Gravity Recovery and Climate Experiment (GRACE) mission data. As such, the glacial isostatic adjustment (GIA) is only corrected in the global MSL record produced by AVISO, which was independently estimated as about −0.3 mm/year for the global ocean surface (Peltier 2001). GIA correction is not applied to the regional data of AVISO as the spatial variations of GIA have not yet been resolved. In this section, to ensure consistency with the regional data, the GIA correction is added back to the global MSL record.

Figure 7a shows the linear trend of MSL in the China seas, for the period 1993–2016. MSL is significantly rising nearly everywhere over this region,

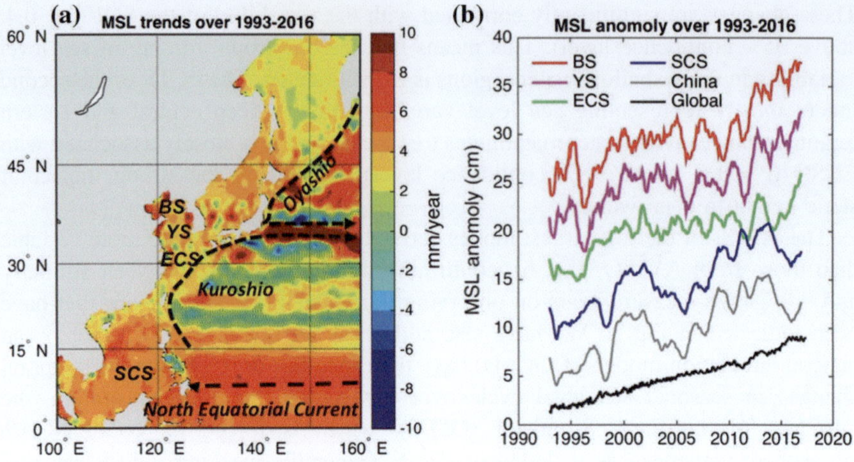

Fig. 7 **a** MSL trends from AVISO over 1993–2016, with black dashed lines showing schematic routes of the key currents and with dashed areas showing the trends are not statistically significant at 95% confidence level. **b** MSL averaged for the Bohai Sea (BS), the Yellow Sea (YS), the East China Sea (ECS), the South China Sea (SCS) and the China seas (China) (all of above), which are geographically indicated in (**a**). Please note that the 12-month moving average is applied in (**b**), and that each series are offset for 5 cm for display. The global MSL from AVISO is also presented in (**b**), with GIA added back

with spatial variations ranging from 1 to 10 mm/year. Larger trends (4–8 mm/year) appear in the centre of the South China Sea, following the footprint of the upper ocean thermal structure via the Equatorial Pacific Current driven by the zonal wind stress (Palanisamy et al. 2015). In the eddy-rich regions, such as east of the Kuroshio and Oyashio Currents, sea level exhibits high variability. Large sea level rises (8–10 mm/year) are also spotted in the north China seas, such as the northwest of the Bohai Sea and the west of the Yellow Sea, probably due to the small-scale impacts of, for example, river discharge. In the west of the East China and the South China Seas, where the Asian monsoons are influential, MSL is rising by 0–3 mm/year.

Averaging the whole area of the China seas, sea level is rising by 3.6 ± 0.3 mm/year over 1993–2016 (Fig. 7b). Specifically, the rising rates are 4.2 ± 0.3 mm/year, 3.5 ± 0.3 mm/year, 2.6 ± 0.2 mm/year, and 3.8 ± 0.3 mm/year for the Bohai, Yellow, the East China, and the South China Seas, respectively. The linear trend of the global MSL record (with GIA added back) is 3.0 ± 0.3 mm/year over 1993–2016, which is smaller than the rates in most of the China seas. The differences relative to the global MSL reflect the regional variations, which are more likely to be related to the dynamical redistribution of heat or water at regional scales.

After 2010, a striking increase of MSL occurs in the north China seas (with rates of 11.0 ± 1.2 mm/year, 8.8 ± 1.5 mm/year and 4.1 ± 1.5 mm/year in the Bohai, the Yellow and the East China Seas, respectively), while the sea level is decreasing in the South China Sea (-1.8 ± 1.4 mm/year) (Fig. 7b). This could be related to phase transitions of large-scale climate variability, such as the ENSO, the PDO or the North Pacific Gyre Oscillation (NPGO) (Di Lorenzo et al. 2008).

Large-scale and low-frequency climate variability is able to induce sea level variability in the China seas (e.g. Cheng et al. 2016c; Zhang and Church 2012). Low-frequency variability in short-term records encompasses the possible linear trend, making it erroneous to interpret the above aliased trends from 24-year altimetry data (Fig. 7a) as the 'aperiodic' linear trend. This is particularly true for the PDO, which is apparently changing on multi-decadal timescales. Studies have shown that a large component of the MSL trends observed in the western tropical Pacific are associated with the intensification of the easterly trade winds in the tropical Pacific since the 1990s (Merrifield 2011). When the effects of this internal variability are removed, the MSL trends over the satellite era could be significantly reduced (Hamlington et al. 2014, 2017). We suspect that the MSL rise rates seen in the China seas (Fig. 7a) are closely related to the dynamical response of sea level to the wind forcing and to the redistribution of water. In the future, subtracting the impacts of internal climate variability from the sea level determined by satellite altimetry is required to understand better the secular trends of sea level in the China seas.

5.2 *Inter-annual Variability*

In the China seas, the inter-annual variability of MSL shows clear regional patterns (Fig. 7b), indicating different mechanisms are driving the inter-annual variability. For example, in 1995–1997 (before El Niño), a rapid sea level fluctuation occurred in the Bohai and the East China Seas, while the MSL tends to be flat in the south. To further reveal the regional patterns, we calculate the correlations of sea level variability with the values at three selected locations: in the East China Sea, in the South China Sea, and in the path of the Kuroshio Current (Fig. 8).

On inter-annual timescales, the sea level is fluctuating homogeneously in the north China seas, and this pattern extends to the west and north of the South China Sea (Fig. 8a). The regional wind forcing (e.g. due to monsoons) plays an important role for changing sea level on the coastal regions of the China seas. In the middle and south of the South China Sea where water is deep, the wind stress curl, which is associated with the ENSO (Cheng et al. 2016c), can drive sea level changes. It also turns out that the sea levels in the above two domains are linked, possibly through the relationships between the Asian monsoons with the ENSO.

Sea level is varying at even larger scales in the South China Sea, and it is well correlated with sea level in the Pacific Warm Pool (Fig. 8b). This means that, compared with the barotropic forcing, the thermal expansion of sea water is more important for sea level variations in the South China Sea. Along the Kuroshio Current, the sea level variability shows a clear footprint of the ocean heat transport (Fig. 8c), indicating significant contributions of the thermal expansion to sea level

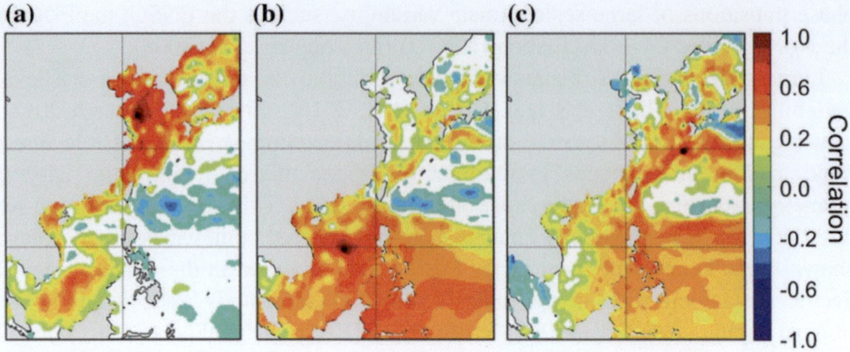

Fig. 8 Correlation coefficients of the detrended MSL anomalies with those at [35°N, 123°E] in the Yellow Sea (**a**), at [15°N, 115°E] in the South China Sea (**b**), and at [30°N and 130°E] in the routine of the Kuroshio Current (**c**), from AVISO over 1993–2016. The selected locations are indicated by black dots. Blank areas show the correlations that are not significant at 95% confidence level

over the west Pacific. However, the ocean heat carried by Kuroshio Current has reduced impacts on sea level in the marginal seas. This, again, reflects the dominance of atmospheric factors in the shallow areas of the China seas.

5.3 Association with Climate Variability

Large-scale climate variability is able to significantly change the MSL on inter-annual timescales. Correlations between the detrended MSL anomaly and the climate indices are shown in Fig. 9. The ENSO is negatively correlated with sea level in the low latitudes, especially in the ocean interior ($r < -0.5$) (Fig. 9a). Lower sea level anomalies are associated with El Niño events, and higher level anomalies with La Niña events. This indicates the important contribution of ocean thermal expansion (steric height) to changing the sea level over these areas. Positive correlations appear in the north China seas and in the middle of the South China Sea, presumably due to the linkage of regional atmospheric forcing with ENSO.

The relationships between MSL and two climate modes in the North Pacific, i.e. the PDO and the NPGO, are examined. The PDO, describing the first principal component of the North Pacific SST anomalies, is an ENSO-like climate variability but happens on longer (multi-decadal) time scales. The NPGO is the second principal component of the North Pacific SSH (sea surface high) anomalies, and it closely tracks the second principal component of the North Pacific SST anomalies (Di Lorenzo et al. 2008). Like the ENSO, the PDO has negative correlations with sea level in the low latitudes, but its impact extends further north to the Sea of Japan, following the path of the Kuroshio Current (Fig. 9b). This implies that the PDO is modulating sea level variability through the large-scale wind-driven ocean

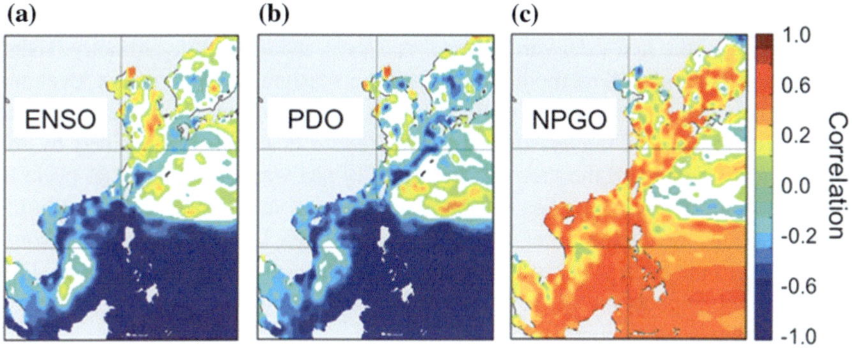

Fig. 9 Correlation coefficients of detrended MSL anomaly from AVISO with climate variability of ENSO (**a**), PDO (**b**) and NPGO (**c**). Blank areas show the correlations that are not significant at 95% level

circulation. Interestingly, the NPGO is a significant predictor of sea level variability ($r > 0.3$) in all of the marginal seas of East Asia (Fig. 9c). The impact of the NPGO is more broadly spread than the correlations with the PDO. Merrifield (2011) revealed that these two climate modes correspond to different large-scale wind patterns in the North Pacific. In comparison, the NPGO can better capture the fluctuations of the easterly trade winds, the mid-latitude westerlies and the northerly winds in the North Pacific basin. Thus, we interpret the NPGO as a better predictor of sea level variations in the China seas possibly because it can explain the oceanic processes that are driven by both local and large-scale winds.

6 Conclusions and Perspectives

Since the 1990s, satellite altimetry has made great contributions to the understanding of ocean surface processes worldwide. This chapter provides some insights into the study of sea levels in the marginal seas of China using satellite altimetry. This chapter is partly based on our recent research on the China seas, but there are also some new aspects such as MSL investigations. We hope our effort can highlight the importance of satellite altimetry for studying sea level in the China seas, and will encourage more work in the near future.

Ocean tides were among the first ocean properties observed by satellite altimetry in the 1990s. Because the Equilibrium tide theory is unable to accurately predict the real tides and the conventional observations are sparse, satellite altimetry with good spatial coverage is critically useful for capturing the tidal features in the China seas. Combined with more along-track sea level data, we have made progress towards resolving more tidal constituents than ever before, and more accurately estimating the main tides.

We systematically investigated the seasonal sea level cycle, the most energetic signal amongst the non-tidal variability of sea level, using satellite altimetry. Using the harmonic regression method, we mapped the seasonal features of sea level and identified the largest seasonal sea level cycle ($\sim$30 cm) in the Bohai Sea. The spatial distribution of the seasonal sea level cycle is mainly determined by the surface air pressure and the thermal expansion of sea water. The seasonal cycle of sea level in the South China Sea exhibits geographical variations of both amplitude and phase lag due to the different mechanisms of sea level changes on seasonal timescales. In the deep water of the South China Sea the oceanic processes produce the dominant impacts on sea level, while in the shallow regions the atmospheric forcing becomes more important.

The instability of the seasonal sea level cycle was also investigated by applying the CSEOF method to the altimetry data. In the north China seas, the seasonal sea level cycle has only one main changing pattern, and its inter-annual variability is correlated with the ENSO. In the South China Sea, there are two independent patterns for sea level seasonality, one located in the shallow water and the other in the deep water. The ENSO has more influences on the temporal variability of latter mode.

Recent MSL change in the China seas was investigated using the latest 24-year altimetry data. MSL is significantly rising nearly everywhere in the China seas in the satellite era, with rates varying from 1 to 10 mm/year. Without the GIA correction, MSL in the China seas is rising by 3.6 $\pm$ 0.3 mm/year over 1993–2016. Regionally, the rising rates are 4.2 $\pm$ 0.3 mm/year, 3.5 $\pm$ 0.3 mm/year, 2.6 $\pm$ 0.2 mm/year, and 3.8 $\pm$ 0.3 mm/year for the Bohai Sea, Yellow Sea, East China Sea, and South China Sea, respectively.

MSL is also greatly varying on inter-annual to decadal timescales. Spatial patterns of MSL variations have been distinguished at regional scales. The MSL fluctuates uniformly in the north China seas and some parts of the South China Sea, indicating the influence of atmospheric forcing. MSL is changing with a large-scale pattern in the South China Sea and western Pacific, showing the strong impact from the ocean interior. Interestingly, the MSL variations have a footprint along the Kuroshio Current. This reflects the impact of steric height regulated by the large-scale ocean heat transport. However, this impact does not impinge upon the shallow China seas, presumably due to the continental shelf barrier. Sea levels across the Kuroshio Current are not apparently associated either. Finally, the roles of climate variability on changing MSL in the China seas were examined. Amongst the climate indices, the NPGO is the best predictor of inter-annual variability of MSL over the whole areas of the China seas.

Satellite altimetry still produces biased estimates of the tides and the seasonal sea level cycle on the Chinese coasts when compared with in situ observations. Similar discrepancies between satellite altimetry and in situ observations were also found in other coastal regions of the world (e.g. Vinogradov and Ponte 2010). These discrepancies can be interpreted as: (i) a combination of data flagging and data filtering in the last 20–30 km from the coasts; (ii) the mapping procedure used in gridding altimeter data tending to smooth the characteristics of the local phenomena of sea level; (iii) radiometer wet tropospheric correction, orbit errors, altimeter parameter instability, etc. For each region, it is vital to identify the uncertainties of satellite altimetry in observing sea levels before using the data in other fields.

Considerable research efforts have been performed to extend the capability of altimetry monitoring coastal regions, for example, by adopting specialized corrections and dedicated processing strategies to compute regional multi-mission altimetry products (Cipollini et al. 2016; Vignudelli et al. 2011). The latest coastal altimetry products reprocessed with improved techniques allow a better representation of sea level variability near the coasts (Passaro et al. 2015). In the future, it will be worth implementing these newly produced data in the China seas.

Acknowledgements We acknowledge the RADS (Radar Altimeter Database System), which enables users to select optimal corrections to altimetry data to decrease the effects of corrections on the retrieval of ocean tides. We also thank AVISO for making the satellite altimeter sea level records and DAC data retrievable (www.aviso.altimetry.fr), and the Permanent Service for Mean Sea Level for making the tide gauge data retrievable (www.psmsl.org). Climate indices are obtained via NOAA Earth System Research Laboratory and www.o3d.org/npgo/. Finally, we thank C. M. Thomas at the University of Reading for improving the paper writing, as well as two reviewers for their comments and suggestions.

References

Carrère L, Lyard F (2003) Modeling the barotropic response of the global ocean to atmospheric wind and pressure forcing-comparisons with observations. Geophys Res Lett 30(6):1275

Cheng Y, Andersen OB (2011) Multimission empirical ocean tide modeling for shallow waters and polar seas. J Geophys Res 116(C11):1–11

Cheng Y, Xu Q, Zhang Y (2016a) Tidal estimation from TOPEX/Poseidon, Jason Primary, and interleaved missions in the Bohai, Yellow, and East China Seas. J Coast Res 320:966–973

Cheng Y, Hamlington B, Plag H-P, Xu Q (2016b) Influence of ENSO on the variation of annual sea level cycle in the South China Sea. Ocean Eng 126:343–352

Cheng X, Xie S-P, Du Y, Wang J, Chen X, Wang J (2016c) Interannual-to-decadal variability and trends of sea level in the South China Sea. Clim Dyn 46(9–10):3113–3126

Cherniawsky JY, Foreman MGG, Kuh Kang S, Scharroo R, Eert AJ (2010) 18.6-year lunar nodal tides from altimeter data. Cont Shelf Res 30(6):575–587

Church JA, White NJ (2011) Sea-level rise from the late 19th to the early 21st century. Surv Geophys 32(4–5):585–602

Cipollini P, Calafat FM, Jevrejeva S, Melet A, Prandi P (2016) Monitoring sea level in the coastal zone with satellite altimetry and tide gauges. Surv Geophys, 1–25

Darwin GH, Adams JC (1883) Report of a committee, consisting of Professors G. H. Darwin and J. C. Adams, for the harmonic analysis of tidal observations. In: British Association for the Advancement of Science Report for 1883, pp 49–117

Di Lorenzo E, Schneider N, Cobb KM, Franks P, Chhak K, Miller AJ, McWilliams JC, Bograd SJ, Arango H, Curchitser E (2008) North Pacific Gyre Oscillation links ocean climate and ecosystem change. Geophys Res Lett 35(8):1–6

Emery KO, Aubrey DG (1986) Relative sea-level changes from tide-gauge records of eastern Asia mainland. Mar Geol 72(1–2):33–45

Emery KO, You F (1981) Sea level changes in the Western Pacific with special emphasis on China. Oceanol Limnol Sin 12(4):297–310

Fang G (1986) Tide and tidal current charts for the marginal seas adjacent to China. Chin J Oceanol Limnol 4(1):1–16

Fang G, Wang Y, Wei Z, Choi BH, Wang X, Wang J (2004) Empirical cotidal charts of the Bohai, Yellow, and East China Seas from 10 years of TOPEX/Poseidon altimetry. J Geophys Res Oceans 109(C11)

Feng X, Tsimplis MN (2014) Sea level extremes at the coasts of China. J Geophys Res Oceans 119(3):1593–1608

Feng X, Tsimplis MN, Woodworth PL (2015a) Nodal variations and long-term changes in the main tides on the coasts of China. J Geophys Res Oceans 120(2):1215–1232

Feng X, Tsimplis MN, Marcos M, Calafat FM, Zheng J, Jordà G, Cipollini P (2015b) Spatial and temporal variations of the seasonal sea level cycle in the northwest Pacific. J Geophys Res Oceans 120(10):7091–7112

Haigh ID, Wahl T, Rohling EJ, Price RM, Pattiaratchi CB, Calafat FM, Dangendorf S (2014) Timescales for detecting a significant acceleration in sea level rise. Nat Commun 5:3635

Hallegatte S, Green C, Nicholls RJ, Corfee-Morlot J (2013) Future flood losses in major coastal cities. Nat Clim Change 3(9):802–806

Hamlington BD, Strassburg MW, Leben RR, Han W, Nerem RS, Kim KY (2014) Uncovering an anthropogenic sea-level rise signal in the Pacific Ocean. Nat Clim Change 4(9):782–785

Hamlington BD, Reager JT, Lo MH, Karnauskas KB, Leben RR (2017) Separating decadal global water cycle variability from sea level rise. Sci Rep 7(1):995

Holgate SJ, Matthews A, Woodworth PL, Rickards LJ, Tamisiea ME, Bradshaw E, Foden PR, Gordon KM, Jevrejeva S, Pugh J (2013) New data systems and products at the permanent service for mean sea level. J Coastal Res 29(3):493–504

Kaula WM (1970) The terrestrial environment: solid earth and ocean physics. Report of the Williams College, Williamstown meeting in August 1969, NASA Report CR-1579

Marcos M, Tsimplis MN, Calafat FM (2012) Inter-annual and decadal sea level variations in the north-western Pacific marginal seas. Prog Oceanogr 105:4–21

Mazzega P, Berge M (1994) Ocean tides in the Asian semienclosed seas from TOPEX/POSEIDON. J Geophys Res Oceans 99(C12):24867–24881

Merrifield MA (2011) A shift in western tropical Pacific sea level trends during the 1990s. J Clim 24(15):4126–4138

Palanisamy H, Cazenave A, Delcroix T, Meyssignac B (2015) Spatial trend patterns in the Pacific Ocean sea level during the altimetry era: the contribution of thermocline depth change and internal climate variability. Ocean Dyn 65(3):341–356

Passaro M, Cipollini P, Benveniste J (2015) Annual sea level variability of the coastal ocean: the Baltic Sea-North Sea transition zone. J Geophys Res Oceans 120(4):3061–3078

Peltier W (2001) Global glacial isostatic adjustment and modern instrumental records of relative sea level history. Int Geophys 75:65–95

Ren M-E (1994) Relative Sea level changes in China over the last eighty years. In: Di Z, Yuan-Bo L, Cheng-Kui Z (eds) Oceanology of China Seas. Springer, The Netherlands, pp 433–444

Vignudelli S, Kostianoy A, Cipollini P, Benveniste J (eds) (2011) Coastal altimetry. Springer, Berlin, p 578

Vinogradov SV, Ponte RM (2010) Annual cycle in coastal sea level from tide gauges and altimetry. J Geophys Res Oceans 115(C4)

Zhang X, Church JA (2012) Sea level trends, interannual and decadal variability in the Pacific Ocean. Geophys Res Lett 39(21)

Part IV
Indian Ocean Marginal Basins

Using SAR Data for an Assessment of the Indonesian Coastal Environment

Martin Gade, Bernhard Mayer, Carolin Meier, Thomas Pohlmann, Mutiara Putri and Agus Setiawan

Abstract This Pilot Study aimed at improving the information on the state of the Indonesian marine environment that is gained from satellite data. More than 2000 historical and actual synthetic aperture radar (SAR) images from ENVISAT ASAR and Sentinel-1A C-SAR, respectively, were used to produce oil pollution density maps of two regions of interest (ROIs) in Indonesian waters. The normalised spill number and the normalised mean polluted area indicate that in general, the marine oil pollution in both ROIs is of different origin: while ship traffic appears to be the main source in the Java Sea, oil production industry causes the highest pollution rates in the Strait of Makassar. In most cases hot spots of marine oil pollution were found in the open sea, and the largest number of oil spills in the Java Sea was found from March to May and from September to December, i.e., during the transition from north-west monsoon to south-east monsoon, and vice versa. This is when the overall wind and current patterns change, apparently making oil pollution detection with SAR sensors easier. In support of our SAR image analyses high-resolution numerical forward and backward tracer experiments were performed. Using the previously gained information we demonstrate that the combination of numerical tracer modelling with (visual) SAR image analyses can be used for an assessment of the marine environment in Indonesian waters, and also helps in better understanding the observed seasonality.

Keywords SAR · Coral Triangle · Indonesia · Oil spill · Numerical modelling Tracer · Marine pollution · Java Sea · Makassar Strait

M. Gade (✉) · B. Mayer · C. Meier · T. Pohlmann
Universität Hamburg, Institut für Meereskunde, Hamburg, Germany
e-mail: Martin.Gade@uni-hamburg.de

M. Putri
Institute Technology Bandung, Bandung, Indonesia

A. Setiawan
Agency for Marine and Fisheries Research and Development, Jakarta, Indonesia

V. Barale and M. Gade (eds.), *Remote Sensing of the Asian Seas*,
https://doi.org/10.1007/978-3-319-94067-0_19

1 Introduction

Indonesian territorial waters cover about three million square kilometres, thereby being larger than the Mediterranean Sea, and are home to more than 3.000 species of fish and more than 500 species of corals. The Indonesian coastline is longer than 99,000 km, and the Indonesian archipelago encompasses more than 17.000 islands. Major ship traffic routes, connecting the economic centres on the South China Sea (and beyond) with Europe, Africa, Australia, or the Persian Gulf, run through Indonesian waters.

Parts of the so-called Coral Triangle, a six million square kilometres area in Pacific Asia, lie in Indonesian territorial waters, where the coral reef area is estimated to be 20,000 km^2 in size. Along with mangrove forests and seagrass meadows on the coasts, these areas are particularly vulnerable to pollutants. Several marine protected areas (MPAs) have been defined in Indonesia, where part or the entire enclosed environment is protected by law or other effective means. Continuous monitoring is of key importance in these areas, but can only be done in an effective manner by taking advantage of state-of-the-art remote sensing and numerical modelling techniques.

The marine ecosystem in Indonesia is under increasing pressure due to social and economic growth in the entire region. Marine fishery has been continuously increasing, making Indonesia the second largest marine capture producer in 2014 (FAO 2016). Consequently, the overall number of fishing vessels has increased, although their vast majority is still small and unpowered (Purwanto 2003).

Although its total oil production has decreased by 25% during the past decade, Indonesia still ranks amongst the top 25 oil producing countries worldwide and is the third-largest oil producer in Pacific Asia (BP 2014). However, not only because of a continuously increasing demand Indonesia, along with its neighbouring countries, has always been importing oil from other countries worldwide.

The joint German-Indonesian Pilot Study IndoNACE (Indonesian seas Numerical Assessment of the Coastal Environment) aimed at improving the information on the state of the Indonesian marine environment that is gained from satellite data. The goal of IndoNACE was to combine spaceborne remote sensing data with a regional hydrodynamical model, in order to meet the monitoring requirements. Synthetic aperture radar (SAR) data were used to assess the overall oil pollution in two dedicated regions of interest (ROIs) in Indonesian waters, namely the Western Java Sea and the Strait of Makassar (Fig. 1).

2 SAR Image Examples and Regions of Interest

Marine mineral oil spills show up on SAR imagery as dark patches (Gade 2006; Gade and Alpers 1999), which however can be confused with other atmospheric or oceanic phenomena such as wind shadowing, biogenic slicks, etc. (Brekke and

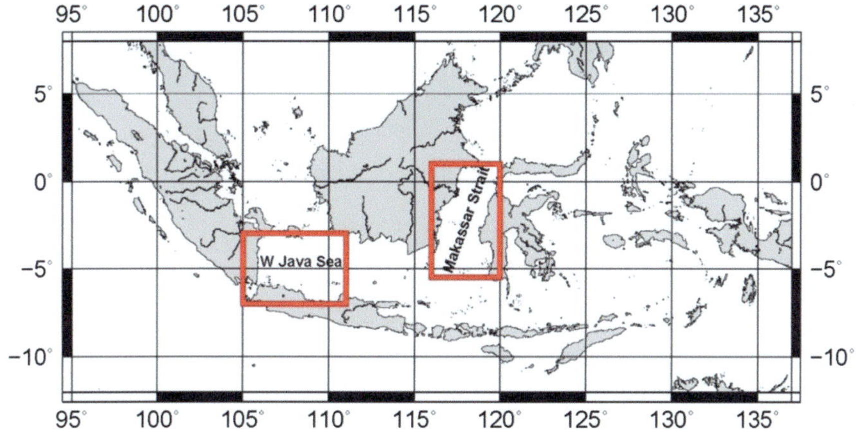

Fig. 1 Two regions of interest (ROI): 'W Java Sea' with borders 105.0°E/111.0°E and 7.0°S/3.0°S, and 'Makassar Strait' with borders 116.0°E/120.0°E and 5.5°S/1.0°N

Solberg 2005). Figure 2 is an ENVISAT ASAR image of the Java Sea, which shows many imprints of marine oil pollution as dark spots. The image was picked arbitrarily and is a good example of the frequent oil pollution in that area.

Both historical and actual SAR data from ESA's ENVISAT archive and Sentinel 1A Rolling Archive, respectively, were used. A visual inspection of all available SAR images of Indonesian waters was performed, along with a manual registration of geo-information on the detected spills (lat/lon, size, etc.) and metadata (wind speed and direction, etc.). Attention was paid to the visual discrimination between anthropogenic (mineral oil) spills and biogenic slicks, since both species tend to cause very similar features on SAR imagery (Gade et al. 1998, 2013).

Figure 3 demonstrates difficulties that may arise during SAR-based oil pollution detection: vast areas in the Makassar Strait, west of Sulawesi, are covered by natural slicks, which show up as elongated narrow dark patches, while larger dark areas are due to low wind speed. However, the major dark patches in the upper left part of the image are likely to be due to marine oil pollution, because of their singular appearance and their stronger radar contrast.

The aim of this study was to gain information on, and to further the knowledge about, the vulnerability of dedicated marine (coastal and offshore) areas in Indonesia to marine oil pollution. Two ROIs were identified (Fig. 1), the western part of the Java Sea, with borders 105.0°E–111.0°E and 7.0°S–3.0°S, and the Makassar Strait, with borders 116.0°E–120.0°E and 5.5°S–1.0°N. The ROIs were chosen because of high economical activities (including oil production industry, ship traffic, and fishery) and a high density of MPAs and coral reefs.

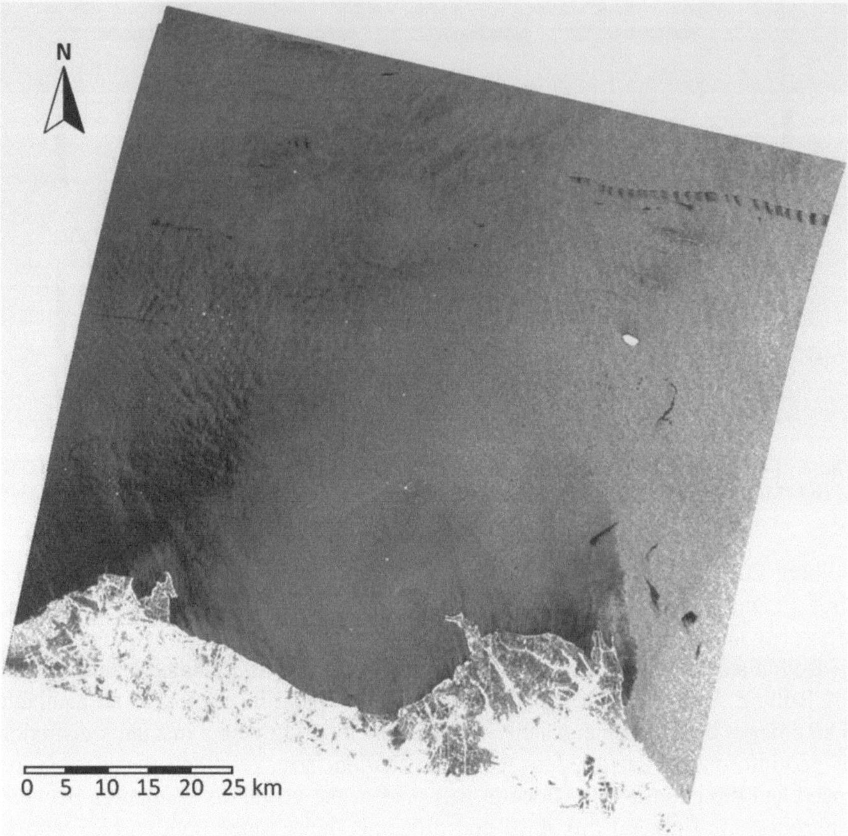

Fig. 2 Envisat ASAR image (82 km × 88 km) of the Java Sea, north off Eretan Bay, Java, acquired on 8 December 2005 at 02:25 UTC. Throughout the image small dark patches can be delineated, which are due to mineral oil spills. However, dark areas on the lower left and off the coast are due to low wind. The upper right part of the image is a Marine Protected Area (MPA)

3 SAR Image Analyses

In total more than 2000 SAR images were available, acquired by ENVISAT ASAR (more than 1600, acquired 2002–2012) and Sentinel-1A (more than 400, acquired from October 2014). Here we mainly concentrate on results obtained through the analysis of the ENVISAT ASAR images, of which approx. 700 images were acquired over the ROI 'W Java Sea' and 900 over the ROI 'Makassar Strait'. Figure 4 shows the total coverage of both ROIs by ENVISAT ASAR, indicating a higher coverage of coastal regions in the Java Sea and in the western Makassar Strait, where the coastal area off the oil production centre in Balikpapan was frequently imaged.

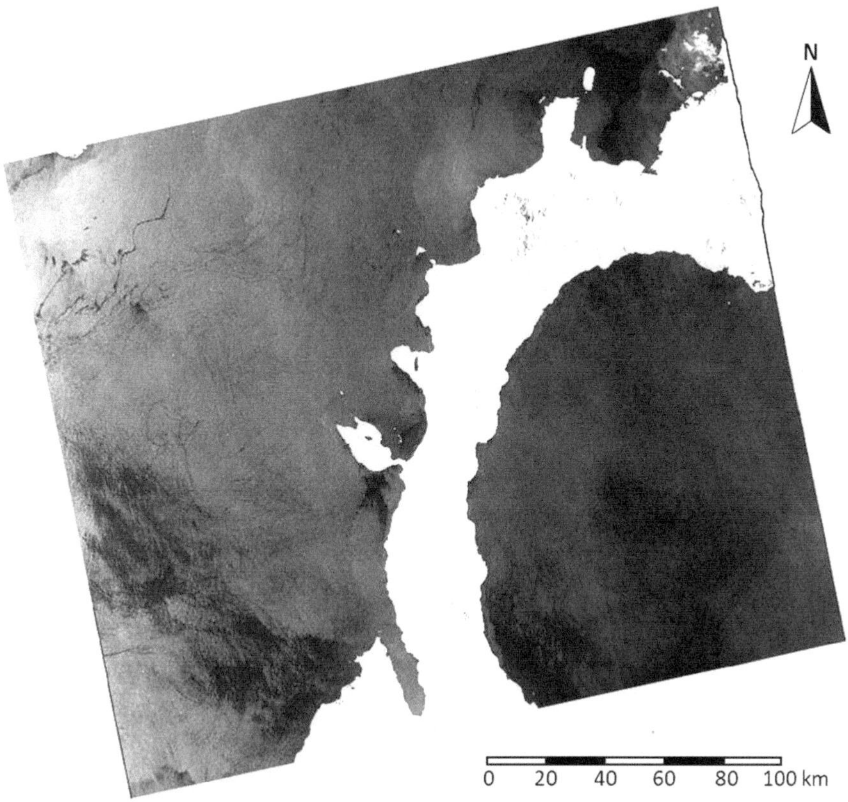

Fig. 3 Sentinel 1A SAR image (214 km × 253 km) of the northern Strait of Makassar, acquired on 18 November 2014 at 10:18 UTC. The north-western part of Sulawesi is reaching through the image. Major dark elongated patches in the upper left part of the image, i.e., at the northern exit of the Strait, are likely to be due to mineral oil spills

In the ROI 'W Java Sea' a total of 2114 oil spills were found (that is, on average, about 3 per image) and in the ROI 'Makassar Strait' a total of 3046 (3.4 per image). Figure 5 shows the geographical locations of all detected oil spills, binned into a 0.05° × 0.05° grid, for both ROIs, 'W Java Sea' (left) and 'Makassar Strait' (right). Clearly visible are areas of higher oil pollution north of western Java, where up to nine spills per grid cell were found, and in the Makassar Strait, off Kalimantan's coast, where the maximum number of oil spills per grid cell well exceeds 30 (for better visualisation the colour coding in both panels is the same, so that high spill numbers in the Makassar Strait, exceeding 9, appear in black).

These areas are marked by high ship traffic ('W Java Sea') and by intense oil production ('Makassar Strait'), both putting the local environment under severe threat. We also note, however, that the areas of maximum detected oil pollution coincide with those areas, where the SAR image coverage is highest (Fig. 4).

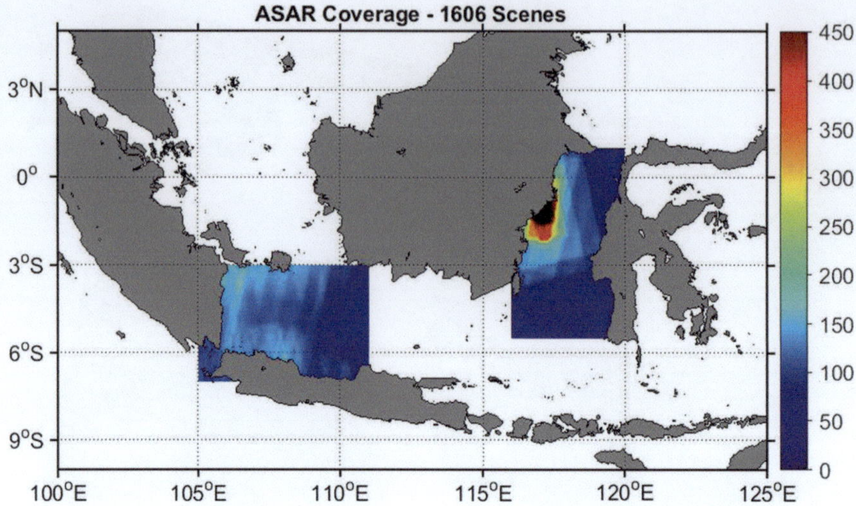

Fig. 4 Coverage of both ROIs by more than 1600 ENVISAT ASAR images of 2002–2012

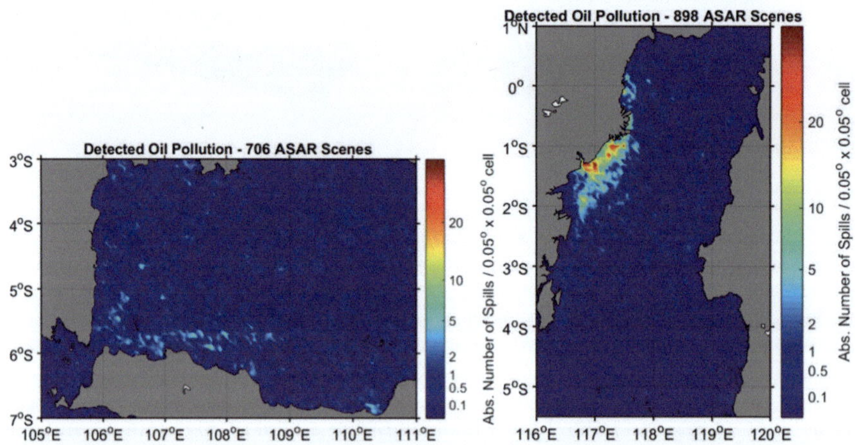

Fig. 5 Number of spills per 0.05° × 0.05° grid cell, as detected in all available ENVISAT ASAR images of each ROI

The length and width of each detected oil spill were protocolled and were used to calculate the spill's size, for simplicity assuming an elliptical shape. We define the normalised mean polluted area as the ratio of the total spill-covered area (i.e., the sum of all spill sizes in a grid cell) and the SAR coverage (Fig. 4), for simplicity given in km^2 for an amount of 100 SAR images. The normalised mean polluted area is rather independent of the heterogeneous data coverage, and assuming a 'critical SAR coverage' at any place in both ROIs, allows comparisons of both

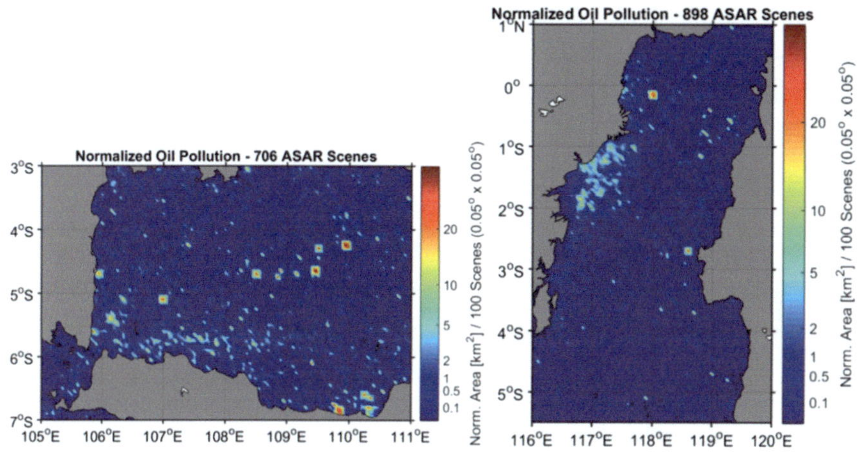

Fig. 6 Normalized mean polluted area per 0.05° × 0.05° grid cell, based on all available ENVISAT ASAR images of each ROI

ROIs and objective conclusions. Normalised polluted areas are shown in Fig. 6, again, binned into a 0.05° × 0.05° grid.

It is interesting to note that most hot spots of marine oil pollution are found in the open sea. Moreover, hot spots in terms of the total amount of detected oil spills (Fig. 5) are no longer areas of highest normalised pollution (Fig. 6). We also note that all coastal industrial areas in both ROIs coincide with regions where the mean oil pollution in a 0.05° × 0.05° grid cell exceeds 5 km^2 per 100 SAR images, i.e., 50,000 m^2 per single SAR image. Statistically, the areal fraction of (detected) pollution in those areas is at least 0.2%, but can easily reach 1.5% in certain locations in the central Java Sea and open Strait of Makassar. According to our analyses of approx. 1600 ASAR scenes this is the fractional areal that is polluted at any time.

Actual SAR data from the Sentinel mission have been used to complement the analyses and results presented above (Meier 2016). A total of 207 Sentinel-1A C-SAR images of the ROI 'W Java Sea' acquired in October 2014–June 2016 (21 months) were analysed. The locations and sizes of all detected oil spills are shown in Fig. 7, where the circles' diameters are proportional to the spill sizes and the colour coding denotes the month in which the oil pollution was imaged by Sentinel-1A.

It is obvious that most pollution was found along the coasts of Java and Sumatra; however, those are the areas that were most frequently imaged. In contrast, no oil pollution was found in the open Java Sea, simply because no Sentinel-1A SAR images of that area were available by the time of our analyses. While no immediate relationship between the amount of oil spills and the months of the year can be seen, the colour coding indicates that less pollution was found in June, July, and August (light green and yellow circles).

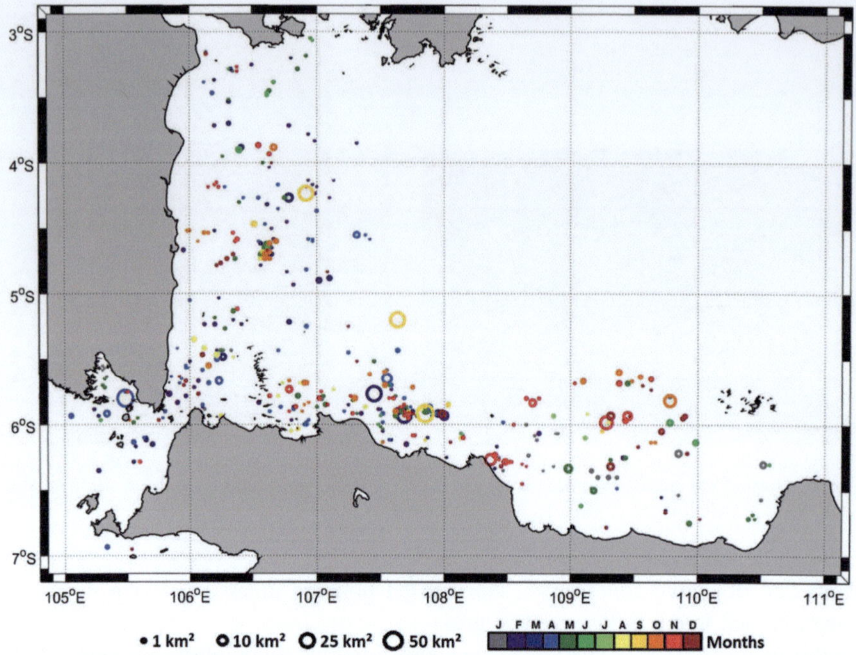

Fig. 7 Map of all oil pollution found on Sentinel-1A SAR imagery of the ROI 'W Java Sea'. Each circle marks an oil spill, whose size is proportional to the circle diameter. The color coding denotes the month of the year, in which the oil pollution was imaged. After Meier (2016)

Spill size distributions for both ROIs are shown in Fig. 8. Added to each histogram are the respective percentile curve and a horizontal dotted line indicating 95%. In both ROIs the largest number of oil spills were of size 0.1–0.3 km^2. However, oil spills found in the ROI 'Makassar Strait' were generally smaller: about 70% of the detected oil spills were smaller than 1 km^2, while it was only 60% in the ROI 'W Java Sea', and about 95% were smaller than 6.6 km^2, while 95% of the oil spills found in the ROI 'W Java Sea' were of sizes up to 10 km^2. Major oil spills found in the open Java Sea mainly contribute to this result.

The seasonality of the observed oil pollution in the Western Java Sea is demonstrated in Fig. 9. The upper panel shows in grey the total number of oil spills detected in each month of the year, and in black the monthly distribution of the approx. 700 ASAR images. The lower panel shows the respective number of oil spills per SAR image of that ROI, i.e., the ratio of the upper two.

On average, between 1.5 and 5 oil spills were found in each ENVISAT ASAR image, with the largest number of oil spills found from March to May and from September to November. We note that these periods mark the transition from north-west to south-east monsoon, and vice versa. During those periods, the overall current pattern in the Java Sea changes (see below) and a greater amount of water

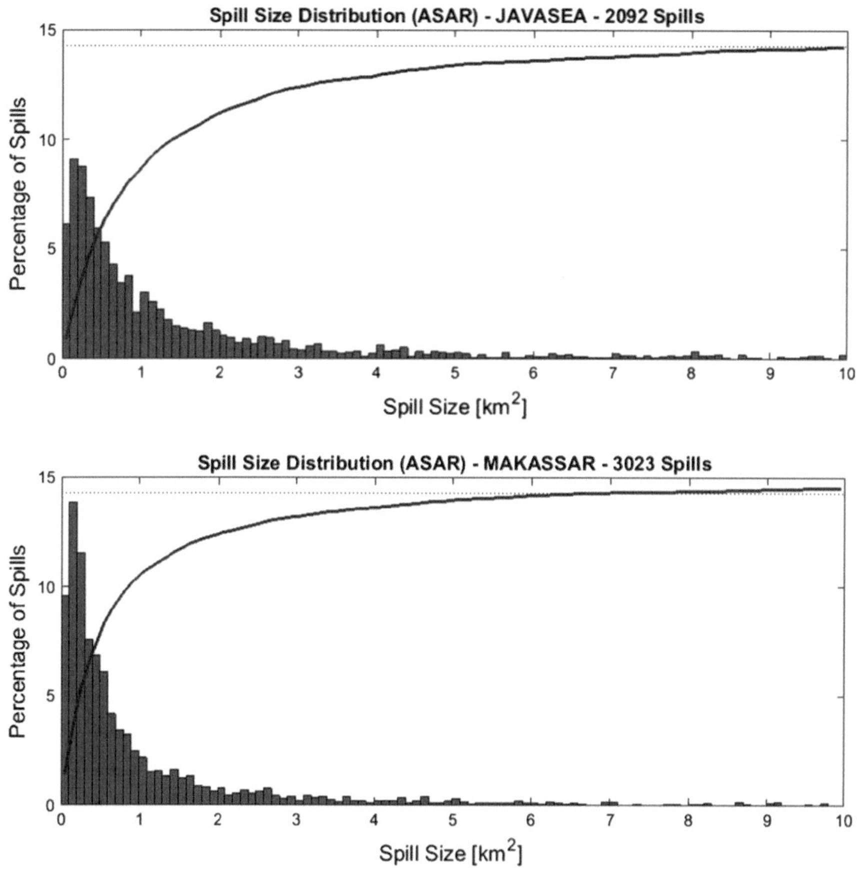

Fig. 8 Histograms of the spill sizes for the ROIs 'W Java Sea' (upper) and 'Makassar Strait' (lower). Added as solid line to each histogram are the percentile curves, with the horizontal dotted line marking 95%

from the inner Java Sea is driven towards south-west, thereby reaching those areas where the highest pollution was found.

The corresponding results for the Makassar Strait are shown in Fig. 10. Here, high pollution rates (average number of spills per SAR image, lower panel) were found in December and January, when the north-west monsoon drives additional water from the Banda Sea in the north into the Makassar Strait, thereby slowing down the overall current speed. Moreover, and similar to the Java Sea, another maximum in the monthly distribution was found in April, May and June, during the transition from north-west to south-east monsoon. The change in the overall wind field, along with the corresponding change in the overall current pattern, appears to trigger the appearance of marine oil pollution in ASAR imagery.

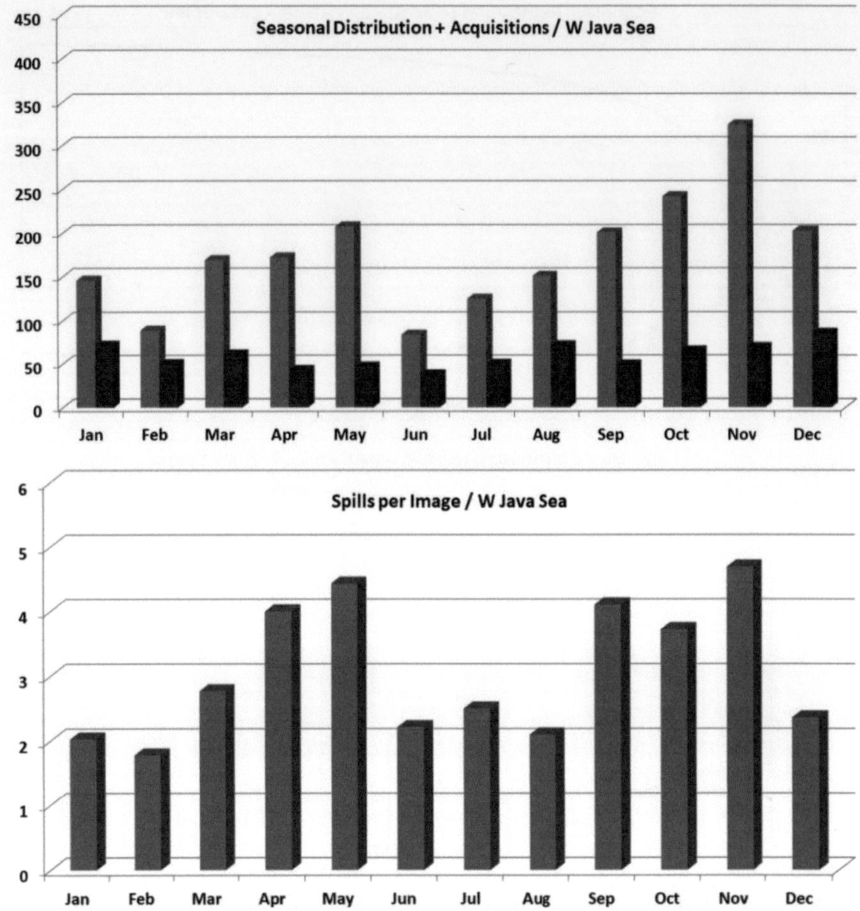

Fig. 9 Seasonal variation of the detected oil pollution in the ROI 'W Java Sea'. The upper panel shows the monthly distribution of all oil spills found (grey) and of the approx. 700 ENVISAT ASAR images (black). The lower Panel shows the respective distribution of the average numbers of oil spills per SAR image

4 Numerical Modelling

In order to get an idea of the origin of potential oil spills at certain locations (e.g. MPAs), a set of numerical models has been applied. The first numerical model is the Hamburg Shelf Ocean Model (HAMSOM) (Backhaus 1985, Mayer and Damm 2012), a three-dimensional regional baroclinic ocean circulation model, here with a horizontal grid resolution of 6′ (approx. 11 km) and a vertical resolution increasing from 6 m at the surface to several hundred meters at greater depths. The meteorological forcing was taken from NCEP/NCAR (Kalnay et al. 1996) and the open boundary conditions (temperature, salinity, sea surface height) were taken from the

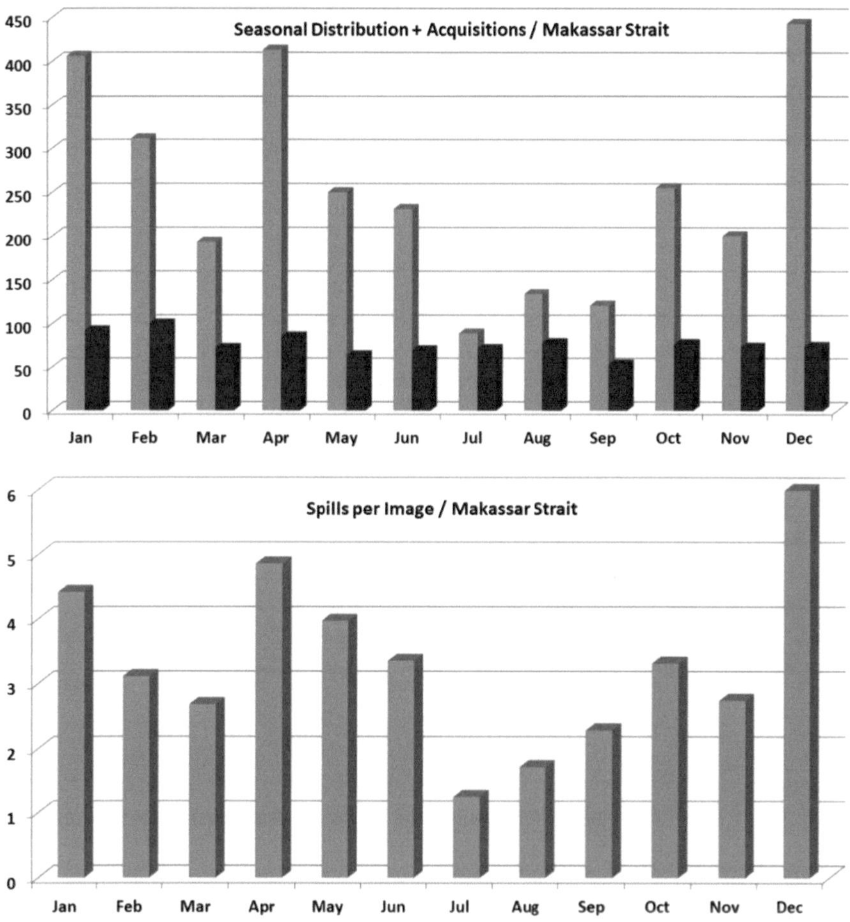

Fig. 10 Same as Fig. 9, but for the ROI 'Makassar Strait'

global circulation model MPIOM (Jungclaus et al. 2013). Results of the model runs were validated using observed velocities from moored current meters of the INSTANT project (Sprintall et al. 2009) at different locations as well as SST and SSS satellite data.

The subsequently applied model is a Lagrangian tracer model (Mayer 1995), which uses the simulated HAMSOM velocities. Depending on the tracer's location within a grid cell it is subject to a spatially interpolated velocity and to acceleration along its path due to spatial velocity changes. This is done for horizontal directions only, because only oil staying in the upper water layer (6 m) is of interest here. Mixing with, and resurfacing from, deeper layers was neglected in this study, because the winds are generally low to moderate in the two ROIs.

As an example, for four dedicated months in the period 2003–2011 backward tracer model runs were performed, each lasting four weeks, from the end until the beginning of a month. Simulation results for oil spills ending up in selected MPAs inside our ROIs are shown in Fig. 11 (W Java Sea) and Fig. 12 (Makassar Strait), for February (fully developed north-west monsoon), April (transition period), August (fully developed south-east monsoon), and October (transition period). Diamonds show the position after one week of backward tracer modelling, i.e., the position of the tracers one week before they would arrive at the MPA.

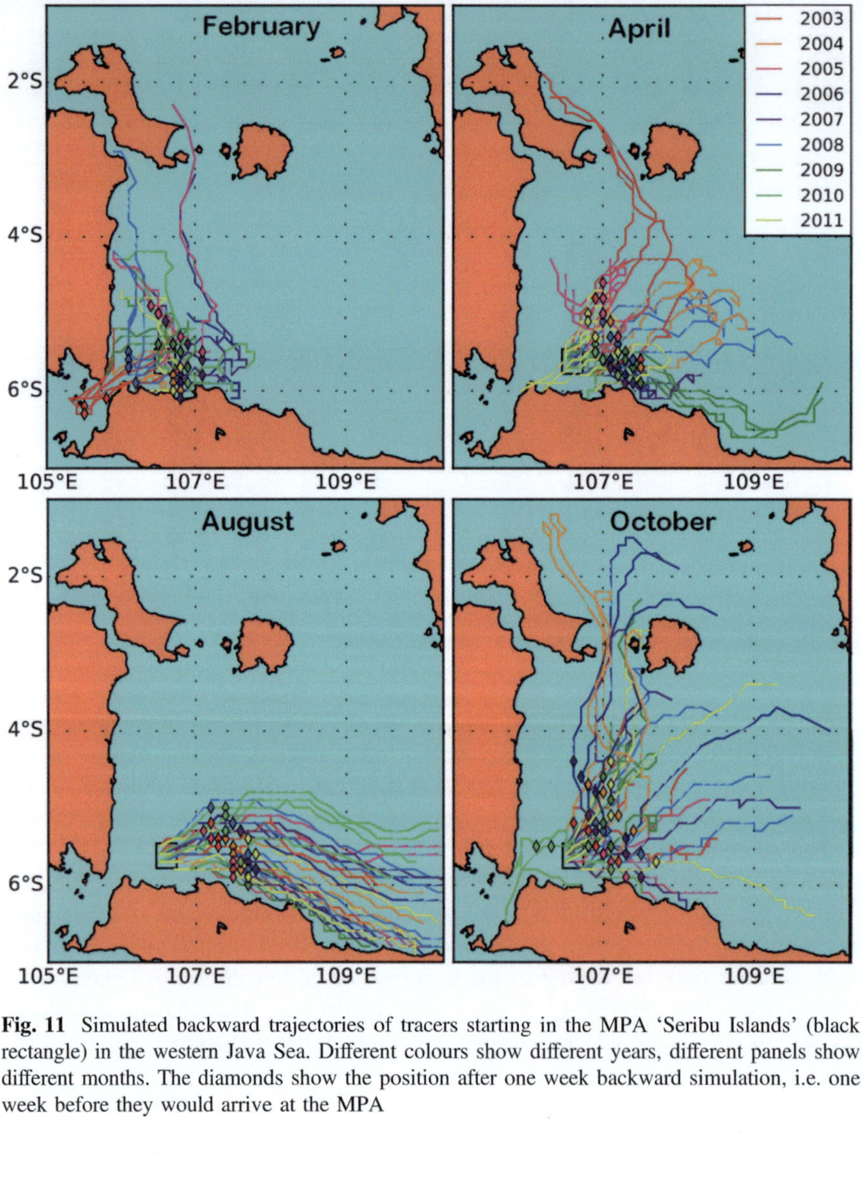

Fig. 11 Simulated backward trajectories of tracers starting in the MPA 'Seribu Islands' (black rectangle) in the western Java Sea. Different colours show different years, different panels show different months. The diamonds show the position after one week backward simulation, i.e. one week before they would arrive at the MPA

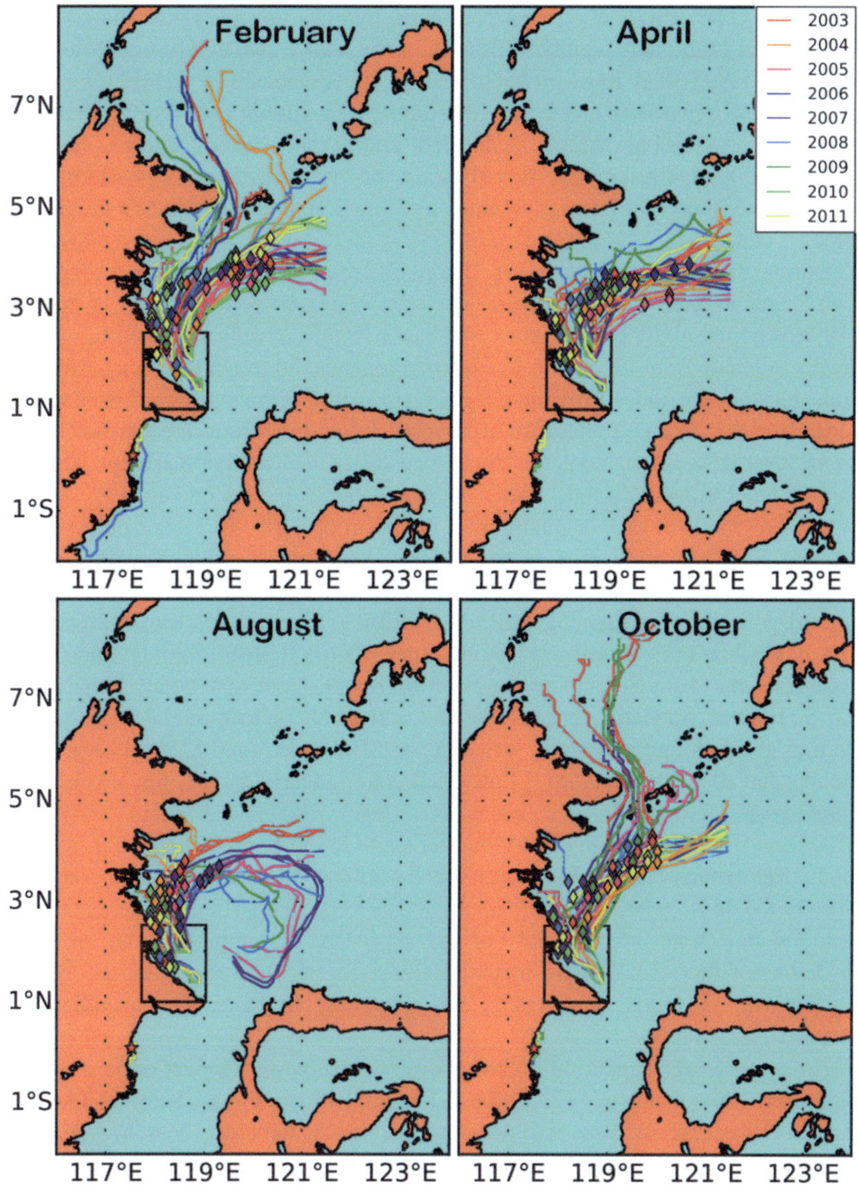

Fig. 12 Same as Fig. 11, but for the MPAs "Berau", northwest of the Makassar Strait (black rectangle), and "Bontang" on the Kalimantan coast in the northwestern Makassar Strait (red star)

For the MPA 'Seribu Islands' (black rectangles at about 5.5°S 106.5°E in Fig. 11) only the August trajectories show a uniform distribution. During dedicated field surveys several oil spills around Pari Island (MPA 'Seribu Islands') were found. The concentration of total petroleum hydrocarbon (TPH) at the southern coast of Pari Island in November was found to be higher than in April. According to our model results it is most probable that oil spills detected in that area originate from the east, regardless of the year. Assuming a maximum age of oil pollution of 4 weeks the possible source is located within an area of approx. 110 km (north-south) by 400 km (west-east) along the northern coast of western Java. During the SE monsoon season, strong surface currents in the Java Sea are clearly directed westward.

During the north-west monsoon (February), the current direction is opposite, with much weaker current speeds. The area of possible oil sources is approx. 110 km (north-south) by 200 km (west-east). Note that only in 2003, the model tracers originate from the Indian Ocean and propagated through the Sunda Strait. In both transition periods, April and October, possible origins are wide-spread and depend on the very year, because of an annually varying transition period.

Another MPA is located in the Berau regency, at the island of Kakaban, which is part of the Derawan Islands (black rectangle in all panels of Fig. 12). At this location oil spills always originate from the north or north-east, due to the general counter-clockwise circulation in the Celebes Sea. Intense oil exploration in the Tarakan Basin (Wight et al. 1993), north of Berau, may therefore be an origin of any oil pollution reaching that MPA and drifting further south into the Strait of Makassar. Only in February and October, (persistent) oil spills might be transported even from the Sulu Sea in the far north to Berau. Note that trajectories east of 121.5 °E were truncated, because of the end of the tracer model grid (tracers starting at this longitude would arrive at the MPA after approx. 9–14 days in all months and all years). In August (lower left panel), a large eddy is obvious in three model years.

Tracers from the Bontang MPA, further south in the Strait of Makassar (red star on the Kalimantan coast in Fig. 12), show hardly any movement, because they stay close to the coast, where currents are obviously very slow. These numerical model results demonstrate the feasibility of our approach, to combine numerical tracer modelling with (visual) SAR image analyses for an assessment of the marine environment in Indonesian waters, and they help in better understanding the observed seasonality.

An example of frequent oil pollution due to offshore oil production is shown in Fig. 13: the five panels are (25.2 km × 19.5 km) excerpts from Sentinel-1A SAR images of the ROI 'W Java Sea' acquired in January, February, March, April, and May 2015 and show dark patches linked to a bright spot at coordinates 5° 54′ 42″S 107° 55′ 35″E. This is an oil rig in the 'ECHO' oil field of the Offshore North West Java (ONWJ) block, from which oil is frequently released, causing oil spills of lengths well exceeding 20 km. In different seasons, the varying currents cause oil drifts of different directions. Such findings can be used as input for the numerical modelling and as validation of the model results.

We used the modelled surface current field to calculate the 'lifetime' of the observed oil spills. Here, the term 'lifetime' refers to the time passed since the (latest) start of the spillage or, in case oil was continuously released for a longer

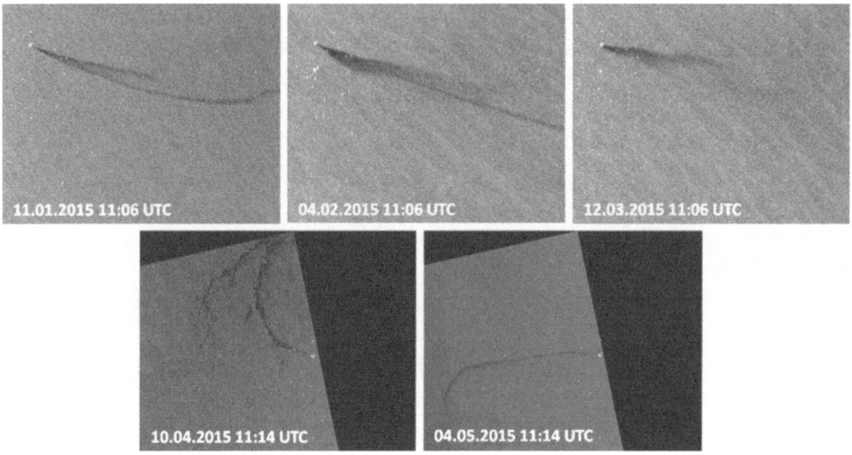

Fig. 13 Case study of frequent oil pollution caused by offshore oil industry and visible on Sentinel-1A SAR imagery (25.2 km × 19.5 km). The bright spot, from which the dark patches in all panels origin, is an oil rig off Java, at 5° 54′ 42″S 107° 55′ 35″E. After Meier (2016)

time, the maximum period, for which the oil spill was visible on SAR imagery. For the particular cases shown in Fig. 13 the ‘lifetime’ is ranging between 8 h (upper right Panel, 12 March 2015) and 33 h (lower left Panel, 10 April 2015). These results are in agreement with observations of an oil spill, which was found in the vicinity of Pari Island and which was detected only for 1–2 days, before it disappeared.

5 Conclusions

We used more than 2000 ENVISAT ASAR and SENTINEL-1A C-SAR images for our statistical analyses and generated oil pollution density maps for two ROIs in Indonesian waters. More oil pollution was found along the Western Java and Kalimantan coasts; however, those are also the areas that were most often imaged by the spaceborne SAR sensors.

The normalised mean polluted area, defined as the ratio of the total polluted area (within a grid cell) and the spatial SAR coverage, is more independent of any heterogeneous data coverage and, assuming that any place in both ROIs is imaged often enough, allows comparisons of both ROIs and objective conclusions.

Our findings indicate that the main source of oil pollution in the two IndoNACE ROIs is different: although major oil production centres are located along the Java coast, the strongest oil pollution (in ENVISAT ASAR imagery) was found in the open Java Sea, indicating that ship traffic is the main source of oil pollution in the Java Sea, where major shipping lanes are located. In contrast, the highest pollution rates in the Strait of Makassar were found in an area of intense oil production, while less pollution was found along the shipping lanes. In comparison with the Java Sea, the main shipping lanes in the Makassar Strait are less polluted.

In most cases local hot spots of marine oil pollution are found in the open sea. We also note, however, that the areas of maximum detected oil pollution coincide with areas of highest SAR image coverage. Our results, therefore, also demonstrate that a larger number of SAR images is necessary for more robust statistical analysis of this kind, particularly for investigations of the seasonal change in regional oil pollution.

The ROI 'W Java Sea' was used to demonstrate the seasonal change in (detected) oil pollution: during the transition times from north-west to south-east monsoon, and vice versa, most pollution was found. Apparently the wind and current conditions during those transition periods are advantageous for SAR-based oil spill monitoring. However, more research is needed here, including more SAR data from the C-SARs aboard Sentinel-1A and 1B.

In summary, our results demonstrate the feasibility of our approach, to combine numerical tracer modelling with (visual) SAR image analyses for an assessment of the marine environment in Indonesian waters, and they help in better understanding the observed seasonality.

Acknowledgements Franziska Leverenz and Anja Berger analysed the ASAR data. The pilot study IndoNACE received funding from the European Space Agency (ESA) under contract ITT AO 1-8176/14/F/MOS. The ENVISAT ASAR imagery was kindly made available by ESA's EO Grid Processing On-Demand (EO-GPOD) Team.

References

Backhaus JO (1985) A three-dimensional model for the simulation of shelf sea dynamics. Deut Hydrogr Z 38:165–187

BP (2014) BP statistical review of world energy, June 2014. http://www.bp.com/content/dam/bp/pdf/Energy-economics/statistical-review-2014/BP-statistical-review-of-world-energy-2014-full-report.pdf. 10 Feb 15

Brekke C, Solberg AHS (2005) Oil spill detection by satellite remote sensing. Remote Sens Environ 95:1–13

FAO (2016) The state of world fisheries and aquaculture 2016. Contributing to food security and nutrition for all. Rome, p 200

Gade M (2006) On the imaging of biogenic and anthropogenic surface films on the sea by radar sensors. In: Gade, M, Hühnerfuss H, Korenowski GM (eds) Marine surface films: chemical characteristics, influence on air-sea interactions and remote sensing, vol 342. Springer, Heidelberg, pp 189–204

Gade M, Alpers W (1999) Using ERS-2 SAR images for routine observation of marine pollution in European coastal waters. Sci Total Environ 237–238:441–448

Gade M, Alpers W, Hühnerfuss H, Masuko H, Kobayashi T (1998) The imaging of biogenic and anthropogenic surface films by a multi-frequency multi-polarization synthetic aperture radar measured during the SIR-C/X-SAR missions. J Geophys Res 103:18851–18866

Gade M, Byfield V, Ermakov S, Lavrova O, Mitnik L (2013) Slicks as indicators for marine processes. Oceanography 26(2):138–149

Jungclaus JH, Fischer N, Haak H, Lohmann K, Marotzke J, Matei D, Mikolajewicz U, Notz D, von Storch JS (2013) Characteristics of the ocean simulations in the Max Planck Institute Ocean Model (MPIOM) the ocean component of the MPI-earth system model. J Adv Model Earth Syst 5:422–446. https://doi.org/10.1002/jame.20023

Kalnay E, Kanamitsu M, Kistler R, Collins W, Deaven D, Gandin L, Iredell M, Saha S, White G, Woollen J, Zhu Y, Leetmaa A, Reynolds R, Chelliah M, Ebisuzaki W, Higgins W, Janowiak J, Mo KC, Ropelewski C, Wang J, Jenne R, Joseph D (1996) The NCEP/NCAR reanalysis 40-year project. Bull Am Meteorol Soc 77:437–471

Mayer B (1995) A three-dimensional numerical SPM transport model with application to the German Bight. In: GKSS Forschungszentrum Geesthacht GmbH (ed) GKSS report 95/E/59. GKSS, p 96 (in German)

Mayer B, Damm PE (2012) The Makassar Strait throughflow and its jet. J Geophys Res 117: C07020. https://doi.org/10.1029/2011JC007809

Meier C (2016) Untersuchungen zur Detektion von mariner Ölverschmutzung in indonesischen Seegebieten mit satellitengestützten Radarsensoren. B.Sc. thesis. Universität Hamburg, Institut für Meereskunde (in German)

Purwanto (2003) Status and management of the Java Sea fisheries, in assessment, management and future directions for coastal fisheries in Asian Countries. In: Silvestre G, Garces L, Stobutzki I, Ahmed M, Valmonte-Santos RA, Luna C, Lachica-Aliño L, Munro P, Christensen V, Pauly D (eds) Proceed. WorldFish center conference 67(1):793–832

Sprintall J, Wijffels SE, Molcard R (2009) Direct estimates of the Indonesian throughflow entering the Indian Ocean: 2004–2006. J Geophys Res 114:1–58

Wight AWR, Hare LH, Reynolds JR (1993) Tarakan Basin, NE Kalimantan, Indonesia: a century of exploration and future potential. Geol Soc Malaysia Bull 33:263–288

Remote Sensing of Coastal Discharge of SE Sumatra (Indonesia)

H. Siegel, M. Gerth, I. Stottmeister, A. Baum and J. Samiaji

Abstract The coastal discharge of Indonesia is driven by high precipitation throughout the year. Monsoon and tides form a highly variable dynamical system that satellite remote sensing is the only method to acquire synoptic information. The large rivers of SE-Sumatra belong to the major tropical carbon and sediment sources for the world ocean. Knowledge of carbon sources is important because of their potential impact on coastal ecosystems and on climate change. Drainage of peatlands enhances the concentration of dissolved organic matter (DOM) and strong tidal currents the suspended particulate matter (SPM) load in the rivers. The high concentrations strongly reduce the available light in the water column of coastal regions. The coloured part of DOM (CDOM) absorbs light in the short wavelength range. Measured very high absorptions changed the colour brownish that the tributaries belong to so-called black water rivers. High SPM concentrations made the water milky and bright. Different compositions of water constituents led to large variations of water colour, which made ocean colour remote sensing to the preferred method for investigating river discharge and coastal transport in different spatial and temporal scales. Sources of different water masses were identified, the estuarine turbidity maximum zone and the spreading of river water in relation to tidal phases were detected. The transport processes in adjacent Malacca and Karimata Straits are described in relation to monsoon phases and ENSO. In addition to available satellite derived level 2 products, CDOM and DOC charts were retrieved using regional algorithms derived from measurements.

H. Siegel (✉) · M. Gerth · I. Stottmeister
Leibniz Institute for Baltic Sea Research Warnemünde,
Seestrasse 15, 18119 Rostock, Germany
e-mail: herbert.siegel@io-warnemuende.de

A. Baum
Leibniz Centre for Tropical Marine Ecology, Fahrenheitstr. 6,
28359 Bremen, Germany

J. Samiaji
University of Riau, Jl. Simpang Panam Km 12.5, Pekanbaru, Riau, Indonesia

V. Barale and M. Gade (eds.), *Remote Sensing of the Asian Seas*,
https://doi.org/10.1007/978-3-319-94067-0_20

Keywords Indonesia · South- East Sumatra · Water constituents Ocean colour · Remote sensing · Coastal discharge · Transport Processes · Tidal and monsoon influence

1 Introduction

In the tropical area off SE-Sumatra (Indonesia), bio-optical satellite and in situ measurements were implemented to study the impact of the main river discharge on the coastal ecosystems and the transport processes in the adjacent seas. The large Indonesian Islands are known as major sources of DOM and SPM (Baum et al. 2007). Satellite derived CDOM absorption and SPM maps show the distribution patterns of coastal runoff. A MODIS standard product combining the absorption of yellow substances and detritus at 443 nm averaged over the entire MODIS Aqua mission (2002–2016) illustrates the high concentrations in the Indonesian coastal waters (Fig. 1). The western part of Indonesia includes the Sunda Shelf surrounded by the Islands of Kalimantan, Java, and Sumatra and the Malaysian Peninsula. The Java Sea is located between Kalimantan and Java. The Karimata Strait between Kalimantan and Sumatra connects the South China Sea in the north with the Java Sea in the south (Fig. 2). The Malacca Strait between Sumatra and Malaysia connects the Andaman Sea and the Indian Ocean in the north with the Karimata Strait in the south. Because of their vast rainforests, high DOM input occurs particularly along the southern and southwestern coast of Kalimantan and along the East and South- East coast of Sumatra. The coastal discharge affects particularly the water of the Malacca and Karimata Straits. Both straits are highly dynamic regions affected by the monsoonal circulation superimposed by various tidal modes that only satellite data can deliver synoptic information about larger areas in different time steps for detailed process studies. Although the Indonesian Rivers drain only 2% of global land area, they transport 20–25% of the total sediment export to the oceans (Milliman et al. 1999). 500 Tg year^{-1} of sediment are only discharged from the east coast of Sumatra. By draining peatlands, the Sumatran coast is with an export of 0.3 Tg C year^{-1} an important C-source for the global ocean (Baum et al. 2007; Rixen et al. 2008; Wit et al. 2015). Ocean colour satellite data (SeaWiFS) were applied for the investigation of Chlorophyll in the Malacca Strait (Tan et al. 2006) and around Java Island (Hendiarti et al. 2004). The South-Sumatran Rivers originate in Barisan Mountains, drain catchment areas from highlands to coastal plains and discharge into the Malacca or Karimata Straits. Malacca Strait is characterized by a general current in north-westerly directions towards the Andaman Sea caused by sea level differences between the south-eastern and north-western entrance with a slope towards Andaman Sea (Wyrtki 1961; Namba and Saadon 2001; Thia-Eng et al. 2000; Siegel et al. 2009). Throughout the year, a salinity gradient in the mixed layer between the southern and northern part generates a small baroclinic sea level gradient inducing transport in north-westerly direction (Wyrtki 1961). In boreal winter (November–March), NE monsoon over the Bay of

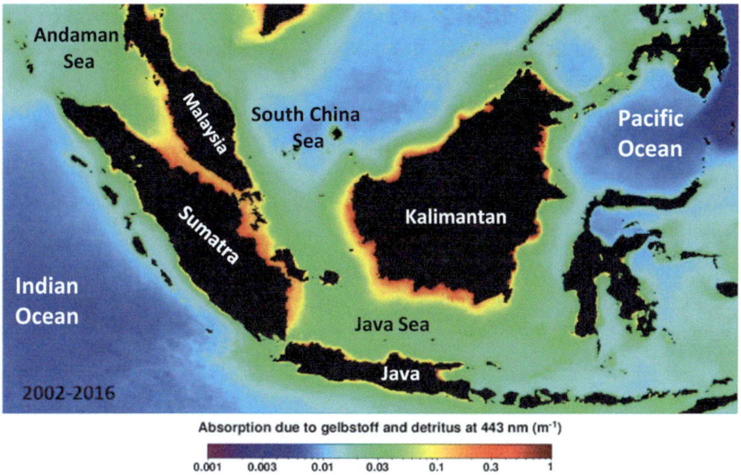

Fig. 1 MODIS-Aqua standard product: absorption of CDOM + detritus at 443 nm; mission mean (2002–2016), for the area of investigation

Fig. 2 The area of Investigation in SE-Sumatra including the main rivers Rokan, Siak, Kampar, Indragiri, and Musi (SE-Sumatra, 2004–2013)

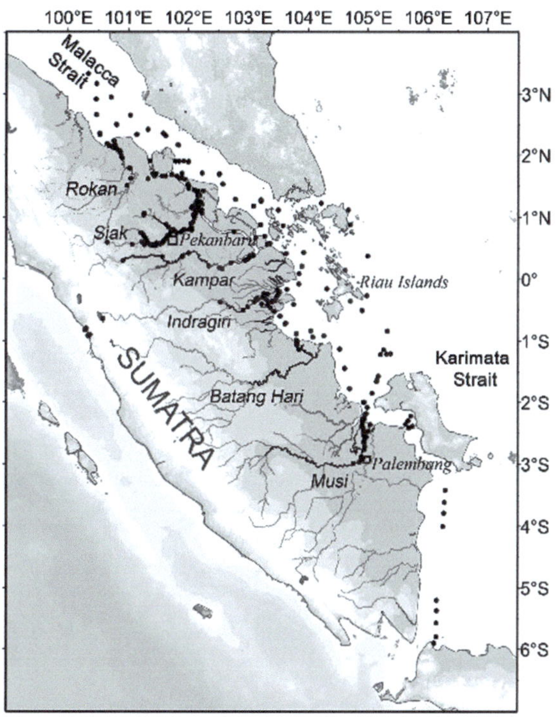

Bengal lowers the sea level in the Andaman Sea, which intensifies the transport as in summer (June–August) the SE monsoon. In the Karimata Strait, the current follows directly the monsoon.

Additionally, tidal currents strongly affect transport and mixing processes (Hatayama et al. 1996; Pang and Tkalich 2003; Siegel et al. 2009). In SE-Asian waters, co-oscillating tides of the Pacific and Indian Oceans modulated by the morphometry of the archipelago yield to a structured tidal pattern (Wyrtki 1961). The tides in the Malacca Strait originate from the northern Indian Ocean with a clear semi-diurnal mode and tidal ranges from 1.6 m to maximum values of 5 m (Thia-Eng et al. 2000). South of the amphidromic point at the southern entrance of the Malacca Strait, regions of mixed tides with prevailing diurnal tides and of full diurnal tides are attached.

2 Study Area, Methods and Methodology

Sources of optically active water constituents are the main river systems and the diffuse coastal discharge of SE-Sumatra. To assess these sources, bio-optical in situ measurements were implemented. SPM and Chlorophyll-a (Chla) concentration, CDOM absorption as well as spectral reflectance and transparency were determined in five rivers in different seasons of the period 2004–2013 (Table 1). These rivers extend over 800 km coastline between 2.5°N and 3°S latitude (Fig. 2). The map contains the main rivers and the adjacent seas, and the sampling stations. The sources of the rivers are located in the Barisan Mountains of West-Sumatra. The rivers are rather large with a length of 350–600 km and a width of up to a few kilometers in the river mouth area. The rivers and their tributaries drain very different landscapes, highlands and coastal plain with peatlands. The Siak River System is the main DOC source in the region, particularly the source river Tapung Kanan and the tributaries Mandau, Siak Kecil, and Bukit Batu (Siegel et al. 2009; Hendiarti et al. 2006). The northern rivers, Rokan, Siak and Kampar, drain into the Malacca Strait. Rokan and Kampar are particularly exposed to tides because of the orientation of their shallow river estuaries. In front of the coast from Kampar to

Table 1 Overview about the campaigns performed in different years and months

Year	Jan	Feb	Mar	Apr	May	Jun	Jul	Aug	Sep	Oct	Nov	Dec
2004			■						■			
2005							■					
2006			■									
2008			■								■	
2010						■						
2012									■			
2013				■								

Batang Hari the Riau Islands are located. Siak and Musi are important waterways connecting the province capitals, Pekanbaru and Palembang, with the open sea. The southern rivers from Indragiri to Musi drain into the Karimata Strait.

Ocean colour satellite data of different spatial and spectral resolutions (MODIS, MERIS, Landsat series, and SPOT) were implemented. Sensors of high spectral resolution are the Moderate-resolution Imaging Spectroradiometer (MODIS) and Medium Resolution Imaging Spectrometer (MERIS). NASA (National Aeronautics and Space Administration) operates MODIS on board the Terra (EOS AM, since 1999) and Aqua (EOS PM, since 2002) satellites. The sensors have a spatial resolution of 1 km and cover the earth every 1–3 days. They contain two channels with a spatial resolution of 250 m.

MERIS was operated from March 2002 until April 2012 by the European Space Agency (ESA) on board the Environmental Satellite (Envisat) covering the earth in 3 days. Full resolution (FR) and reduced resolution (RR) products had spatial resolutions of 300 and 1200 m, respectively. MODIS and MERIS level 2 products with 1 km resolution were used to investigate the distribution of sea surface temperature (SST), Chla, SPM, and CDOM absorption. True colour images were derived from MODIS and MERIS high-resolution data (250 and 300 m spatial resolution). MODIS level 3 products with a 4 km $\times$ 4 km spatial resolution were included for general applications.

Broadband data of higher spatial resolution of Landsat and SPOT series were applied for detailed case studies of river plumes and estuaries. The Landsat series operates since 1982 and the SPOT series since 1986. The visible channels had a spatial resolution of 30 and 20 m, and repetition rates of 16 and 23 days only.

Wind data of Topex/Poseidon (1992–2006), TMI (TRMM Microwave Imager, TRMM—Tropical Rainfall Measuring Mission, 1997–), AMSR-E (Advanced Microwave Scanning Radiometer—Earth Observing System) on MODIS Aqua or QuikSCAT (Quick Scatterometer, 1999–2009) were used for climate studies.

The in situ methods and the theoretical background are described in detail in Siegel et al. (2005, 2009). Two radiation sensors RAMSES (Radiation Measurements Sensor with Enhanced Spectral Resolution) of the German Company TriOS were applied to measure upward radiance $L_u(\lambda)$ just beneath and downward irradiance $E_s(\lambda)$ above the sea surface at 255 channels (320–950 nm) to calculate the total internal spectral reflectance $\Re(\lambda)$ (Eq. 1).

$$\Re(\lambda) = \frac{\pi L_u(\lambda)}{E_s(\lambda)} \tag{1}$$

$\Re(\lambda)$ is proportional to the water reflectance $R_w(\lambda)$, which is the ratio of the total backscattering $b_b(\lambda)$ and total absorption coefficients $a(\lambda)$ (Kirk 1994).

$$R_w(\lambda) = 0.33 \frac{b_b(\lambda)}{a(\lambda) + b_b(\lambda)} \tag{2}$$

Total spectral absorption and backscattering coefficients are the sum of the contributions of clear water and optically active water constituents. The backscattering coefficient $b_b(\lambda)$ consists of contributions by water molecules, chlorophyllous, and non-chlorophyll particles. The absorption coefficient $a(\lambda)$ contains contributions of water, phytoplankton, detritus, CDOM, and inorganic suspended matter (Morel and Prieur 1977). Scattering by particles increases the reflectance in the entire spectral range and makes the water turbid and bright. Changes in water colour are dominated by selective spectral absorption of pure water, Chla, and CDOM (Siegel et al. 2005). The absorption of water has a minimum in the blue spectral range and a maximum in the red and leads to the dark blue colour of clear ocean water. Chla absorption has two maxima near 443 and 670 nm, which can be used for the determination of Chla from satellite data. Spectral absorption of CDOM increases exponentially to shorter wavelength, ultraviolet to green light.

Chla, SPM, and CDOM were measured on water samples taken just beneath the sea surface. SPM was determined gravimetrically according to Doerffer (2002). Chla was measured according to Lorenzen and Jeffrey (1980) after ethanol extraction and CDOM-absorption on Whatman GFF filtered water (Højerslev 1980). The photometric measurements were performed on site using a spectral photometer PC-Spec (380–700 nm). Transparency was measured in terms of Secchi disc depth. Dissolved organic carbon (DOC) an important component in the carbon cycle was measured according to Baum et al. (2007).

3 Bio-optical Characterisation of Coastal Discharge

Chla, SPM, and CDOM distributions of SE-Sumatra measured in different seasons in the period 2004–2013 document highly variable concentration and composition of optically active water constituents (Fig. 3). These variations are also visible in

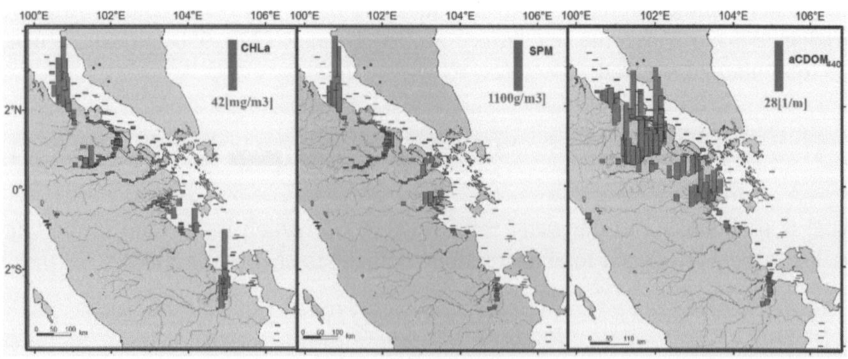

Fig. 3 Concentration of Chla and SPM, and CDOM absorption at 440 nm of the main rivers Rokan, Siak, Kampar, Indragiri, and Musi (SE-Sumatra, 2004–2013)

Table 2 Mean Chla, SPM, and $aCDOM_{440}$ and the variation range in the freshwater part of the main rivers (salinity = 0) and in the estuaries and adjacent seas (salinity > 0)

Location	Chla (mg/m^3)	max	min	SPM (g/m^3)	max	min	$aCDOM_{440}$ (m^{-1})	max	min
Rokan	8.1	19.3	1.4	274.8	776.3	16.6	14.8	18.4	8.1
Siak river	1.8	10.4	0.1	41.9	208.0	2.75	33.6	51.8	6.5
Kampar	1.8	3.0	1.2	38.2	49.4	26.1	10.9	15.8	6.3
Indragiri	1.9	3.6	0.5	78.2	114.4	32.9	13.1	16.8	5.9
Musi	1.0	2.7	0.2	54.1	101.0	22.1	3.8	5.1	1.8
Rokan estuary	35.7	77.6	11.1	668.6	2211.7	106.3	6.2	15.2	1.5
Siak estuary	2.4	4.8	0.3	49.8	82.5	18.6	22.8	25.1	1.1
Kampar estuary	2.2	3.9	0.9	114.0	348.4	29.9	10.5	15.0	3.3
Indragiri estuary	1.2	3.0	0.2	78.5	216.8	22.9	11.4	18.4	6.7
Musi estuary	6.1	13.9	0.2	59.3	84.2	21.9	2.5	3.7	1.4
Malacca Strait	1.3	1.5	1.0	5.9	11.7	2.2	0.4	0.5	0.2
Karimata Strait	0.6	0.9	0.3	2.3	6.1	0.9	0.2	0.34	0

Table 2, where mean values and variation ranges of the unique data set are summarized for the rivers (salinity = 0), estuaries and adjacent seas.

Chla, an important indicator for phytoplankton activity, varied in freshwater on average between 8 mg m^{-3} in the Rokan and 1–2 mg m^{-3} in the other rivers. The minimum of 0.1 mg m^{-3} was determined in the industrial area of the Siak with concomitant low values of oxygen, and the maximum of 19.3 mg m^{-3} in the Rokan. Chla maxima exist in the coastal plume of Rokan and Musi with 77.6 and 13.9 mg m^{-3} due to nutrient input.

SPM provides information on erosion or soil leaching with highest mean values in the Rokan (275 g m^{-3}), followed by Indragiri and Musi. The highest SPM concentration occurred in the Rokan with 776 g m^{-3} and in the Rokan estuary with 2200 g m^{-3} due to resuspension caused by strong tidal currents in the shallow north-westwards directed estuary. The high concentrations can be also observed by satellite data. Increased values occurred in the shallow estuaries of Kampar and Indragiri with 348 and 216 g m^{-3}.

The CDOM load, detected by CDOM absorption at 440 nm ($aCDOM_{440}$), delivers information about catchment areas. High mean CDOM absorption of 33.6 m^{-1} in the Siak followed by Rokan (14.8 m^{-1}), and Indragiri (13.1 m^{-1}) reflect the drainage of peat areas. Maximum value of 51.8 m^{-1} measured in Siak Kecil exceeded the known global maximum (Siegel et al. 2009). These rivers are globally a significant source of DOC and their source rivers and tributaries can truly be called “black-water Rivers”.

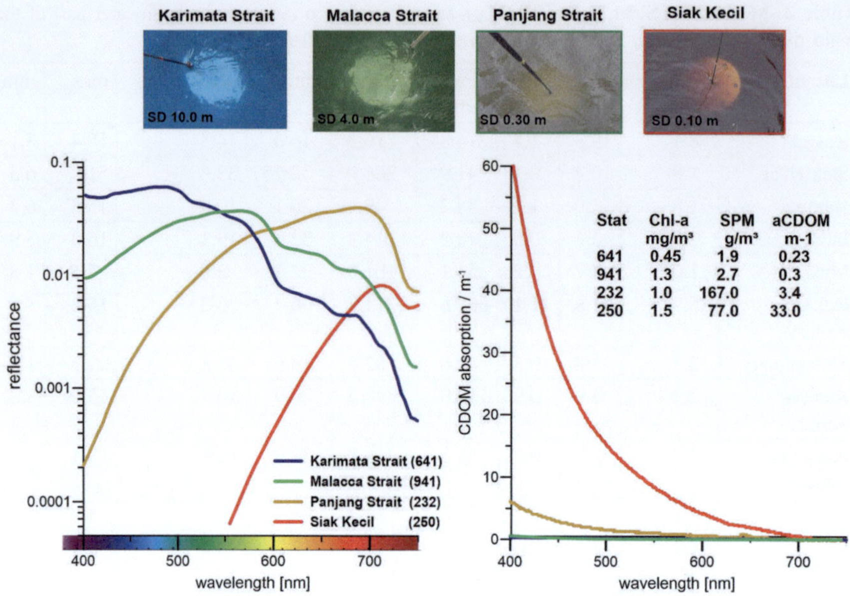

Fig. 4 Photographs and data from four different water masses (Karimata Strait, Malacca Strait, Panjang Strait, and Siak Kecil) illustrate the variations of ocean colour. Changes in water colour (left diagram) are dominated by strong CDOM absorption (right diagram)

Optically active water constituents and their inherent properties determine the water colour in terms of spectral reflectance (Siegel et al. 1994, 1997, 2005). Figure 4 presents spectral reflectances, CDOM absorption and water constituents for four different regions, Karimata Strait, Malacca Strait, Panjang Strait and Siak Kecil from the Siak River System. Photographs of Secchi discs and the transparency in terms of Secchi disc depth (SD) illustrate the variation range of water colour and transparency particularly caused by CDOM absorption and SPM.

In contrast to the rather blue water of the Karimata Strait with very low concentrations of water constituents and a reflectance maximum between 450 and 500 nm, the reflectance maximum is shifted to 530–550 nm in the Malacca Strait. In the Panjang Strait, high SPM concentrations and a high background of CDOM led to highest reflectances in the range of 550–680 nm and a strong slope in the short wavelength range reflecting turbid yellow-brownish water. The peat-draining river Siak Kecil was selected as a "black-water river" where very high CDOM absorption shifted the reflectance maximum to 700 nm. The transparency varied from 10 m in the Karimata Strait to 10 cm in the black water river.

For the area of investigation, we derived from reflectance and CDOM measurements a regional algorithm to retrieve CDOM absorption $aCDOM_{440}$ at 440 nm from MODIS data. Regression analyses were performed between $aCDOM_{440}$ and ratios of reflectances of different wavelength representing MODIS channels. Best results represents the equation ($aCDOM_{440} = -3.12 * R488/R555-1.88$, $R^2 = 0.83$). Based on DOC and CDOM absorption measurements performed at the campaign an empirical algorithm ($DOC = 1.1435\ ay + 3.399$, $N = 55$, $R^2 = 0.906$) was developed, that allows deriving DOC concentration from satellite data.

4 Coastal Transport Processes Caused by Tides and Monsoon

The transport processes in SE Sumatran waters are forced by monsoon and superimposed by tides. Detailed studies on tidal influence were performed in the Siak River System (Siegel et al. 2009), where the rivers Siak, Siak Kecil and Bukit Batu discharging into the Bengkalis Strait (Fig. 5). The Strait connects the rivers with the Malacca Strait. The distribution of river water strongly depends on tidal phases. During low tide, the Siak River water propagates northward, mixes with Siak Kecil and Bukit Batu waters and enter the Malacca Strait as shown in the quasi-true colour image of Landsat 8 OLI from 12 July 2016.

During increasing tide, the Siak discharge stops, estuarine water may enter the Siak, and the water of the last ebb phase is transported further southward into the

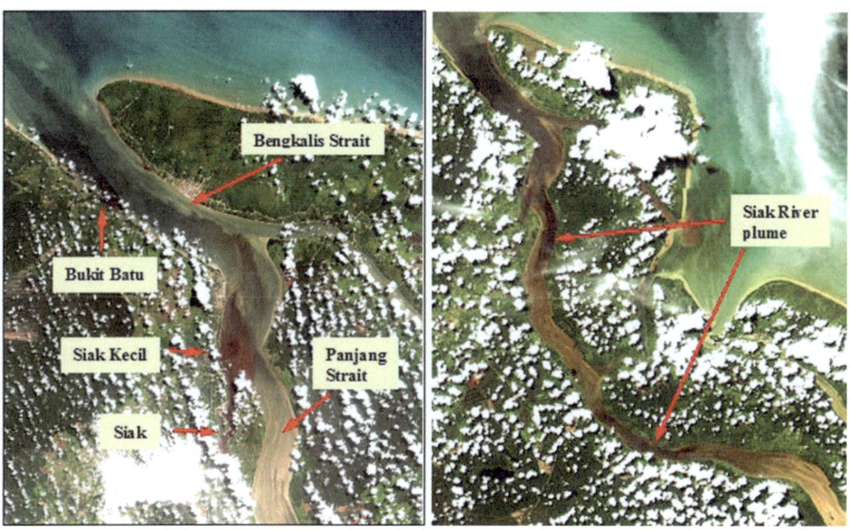

Fig. 5 Left panel: Landsat 8 OLI quasi-true colour image on 12 July 2016 of the Siak River, the estuary and the Malacca Strait, showing the wide range of water colour with the river plumes of Siak, Siak Kecil and Bukit Batu in the estuary. Right panel: During increasing tide on 29 May 2006 two plumes are identified in the Selat Panjang, the actual plume just outside the Siak River mouth and that of the low tide phase before (high tide 6.06 UTC, tidal range 3 m)

Panjang Strait, about 80–85 km until the next ebb phase. In the image from 29 May 2006, these two plumes are visible in the Panjang Strait, the actual plume just outside the Siak River mouth and that of the low tide phase before. Only satellite data deliver snapshots of these highly dynamic processes.

Detailed in situ measurements and remote sensing studies in the Siak River have shown the occurrence of a permanent estuarine turbidity maximum (ETM) zone in the saline front due to high turbulence during increasing and decreasing tide resulting in high SPM concentrations and higher reflectances in the satellite images (Siegel et al. 2009).

Monsoon describes seasonal changes in atmospheric circulation consisting of reversing wind and upper ocean current and changes in precipitation rate. A comparison between the regions based on satellite derived wind, precipitation, and SST is shown in Fig. 6 for the period 2004–2009. Shaded area show periods of NE monsoon from December to March. SE monsoon occurs from May to September and in the inter-monsoon periods in April and Oct/Nov, the wind conditions are highly variable. In Malacca Strait, the wind speed has no significant seasonal cycle, slight maxima occurred at the end of the inter-monsoon period and beginning of the NE monsoon in Nov/Dec. In the Karimata Strait, the seasonal development of wind speed is characterized by minima in Mar/Apr and Oct/Nov in the inter-monsoon periods and maxima in Dec/Jan and Jul/Aug in the monsoon phases. Monthly precipitation in South-East Sumatra and the Malacca Strait indicates dryer SE and wetter NE monsoonal season. Tropical climate with heavy and frequent rainfall occurs during all seasons. In the Malacca Strait, the precipitation maximum correlates with the transition to NE monsoon and increasing winds, the minimum appears during the SE monsoon. In the Karimata Strait, an additional precipitation minimum exists during the SE monsoon. The development of SST is very different in the two regions. In the Malacca Strait, the seasonal cycle consists of a minimum in Dec/Jan and a maximum in Mar/Apr. The increase from minimum (28 °C) to maximum (32 °C) occurs very quickly during the low wind period in Jan/Feb. After the maximum, SST decreases slowly during the SE monsoon and stronger than in the following inter-monsoon period of maximum wind speed. The maxima are higher than in the Karimata Strait where a clear correlation exists between decreasing winds and increasing SST. In the Karimata Strait, SST minima occur in Dec/Jan and Aug/Sep and the maxima in Apr/May and Oct/Nov. With increasing wind speed, wind mixing increases and SST decreases. The NE monsoon transports cold water from the South China Sea into the Karimata Strait that leads to lower SST (SST minimum).

For effects on the ecosystem, Chla and SST were compared for both regions. Chla maxima occur in Dec/Jan and minima in Mar to Aug in the Malacca Strait. The maxima correspond to the SST minimum and the SST maxima in Mar/Apr to the Chla minima. In the Karimata Strait, the annual cycles of SST consists of two periods and Chla follows in the opposite direction. High wind speed generates strong mixing with decreasing SST and transport of nutrients into the euphotic zone inducing phytoplankton development. Low wind means little mixing, strong stratification, high SST, nutrient depletion in the surface layer and reduced Chla.

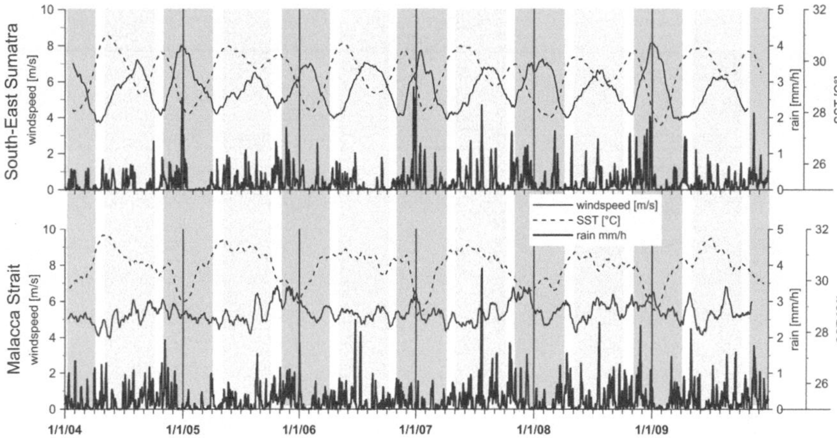

Fig. 6 Wind speed, SST and precipitation for the Malacca and Karimata Straits (South-East Sumatra) for the period 2004–2009. Shaded area show the period of NE monsoon from Dec to Mar. SE monsoon occurs from May to Sep and in the inter-monsoon periods Apr and Oct/Nov the wind is highly variable

MODIS true colour images deliver distribution patterns of water masses and information about main transport direction in relation to monsoon phases (Fig. 7). The water of the northern rivers (Rokan, Siak, and Kampar), discharged into the Malacca Strait, follows the general north-westward transport during all monsoon phases (Wyrtki 1961; Saadon and Namba 2001). The baroclinic pressure gradient between both entrances of the Malacca Strait forces the transport supported by monsoon. Clearer water from Karimata Strait enters the Malacca Strait and improves the water quality. The spreading of the southern rivers (Indragiri, Musi) in the Karimata Strait is strongly related to monsoon phases.

On 16 January, the Indragiri water was spreading southward during the northerly wind. The image from 27 April 2008 shows the entire southern region for south-easterly winds with a northward directed transport of Indragiri water.

Figure 8 presents examples from the Musi river region in the south. The MODIS image from 26 March 2008 shows the distribution of Musi water and adjacent small brownish peat draining rivers during northerly winds. The water of the small rivers north of the Musi River mouth is transported south- and eastward in the direction of Bangka Strait. During southerly winds on 26 June 2007, the coastal discharge including the Musi water is spreading northward.

MERIS derived products from 4 Nov 2008 show the distribution of Chla, SPM, CDOM absorption and DOC (Fig. 9). CDOM and DOC were generated using the derived empirical algorithms (third paragraph). The scene represents a situation during north-easterly winds. Strong gradients, particularly in CDOM with very high values in the river mouth area and in a coastal strip along the northern rivers and low values in the coastal area near the Musi river mouth. In DOC, this distribution

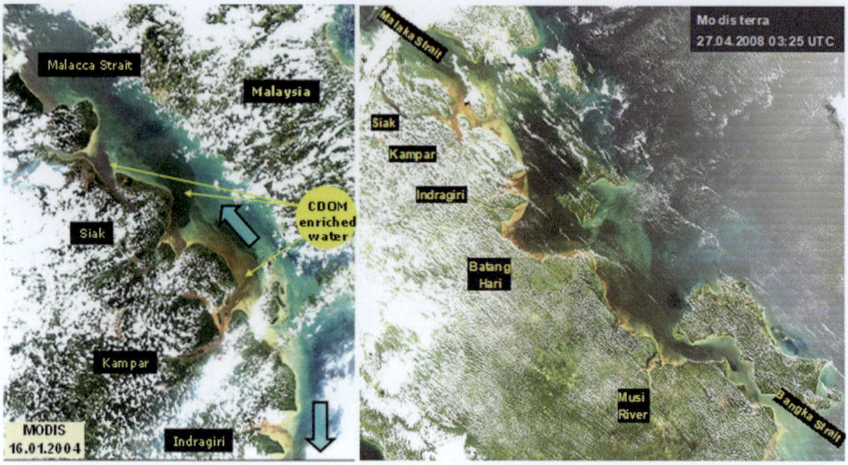

Fig. 7 MODIS quasi-true colour images on 16 January 2004 showing north-western transport in the Malacca Strait and southward transport of Indragiri water during northerly winds and on 27 April 2008 northward-directed transport during south-easterly winds

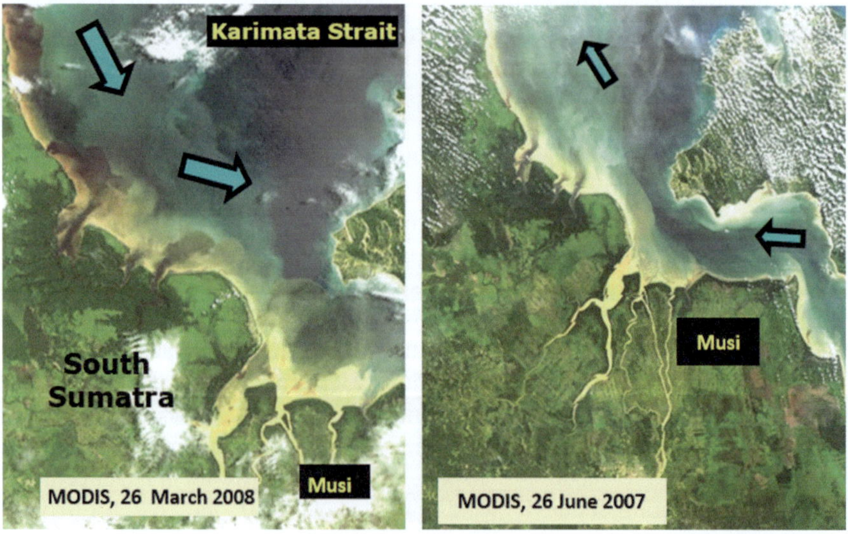

Fig. 8 MODIS quasi-true colour images from 26 March 2008 and from 26 June 2007 showing distribution patterns of Musi River water and adjacent small rivers during northerly and southerly winds. During northerly wind, the water is transported south- and eastwards and during southerly winds, the river water is spreading northward

is even stronger pronounced. Distribution of SPM is different to the distribution of Chla, $aCDOM_{440}$, and DOC.

MERIS products of 31 July 2008 and 22 January 2009 in Fig. 10 show the general distribution patterns during the monsoon phases. During both phases CDOM, Chla and TSM were high in the coastal strip of Sumatra with a larger lateral spread during the SE monsoon. Year-round, cells of clear water occur in the basin between Kampar, Indragiri and offshore islands. These low concentrations are also reflected in measured concentrations in Fig. 3.

In the Malacca Strait, the concentration gradients are more pronounced during NE than during SE monsoon. The concentrations are lower on the Malaysian side than on the Indonesian, because the NE monsoon transports the surface water towards the Indonesian coast and creates upwelling on the northern side. During the SE monsoon, the nutrient-rich coastal runoff of Sumatra is more distributed in the Malacca Strait and higher concentrations are observed, especially in Chla.

In Fig. 11, seasonal means of MODIS standard products, absorption of Gelbstoff (CDOM) + detritus, Chla, and SST, are shown for SE monsoon in boreal summer

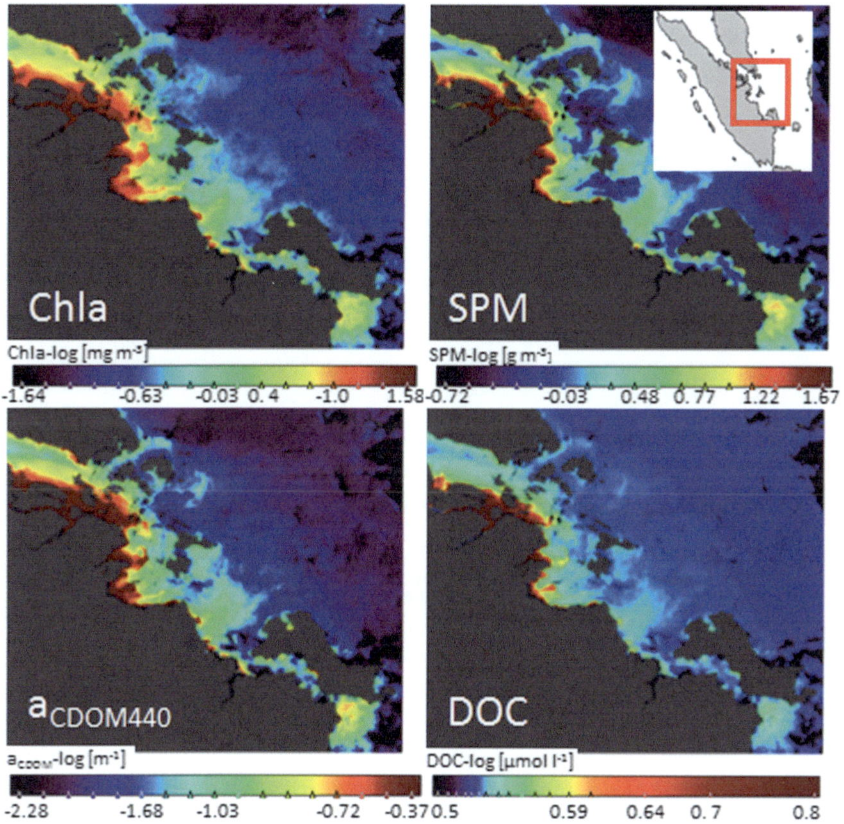

Fig. 9 MERIS derived Chla, SPM, $aCDOM_{440}$, and DOC on 4 Nov 2008 during NE monsoon

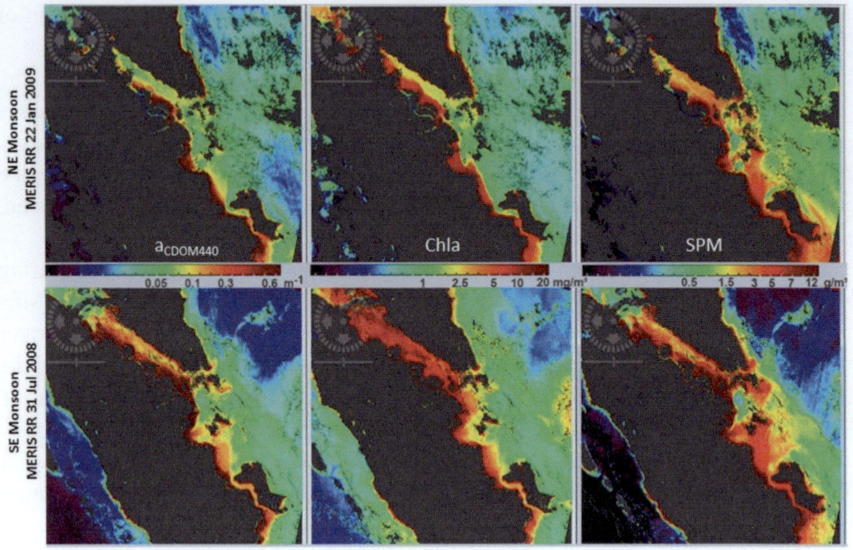

Fig. 10 MERIS derived CDOM absorption, Chla, and SPM on 22 January 2009 and 31 July 2008 showing the strong differences between NE and SE monsoon

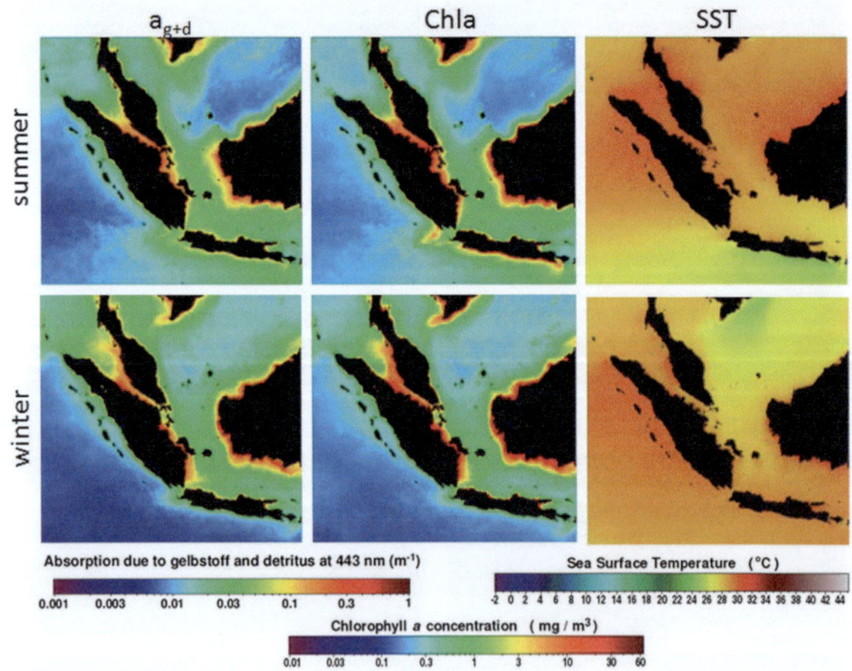

Fig. 11 Seasonal means of MODIS standard products for the period 2002–2015/16: ag + d—absorption of Gelbstoff (CDOM) + detritus, Chla, and SST during SE monsoon in boreal summer and NE monsoon in boreal winter

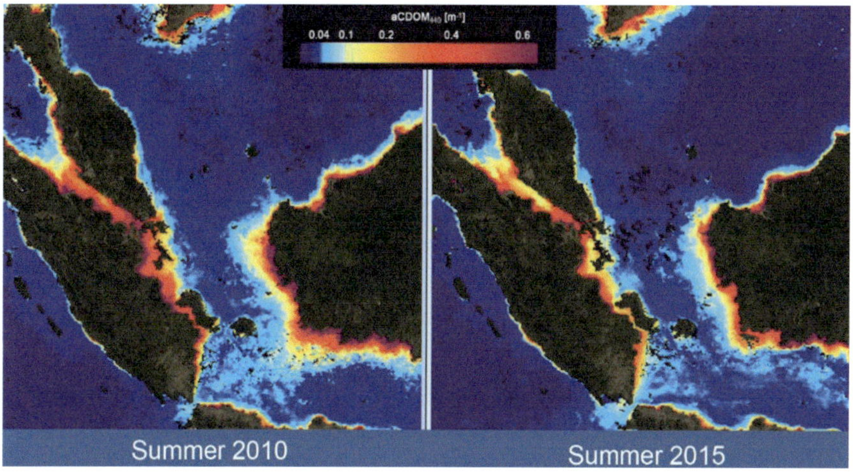

Fig. 12 CDOM absorption derived from MODIS data for the La Niña year 2010 and the El Niño year 2015

and NE monsoon in boreal winter. Monthly precipitation in Karimata and Malacca Straits indicate a dryer SE and a wetter NE monsoon season. Higher precipitation enhances coastal discharge and transport of water constituents into coastal seas.

During NE monsoon, currents prevent offshore propagation of coastal discharge. During SE monsoon, weaker winds support offshore propagation of water despite the lower precipitation rates. During SE monsoon, economic sources like coral reefs or touristic areas at Riau Islands between Malacca und Karimata Straits are more influenced by coastal discharge. During NE Monsoon, water with high concentrations of CDOM and Chla propagates along the coasts of SW-Malaysia in the Malacca Strait and SE Sumatra. During NE monsoon the higher precipitation intensifies coastal discharge, but the north-easterly currents brings clear water from South-China Sea preventing this propagation. Between the southern entrance of Malacca Strait and the Riau Islands clear South China Sea water fill the gap during the north-westward transport in Malacca Strait and southward transport in the Karimata Strait.

The archipelago of Indonesia plays a central role as an atmospheric source of heat in the coupled ocean-atmosphere circulation system of the tropical Pacific, El Niño-Southern Oscillation (ENSO). ENSO has three phases, the normal state, El Niño and La Niña. El Niño (2015) reduced the rainfall and La Niña (2010) brings stronger Pacific trade winds, higher SST and precipitation. Higher precipitation strengthened coastal discharge and transport of CDOM into the sea. The NE monsoon prevents the offshore spreading of coastal discharge. Figure 12 shows MODIS derived CDOM absorption for the boreal summer (SE monsoon) of La Niña year 2010 and El Niño year 2015. SE-winds are weaker, recognizable by higher CDOM values in the coastal area, especially during the La Niña year 2010, and distribute the coastal outflow of Sumatra more in the Malacca Strait. La Niña

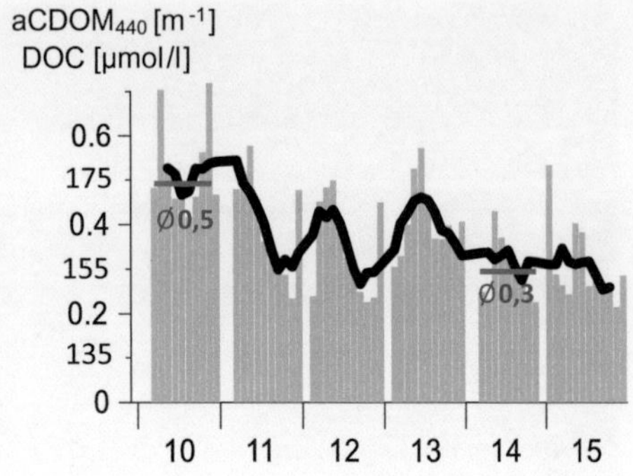

Fig. 13 Monthly mean CDOM absorption and DOC content derived from MODIS for Malacca Strait showing the highest values in the La Niña year 2010 with a negative trend until 2015

causes frequent rainfall. In winter, the northeast monsoon presses the water masses towards the coast, a gradient in Malacca Strait is more pronounced than during the SE monsoon. On the Malaysian side, the concentration is lower than on the Indonesian side, because the surface water is transported towards the Indonesian coast and upwelling of clearer water develops.

The development of the CDOM absorption and the DOC content in Fig. 13 shows for the period 2010–2015 in the Malacca Strait a decrease. The highest monthly values were recorded in the La Niña year of 2010.

5 Conclusions

The application of synoptic satellite data supports the investigation of coastal discharge in tropical regions of Indonesia, where its distribution is strongly influenced by monsoon overlaid by tides. River run-off from SE Sumatra is among the major carbon and sediments sources for the world ocean. The northern rivers, Rokan, Siak, and Kampar, are discharging into the Malacca Strait, and Indragiri, Batang Hari and Musi in the western Karimata Strait. Different catchment areas like peatland and soil leaching area determine the composition of optically active water constituents in the rivers.

A unique data set of optical in situ measurements provides an extremely high variation range of all optically active water constituents. CDOM absorption coefficient at 440 nm in the Siak Kecil (51.8 m^{-1}) exceeded the published maximum. Source Rivers and tributaries of the Siak River system are true "black Water Rivers". SPM concentration was highest in Rokan, with 275 g m^{-3} in the freshwater part, 776 g m^{-3} in the estuarine turbidity maximum, and up to 2200 g m^{-3} in the estuary caused by strong tidal resuspension. The north-west directed Rokan estuary allows a direct inflow of tidal waves from Indian Ocean, which supports the resuspension. Strong variations of water constituents produce changes in ocean colour from clear black to turbid, milky waters. In the high dynamic system, the high spatial and temporal variations can be recorded only by satellite remote sensing providing synoptic information from large areas. High content of water constituents strongly influence the coastal ecosystems, i.e. reducing photosynthetically available radiation in the water column. In extreme cases, the Secchi disc depths were lower than 10 cm.

Satellite data are suitable to record large areas at different temporal scales, from short-term distribution patterns of river plumes during different tidal phases, over the transport in coastal and adjacent seas during different monsoon phases, to climatological aspects concerning economically important areas. Data obtained from satellite remote sensing and its interpretation can provide important information for local authorities.

Acknowledgements The authors thank scientists and students of the University of Riau in Pekanbaru and crews of the vessels for the support during the cruises. MODIS data were received from NASA Rapid Response System and MERIS data from ESA. The study was funded by the Federal German Ministry for Education, Science, Research and Technology (Grand No. 03F0473E-IOW). The authors thank the editors and reviewers for the helpful comments and constructive hints.

References

Baum A, Rixen T, Samiaji J (2007) Relevance of peat draining rivers in central Sumatra for the riverine input of dissolved organic carbon into the ocean. Estuar Coast Shelf Sci 73(3–4): 563–570

Doerffer R (2002) Protocols for the validation of MERIS water products, no. PO-TN-MEL-GS-0043. ESA

Hatayama T, Awaji T, Akitomo K (1996) Tidal currents in the Indonesian seas and their effect on transport and mixing. J Geophys Res C Oceans 101(C5):12353–12373

Hendiarti N, Siegel H, Ohde T (2004) Investigation of different coastal processes in Indonesian waters using SeaWiFS data. Deep Sea Res Part 2 51:85–97

Hendiarti N, Siegel H, Frederik MCG, Reissmann J, Andiastuti R (2006) Remote sensing investigation of coastal discharge of Siak estuary and nearby Jakarta Bay. In: Proceedings of the 16th APEC workshop on ocean models and information system for the APEC region. OMISAR Project Publication, Ho Chi Minh City, 2005, p 5

Højerslev NK (1980) On the origin of yellow substances in marine environments, vol 42. Report Institute Physical Oceanography of University of Copenhagen, pp 39–56

Kirk JTO (1994) Light and photosynthesis in aquatic ecosystems, 2nd edn. Cambridge University Press, New York, p 509

Lorenzen CJ, Jeffrey SW (1980) Determination of chlorophyll in seawater. UNESCO Techn Pap Mar Sci 35:20

Milliman JD, Farnsworth KL, Albertin CS (1999) Flux and fate of fluvial sediments leaving large islands in the East Indies. J Sea Res 41:97–107

Morel A, Prieur L (1977) Analysis of variations in ocean colour. Limnol Oceanogr 22:709–722

Namba T, Saadon MN (2001) Water characteristics of the Malacca Strait and the adjacent sea. In: International conference on international oceanographic data and information exchange in the Western Pacific (IODE-WESTPAC) 1999, Langkawi, Malaysia

Pang WC, Tkalich P (2003) Modeling tidal and monsoon driven currents in the Singapore Strait. Singap Marit Port J, 151–162

Rixen T, Baum A, Pohlmann T, Balzer W, Samiaji J, Jose C (2008) The Siak, a tropical black water river in central Sumatra on the verge of anoxia. Biogeochemistry 90:129–140

Siegel H, Gerth M, Beckert M (1994) The variation of optical properties in the Baltic Sea and algorithms for the application of remote sensing data. SPIE, vol 2258. Ocean optics XII, pp 894–905

Siegel H, Gerth M, Beckert M (1997) Variation of specific optical properties and their influence on measured and modelled spectral reflectances in the Baltic Sea. In: Ackleson SG, Frouin R (eds) Ocean optics XIII. Proceedings of the SPIE, vol 2963, pp 526–531

Siegel H, Gerth M, Ohde T, Heene T (2005) Ocean colour remote sensing relevant water constituents and optical properties of the Baltic Sea. Int J Rem Sens 26:315–330

Siegel H, Stottmeister I, Reißmann J, Gerth M, Jose C, Samiaji J (2009) Siak river system—East-Sumatra, characterisation of sources, estuarine processes, and discharge into the Malacca Strait. J Mar Syst 77:148–159

Tan CK, Ishizaka J, Matsumura S, Yusoff FM, Mohamed MH (2006) Seasonal variability of SeaWiFS chlorophyll a in the Malacca Straits in relation to Asian monsoon. Cont Shelf Res 26(2):168–178

Thia-Eng C, Gorre IRL, Ross SA, Bernad SR, Gervacio B, Ebarvia MC (2000) The Malacca Straits. Mar Pollut Bull 41(1–6):160–178

Wit F, Muller D, Baum A, Warneke T, Pranowo WS, Muller M, Rixen T (2015) The impact of disturbed peatlands on river outgassing in Southeast Asia. Nat Commun 6:10155

Wyrtki K (1961) Physical oceanography of the Southeast Asian waters. Naga report 2. Scripps Institution of Oceanography, San Diego

Using WorldView-2 Imagery to Estimate Mangroves Density in the Porong Estuary

Agus Setiawan, Bernadinus Realino, Iis Triyulianti, Faisal Hamzah, Ari Murdimanto, Mutiara Rachmat Putri and Dwiyoga Nugroho

Abstract The Porong Estuary is an estuarine ecosystem influenced by the Porong River's interaction with the Java Sea. This estuary plays a significant ecological role, acting as source that provides nutrients and organic materials that are then transported through the river and mixed in with the tidal currents. Some aquatic biota, for example crabs (*Scylia serrata*), oysters (*Crassostrea cucullata*), shrimps, and fishes utilize this estuary as a shelter, a nursery ground, and a source of food. Furthermore, because it is surrounded by numerous fish and shrimp ponds, this estuary is also relied upon by thousands of people. In response to the Sidoarjo mudflow disaster that occurred on May 29, 2006, the Porong River has been used as a channel diverting the mudflow to the Java Sea since 2007. This channelling activity will potentially increase the sedimentation rate and decrease the water quality of the Porong Estuary. Therefore, to investigate the potential impact of mudflow channelling activity and provide time series data, the Institute for Marine Research and Observation (IMRO) has been monitoring the ecological condition of the Porong Estuary regularly since 2009; this includes estimating the mangroves density. The estimation of mangroves density was carried out based on remote sensing data. In 2003, 2007, and 2009, the data used was Landsat ETM+. Meanwhile, in 2010, the estimation was done based on WorldView-2 imagery. By comparing the results gathered between 2003 to 2007, before being used as a channel to divert the mudflow, the surface area of mangrove forests in the Porong Estuary decreased significantly, i.e., 666.709–435.365 ha. This generally occurred to mangroves with high and moderate density. On the other hand, from 2007 to

A. Setiawan (✉) · B. Realino · D. Nugroho
Marine Research Centre, Agency for Marine and Fisheries Research and Human Resources, Ministry of Marine Affairs and Fisheries, Jakarta, Indonesia
e-mail: setiawan.agus@kkp.go.id

I. Triyulianti · F. Hamzah · A. Murdimanto
Institute for Marine Research and Observation, Agency for Marine and Fisheries Research and Human Resources, Ministry of Marine Affairs and Fisheries, Bali, Indonesia

M. R. Putri
Study Program of Oceanography, Faculty of Earth Science and Technology, Institute Technology of Bandung, Bandung, Indonesia

V. Barale and M. Gade (eds.), *Remote Sensing of the Asian Seas*,
https://doi.org/10.1007/978-3-319-94067-0_21

2009 the mangrove area started to increase again, going from 435.4 to 550.1 ha. By 2010, its area continued to increase reaching 654.2 ha. In general, from 2007 to 2010, mangroves with moderate and high density increased significantly from 33.964 to 256.81 ha and from 26.37 to 344.27 ha, respectively. These results show that channelling activity of the Sidoarjo mudflow to the Porong River has provided more substrate to the Porong Estuary area. Moreover, the coastal dynamics around the Porong Estuary give mangroves an opportunity to grow, especially in areas where sediment transport is dominant.

Keywords Porong Estuary · Sidoarjo mudflow disaster · Mangrove density WorldView-2 imagery

1 Introduction

An estuary is "a semi-enclosed coastal body of water that extends to the effective limit of tidal influence, within which sea water entering from one or more free connections with the open sea, or any other saline coastal body of water, is significantly diluted with fresh water derived from land drainage, and can sustain euryhaline biological species from either part or the whole of their life cycle" (Perillo 1995). Similar to a coastal wetland, an estuary is a "critical transition zone (CTZ) that links land, freshwater habitats, and the sea that provide essential ecological functions, including decomposition, nutrient cycling, and nutrient production, as well as regulation of fluxes of nutrients, water, particles, and organisms to and from land, rivers, and the ocean" (Levin et al. 2001).

Mangrove forests are the major ecosystems usually found in the CTZs, particularly in tropical regions. Mangroves play an important role in producing dead plant materials, which are essential for decomposition and nutrient recycling (Levin et al. 2001). They export the nutrients, organic materials, and organisms to the environment (Alongi 2009). Mangroves also have a specific vegetation composition. Their shapes are comprised by various species of plants that have physiological adaptability to their environment, such as salinity, soils types, and tides (Walter 1971).

As intertidal plant communities spread across tropical accreting shorelines, established when sediments accumulate at elevations close to the mean sea level (Sidik et al. 2016), mangroves can trap and stabilize sediments with their above- and below-ground roots system (Furukawa and Wolanski 1996). However, in some cases, the increasing rates of sediment loading tend to trigger a widespread and detrimental effect to the estuarine and coastal ecosystems; this includes the mangrove communities, particularly because of their sensitivity to environmental changes (Thrush et al. 2004). The various studies regarding the soils which are suitable for the growth of mangroves indicate that the soil's characteristics are a major limiting factor for the mangroves growth rate and distribution and organisms (Gledhill 1963; Aksornkoae 1975; Giglioli and King 1966; Clarke and Hannon 1967; Aksornkoae et al. 1985; Sanders 1958; Frith et al. 1976).

After an unexpected eruption of mud and gases from hydrocarbon exploration well near Sidoarjo, East Java on May 29, 2006, Lapindo Brantas Inc. started channelling the mudflow into the Porong River in 2007 to reduce the ongoing damage to the local communities and infrastructure in the area. In 2009, the Recovery Agency for Sidoarjo Mudflow (BPLS) started to take over the mud channelling activity and continued to pump the slurry (a mixture of mud and water) from the mud ponds, disposing of it in the Porong River. This chosen action helped decrease the volume of mudflow accumulation in the mud ponds with limited capacity. At the same time, the eruption rate remained high until 2010, approximately 1000–76,000 m^3 day^{-1} (Kure et al. 2014).

Figure 1 shows the situation map of Sidoarjo mudflow ponds and the Porong River and its estuarine ecosystem. The Porong Estuary is part of the Java Sea, which is influenced mainly by fluvial and tidal processes, and located about 20 km eastward from the mud ponds. The amount of sediment load from the Porong River to the Porong Estuary is quite high, especially during the wet season. This results in the formation of a strong prograding delta (Jennerjahn et al. 2004).

By channelling the Sidoarjo mudflow into the Porong River, the transport of sediment to the Porong Estuary increases significantly. Moreover, the characteristics of sediment from the mudflow are different than those of the sandy bed load that previously dominated the transport (Hoekstra and Tiktanata 1988); for example, predominantly clay (81.5%) with porosity of approximately 30% and dry bulk density ranging from 1.24 to 1.37 g cm^{-3} (Karyadi et al. 2012). Based on data from

Fig. 1 Map of study area, the situation of Sidoarjo mudflow ponds, and the position of pipeline for channelling the slurry to the Porong River (Google Earth). Insert figure shows the study area (indicated by black rectangle) and its surrounding area

BPLS in 2014, slurry that has already been disposed off through the Porong River in 2010 is approximately 31 million m^3. To routinely monitor the impact of the mudflow channelling the Porong Estuary, the Institute for Marine Research and Observation (IMRO), supported by the BPLS, has conducted comprehensive studies since 2007. The research presented in this paper comes, in part, from those studies. Ultimately, these studies aimed to assess the changes of the mangrove forest's ecosystem near the Porong Estuary.

2 Environmental Conditions of the Porong Estuary

The water depth at the Porong Estuary is relatively shallow, with a gentle slope around the river mouth to the north and south. Up to 1 km eastward from the coastline, water depth is still below 5 m and starts to increase significantly afterwards. Comparing the bathymetric maps of 2008 and 2010, one can observe that some bathymetric changes occurred. In general, bathymetry along the eastern portion of the Porong River was shallower due to sedimentation while, closer to the mouth of the river, a small mud island had been created with materials taken from dredged sediment around the Porong Estuary (see Fig. 2).

The Porong Estuary and the adjacent waters have mixed predominantly semi-diurnal tides with a range of 1.1 m at neap tides and 2.6 m at spring tides. In the wet season, due to a high discharge from the Porong River, fresh water is

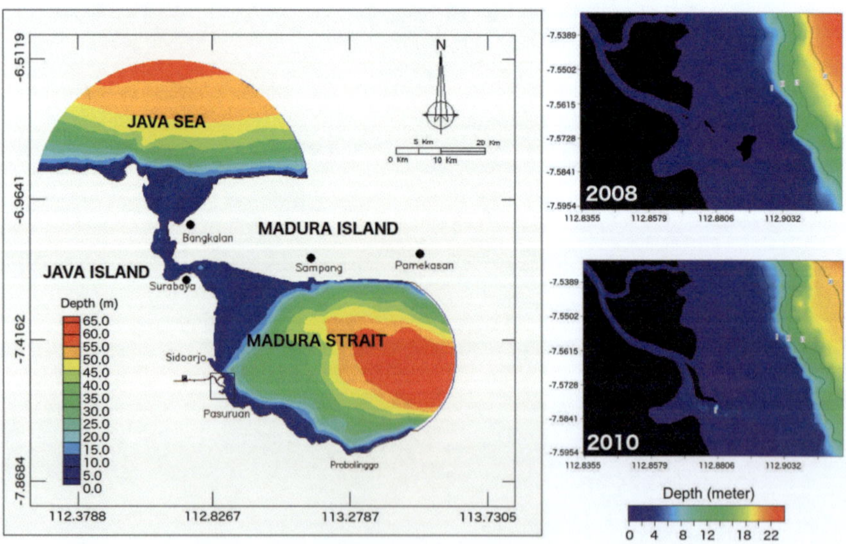

Fig. 2 Bathymetry of Porong Estuary and the adjacent waters (left panel) and a close-up of bathymetry in the Porong Estuary (indicated by black box in figure at left panel) measured in 2008 (top right panel) and 2010 (bottom right panel)

observed as far as 4 km offshore. On the other hand, when the discharge from the Porong River is almost zero during the dry season, the saline water penetrates up to 12 km upstream (MMAF 2010) and introduces mangrove detritus on the riverside. Discharge from the Porong River is highest during the wet season (approximately 450 m^3 s^{-1}) and declines significantly during the dry season (June to October) to approximately 50 m^3 s^{-1} (Jennerjahn et al. 2013). Rates of sedimentation in the Porong Estuary are approximately 20 and 5 cm $year^{-1}$ during wet and dry seasons, respectively (Sidik et al. 2016).

Porong River's biogeochemistry in the area where the mudflow channels is characterized by a high concentration of total suspended matter (approximately 2298.9 mg l^{-1}), particulate organic carbon (1711 μM), ammonium (317 μM), and inorganic nitrogen (145 μM); however, these figures decrease significantly in the estuary. In contrast, dissolved oxygen (DO) is low in the mudflow channeling area (1 mg l^{-1}) and higher in the estuary (8.6 mg l^{-1}) (MMAF 2010). Annual supply of particulate organic carbon (POC) as fertilizer and source of nutrition for mangroves growth increases significantly from upstream to downstream, from 15,300 to 36,500 t $year^{-1}$ (Jennerjahn et al. 2013). Mangrove forests in the Porong Estuary were characterized as having low diversity, dense mangrove trees (dominated by *Avicennia marina*), and, lastly, of being comprised of small trees growing at high densities (Sidik and Lovelock 2012).

3 Image Processing Methods

To achieve the above objective, the use of satellite images has appeared to be a promising solution for mapping the Porong Estuary and the surrounding area, particularly the mangroves vegetation area. The various Very High Spatial Resolution (VHSR) satellite images, such as Quickbird (Laba et al. 2008; Belluco et al. 2006; Harris and Bryant 2009) and WorldView-2 (Rapinel et al. 2014) have been used to map vegetation. In this study, the WorldView-2 satellite has been chosen since its sensor can provide VHSR images in 8 spectral bands, ranging from blue to near infrared, including additional coastal-blue, yellow, and red-edge bands. It has also has proven to be an effective tool for identifying and mapping natural vegetation on coastal sites (Rapinel et al. 2014).

WorldView-2 Satellite was launched on October 8, 2009. It operated on the orbit altitude of 770 km, with an inclination of 97.2° for a maximum orbital period of 100 min. It also has a spatial resolution of 0.46 m for panchromatic channel and 1.8 m for multispectral channel. With a temporal resolution of 1.1 day, WorldView-2 is capable of recording up to 975,000 km^2 per day with higher performance on monitoring and mapping in comparison to the previous series. With its 4 standard red, green, blue, and infrared channels and 4 additional coastal, yellow, red edge, and near-infrared 2 channels, images from WorldView-2 fit our purposes in identifying and estimating the density of mangrove communities near the Porong Estuary.

Fig. 3 The natural colour of WorldView-2 imagery around the Porong Estuary and positions of GCPs, GC, and transect (indicated by ▲, ⬟, and ●, respectively) used for geometric corrections, supervised classification, and mangroves density analysis

The WorldView-2 satellite imagery used in this study consists of panchromatic and multispectral images from three different acquisition times in 2010, specifically February 13, June 22, and September 4. Those images were then combined and pre-processed to generate a map with natural colour and a spatial resolution of 0.5 m, as shown in Fig. 3. Afterwards, this map was used to define the field surveys.

The field surveys have been conducted to collect Ground Control points (GCPs) and to carry out a field check, or Ground Check (GC), of objects or appearances found on the satellite images. The GCPs survey aims to determine control points on earth that will be used as reference points to perform geometric corrections on WorldView-2 satellite imagery. The control points used are usually objects on earth that are easily recognized, permanent, and show up clearly in the images. Some examples include bridges, crossroads, or monuments. Meanwhile, GC surveys are necessary to obtain an accurate image interpretation. This GC surveys would provide an input for supervised classification in interpreting the land cover. The Global Positioning System (GPS) has been used to determine the coordinate positions of those GCPs and GCs.

In order to obtain complete information on GCPs and GC, a standard survey form to be used by the surveyors has been provided. The information for this form should be collected at the time the observations are made; such information includes type of survey, survey time, name of surveyor, geographical position, name of the location, type of land use, and aerial photography taken to the north, west, south, and east directions of the objects.

GCPs for geometric corrections have been collected from 15 points, while GC for training areas used in supervised classification have been collected from 46 points. Those point locations are defined before conducting field survey activities. Aside from the two field surveys, a systematic sampling of transects activity used to identify mangrove at 11 locations has been carried out (see Fig. 3). This transects activity (see Fig. 4) aimed to collect some information related to the mangroves characteristics, for instance, type, number, and size of mangrove trees that will be used afterwards to estimate the importance value index (IVI) of any mangrove species. IVI is an index to express the dominance and ecological success of any species, which has value ranging from 1 to 300 (Bengen 2001). This value is obtained by summing up the mangroves species relative density (*RDi*), relative frequency (*RFi*), and relative

Fig. 4 Field survey for mangrove identification, which collects information on species, number, and size of mangrove trees at 4 different locations: Station 1 (left top panel), station 2 (right top panel), station 5 (left bottom panel), and station 9 (right bottom panel)

closure (*RCi*) of mangrove species that calculated from results of transect survey. Usually, IVI is divided into 3 classes, i.e. low (1 < IVI < 100), moderate (100 < IVI < 204), and high (204 < IVI < 300) roles, respectively.

The method used for mangroves transect is point centre quarter or quadrant method (PCQM) (Satyanarayana et al. 2001). This method didn't use any sample plot and was very efficient in estimating mangrove community with tree and pole shapes in a dense forest. With this method, transect locations from the outermost to the innermost zones are arranged to be perpendicular to the coastline. The density values obtained are expected to reflect the distribution of mangrove class and the representation of each zoning in the study area. This method is considered appropriate when generating accurate density values (Dahdouh-Guebas and Koedam 2006), and allows for quick and efficient measuring in a survey of mangrove ecosystems.

Figure 5 shows the image-processing steps applied in this study. Since the images used in this study consist of three different acquisition times, a mosaicking process is performed to overlay all images. The images used for this purpose are already corrected geometrically, making use of GCPs obtained from the field

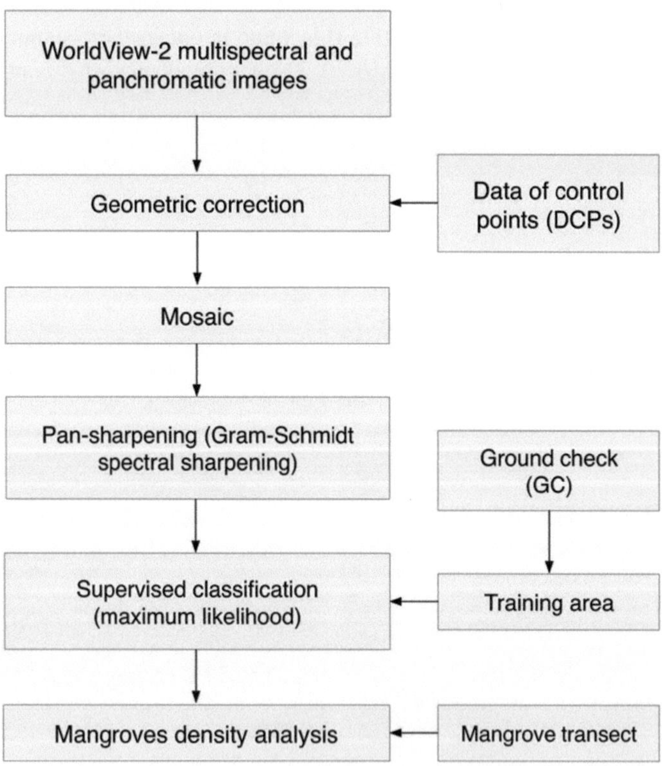

Fig. 5 The image-processing steps and field data used to obtain a map of mangroves' density around Porong Estuary

surveys. Afterwards, to improve the spatial resolution of multispectral bands, a pan-sharpening process of Gram-Schmidt (Laben and Brower 2000) is performed. In the ENVI image analysis software, which is used in this study, this process is divided into 4 steps. In the first step, a panchromatic channel is simulated from the lower spatial resolution spectral bands. In the second, a Gram-Schmidt transformation is performed on the simulated panchromatic and spectral bands, where the simulated panchromatic band is used as the first band. In the third step, the high spatial resolution panchromatic band was swapped with the first Gram-Schmidt band. In the last and final step, an inverse Gram-Schmidt transformation is applied to form the pan-sharpened spectral bands.

To process a supervised classification, training areas with 7 classes of land use, namely settlement, rice field, cultivated area, shrimp/fish ponds, mangroves, bush, and barren land, need to be set. These 7 classes of land use were the most dominant land use found in the study area, where the spatial information about them has already been figured out in the map of spatial planning. GC from 46 locations, all chosen because of their specificity and ability to represent each class effectively, were used in this process. The maximum likelihood algorithm (Richards 1999) was used in this process. This classification assumes that the statistics for each class in each band is normally distributed. To implement this classification, the discriminant functions for each pixel in the image are calculated using (Richards 1999):

$$g_i(x) = \ln p(\omega_i) - \frac{1}{2}\ln\left|\sum_i\right| - \frac{1}{2}(x - m_i)^T \sum_i^{-1}(x - m_i) \tag{1}$$

where i indicates the class, x indicates n-dimensional data with n number of bands, $p(\omega_i)$ indicates probability that class ω_i occurs in the image and is assumed the same for all classes, $|\Sigma_i|$ and Σ_i^{-1} indicate determinant of the covariance matrix of the data in class ω_i and its inverse matrix, and m_i indicates the mean vector.

From those processes, a map of land cover will be obtained and can be utilized for further analysis to estimate the mangroves density. A Normalized Difference Vegetation Index (NDVI) was derived from two bands that are most sensitive to vegetation information, which are red (0.6–0.7 μm) and infrared (0.7–1.3 μm). These bands are known to have a strong relationship with vegetation chlorophyll absorption electromagnetic spectrum that produced because of reaction between vegetation cover and the radiation above. NDVI indicates the mass of bio-vegetation and can be calculated with:

$$NDVI = \frac{(NIR - Red)}{(NIR + Red)} \tag{2}$$

where Red and NIR stand for the spectral reflectance measurement acquired in the red and near-infrared regions, respectively.

Based on this NDVI, the mangrove density can be classified into three equal-interval classes as shown in Table 1, and mangrove density analysis can be

Table 1 NDVI classification used for mangroves density analysis

Class of mangroves density	NDVI value
1. Low	0.200001–0.4
2. Moderate	0.400001–0.6
3. High	0.600001–1.0

performed to estimate the density level. Figure 5 shows the comprehensive image processing steps conducted in this study to produced maps of land cover and the density of mangrove communities.

4 Results and Discussions

The transect survey results from 11 locations (Table 2) shows that *Avicennia* spp. and *Sonneratia* spp. are the two species most widely distributed in the Porong Estuary area. The minimum and maximum density of *Avicennia* spp. and *Sonneratia* spp. are 1700 and 3900 trees ha^{-1} and 67 and 1033 trees ha^{-1}, respectively. Based on the IVI value, *Avicennia* spp. is the highest at all stations (ranging from 125 to 275), followed by *Sonneratia* spp. that also dominates at stations 1–7 (Fig. 6). Those results indicate that *Avicennia* spp. and *Sonneratia* spp., as species commonly associated with accreting shorelines, are the most dominant and have a moderate to high role on the structure of mangrove communities. Since *Avicennia* is one of the mangrove genera with a high adaptive capacity to increase sedimentation rate (Vaiphasa et al. 2007), this conclusion seems rational. In addition, the structure of its pneumatophores may enable this tree to acclimate low oxygen conditions (Ball 1988).

Table 2 Estimation of mangrove density of five species based on the transects data taken from 11 stations around the Porong Estuary; the size of each transect station is 10 m × 10 m

Station ID	Mangrove density (trees ha^{-1})				
	Avicennia spp.	*Bruguiera* spp.	*Derris trifoliata*	*Rhizophora* spp.	*Sonneratia* spp.
1	3200	0	0	0	767
2	3000	0	0	267	1033
3	3300	300	0	0	667
4	1967	0	0	0	533
5	3633	0	0	67	533
6	1700	0	0	0	500
7	3900	0	0	0	267
8	3700	0	0	0	67
9	2700	0	167	0	100
10	2233	0	0	533	100
11	2333	0	567	0	67

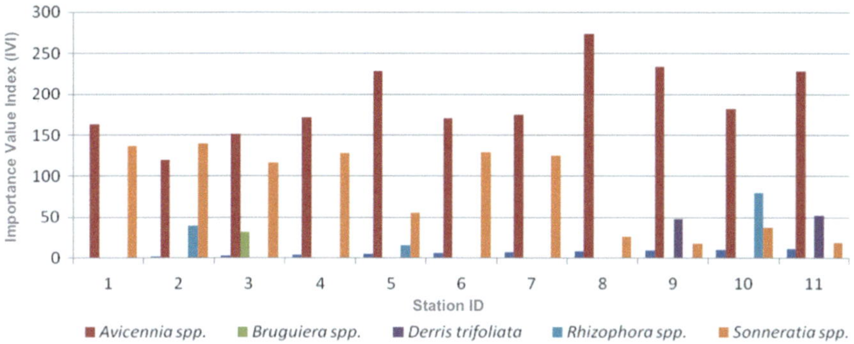

Fig. 6 The importance value index (IVI) of five mangrove species estimated from transect data at 11 stations around the Porong Estuary

Table 3 Surface area of existing land covers around the Porong Estuary based on supervised classification of WorldView-2 images making use of 46 GCs for the training area

Land cover	Surface area (ha)
1. Settlement	782.63
2. Rice field	744.34
3. Cultivated area	3.33
4. Shrimp/fish ponds	14,385.28
5. Mangroves	654.19
6. Bush	62.07
7. Barren land	58.02

The supervised classification of the images, making use of GC data from 46 stations, shows that the study area can be classified into seven land covers, namely settlement, rice field, shrimp/fish ponds, cultivated area, mangroves, bush, and barren land. The estimation to their surface area shows that the shrimp/fish ponds occupy the largest area (14,385.28 ha), while the mangroves have a surface area of approximately 654.19 ha (see Table 3).

In this study, the northern and southern parts of the Porong Estuary are included when estimating land use and determining mangrove density. This is because, based on the results from the sediment transport model (shown in Fig. 7, left panel), the sediment is also transported to those areas. Those sediment transports can also be seen clearly in the WorldView-2 imagery used by this study (see Fig. 7, right panel). Figure 8 shows the map produced from supervised classification of WorldView-2 images, where most of the study area is dominated by shrimp/fish ponds that are operated actively by local people, with milkfish and shrimps being major products. The surface area of shrimp/fish ponds is rather smaller compared to statistical data from Sidoarjo Regency's Spatial Plans 2009–2029, i.e. 18,672 ha. However, since our study area did not cover all Sidoarjo Regency region, this variance is reasonable.

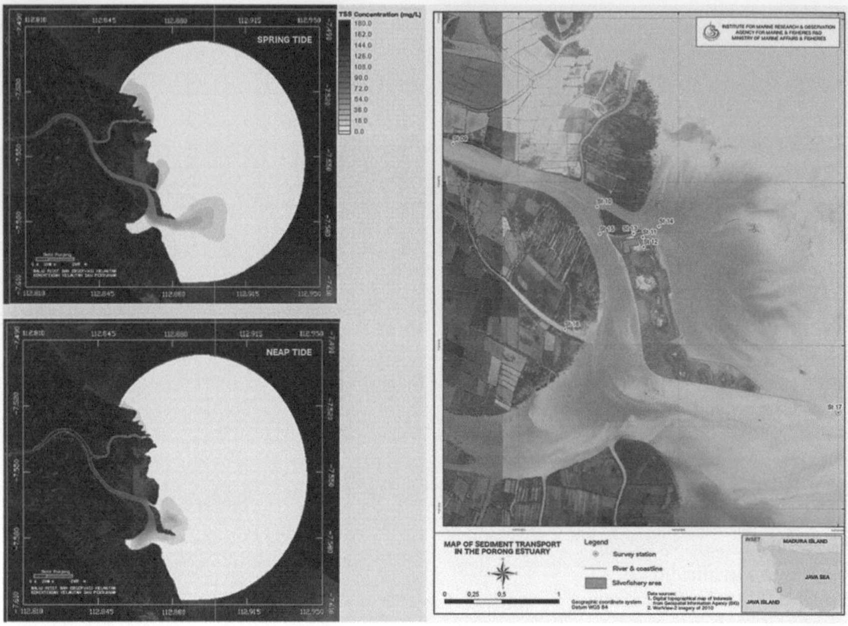

Fig. 7 Sediment transport in the Porong Estuary during spring and neap tides based on the numerical model (left panel) and captured by WorldView-2 imagery (right panel)

The existing mangrove forests were mostly found around the southern part of Porong Estuary (region A in Fig. 8), some parts along the coastline of Delta Porong (region B), and along the coastline of the Java Sea in the southern (region C) and northern (region D) areas, respectively. After comparing the results obtained from this study with those obtained in previous ones, one can conclude that, despite a significant decrease in 2007, the total surface area of mangrove forests around the Porong Estuary has increased. This finding indicates that, after the mudflow disaster and slurry channelling through the Porong River, the mangrove ecosystem around the Porong Estuary can adapt to the changing conditions. Interestingly, in the spoil bank—a small island created from the dredged sediment around the Porong Estuary—it is hard to witness growth among mangroves. The exception is a small area in the northern part. This condition results from the high salinity in the spoil bank's soil, where water can neither flow in or out to flush out the salt trapped in the soil.

By using Landsat ETM+ images from 2003, 2007, and 2009, and WorldView-2 images from 2010, some changes in the mangrove forest's surface area and distribution can be clearly identified (see Fig. 9). Quantitatively, from 2003 to 2007, the mangroves forests experienced deterioration. Its total surface area of 666.709 ha decreased significantly to 435.365 ha. Meanwhile, from 2007 to 2009 there was an increase from 435.365 ha to 550.08 ha, and another to 654.19 ha in 2010 (see Table 4).

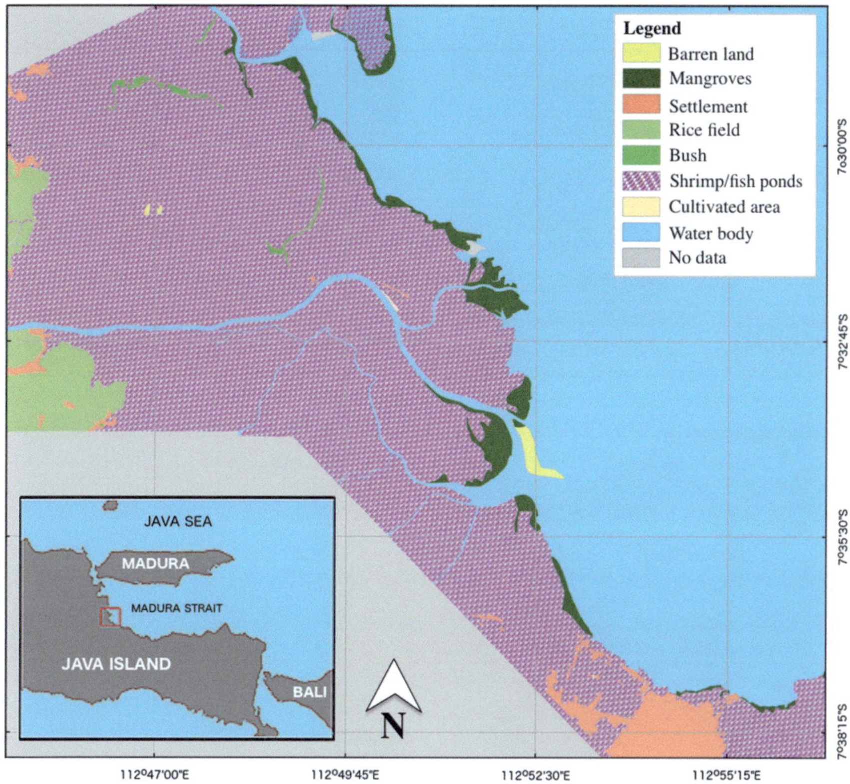

Fig. 8 Map of land cover in the study area estimated from supervised classification of WorldView-2 images used in this study making use of 46 GCs for the training area

In 2010, mangroves with moderate and high density covered 39 and 53% of total mangroves surface area, respectively. Mangroves with moderate density were mostly found in regions A and C, while regions B and D were dominated by mangroves with high density. These conditions are reasonable after observing the mangrove density map of 2009 (see Fig. 9), which showed regions B and D were already covered in high-density mangroves. Interestingly, high-density mangrove forests started forming in the northern part of the spoil bank. Apparently, this condition occurs because of high sedimentation in this area, which also spreads mangrove seeds naturally.

Mangroves with moderate density are generally located in the accretion areas. This suggests that the existing mangrove trees tend to trap the sediment load from the Porong River. The tidal current and the spreading of mangrove seeds also play an important role in the sedimentation processes around the Porong Estuary. Since the sediment transported there is rich with particulate organic carbon as fertilizer and nutrition for mangroves to growth, the surface area of mangrove communities tends to increase continuously downstream, inline with the increasing mainland.

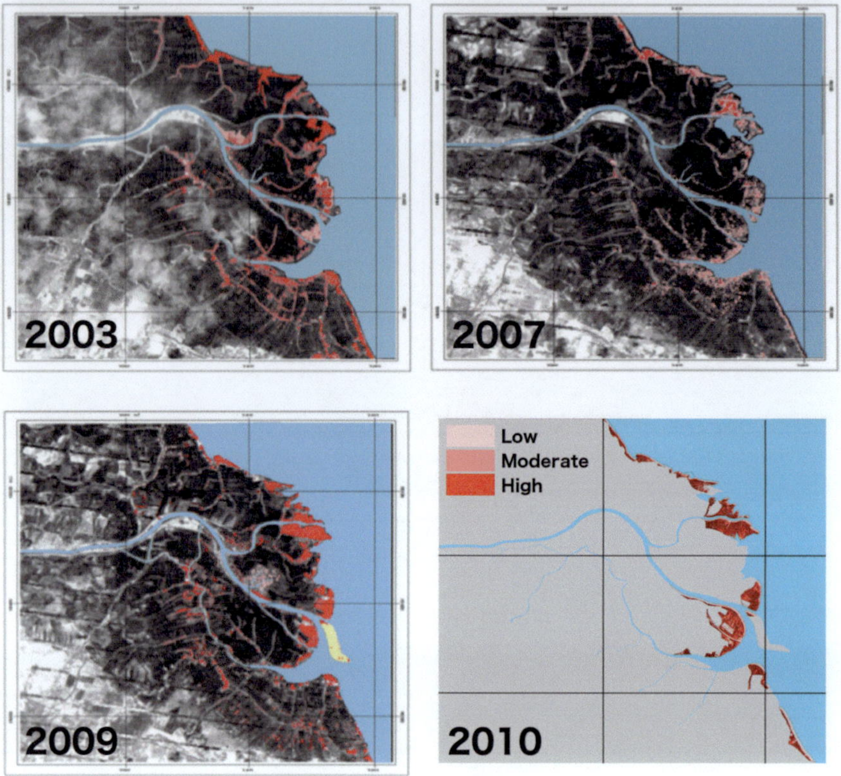

Fig. 9 Changes of mangroves forests distribution and their density around the Porong Estuary estimated from Landsat 7 ETM+ images of 2003, 2007, and 2009 and WorldView-2 image of 2010. Maps for 2003, 2007, and 2009 are taken from MMAF (2009)

Table 4 Mangroves surface area around Porong Estuary based on their density for the years 2003, 2007, 2009, and 2010

Mangroves density	Surface area (ha)			
	2003	2007	2009	2010
1. High	157.063	26.370	323.730	376.32
2. Moderate	445.191	33.964	180.900	256.81
3. Low	64.455	375.031	45.450	21.06
Total surface area	666.709	435.365	550.080	654.19

Pahlevi and Wiweka, who analysed the coastline changes and sedimentation rate around the Porong Estuary based on ASTER, Landsat 5 TM, and Landsat 7 ETM+ images, found that in average the sedimentation rate around Porong Estuary is 59 ha $year^{-1}$ (Pahlevi and Wiweka 2010). The highest sedimentation rate occurred between 2006 and 2007 at 93.3 ha $year^{-1}$. This could be the reason why, in 2009, the mangrove surface area in the Porong Estuary started increasing despite undergoing significant decreases in 2007.

5 Conclusions

From time to time, the Porong Estuary experiences a load of sediment, both from anthropogenic and Sidoarjo mudflows. Several studies suggest that the load of sediment from the Sidoarjo mudflow will ensue for several years to come (Jennerjahn et al. 2013). Due to high accretion rates with the potential to decrease the mangrove growth rate, this could negatively impact the mangrove ecosystem (Sidik et al. 2016). Although *Avicennia* spp., the most dominant mangrove trees in the Porong Estuary, can generally adapt very well to the increasing accretion rate, periodic assessment using satellite imagery needs to be done. This is because, in addition to the Sidoarjo mudflow, human modifications on the upstream and coastal zones tend to increase with a rise in population.

In general, from this study, the powerful of remote sensing technology on assessing the dynamics changing in estuary can be seen clearly. Comparing results of remote sensing data over a certain periods, even though from different satellite mission, could give clear description on the dynamics of the estuary and its relation to the ecosystem changes, particularly mangroves. This study also demonstrates that making use of WorldView-2's detailed imagery to assess the changes in the mangroves ecosystem is both easy and efficient. Its accuracy in identifying the land use and producing a map of natural vegetation suggests that WorldView-2 imagery can be used further for more complex classification studies.

Acknowledgements This study was conducted in the framework of cooperation between Badan Penanggulangan Lumpur Sidoarjo (BPLS) and the Institute for Marine Research and Observation (IMRO)—Ministry of Marine Affairs and Fisheries (MMAF). We gratefully acknowledge financial support by the BPLS for this study. We also thank IMRO—MMAF for supporting us in field measurements and laboratory analysis. We also thank to anonymous reviewers for providing comments on the manuscript. We especially acknowledge Mr. Soegiarto and Mr. Karyadi for their tremendous efforts for the success of this study.

References

Aksornkoae S (1975) Structure, regeneration, and productivity of mangroves in Thailand. Ph.D thesis. Michigan State University, USA

Aksornkoae S, Srisawasdi W, Panichsuko P (1985) Productivity and mortality of mangrove plantation in an abandoned mining area for coastal zone development in Thailand. BAKAWAN 6(3):6–8

Alongi DM (2009) The energetics of mangrove forests. Springer, The Netherlands

Ball MC (1988) Ecophysiology of mangroves. Trees-Struct Funct 2:129–142

Belluco E, Camuffo M, Ferrari S, Modenese L, Silvestri S, Marani A, Marani M (2006) Mapping salt-marsh vegetation by multispectral and hyperspectral remote sensing. Remote Sens Environ 105:54–67

Bengen DG (2001) Technical guideline on introduction and management of mangroves ecosystem. Pusat Kajian Sumberdaya Pesisir dan Lautan IPB, Bogor (in Indonesia)

Clarke LD, Hannon NJ (1967) The mangrove swamp and salt marsh communities of the Sydney district. I. Vegetation, soils and climate. J Ecol 55:753–771

Dahdouh-Guebas F, Koedam N (2006) Empirical estimate of the reliability of the use of the Point-Centred Quarter Method (PCQM): solutions to ambiguous field situations and description of the PCQM + protocol. For Ecol Manage 228:1–18

Frith DW, Tantanasiriwong R, Bhatia O (1976) Zonation and abundance of macrofauna on a mangrove shore, Phuket Island. Phuket Mar Biol Cent Phuket Thail Res Bull 10:1–37

Furukawa K, Wolanski E (1996) Sedimentation in mangrove forests. Mangrove Salt Marshes 1:3–10

Giglioli MEC, King DD (1966) The mangrove swamps of Keneba, the lower Cambia River Basin. III. Seasonal variation in the chloride and water content of swamp soils, with observation on the water level and chloride concentration of free soil water under a barren mudflat during the dry season. J Appl Acol 3:1–19

Gledhill D (1963) The ecology of the Aberdeen Creek mangrove swamp. J Ecol 51:639–703

Harris A, Bryant RG (2009) A multi-scale remote sensing approach for monitoring northern peatland hydrology: present possibilities and future challenges. J Environ Manage 90:2178–2188

Hoekstra P, Tiktanata (1988) Coastal hydrodynamics, geomorphology and sedimentary environments of two major Javanese river deltas. Program and preliminary results from the Snellius-II expedition (Indonesia). J SE Asian Earth Sci 2:95–107

Jennerjahn TC, Ittekot V, Klöpper S, Adi S, Nugroho SP, Sudiana N, Yusmal A, Prihartanto P, Gaye-Haake B (2004) Biogeochemistry of a tropical river affected by human activities in its catchment: Brantas River estuary and coastal waters of Madura Strait, Java, Indonesia. Estuar Coast Shelf Sci 60:503–514

Jennerjahn TC, Jänen I, Propp C, Adi S, Nugroho SP (2013) Environmental impact of mud volcano inputs on the anthropogenically altered Porong River and Madura Strait coastal waters, Java, Indonesia. Estuar Coast Shelf Sci 130:152–160

Karyadi, Soegiarto, Harnanto A (2012) Pengaliran lumpur Sidoarjo ke laut melalui Kali Porong. Bayumedia Publishing, Malang

Kure S, Winarta B, Takeda Y, Udo K, Umeda M, Mano A, Tanaka H (2014) Effects of mud flows from the LUSI mud volcano on the Porong River Estuary, Indonesia. In: Proceedings of the 13th international coastal symposium. J Coast Res Spec 70:568–573

Laba M, Downs R, Smith S, Welsh S, Neider C, White S, Richmond M, Philpot W, Baveye P (2008) Mapping invasive wetland plants in the Hudson River National Estuarine Research Reserve using quickbird satellite imagery. Remote Sens Environ 112:286–300

Laben CA, Brower BV (2000) Process for enhancing the spatial resolution of multispectral imagery using pan-sharpening. Google Patents

Levin LA, Boesch DF, Covich A, Dahm C, Erséus C, Ewel KC, Kneib RT, Moldenke A, Palmer MA, Snelgrove P, Strayer D, Weslawski JM (2001) The function of marine critical transition zones and the importance of sediment biodiversity. Ecosystems 4:430–451. https://doi.org/10.1007/s10021-001-0021-4

MMAF (2009) Assessment of the impact of Lumpur Sidoarjo on sediment dispersion and coastal change. Ministry of Marine Affairs and Fisheries, Indonesia

MMAF (2010) Assessment on the utilization and development of Porong Estuary. Ministry of Marine Affairs and Fisheries, Indonesia

Pahlevi AM, Wiweka (2010) Sedimentation analysis in Porong estuary due to channeling mud of Lapindo using ASTER satellite imagery. J Ilmiah Geomatika 16(2):23–42 (in Indonesia)

Perillo GME (1995) Geomorphology and sedimentology of estiaries. In: Developments in sedimentology, vol 53. Elsevier, Amsterdam

Rapinel S, Clément B, Magnanon S, Sellin V, Hubert-Moy L (2014) Identification and mapping of natural vegetation on a coastal site using a Worldview-2 satellite image. J Environ Manage 144:236–246

Richards J (1999) Remote sensing digital image analysis. Springer, Berlin

Sanders HL (1958) Benthic studies in Buzzards Bay. I. Animal-sediment relationships. Limnol Oceanog 3:245–358

Satyanarayana B, Thierry B, Lo-Seen D, Raman AV, Muthusankar G (2001) Remote sensing in mangrove research—relationship between vegetation indices and dendrometric parameters: a case for Coringa, East Coast of India. In 22nd Asian conference on remote sensing, 5–9 Nov 2001

Sidik F, Lovelock CE (2012) Mitigating disaster with mangroves: lessons learned from the Indonesian mud volcano risk assessment and management process. In: Proceeding of the 1st ASEAN congress on mangrove research and development, 3–7 Dec 2012, pp 32–44

Sidik F, Neil D, Lovelock CE (2016) Effect of high sedimentation rates on surface sediment dynamics and mangrove growth in the Porong River, Indonesia. Mar Pollut Bull http://dx.doi.org/10.1016/j.marpolbul.2016.02.048

Thrush S, Hewitt J, Cummings V, Ellis J, Hatton C, Lohrer A, Norkko A (2004) Muddy waters: elevating sediment input to coastal and estuarine habitats. Front Ecol Environ 2:299–306

Vaiphasa C, De Boer WF, Panitchart S, Vaiphasa T, Bamrongrugsa N, Santitamnont P (2007) Impact of solid shrimp pond waste materials on mangrove growth and mortality: a case study from Pak Phanang, Thailand. Hydrobiologia 591:47–57

Walter H (1971) Ecology of tropical and subtropical vegetation. Oliver and Boyd, New York

Internal Waves in the Andaman Sea

Werner Alpers and Vasiliy Vlasenko

Abstract The Andaman Sea of the Indian Ocean is one of the sites in the World's ocean where extraordinarily large internal waves are encountered. For centuries, seafarers have noticed in the Andaman Sea often bands of strongly increases sea surface roughness extending from horizon to horizon. Now we know that they are sea surface manifestations of large amplitude internal waves (LAIWs), which are waves in the interior of the ocean. They can exist when the water body is stratified, i.e., when the water body consists of layers of different density. The LAIWs in the Andaman Sea are nonlinear waves which are generated by tidal flow over shallow bottom topography, like shallow ridges or sea mounts. Synthetic aperture radar (SAR) images have been very instrumental in studying the generation and propagation of internal waves in the Andaman Sea. The predominantly eastward propagating internal waves finally shoal on the shelf at the west coast the Malay Peninsula, where they usually break and generate turbulence. This causes transport of nutrient-rich water from deeper water layers to near-surface layers and gives rise to increased plankton growth. The LAIWs are associated with strong currents at depths and are a potential threat to oil drilling operations at sea.

Keywords Internal waves · Solitary waves · Andaman Sea · Synthetic aperture radar · Tidal flow · Dreadnought Bank

1 Introduction

The Andaman Sea of the Indian Ocean is known to be one of the sites in the World's ocean where extraordinarily large internal waves are encountered. They are waves in the interior of the ocean, but they are, to first order, not associated with an

---

W. Alpers (✉)
Institut für Meereskunde, Universität Hamburg, Hamburg, Germany
e-mail: werner.alpers@uni-hamburg.de

V. Vlasenko
School of Marine Science and Engineering, Plymouth University, Plymouth, UK

V. Barale and M. Gade (eds.), *Remote Sensing of the Asian Seas*,
https://doi.org/10.1007/978-3-319-94067-0_22

elevation or depression of the sea surface like the surface waves. They become visible on the sea surface indirectly by variations of the sea surface roughness. Seafarers passing through the Strait of Malacca on their journeys between India and the Far East have often noticed in the Andaman Sea between the Nicobar Islands and the north east coast of Sumatra bands of choppy waters or ripplings reaching from horizon to horizon. A description of such bands of choppy waters observed in the Andaman Sea can be found in the book of Mauray which was published in 1861 and which is quoted in Osborne and Burch (1980): "The ripplings are seen in calm weather approaching from a distance, and in the night their noise is heard a considerable time before they come near. They beat against the sides of a ship with great violence, and pass on, the spray sometimes coming on deck; and a small boat could not always resist the turbulence of these remarkable ripplings". Now we know that these ripplings were sea surface manifestations of nonlinear internal waves, which are also called internal solitary waves, internal solitons, or large amplitude internal waves (LAIWs). Perry and Schimke (1965) were the first to show by oceanographic measurements carried out from a ship that these bands of choppy water in the Andaman Sea are associated with LAIWs. Later Osborne and Burch (1980) analyzed oceanographic data collected by the Exxon Production Research Company in the southern Andaman Sea with the aim to assess the impact of underwater current fluctuations associated with oceanic internal waves on drilling operations carried out from a drill ship. Their oceanographic measurements showed that the time interval between the first solitons in a wave packet was typically 40 min and then decreased towards the end. In one event, the amplitude (peak-to-trough distance) of the foremost soliton was estimated to be 60 m, i.e., warm water in the upper water layer was pushed down by the internal soliton by 60 m. The roughness bands associated with one of the internal soliton packets extended from horizon to horizon and were 600–1200 m wide. They concluded that the visually observed roughness bands are caused by internal solitary waves which can be described by the soliton theory developed by Korteweg and de Vries (1895).

Internal waves in the Andaman Sea were first detected from space on optical images (Apel et al. 1985) and later on synthetic aperture radar (SAR) images acquired from several satellites, like the ERS-1, ERS-2, Radarsat-1, Radarsat-2, Envisat, and AlOS-1 satellites, which were launched in 1991, 1995, 1995, 2001, 2006, respectively. With these SAR images it was possible for the first time to study systematically the spatial distribution of internal solitons in the Andaman Sea and thus obtain information on their generation and propagation characteristics (Alpers et al. 1997; Jackson 2004a).

The LAIWs can reach in the Andaman Sea amplitudes larger than 80 m, wavelengths larger than 15 km, and propagation speeds larger than 1.5 m/s. They are potentially hazardous to offshore oil and gas drilling operations and have already caused several costly and dangerous incidents in Southeast Asian waters (Hyder et al. 2005). Strong currents and shears associated with internal solitons may also exert extreme stresses on drill pipes during offshore operations, which is the reason why oil companies engaged in offshore explorations are interested in internal waves. Furthermore, internal waves affect submarine navigation and underwater sound propagation, which is the reason why the military is interested in internal waves. Also marine biologists interested in them because breaking internal waves in shelf regions cause transport of nutrient-rich water from deeper water layers to near-surface layers giving rise to increased plankton growth. Furthermore, they nourish coral reefs (Schmidt et al. 2012; Wall et al. 2015) and pilot whale population that forage in their wakes (Moore and Lien 2007).

Since the LAIWs are generated at different locations in the Andaman Sea, the solitary wave packets often interact with each other giving rise to phase shifts and redistribution of energy. They also can generate secondary wave packets when they interact with shallow bottom topographic features (Vlasenko and Alpers 2005). Most of the internal waves in the Andaman Sea propagate eastward, where they shoal at the shelf of the Malay Peninsula and usually break.

In the following chapters, we first give a short description of the Andaman Sea, of internal waves, and of the SAR imaging mechanism of internal waves, and then present several SAR images and one optical image acquired over the Andaman Sea showing sea surface manifestations of internal wave packets. Furthermore, we present two SAR images showing the interaction of two internal solitary wave packets and then a SAR image showing the interaction of a strong internal wave packet with an underwater bank, which causes generation of a secondary wave packet. Then we present a simulation showing the shoaling of a solitary internal wave at the shelf of the Malay Peninsula, which ends up in wave breaking. Finally, we give an outlook on other areas of the Asian Seas, where also LAIWs are encountered and recall how first internal waves were discovered.

2 Basics

2.1 The Andaman Sea

The Andaman Sea is a marginal sea of the northeastern Indian Ocean extending approximately from 6 to 14°N in latitude and from 93 to 90°E in longitude. It is bounded to the north by the Irrawaddy River delta of Myanmar (Burma), to the east by the Malay Peninsula, to the south by the Indonesian island of Sumatra and the Strait of Malacca, and to the west by the Indian Andaman and Nicobar Islands. A bathymetric map of the Andaman Sea is depicted in Fig. 1a. The Andaman and Nicobar Islands on the western side of the Andaman Sea are volcanic in origin. To the east of these islands, and also northwest of Sumatra, there are several shallow areas, ridges, and seamounts which serve as birthplaces for internal waves. The center of the Andaman Sea is quite deep, reaching 3075 m in the Central Andaman Trough. To the east, there is a 250 km wide shelf west of the Malay Peninsula. The inner shelf off Phuket (Thailand) is 35 km wide and has a depth of 100 m. Close to the coast, the bathymetry is very inhomogeneous causing varying interactions with shoaling internal waves. Near the Andaman and Nicobar islands and along the north/northwest coast of Sumatra, the tidal range during spring tide is 1–2 m, at the east

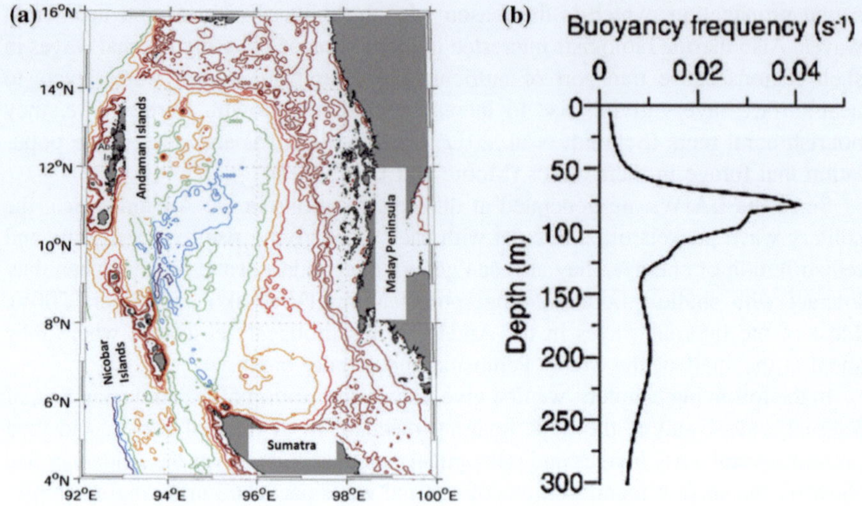

Fig. 1 **a** Bathymetry map of the Andaman Sea derived from Smith and Sandwell version 8.2 (courtesy: Ch. Jackson). **b** Typical buoyancy profile of the Andaman Sea waters in winter (Vlasenko and Stashchuk 2007)

coast of Thailand it is around 3 m, and at the Mergui Archipelago of Myanmar it is >5 m. The water column of the Andaman Sea is strongly layered all year round, i.e., its density varies strongly with depth due to variations of water temperature and salinity, such that it can support internal waves. The layering of the water column in the Andaman depends strongly on location and time of the year. When studying internal waves, the layering enters in the form of the buoyancy frequency (or Brunt-Vaisala frequency) N, which is a function of the vertical gradient of the water density ρ:

$$N(z) = \sqrt{-\frac{g}{\rho}\frac{\partial \rho}{\partial z}} \qquad (1)$$

where g denotes the gravitational constant and z is the vertical coordinate (positive upwards).

A typical depth profile of the buoyancy frequency of the Andaman Sea in winter is shown in Fig. 1b. This profile was used by Vlasenko and Stashchuk (2007) in their study on shoaling of LAIWs on the inner shelf of Malay Peninsula (see Sect. 9).

2.2 Internal Waves

Internal waves can exist in the interior of the ocean when the density of the water body varies with depth. This difference in water density is mostly due to a difference in water temperature, but can also be due to a difference in salinity or both, temperature and salinity. Often the density stratification of the ocean can be approximated by two layers. Like the well-known ocean surface waves, which are waves at the interface of two media of different density, i.e., of water and air, the internal waves are also waves at the interface between two water layers of different density. This interface is called pycnocline (from Greek: pycnotes = density, klinein = to slope) or thermocline (from Greek: thermos = hot) because the density difference is often caused mainly by the difference in water temperature. In the case of surface waves as well as in the case of internal waves, the restoring force is gravity, which is the reason why both waves sometimes are called gravity waves. These waves are generated when the interface is disturbed. In the case of surface waves, the disturbance can be caused by a stone being thrown into the water or by wind blowing over the water surface. In the case of internal waves, the disturbance is usually caused by tidal flow pushing a layered water body over shallow underwater obstacles, e.g., over shallow sills or shallow ridges (Maxworthy 1979; Helferich and Melville 1986). Internal waves in the ocean typically have wavelengths from hundreds of meters to tens of kilometers and periods from tens of minutes to several hours. Their amplitude (crest-to-trough distance) often exceeds 50 m. Associated with internal waves are orbital motions of the water particles as depicted in Fig. 2 (the dashed circular lines). The radius of the circular motion of the water particles is largest at the pycnocline (or thermocline) and decreases downwards as well as upwards. At the sea surface, the water motion give rise to variable (horizontal) surface currents causing convergent and divergent flow region as depicted in Fig. 2. The variable surface current interacts with the surface waves and modulates the sea surface roughness.

2.3 SAR Imaging Mechanism of Internal Waves

As mentioned already in the Introduction, LAIWs can be detected remotely from satellites and aircrafts only indirectly via variations of the sea surface roughness. The internal waves are associated with variable surface currents which modulate the sea surface roughness. A theoretical model describing the modulation of short-scale sea surface roughness by variable surface currents has been developed in the framework of weak hydrodynamic interaction theory by Alpers (1985). When using this theory together with Bragg scattering theory, which relates spectral values of the ocean surface waves to the normalized radar cross section (NRCS), the relationship between the NRCS and surface current gradient $\frac{dU_x}{dx}$ can be approximated by

$$\sigma = \sigma_0 \left(1 - A \frac{dU_x}{dx} \right) \quad (2)$$

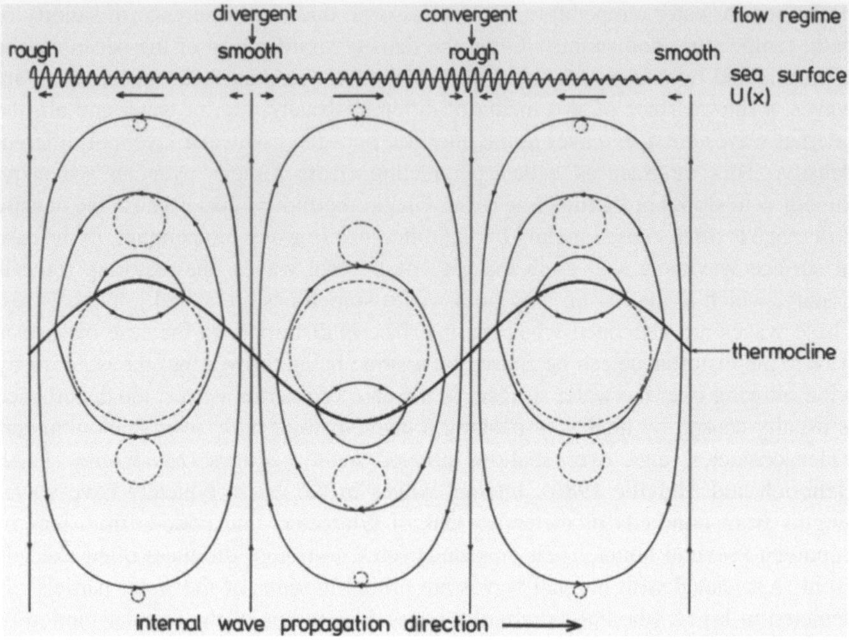

Fig. 2 Schematic plot of processes associated with the passage of a linear oceanic internal wave. Deformation of the thermocline (heavy solid line), orbital motions of the water particles (dashed lines), streamlines of the velocity field (light solid lines), surface current velocity vectors (arrows in the upper part of the image), and variation of the amplitude of the Bragg waves (wavy line at the top). Reproduced from Alpers 1985

Here σ denotes the total NRCS, σ_0 the NRCS of the background, x the coordinate in the look direction of the SAR antenna projected onto the horizontal plane, and A denotes a constant that depends on radar wavelength, incidence angle, and relaxation rate. The relaxation rate is quite variable, depending, among others, on wind speed and direction. The gradient $\frac{dU_x}{dx}$ is related to internal wave and upper ocean parameters such as amplitude and wavelength of the internal wave, as well as to the stratification of the water column. The larger the amplitude of the internal wave, the larger $\frac{dU_x}{dx}$ becomes. From Eq. (2) we see that the NRCS becomes larger in convergent flow regions $\left(\frac{dU_x}{dx} < 0\right)$ and smaller in divergent flow regions $\left(\frac{dU_x}{dx} > 0\right)$ producing the bands of increased and decreased image intensity relative to a background, respectively. Other, more advanced, weak hydrodynamic interaction theories have been proposed (Romeiser and Alpers 1997) and yield qualitatively similar results. In these theories, the variation of the NRCS or image intensity (also called the modulation depth) depends strongly on wind speed and direction (Brandt et al. 1999) and usually underestimates the strength of the NRCS variation because it does not include wave breaking. In addition, the theories require wind speeds above the threshold for Bragg wave generation (approximately 2–3 m/s) and below

8–10 m/s because at high wind speeds the internal wave—induced roughness is masked by the wind generated roughness.

Usually, internal waves visible on SAR images are not linear waves, but are strongly nonlinear and appear in wave packets. In this case they have the shape shown in Fig. 3. The distance between the solitons in a wave packet and also the amplitude of the solitons decreases from the front to the rear. Depicted is in Fig. 3 an internal wave packet consisting of three soltions of depression, which means they are associated with a depression of the pycnocline. In this case, the front of the soliton is associated with a convergent surface current region (causing enhancement of the Bragg wave amplitude) and the rear with a divergent region (causing a decrease of the Bragg wave amplitude). As a consequence, the radar image intensity is increased in front and decreased in the rear as shown in Fig. 3. Thus an internal soliton of depression is imaged by SAR always as band consisting of a bright sub-band in front followed by a dark sub-band. However, for an internal soliton of elevation the radar signature is reversed. An internal wave of elevation is a soliton, where the pycnoline is elevated. Soliton theory predicts that solitons of elevation exist when the depth of the upper water layer is smaller than the depth of the lower layer (Klymak and Moum 2003). The reversal of the radar signature (dark band in front followed by a bright band in the rear) has been used for identifying solitons of elevation on SAR images (Liu et al. 1998).

However, often the radar image intensity pattern associated with LAIWs does not have the form shown in Fig. 3 or a reversed form (dark-bright) as in the case of solitons of elevation. Often the radar signature of a LAIW consists only of bright bands on a grey or dark background (when the wind speed is below the Bragg generation threshold). This then can be explained by [specular]reflection from breaking surface waves and Bragg scattering from short surface waves generated by wave breaking (Kudryavtsev et al. 2005).

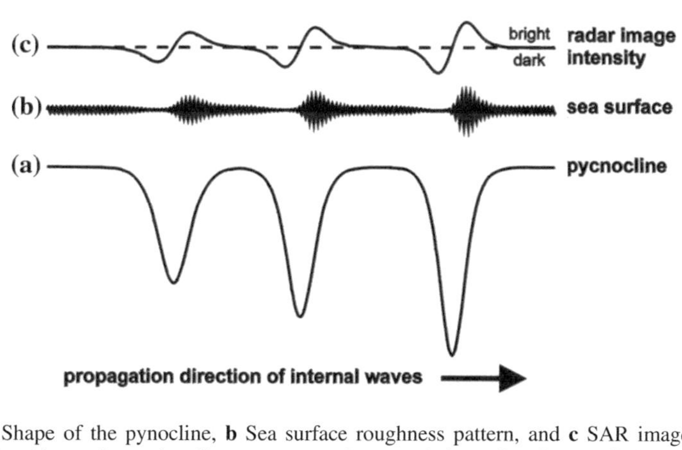

Fig. 3 **a** Shape of the pynocline, **b** Sea surface roughness pattern, and **c** SAR image intensity associated with an internal solitary wave packet consisting of solitons of depression with amplitudes decreasing from front to rear. Reproduced from https://www.earth/esa.int/web/guest/-/ers-sar-tropical-5883

3 Satellite Remote Sensing of Internal Waves

3.1 SAR Images of Internal Waves in the Andaman Sea

LAIWs are generated in the Andaman Sea by tidal flow over shallow ridges and seamounts that abound in the western and southwestern section of the Andaman Sea. The SAR image depicted in Fig. 4a was acquired by the European Remote Sensing Satellite-2 (ERS-2) over the southern Andaman Sea together with a scan through the pattern showing the variation of the NRCS (Fig. 4c). It shows two internal wave packets propagating eastward, which have an inter-packet separation of approximately 75 km. They were generated at two successive cycles of the semi -diurnal tide having a period of 12.4 h. From the distance and the time separation we obtain the propagation speed of 1.7 m/s, which is a typical propagation speed of

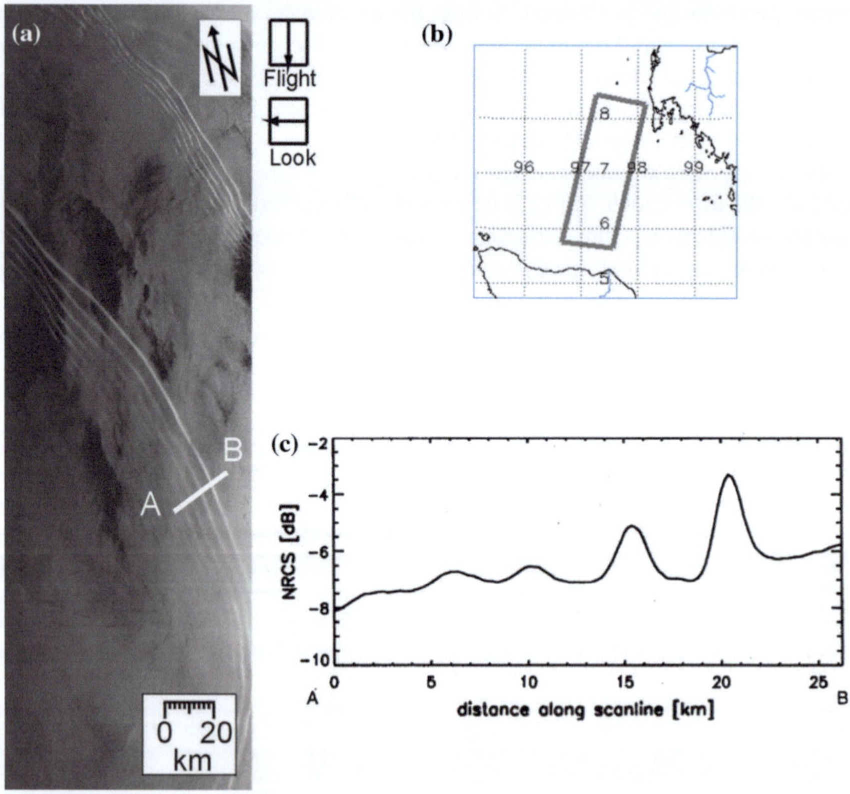

Fig. 4 **a** ERS-2 image acquired on 04 May 1996 at 03:52 UTC over the southern Andaman Sea showing sea surface manifestations of two internal wave packets generated at successive semi-diurnal tidal cycles. **b** Location of the imaged scene. **c** Variation of the NRCS along the profile A-B inserted in (**a**). Reproduced from: https://earth.esa.int/web/guest/-/ers-sar-tropical-5883

LAIWs in the Andaman Sea. This image shows further, that the wavelength within a wave packet decreases from front to rear, and that the wave packet gets weaker the further it has propagated from its source.

Figure 5 shows two SAR images of the southern Andaman Sea, one acquired by the ERS-2 SAR with a swath width of 100 km and the other acquired by Radarsat-1 with a swath width of 500 km. On both images, a plethora of wave packets can be

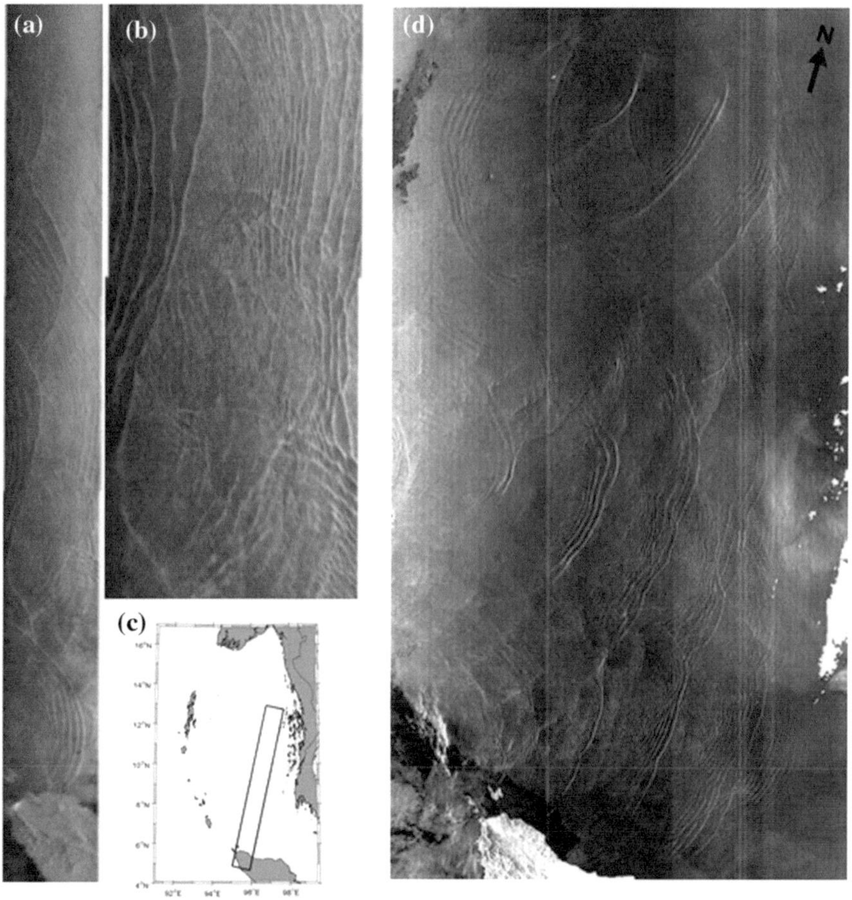

Fig. 5 **a** ERS-2 SAR image acquired on 11 February 1997 at 03:59 UTC over the southern Andaman Sea showing a large number of internal wave packets. The swath width is 100 km. **b** Enlargement of the lower middle section of this SAR image. The imaged area is 235 km × 100 km. **c** Location of the SAR image depicted in (**a**). **d** Radarsat-1 SAR image acquired on 17 December at 11:49 UTC over the Andaman Sea showing a plethora of radar signatures of internal wave packets. The image was acquired in the ScanSAR Wide mode, which has a swath width of 500 km. Reproduced from Jackson et al. (2013)

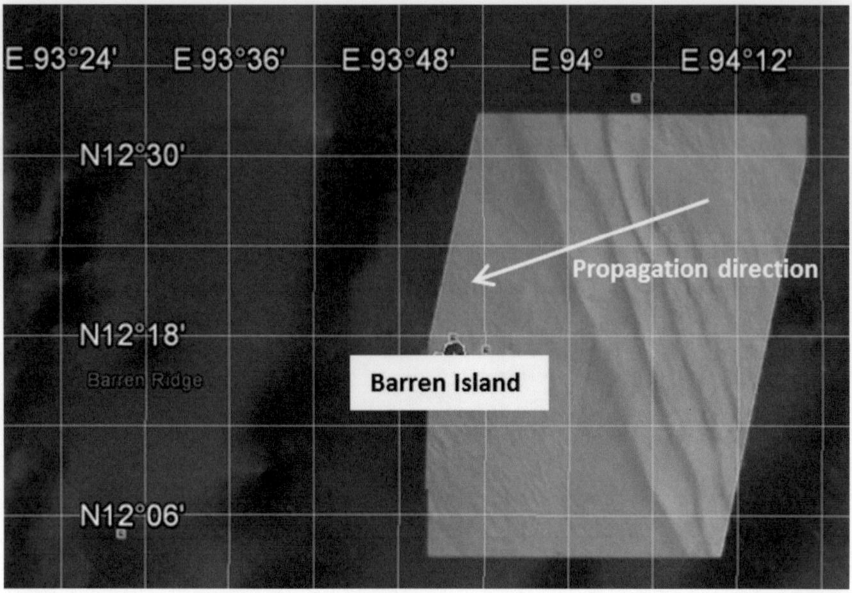

Fig. 6 **a** Section of an optical image acquired on 6 March 2007 by the Advanced Land Imager (ALI) onboard the Earth Observing-1 (EO-1) satellite over the western section of the Andaman Sea near Barren Island showing a south-westward propagating internal wave packet. The image is inserted in Google Earth and has the dimension 36 km × 19 km. *Image credit* https://earthobservatory.nasa.gov/IOTD/view.php?id=44567. Image created by Jesse Allen and Robert Simmon

delineated, on the Radarsat-1 image even more than 20. Most of them propagate eastward as evidenced by their curvature and their bright line in front. However, on the Radarsat-1 image also a westward propagating wave packet is visible.

3.2 Optical Image of an Internal Wave Packet in the Andaman Sea

Roughness patterns associated with LAIWs can also be delineated on optical images when the illumination conditions are favorable. The best images are obtained under "sunglint" conditions, i.e., when the sensor looks at the sea surface approximately with the same angle as the sun from the opposite side (Jackson and Alpers 2010). Figure 6 shows such an example, where the wave packet propagates in a southwestward direction.

4 Dynamical Processes of Internal Waves

4.1 Interaction of Internal Solitary Waves

A theory on the interaction of small-amplitude shallow water solitary wave was first developed by Miles (1977a, b), but theories on the interaction of LAIWs waves are still in the infancy stage. Studies on this interaction have been motivated mainly by SAR images on which often complex patterns are visible which apparently are caused by interactions of solitary internal waves (Wang and Pawlowicz 2012). One of the most noticeable features often visible on SAR images is that there is a phase shift when two internal solitons cross each other and that in the interaction region the SAR image intensity is increased which suggests that here the amplitude is increased, see Fig. 7a. Theoretical investigation with the aim of explaining these phenomena were carried out by Chen et al. (2011) and Xue et al. (2012) by using the Kadomtsev-Petviashvili equation. LAIWs also interact with small-amplitude internal waves, see Fig. 7b. This SAR image, as many other SAR images, suggests that LAIWs sweep clean small-scale and weak internal waves (in this case the tail of an internal wave packet from the previous tidal cycle).

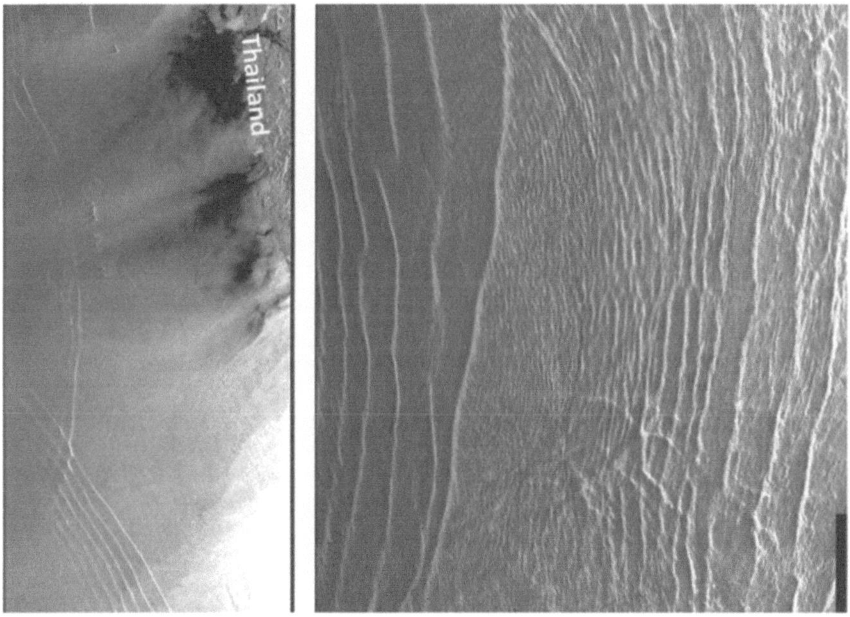

Fig. 7 **a** Envisat ASAR image acquired on 18 November 2006 at 03:22 UTC showing sea surface manifestations of two interacting packets of internal solitary waves propagating in different directions causing a phase shift and an increase of intensity in the interaction region. **b** ERS-2 SAR image acquired on 1 February 1997 at 03:58 UTC over the southern central section of the Andaman Sea (8° 48′N, 96° 22′E). The front of a packet of LAIWs (left section of the image) interacts with the tail of another internal solitary wave packet (right-hand section of the image)

4.2 Interaction of Internal Solitary Waves with the Dreadnaught Bank

Internal solitary waves can also interact with shallow bathymetric features and generate secondary internal waves. The ERS-2 SAR image depicted in Fig. 8a, b shows sea surface manifestation of a packet of strong internal solitary waves in which a smaller-scale weak internal wave packet with a semi-circular shape is embedded. Comparison with a bathymetric map reveals that the focal point of this small-scale wave pattern coincides with the shallowest point of the Dreadnought Bank, which has a depth of 241 m. Simulations carried out with a 2-dimensional hydrodynamic model (Vlasenko and Alpers 2005) shows that the secondary internal wave packet was generated by the interaction of a large-amplitude leading internal

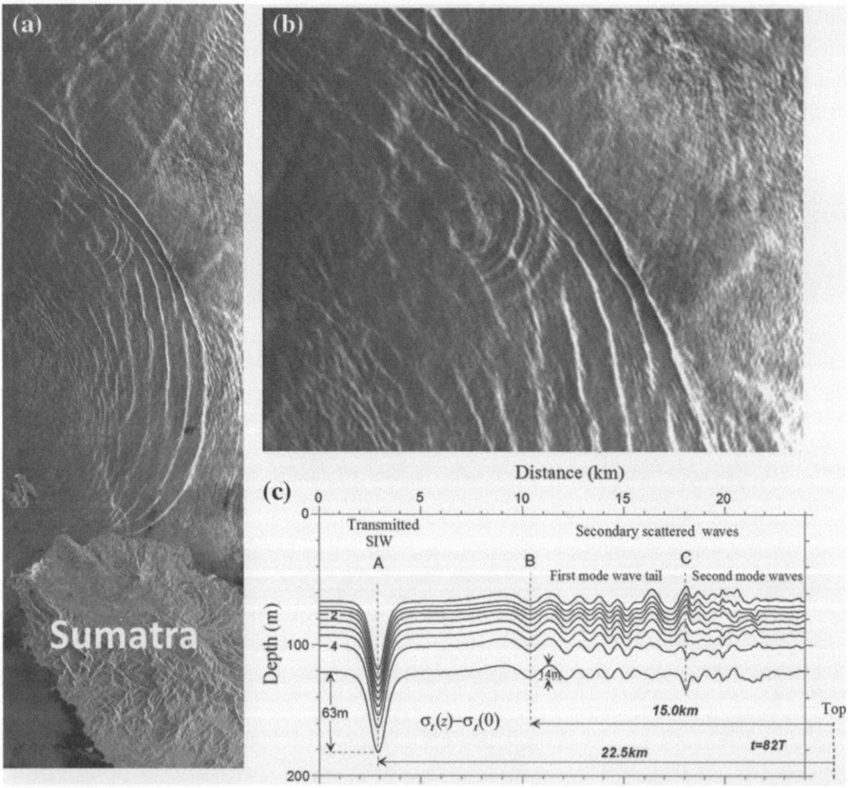

Fig. 8 **a** ERS-2 SAR strip acquired over the Andaman Sea on 11 February 1997 at 03:59 UTC showing the generation of secondary wave packet by the interaction of a strong leading soliton in a wave packet with an underwater bank. The imaged area is 100 km × 300 km. **b** Zoom of the ERS-2 SAR image on the interaction area. **c** Density field obtained from simulations carried out with a 2-dimensional hydrodynamic model showing the density field at the time of the SAR data acquisitions. Reproduced from Vlasenko and Alpers (2005)

soliton in an internal wave packet with the Dreadnought Bank, see Fig. 8c. The simulations show further that this can happen only when the amplitude of the impinging soliton is sufficiently large (at spring tide) such that it can overturn and break.

4.3 Shoaling of Internal Waves in the Shelf Region of the Malay Peninsula

Most of the internal waves propagate eastward, where they enter the shallow shelf region at the west coast the Malay Peninsula and shoal. The shoaling process can take different forms depending on the bottom topography and the strength of the LAIW. The shoaling process in the northern part of Andaman Sea has been studied by Vlasenko and Stashchuk (2007) using a fully nonlinear nonhydrostatic numerical model and using climatic data for the stratification, see Fig. 1b. The shoaling process can take different forms depending on the bottom topography in the shoaling area and the strength of the LAIW.

When the bottom inclination is small, the solitary wave of depression transforms into a solitary wave of elevation, and finally disintegrates into a packet of secondary short waves (Liu et al. 1998). When the bottom inclination is large, the solitary wave of depression first transforms into a soliton of elevation, then breaks down

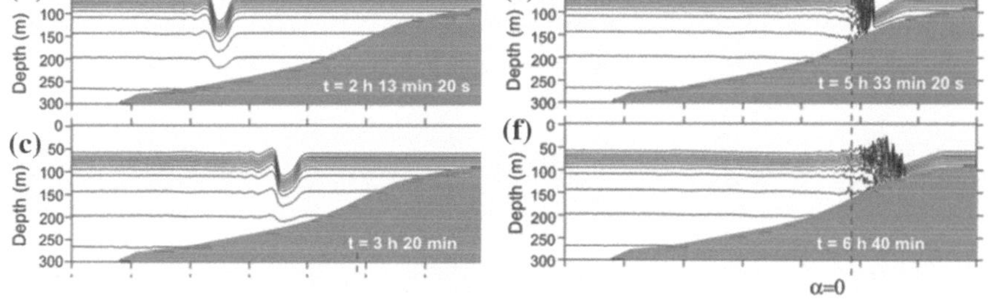

Fig. 9 **a–f** Time evolution of an internal soliton approaching a shallow shelf as simulated by a fully nonlinear nonhydrostatic numerical model. The internal soliton finally breaks near the shore where it generates turbulence. The contour lines represent the density field $\rho(z) - \rho(0)$ with a spacing of 0.5 kg m^{-3}. Vertical dashed line represents the position of the "turning point" where the weakly nonlinear soliton changes its characteristics (see Fig. 8 of Vlasenko and Stashchuk 2007)

producing multiple soliton-like waves of elevation (turbulent surges or boluses) and finally generates turbulence. In Fig. 9 is shown a time series of such an evolution as simulated by a numerical model (Fig. 8 of Vlasenko and Stashchuk (2007)). The turbulence gives rise to transport of nutrient-rich water from deeper water layers to near-surface layers causing increased plankton growth.

5 Outlook

LAIWs are not only encountered in the Andaman Sea, but also in many other parts of the World's ocean (for a review, see the internal wave atlas compiled by Jackson (2004b)). Probably the sea area, where the most powerful internal waves exist, is the South China Sea, where internal waves generated in the Luzon Strait can attain (crest-to-trough) amplitudes exceeding 100 m (Liu et al. 1998; Li et al. 2013; Alford et al. 2015). Other prominent areas with LAIWs are the Sulu Sea (Liu et al. 1985; Zeng and Alpers 2004) and the Macarene Ridge in the Indian Ocean (da Silva et al. 2011). Internal waves are also generated in straits: in Europe, in the Straits of Gibraltar and Messina and in Southeast Asia, in the Lombok Strait (between the Indonesian Islands Bali and Lombok) (Mitnik et al. 2000; Susanto et al. 2005).

However, internal waves are not only generated by tidal flow over shallow underwater bottom topography, but also by the action of the wind or by atmospheric pressure fluctuations. Also ships may generate internal waves, e.g., in fjords, where a layer of fresh or brackish water (from snowmelt) rests on top of denser salt water. Historically, it was the generation of internal waves by ships that led to the discovery of internal waves. These waves, invisible at the sea surface, exert an extra drag on a ship which leads to a loss of steering power and reduction of cruising speed. Mariners called this phenomenon "dead water", which was known already in ancient times (Mercier et al. 2011). In modern times, the dead water phenomenon was first described in detail by the Norwegian scientist and explorer Nansen (1897) and later explained by the Swedish scientist Ekman (1904), who attributed it to interfacial waves between the two layers interacting with the boat. Later these interfacial waves were termed internal waves.

References

Alford MH et al (2015) The formation and fate of internal waves in the South China Sea. Nature 521:65–69. https://doi.org/10.1038/nature14399

Alpers W (1985) Theory of radar imaging of internal waves. Nature 314:245–247

Alpers W, Wang-Chen H, Hock L (1997) Observation of internal waves in the Andaman Sea by ERS SAR. Paper presented at the third ERS symposium on space at the service of our environment, Space Agency, Florence, Italy, pp 1518–1520. https://doi.org/10.1109/igarss.1997.608926

Apel JR, Thomson DR, Tilley DG, van Dyke P (1985) Hydrodynamics and radar signatures of internal solitons in the Andaman Sea. John Hopkins APL Tech Dig 6(4):3330–3337

Brandt P, Romeiser R, Rubino A (1999) On the determination of characteristics of the interior ocean dynamics from radar signatures of internal solitary waves. J Geophys Res 104:30039–30045. https://doi.org/10.1029/1999JC900092

Chen GY, Liu CT, Wang YH, Hsu MK (2011) Interaction and generation of long-crested internal solitary waves in the South China Sea. J Geophys Res 116:C06013

da Silva JCB, New AL, Magalhaes JM (2011) On the structure and propagation of internal solitary waves generated at the Mascarene Plateau in the Indian Ocean. Deep Sea Res Part 1 58:229–240. https://doi.org/10.1016/j.dsr.2010.12.003

Ekman VW (1904) On dead water, Norw. N. Polar Exped. 1893–1896, Sci. Results, XV, Christiana. Ph.D. thesis, 1904

Helfrich KR, Melville WK (1986) On long nonlinear internal waves over slope-shelf topography. J Fluid Mech 167:285–308

Hydera P, Jeans DRG, Cauquil E, Nerzi R (2005) Observations and predictability of internal solitons in the northern Andaman Sea. Appl Ocean Res 27:1–11

Jackson CR (2004a) An atlas of internal solitary-like waves and their properties. Report: Global Ocean Associates, Alexandria, VA. Available at http://www.internalwaveatlas.com/Atlas2_PDF/IWAtlas2_FrontMatter.pdf

Jackson CR (2004b) An atlas of internal solitary-like waves and their properties, 2nd ed, 560 pp, Global Ocean Associates, Alexandria, VA. Available online: http://www.internalwaveatlas.com/Atlas2_index.html

Jackson CR, Alpers W (2010) The role of the critical angle in brightness reversals on sunglint images of the sea surface. J Geophys Res 115:C09019. https://doi.org/10.1029/2009JC006037

Jackson CR, da Silva JCB, Jeans G, Alpers W, Caruso MJ (2013) Nonlinear internal waves in synthetic aperture radar imagery. Oceanography 26(2):68–79. https://doi.org/10.5670/oceanog.2013.32

Klymak JM, Moum JN (2003) Internal solitary waves of elevation advancing on a shoaling shelf. Geophys Res Lett 30:2045. https://doi.org/10.1029/2003GL017706

Korteweg DJ, de Vries G (1895) On the change of long waves advancing in a rectangular canal and a new type of long stationary waves. Phil Mag 5:422

Kudryavtsev V, Akimov D, Johannessen JA, Chapron B (2005) On radar imaging of current features: 1. Model and comparison with observations. J Geophys Res 110:C07016. https://doi.org/10.1029/2004JC002505

Li X, Jackson CR, Pichel WG (2013) Internal solitary wave refraction at Dongsha Atoll, South China Sea. Geophys Res Lett 40:3128–3132. https://doi.org/10.1002/grl.50614

Liu AK, Chang YS, Hsu MK, Lang NK (1998) Evolution of nonlinear internal waves in the East and South China Seas. J Geophys Res 103:7995–8008

Lui AK, Holbrook JR, Apel JR (1985) Nonlinear internal wave evolution in the Sulu Sea. J Phys Oceanogr 15:1613–1624

Maxworthy T (1979) A note on the internal solitary waves produced by tidal flow over a three-dimensional ridge. J Geophys Res 84:338–346

Mercier MJ, Vasseur R, Dauxois T (2011) Resurrecting dead-water phenomenon. Nonlin Process Geophys, 18:193–208. www.nonlin-processes-geophys.net/18/193/2011/, https://doi.org/10.5194/npg-18-193-2011

Miles JW (1977a) Obliquely interacting solitary waves. J Fluid Mech 79:157–169

Miles JW (1977b) Resonantly interacting solitary waves. J Fluid Mech 79:171–179

Mitnik L, Alpers W, Lim H (2000) Thermal plumes and internal solitary waves generated in the Lombok Strait studied by ERS SAR. ERS-envisat symposium, Gothenburg, Sweden, 16–20 Oct 2000, ESA Publication SP-461, 1-9.9

Moore S, Lien RC (2007) Pilot whales follow internal solitary waves in the South China Sea. Mar Mann Sci 23:193–196

Nansen F (1897) Farthest North: the epic adventure of a visionary explorer. Skyhorse Publishing, New-York, p 1897

Osborne AR, Burch TL (1980) Internal solitons in the Andaman Sea. Science 208:451–460. https://doi.org/10.1126/science.208.4443.451

Perry BR, Schimke GR (1965) Large-amplitude internal waves observed off the northwest coast of Sumatra. J Geophys Res 70:2319–2324

Romeiser R, Alpers W (1997) An improved composite surface model for the radar backscattering cross section of the ocean surface: part 2. Model response to surface roughness variations and the radar imaging of underwater bottom topography. J Geophys Res 102:25251–25267. https://doi.org/10.1029/97JC00191

Schmidt GM, Phongsuwan N, Jantzen C, Roder C, Khokittiawong S, Richter C (2012) Coral community composition and reef development at large amplitude internal wave affected coral reefs in the Andaman Sea. Mar Ecol Prog Ser 456:113–126. https://doi.org/10.3354/meps09682

Susanto RD, Mitnik L, Zheng Q (2005) Ocean internal waves observed in the Lombok Strait. Oceanography 18:80–87. https://doi.org/10.5670/oceanog.2005.08

Vlasenko V, Alpers W (2005) Generation of secondary internal waves by the interaction of an internal solitary wave with an underwater bank. J Geophys Res 110(C2):C02019. https://doi.org/10.1029/2004jc002467

Vlasenko V, Stashchuk N (2007) Three-dimensional shoaling of large-amplitude internal waves. J Geophys Res 112:C11018. https://doi.org/10.1029/2007JC004107

Wall M, Putchim L, Schmidt GM, Jantzen C, Khokiattiwong S, Richter C (2015) Large-amplitude internal waves benefit corals during thermal stress. Proc R Soc B 282:20140650. http://dx.doi.org/10.1098/rspb.2014.0650

Wang C, Pawlowicz R (2012) Oblique wave-wave interactions of nonlinear near-surface internal waves in the Strait of Georgia. J Geophys Res 117:C06031. https://doi.org/10.1029/2012JC008022

Xue J, Lund B, Graber HC, Romeiser R (2012) A case study of internal wave-wave interaction during the shallow water 2006 experiment. Proc IGARSS 2012:2845–2848

Zeng K, Alpers W (2004) Generation of internal solitary waves in the Sulu Sea and their refraction by bottom topography studied by ERS SAR imagery and a numerical model. Int J Remote Sens 25:1277–1281. https://doi.org/10.1080/01431160310001592175

Remote Sensing Observations of Ocean Surface Chlorophyll and Its Controlling Parameters in Southern Peninsula Indian Waters

R. K. Sarangi

Abstract The analysis of chlorophyll, sea surface temperature (SST), wind speed and nitrate data have been carried out in the monthly, seasonal and inter-annual scales during 1999–2004. The monthly averaged chlorophyll concentration indicates high chlorophyll concentration (0.50–2.0 mg/m^3) in the southern peninsula around the tip of India. The movement of chlorophyll from the Arabian Sea and Gulf of Mannar region towards the east via Sri Lankan region has been observed during the southwest monsoon (SWM) season. The algal bloom has been observed both during southwest and northeast monsoon (NEM) period. The SST has been observed to be high (29–31 °C) during the spring inter monsoon (SIM) during March–May and low during SWM and NEM ($\sim$27 °C). Wind speed has been observed to be very high (8–12 m/s) during the SWM and NEM periods. The relationship between in situ nitrate and temperature has been established with R^2 value 0.912 with 1537 data points. The nitrate concentration has been observed to be high (0.20–0.50 μmol/l) during SWM due to the upwelling process. Relationship has been established between chlorophyll, SST, wind speed and nitrate covering the seasonal averaged data over five-year period. The increase in wind speed may be causing upwelling and mixing phenomenon, bringing up the nutrient rich bottom water to surface and mixes up the water column and hence the decrease in SST. So, the enhancement in productivity/phytoplankton chlorophyll biomass has been observed. The interrelationship of the parameters in the Arabian Sea, Indian Ocean and the Bay of Bengal subsets has been derived. The chlorophyll in the Indian Ocean has been observed to be primarily dependent on nitrate ($R^2 = 0.39$) and SST ($R^2 = 0.38$) and to a lesser extent wind speed ($R^2 = 0.11$). But, the Arabian Sea and Bay of Bengal chlorophyll has been observed to be more dependent on SST ($R^2 = 0.43$ and 0.52) followed by nitrate ($R^2 = 0.30$ and 0.27) and wind speed ($R^2 = 0.18$ and 0.11), obtained from regression analysis.

Keywords Chlorophyll · Sea surface temperature (SST) · Nitrate
Wind speed · Bay of Bengal · Arabian Sea · Northern Indian Ocean

---

R. K. Sarangi (✉)
Marine Ecosystem Division, Space Applications Centre (ISRO), Ahmedabad, India
e-mail: sarangi@sac.isro.gov.in

V. Barale and M. Gade (eds.), *Remote Sensing of the Asian Seas*,
https://doi.org/10.1007/978-3-319-94067-0_23

1 Introduction

Satellite observations are the only means to measure the phytoplankton concentration on global scale and also to sample with sufficient spatial and temporal resolution to avoid aliasing the long-term and large-scale estimates. Satellite observations are also the means to "revisit" large areas of the oceans on a precise scale to monitor any changes in the pigment content, whether influenced by biological process such as growth and grazing, or physical transport by surface winds and ocean currents. In the oceans, the physical, chemical and biological processes are linked in an intimate manner. The biological and chemical-flux, to a certain extent, are also affected by various physical forcing like winds, currents, mixed layer depth and temperature. The present availability of satellite ocean colour data offers an unprecedented opportunity to force biogeochemical models to estimate the pigment distributions. Phytoplankton pigments, principally chlorophyll-a, are of intrinsic interest as an estimate of phytoplankton biomass in regional or seasonal contexts, including the spatial and temporal occurrence of blooms.

Such images are also one of the important data sources for understanding how physical processes affect biological distributions on ocean basins and global scales. The CZCS imagery and that from latter sensors (POLDER/GLI, IRS-P3 MOS, SeaWiFS, IRS-P4 OCM-1, MODIS Terra and Aqua, MERIS, OCM-2, etc.) have been extensively used. Ocean margin waters have particularly strong signatures in chlorophyll-a images and the imagery has been effectively used to study a wide range of environments, including coastal upwelling systems and river plumes. Chlorophyll-a imagery is the only tool for observing the mean and time-varying components of a biological variable at ocean basin scale with the influence of physical and chemical parameters. The concentration of phytoplankton in the ocean depends on the availability of nutrients dissolved in seawater, and sunlight. In the tropics, the growth of phytoplankton is generally nutrient-limited, so that the supply of nutrients into the euphotic zone can trigger phytoplankton blooms. Therefore, physical processes that inject nutrients into the euphotic zone from below are of prime importance. Examples of such processes are coastal upwelling, Ekman pumping in the ocean and entrainment of nutrient-rich water from deeper depths into the mixed layer (McCreary et al. 1996).

Several studies have been carried out to measure the ocean colour parameters in Indian waters (Dwivedi and Narain 1987; Mohan et al. 1998; Namita et al. 2000; Nayak et al. 2001; Chauhan et al. 2002, 2003; Sarangi et al. 2001a, b, 2005; Tang et al. 2002; Dey and Singh 2003; Vinaychandran et al. 2004; Luis and Kawamura 2004). Algorithms have been developed to derive indicators such as chlorophyll, SST and other ocean surface parameters like surface wind vectors. The biological production is affected by the physical processes altering the vertical flux of nutrients in the mixed layer (Banse and English 2000). The influence of northeasterly trade winds on intensity of winter bloom has been studied in the northern Arabian Sea (Dwivedi et al. 2006). The changes in the oceanic surface chlorophyll and physical parameters have been studied in the Arabian Sea off Gujarat (Singh et al. 2001) and in the southern peninsular region (Sarangi et al. 2005, 2008). The continental shelf in this part of the Arabian Sea is much shallower compared to the shelf in the southern Bay of Bengal (Dey and Singh 2003). Using numerical model experiment results, Sathyendranath et al. (1991) demonstrated a significant biological impact on the Arabian Sea's SST. Efforts have been made to study the chlorophyll distributions in the Arabian Sea (Tang et al. 2002) and its influence on the sea surface temperature (Sathyendranath et al. 1991). The variability of phytoplankton chlorophyll and relationship with SST has been studied in the Arabian Sea using SeaWiFS and NOAA-AVHRR data, respectively (Chaturvedi and Narain 2003). Synergistic application based on ocean color and SST information has been carried out to retrieve the potential fishing zones in the eastern Arabian Sea in Indian water (Solanki et al. 2001). SeaWiFS images of the summer monsoon of 1998 (Murtugudde et al. 1999) indicate a seasonal bloom south of Sri Lankan, in the northern Indian Ocean. The Bay of Bengal has several distinguishing features, which make it a particularly unique and dynamic area of study. Like the extensively studied Arabian Sea to its west is semi-enclosed basin in the north Indian Ocean is forced by seasonally reversing monsoons (Shetye et al. 1991, 1993). Its major distinctive feature however, is the tremendous fresh water flux that it receives from rivers. The fresh water influx together with the monsoon winds exerts a strong influence on the surface water circulation and stratification (Potemra et al. 1991). Specifically, the strong stratification in Bay of Bengal due to the barrier layer impedes the transfer of nutrients into the euphotic zone from below (Gomes et al. 2000). The Bay of Bengal is usually considered less productive compared to the Arabian Sea. Although many major river systems bring in large quantities of nutrients, narrow shelf and persistent cloud cover and low light penetration are responsible for low productivity (Narvekar and Prasanna Kumar 2006; Prasanna Kumar et al. 2010). However, recent studies show that episodic events such as cyclones and mesoscale features such as cold-core eddies are capable of enhancing biological productivity in Bay of Bengal (Vinaychandran and Mathew 2003; Srinivasa et al. 2009). The chlorophyll-a rich water from the Indian coast advects along with the southwest monsoon current (SMC) towards Sri Lankan from the Indian southern peninsula and in the northern Indian Ocean around Indian tip (Vinayachandran et al. 2004).

The present paper emphasizes the observation of surface chlorophyll, SST, wind speed and nitrate concentration in the southern Indian water (Longitude: 75–85°E and Latitude: 5–12°N) covering portions from three ocean regimes; southwest Bay of Bengal, northern Indian Ocean adjacent to the southern Indian tip and southeast Arabian Sea and their relationship on seasonal scale. The inter-relationship between these parameters can be utilized for ecological modeling and forecast development. Attempts can be made to understand the possible causes and mechanisms regulating the concentration and magnitude of above parameters and the inter-relationship of parameters during five-year study period.

2 Study Area and the Essence of the Study

The present study has been carried in the southern peninsular Indian waters that include a part of the Arabian Sea (AS), Bay of Bengal (BoB) and Indian Ocean (IO) (Fig. 1). The study area is an interesting area, because it consists of three different water types with their own productivity potential and other environmental features. The variability of phytoplankton production could be established with respect to alongshore and offshore wind stress, seasonal upwelling, nutrient enrichment, temperature variability, currents, eddies and other physical oceanographic parameters on different time scales. The southern tip of India is associated with the IO. Mixing and convergence takes place between the waters of the AS and the BoB. The shelf flow also determines the productivity pattern of the IO. The study of SST, wind stress and surface nitrate concentration on seasonal scale along with phytoplankton production will help to predict potential bio-ocean productivity on temporal scales and factors influencing the phytoplankton production in oceanic water. The BoB is activated by the cyclonic winds, storms, gyres and sea surface is rough compared to the AS. The surface primary productivity is controlled by the oceanic features, which can vary from season to season.

The BoB is a semi-enclosed tropical basin in the northern IO. It experiences seasonal changes in circulation and weather due to the seasonally reversing monsoons. The riverine inputs are a major potential source of nutrients such as nitrate, phosphate and silica to the Bay. Also, one of the major suppliers of the nutrients to

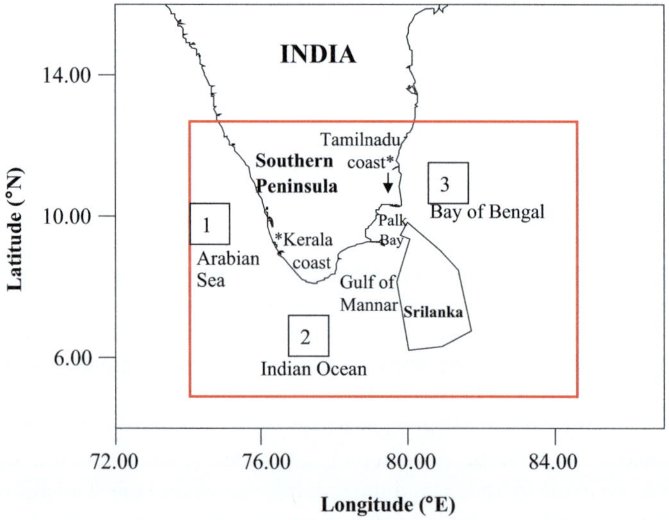

Fig. 1 Study area [Latitude: 5.5–12°N and Longitude: 75–84°E] indicating the Arabian Sea, Indian Ocean and the Bay of Bengal with respective boxes (1, 2, 3)

the surface in BoB is mixing due to cyclones, frequent during post-monsoon season (Madhu et al. 2002). The formation of localized intense blooms and also the intensification of bloom generated by anti-cyclonic gyre is known due to injection of nutrients by cyclonic activity (Vinayachandran and Mathew 2003). Biological productivity in the BoB is believed to be considerably less than the AS, particularly during the summer monsoon (Prasanna Kumar et al. 2002; Madhupratap et al. 2003).

3 Data and Methodology

In the present study dataset has been utilized to retrieve ocean surface chlorophyll-a and interpret with processed sea surface temperature (SST), ocean surface wind speed/vector and surface water nitrate datasets for a period of five years during July 1999 to June 2004. This study focuses on the estimation of phytoplankton pigment concentration using IRS-P4 OCM data relating to above area in order to assess the influence of oceanographic features on chlorophyll variability. The IRS-P4 (Oceansat-1) OCM sensor data has been used to retrieve surface ocean chlorophyll-a images. The details of the OCM, technical specifications have been described in the published literature (Navalgund and Kiran Kumar 1999). NOAA-AVHRR pathfinder data has been used to generate SST maps. NASA Quickscat scatterometer data has been used to generate wind speed maps. The in situ nitrate and SST based modeled nitrate maps are generated for the study area. The monthly images were averaged to seasonal composite images as follows; (1) Southwest monsoon (SWM), June–September, (2) Fall inter-monsoon (FIM), October–November, (3) Northeast monsoon (NEM), December–March and the (4) Spring inter-monsoon (SIM), April–May (JGOFS 1996).

The retrieval of ocean colour parameters such as phytoplankton pigment chlorophyll-a involves two major steps, (i) Atmospheric correction of visible channels to obtain normalized water leaving radiances (nLw) and (ii) Retrieval of water constituents, such as chlorophyll, suspended sediments, etc. The procedures adapted for atmospheric correction (to remove Rayleigh and aerosol scattering etc.) has been mentioned in published literatures (Gordon and Clarke 1980; Mohan et al. 1997). The water leaving radiances derived from atmospheric correction procedure are used to compute chlorophyll-a concentration.

3.1 Chlorophyll Algorithm

The Ocean Chlorophyll 2 or OC2 algorithm with five coefficients (O'Reilly et al. 1998) applied for chlorophyll retrieval and has following mathematical form:

$$C = -0.040 + 10^{0.341 - 3.001X + 2.811X^2 - 2.041X^3} \quad (1)$$

where C is chlorophyll concentration in mg/m^3 and $X = \log_{10}[R_{rs}490/R_{rs}555]$, $R_{rs}490$ and $R_{rs}555$ are remote sensing reflectance for 490 and 555 nm wavelength channels. While comparing all available algorithms for Indian waters, this algorithm provided best results for chlorophyll retrieval and shown $\pm$30% error accuracy in open ocean (Chauhan and Mohan 2000). The chlorophyll images were generated for July 1999 to June 2004. The IRS-P4 OCM data has the resolution of 360 m. The above images were geometrically corrected and gridded and monthly and seasonal binned images were generated.

3.2 Sea Surface Temperature (SST)

The NOAA-AVHRR Pathfinder based monthly averaged SST ascii data with latitude, longitude and SST columns were accessed from the JPL Physical Oceanography Distributed Active Archive Center (PO.DAAC) data distribution archive[1] for the period July 1999 to June 2004 and utilized for generating SST maps over the study area. The Pathfinder data has the resolution of 4 km and gives SST accuracy of 0.2 °C. The monthly and seasonal binned images were processed covering the study area (Fig. 1), Latitude: 5.5–12°N and Longitude: 75.5–85°E in the Indian southern peninsula water.

3.3 Wind Speed Maps

Quickscat scatterometer data was utilized to acquire all-weather, high-resolution measurements of near-surface winds over global oceans. A 1800-kilometer swath during each orbit provides approximately 90-percent coverage of Earth's oceans every day. Wind speed measurements of 3–20 m/s range, with an accuracy of 2 m/s with its direction, with an accuracy of 20° with wind vector resolution of 25 km has been the target of retrieval. Scatterometer data products have been archived through the JPL Physical Oceanography Distributed Active Archive Center (PO.DAAC) data distribution site.

[1]http://podaac.jpl.nasa.gov.

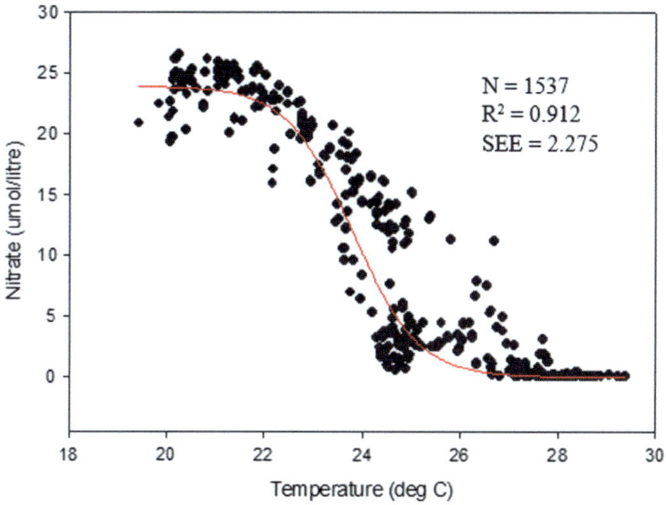

Fig. 2 Sigmoid relationships between Temperature and Nitrate have been displayed for the Arabian Sea and Bay of Bengal data collected from JGOFS and Indian cruises (Sarangi 2011)

3.4 Nitrate Retrieval Algorithm

In this study attempt was made to find the relationship between the nitrate and temperature from the in situ data. As earlier, linear relationship has been established in the European waters (Morin et al. 1993). So, for Indian water attempt was made to produce relationship between nitrate and SST from the US JGOFS dataset for the Arabian Sea (1994–1997) and Indian cruise datasets (2000–2006) covering the Arabian Sea and Bay of Bengal (Fig. 2) and the relation found (Sarangi 2011) is as follows:

$$\begin{array}{c} \text{Nitrate}\,(\text{NO}_3) = 23.924/[1 + \text{e}^{\{(\text{SST}-23.793)/(0.662)\}}] \\ \text{R}^2 = 0.912,\ \text{Standard Error of Estimate}(\text{SEE}) = 2.2755 \end{array} \quad (2)$$

The pockets/boxes data were retrieved from three different Oceans; (1) Arabian Sea (Latitude: 8.95–9.95 N and Longitude: 74.6–75.6 E), (2) Indian Ocean (6.24–7.24 N and 77.1–78.1 E) and (3) Bay of Bengal (10.2–11.2 N and 80.4–81.4 E) (Fig. 1). The three pockets are off the coast and covering the water in the depth ranges to 1000–3000 m (Fig. 1).

4 Results and Discussion

Seasonal upwelling in the southwest monsoon around the Kerala coast and southeast Arabian Sea has been observed to be associated with high chlorophyll (1.0 mg m^{-3}), low SST (25–27 °C) and high nitrate (~0.50 µmol/l) in respective images (Figs. 3, 4 and 5). In the Gulf of Mannar and the Palk Bay region, the chlorophyll biomass observed consistently very high (1.0–10.0 mg/m^3), where the churning of water in the shallower depth (<200 m) takes place in the semi-enclosed

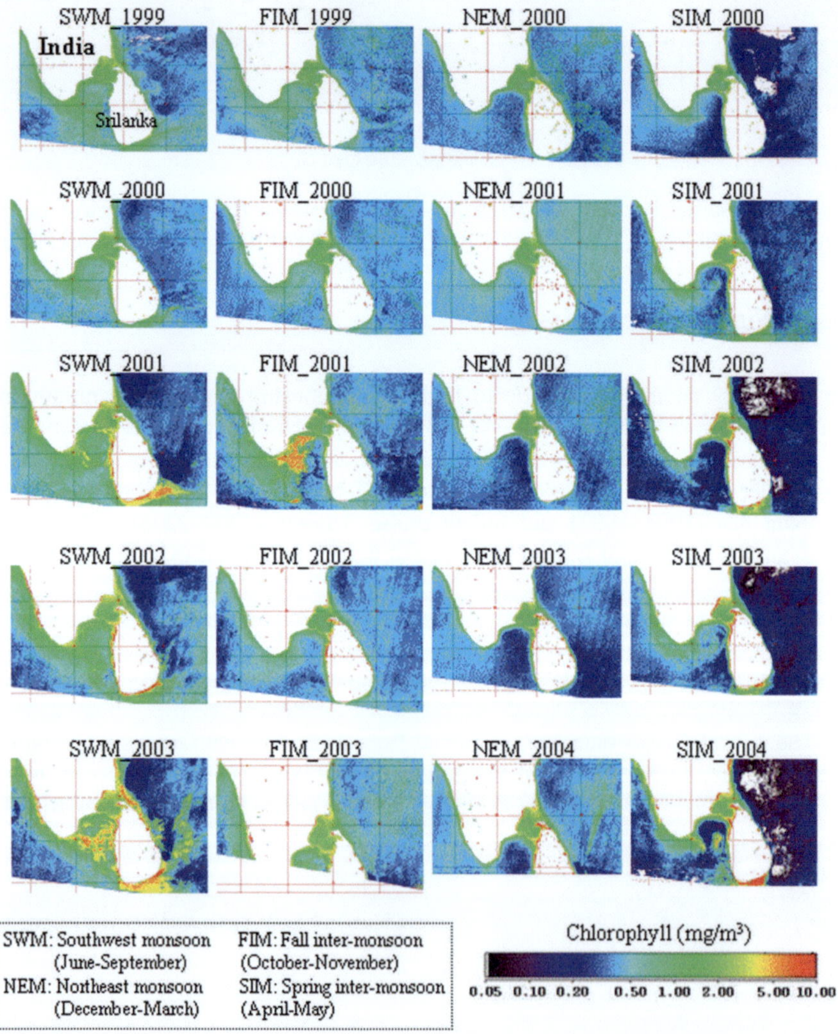

Fig. 3 IRS-P4 OCM derived seasonal composite chlorophyll images during 1999–2004

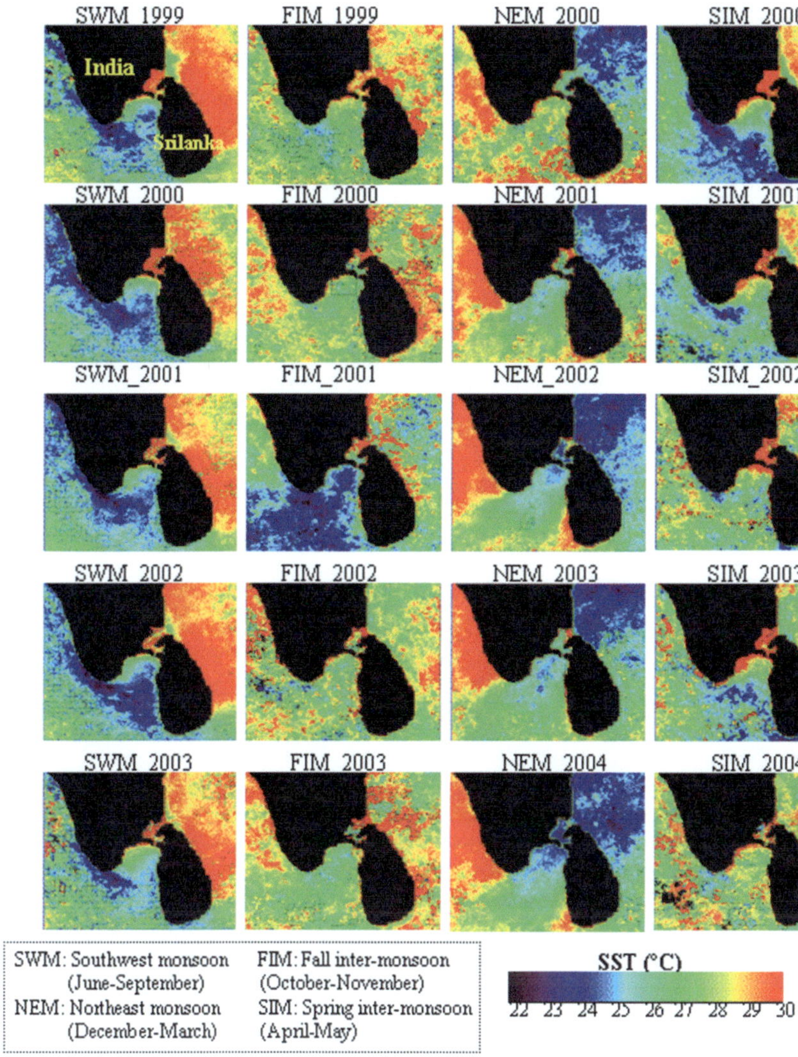

Fig. 4 NOAA-AVHRR derived seasonally averaged SST images during 1999–2004 in the southern peninsular Indian water

Gulf water. This region has been predicted to be a region of high fish catch throughout the year. Even the biogeochemical cycling rate of carbon and nutrients would be high.

The impact of East Indian Coastal Current (EICC) is seen as tracer in the chlorophyll images with the movement of high chlorophyll concentration (0.5–2.0 mg m^{-3}) patches in current direction during the months July–August off Tamilnadu coast in the Bay of Bengal water (Fig. 3). The West India coastal current

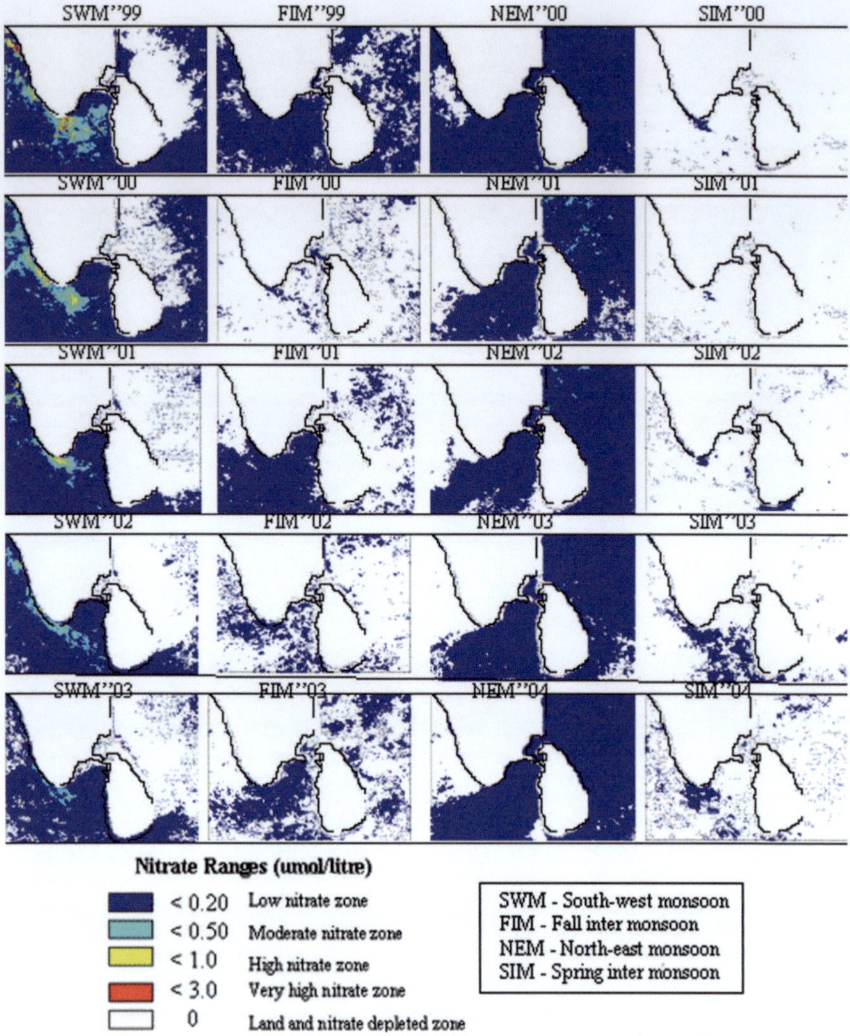

Fig. 5 In-situ nitrate and temperature based five-year seasonal composite modeled nitrate images during July 1999 to June 2004 retrieved from NOAA-AVHRR pathfinder data

(WICC) is seen in the Arabian Sea from the chlorophyll images associated with the movement of features/patches of chlorophyll in northwest direction. The other currents like southwest monsoon current (SMC) and northeast monsoon current (NMC) has been observed in the chlorophyll and SST images around southern part of Sri Lankan coast during southwest and northeast monsoons. The southeasterly and northwesterly of the currents have been understood with the chlorophyll images pattern.

The movement of algal bloom patches has been observed. The full image of the study area (covering 3 oceanic sub-regions) indicated that the relationship of chlorophyll is more dependent on wind speed than followed by nitrate and SST (Fig. 7). It shows that for the southern peninsular Indian water, the wind speed is playing a major role throughout the year, compared to nitrate and SST.

In the Arabian Sea, the chlorophyll is observed to be more dependent on SST than nitrate and wind speed. This is confirmed from our study of regional pockets and their correlations. Similarly, the northern Indian Ocean chlorophyll is more dependent on nitrate than SST and wind speed. The Bay of Bengal like Arabian Sea indicated similar dependency parameter as the chlorophyll concentration is depending more on SST than nitrate and wind speed.

Observation of algal bloom around Sri Lankan east coast has been triggered by eddies around Sri Lankan dome and it is a regular phenomenon. During southwest monsoon (June–September) the Indian southern tip and water around Sri Lankan southern part indicates the algal bloom of high chlorophyll concentration. Western boundary current (WBC) and its impact on chlorophyll images is observed. The bloom of the year 2003, showed very high chlorophyll concentration patches during the southwest monsoon, supported by low SST, high nitrate and high wind speed. The Indian Ocean region around the southern tip of India can be recognized as the most potential region for the fishery and high biogeochemical fluxes on seasonal and annual scale, where even the nitrate concentration observed to be higher ($\sim$0.50 μmol/l) during both the major monsoons, SWM and NEM periods. The seasonal average chlorophyll concentration was highest in the Indian Ocean region i.e. 0.7 mg m^{-3} (Fig. 4). The seasonal average SST was highest in year 2003, i.e. 29.4 °C in the Arabian Sea (Fig. 4). The seasonal average wind speed was highest for the year 2003 in the Indian Ocean, i.e. 7 m/s (Fig. 6) and the seasonal averaged nitrate was highest in the Arabian Sea, i.e. 0.08 μmol/l (Fig. 5).

There has been high chlorophyll concentration (1.0–2.0 mg/m^3) algal bloom observed in the western Bay of Bengal off the Indian east coast during the northeast monsoon, December–January. Along with high chlorophyll, the nitrate concentration observed high (0.2–2.0 μmol/l) in the BoB. The SST observed to be low (24–26 °C) and wind speed moderate (4–6 m/s). It seems to be occurrence of upwelling in the western BoB during northeast monsoon months. Five-year seasonal average images indicate the high chlorophyll trend as Indian Ocean > Arabian Sea > Bay of Bengal (Table 1); The SST trend as Arabian Sea > Bay of Bengal > Indian Ocean; The wind speed trend as Indian Ocean > Bay of Bengal > Arabian Sea and the nitrate trend as Arabian Sea > Indian Ocean > Bay of Bengal.

Out of the three pockets of the study area in the Arabian Sea, Indian Ocean and the Bay of Bengal, the Indian Ocean region found to be highly productive during most of the period of the year with average chlorophyll concentration, 0.5–2.0 mg/m^3, covering the coastal and offshore waters, followed by the Arabian Sea and the Bay of Bengal region.

There is indication of strong upwelling, Ekman pumping and convective processes, which brings up nutrients from water column and bottom to surface, which is also observed in the wind speed information. The nutrient nitrate has acted as

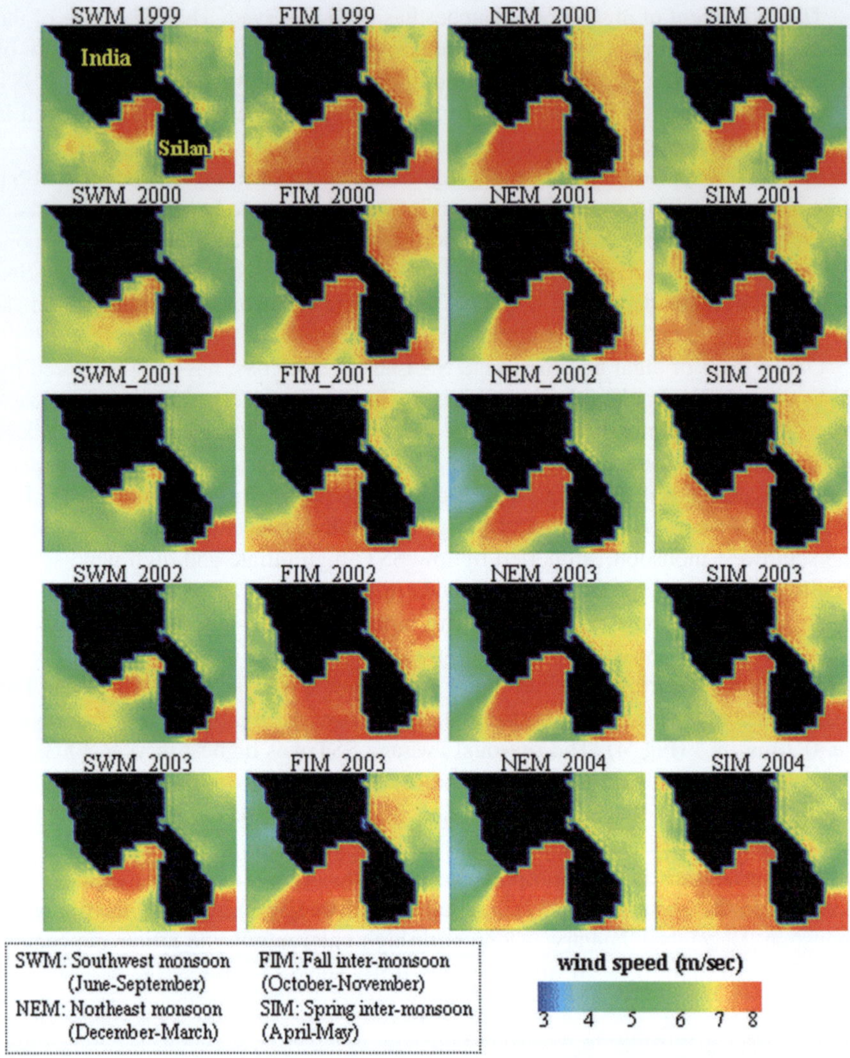

Fig. 6 Quickscat scatterometer retrieved seasonally averaged wind speed images during 1999–2004 in the southern peninsular Indian water

Table 1 Five years seasonally averaged chlorophyll concentration (mg/m^3)

Seasons/study region	Arabian Sea	Indian Ocean	Bay of Bengal
Southwest monsoon	0.508	0.564	0.268
Fall inter-monsoon	0.436	0.62	0.442
Northeast monsoon	0.374	0.397	0.414
Spring inter-monsoon	0.344	0.392	0.128

Table 2 Seasonal trend of nitrate variability

Study area	High nitrate/months
Arabian Sea	June–September
Indian Ocean	June–September and December–February
Bay of Bengal	December–February

April–May showed fully NO_3 depleted zones throughout the study area

limiting parameter to the chlorophyll biomass in the Indian Ocean region. The chlorophyll biomass and variability in the AS is highest during the SWM period. In the BoB the chlorophyll biomass and variability is influenced by the NEM period.

The Indian Ocean region is influenced by both the SWM and NEM period, even the impact during the transition/inter-monsoon periods has been observed. The nitrate depleted zones (NDZ) have been found in the month of April and May along with low chlorophyll and high SST, void of vertical processes in the study region (Table 2).

All the four parameters (chlorophyll, SST, wind speed, nitrate), four-year composite indicates that with the impact of higher temperature there has been reduction in nitrate concentration as no vertical movement of water and any advection or flux of water, so low chlorophyll and low wind speed is observed in the Bay of Bengal off Sri Lankan east coast and off Indian southeast coast. In the southern tip of India, the wind speed has been very high (>8 m/s), lower temperature (24–27 °C SST) and high nitrate ranged between (0.1–0.3 μmol/l) and high chlorophyll 1.0–2.0 mg/m^3 is observed (Figs. 3, 4, 5 and 6). The surface nitrate concentration was highest in the Arabian Sea during June–September (SWM) period, in the BoB the nitrate was highest during December–February (NEM) and the Indian Ocean exhibited high nitrate during both the seasons SWM and NEM.

Upwelling during the SWM has good link to the high nitrate concentration, enhancing the chlorophyll concentration in the Arabian Sea and around southern tip of India (Indian Ocean). The relationship between chlorophyll with SST, wind and nitrate indicate the interdependency of parameters (Fig. 7a–l). But the situation is different in the case of different seas; their chlorophyll biomass and dependency parameters (Table 3). The northern Indian Oceans chlorophyll is more dependent on nitrate (R^2 = 0.39) than SST and wind speed, due to the seasonal upwelling phenomenon. Observation oceanic feature eddies also play a role to indicate high chlorophyll triggered by the vertical nitrates churning. The Bay of Bengal (BoB) like Arabian Sea (AS) indicated similar dependency parameter as the chlorophyll concentration is depending more on SST (R^2 = 0.52 and 0.43 respectively) than nitrate and wind speed (Fig. 7a–l). Observation of algal bloom around Sri Lankan east coast has been triggered by eddies around Sri Lankan dome and is known to be a regular phenomenon.

The result depicts the inference that the Arabian Sea undergoes upwelling (Prasanna Kumar et al. 2002; Subrahmanyam et al. 2002) and convective eddies

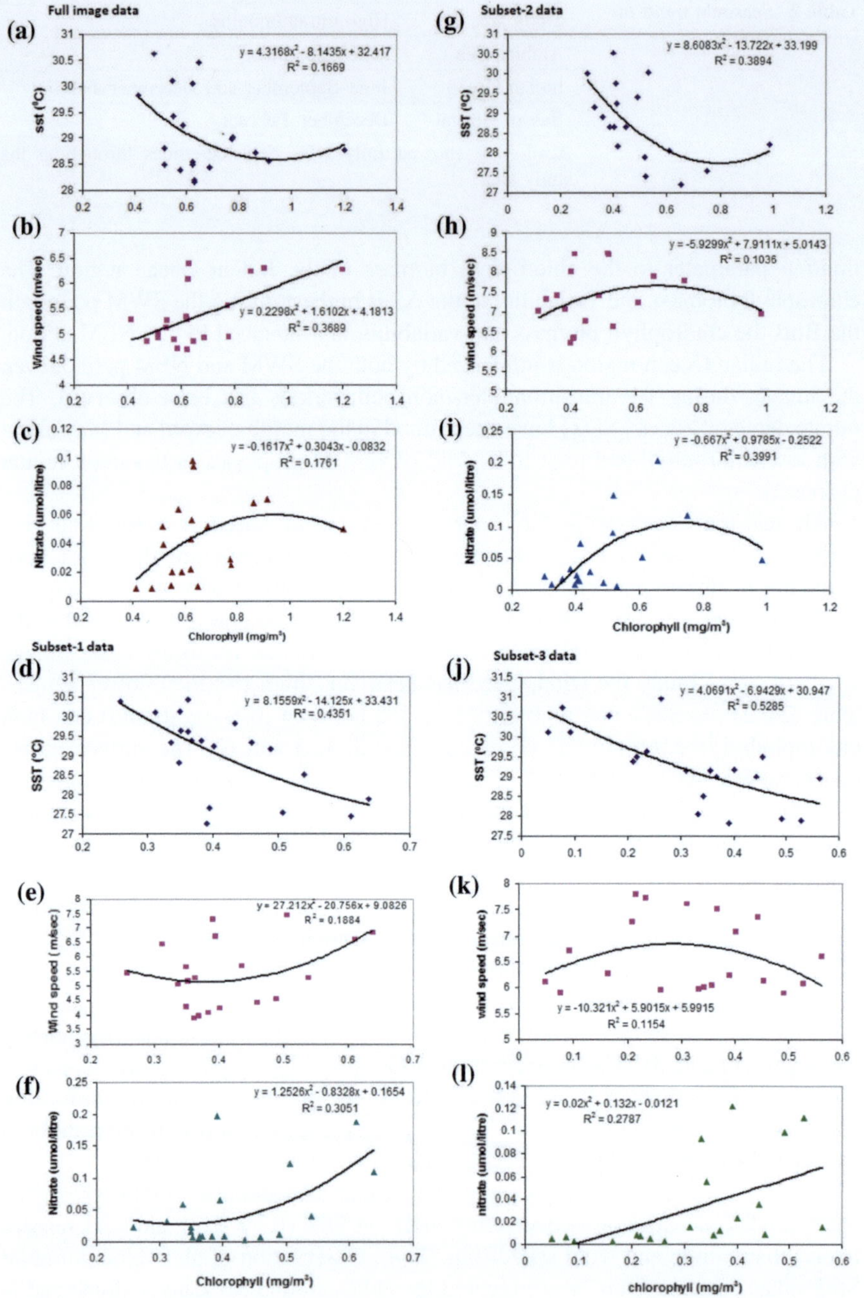

Fig. 7 Inter-relationship of OCM derived chlorophyll with NOAA-AVHRR SST, Quickscat scatterometer wind speed and modeled nitrate for the study area depicting (full image, **a–c**; Arabian Sea/subset-1, **d–f**; Indian Ocean/subset-2, **g–i** and Bay of Bengal/subset-3, **j–l**) on seasonal scale data (1999 July to 2004 June)

Table 3 Relationship (R^2 value) of chlorophyll with regulating parameters

Study area (Fig. 1) (Subsets & full image)	Chl versus SST	Chl versus wind speed	Chl versus nitrate
Arabian Sea (AS)	0.435	0.188	0.305
Bay of Bengal (BoB)	0.528	0.115	0.278
Indian Ocean (IO)	0.389	0.103	0.399
Full image (AS, BoB and IO)	0.166	0.368	0.176

which cools the surface water and brings up nutrient rich water and similar case was evident for the NIO pocket. The high wind speed enhances the surface NIO water and the chlorophyll concentration by bringing nutrients up.

BoB with higher SST diminishes the chlorophyll concentration as compared to AS and NIO region. Even more rainfall and fresh water flux from rivers may be playing a predominant role for decrease in chlorophyll concentration, even if nitrate availability from river sources used to be there.

5 Conclusions

The inter-annual, intra-annual and intra-seasonal variability of chlorophyll and corollary parameters, SST, wind speed and derived nitrate, in the study areas Arabian Sea (AS), north Indian Ocean (NIO) and Bay of Bengal (BoB) has been studied during a five-year period 1999–2004 using Oceansat-1 OCM data. The impact of southwest monsoon (SWM) and northeast monsoon (NEM) on the study area has been studied, along with the role of nutrient nitrate on ocean chlorophyll with associated ocean processes like upwelling, Ekman pumping, convection etc. The enhancement of chlorophyll has been observed during both the monsoon (SWM and NEM) phases. The AS productivity is higher during the SWM and the BoB productivity during the NEM. The correlation of chlorophyll and nitrate was found as R^2 = 0.305, 0.278 and 0.399 for AS, BoB and NIO, respectively, and with wind speed as 0.188, 0.115 and 0.103, respectively. For the entire study area covering these 3 sub-regions, the correlation of chlorophyll and wind speed was even better, R^2 = 0.368. This study shows that SST is the most important parameter for the BoB and AS, while nitrate is important in regulating the chlorophyll variability. The role of wind speed is vital for the entire region and can be considered as third parameter for the BoB and AS productivity. For larger regions the wind speed role would be vital for this type of study. Basing on this study data on SST, nitrate and wind speed along with chlorophyll and productivity data needs to be correlated for larger basins and ocean regions in climatic time scales. The inference can be modeled to predict the ocean productivity with help of multiple parameters regression in different regional seas and time scales and can be attempted using remote sensing based datasets. So, this study is very interesting for understanding the causative factors and even the phenomenon which regulates the ocean chlorophyll and productivity in the southern peninsular Indian water over five-year time period using Indian satellite Oceansat-1 OCM and other satellite datasets.

Acknowledgements The author mentions thanks to Dr. Prakash Chauhan, GD/BPSG, Dr. J. S. Parihar, former Deputy Director, EPSA and Director, Space Applications Center, Ahmedabad for necessary guidance and facilities for carrying out the work.

References

Banse K, English DC (2000) Geographical differences in seasonality of CZCS-derived phytoplankton pigment in the Arabian Sea for 1978–1986. Deep Sea Res 47:1623–1677

Chaturvedi N, Narain A (2003) Chorophyll distribution pattern in the Arabian Sea: seasonal and regional variability, as observed from SeaWiFS data. Int J Remote Sens 24(3):511–518

Chauhan P, Mohan M (2000) Technical report on IRS-P4 OCM Special Product Generation Software, (SAC, Ahmedabad), 4–6, 2000

Chauhan P, Mohan M, Sarangi RK, Kumari Beena, Nayak SR, Matondkar SGP (2002) Surface chlorophyll-a estimation in the Arabian Sea using IRS-P4 ocean colour monitor (OCM) satellite data. Int J Remote Sens 23(8):1663–1676

Chauhan P, Mohan M, Nayak SR (2003) Comparative analysis of ocean color measurements of IRS-P4 OCM and SeaWiFS in the Arabian Sea. IEEE Trans Geosci Remote Sens 41(4): 922–926

Dey S, Singh RP (2003) Comparison of chlorophyll distributions in the northeastern Arabian Sea and southern Bay of Bengal using IRS-P4 ocean color monitor data. Remote Sens Environ 85:424–428

Dwivedi RM, Narain A (1987) Remote sensing of phytoplankton. An attempt from the landsat thematic mapper. Intl J Rem Sens 8:1563–1569

Dwivedi RM, Raman M, Parab S, Matondkar SGP, Nayak S (2006) Influence of northeasterly trade winds on intensity of winter bloom in the northern Arabian Sea. Curr Sci 90(10):1397–1406

Gomes HR, Goes JI, Saino T (2000) Influence of physical processes and freshwater discharge on the seasonality of phytoplankton regime in the Bay of Bengal. Cont Shelf Res 20:313–330

Gordon HR, Clark DK (1980) Atmospheric effects in the remote sensing of phytoplankton pigments. Bound-Layer Meteorol 18:299–313

JGOFS (1996) Protocols for the Joint Global Ocean Flux Study (JGOFS) core measurements. Report No. 19, Intergovernmental Oceanographic Commission, Bergen, Norway. 170 pp

Luis AJ, Kawamura H (2004) Air-sea interaction, coastal circulation and primary production in the eastern Arabian Sea: a review. J Oceanogr 60:205–218

Madhu NV, Maheswaran PA, Jyotibabu R, Sunil V, Ravichandran C, Balasubramaniam T, Gopalakrsishnaan TCAND, Nair KKC (2002) Enhanced biological production off Chennai triggered by October 1999 super cyclone (Orissa). Curr Sci 82(12):1472–1479

Madhupratap M, Gauns M, Ramaiah N, Kumar SP, Muraleedharan PM, de Souza SN, Sardesai S, Muraleedharan U (2003) Biogeochemistry of the Bay of Bengal: physical, chemical and primary productivity characteristics of the central and western Bay of Bengal during summer monsoon 2001. Deep Sea Res Part II, 50:881–896

McCreary JP, Han W, Shankar D, Shetye SR (1996) Dynamics of East India Coastal current, numerical solutions, J Geophys Res Part 2, 101:13993–14010

Mohan M, Chauhan P, Mathur A, Dwivedi RM (1998) Atmospheric correction of MOS-B data using long wavelength and PCI approaches: IRS-P3 MOS validation experiment: ocean applications. SAC Report No. SAC/RESA/MWRD/ IRSP3/SN/02/98, pp 14–22

Mohan M, Chauhan P, Raman M, Solanki HU, Mathur AK, Nayak SR, Jayaraman A, Satheesh SK, Krishnan SK (1997) Initial results of MOS validation experiment over the Arabian Sea, atmospheric correction aspects. In: Proceedings of the first international workshop on MOS-IRS and ocean colour, DLR, Berlin

Morin P, Wafar MVM, LeCorre P (1993) Estimation of nitrate flux in a tidal front from satellite derived temperature data. J Geophys Res 98:4689–4695

Murtugudde RG, Signorini SR, Christian JR, Busalachhi AJ, McClain CR, Picaut J (1999) Ocean color variability of the tropical Indo-Pacific basin observed by the SeaWIFS. J Geophys Res 104:18351–18366

Namita AJ, Rao LVG, Panigrahi RC (2000) Pigment patterns in the Arabian Sea: spatiotemporal variability as observed by ADEOS. PORSEC Proc 1:50–54

Narvekar J, Prasanna Kumar S (2006) Seasonal variability of the mixed layer in the central Bay of Bengal and associated changes in nutrients and chlorophyll. Deep-Sea Res I, 53:820–835

Navalgund RR, Kiran Kumar S (1999) IRS-P4 Ocean color monitor (OCM). IOCCG web page (http://www.ioccg.org/ocm/ocm.htm)

Nayak SR, Sarangi RK, Rajawat AS (2001) Application of IRS-P4 OCM data to study the impact of cyclone on coastal environment of Orissa. Curr Sci 80(10):101–106

O'Reilly JE, Maritorena S, Mitchell BG, Siegel DA, Carder KL, Garver SA, Kahru M, McClain CR (1998) Ocean color chlorophyll algorithms for SeaWiFS. J Geophys Res 103:24937–24953

Potemra JT, Luther ME, Brien JJ (1991) The seasonal circulation of the upper ocean in the Bay of Bengal. J Geophys Res 96:12667–12683

Prasanna Kumar S, Muraleedharan PM, Prasad TG, Gauns M, Ramaiah N, De Souza SN, Sardesai S, Madhupratap M (2002) Why is the Bay of Bengal less productive during summer monsoon compared to the Arabian Sea? Geophys Res Lett 29(24):2235. https://doi.org/10.1029/2002gl016013

Prasanna Kumar S, Narvekar J, Nuncio M, Kumar A, Ramaiah N, Sardesai S, Gauns M, Fernandes V, Paul J (2010) Is the biological productivity in the Bay of Bengal light limited? Curr Sci 98:1331–1339

Sarangi RK, Chauhan P, Mohan M, Nayak SR, Navalgund RR (2001a) Phytoplankton distribution in the Arabian Sea using IRS-P4 OCM satellite data. Int J Remote Sens 20(15):2863–2866

Sarangi RK, Chauhan P, Nayak SR (2001b) Phytoplankton bloom monitoring in the offshore water of Northern Arabian Sea using IRS-P4 OCM Satellite data. Indian J Mar Sci 30(4):214–221

Sarangi RK, Chauhan P, Nayak SR (2005) Inter-annual variability of phytoplankton blooms in the northern Arabian Sea during winter monsoon period (February–March) using IRS-P4 OCM data. Indian J Mar Sci 34(2):163–173

Sarangi RK, Nayak SR, Panigrahy RC (2008) Monthly variability of chlorophyll and associated physical parameters in the southwest Bay of Bengal water using remote sensing data. Indian J Mar Sci 37(3):256–266

Sarangi RK (2011) Remote-sensing-based estimation of surface nitrate and its variability in the Southern Peninsular Indian waters. Intern J Oceanogr 2011:16, Article ID 172731. https://doi.org/10.1155/2011/172731

Sathyendranath S, Gouveia AD, Shetye SR, Platt T (1991) Biological controls of surface temperature in the Arabian Sea. Nature 349:54–56

Shetye SR, Shenoi SSC, Gouveia AD, Michael GS, Sundar D, Nampoothiri G (1991) Wind-driven coastal upwelling along the western boundary of the Bay of Bengal during the southwest monsoon. Cont Shelf Res 11:1397–1408

Shetye SR, Gouveia AD, Shenoi SSC, Sundar D, Michael GS, Nampoothiri G (1993) The western boundary current of the seasonal subtropical gyre in the Bay of Bengal. J Geophys Res 98(C1):945–954

Singh RP, Bhoi S, Sahoo AK (2001) Significant changes in the ocean parameters after the Gujarat earthquake. Curr Sci 80(11):1376–1377

Solanki HU, Dwivedi RM, Nayak SR, Jadeja JV, Thaker DB, Dave HB, Patel MI (2001) Application of ocean colour monitor chlorophyll and AVHRR SST for fishery forecast: preliminary validation result off Gujarat Coast, northwest coast of India. Indian J Mar Sci 30:132–138

Srinivasa KT, Nayak SR, Mupparthy RS, Nagaraja KM (2009) Phytoplankton bloom due to Cyclone Sidr in the central Bay of Bengal. J Appl Remote Sens 3(1):033547

Subrahmanyam B, Rao KH, Srinivasa Rao N, Murty VSN, Sharp RJ (2002) Influence of a tropical cyclone on Chlorophyll-a concentration in the Arabian Sea. Geophys Res Lett 29(22):1–4

Tang DL, Kuwamura H, Luis AJ (2002) Short term variability of phytoplankton blooms associated with a cold eddy in the northwestern Arabian Sea. Remote Sens Environ 81:82–89

Vinayachandran PN, Mathew S (2003) Phytoplankton bloom in the Bay of Bengal during the northeast monsoon and its intensification by cyclones. Geophys Res Lett 30(11):1572. https://doi.org/10.1029/2002GL016717

Vinayachandran PN, Chauhan P, Mohan M, Nayak SR (2004) Biological response of the sea around Sri Lankan to summer monsoon. Geophys Res Lett 31:L01302. https://doi.org/10.1029/2003GL018533

Ocean Surface Vector Wind Observations

Ad Stoffelen, Raj Kumar, Juhong Zou, Vladimir Karaev, Paul S. Chang and Ernesto Rodriguez

Abstract Ocean surface vector winds (OSVW) play a fundamental role in the Asian Seas through air-sea interaction; this applies to the modest winds in the trades, the winds associated with the extensive areas of tropical convection, sea and land breezes and, of most direct human relevance, the winds associated with hurricane-force typhoons. Predicting the air-sea exchanges in the cold polar seas and the atmospheric dynamics of tropical mesoscale convective systems or the strength and track of typhoons remains equally a challenge, but is of fundamental importance for weather forecasting and climate change studies. It is briefly described how wind vector information is obtained from satellite microwave active and passive measurements off the wind-roughened ocean surface, and subsequently an evaluation of the wind vector product services, for example, in coastal areas, is provided. India, China, Russia and Japan, inter alia, have been, are, or will be contributing to a global virtual constellation of scatterometers that provide increasing temporal coverage of ocean surface vector wind information.

A. Stoffelen (✉)
Royal, Netherlands Meteorological Institute (KNMI),
de Bilt, The Netherlands
e-mail: Ad.Stoffelen@knmi.nl

R. Kumar
Indian Space Research Organisation (ISRO), Ahmedabad, India

J. Zou
National Space Ocean Application Service (NSOAS), Beijing, China

V. Karaev
Institute of Applied Physics, Russian Academy of Sciences,
Nizhny Novgorod, Russia

P. S. Chang
National Ocean and Atmosphere Administration (NOAA), Washington,
DC, USA

E. Rodriguez
National Aeronautics and Space Administration (NASA), Pasadena, USA

V. Barale and M. Gade (eds.), *Remote Sensing of the Asian Seas*,
https://doi.org/10.1007/978-3-319-94067-0_24

The application of scatterometer winds for weather nowcasting, for mesoscale and global numerical weather prediction and for oceanography and climate studies is highlighted.

Keywords Scatterometer · Ocean surface vector winds · Virtual constellation Air-sea interaction · Typhoons · Moist convection · Mesoscale Nowcasting · Oceanography · Climate

1 Introduction

Scatterometer winds have proven to be very relevant in driving ocean circulation and air–sea interactions (Chelton et al. 2004; Tokmakian 2005; Blanke et al. 2005; Liu and Xie 2006), which in turn play a major role in the climate system (Xie et al. 2007; Tokinaga et al. 2009; Shimada and Kawamura 2006) and marine biology (Moore and Renfrew 2005). OSVW and wind stress force the ocean in many ways, where most uncertainty exists in these interaction processes at mesoscales and at extreme winds. High winds play a disproportionately large role in Earth's weather and climate. Mid and high latitude, high wind events (cold air outbreaks) lasting several days, can remove what at typical wind speeds would be a month's worth of the ocean's heat and moisture, leading to the formation of "deep water" that helps drive global ocean circulation patterns. High winds also help exchange disproportionately large amounts of carbon dioxide. Moreover, long-term stable instrument records are essential to build confidence in the fidelity and limitations of model reanalyses and guide their application in trend analyses.

Understanding the peculiarities of such a large and varied collection of marine and coastal atmospheric dynamics requires the adoption of integrated systems of earth observation (EO), in which microwave remote sensing by satellite instruments, such as scatterometers (Bourassa et al. 2009), Synthetic Aperture Radars (SAR) and radiometers, plays an essential role.

This chapter reviews the current potential of EO wind in the assessment of the many Asian seascapes, using both passive and active microwave techniques, measuring emitted or scattered microwaves. An in-depth evaluation of the available microwave wind observation techniques, as well as of novel methods, ensures that suitable instruments are indeed accessible for managing and exploiting the wealth of resources that the Asian Seas have to offer. First, the physical regime of mesoscale winds is described to lead on to the application-based requirements and the EO-based capabilities for OSVW. The relationship between EO measurements, OSVWs and wind stress will be elaborated, before discussing the application of EO winds in the last section, including extreme winds (Yu et al. 2015).

2 Ocean Vector Winds and Stress

Though essential and relevant for a wide range of applications in economy and society, such as coastal protection, ship routing, off-shore wind energy, climate-scale circulation, water cycle and ocean forcing, only sparse coverage of the ocean surface by conventional anemometers exists. In particular, spatial gaps of several 1000 km exist, which clearly limit the capability to obtain detailed information on the winds and wind stress over the world oceans. The advent of EO satellites provides a leap forward to fulfil the requirements for wind and stress measurements over the oceans. On the large scales, wind measurements are mainly relevant in the tropics, since elsewhere large-scale wind circulations may be determined from the atmospheric mass field. However, as meteorological analyses more and more focus on mesoscale processes, wind measurements are needed globally as further discussed in the first subsection. Based on this basic need, user requirements are subsequently formulated in the different basic application areas and the associated EO capability on ocean winds. Finally, the relationship between these EO ocean measurements and both winds at 10 m height, which is a World Meteorological Organisation (WMO) standard height, and wind stress, which is the most relevant input parameter for ocean forcing, is discussed.

It is clear that in the next decades focus will be put on mesoscale weather phenomena and coastal dynamics, as these will be required in many off-shore and coastal applications and increasingly resolved by Numerical Weather Prediction (NWP) and ocean models.

2.1 3D Turbulence and Convection

Winds are associated with turbulence and convection, but also with mixing of air, dispersion, fluxes of momentum, heat and moisture, and cloud processes. For example, important exchanges appear due to downbursts of relatively dry and cool air near moist convection, which is widespread in tropical areas in particular. Given these important atmospheric and ocean exchange processes, detailed knowledge of wind fields and their dynamical evolution appears critical to understanding the earth system, since most of the earth surface is covered by water.

Nastrom and Gage (1987) supplemented large-scale experimental evidence to the ideas of Kolmogorov (1941) on atmospheric turbulence, as depicted in Fig. 1. For scales larger than 500 km, the spectral signature designates two-dimensional (2D) turbulence and geostrophic flow, while for smaller scales 3D turbulence appears. Although the depth of the troposphere is order 10 km, 3D turbulence starts to dominate on scales below about 500 km. So, the horizontal-to-vertical aspect ratio of the associated turbulent structures is about 100, with, for example, vortices of 1 km depth being spread 100 km horizontally.

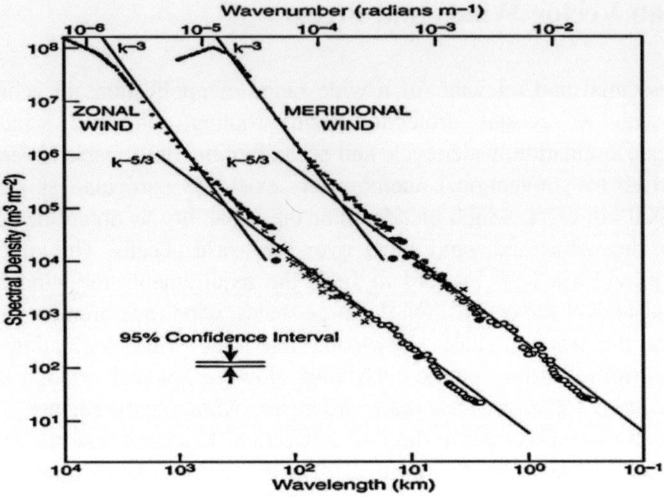

Fig. 1 Spectral wind energy density as a function of wave length (bottom) and wave number (top) for measured upper air zonal and meridional winds, where the latter is shifted right by an order of magnitude for clarity. Nastrom and Gage (1985). © AMS

The spectra in Fig. 1 are determined statistically and the actual two-way spread (2σ) around the climatological mean at small scales is about an order of magnitude in spectral energy. Meanwhile, many other studies confirmed these findings (e.g., Vogelzang et al. 2011). Note that the spectral turbulent energy in the meridional and zonal components in Fig. 1 is very similar and that mesoscale wind variations tend to be statistically isotropic, i.e., variations along the mean large-scale wind tend to be similar to variations across (Vogelzang et al. 2015).

The large-scale heating of the atmosphere results in large-scale circulation patterns, broken down to smaller scales (downscale propagation), until eventual dissipation on the molecular scale. On the other side, atmospheric instability and (moist) convection processes cause the generation of wind variability on the smaller scales, which leads in turn to so-called upscale propagation (King et al. 2015). Note that the dynamics of these small scales are steered by the local winds and that wind measurements are most effective to capture their evolution (Stoffelen et al. 2006).

Moist convection is widespread in the tropics and causes rapid changes in the ocean surface vector winds. Moist convective cells are identified by Meteosat Second Generation (MSG) geostationary imagery, notably the day-time KNMI MSG rain product. Collocated ASCAT winds show characteristic patterns in the wind convergence and curl, as illustrated in Fig. 2. The two panels are centred on the same location and the intense red blob in the left panel of Fig. 2 is closely overlapping the MSG rain pattern and designates a wind downburst due to rapidly descending rain-cooled air aloft. It may be clear that moist convection invigorates mixing of air, dispersion, air-sea fluxes of momentum, heat and moisture, and further lifting of air and cloud processes.

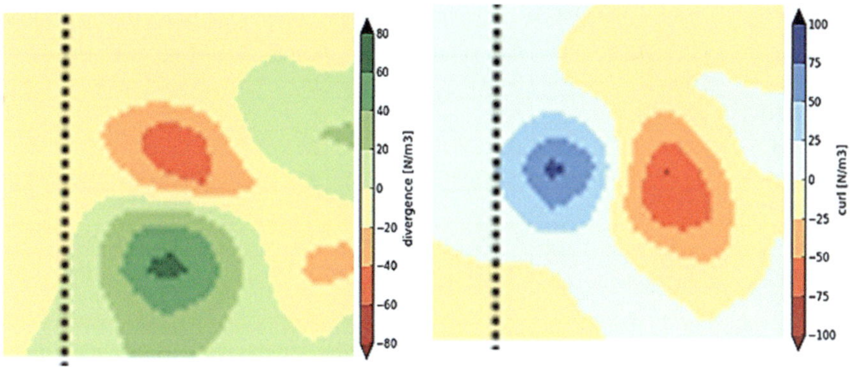

Fig. 2 ASCAT wind convergence (left) and curl (right) around a deep moist convective cell in the tropics (as identified by KNMI MSG rain). The wind convergence and divergence (downburst) are clearly visible, depicting intense air-sea interaction. Moist convection is widespread in the tropics

Although advancing meteorological applications to resolve smaller scales appears natural, the interest in measuring mesoscale and convective-scale wind structures poses daunting challenges:

- Spatial resolution and coverage must be sufficient to capture these scales;
- Accuracy must be sufficiently high to measure the small amplitudes of the small scales (see Fig. 1);
- Temporal sampling must be enhanced to match convective time scales (30 min);
- Observation timeliness must be commensurate this time scale and also the warning horizon needed to take action to avoid societal impacts.

2.2 *Wind Requirements*

The WMO Observing Systems Capability Analysis and Review[1] tool is maintained by application experts, who specify consensus requirements for geophysical variables in diverse application areas, such as climate applications, global and regional NWP, (high-resolution) weather nowcasting, ocean applications, meteorological research (GEWEX, SOLAS), etc. Moreover, these basic meteorological requirements serve downstream applications, such as ocean wave and storm surge predictions, or more novel applications, such as energy production in off-shore wind farms (Dong et al. 2013).

The geophysical variable of interest here is the horizontal wind vector over water surfaces defined by WMO as: “Horizontal (2D) vector component of the 3D wind vector, conventionally measured at 10 m height”. Table 1 provides a brief summary

[1]WMO OSCAR: www.wmo-sat.info/oscar.

Table 1 Summary of WMO OSCAR requirements defined for 2D wind vector over the surface

Application area	Uncertainty (ms^{-1})	Spatial resolution (km)	Observation cycle (h)	Timeliness (h)
Global NWP	2	100	6	1
Regional NWP	1	10	3	½
Nowcasting	2	10	½	½
Ocean	2	5–25	3	1–24
Climate	1–2	5–100	3–18	1–1000

of the so-called breakthrough requirements for the horizontal wind as provided in the OSCAR data base. This "breakthrough" would, if achieved, result in a significant improvement for the targeted application and is an intermediate level between a threshold requirement and an ultimate goal. The breakthrough level may be considered as an optimum from a cost-to-benefit point of view, when planning or designing observing systems. Since multiple entries exist in the data base in an application area, ranges are specified when needed. In Table 1 uncertainty is meant as vector error, i.e., the modulus of the vector difference between the observed wind vector and the true vector. Since the two horizontal wind vector components are assumed equally accurate, the wind vector component uncertainty requirement is a factor $\sqrt{2}$ more stringent. Moreover, the "uncertainty" characterizes the estimated range of observation errors on the given variable, with a 68% confidence interval (1 σ).

All application areas require 1−2 ms^{-1} vector wind quality. Biases are assumed negligible and, therefore, accurate (inter)calibration of instruments is necessary. The requirements for spatial resolution are more diverse, even within an application area. In this regard it is noted that wind spectra fall off exponentially and smaller scales contain less wind variance. Therefore, when small scale wind features are targeted, small uncertainty must be required, typically better than the integrated climatological wind variance on scales smaller than the targeted horizontal resolution. Similarly, small-scale wind structures evolve must faster than large-scale atmospheric structures and therefore a high spatial resolution generally implies a high observation frequency in order to be able to capture all small-scale variations. Finally, timeliness is critical for atmospheric, ocean wave and storm surge hazard forecasting applications, since it is generally directly compromising the timeliness of alerting economy and society.

Although WMO OSCAR also provides requirements on wind speed in a similar fashion, it may be clear, however, that wind vector information is generally more desirable, for example to obtain spatial wind derivatives, such as wind stress curl and divergence.

In addition to the quality parameters in Table 1, the WMO Global Climate Observing System section formulated a requirement on the stability of essential climate data records (GCOS 2011). In a report on Systematic Observation Requirements for Satellite-based Products for Climate, supplemental details to the satellite-based component of the Implementation Plan for the GCOS are provided and a stability of 0.1 m/s per decade for observational wind records is requested.

2.3 Virtual Ocean Vector Wind Constellation

Scatterometer instruments are part of so-called virtual constellations through international collaboration. The Committee on Earth Observation Satellites Ocean Surface Vector Wind Virtual Constellation (CEOS OSVW VC) interagency working group is making steps to align requirements and product standards to build a global constellation of scatterometers. Given the above-mentioned daunting challenges on spatial and temporal sampling, accuracy and timeliness, international collaboration appears extremely profitable for all space agencies involved. With more agencies now planning scatterometers (e.g., ISRO on ScatSat and OceanSat, the Chinese Meteorological Agency on the FY series of satellites, the Chinese State Ocean Administration on the HY series and CFOSAT[2] and Russia on the Meteor series) and scatterometers likely available from three[3] EUMETSAT Metop satellites at the same time, a virtual constellation is being built to address spatio-temporal coverage needs.

Scatterometers are usually mounted on polar satellites in sun-synchronous orbits, i.e., they pass a given location at approximately the same time every day, defined by the satellite's Local Time of equator Acending Node (LTAN). By coordinating the LTAN of the different satellites, the CEOS OSVW VC may effectively cover all times of day and therefore for example the diurnal cycle of the ocean winds, which is particularly substantial near land masses.

Due to the polar orbits and different LTAN, close collocations between scatterometer measurements are relatively infrequent and mainly situated near the poles, where orbits cross. The NASA RapidScat scatterometer, mounted on the International Space Station, is neither sun-synchronous nor polar and provides scatterometer winds at different times of day. Moreover, in due time, close collocations with other operational scatterometers, notably ASCAT-A and –B and NASA's QuikScat (now only pointing at one fixed azimuth angle), which favours accurate intercalibration and geophysical interpretation (Wang et al. 2017).

In addition, the close RapidScat and ASCAT collocations provide an excellent data base to further improve the wind retrieval from scatterometer data, both at Ku and C band, in terms of quality control, wind speed scale, wind direction biases, and ambiguity removal (Stoffelen et al. 2017) and moreover fosters the opportunity of international collaboration towards wind scatterometer retrieval improvement as elaborated in the International Ocean Vector Winds Science Team (IOVWST).

A new asset to the virtual constellation is the recently launched ScatSat mission by ISRO, which has higher incidence angles than RapidScat and will initially provide abundant collocations with ASCAT to provide further intercalibration and processing advancements of C- and Ku-band scatterometers. Figure 3 shows a depiction of a typhoon by ASCAT.

---

[2]Chinese-French Ocean Satellite SCAT wind scatterometer.

[3]Scatterometers (ASCAT) on board the European Organisation for the Exploitation of Meteorological Satellites (EUMETSAT) Metop satellites A, B and C.

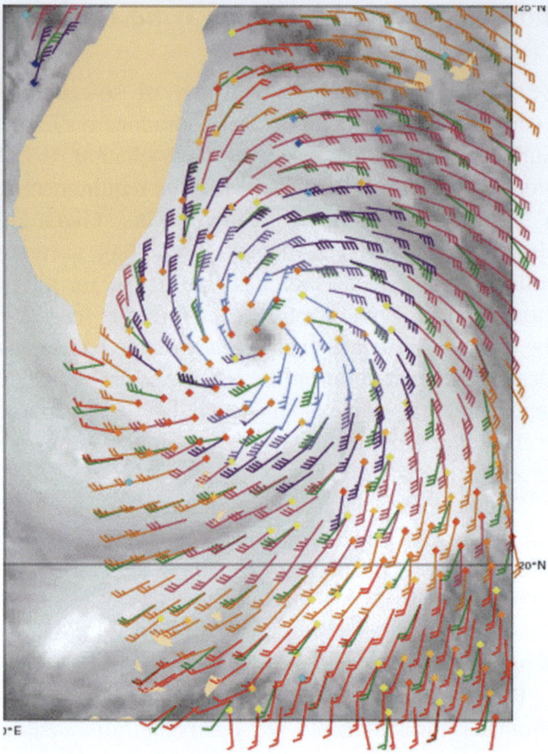

Fig. 3 Acquisition of typhoon Nepartak on the 8th of July 2016 before hitting Taiwan; the typhoon shows 25-km scale eye wall winds up to 35 ms^{-1} in ASCAT (blue and red), but, while the 12-hour forecast position is excellent, considerable weaker winds in the collocated ECMWF winds (green)

2.4 *Stress-Equivalent Winds*

Scatterometer, SAR and radiometer measurements relate to the ocean surface roughness and kinematic wind stress, while their wind products are represented by a wind vector at 10 m height over a Wind Vector Cell (WVC). How can this be consolidated? It is important to realize that in the approach followed, the radar backscatter measurement σ^0 is related to the wind at 10 m height above the ocean surface, simply because such measurements are widely available for validation. This implies, however, some geophysical error. In particular, air stability and mass density are not directly sensed by the wind scatterometer, SAR nor radiometer and should be compensated in the derived 10 m wind, which is affected by these air properties.

Fortunately, in moored buoy and NWP model validation sources, the 10 m wind may be corrected for air stability and mass density with good precision. Therefore, scatterometer winds are validated and compared as so-called stress-equivalent 10 m winds, abbreviated as $\mathbf{U}_{10S}$. To avoid atmospheric stability effects, the 10-m equivalent-neutral wind vector, $\mathbf{U}_{10N}$, has been defined in the past, to replace the actual 10-m wind vector, $\mathbf{U}_{10}$. Using Monin-Obukhov similarity scaling, the equivalent-neutral wind vector amplitude is simply given by

$$U_{10N} = \frac{u_*}{\kappa} \ln\left(\frac{10}{z_0}\right) \tag{1}$$

where z_0 is the aerodynamic roughness length, the friction velocity is defined by the equation for the kinematic wind stress $u_*^2 = \tau/\rho$ and κ is von Karman's constant. The aerodynamic roughness length

$$z_0 = \frac{0.11\nu}{u_*} + \alpha \frac{u_*^2}{g} \tag{2}$$

is approximated from the geophysical variables ν (kinematic viscosity of air, 1.5×10^{-5} m^2/s), α (dimensionless Charnock parameter, see Charnock 1955), and g (gravitational constant, 9.81 m/s^2). The Charnock parameter, which depends on sea-state, varies substantially for different surface layer schemes, from 0.011 to around 0.018, and affects wind stress τ, while it does not much affect U_{10N} (Portabella and Stoffelen 2009), such that U_{10N} may be used as a standard. To obtain τ from U_{10N} one needs information on the stability of the atmospheric boundary layer, which may be obtained from moored buoys or NWP models with sufficient precision, and apply it in one's favorite atmospheric surface layer scheme.

For the same $\mathbf{U}_{10N}$, cold heavy air will produce more stress (and roughness) than lighter (warmer or dryer) air. This effect is expressed by the surface stress equation. The surface wind stress $\boldsymbol{\tau} = \rho\, u_*\, \mathbf{u}_*$ depends on the air mass density ρ and NWP and buoy winds may be converted to so-called stress-equivalent winds

$$U_{10S} = \sqrt{\frac{\rho}{\langle\rho\rangle}} U_{10N} \tag{3}$$

where U_{10S} is the 10 m-height stress-equivalent NWP or buoy winds that may be compared to scatterometer wind retrievals and $\langle\rho\rangle$ is the average air density as defined in a standard atmosphere (1.225 kg/m^3). ρ variations, which depend on surface pressure, air temperature, and humidity, are generally small (1–2%) and can exceptionally increase locally in cases such as cold air outbreaks. These straightforward corrections for air stability and mass density reduce geophysical errors when validating and applying scatterometer, SAR and radiometer wind retrievals. Equation (3) suggests that the global mean U_{10S} and $U_{10\mathrm{N}}$ are equal, but differ locally. It implies further that validated scatterometer products of $U_{10\mathrm{N}}$ can be interpreted as U_{10S} with improved geophysical accuracy (de Kloe et al. 2017).

Kinematic wind stress may now be easily obtained from $\mathbf{U}_{10S}$, since

$$\boldsymbol{\tau} = \langle\rho\rangle.\left(\kappa/\ln\left(\frac{10}{z_0}\right)\right)^2 U_{10S}\cdot\mathbf{U_{10S}} = \langle\rho\rangle.C_{DN}U_{10S}\cdot\mathbf{U_{10S}} \tag{4}$$

where it is noted that $\mathbf{U}_{10S}$ already includes the effect of varying air mass density on the stress exerted on the ocean surface to create roughness. C_{DN} is the neutral air drag coefficient.

However, wind scatterometers may be sensitive to more parameters than only those that relate (correlate) with $\mathbf{U}_{10S}$. For example, the appearance of surface slicks, suppressing the amplitude of small-scale ocean waves and thus microwave-sensed roughness, is associated with low winds and may, to some degree, be fitted by a GMF (Stoffelen 1998; Chap. I). However, abundant surfactants, generated by natural or human causes, may render the nominal wind-to-backscatter relationship, as captured by the GMF, less accurate. Other variable effects, such as rain (mainly for Ku-band scatterometers), extreme wind variability, complex sea states, SST, etc., may affect GMF accuracy too, but only to a small degree.

2.5 Accuracy

Scatterometer winds are averaged over a WVC and do not spatially represent buoy nor NWP winds, which complicates calibration. Stoffelen (1998; Chap. IV) discusses a unique method to determine the accuracy of scatterometer, buoy, and NWP model $\mathbf{U}_{10S}$ winds: triple collocation. The method takes into account the different spatial representations of local winds from moored buoys, scatterometer (25 km WVCs) and global NWP winds (>100 km) (Vogelzang et al. 2011).

Table 2 provides some typical wind error standard deviations (SDE) of scatterometer, buoy and ECMWF winds for the same collocation criteria, though (necessarily) over different years. All verification is naturally provided on the spatial aggregation scale of the scatterometer, which here is either 12.5, 25 or 50 km. As may be expected, a larger aggregation to a 50 km scale, suppresses noise and slightly improves the SDE. However, at the same instance, the buoy SDE

Table 2 Triple collocation results for KNMI scatterometer winds in subsequent months of November, December and January, starting in 2007 for ASCAT, 2015 for RapidScat, and 2009 for OSCAT and QuikScat

Product	Scatterometer		Buoys		ECMWF	
	σ_U	σ_V	σ_U	σ_V	σ_U	σ_V
	[m/s]		[m/s]		[m/s]	
12.5 km ASCAT	0.7	0.8	1.2	1.2	1.7	1.7
25 km ASCAT	0.6	0.7	1.2	1.2	1.6	1.6
25-km QuikScat	0.6	0.5	1.4	1.4	1.3	1.3
25-km RapidScat	0.6	0.7	1.3	1.4	1.2	1.2
50-km RapidScat	0.6	0.5	1.4	1.5	1.1	1.1
50-km OSCAT	0.7	0.5	1.5	1.6	1.0	1.1

The zonal (u) and meridional (v) wind component error standard deviations are provided on the scatterometer scale (e.g., Verhoef et al. 2015)

slightly increases at 50 km, indeed suggesting that the scatterometer 50-km products are a worse representation of the local buoy winds than their 25-km equivalents. Note that the 50-km products are a better representation of the ECMWF winds (ECMWF has lower SDE on the 50-km scale), which do not resolve 50-km scales.

It is clear from these results that moored buoys are not a good validation reference for scatterometer winds, due to their relatively large SDE on the scatterometer scale. However, the triple collocation method takes into account the SDE in estimating the calibration of scatterometers with respect to buoys and thus enables to exploit the main asset of moored buoys, which is their absolute calibration. As such, scatterometers are calibrated with respect to moored buoys and thus moored buoys are absolutely critical in scatterometry (Stoffelen et al. 2015).

3 Applications of Stress-Equivalent Vector Winds

Satellite stress-equivalent vector winds are used in a wide variety of applications from weather and climate to oceanography, as outlined below.

3.1 Global NWP

Scatterometer winds have proven to be very useful for the forecasting of dynamic weather (Isaksen and Stoffelen 2000) using the 4D-Var data assimilation system, which is capable of assimilating winds with dual ambiguity (Stoffelen and Anderson 1997) and propagating the scatterometer surface information vertically (Isaksen and Janssen 2004). Increased coverage clearly improves forecasts of extreme events (Stoffelen et al. 2013). Severe storms often originate from ocean areas, where only sparse meteorological observations are available. Consequently, the initial stages of severe storms are often poorly analyzed and their development poorly predicted (ESA 1999) and occasional devastating ocean or coastal wind and wave conditions remain a major challenge for global NWP. Due to cut-off times in NWP, particularly in cases of fast weather developments, which often have large societal impacts, the timely use of satellite observations is complicated.

Figure 4 depicts regular monitoring statistics in the EUMETSAT NWP SAF. Note that larger biases and standard deviations of the differences between scatterometer and NWP model appear in the same regions: near the ITCZ, in regions with ocean currents, and near land masses. These areas correspond to areas with large atmospheric dynamics and substantial speed and systematic wind direction biases (Sandu et al. 2011).

Similar NWP biases occur in most global NWP models and are due to fast or mesoscale processes resolved by the scatterometer, but not by models, such as

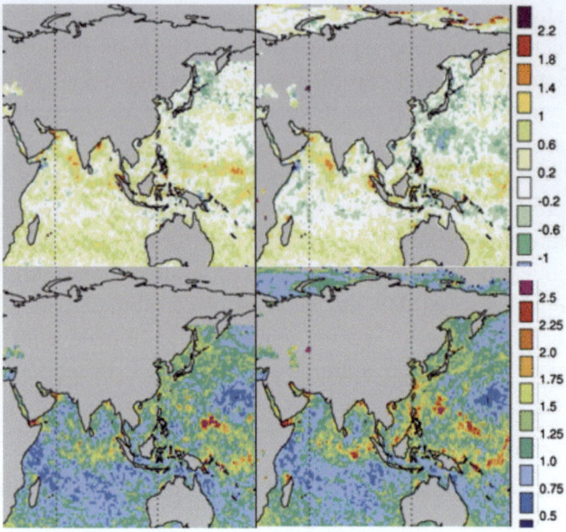

Fig. 4 Speed bias (top) and SD (bottom) of differences (O-B) of scatterometer winds (O) with respect to the Met. Office global model (B) for RapidScat on ISS (left) and ASCAT-A on MetOp (right) for July 2015 (note different legends). Different scatterometers produce similar local mean and variability of differences with respect to weather models. © EUMETSAT NWP SAF

(moist) convection (Lin et al. 2015), atmospheric turbulence (King et al. 2017), gravity waves and systematic errors in boundary layer parameterisations or the lack of ocean currents. These biases seasonally evolve, but tend to persist over time. The occurrence of these biases is however ignored in data assimilation and their existence prevents the correct assimilation of observed dynamical weather features. Whereas variational bias correction is quite common for satellite observations, the local model biases of Fig. 4 are not corrected yet, but may be estimated and prevented to a large extent, thus facilitating improved dynamical wind data assimilation.

All aspects of global NWP apply to regional NWP as well, but in regional NWP higher spatial and temporal resolution is attempted, as well as faster cycling.

3.2 *Regional NWP*

Although for regional NWP similar objectives hold as for global NWP, these applications will need faster observations, higher resolution, and improved coverage. Nevertheless, insufficient observations exist to initialise the fast-evolving turbulent scales. Therefore, while these mesoscale wind structures perhaps may be realistic, they are not real. Over the ocean and in the upper air they are generally not in phase with the real structures as observed by, for example, a scatterometer. This leads to a so-called double penalty in the scatterometer minus first guess departures and thus to larger departures than in global NWP.

Such large departures, dispersed over a range of spatial scales, are challenging in data assimilation and weather-dependent background error covariances will be needed to obtain appropriate and balanced distributions of the innovations over the broad spectrum of background error scales. Such background error covariances may be provided by ensemble techniques, but also scatterometer OSVW provide information on NWP model error structures (Lin et al. 2016). Moreover, improved initialisation may be achieved by assimilating a constellation of wind scatterometer instruments (see also, e.g., Žagar et al. 2016).

3.3 Nowcasting

Scatterometer winds are widely used for nowcasting, e.g., to depict the evolution of typhoons, tropical cyclones or mesoscale complex systems (MCS), but also in case of marine and coastal forecasting for traffic, off-shore activities, wind energy, etc. Scatterometer data are furthermore used for typhoon warnings over extratropical seas (Von Ahn et al. 2006; Chang et al. 2009). Moreover, also the depiction of coastal features and moist convection warrants high spatio-temporal resolution.

Recently, further progress has been made in the quality assessment of ASCAT winds near rain (Portabella et al. 2012) and in associating the inversion residual, called MLE and as depicted in Fig. 3, with local wind variability (Lin et al. 2015). It is very useful when scatterometer winds and wind variabilities should be part of advanced NRT integrated visualization in meteorological work stations, where mesoscale weather information from satellites, in situ observations and models can be compared. Furthermore, on-the-fly warning systems may be developed to effectively focus attention on areas where large discrepancies exist between observed and modelled weather phenomena.

Typhoons cause much havoc in the Asian waters, where its winds are devastating and occasionally cause water levels to surge to catastrophic heights. A particularly pressing requirement at most participating CEOS OSVW agencies is therefore to obtain extreme winds in hurricanes (>30 m/s) from scatterometers, since extreme weather classification, surge and wave forecasts for societal warning are a high priority in nowcasting (van Zadelhoff et al. 2014).

3.4 Extreme Winds, Waves and Storm Surges

However, measuring extreme winds is challenging as vicarious calibration is needed. The NOAA hurricane hunters go into hurricanes to drop sondes, and thus obtain wind profiles in the lowest few kilometers of hurricanes, and mount dedicated microwave instrumentation on aircraft, to obtain detailed wind patterns in hurricanes, such as the Stepped-Frequency Microwave Radiometer (SFMR). Ideally, local dropsonde winds may be statistically used to calibrate SFMR, which in turn is used to calibrate satellite scatterometers and radiometers in overflights. Given the scale of typhoons and the footprint of radiometers and scatterometers,

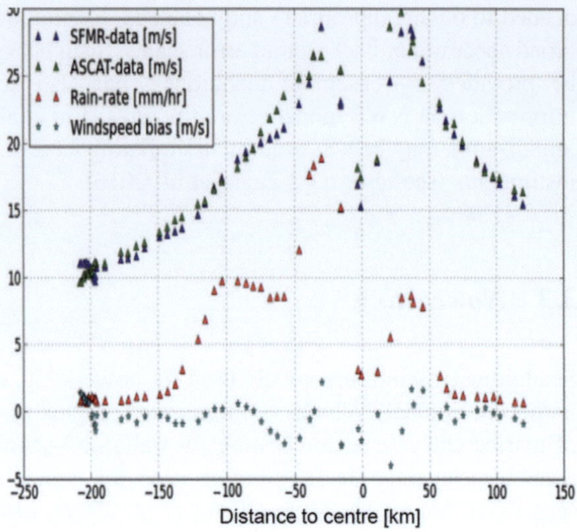

Fig. 5 Illustration of modified (2012) SFMR wind speeds (blue), rain rate (red), collocated ASCAT speed (green) and SFMR-ASCAT speed difference (light blue). SFMR speeds have been ad hoc modified to match ASCAT over a large number of typhoons by subtraction of ln(RR) and ln(V) and provides an estimate of current typical calibration uncertainty. The ASCAT depiction of typhoon structure appears faithful and not much affected by rain

satellite typhoon speeds are expected to be less extreme than in situ measured extremes, due to the implicit spatial aggregation of the satellite-based winds. The calibrated spatial wind patterns measured by SFMR may however be averaged to approximately match the satellite footprints and thus provide a calibration reference and an estimate of local wind variability (see Fig. 5). Using similar tracking and collocation procedures, van Zadelhoff et al. (2014) used SFMR to obtain a C-band cross-polarisation geophysical model function (GMF).

Although this approach is credible in principle and physically more consistent than any other global method to obtain maximum winds in typhoons, research is ongoing to understand the exact physical interpretation of dropsonde winds and SFMR in the inherently extremely variable conditions in typhoons and to obtain more hurricane flight data to relate to satellite scatterometer overpasses.

4 Wind Energy and Other Coastal Applications

Offshore wind energy is a sustainable technique in development in an attempt to reduce the burning of fossil fuel and temper anthropogenic climate change. It however results in a demand for observed off-shore wind information to

1. estimate long-term wind production,
2. aid in the design of the wind mills,
3. plan the occurrence of maintenance time windows and
4. continuously forecast the electrical input to the grid of available wind farms.

Scatterometer winds are helpful in all these aspects since they are being carefully intercalibrated (e.g., KNMI Wind Atlas). For points 2 and 3 wave or storm surge information is relevant, which is usually based on wind vector input to wave or storm surge models.[4]

Not only wind energy, but also shipping, e.g., on the northern route from China to Europe, and other economic activities in coastal areas will increase as the world population grows, as well as coastal protection. Improved scatterometer wind and sea ice (Belmonte Rivas and Stoffelen 2011) processing in coastal areas will thus remain a high priority in the years to come. The developments in regional NWP, Nowcasting and Wind Energy above, all apply to the coastal area in particular.

4.1 Climate Data Records

Wind and stress climatologies, based on a long heritage of scatterometer measurements, are considered essential climate variables (ECV) and used for the design of structures at sea, by policy-makers, in wind energy applications, climate change adaptation, etc. In particular, improvements in the coastal region brought by increased resolution will be beneficial for many climatological applications.

Alternatively, model reanalyses are conveniently used for wind climatologies at regular and high temporal resolution. However, the observational inputs to these reanalyses vary in time, which may affect the climatology artificially. An example is provided in Fig. 6 for the ECMWF reanalysis ERA. Here, a detection algorithm has been used to find typhoons in ERS-2 and ASCAT scatterometer winds and, subsequently, another detection algorithm uses a radius of 800 km to search for that typhoon in ERA. While the statistical characteristics of the typhoons found in ERS-2 and ASCAT are quite similar (e.g., right panel of Fig. 6), the ERA hit rate is quite different (left panel). A plausible reason for this difference may be that ERS-2 winds have been assimilated in ERA, while ASCAT winds were not assimilated. Although the skill is generally low, one may conclude that a change of observational input, clearly has changed the skill of ERA to represent typhoons over time. Climate data records of essential climate variables, such OSVW or wind stress, based on long-term stable instrument records are thus essential to build confidence in the fidelity and limitations of reanalyses based on general circulation models, and identify and possibly correct their time-dependent biases.

The construction of OSVW ECV CDRs consists in the reprocessing of scatterometer data, which profits from the strict automated monitoring developed in near-real time (NRT), since very similar procedures are implemented for reprocessing. In addition, ocean calibration methods and triple collocation are used for long-loop instrument and parameter performance monitoring (e.g., calibration stability) on a monthly and yearly basis. Triple collocation (TC) with moored buoys, scatterometers and GCMs is performed to establish the accuracy and intercalibration of the scatterometer winds and the GCMs at the available moored buoy

[4]eSurge: www.storm-surge.info/.

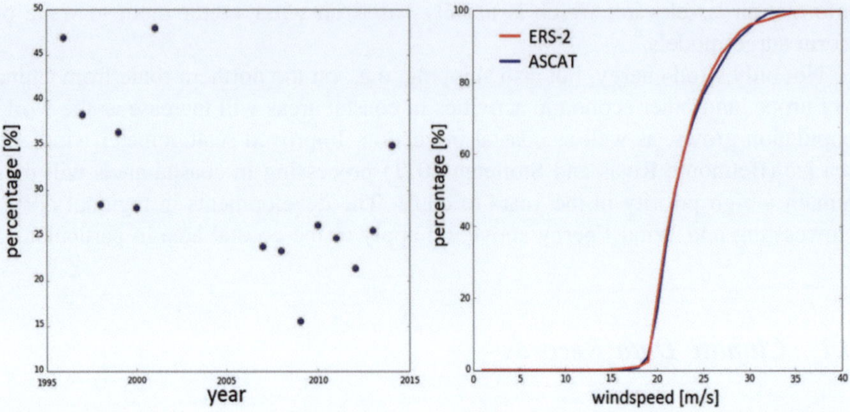

Fig. 6 Left: Tropical hit rate of typhoons in the ECWWF reanalysis (ERA) for cases where a typhoon has been detected in ERS or ASCAT scatterometer winds. Right: PDF of typhoon winds for ERS-2 from 1995 to 2000 and for ASCAT from 2007 to 2015. Although ERS and ASCAT typhoon detection behaves very similar, the ERA typhoon representation appears quite different in the 1990s and 2000s

positions. By physical inference, it is assumed that the spatial sample of buoys is sufficient to obtain a globally representative absolute calibration for scatterometers. This can obviously not be proven, as no globally representative in situ wind network is available (Stoffelen et al. 2015). However, given such plausible inference, it appears possible to reach the 0.1 m/s per decade stability in a representative global metric from scatterometers (Verhoef et al. 2015).

Recent publications indeed suggest that observation of OSVW variability in the tropics is quite relevant, e.g., Sherwood et al. (2014) or Sandu et al. (2011), suggesting that spread in climate model sensitivity and model bias can be related to subtle dynamical model aspects, such as moist convection. More work is needed on dynamical meteorological atmospheric and oceanographic interaction processes, relevant for the realism of climate models, and may be addressed by measurement capability in the satellite era. Scatterometer OSVW can play a relevant role here (Lin et al. 2015; King et al. 2017).

5 Conclusions

An overview of satellite Ocean Surface Vector Winds, physical phenomena, observation requirements, capabilities and applications is provided. OSVWs contribute to weather and climate analyses and forecasts, which in turn serve many applications in economy and society, related to coastal hazard warnings, off-shore marine applications, such as shipping, exploration, infrastructure, wind energy and tourism. Besides these applications of winds, sea waves and storm surges, near-surface wind conditions drive the ocean circulation, which in turn plays a major role in the climate system and in marine ecosystems (e.g., fisheries).

OSVW play a fundamental role in the Asian Seas through air-sea interaction. This applies to the modest winds in the trades, the winds associated with the extensive areas of tropical convection, sea and land breezes and, of most direct human relevance, the winds associated with hurricane-force typhoons.

Accurate wind vector information is obtained from satellite microwave active measurements off the wind-roughened ocean surface, and wind speed information from passive microwave sensors. India, China, Russia and Japan, inter alia, have been, are, or will be contributing to a global virtual constellation of scatterometers that provide increasing temporal coverage of ocean surface vector wind information. This much facilitates the application of scatterometer winds for weather nowcasting, for mesoscale and global numerical weather prediction and for oceanography and climate studies.

Acknowledgements The views expressed in this chapter have been developed through discussions in a wide international forum, among which the IOVWST and the International Winds Working Group (IWWG). Figure 2 was provided by student Patrick Bunn. Colleagues at KNMI provided much of the background material used, through the EUMETSAT OSI and NWP Satellite Application Facilities and the Copernicus Marine Environment Monitor. Service (CMEMS).

References

Belmonte Rivas M, Stoffelen A (2011) New Bayesian algorithm for sea ice detection with QuikSCAT. IEEE Trans Geosci Remote Sens 2(49). https://doi.org/10.1109/tgrs.2010.2101608

Blanke B et al (2005) Modelling the structure and variability of the southern Benguela upwelling using QuikSCAT wind forcing. J Geophys Res 110(C). https://doi.org/10.1029/2004JC002529

Bourassa M et al (2009) Remotely sensed winds and wind stresses for marine forecasting and ocean modelling. In: Proceedings of OceanObs'09, https://coaps.fsu.edu/scatterometry/reports/docs/OceanObs09_Winds_White_Paper.pdf

Chang P, Jelenak Z, Sienkiewicz J, Knabb R, Brennan M, Long D, Freeberg M (2009) Operational use and impact of satellite remotely sensed ocean surface vector winds in the marine warning and forecasting environment. Oceanography 22(2):194–207

Charnock H (1955) Wind stress on a water surface. Quart J Royal Meteorol Soc 81(350):639–640

Chelton DB, Schlax MG, Freilich MH, Milliff RE (2004) Satellite measurements reveal persistent small-scale features in ocean winds. Science 303:978–983

de Kloe J et al (2017) Improved use of scatterometer measurements by using stress-equivalent reference winds. IEEE JSTARS. Accepted, JSTARS-2016-00580

Dong J et al. (2013) Evaluating the spatio- temporal variation of China's offshore wind resources based on remotely sensed wind field data. Renew Sust. Energy Rev 24:142–148. https://doi.org/10.1016/j.rser.2013.03.058

ESA (1999) Atmospheric dynamics mission: core earth explorer mission selection report. ESA SP-1233(4), European Space Agency, Noordwijk, The Netherlands

GCOS (2011) Systematic observation requirements for satellite-based products for climate, 2011. GCOS Report 154. www.wmo.int/pages/prog/gcos/Publications/gcos-154.pdf

Isaksen L, Stoffelen A (2000) ERS-scatterometer wind data impact on ECMWF's tropical cyclone forecasts. IEEE Trans Geosci Remote Sens 38(4):1885–1892

Isaksen L, Janssen PAEM (2004) Impact of ERS scatterometer winds in ECMWF's assimilation system. Q J R Meteorol Soc 130(600):1793–1814

King G P, Portabella M, Lin W, Stoffelen A (2017) Correlating extremes in wind and stress, EUMETSAT ocean and sea ice SAF, Scientific Report OSI_AVS_15_02, https://www.osi-saf.org

King GP, Vogelzang J, Stoffelen A (2015) Upscale and downscale energy transfer over the tropical Pacific revealed by scatterometer winds. J Geophys Res. https://doi.org/10.1002/2014JC009993

Kolmogorov AN (1941) Dissipation of energy in the locally isotropic turbulence. Proc USSR Acad Sci 32:16–18 (in Russian). English by Kolmogorov AN (1991) The local structure of turbulence in incompressible viscous fluid for very large Reynolds numbers. Proc R Soc A 434 (1980):15–17

Lin W, Portabella M, Stoffelen A, Verhoef A, Turiel A (2015) ASCAT wind quality control near rain. IEEE Trans Geosci Remote Sens 53(8):4165–4177

Lin W, Portabella M, Stoffelen A, Vogelzang J, Verhoef A (2016) On mesoscale analysis and ASCAT ambiguity removal. Quart J R Meteorol Soc QJ-15-0247.R2 (in press)

Liu WT, Xie X (2006) Measuring ocean surface wind from space. In: Gower J (ed) Remote sensing of the marine environment, 3rd edn, vol 6. ISBN1570830800

Moore GW, Renfrew IA (2005) Tip jets and barrier winds: a QuikSCAT climatology of high wind speed events around Greenland. J Climate 18:3713–3725

Nastrom GD, Gage KS (1985) A climatology of atmospheric wavenumber spectra of wind and temperature observed by commercial aircraft. J Atmos Sci 42:950–960

Portabella M, Stoffelen A (2009) On scatterometer ocean stress. J Atm Oceanic Technol 2 (26):368–382. https://doi.org/10.1175/2008JTECHO578.1

Portabella M, Stoffelen A, Verhoef A, Verspeek J (2012) A new method for improving scatterometer wind quality control. IEEE Geosci Remote Sens Lett 9(4):579–583

Sandu I et al (2011) Why is it so difficult to represent stably stratified conditions in numerical weather prediction (NWP) models? J Adv Model Earth Syst 5:117–133

Sherwood SC, Bony S, Dufresne J-L (2014) Spread in model climate sensitivity traced to atmospheric convective mixing. Nature 505:37–42. https://doi.org/10.1038/nature12829

Shimada T, Kawamura H (2006) Satellite observations of sea surface temperature and sea surface wind coupling in the Japan Sea. J Geophys Res 111(C). https://doi.org/10.1029/2005jc003345

Stoffelen A (1998) Scatterometry. PhD thesis at the University of Utrecht. ISBN 90-393-1708-9

Stoffelen A, Anderson D (1997) Ambiguity removal and assimilation of scatterometer data. Quart J R Meteorol Soc 123:491–518

Stoffelen A, Aaboe S, Calvet J-C, Cotton J, De Chiara G, Saldana JF, Mouche AA, Portabella M, Scipal K, Wagner W (2017) Scientific developments and the EPS-SG scatterometer. IEEE J Sel Topics Appl Earth Observ Remote Sensing 10(5):2086–2097

Stoffelen A et al (2006) ADM-Aeolus doppler wind lidar observing system simulation experiment. Quart J R Meteor Soc 132:1927–1947. https://doi.org/10.1256/qj.05.83

Stoffelen A et al (2013) Research and development in Europe on global application of the OceanSat-2 Scatterometer Winds. NWP SAF report NWPSAF-KN-TR-022. http://www.knmi.nl/publications/fulltexts/oceansat_cal_val_report_final_copy1.pdf

Stoffelen A, Vogelzang J, Lin W (2015) On buoys, scatterometers and reanalyses for globally representative winds. EUMETSAT NWP SAF, Document NWPSAF-KN-TR-024, V1.1

Tokinaga H et al (2009) Ocean frontal effects on the vertical development of clouds over the Western North Pacific: in situ and satellite observations. J Climate 22. https://doi.org/10.1175/2009jcli2763.1

Tokmakian R (2005) An ocean model's response to scatterometer winds. Ocean Model 9:89–103

Verhoef A, Vogelzang J, Stoffelen A (2015) SeaWinds wind climate data record validation report. www.knmi.nl/scatterometer/publications/pdf/seawinds_cdr_validation.pdf

Vogelzang J, King GP, Stoffelen A (2015) Spatial variances of wind fields and their relation to second-order structure functions and spectra. J Geophys Res. https://doi.org/10.1002/2014JC010239

Vogelzang J, Stoffelen A, Verhoef A, Figa-Saldaña J (2011) On the quality of high-resolution scatterometer winds. J Geophys Res 116(C10033). https://doi.org/10.1029/2010jc006640
Von Ahn J, Sienkiewicz J, Chang C (2006) Operational impact of QuikSCAT winds at the NOAA ocean prediction center. Weather Forecast 21:523–539
Wang Z, Stoffelen A, Zhao C, Vogelzang J, Verhoef A, Verspeek J, Lin M, Chen G (2017) A SST-dependent Ku-band geophysical model function for RapidScat. Accepted for J Geoph Res Oceans. https://doi.org/10.1002/2016jc012619
Xie S-P et al (2007) Intraseasonal variability in the summer South China Sea: wind jet, cold filament, and recirculations. J Geophys Res 112:C10008. https://doi.org/10.1029/2007JC004238
Yu Y et al (2015) Assimilation of HY-2A sea surface wind data in a 3DVAR DAS—a case study of Typhoon Bolaven. Front Earth Sci 9:192. https://doi.org/10.1007/s11707-014-0461-8
van Zadelhoff G-J et al (2014) Retrieving hurricane wind speeds using cross-polarization C-band measurements. Atmos Meas Technol 7(2):437–449. https://doi.org/10.5194/amt-7-437-2014
Žagar N et al. (2016) Mesoscale data assimilation and the role of winds in limited-area NWP. http://meteo.fmf.uni-lj.si/sites/default/files/MesoWindsWorkshopLjubljana2016_Summary.pdf

Using Remote Sensing to Study Phytoplankton Biomass and Its Influence on Herbivore Fishery in the South-Eastern Arabian Sea

Ammamkuzhiyil Smitha, Sankar Syam, N. Nandini Menon and Lasse H. Pettersson

Abstract In this chapter, satellite remote sensing data are analysed to study the physical forcing that favors coastal upwelling and the variability in phytoplankton biomass in the South-Eastern Arabian Sea for a time period of 14 years. Analysis of monthly binned measurements of Sea Surface Temperature (SST) coupled with reanalysis wind data clearly demonstrate the dominating influence of the strong southwest monsoon in cooling the SST, breaking down the surface water warm pool, and the subsequent development of wind induced upwelling off the southwest coast of India. During southwest monsoon (June–September), the depth of the 20 °C isotherm shoals from typically 140 m to about 80 m. The development of negative Sea Level Anomaly (SLA) along with cyclonic eddies during summer monsoon season prove the occurrence of divergence and upwelling. Ekman mass transport computed using the monthly reanalysis wind data show strong negative values during the southwest monsoon, which indicate strong upwelling along the coastal regions of southwest India, also discernible from the high chlorophyll-*a* (Chl-*a*) concentrations (≥ 1 mg m^{-3}) during southwest monsoon season, which decrease to about 0.2–0.5 mg m^{-3} in non-monsoon months. Increased primary production triggers higher catches of Indian oil sardine (*Sardinella longiceps* Valenciennes), and the analysis of fish landing data for the State of Kerala, India show that the sardine catch follows Chl-*a* peaks with a lag of one season with significant positive correlation.

Keywords Arabian Sea · Coastal upwelling · Chlorophyll-*a* · Indian oil sardine Herbivore fishery · Ekman mass transport

---

A. Smitha (✉) · S. Syam · N. N. Menon
Nansen Environmental Research Centre (India), Kochi, India
e-mail: smitha.a14@gmail.com

L. H. Pettersson
Nansen Environmental and Remote Sensing Center, Bergen, Norway

V. Barale and M. Gade (eds.), *Remote Sensing of the Asian Seas*,
https://doi.org/10.1007/978-3-319-94067-0_25

1 Introduction

Wind driven coastal upwelling is a major feature during southwest monsoon in the coastal Arabian Sea—a typical eastern boundary coastal upwelling system. Upwelling is less prominent in the South-Eastern Arabian Sea (SEAS) compared to the northern and northwestern coastal waters of the Arabian Sea. Nevertheless, upwelling that brings nutrient rich water to the surface supports the primary productivity of SEAS and in turn the coastal fisheries, especially that of the State of Kerala, located in the southwest tip of India. Pelagic fisheries in the SEAS are dominated by planktivorous fish species such as oil sardines, mackerels and anchovies. Figure 1 illustrates the bathymetry map of SEAS (72–78°E, 7–13°N) defined in this chapter.

The SEAS forms part of the northern Indian Ocean where seasonally reversing monsoon winds force a seasonally reversing circulation in the upper ocean. A schematic view of the ocean surface circulation in the northern Indian Ocean is shown in Fig. 2 (Shankar et al. 2002). The circulation map resolves the two monsoon currents, namely, Summer Monsoon Current (SMC) which flows eastward during summer monsoon and the Winter Monsoon Current (WMC) which flows westward during winter monsoon. In addition to these monsoon currents, SEAS also comes under the influence of West India Coastal Current (WICC) (Shankar et al. 2002). In its simplest form, WICC is a superposition of annual and semi-annual coastally-trapped Kelvin waves (Shetye 1998). The Lakshadweep High/Low forms when Kelvin waves propagate northward along the west coast of India and radiate Rossby waves (Shankar and Shetye 1997; Shetye 1998). Northern Indian Ocean becomes the warmest area of the world oceans prior to the onset of the southwest monsoon (Shenoi et al. 1999; Joseph 1990).

The coastal upwelling processes along the west coast of Kerala and its impact on regional pelagic fisheries was first addressed by Johannessen et al. (1987). Using the Miami Isopycnic Coordinate Ocean Model (MICOM), Haugen et al. (2002a) studied the seasonal circulation and coastal upwelling along the southwest coast of India. Their model simulations resolved that upwelling during the southwest

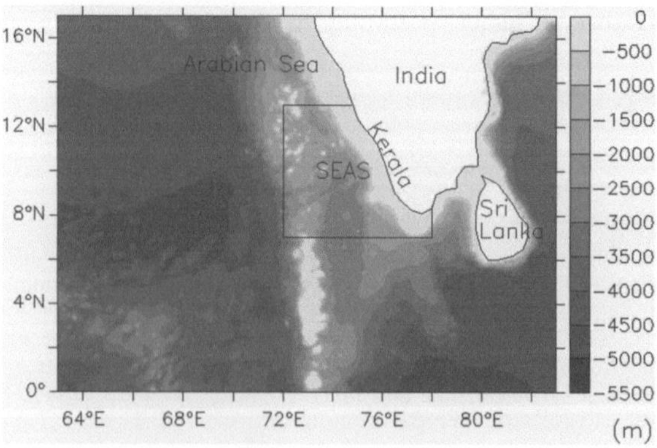

Fig. 1 Bathymetry of the South-Eastern Arabian Sea (SEAS) and adjacent waters. Data—5 min gridded global relief data (ETOPO5) from National Oceanic and Atmospheric Administration National Geophysical Data Centre (NOAA NGDC)

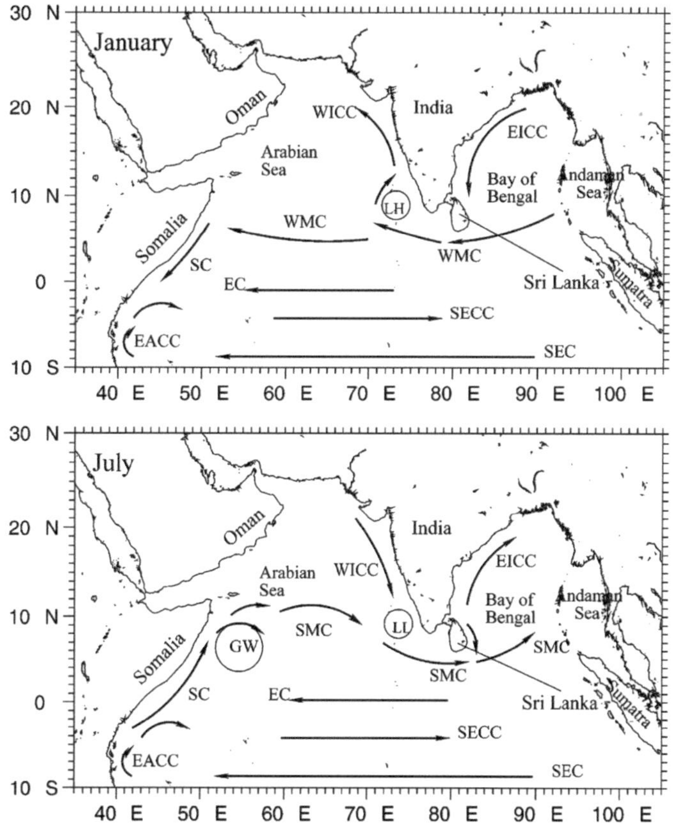

Fig. 2 Ocean surface circulation in the Indian Ocean during January (Upper panel) and July (Lower panel) from Shankar et al. (2002). Reprinted from Shankar et al. (2002), Copyright (2002), with permission from Elsevier. SC: Somali Current; EC: Equatorial Current; SMC: Summer Monsoon Current; WMC: Winter Monsoon Current; EICC: East India Coastal Current; WICC: West India Coastal Current; SECC: South Equatorial Counter Current; EACC: East African Coastal Current; SEC: South Equatorial Current; LH: Lakshadweep high; LL: Lakshadweep low; and GW: Great Whirl

monsoon was strongest in the waters off Cochin and Quilon in Kerala and that the upwelled cold water was less saline. The model simulations also, covering the period from 1958 to 1998, showed a reduction in upwelling off the southwest coast of India during El Nino years (Haugen et al. 2002b).

Satellite remote sensing data enable us to consistently monitor physical and biological oceanographic parameters at regional to local scales over large time periods. Accordingly, in this chapter, we have used satellite remote sensing and model data to analyse the dominant physical forcings in the coastal SEAS and its influence on phytoplankton biomass for a period of 14 years from 1998 to 2011. Variations in environmental conditions and marine primary production affect the recruitment, distribution, abundance and availability of fishery resources. The influence of varying oceanographic forcings on the fishery of Indian oil sardine (*Sardinella longiceps)*, a major pelagic herbivore, is also analysed in this study.

Monthly to seasonal variability of Sea Surface Temperature (SST), Sea Level Anomaly (SLA), 20 °C isotherm depth (D20) and Ekman mass transport are presented in Sects. 2–5. Section 6 explains the seasonal and monthly variability of chlorophyll-*a* (Chl-*a*) concentration in the SEAS during the study period. The relationship of seasonal sardine landing of Kerala to Chl-*a* concentration is examined in Sect. 7. The results are discussed and summarized in Sect. 8.

2 Seasonal Characteristics of Surface Winds and SST

The surface winds over the northern Indian Ocean above 10°S reverse direction seasonally in relation to the monsoon (Shankar et al. 2002). The surface waters of the SEAS are seasonally characterized by strong variability from high surface winds, upwelling and strong vertical mixing during southwest monsoon and relatively weak surface winds during northeast monsoon. Accordingly, the seasonal variability in the SST of the Arabian Sea (Jaswal et al., 2012) is distinct, which is marked by a cooling from May to August when rest of the oceans in the northern hemisphere is in a "summer time" warming period.

European Centre for Medium-Range Weather Forecasts (ECMWF) ERA-Interim monthly mean reanalysis wind data at $0.25° \times 0.25°$ resolution and Advanced Very High Resolution Radiometer (AVHRR) monthly mean SST data at 4.0 km resolution are used to analyse the seasonal variability in surface winds and SST of the SEAS. The mean winds superimposed with SST over the SEAS for the northeast monsoon (December, January, February—DJF), pre-monsoon (March, April, May—MAM), southwest monsoon (June, July, August, September—JJAS) and post-monsoon (October, November—ON) are shown in Fig. 3a–d. The well-developed warm pool over the SEAS with SSTs above 29 °C during pre-monsoon season (Fig. 3b) is clearly observed. This warm pool breaks down quite rapidly under the influence of strong surface winds (about 6–7 ms^{-1}) with the onset of southwest monsoon in June. The shift in the prevailing wind direction results in the development of an upwelling zone off the southwest coast of India. This is corroborated by the decrease in the average SST to about 27.5 °C near the coast during this season (Fig. 3c). With the withdrawal of monsoon towards post-monsoon season, the upwelling region off the coast disappears (Fig. 3d). The surface winds are generally weak during the transition periods, pre-monsoon and post-monsoon seasons (Fig. 3b, d). During northeast monsoon, northwesterlies dominate the southern parts of the SEAS, but its magnitude is less than half of that during southwest monsoon. This is also proved by the variation noted in SST, which from a high value of about 28.7–29 °C during post-monsoon (Fig. 3d) decreases slightly to less than 28.5 °C during northeast monsoon season.

SST is suggested to be the most reliable indicator of coastal upwelling in the Arabian Sea (Prell and Curry 1981). The cooler surface temperatures are checked for signatures of upwelling and the corresponding changes that it would bring about in phytoplankton biomass are discussed in the following sections.

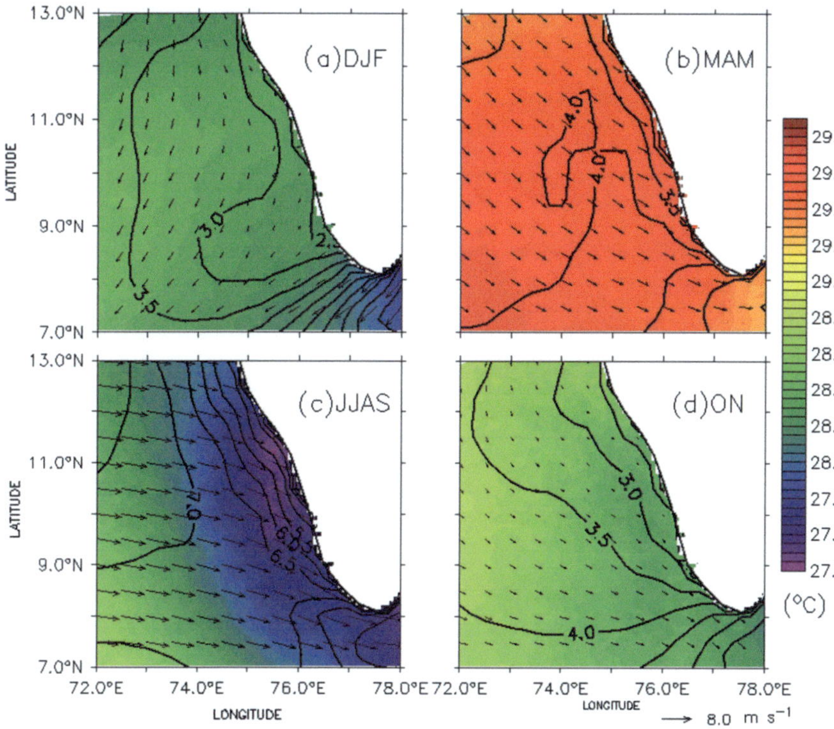

Fig. 3 Seasonal mean SST (°C) derived from AVHRR data, superimposed with seasonal winds (m s^{-1}) from ERA-Interim reanalysis, over the SEAS for **a** December–February (DJF), **b** March–May (MAM), **c** June–September (JJAS), and **d** October–November (ON)

3 Variability in the Depth of 20 °C Isotherm

Upwelling results in the shoaling of thermocline of the water column. In tropical waters the 20 °C isotherm is typically located near the centre of the main thermocline. Therefore, the depth of 20 °C isotherm from the surface, known as D20 is broadly used as the thermocline depth for tropical ocean studies (Meyers 1979; Vialard and Delecluse 1998; Meinen and McPhaden 2000; Yu 2003). In order to understand the temporal variability in the shoaling of thermocline and to monitor the evolution of upwelling phenomenon as a whole, monthly mean D20 is analysed (Fig. 4). The analysis is carried out using subsurface water temperature data at 1° × 1° resolution by Ishii et al. (2005, 2006). The average D20 in the SEAS shows seasonal variability dominated by the influence of southwest monsoon. Strong alongshore winds during southwest monsoon are expected to cause intense upwelling in the SEAS. As a result, D20 shoals to about 80 m depth from about 140 m depth observed during the winter season. The withdrawal of monsoon in October accompanied by weak and variable surface winds with little wind induced mixing, causes the 20 °C isotherm to deepen and reach its maximum depth in

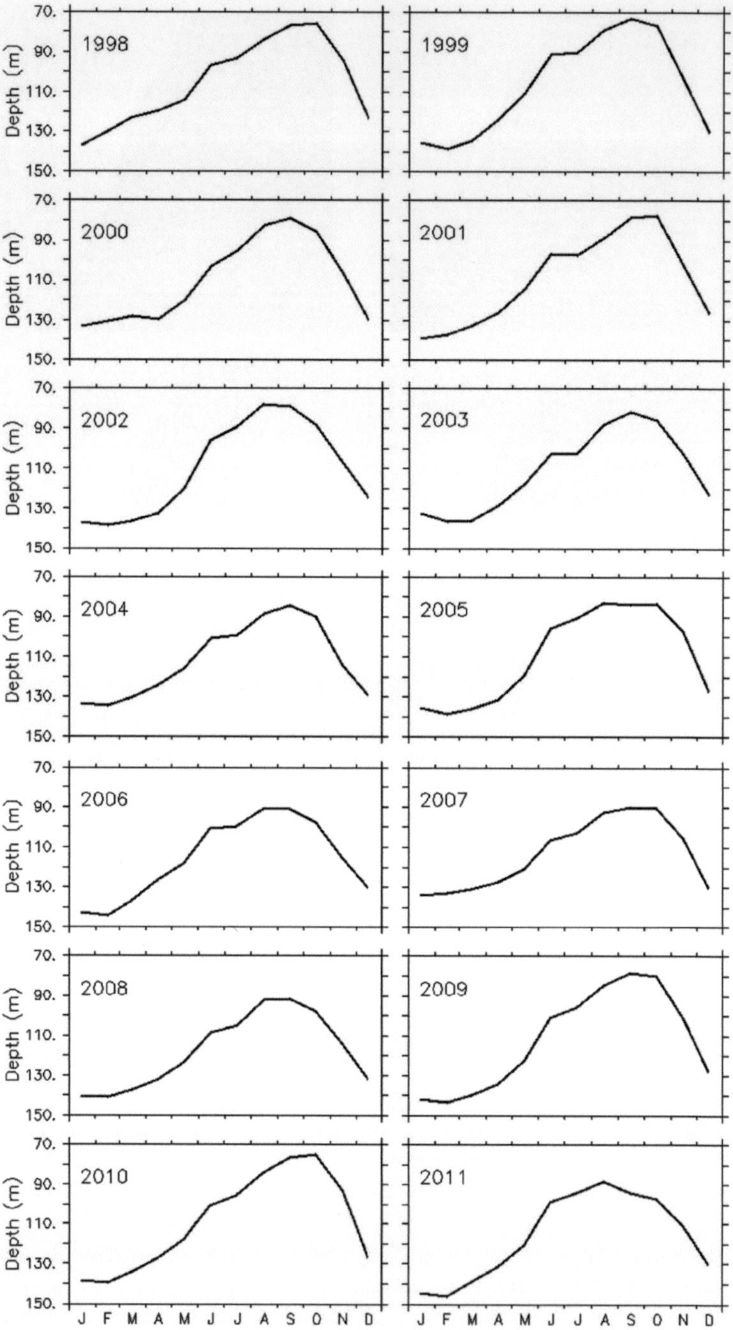

Fig. 4 Depth of the 20 °C isotherm (D20) averaged over SEAS (72–78°E; 7–13°N) depicting the shoaling of thermocline during southwest monsoon season (JJAS) for the years 1998–2011. *Data* —Subsurface temperature by Ishii et al. (2005, 2006)

winter. In addition to seasonal variability, D20 also exhibits inter-annual variability during the study period. The large seasonal variability of D20 in the SEAS can be attributed to coastal dynamics associated with boundary currents, upwelling and propagating Kelvin and Rossby waves (Rao and Sivakumar 1996).

4 Variability of the Sea Level Anomaly

SLA reflect the changes on ocean surface with sea level dropping in response to coastal upwelling induced by alongshore winds in the SEAS. Monthly climatology of SLA at 0.25° × 0.25° resolution computed from merged data by Archiving, Vaidation and Interpretation of Satellite Oceanographic Data (AVISO) which is produced using up-to-date datasets with up to four satellite radar altimeters at a given time from Jason-2/Jason-1/Envisat/Topex/Poseidon/GFO is used to study the variability in sea level in the SEAS. The surface geostrophic currents computed from SLA are overlaid on monthly mean SLA (Fig. 5).

During January–February, the well-developed Lakshadweep High (as resolved in Fig. 2—Upper panel) is clearly present in the SEAS, reflecting a positive SLA as

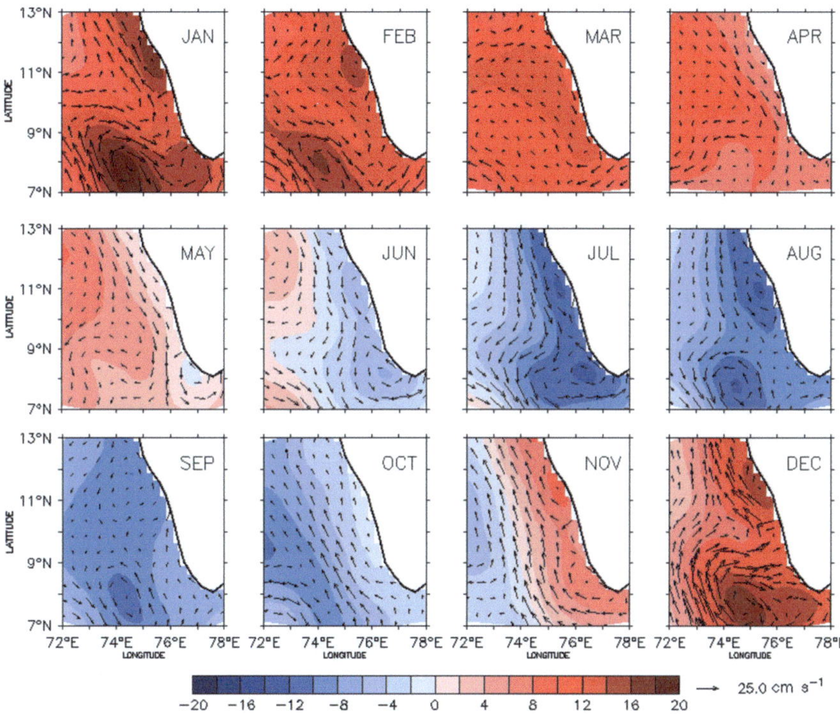

Fig. 5 Monthly average (period from 1998 to 2011) of Sea Level Anomalies (SLA in cm) superimposed with geostrophic currents (cm s^{-1}) in the SEAS. The data source is AVISO

well as anti-cyclonic surface current pattern. The anti-cyclonic flow around the Lakshadweep High and WICC flowing northwards are also depicted in the monthly climatology of SLA for December, January and February (Fig. 5) as also observed by Raj (2017). From March through May, Lakshadweep High weakens considerably and the geostrophic currents become weaker and variable. With the onset of monsoon in June, SLA over the SEAS gradually becomes negative and the anti-cyclonic currents are replaced by cyclonic currents. The presence of cyclonic eddies is an indication of divergence of the sea surface that favour upwelling. The Lakshadweep Low starts developing in June and reaches its maximum intensity in July–August. Signatures of strong upwelling off the west coast are manifested as regions of negative SLA during southwest monsoon. By September, the Lakshadweep Low weakens and the surface currents become weak and variable. During the post-monsoon season, Lakshadweep Low propagates westward as a Rossby wave and the currents start to reverse direction again. By November–December the SEAS again develops into a region of positive SLA with northward boundary currents and anticyclonic currents around the well developed Lakshadweep High. According to Shetye et al. (1991), positive SLA during northeast monsoon develops when the East India Coastal Current (EICC) along the periphery of Bay of Bengal moves southward towards equator, around Sri Lanka and the southern tip of India. Accordingly, the WICC flows northward along the west coast of India during the winter period (Fig. 2—Upper panel). Our study confirms the view of Hareesh Kumar and Anand (2016) that SLA has an annual cycle for this part of the ocean and the upwelling signals first appear at southern latitudes and proceed towards north along the coast.

5 Ekman Mass Transport

The intensity of upwelling is estimated from Ekman mass transport, which is perpendicular to the prevailing wind direction in the region and is given by

$$M_e = \tau_y / f \tag{1}$$

where M_e is the Ekman mass transport, τ_y is the meridional component of wind stress, and f is the Coriolis parameter, $2\Omega \sin \varphi$, with Ω being the Earth's angular frequency and φ the latitude.

Ekman mass transport is computed over the SEAS using the monthly ECMWF reanalysis wind data (Dee et al. 2011) to describe and understand the coastal upwelling in this region. Negative Ekman mass transport indicates offshore transport of surface waters from near the coast leading to an upwelling condition. The seasonal and inter-annual variability of Ekman mass transport in the SEAS is illustrated using Hovmoller plots between 7 and 13°N averaged over 72–78°E for the years from 1998 to 2011 (Fig. 6).

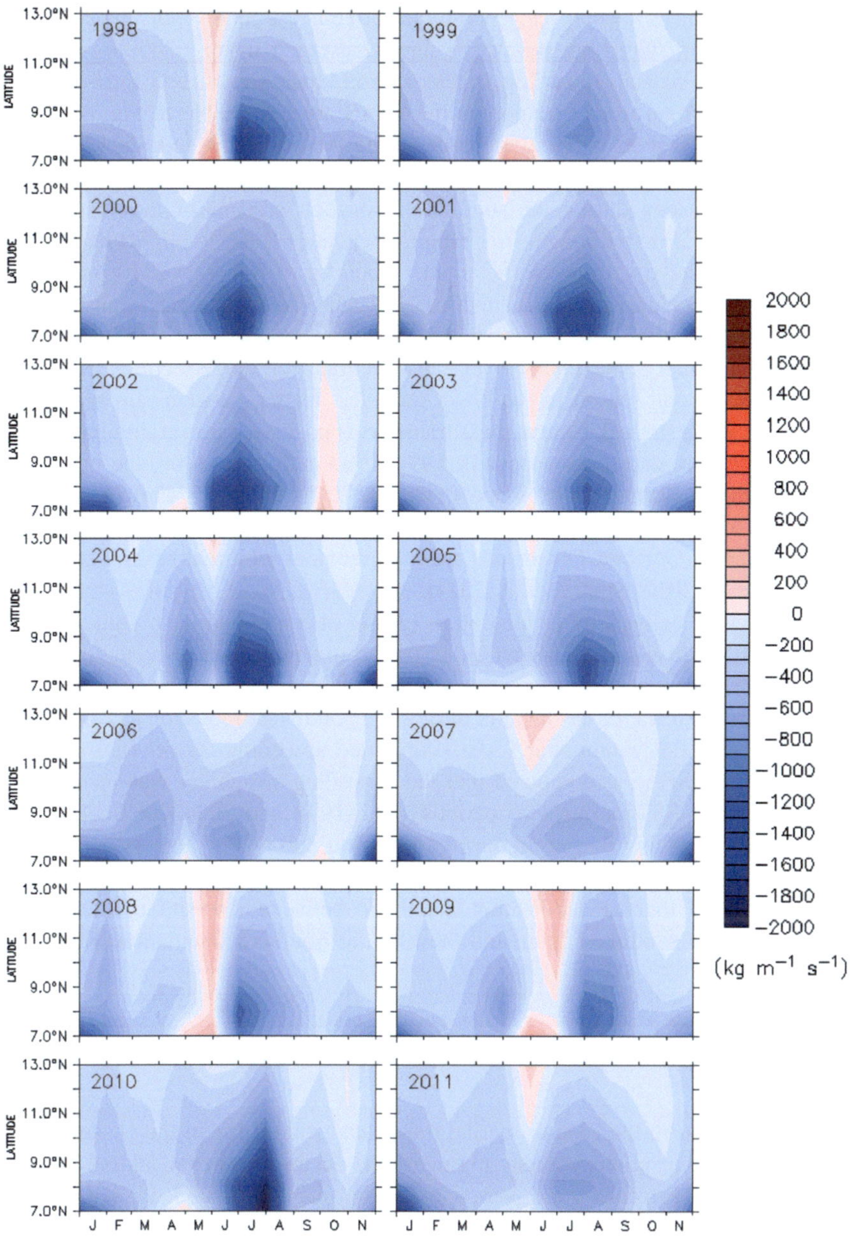

Fig. 6 The seasonal and inter-annual variability (1998–2011) of the Ekman mass transport (kg m^{-1} s^{-1}) in the SEAS between 7 and 13°N averaged over 72–78°E, illustrated by Hovmoller plots

The Ekman mass transport is generally negative in the SEAS indicating offshore water movement, with high negative values that reach up to −2000 kg m^{-1} s^{-1} responsible for coastal upwelling in this region. Inter-annual and intra-seasonal variations in Ekman mass transport in the SEAS are discernible in Fig. 6. Strong offshore mass transport is found to occur during the southwest monsoon in all the years mostly up to 10°N, with inter-annual variability in intensity. The Ekman mass transport was strong during the southwest monsoon seasons of the years 1998, 2000–2002 and 2004 with intensities ranging between −1400 to −1500 kg m^{-1} s^{-1} and reached a maximum value of about −2000 kg m^{-1} s^{-1} in 2010, indicating strong upwelling along the coastal regions of SEAS. The Ekman mass transport was weak (~−600 kg m^{-1} s^{-1}) during the southwest monsoon of 2006. During November–February, the intensity of offshore transport is less compared to the southwest monsoon season in the lower latitudes. In the pre-monsoon seasons of most of the years studied, Ekman mass transport is positive with varying intensities.

During the pre-monsoon periods of 1998–1999 and 2008–2009, the high positive Ekman mass transport extended from 7 to 13°N in SEAS showing onshore transport and subsequent coastal downwelling. In 2002, positive Ekman transport occurred during October. Our results are in agreement with that of similar studies (Jayaram et al. 2010, Smitha et al. 2014).

A combined analysis by Smitha et al. (2014) of wind data from QuikSCAT II scatterometer, Chl-*a* and SST, found that the prominent region of upwelling along the southwest coast of India lies between 8 and 14°N. They observed strong Ekman mass transport off the coast of India during the southwest monsoon season due to favourable winds. Jayaram et al. (2010) computed upwelling indices for the period 1988–2007 for SEAS and showed that the upwelling variability was inconsistent with the rise in SST during two periods 1992–1998 and 2004–2007. They also showed that upwelling decreased during the period 2005–2007 and also during the years 1991–92 and 1997–98 when El Nino and Indian Ocean Dipole (IOD) appeared together, indicating a linkage between intensity of the coastal upwelling along southwest India and the El Niño-Southern Oscillation (ENSO).

6 Surface Chlorophyll-*a* Concentration in the SEAS

Phytoplankton form the base of marine food webs and all life in the oceans. Since Chl-*a* is the major photosynthetic pigment that occurs in phytoplankton, its concentration serves as a convenient parameter of phytoplankton abundance and biomass (Ryther and Yentsch 1957; Pettersson and Pozdnyakov 2013), which can be measured from spaceborne optical sensors. Variability in the phytoplankton biomass in the SEAS is studied using the merged Chl-*a* concentration data provided by the Ocean Colour—Climate Change Initiative (OC-CCI). The OC-CCI products are provided by the European Space Agency (ESA) under CCI programme and contains high quality ocean colour data at 4.0 km spatial resolution processed with retrieval algorithms tuned for offshore case 1 waters (Product User Guide—ESA CCI 2015).

In this study, the seasonal average Chl-*a* concentration for the SEAS is retrieved from the monthly mean data over the period 1998–2011 (Fig. 7). The seasonal variability in the spatial distribution and magnitude of Chl-*a* abundance is clearly identified. Chl-*a* concentration near the coast is lowest during the northeast monsoon (~0.2–0.5 mg m^{-3}) followed by post-monsoon season. Chl-*a* concentration reaches as high as 8 mg m^{-3} along the coast during the southwest monsoon season, with high Chl-*a* concentration in the range of 2–5 mg m^{-3} extending offshore from the coast. During pre-monsoon season, high chlorophyll is observed in the southern parts of SEAS.

Hovmoller plots of Chl-*a* (Fig. 8) averaged over 7–13°N along 72–78°E illustrates the inter-annual and intra-seasonal variations in Chl-*a* concentration in the SEAS domain for the time period 1998–2011. In agreement with the findings of Smitha et al. (2008), it is found that even though upwelling in the region started well before the onset of southwest monsoon, increased surface Chl-*a* is observed only after the southwest monsoon has set in. Chl-*a* concentration increases to about 1 mg m^{-3} during June, with the highest value (~8 mg m^{-3}) observed during July–August. High Chl-*a* concentrations sometimes extend also into October. During the rest of the year, Chl-*a* concentration is generally low at about 0.5 mg m^{-3}.

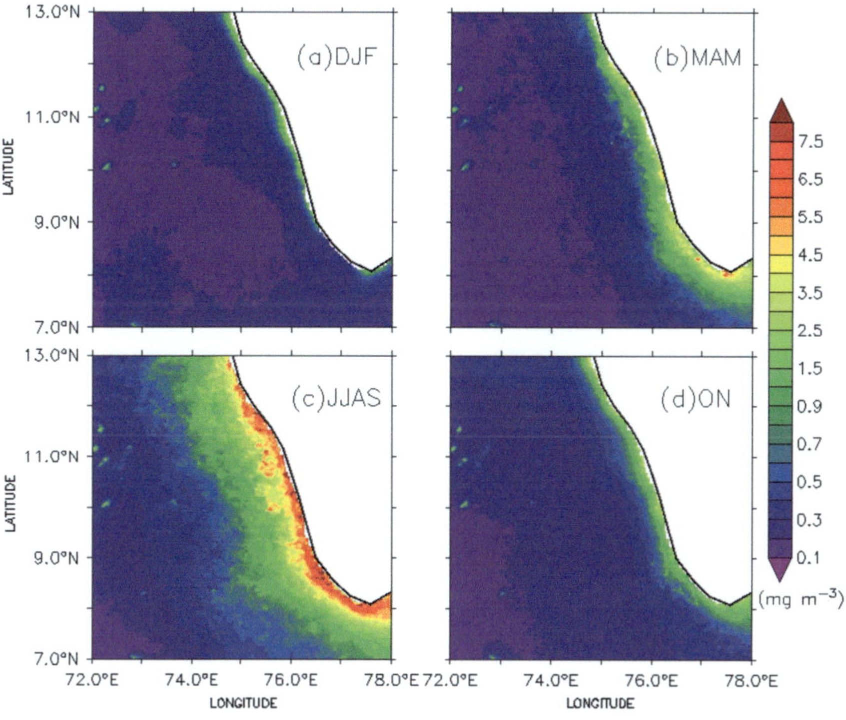

Fig. 7 The spatial distribution of seasonal Chl-*a* concentration (mg m^{-3}) in the SEAS for **a** northeast monsoon (DJF), **b** pre-monsoon (MAM), **c** southwest monsoon (JJAS), and **d** post-monsoon season (ON). The data source is OC-CCI

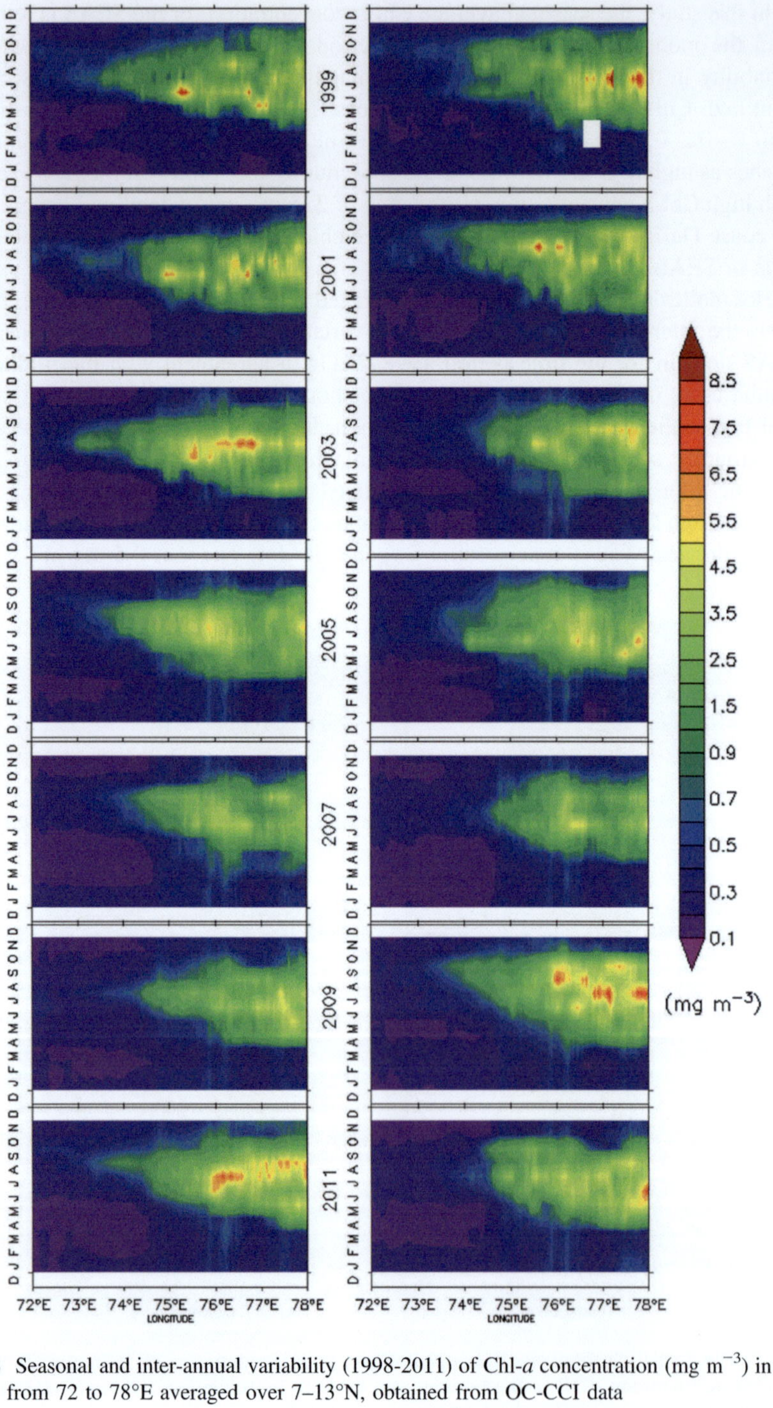

Fig. 8 Seasonal and inter-annual variability (1998-2011) of Chl-*a* concentration (mg m^{-3}) in the SEAS from 72 to 78°E averaged over 7–13°N, obtained from OC-CCI data

The offshore extent of high chlorophyll waters varies slightly from year to year. The increased offshore extent of high chlorophyll concentration (>3 mg m^{-3}) during the southwest monsoon season is seen (Fig. 8) with inter-annual variability in location and magnitude. The highest offshore extent occurred during 2002 (up to 73°E), which was a deficient monsoon year, followed by the years 1998, 2004, 2009 and 2010. These four years (1998, 2002, 2004 and 2010) with the exception of 2009 experienced high offshore mass transport in the SEAS (Fig. 6).

During 2007, which was an excess monsoon year, the offshore extent was less and the concentrations of Chl-*a* were significantly low compared to other years. From 2005 to 2008, there was a decrease in the peak magnitude of Chl-*a* concentration as well as its offshore extent during the southwest monsoon, irrespective of the intensity of monsoon. Jayaram et al. (2013) in their study using SeaWiFS data reported a decrease in Chl-*a* during 1997–1998 El Nino and also during 2005–2007 in the SEAS. They attributed the decrease in Chl-*a* during 2005–2007 to the decrease in the alongshore component of wind that resulted in reduced upwelling intensity. Our observations on wind and Ekman transport also corroborate this view.

7 Seasonal Fish Landing in the Southwest Coast of India

Out of the fourteen species of sardines that occur in the Indian waters, *Sardinella longiceps* Valenciennes is the major single species fishery which contributes to about 15% of the total marine fish production in the country. Fishery of *S. longiceps* is important as the fish is restricted to a narrow coastal belt of about 15 km offshore (Mohanty et al. 2005) and is important for artisanal and coastal fisheries in India. Oil sardine is a pelagic herbivore whose presence and accordingly fishery is directly related to the phytoplankton abundance and concentration in the coastal waters. Sardine fishery shows large seasonal, annual and decadal fluctuations (Pillai et al. 2003), the reason for which has not yet been clearly deciphered.

In the present study, we attempt to link the Chl-*a* variability during the period 1998–2011 to that of *S. longiceps* landings in southwest India. For this, monthly oil sardine landing data from Central Marine Fisheries Research Institute (CMFRI) for the Kerala coast is used. Cross-correlation is done between seasonal Chl-*a* anomaly and seasonal oil sardine landing (derived from monthly landing data) anomaly with a lag of one season. This is done because in the case of *S. longiceps,* which has a zero year fishery in India, spawning usually occurs during May–September. Larval development is rapid during the first few months and the earliest spawned and survived individuals are recruited to the fishery in the post-monsoon months. This in turn determines the yearly landings. The standardized anomalies of seasonal Chl-*a* concentration and sardine landings with a time lag of one season for the sardine are presented in Fig. 9. The linear correlation coefficient (r) between these parameters is 0.56 statistically significant at the 99% confidence level.

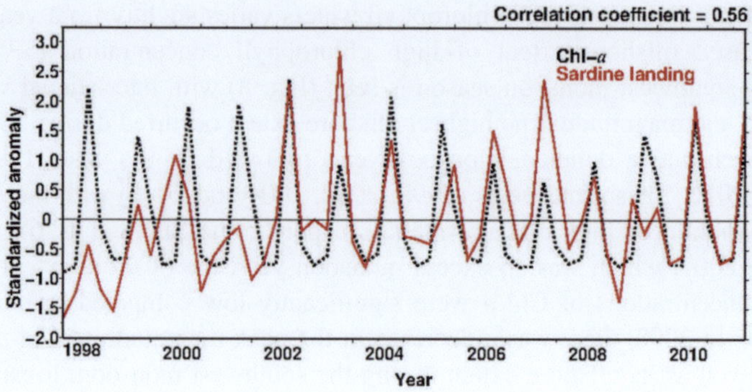

Fig. 9 Standardized anomalies of seasonal Chl-*a* concentration (black dotted line) and oil sardine landing (red line) for the SEAS with a time lag of one season for the sardine for the period 1998–2011

It is found that the period of high Chl-*a* concentration matches with the active breeding period of sardine i.e., southwest monsoon, in most of the years studied. Mean chlorophyll values in the SEAS during the southwest monsoon season in all the years studied are >1 mg m^{-3}. This is based on the approach of George et al. (2012) in which they fixed a threshold level of above 1 mg m^{-3} as the average chlorophyll concentration that determines the active breeding phase of oil sardine. The high positive correlation (linear correlation coefficient of 0.56) between Chl-*a* concentration and sardine landings indicates that in most of the years studied, increased food availability for the larvae during the southwest monsoon season resulted in high landings of sardine in that particular year. This is in agreement with the finding of Madhupratap et al. (1994) that, fish itself times its appearance to exploit the productive southwest monsoon period. As the chlorophyll rich waters move northwards with the progression of southwest monsoon, proportional change is reflected in the quarterly sardine landings in each maritime states on the west coast of India (Madhupratap et al. 1994).

Figure 10 shows the variation of mean Chl-*a* concentration during the southwest monsoon in the SEAS against the changes in annual sardine landings of Kerala. It shows that Chl-*a* is not the sole factor determining the sardine fishery in Kerala. Naturally there will be a difference between the fish landings/catch and their actual abundance due to environmental conditions and fishing intensity. Krishnakumar and Bhat (2008) showed that environmental variables like SST, surface salinity, surface dissolved oxygen, pH, nutrients, zooplankton, rainfall, ENSO index and coastal upwelling index have a decisive role in explaining the sardine catch from Malabar upwelling area. Using Chl-*a* as an index, this study analysed only the direct trophic link to sardine landings.

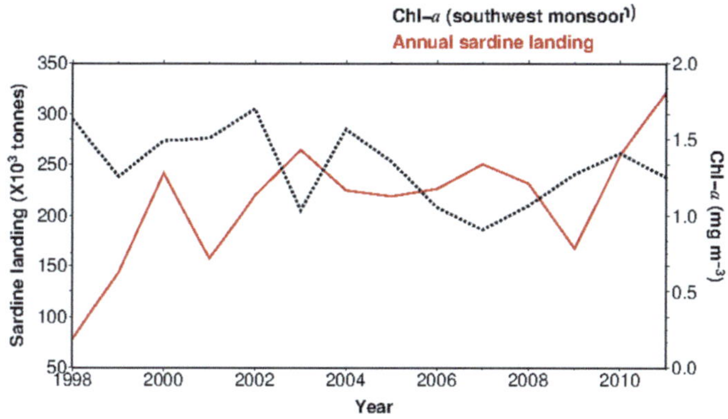

Fig. 10 Mean Chl-*a* concentration (black dotted line) during southwest monsoon and the annual sardine landings (red line) of Kerala during 1998–2011

8 Summary and Conclusions

The influence of important physical forcings in the South-Eastern Arabian Sea (SEAS) that affect the phytoplankton biomass in the surface waters are studied using satellite remote sensing data. Southwest monsoon is the season when major variations are brought about in the physical forcings like wind, temperature, upwelling and SLA, which in turn influence the phytoplankton biomass as reflected in the variations of Chl-*a* concentration in the surface waters. Among the physical drivers, seasonal variability is more pronounced for D20 as well as SLA. The shoaling of D20 and negative SLA maximum occur during southwest monsoon indicating coastal upwelling. Strong offshore mass transport is found to occur in all the years studied during the southwest monsoon season, mainly up to 10°N with inter-annual variability in intensity. Accordingly, Chl-*a* concentration is high during southwest monsoon compared to non-monsoon months when waters are typically oligotrophic with Chl-*a* concentrations ≤ 0.5 mg m^{-3}. However, maximum offshore extent of Chl-*a* concentration is observed during deficient monsoon years. More detailed study into this aspect considering the inter-annual variation in precipitation and related factors is necessary to understand the primary production dynamics of SEAS. Spatial and temporal variations in the Chl-*a* concentration are reflected in the landings of *S. longiceps*, a major herbivorous pelagic fish in the SEAS. Variability in Chl-*a* concentration during the southwest monsoon season is reflected strongly on the catch of oil sardines with a lag of one season. This proves beyond doubt that the availability of phytoplankton during the developmental stages of sardines has a decisive role in the success of the coastal pelagic fishery. The fluctuation of fishery resource in relation to the variation of primary production in the SEAS will provide an opportunity to support planning and management of fishery resources in future years.

Acknowledgements This work was initiated under the project *INDO-European Research Facilities for Studies on MARine Ecosystem and CLIMate in India* (*INDO-MARECLIM*, GA#295092) supported by the European Commission under the Seventh Framework Programme (INCO-LAB). Authors gratefully acknowledge the financial support provided by Nansen Environmental and Remote Sensing Center (NERSC) in Bergen, Norway. Authors acknowledge the use of OC CCI dataset, Version [3.1], ESA, available online at http://www.esa-oceancolour-cci.org/. The authors thank Frontier Research System for Global Change, Japan Marine Science and Technology Center/Japan. 2005 for the *Subsurface Temperature Analyses by Ishii* et al., Research Data Archive at the National Center for Atmospheric Research, Computational and Information Systems Laboratory, http://rda.ucar.edu/datasets/ds285.3/. Authors acknowledge CMFRI, Kochi for providing the fish landing data. Authors acknowledge Dr. Ajith K. Joseph for his comments during the final discussions of this paper.

References

Dee DP, Uppala SM, Simmons AJ, Berrisford P, Poli P, Kobayashi S, Andrae U, Balmaseda MA, Balsamo G, Bauer P, Bechtold P, Beljaars ACM, van de Berg L, Bidlot J, Bormann N, Delsol C, Dragani R, Fuentes M, Geer AJ, Haimberger L, Healy SB, Hersbach H, Hólm EV, Isaksen L, Kållberg P, Köhler M, Matricardi M, McNally AP, Monge-Sanz BM, Morcrette JJ, Park B-K, Peubey C, de Rosnay P, Tavolato C, Thépaut J-N, Vitart F (2011) The ERA Interim reanalysis: configuration and performance of the data assimilation system. QJR Meteorol Soc 137:553–597. https://doi.org/10.1002/qj.828

George G, Meenakumari B, Raman M, Verlecar X (2012) Remotely sensed chlorophyll: a putative trophic link for explaining variability in Indian oil sardine stocks. J Coastal Res 28(1A): 105–113

Hareesh Kumar PV, Anand P (2016) Coastal upwelling off the southwest coast of India: observations and simulations. Int J Digit Earth 9(12):1256–1274

Haugen VE, Johannessen OM, Evensen G (2002a) Mesoscale modelling of oceanographic conditions off southwest coast of India. Proc Indian Acad Sci (Earth Planet Sci) 111(3): 321–337

Haugen VE, Johannesen OM, Evensen G (2002b) Indian Ocean: validation of the Miami Isopycnic Coordinate Ocean Model and ENSO events during 1958–1998. J Geophys Res 107:3043. https://doi.org/10.1029/2000JC000330

Ishii M, Shouji A, Sugimoto S, Matsumoto T (2005) Objective analyses of SST and marine meteorological variables for the 20th century using COADS and the Kobe Collection. Int J Climatol 25:865–879

Ishii M, Kimoto M, Sakamoto K, Iwasaki SI (2006) Steric sea level changes estimated from historical ocean subsurface temperature and salinity analyses. J Oceanogr 62(2):155–170

Jaswal AK, Singh V, Bhambak SR (2012) Relationship between sea surface temperature and surface air temperature over Arabian Sea, Bay of Bengal and Indian Ocean. J Ind Geophys Union. 16(2):41–53

Jayaram C, Chacko N, Joseph KA, Balchand AN (2010) Interannual variability of upwelling indices in the southeastern Arabian Sea: a satellite based study. OceanSci J 45(1):27–40. https://doi.org/10.1007/s12601-010-0003-6

Jayaram C, Joseph KA, Balchand AN (2013) Interannual variability of chlorophyll-*a* concentration along the southwest coast of India. Int J Remote Sens 34(11):3820–3831

Johannessen OM, Subbaraju G, Blindheim J (1987) Seasonal variations of the oceanographic conditions off the southwest coast of India during 1971–1975. Fisk Dir Skr Hav Unders 18:247–261

Joseph PV (1990) Warm pool over the Indian Ocean and monsoon onset. Trop Ocean-Atmos Newslett 53:1–5

Krishnakumar PK, Bhat GS (2008) Seasonal and interannual variations of oceanographic conditions off Mangalore coast (Karnataka, India) during 1995–2004 and their influences on pelagic fishery. Fish Oceanogr 17:45–60

Madhupratap M, Shetye SR, Sreekumaran Nair KNV (1994) Oil sardine and Indian mackerel: their fishery, problems and coastal oceanography. Curr Sci 66:340–348

Meinen CS, McPhaden MJ (2000) Observations of warm water volume change in the equatorial pacific and their relationship to the El-niño and La-niña. J Climate 13:3551–3559

Meyers G (1979) On the annual Rossby wave in the tropical North Pacific Ocean. J Phys Oceanogr 9:663–674

Mohanty PK, Khora SS, Panda US, Mohapatra GN, Mishra P (2005) An overview of sardines and anchovies fishery along the Indian coasts. Department of Marine Sciences, Berhampur University, Berhampur, Orissa, India

Pettersson LH, Pozdnyakov D (2013) Monitoring of harmful algal blooms. Springer Praxis books. https://doi.org/10.1007/978-3-540-68209-7. ISBN: 978-3-540-22892-9

Pillai NGK, Ganga U, Jayaprakash AA (2003) Indian oil sardine. In: Status of exploited marine fishery resources of India. CMFRI Cochin, pp. 18–24. ISBN 81-901219-3-6

Prell WL, Curry WB (1981) Faunal and isotopic indices of monsoonal upwelling: Western Arabian Sea. Oceanol Acta 4:91–95

Product User Guide v. 2.0.5. D3.4 PUG (2015) Ocean Colour Climate Change Initiative (OC_CCI)—Phase Two. 28 April

Raj RP (2017) Surface velocity estimates of the North Indian Ocean from satellite gravity and altimeter missions. Int J Remote Sens 38(1):296–313

Rao RR, Sivakumar R (1996) Seasonal variability of near-surface isothermal layer and thermocline characteristics of the tropical Indian Ocean. Meteorol Atmos Phys 61:201–212

Ryther JH, Yentsch CS (1957) The estimation of phytoplankton production in the ocean from chlorophyll and light data. Limnol Oceanogr 2:281–286

Shankar D, Shetye SR (1997) On the dynamics of the Lakshadweep high and low in the southeastern Arabian Sea. J Geophys Res Oceans. 102(C6):12551–12562

Shankar D, Vinayachandran PN, Unnikrishnan AS (2002) The monsoon currents in the north Indian Ocean. Prog Oceanogr 52:63–120

Shenoi SSC, Shankar D, Shetye SR (1999) On the sea surface temperature high in the Lakshadweep Sea before the onset of the southwest monsoon. J Geophys Res. 104:15,703-15,712

Shetye SR (1998) West India coastal current and Lakshadweep high/low. Sadhana 23(5–6):637–651

Shetye SR, Shenoi SSC, Gouveia AD, Michael GS, Sundar D, Nampoothiri G (1991) Winddriven coastal upwelling along the western boundary of the Bay of Bengal during the southwest monsoon. Cont Shelf Res 11:1397–1408

Smitha A, Joseph KA, Jayaram C, Balchand AN (2014) Upwelling in the southeastern Arabian Sea as evidenced by Ekman mass transport using wind observations from OCEANSAT–II Scatterometer. Ind J Mar Sci 43(1):111–116

Smitha BR, Sanjeevan VN, Vimalkumar KG, Revichandran C (2008) On the upwelling off the southern tip and along the west coast of India. J Coast Res 24:95–102

Vialard J, Delecluse P (1998) An OGCM study for the TOGA decade. Part II: barrier layer formation and variability. J Phys Oceanogr 28:1089–1106

Yu L (2003) Variability of the depth of the 20 °C isotherm along 6°N in the Bay of Bengal: its response to remote and local forcing and its relation to satellite SSH variability. Deep Sea Res Part II 50:2285–2304

Remote Sensing of Upwelling in the Arabian Sea and Adjacent Near-Coastal Regions

K. Ajith Joseph, Chiranjivi Jayaram, Archana Nair, Mary Swapna George, A. N. Balchand and Lasse H. Pettersson

Abstract Upwelling is a dominant mechanism in the Arabian Sea that occurs annually during southwest monsoon summer season. This results in abundance of phytoplankton and zooplankton in the region and has profound influence on the coastal fisheries. During the southwest monsoon, an intense low-level wind jet blows diagonally across the Arabian Sea generating coastal upwelling along the coasts of Somalia, Oman and the southeastern Arabian Sea. In this study, a synergy of different parameters like sea surface winds, chlorophyll (chl-*a*), sea surface temperature (SST) and sea level anomaly (SLA) retrieved from remote sensing were used to make a more detailed analysis on upwelling features for the summer seasons of the years 1982–2015. From the analysis, it is observed that upwelling in the Arabian Sea is not homogeneous across the basin despite being driven by monsoon winds. During the study period, Ekman transport, SLA and SST anomaly showed positive trend, whereas chlorophyll showed negative trend of varying strengths. Increased Ekman transport has not generated increased productivity indicating the role of other governing mechanisms on the availability of nutrients in the region.

Keywords Coastal upwelling · Arabian sea · Sea surface temperature (SST) Sea level anomalies (SLA) · Ekman transport

---

K. Ajith Joseph (✉) · M. S. George
Nansen Environmental Research Centre (India), Kochi, India
e-mail: ajith_jk@nerci.in

C. Jayaram
RRSC-East, NRSC/Indian Space Research Organisation, Kolkata, India

A. Nair
SEOCS, Indian Institute of Technology Bhubaneswar, Bhubaneswar, India

A. N. Balchand
Cochin University of Science and Technology, Kochi, Kerala, India

L. H. Pettersson
Nansen Environmental and Remote Sensing Center, Bergen, Norway

Present address
Centre for Earth Research and Environment Management, Kochi, India

V. Barale and M. Gade (eds.), *Remote Sensing of the Asian Seas*,
https://doi.org/10.1007/978-3-319-94067-0_26

1 Introduction

Coastal upwelling systems are major economic sources, which contribute to 20% of the global fish production for less than 3% of the world oceans' surface (Bennazzouz et al. 2014). Environmental factors like wind, productivity and the gas exchange make upwelling ecosystems sensitive to anthropogenic influence and global warming (Freon et al. 2009). Within the northern Indian Ocean, upwelling is a dominant mechanism especially pertaining to the Arabian Sea (AS) that occurs during the summer time southwest monsoon enhancing the productivity of the region. The significant event during the summer monsoon is, upwelling induced Ekman dynamics and a subsequent shoaling of the nutracline (*euphotic zone?*) in the northern Arabian Sea (Bauer et al. 1991). The combined effect of upwelling and the surface layer deepening results in the latitudinal gradients of phytoplankton biomass. The low-level monsoon jet stream (so called Findlater jet) (Joseph and Raman 1966) blows diagonally across the Arabian Sea resulting in upwelling along the coasts of Somalia (R3), Oman (R2) and southeast Arabian Sea (SEAS), India (R1—see Fig. 1) (Wyrtki 1973). The upwelling regime of the AS during the southwest monsoon (SWM) is a combination of coastal upwelling driven by the

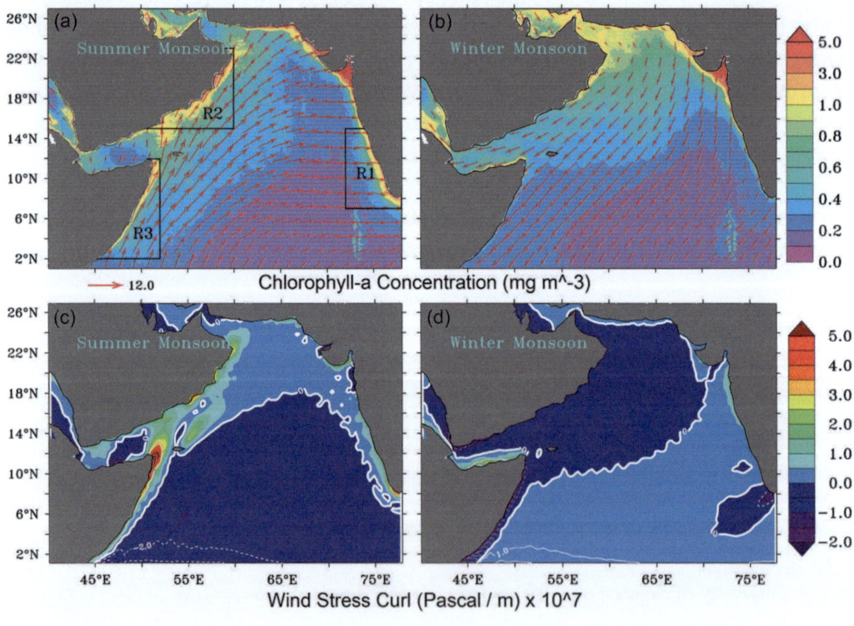

Fig. 1 Surface chlorophyll-*a* concentration (mg m^{-3}) and wind direction during: **a** summer southwest monsoon and **b** winter northeast monsoon. Wind stress curls (Pascal/m) × 10^7 during: **c** summer monsoon and **d** winter monsoon. The boxes in (**a**) indicate the predominant upwelling regions in the Arabian Sea: R1—Southwest coast of India (SEAS) (Lat: 7–15°N, Lon: 70–78°E), R2—Oman coast (Lat: 16–23°N, Lon: 52–60°E) and R3—Somalia coast (Lat: 2–11°N, Lon: 45–52°E)

alongshore wind stress as well as the wind stress curl driven offshore upwelling that occurs inshore of the Findlater jet (Findlater 1969; Bauer et al. 1991; Prasanna Kumar et al. 2001). The Indian summer monsoon rainfall (ISMR) exhibits variability with low frequency component that alternates between epochs of above normal (wet) and below normal (dry) rainfall, each about three decades long (Krishnamurthy and Goswami 2000). During the summer monsoon season with lower (higher) than normal rainfall, the warm upper surface of the ocean is thicker (thinner) and thereby resulting in weak (strong) upwelling (Dube et al. 1990). Therefore, it could be inferred that the upwelling is relatively weak during weaker monsoon with below normal rainfall years and vice versa. The significance of the upwelling in AS is the governing mechanisms that are prevalent in the different coastal regions.

An overview of upwelling phenomena in the Arabian Sea has been detailed by Kamph and Chapman (2016) and the characteristic features of the seasonal wind driven coastal upwelling systems in this region is available in the references there in. Upwelling along the Somali coast is considered as the fifth major upwelling in the world in terms of volume of upwelled water and it is a western boundary upwelling system (de Castro et al. 2016). One of the main differences between the Somali upwelling and other major upwelling regions is the short residence time of upwelled surface waters in the coastal upwelling area due to rapid offshore transport (Hitchcock et al. 2000). Off the Oman coast, intense upwelling is associated with varying currents that are dominated by eddies. The mean upwelling volume transports at a depth of 100 m within an area encompassing the region 260 km offshore and 600 km alongshore (Shi et al. 2000). The cooler SST pattern is defined by upwelling as well as flow distribution in the upwelling region and it is observed approximately up to one month after the collapse of Southwest monsson (SWM). Upwelling along the SW coast of India (SEAS) is prevalent during SWM (Jayaram et al. 2010). Upwelling in this region is a combination of wind and remote forcing, which spreads northward with time since the onset of the SWM propagating from southern tip of India upward along the Karnataka coast. Wind speed gradually increases from the middle of May and reaches its peak by July/August, and in September, the wind speed tapers away signaling the withdrawal phase of summer monsoon (Johannessen et al. 1987; Smitha et al. 2014; George et al. 2017). The enhanced productivity due to upwelling makes AS one of the most productive regions of the world ocean and has profound impact on the local fishery (Smitha et al. 2008 and Chapter 25).

Mechanisms that govern upwelling, originate from large scale air-sea coupling, wherein, the amplitude and timing of the upwelling conducive winds are affected by climate variability and global warming (Bakun et al. 2015; García-Reyes et al. 2015). Short term oscillations in the strength of upwelling and the resultant productivity changes, influence the abundance of fishery resources and pose a serious impact on the socio-economic conditions of the coastal communities in the region. However, consistency of the hypothesis on long term variability remains ambiguous owing to various factors like the spatial resolution and length of data records, source of data etc., (Cropper et al. 2014). In this context, utilization of satellite remote sensing observations to assess the extent and variability of upwelling in the Arabian Sea over a time period is prudent as the satellite data products are capable of monitoring oceans over a large scale and with consistent methodology over longer time-periods.

The emphasis of the present study is on the upwelling phenomenon prevalent in the major upwelling zones of the Arabian Sea viz., central Arabian Sea (open ocean), southwest coast of India (eastern boundary), Oman coast (northern boundary) and Somalia coast (western boundary). Long term satellite data records over the last several decades pertaining to the parameters like sea surface winds, sea surface temperature, sea level anomaly and chlorophyll-*a* concentration (chl-*a*) is analyzed to identify the predominant trends in their variability and thereby quantifying the strength of the coastal upwelling.

2 Data and Methods

The present study is carried out using the available long term satellite data products merged from different satellites and sensors pertaining to the signatures of upwelling in terms of Ekman transport (UI_{wind}) derived from sea surface winds, negative sea level anomaly, cooler sea surface temperature and higher chl-*a*. The data products for each of the parameter were selected in such a way that the entire time series is generated using uniform merging and quality. The time series plots are prepared after removing the regression estimate (trend) from the data in order to avoid the influence of the global signatures on the upwelling.

2.1 Sea Surface Winds

Wind speed is one of the important factors in determining the strength of upwelling. Wind based coastal upwelling index is interpreted as the water flux transported offshore by the wind stress from the entrainment of cooler coastal water assuming an infinite ocean (Bennazzouz et al. 2014). The Cross-calibrated, Multi-Platform Ocean Surface Wind Velocity project (CCMP) Version 2.0, is used in this study for the period 1987 – 2015 obtained from www.remss.com. This data is obtained by applying variational analysis on wind observations from different scatterometers viz., QuikSCAT, ADEOS-II, NSCAT and radiometers like AMSR-E, TRMM, SSM/I and SSMIS (Atlas et al. 2011). The data from scatterometers and radiometers were evaluated with in situ observations and the final product is obtained at a spatial resolution of 25 km $\times$ 25 km and a temporal interval of one day. Monthly data was created from the daily data. Wind stress and UI_{wind} were computed along the coasts of the three selected regions following Varela et al. (2015). The alongshore wind stress is calculated as follows:

$$\tau_l = \rho_a C_d V_l \left(u^2 + v^2\right)^{1/2} \tag{1}$$

where, ρ_a is the air density, C_d is the dimensionless drag coefficient, V_l is the upwelling favorable wind, calculated as follows:

$$V_l = -\left(\frac{lat}{abs(lat)}\right)\left[u\ \cos\left(\theta - \frac{\pi}{2}\right) + v\ \sin\left(\theta - \frac{\pi}{2}\right)\right] \tag{2}$$

where, *lat* is the latitude, *u* is the zonal wind component, and *v* is the meridional wind component, θ is the angle defined by a unitary vector normal to the shoreline and pointing seaward. Thus, the Ekman transport, UI_{wind}, is obtained from the alongshore wind stress following Smith (1968), as given below:

$$UI_{wind} = \frac{\tau_i}{\rho_w f} \tag{3}$$

where ρ_w is the water density and f is the Coriolis force, $2\Omega\ \sin\Phi[s^{-1}]$.

2.2 Sea Surface Temperature

NASA's optimally interpolated sea surface temperature at 25 km × 25 km resolution (OISST v2, Reynolds et al. 2007) for the period 1982–2015 was used in this study. This data is a combination of measurements from satellites, ships and buoys. OISST gains in terms of better coverage and ability to obtain SST during clouds by merging different products. The sensitivity of OISST to seasonal factors is less than 0.3 °C with mean bias error of 0.04 °C and standard deviation of 0.66 °C and provides one of the most reliable global SST products (Marullo et al. 2014). The daily OISST data is averaged to create monthly data and monthly SST anomalies were obtained by subtracting the long term mean of the corresponding month. Anomalies were computed to ascertain the variability of SST over the study period in the region of upwelling.

2.3 Sea Level Anomaly

Daily sea level anomaly data produced from satellite altimetry available at 25 km × 25 km resolution for the period 1993–2015 is obtained from the French CNES Archiving, Validation and Interpretation of Satellite Oceanographic data (AVISO). The merged satellite radar altimetry product was generated from Topex/Poseidon, Jason, ERS and Envisat satellites and the reference period for computing SLA was considered as 1993–2012 (SSALTO/DUACS 2015). The daily data is used to generate monthly mean values of SLA for the study region.

2.4 Chlorophyll-a Concentration

Chlorophyll-*a* concentration data used in this study is obtained from the European Space Agency's (ESA) Ocean Colour CCI (OC_CCI, version 1, CCI 2011) project. The merged product is obtained from combining data from Sea-viewing Wide Field-of-view Sensor (SeaWiFS) and Moderate Resolution Imaging Spectroradiometer (MODIS) onboard Aqua and Medium Resolution Imaging Spectrometer (MERIS) (Nieto and Mélin 2017). The chl-*a* was computed using maximum band ratio algorithm OC4v6 (O'Reilly et al. 2000). The merged product has twice the mean global coverage from individual sensors mapping (Maritorena et al. 2010). The spatial resolution of the data is 4 km $\times$ 4 km with a temporal interval of 8-days. Monthly mean chl-*a* for the study region is obtained from the 8-day data.

3 Results and Discussion

3.1 Central Arabian Sea

Upwelling is prevalent in the Arabian Sea, during the summer monsoon season (June – September) with varied intensities and temporal distribution along the southwest coast of India (R1, Fig. 1), Oman coast (R2) and Somalia coast (R3). Central Arabian Sea is distinguished by two different mechanisms to the north and south of the axis of the Findlater Jet. Cyclonic wind stress curl (positive) in the northern part stimulates a divergent Ekman transport in the upper layers of the ocean and drives open-ocean upwelling through upward Ekman pumping (Bauer et al. 1991). The upward movement and the entrainment of nutrients that are driven by basin wide winds, together with advection of the upwelled waters from the Arabian coast enhance the productivity of the region. Also, the upwelling induced Ekman dynamics shoals the nutracline of the northern Arabian Sea. The productivity in the southern region is enhanced by the advected nutrients from the Somalia upwelling zone (Prasanna Kumar et al. 2001). As a result of this, during the upwelling period, the amount of nutrients upwelled in to the euphotic zone and the thickness of the mixed layer will be controlled by the relative strength of wind-driven mixing (entrainment) and Ekman pumping (Bauer et al. 1991).

Wind direction during both the monsoon seasons along with the surface chlorophyll-*a* concentration, wind stress curl and the Ekman pumping velocity are shown in Fig. 1. From the figure, it is inferred that during the summer monsoon, upwelling induced productivity is abundant in the regions marked R1 – R3, which is substantiated by positive wind stress curl and Ekman pumping velocity. However, during the winter monsoon, the wind direction is reversed and it is not favorable for upwelling. The regions where higher chlorophyll prevailed during summer monsoon are devoid of higher chlorophyll; indicating the lack of upwelling in this season.

During summer monsoon, wind stress curl and Ekman pumping velocities are observed to be positive along the coast and also between 17 and 23° N in the central AS (Fig. 1) indicating the presence of upwelling. Maximum intensity of curl is observed along the Somalia coast of the order of 4 x 107 Pascal / m, followed by Oman coast (2 – 3 x 107 Pascal/m). The offshore spread of upwelled waters up to a distance of ~200 km off the Oman coast is also observed. Positive curl (0 – 1 x 107 Pascal/m) is also observed along the southwest coast of India though not of the order of magnitude observed along the Somali and Oman coasts. From the figure, the upwelling and sinking mechanism in terms of both wind stress curl and Ekman pumping velocities are observed to be delineated by the axis of the *Findlater Jet*, concurring with the ship borne observations of Bauer et al. (1991) and Muraleedharan and Prasanna Kumar (1996), wherein it was observed that the mixed layer depth shoaled from 110m at 11°N to 45m at 17°N. This shoaling of thermocline results in the entrainment of nutrients in to the euphotic zone and thereby the latitudinal gradient of biomass.

3.2 *Southwest Coast of India*

During summer monsoon, along the southwest coast of India, upwelling is driven by both wind and remote forcing (George et al. 2017). Equatorward alongshore wind stress drives the coastal surface waters away from the coast resulting in the upwelling. Apart from winds, remote forcing by the way of Kelvin waves propagating along the equator and the east coast of India enter the southeastern Arabian Sea by planetary waves also contribute to the upwelling phenomenon in the region (Rao et al. 2010). Upwelling along the southwest coast of India is noticeable at around 100 m depth between January and March and reaches the surface by May and strengthens during June–July (the southwest monsoon period). The phenomenon ceases to completely by October (Gupta et al. 2016). The upwelling intensity and variability along the southwest coast of India were assessed in terms of Ekman transport, SLA, SST anomaly and chl-*a* during the summer monsoon (June–September) for the entire period of study. Figure 2(a–d) shows the long term seasonal mean of the upwelling signatures for these parameters, respectively. Since the offshore transport as a result of upwelling is from east towards west, the sign of UI_{wind} (Ekman transport) is negative. Wind direction in this region along the coast is equatorward and oriented in northwest direction. From the Fig. 2(a), intensity of upwelling is stronger to the southern part of the region between (8–9°N) with higher UI_{wind} and lower SLA (Fig. 2b) coinciding with negative SST anomaly

(Fig. 2c) and higher chl-*a*. There appears a distinct synergy between wind induced UI_{wind} pattern and SLA with their westward propagation; while the SST anomalies and chl-*a* are oriented all along the coast with an offshore extent of 100 to 150 km.

Temporal variability in terms of seasonal southwest summer mean (June–September) of the entire region is calculated for each of the year in the period 1982–2015, depending on the actual available data record for each parameter, for the southwest coast of India and it is shown in the Fig. 3. It is observed that there exists nominal increase in Ekman transport UI_{wind}, SLA and SST anomaly and a negative trend observed for chl-*a* over the analyzed time period. Jayaram et al. (2010) showed that during 2005, 2006 and 2007 when the meridional component of wind speed was less, upwelling intensity was less and they confirmed this hypothesis with variability of the available chl-*a* data. In the present study, the decrease in the upwelling intensity for the period up to 2007 as observed in Jayaram et al. (2010) is established and thereafter a slight increase in the UI_{wind} is noticed. Though the length of the data record for each parameter is different, the signatures of all parameters are varying in tandem. Upwelling is very less during the strong ENSO years of 1992, 1997–1998 and 2015 as depicted in the UI_{wind} and SLA signals.

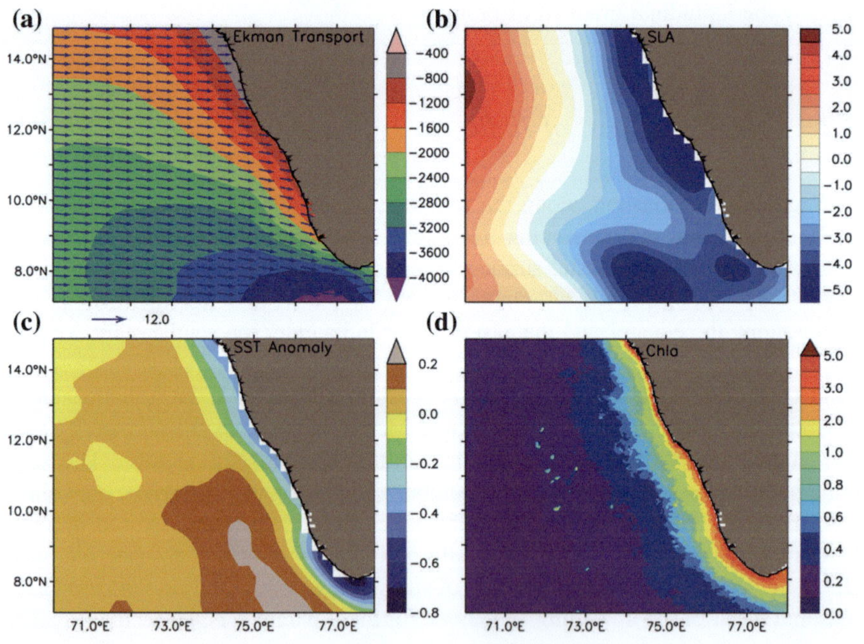

Fig. 2 Long term mean (binned over years 1982–2015, pending on the actual available data record for each parameter) of: **a** Ekman transport quantified by UI_{wind} ($m^2\ s^{-1}$) overlaid by the prevalent wind direction. **b** Sea Level Anomaly (cm). **c** Sea Surface Temperature Anomaly (°C) and **d** chl-*a* (mg m^{-3}) during the upwelling period (June–September); along the southwest coast of India

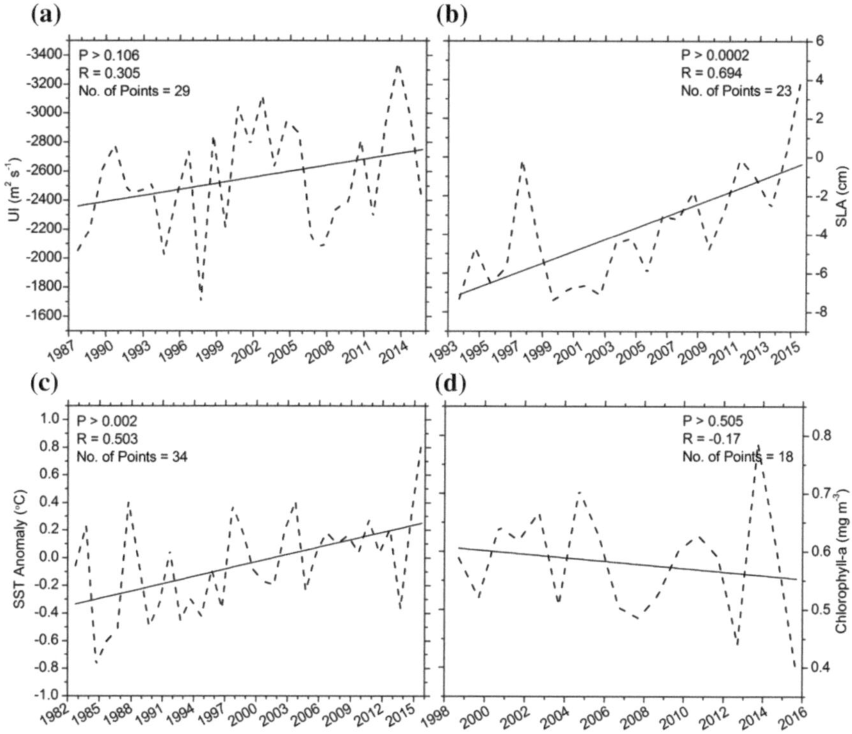

Fig. 3 Variability on the southwest Indian coast: **a** Ekman transport quantified by UI_{wind} (m^2 s^{-1}). **b** Sea Level Anomaly (cm). **c** SST Anomaly (°C) and **d** chl-*a* (mg m^{-3}); during the upwelling period (June–September); in the southwest coast of India (R1). Data are for 1982–2015, depending on the actual available data record for each parameter (see each x-axis). The data is detrended by removing the regression estimate from the time series

Similar pattern was reported by Currie et al. (2013) and Roxy et al. (2016) from their studies based on analyses of numerical ocean model results and satellite observations. There is a consistent raise in the SLA record between 1999 and 2015, which is the most prominent trend among all the parameters. SST anomaly also showed an increasing tendency from 1988 with varying intensities. All the parameters resolve strong upwelling during the year 2014. From the time series Figures, it is evident that all the upwelling signatures are not following uniform trend and represent a distinct inter annual variability.

3.3 Oman Coast

Upwelling along the Oman coast (R2) has garnered the attention of researchers based on model simulations to gauge the trend in the global warming scenario (de Castro et al. 2016; Praveen et al. 2016). From these model simulation studies, they concluded that the coastal upwelling along the northern Oman coast might intensify in future due to the northward shift in the low-level jet. However, satellite observations and coarse resolution climate model based studies by Varela et al. (2015) and Roxy et al. (2016) indicate declining strength of upwelling in the region over the resent decades. In this context, the present study aims to discuss the distribution of upwelling signatures in the spatial domain as well as their temporal variability. Figure 4 shows the spatial pattern of the Ekman transport (UI_{wind}), prevailing wind direction, SLA, SST anomaly and chl-*a* in the Oman coast during the upwelling favorable summer monsoon season (June–September) averaged over the entire study period. Since the offshore movement of water is from west to east, positive sign indicates upwelling for Oman coast. UI_{wind} gained in strength progressing from the coast that is corroborated by the eastward moving SLA anomalies especially between 17°N and 19°N. There exist two different zones of negative SST anomalies between 17°N and 20°N and again from 21°N to 22°N. Similarly, higher chl-*a* is observed from 17°N to 20°N. Spatial distribution of the parameters are considered for this study and it is understood that the upwelling along the Oman coast is predominant between 17°N and 20°N.

Temporal variability of UI_{wind}, SLA, SST anomaly and chl-*a* along the Oman coast during the upwelling period (June–September) is shown in Fig. 5. The strength of Ekman transport is greater in this region than the southwest coast of

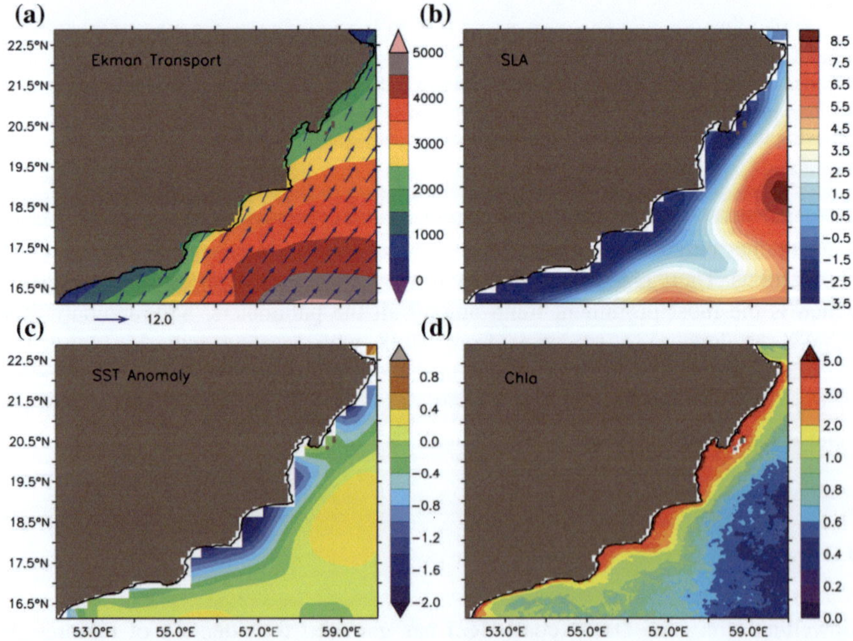

Fig. 4 Same as Fig. 2 but for the Oman Coast (R3)

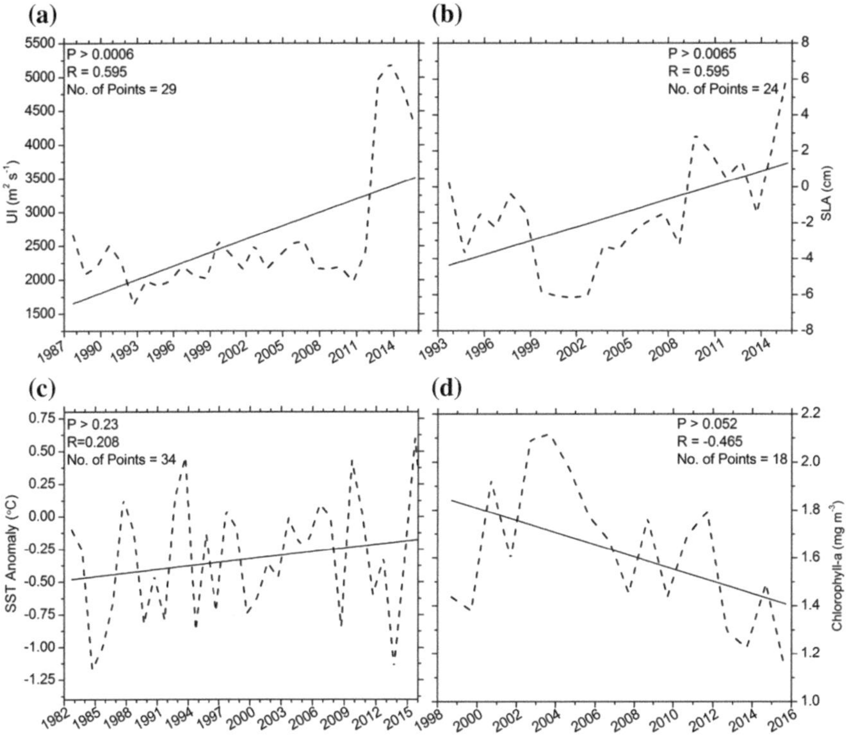

Fig. 5 Same as Fig. 3, but for the Oman coast

India (R1) supporting the fact that upwelling in the Arabian Sea is spatially not homogeneous between the different coastal boundary regions. Consistent increase in Ekman transport UI_{wind} is observed between 1993 and 2007 followed by decreasing tendency during 2008–2011. Similar to southwest coast of India (R1), intense Ekman transport is also observed in 2014 along Oman coast. SLA and SST anomaly showed increasing tendency during the entire study period while chl-*a* have decreased over the last decades. The enhanced Ekman transport despite increase in the SST anomaly infers to the intensification of upwelling in the warming scenario as hypothesized by Bakun (1990) and also simulated by Praveen et al. (2016) for the Oman coastal waters. During the year 2015 all the parameters indicated a subdued upwelling in this region with positive SLA and SST anomaly and negative chl-*a*. Despite an increase in the Ekman transport, chl-*a* showed decreasing tendency indicating the role of other factors that may influence the availability of nutrients in the region or other factors causing a decline in the marine primary production. However, the quantification of the impacts on upwelling under warming scenario is influenced by the length of the analyzed data record and also the resolution of the model domain (Praveen et al. 2016).

3.4 Somalia Coast

During summer monsoon, the Somalia current changes its direction from south to northward favoring upwelling along the coast and a sub-surface current at around a depth of 700 m in the opposite direction (Praveen et al. 2016). In the Arabian Sea, upwelling along the Somalia coast is the most intense in terms of the volume of upwelled water and the rapid rate at which the offshore water is exported. The presence of the great whirl near to the upwelling zone results in short residence period of the upwelled waters in the upwelling area (Santos et al. 2015). The mean intensity of Somalia upwelling is stronger due to its occurrence in the lower latitudes. Long term seasonal mean of Ekman drift quantified by UI_{wind}, SLA, SST anomaly and chl-*a* during the summer monsoon is shown in Fig. 6. The Ekman transport is very strong towards the southern part of the coast and tapered its intensity towards north. Negative SLA and SST anomaly are prevalent all along the coast as is the nature of an upwelling region. The intensity in offshore extent of chl-*a* varied along the coast with increasing from south to north. However, the spatial distribution might change due to the change in the intensity of monsoon winds and land-ocean thermal gradient.

The temporal variability for the summer monsoon (June–September) of the upwelling signatures is shown in Fig. 7. The Ekman transport quantified by UI_{wind}

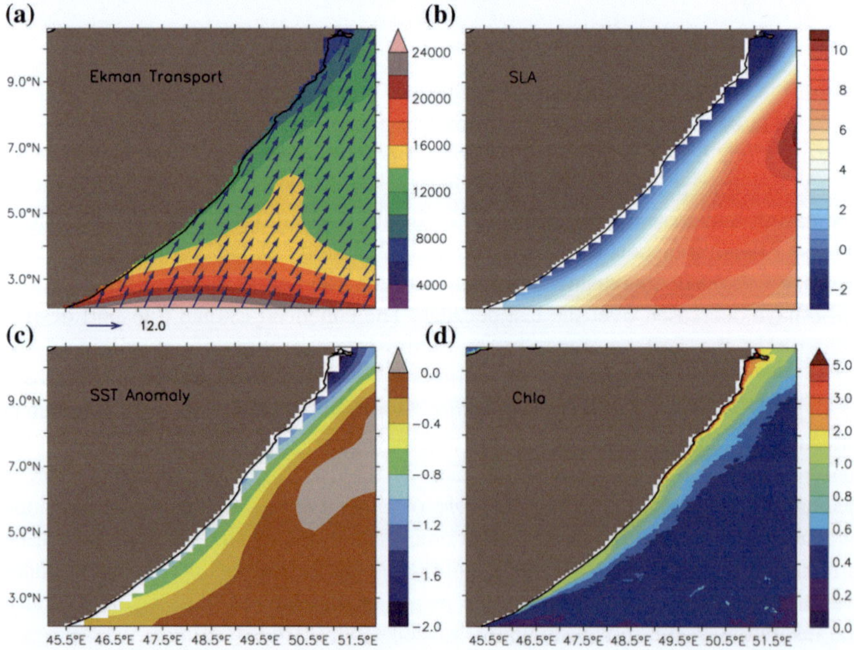

Fig. 6 Same as Fig. 4, but for the Somali coast (R3)

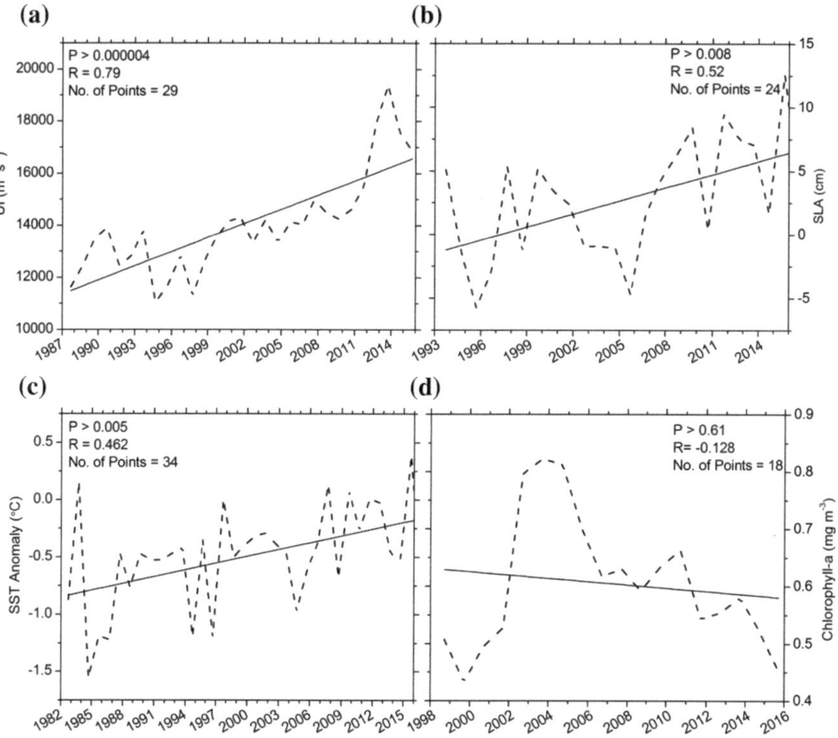

Fig. 7 Same as Fig. 3, but for the Somali coast (R3)

increased continuously/linear since 1998 with significant increase during 2014. Unlike the other two regions, SLA and SST anomaly varied inconsistently over the study period with an overall increasing tendency for the coastal waters off Somalia. Chl-*a* showed a steep increasing tendency from 1998 to 2004 followed by a decline after the peak in 2004. However, the decline in chlorophyll in the region could be either due to the increase in the sea level as opined by Prakash et al. (2012) or could be due to raise in SST observed from the studies of Roxy et al. (2016). Both factors may influence the marine primary productivity. The change in the seasonal mean values observed between the present study and by Prakash et al. (2012), may be due to the fact that this analysis is covering a larger domain to including the area of higher chl-*a* and production off the Horn of Africa. Despite the negative trend of chlorophyll, Ekman transport showed positive trend indicating intensification of upwelling along the Somalia coast similar to the other regions around the Arabian Sea. Overall, the synergy of the upwelling signatures observed over the other regions R1 (India) and R2 (Oman) is not observed in the Somalia coast (R3) indicating the scope of other mechanisms and constituents that are influencing the upwelling related productivity of the coastal waters off Somalia, which is beyond the scope of the present study.

4 Summary and Conclusions

Upwelling in the Arabian Sea during summer southwest monsoon is analyzed by applying satellite remote sensing data records for the last couple of decades. The signatures and variability of coastal upwelling in different regions are analyzed through the Ekman transport derived from sea surface winds measured using scatterometers, the sea surface height derived from radar altimeter observations, the SST measured by radiometers and the chl-*a* derived from ocean color observations. The data records available to this study varied for the different parameters viz., sea surface winds (1987–2015), SLA (1993–2015), SST (1982–2015) and chl-*a* (1998–2015). In the present study, these signatures are discussed based on their spatial distribution and temporal interannual variability. The major regions of coastal upwelling in the Arabian Sea are along the southwest coast of India (R1), Oman coast (R2) and the Somalia coast (R3). The analysis is carried out exclusively for the summer (southwest monsoon) season, when upwelling is prevalent in the region, in order to delineate the variability in the strength of upwelling in each region. Ekman transport is estimated for each region from the upwelling favorable wind direction and alongshore wind stress based on the angle of orientation of the coastline, while SST anomalies were obtained after removing the long term monthly mean from the corresponding months. Monthly average of sea level anomaly (SLA) and chl-*a* is derived from daily and 8-day temporal data, respectively; and seasonal mean is extracted from monthly data to ascertain the variability of the upwelling signatures.

Along the southwest coast of India (R1), upwelling commences in the southern region at the annual onset of the summer monsoon (in June) and propagate northward with the development of southwest monsoon attaining peak strength by July/August. Intensity of upwelling is stronger between 8°N and 9°N with higher Ekman transport quantified by UI_{wind}. Synergy of westward propagation is observed between UI_{wind} and SLA. The influence of upwelling is observed to an extent of approximately 150 km off from the coast through the negative SST anomaly and higher chl-*a*. From the temporal analysis, positive trend is observed for UI_{wind}, SLA and SST anomalies for the entire studies data record with episodic intra-annual fluctuations as observed by Jayaram et al. (2010). However, this homogeneous trend is not reflected in the variability of the chl-*a* for the region. The year 2014 is observed to be the strongest upwelling season, which is reflected in all the parameters (including chl-*a*), while weak upwelling is observed during the ENSO years of 1992, 1997–98 and again in 2015. Continuously/linear increase in SLA is observed from 1999 to 2015 in the region, whereas SST anomaly does not resolve a similar secular tendency.

Ekman transport along the Oman coast increased in strength expanding offshore from the coast indicating the combined effect of upwelled waters being carried offshore due to the prevalent currents and wind as observed from the eastward propagating SLA contours. Strong upwelling is observed along the southern part of the coast between 17°N and 20°N, considering the signatures of upwelling retrieved in all analyzed parameters. Ekman transport quantified by UI_{wind} showed increasing tendency between 1993 and 2007 and reduced during 2008–2011 attaining peak value in 2014. SLA and SST anomaly showed positive increasing trends during the study period with minor intra annual oscillations, while chl-*a* showed very minor negative trend over the analyzed period. Similar to the Indian and Oman coastal regions also weak upwelling was identified in the Somali waters during 2015.

Somalia coastal upwelling is the strongest upwelling zone in the Arabian Sea as retrieved from the strength of the Ekman transport compared to the other regions (R1 and R2). Ekman transport and chl-*a* varied in strength and offshore extent from south to north along the coast of Somalia. Gradients in both SLA and SST anomalies are oriented perpendicular to the coast indicating the offshore extent of the of the upwelling region. Ekman transport showed positive trend from 1998 to 2014, while SLA and SST anomaly varied inconsistently. Chl-*a* showed an increasing tendency from 1998 to 2004 and decreased drastically thereafter, which is in concurrence with the earlier studies on the marine productivity of this region. The major difference observed in the Somalia region is the non-secular variability of SLA and SST anomaly, which showed relatively consistent increasing trend during the entire study period.

In the Arabian Sea the spatial-temporal distribution and variability of upwelling is predominantly influenced by the strength of monsoon winds and land-ocean thermal gradient. From the analysis of three different coastal upwelling regions in the Arabian Sea it is concluded that the variability of the coastal upwelling phenomenon is not homogeneous across the basin despite the monsoon winds being the triggering mechanism. Throughout the Arabian Sea, Ekman transport, SLA and SST anomaly showed positive trend, whereas chl-*a* showed negative trend of varying strengths. The noteworthy point here is that the increase in Ekman transport, which is an indicator of coastal upwelling, is not reflected in terms of the marine primary productivity, when analyzing the entire study period. However, there exists some sort of compatibility over short durations. This inconsistent relationship expands the scope of research on the importance of governing mechanism on the availability of nutrients one factor controlling the marine primary production. The enhanced Ekman transport and increasing SST anomaly concurs with the upwelling intensification in a warming climate hypothesis subject to the length and resolution of the data records as well as the quality of the models used to simulate upwelling variability.

Acknowledgements The authors thank the NASA Cross-calibrated, Multi-Platform Ocean Surface Wind Velocity project (CCMP) for providing sea surface winds ESRL, NOAA is acknowledged for providing the Optimum Interpolation SST V2 (OISST v2). Sea Level Anomaly data is obtained from the French AVISO and Ocean Colour Climate Change Initiative (OC_CCI) of the European Space Agency (ESA) provided merged chlorophyll data. This work was initiated under the project INDO-European Research Facilities for Studies on MARine Ecosystem and CLIMate in India (INDO-MARECLIM, GA#295092) supported by the European Commission under the Seventh Framework Programme (INCO-LAB). AJK, MSG and LHP are grateful for the financial support under the institutional basic grant from the Nansen Center in Bergen, Norway. CJ thanks the Head and General Manager, RRSC-East for the support, ANB acknowledges CUSAT for the facilities. Authors thank the anonymous reviewers for their constructive comments to improve the manuscript.

References

Atlas R, Hoffman RN, Ardizzone J, Leidner SM, Jusem JC, Smith DK, Gambos D (2011) A cross-calibrated multiplatform ocean surface wind velocity product for meteorological and oceanographic applications. Bull Am Meteor Soc 92:157–174

Bakun A (1990) Global climate change and intensification of coastal ocean upwelling. Science 247:198–201

Bakun A, Black BA, Bograd SJ, García-Reyes M, Miller AJ, Rykaczewski RR (2015) Anticipated effects of climate change on coastal upwelling ecosystem. Curr Climate Change Rep 1:85–93. https://doi.org/10.1007/s40641-015-0008-4

Bauer S, Hitchcock GL, Olson DB (1991) Influence of monsoonally-forced Ekman dynamics upon surface layer depth and plankton biomass distribution in the Arabian Sea. Deep Sea Res Part A Oceanogr Res Pap 38(5):531–553

Benazzouz A, Mordane S, Orbi A, Chagdali M, Hilmi K, Atillah A, Pelegri JL, Demarcq H (2014) An improved coastal upwelling index from sea surface temperature using satellite-based approach—the case study of the Canary Current upwelling system. Cont Shelf Res 81:38–54

CCI (2011) Pixel identification, OC-CCI algorithm theoretical basis document. http://www.esa-oceancolour-cci.org/, 37pp

Cropper TE, Hanna E, Bigg GR (2014) Spatial and temporal seasonal trends in coastal upwelling off Northwest Africa, 1981–2012. Deep Sea Res-I 86:94–111

Currie JC, Lengaigne M, Vialard J, Kaplan DM, Aumont O, Naqvi SWA, Maury O (2013) Indian Ocean dipole and El Nino/southern oscillation impacts on regional chlorophyll anomalies in the Indian Ocean. Biogeoscience 10:6677–6698

de Castro M, Sousa MC, Santos F, Dias JM, Gómez-Gesteria M (2016) How will Somali coastal upwelling evolve under future warming scenarios? Nat Sci Rep 6:30137. https://doi.org/10.1038/srep30137

Dube SK, Luther ME, O'Brien JJ (1990) Relationships between interannual variability in the Arabian Sea and Indian summer monsoon rainfall. Meteor Atmos Phys 44:153–165

Findlater J (1969) A major low-level air current near the Indian Ocean during the northern summer. Q J R Meteor Soc 95:362–380

Freon P, Barange M, Arisegui J (2009) Eastern boundary upwelling ecosystems: integrative and comparative approaches. Progr Ocean 83:1–14

García-Reyes M, Sydeman WJ, Schoeman DS, Rykaczewski RR, Black BA, Smit AJ, Bograd SJ (2015) Under pressure: climate change, upwelling, and eastern boundary upwelling ecosystems. Front Mar Sci 2:109. https://doi.org/10.3389/fmars.2015.00109

George MS, Joseph PV, Ajith Joseph K, Bertino Laurent, Johannessen OM (2017) The cold pool of the Bay of Bengal and its association with the break phase of the Indian summer monsoon. Atmos Ocean Sci Lett. https://doi.org/10.1080/16742834.2017.1294017

Gupta GVM, Sudheesh V, Sudharma KV, Saravanane N, Dhanya V, Dhanya KR, Lakshmi G, Sudhakar M, Naqvi SWA (2016) Evolution to decay of upwelling and associated biogeochemistry over the southeastern Arabian Sea Shelf. J Geophys Res Biogeosci 121:159–175

Hitchcock GL, Key EL, Masters J (2000) The fate of upwelled waters in the great whirl, August 1995. Deep Sea Res II 47:1605–1621

Jayaram C, Chacko N, Ajith Joseph K, Balchand AN (2010) Interannual variability of upwelling indices in the southeastern Arabian Sea: a satellite based study. Ocean Sci J 45:27–40

Johannessen OM, Subbaraju G, Blindheim J (1987) Seasonal variation of oceanographic conditions off the southwest coast of India during 1971–1975. Fisk Dir Skr Ser Hav Unders 18:247–261

Joseph PV, Raman PL (1966) Existence of low level westerly jet stream over peninsular India during July. Indian J Meteor Hydr Geophys 17:407–410

Kamph J, Chapman P (2016) Upwelling systems of the world: a scientific journey to the most productive marine ecosystems. Springer, Berlin, 425pp

Krishnamurthy V, Goswami BN (2000) Indian monsoon-ENSO relationship on interdecadal timescale. J Climate 13:579–595

Maritorena S, D'Anton OHF, Mangin A, Siegel DA (2010) Merged satellite ocean color data products using a bio-optical model: characteristics, benefits and issues. Remote Sens Environ 114:1791–1804

Marullo S, Santoleri R, Ciani D, Borgne PL, Péré S, Pinardi N, Tonani M, Nardone G (2014) Combining model and geostationary satellite data to reconstruct hourly SST field over the Mediterranean Sea. Remote Sens Environ 146:11–23

Muraleedharan PM, Prasanna Kumar S (1996) Arabian Sea upwelling – A comparison between coastal and open ocean regions. Curr Sci 71:842–846

Nieto K, Mélin F (2017) Variability of chlorophyll-a concentration in the Gulf of Guinea and its relation to physical oceanographic variables. Progr Ocean 151:97–115

O'Reilly JE, Maritorena S, Siegel DA, O'Brien MC, Toole DA, Mitchell BG, Kahru M, Chavez F P, Strutton P, Cota GF, Hooker SB, McClain CR, Carder KL, Mueller-Karger F, Harding L, Magnusson A, Phinney D, Moore GF, Aiken J, Arrigo KR, Letelier R, Culver M (2000) Ocean color chlorophyll a algorithms for SeaWiFS, OC2 and OC4: version 4. In: Hooker SB, Firestone ER (eds) NASA Technical Memorandum 2000-206892, 20. NASA-GSFC, Greenbelt, Maryland, pp 9–23

Prasanna Kumar S, Madhupratap M, Dileep Kumar M, Muraleedharan PM, de Souza SN, Gauns M, Sarma VVSS (2001) High biological productivity in the central Arabian Sea during the summer monsoon driven by Ekman pumping and lateral advection. Curr Sci 81:1633–1638

Praveen V, Ajayamohan RS, Valsala V, Sandeep S (2016) Intensification of upwelling along the Oman coast in a warming scenario. Geophys Res Lett 43. https://doi.org/10.1002/2016gl069638

Prakash P, Prakash S, Rahaman H, Ravichandran M, Nayak S (2012) Is the trend in chlorophyll-a in the Arabian Sea decreasing? Geophys Res Letters 39:L23605. https://doi.org/10.10290/2012GL054187

Rao RR, Girishkumar MS, Ravichandran M, Rao AR, Gopalakrishna VV, Thadatil P (2010) Interannual variability of Kelvin wave propagation in the wave guides of the equatorial Indian Ocean, the coastal Bay of Bengal and the southeastern Arabian Sea during 1993–2006. Deep Sea Res I 57:1–13

Reynolds RW, Smith TM, Liu C, Chelton DB, Casey KS, Schlax MG (2007) Daily high resolution blended analyses for sea surface temperature. J Climate 20:5473–5496

Roxy MK, Modi A, Murthugudde R, Valsala V, Panickal S, Kumar SP, Ravichandran M, Vichi M, Levy M (2016) A reduction in marine primary productivity driven by rapid warming over the tropical Indian Ocean. Geophys Res Lett 43:826–833

Santos F, Gómez-Gesteria M, de Castro M, Dias JM (2015) A dipole-like SST trend in Somalia region during the monsoon season. J Geophys Res 120:597–607

Shi W, Morrison JM, Böhm E, Manghnani V (2000) The Oman upwelling zone during 1993, 1994 and 1995. Deep Sea Res II 47:1227–1247

Smith RL (1968) Upwelling. Oceanogr Mar Biol Ann Rev 6:11–46

Smitha BR, Sanjeevan VN, Vimalkumar KG, Revichandran C (2008) On the upwelling off the southern tip and along the southwest coast of India. J Coastal Res 24(4C):95–102

Smitha A, Ajith Joseph K, Jayaram C, Balchand AN (2014) Upwelling in the southeastern Arabian Sea as evidenced by Ekman mass transport using wind observations from OCEANSAT–II Scatterometer. Ind J Geomarine Sci 43(1):111–116

SSALTO/DUACS User Handbook (2015) (M)SLA and (M)ADT near-real time and delayed time products. Reference: CLS-DOS-NT-06-034, 72pp

Varela R, Alvarez I, Santos F, de Castro M, Gómez-Gesteira M (2015) Has upwelling strengthened along worldwide coasts over 1982–2010? Nature Sci Rep 5:10016. https://doi.org/10.1038/srep10016

Wyrtki K (1973) Physical oceanography of the Indian Ocean. In: Zeitzschel B et al (eds) The biology of Indian Ocean, vol 3. Springer, New York, pp 18–36

Remote Sensing of Phytoplankton Variability in the Arabian/Persian Gulf

Igor Polikarpov, Faiza Al-Yamani and Maria Saburova

Abstract The Arabian/Persian Gulf [hereafter the Gulf (Sheppard et al. (Mar Pollut Bull 60:13–38, 2010) mentioned that fourteen historical variants of the name of the Arabian/Persian Gulf are known. Here the name 'Gulf' is used, as is the case in several preceding scientific papers.)] is a marginal sea of the Indian Ocean connected with the Gulf of Oman through the Strait of Hormuz. Remote sensing approaches to the studies of phytoplankton biomass variability within this very productive and hydrographically and optically complex area are reviewed and analyzed. The remote-sensing reflectance of the Gulf surface is significantly affected by bottom reflection due to the Gulf's shallowness. Another crucial factor is the deposition of aeolian dust transported through the atmosphere from the adjacent deserts. Spatial and temporal variability in phytoplankton biomass estimated as remotely sensed chlorophyll concentrations together with physical factors are analyzed with a special emphasis on high biomass and toxic phytoplankton blooms.

Keywords Arabian Gulf · Chlorophyll · Harmful Algal Blooms
Phytoplankton · Remote sensing

1 Introduction

The Arabian Gulf, also known as the Persian Gulf or the Inner ROPME[1] Sea Area (Hamza and Munawar 2009) is a remarkable West Asian Mediterranean-type marginal sea of the Indian Ocean (Tomczak and Godfrey 2003), located in a region of the Middle East (Fig. 1) with subtropical hyper-arid climate, and playing

[1]ROPME—Regional Organization for the Protection of the Marine Environment.

I. Polikarpov (✉) · F. Al-Yamani · M. Saburova
Ecosystem-Based Management of Marine Resources Program, Environment and Life Sciences Research Center, Kuwait Institute for Scientific Research, P.O. Box 1638, Salmiya 22017, Kuwait
e-mail: igor.polikarpov@gmail.com

V. Barale and M. Gade (eds.), *Remote Sensing of the Asian Seas*,
https://doi.org/10.1007/978-3-319-94067-0_27

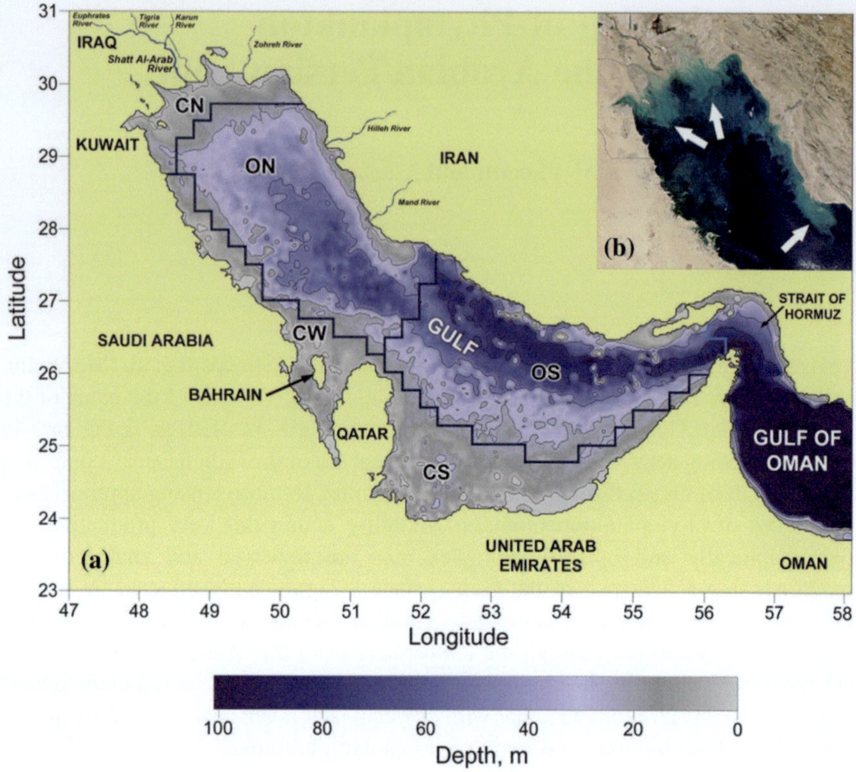

Fig. 1 **a** The Gulf's coastline, bathymetry, and sub-basin regions: northern (ON) and southern (OS) open regions, coastal northern zone of Shatt Al-Arab plume (CN), coastal western zone along the Saudi coast (CW), and shallow area between Qatar and the UAE coast (CS) (after Nezlin et al. 2010, modified); **b** NASA Terra MODIS true colour image from November 11, 2002, showing river plums as areas of high turbidity (*arrows*) in the north-western Gulf

noticeable role in the world economy and geopolitics (e.g. Hamza and Munawar 2009; Sheppard et al. 2010).

In ancient times, the Gulf was the natural corridor between Mesopotamia and India. This maritime route was extended up to the Far East. Famous Muslim geographer Ahmad al-Ya'qubi, who died in 897, described the Gulf (Sea of Fars) in the following way: "*Whoever wants to go to China must cross seven seas, each one with its own color and wind and fish and breeze, completely unlike the sea that lies beside it. The first of them is the Sea of Fars*" (Lunde 2005).

The Gulf supports highly productive marine ecosystems (Sheppard et al. 2010) including economically important species of fish, crustaceans and cephalopods. Over the past decades, marine environment of the Gulf has been changed rapidly under variety of anthropogenic and natural stresses resulted in ecosystems degradation (Hamza and Munawar 2009; Sheppard et al. 2010). Problems associated with intensive blooms of marine microalgae, known as Harmful Algal Blooms (HABs), or "red tides", are global and increasing in severity and extent (e.g. Anderson et al. 2012). Within the Gulf area, HABs have many economic and ecological impacts including mass mortalities of marine biota (e.g. Richlen et al. 2010). The regional monitoring program is essential to develop early warning systems aimed correct assessment of the red tides (e.g. Al-Yamani et al. 2012).

Remote sensing of ocean colour is a well-known tool to detect and observe elevated chlorophyll concentrations with the synoptic coverage of the entire basin. For the Gulf, this tool is rather new, but different approaches for analysis of remotely-sensed data have been tested recently (e.g. Nezlin et al. 2010; Sheppard et al. 2010; Hamzehei et al. 2013; Olalekan and Malik 2015; Zhao et al. 2015; Ghanea et al. 2016). In this chapter, most promising recent approaches used to evaluate the chlorophyll concentration and HABs discrimination from satellite ocean colour data in the Gulf and adjacent waters are reviewed.

2 Study Area

The Arabian Seas Marine Region consists of marine areas from Djibouti to Pakistan, including the northern part of Somalia, the Red Sea, the Gulf, and some parts of the Arabian Sea. The Gulf has been regionalized on the basis of temperature, salinity, and circulation patterns (Chiffings 1995). This marginal sea is located between latitudes 24° and 30°N and longitudes 48° and 57°E (Fig. 1). The Gulf is nearly 990 km long and has a maximum width of 370 km, and occupies the surface area of 239 thousand km^2. The narrow Strait of Hormuz restricts water exchange between the Gulf and the northern Indian Ocean. The Gulf's bathymetry is basically asymmetric along the main northwest-southeast axis with an average depth of 36 m. This axis separates a broad, shallow ($\sim$10 m), gradually sloping shelf of the Arabian Peninsula from Kuwait to the United Arab Emirates (UAE) from the deeper waters (deepest trough by 120 m) off the Iranian coast (Fig. 1a).

The shallowness and very high net evaporation rate, estimated by different authors as 1.44–1.68 m year^{-1} (e.g. Johns et al. 2003) coupled with very limited freshwater river runoff and negligible rainfall result in hypersaline water mass production making the Gulf one of the most saline basin on Earth. Throughout the Gulf, salinities generally range from about 38–40, increasing up to 50–70 in some embayments along the shallow south-eastern shelf, but falling to around 37 towards the Strait of Hormuz (John et al. 1990; Reynolds 1993).

The main freshwater supply of the Gulf comes from the Shatt Al-Arab deltaic system composed of Tigris, Euphrates and Karun rivers (Fig. 1a), with the maximum discharge occurs in late spring—early summer (Al-Yamani et al. 2004). Other major rivers are the Zohreh, the Hilleh, and the Mand along the north-western Iranian coast.

Deep part of the Gulf is characterized by pronounced haline stratification, but water column in the shallow (10–15 m) parts is well mixed as a result of wind stress and tidal turbulence. Persistent north-western winds produce south-eastward coastal currents resulting in upwelling along the Iranian and downwelling along the Arabian coasts. Due to Gulf's shallowness, winds strongly affect the water circulation, mixing processes, and the sediment transport and resuspension, particularly in the northern part above Qatar (Reynolds 1993).

The region of the Gulf occupies the significant part of the "dust belt" area and represents one of the major sources of aeolian dust in the world (e.g. Husar et al. 1997), so strong wind events in this region are often associated with dust storms, with the maximum in the late spring and summer and is at the minimum in the winter. Dust is an important factor limiting the remote sensing in the Gulf area (e.g. Nezlin et al. 2010; Zhao and Ghedira 2014). Sea surface temperatures are markedly fluctuated between winter minima (down to 12 °C) and summer maxima (up to 38 °C), ranging from 15 to 36 °C on the average (e.g. John et al. 1990; Al-Yamani et al. 2004).

In shallow parts of the Gulf, euphotic layer often extends to the bottom and is well mixed (Subba Rao and Al-Yamani 1999). Weak correlations between chlorophyll and photosynthetically available radiation (PAR) indicate that generally the algal growth in the Gulf is not light-limited (Nezlin et al. 2010). The field-based measurements of the phytoplankton biomass given as a chlorophyll-*a* are scarce and widely ranged between 0.01 and 10 mg m^{-3} (Sheppard et al. 2010), but under bloom conditions could reach from 55.4–262.7 up to 4525 mg m^{-3} in Kuwait's waters (Al-Yamani et al. 2012).

The biological productivity of the Gulf is highly variable and closely limited to the nutrient distribution throughout the basin. In the Gulf, lower concentrations of nutrients were recorded in both surface and near-bottom waters when compared to the adjacent Gulf of Oman, in which the productivity is largely governed by the monsoon cycle and related upwelling (e.g. Jones et al. 2002; Polikarpov et al. 2016). The lack of the favorable upwelling conditions results in nutrient limitation in most of the offshore Gulf waters (Jones et al. 2002), which, in turn, determines the low to moderate productivity of the basin. Conversely, some coastal areas, especially influenced by river discharge (Fig. 1b), are much more productive (e.g. Al-Yamani et al. 2004; Sheppard et al. 2010).

3 Chlorophyll Remote Sensing in the Gulf

3.1 Interactive Visualization and Analysis Tools

To date, there are several information systems for accumulation, processing and dissemination of the remote sensing of environmental data. Some of them have free of charge, open-access interactive Internet-based applications for Earth observing satellite data analysis. One of such instruments was developed by the NASA Goddard Earth Sciences Data and Information Services Center (GES DISC). Its acronym is GIOVANNI—Geospatial (Goddard) Interactive Online Visualization and Analysis Infrastructure.[2] It provides a simple and intuitive way to visualize, analyze, and access massive amounts of Earth remote sensing data (Acker and Leptoukh 2007) collected by many sensors based on several Earth orbiting satellites launched by NASA.

For the entire Gulf, the first known application of GIOVANNI was the visualization of surface chlorophyll distribution derived from the data collected by MODIS-Aqua spectroradiometer with the 9-km resolution during May–August 2007 and December 2007–February 2008 (Fig. 13 in Sheppard et al. 2010). The extensive areas with low chlorophyll concentrations in offshore waters and hot spots near the river mouths were revealed in comparison between two different seasons. While synoptic-scale analysis of remotely-sensed chlorophyll concentrations was of little use to detect the ephemeral phytoplankton outbreaks within the Gulf, GIOVANNI has been shown to be valuable tool for description of widespread and progressive harmful algal bloom dominated by ichthyotoxic dinoflagellate *Cochlodinium polykrikoides* in 2008–2009 (Al-Azri et al. 2014; Olalekan and Malik 2015; see also Fig. 3b).

Another excellent example of the similar web-based tool for mapping of remotely-sensed chlorophyll is the Global Marine Information System[3] (GMIS) geo-portal application (Hoepffner et al. 2014; see also Fig. 2).

3.2 *Chlorophyll Variability Along the Gulf*

In their comprehensive study of the factors affecting phytoplankton abundance within the Gulf, Nezlin et al. (2010) have been focused on the analysis of seasonal and interannual variabilities of remotely-sensed chlorophyll concentration in conjunction with the environmental factors potentially related to phytoplankton distribution. This study was based on dataset derived from Seaviewing Wide Field-of-view Sensor (SeaWiFS) and MODIS satellite instruments.

Based on bathymetry and known water circulation pattern (Reynolds 1993), the entire Gulf area was subdivided into five non-rectangular regions (Fig. 1a). In the open Gulf's waters (ON, OS), seasonal cycle of the surface chlorophyll variability was characterized by winter maximum and summer minimum (see also Fig. 2). In the shallow northern part of the Gulf (CN, the Shatt Al-Arab area), chlorophyll concentrations were much higher compared to other regions year round, and decreased in summer. However, in shallow and extremely highly turbid the Shatt Al-Arab plume zone, the remotely-sensed chlorophyll calculated by standard OC3 algorithm developed for Case 1 waters (O'Reily et al. 1998) might be overestimated through bottom reflection or by turbidity influence. In shallow regions (CS, WS), the minimum of chlorophyll concentrations was observed in spring (February–

[2]http://giovanni.gsfc.nasa.gov/. Accessed 2 Feb 2017.

[3]http://gmis.jrc.ec.europa.eu. Accessed 25 Dec 2016.

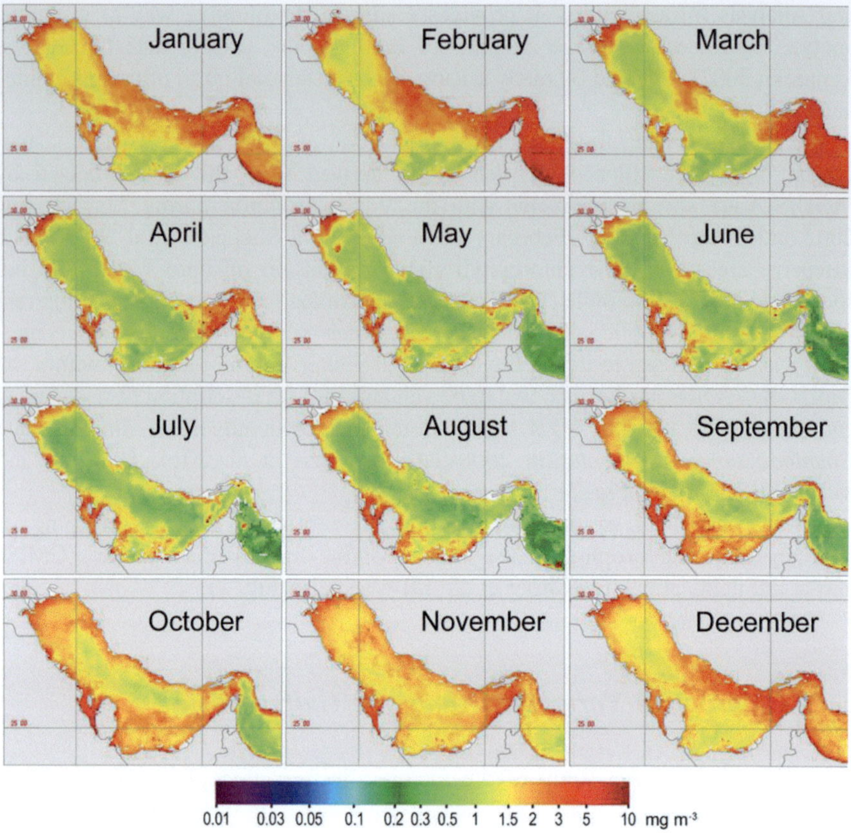

Fig. 2 Monthly climatologies of surface chlorophyll concentration (mg m^{-3}) in the Gulf and nearby waters. Data from MODIS Aqua spectrometer (2002–2015). Images were extracted from Global Marine Information System (GMIS): http://gmis.jrc.ec.europa.eu

April), whereas the maximum in late summer and autumn (August–October) coincided with the seasonal minimum of wind mixing over the Gulf suggesting low turbidity and increased bottom reflectance in shallow waters (Nezlin et al. 2010).

3.3 *Chlorophyll Climatology in the Gulf*

Monthly sea surface chlorophyll-*a* concentrations with 9-km spatial resolution retrieved by MODIS-Aqua sensor during 2002–2015 and obtained from GMIS web-based service were utilized to show the mean climatological annual cycle of chlorophyll within the Gulf and the adjacent waters (Fig. 2). Over the entire Gulf's area, a remarkable seasonal cycle of chlorophyll was characterized by maximum in cold season from November to February followed by minimum in warm season from April to August, consistent with the results of Nezlin et al. (2010). Monthly climatology of chlorophyll in synoptic scale shows the distinct and persistent spatial pattern. Within this pattern, the marginal coastal and shelf areas, in particular, river mouths with very high chlorophyll concentration are clearly distinguished from the central areas with reduced chlorophyll level. Although seasonal variability of chlorophyll concentration is clearly important, the dominant spatial pattern is persistent over an annual cycle and corresponds well to the water circulation pattern throughout the Gulf according to Reynolds (1993). Chlorophyll climatology pattern clearly visualizes the inflow of surface waters with lower salinity from the Gulf of Oman through the Strait of Hormuz (Fig. 2) along the Iranian coast.

Monthly chlorophyll climatology of the Gulf clearly separates this shallow basin from the nearby deep waters of the Strait of Hormuz and the western Gulf of Oman (Fig. 2) based on differences in monthly climatological means and the range of chlorophyll variations, and in timing of the seasonal chlorophyll blooms between these basins. The distinct peak of the chlorophyll bloom (more than 5 mg m^{-3}) occurs later in the Strait of Hormuz and the western Gulf of Oman (in February and March) and is much abruptly than in the Gulf's waters (Fig. 2).

Westward from the Strait of Hormuz, biological processes in water column are under the strong influence of seasonal phenomena such as Indian Ocean monsoon and forced cyclic upwelling, whereas the Gulf is affected mainly by the extra-tropical weather systems from the north-west and subject to strong winds influence on the water circulation and mixing processes (Reynolds 1993). Noticeable distinctions in climatic conditions between these two adjoining basins reflect monthly, seasonal and interannual differences in chlorophyll variability and in the timing, amplitude and duration of the phytoplankton blooms. The reported seasonal variations in monthly satellite-derived chlorophyll concentrations ranged from 0.85 to 2.92 mg m^{-3} in the south-eastern Gulf compared to 0.37–4.96 mg m^{-3} in the north-western Gulf of Oman, and were negatively correlated with sea surface temperatures (Kwarteng and Mozumder 2016).

Monthly climatology of chlorophyll in synoptic scale shows the baseline of the phytoplankton development in the basin. Based on range of mean chlorophyll concentrations, the Gulf's waters can be classified to be either mesotrophic (0.1–1 mg m^{-3}) or eutrophic (>1 mg m^{-3}) (according to Antoine et al. 1996). Over the annual cycle, the offshore deep (>30 m) Gulf's waters shift progressively from mesotrophic during warm season from March to September to eutrophic from October to February, whereas eutrophic status of the coastal waters and river mouths is persistent year round (Fig. 2). Presented here monthly climatological means of satellite-retrieved chlorophyll concentration are broadly consistent with the range of the reported field-based measurements, at least in case of the most studied Shatt Al-Arab area (0.2–13.9 mg m^{-3}; Subba Rao and Al-Yamani 1998).

3.4 Harmful Algal Blooms in the Gulf and Adjacent Waters

HABs and related fish kills are relatively new phenomena in the Gulf area except the sporadic red tide events caused by dinoflagellate *Noctiluca*, autotrophic ciliate *Myrionecta*, haptophyte *Phaeocystis*, cyanobacterium *Trichodesmium* and some other relatively harmless species of plankton (e.g. Subba Rao and Al-Yamani 1998; Jones et al. 2002; Al-Yamani et al. 2012). The extensive phytoplankton bloom resulted in massive mortality of wild and aquaculture-related fish in the north-western Gulf's waters was recorded in September and October 1999 in Iranian and Kuwaiti coastal waters (Heil et al. 2001). Fish kills in Kuwait Bay with an estimated economic loss of US $7 million were attributed to a massive bloom of toxigenic dinoflagellates *Karenia selliformis* and *Prorocentrum rhathymum* (Heil et al. 2001; Al-Yamani et al. 2004).

Recently, the diverse assemblages of harmful phytoplankton species have been outlined in the Gulf and adjacent waters (e.g. Al Gheilani et al. 2012; Al-Yamani et al. 2012). Moreover, over the past decade notably, the occurrence of algal blooms in the Gulf and nearby waters has increased both in frequency and severity (Richlen et al. 2010; Al Gheilani et al. 2012; Al-Yamani et al. 2012) that is in compliance with reported global HABs expansion (e.g. Anderson et al. 2012).

3.5 Case-Study on Massive Algal Blooms in the Gulf

The continuity, global coverage, and high temporal and spatial resolution of satellite data make it an important tool for monitoring of the phytoplankton blooms worldwide. Since the remote sensing tools have been applied to chlorophyll monitoring across the Gulf, two intensive and large-scale algal blooms were detected.

Based on interannual variations of seasonal anomalies SeaWiFS-derived chlorophyll concentration (Nezlin et al. 2010), exceptional high chlorophyll concentrations were revealed along the Gulf in summer 2000, when chlorophyll exceeded the climatic concentration by a factor of 1.5. The bloom was characterized by elevated chlorophyll level over the majority of the Gulf area (Fig. 3a) as well as in the adjacent Arabian Sea (Patra et al. 2007). Satellite-based observations of the summer bloom in 2000 are verified by field-based measurements of the phytoplankton biomass in the north-western Gulf off Kuwait (Subba Rao et al. 2003). In Kuwait's waters, the intensive bloom of filamentous cyanobacterium *Trichodesmium erythraeum* in May 2000 followed by multispecies phytoplankton bloom in July. During bloom incidents, the sea surface water was strongly discoloured and characterized by extremely high chlorophyll content ranged from more than 900 mg m^{-3} in May up to 2906.8–4525.5 mg m^{-3} in July (Subba Rao et al. 2003).

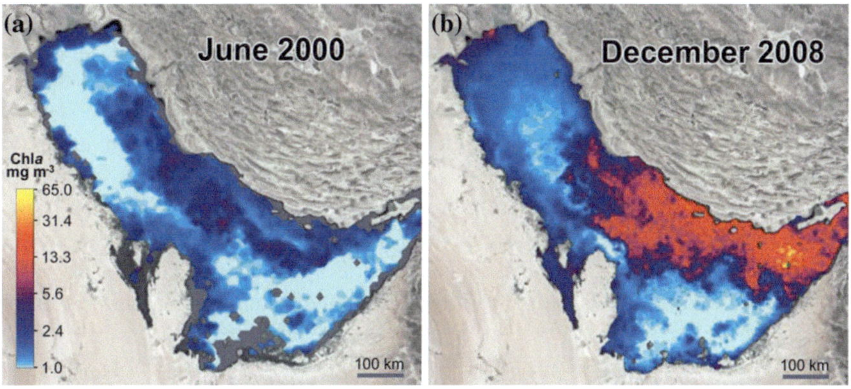

Fig. 3 Time average maps of chlorophyll-*a* concentration (mg m^{-3}) observed within the Gulf during phytoplankton blooms in June 2000 and December 2008 plotted using GIOVANNI v.4.21.6: **a** SeaWiFS_L3m_CHLv2014_9 km, **b** MODISA_L3m_CHLv2014_4 km

In August 2008, an extensive bloom of ichthyotoxic dinoflagellate *Cochlodinium polykrikoides* affected the Gulf, Strait of Hormuz and the Gulf of Oman (Fig. 3b). The bloom was notable for its coverage, intensity, and atypical duration during August 2008 through May 2009. It caused massive fish kills in Iran, Oman and UAE, damaged coral reefs, restricted fishing activities, and forced desalination plants in Oman and UAE (e.g. Richlen et al. 2010; Hamzehei et al. 2013). This was the first HAB event infected the significant part of the Gulf and the adjacent waters and affected more than 1200 km of coastline in the region, and the first to which a remote sensing approaches were applied intensively. Owing to its duration, severity and catastrophic ecological and economic impact to the Gulf and the nearby waters, this HAB incident has become the subject of thorough multidisciplinary studies and can be considered to be an opportunity to obtain coincident data sets of satellite and in situ observations available to assess the quality of the remotely-sensed data in the Gulf.

The development of *C. polykrikoides* bloom in 2009–2009 was monitored using a range of remote sensing tools. Chlorophyll anomalies observed over the Gulf area by SeaWiFS (Nezlin et al. 2010) were found to be 2–8-fold higher than the climatic concentration during November 2008–January 2009 (see Fig. 3b for December 2008). Satellite imagery from MODIS, MERIS, and SeaWiFS was used to track the outbreak and evolution of the red tide event by Zhao and Ghedira (2014). At the bloom culmination from December 2008 through January 2009, the monthly averaged SeaWiFS-derived chlorophyll concentrations exceeded 7 mg m^{-3} (Sharifinia et al. 2015). Using MODIS Aqua satellite-derived data, the progression of *C. polykrikoides* bloom from the Gulf of Oman through the Strait of Hormuz alongside the Iranian coast in the Gulf was tracked (Hamzehei et al. 2013).

GIOVANNI online tool was applied to monitoring of long-lived *C. polykrikoides* bloom by Olalekan and Malik (2015). Based on remotely-sensed chlorophyll concentrations from 4 km resolution data, the extremely high compared to background chlorophyll levels in the Gulf's waters were shown from September 2008 to January 2009 with the distinct peak between December 2008 and February 2009. Additionally, the authors applied the visualization features of GIOVANNI to show the origin and gradual changes of *C. polykrikoides* bloom using the animation tool. Starting from September 2008 in the north-western Gulf of Oman, the anomalous chlorophyll concentrations related to HAB of *C. polykrikoides* subsequently moved into the Gulf through the Strait of Hormuz, spreading to coastal waters of the UAE, Qatar, and Iran and southwards along the east coast of the UAE, generally following known spatial hydrological features, and gradually disappeared after February 2009. Similar way of this bloom development was shown by Hamzehei et al. (2013), Al-Azri et al. (2014), Zhao and Ghedira (2014), Sharifina et al. (2015) and Kwarteng and Mozumder (2016) based on satellite imagery derived from various platforms.

Presently, anthropogenic nutrient enrichment is generally accepted as a principal causative factor of a global HAB expansion in the coastal waters worldwide (e.g. Anderson et al. 2012). The bloom of *C. polykrikoides* in 2008–2009 has been presumably attributed to nutrient enrichment of the Gulf's marine environment, potentially from aquaculture activities as well as industrial and sewage inputs (Richlen et al. 2010; Sharifinia et al. 2015), whereas the initial development of bloom in Omani waters has been linked to strong upwelling along the Iranian and northern Omani coasts during the southwest monsoon in late summer (Al-Azri et al. 2014). However recently, anthropogenic nutrient-HAB link has been shown to be far from universal, and the contribution of anthropogenic nutrients to HAB promoting has been found to be rather site-specific (Davidson et al. 2014).

The Gulf's area is largely affected by dust storms occurring frequently, in particular in May–July. The roughly estimated total atmospheric dust input into the Gulf's waters is about 5.5×10^6 tons per year, assuming a 10% uniform deposition (Hamza et al. 2011). Nezlin et al. (2010) underlined that both 2000 and 2008 years were characterized by very low rain precipitation coupled with high concentration of large aerosols, probably desert dust, transported by wind over the Gulf and followed by intensive and wide-spread phytoplankton blooms. The observed low precipitation might result in higher than normal dust transport over the Gulf. Aeolian erosion in the desert areas surrounding the Gulf was high due to lower than normal soil moisture (Mahowald et al. 2005). The hypothesized significant contribution of aeolian dust-induced deposition (supposedly associated with iron micronutrient) to the enhancement of marine productivity in the Gulf (Nezlin et al. 2010; Hamza et al. 2011) is consistent with experimental studies of the phytoplankton growth under dust-enriched conditions (Subba Rao et al. 2003) and with direct observations of phytoplankton blooms following dust storms in the north-western Gulf (Subba Rao et al. 2003; Nezlin et al. 2010) and elsewhere (e.g. Meskhidze et al. 2005).

Apart from dust-induced fertilization of the surface oceanic waters, in the presence of strongly absorbing dust-produced aerosols, the standard atmospheric correction algorithms often underestimate water-leaving radiance values in the violet and blue spectrum resulting in challenges of accurately retrieving surface phytoplankton pigment concentrations using satellite tools (e.g. Nobileau and Antoine 2005). While over past decade various approaches to dust correction of satellite-derived ocean colour data have been developed and validated by in situ observations, in particular in cases of Saharan dust over the Mediterranean Sea and African coastal waters (Banzon et al. 2009; Mallet et al. 2009) and Asian dust over the northern Pacific waters (Shanmugam and Ahn 2007), dust events over the Arabian Seas, and over the Gulf in particular, are poorly characterized so far (Nezlin et al. 2010; Hamza et al. 2011).

3.6 Fluorescence Remote Sensing Products for the Gulf

Besides empirical algorithms with the blue-green ratio, the algorithms based on fluorescence has been found to be valid methods for retrieving chlorophyll-*a* concentration, in particularly for Case 2 waters and for algal bloom monitoring worldwide (e.g. Gower et al. 2005; Blondeau-Patissier et al. 2014).

To evaluate a potential of high resolution MERIS data in detecting and monitoring of algal blooms under the specific conditions of the Gulf region, Zhao et al. (2015) tested applicability of fluorescence products for the HAB characterization and delineation of toxic species from harmless. In this study, the satellite-measured fluorescence signal was extracted using two subtraction-based algorithms, fluorescence line height (FLH) and maximum chlorophyll index (MCI), to map *C. polykrikoides* bloom in 2008 in the Gulf area and nearby waters and to assess the range of applicability for these conventional MERIS fluorescence products along the different chlorophyll concentrations. MCI was found to be less efficient in detecting blooms with low chlorophyll concentration compared to FLH.

With aim to effectively quantify algal blooms and address the shortcomings of FLH and MCI, when each of them is used separately, a new index, termed as modified FLH (MFLH), was developed in this study as a surrogate of chlorophyll to describe HAB event in the Gulf area in 2008 (Zhao et al. 2015). The higher sensitivity of MFLH to the gradients between bloom and non-bloom waters compared to FLH and MCI was corroborated by in situ chlorophyll measurements. This newly proposed index is immune to coloured dissolved organic matter (CDOM), but can be sensitive to high concentration of suspended sediments and the dust, producing a false signal. Moreover, the application of MFLH images coupled with some other satellite derived data allowed to distinguish *Cochlodinium*-dominated bloom from other commonly observed in the region phytoplankton blooms.

The combination of high MFLH values with dark colours in enhanced Red-Green-Blue (ERGB) images was successfully applied to delineate the bloom-affected areas, whereas the ratio of particulate backscattering (b_{bp}) to MFLH (b_{bp}/MFLH) coupled with true colour images were found to be effective to identify species-specific optical signature of bloom-forming phytoplankton species. In bloom conditions, both dinoflagellate species, *Cochlodinium* and *Noctiluca*, were characterized by low b_{bp}/MFLH, but were discriminated by true colour images resulted from the difference in the pigment composition of these two dinoflagellate species. In contrast with reddish-brown colour of the surface waters caused by the *C. polykrikoides* bloom, the bloom of symbiont-bearing form of *Noctiluca* resulted in green water discolouration. Phylogenetically distant diatom-dominated blooms were distinguished by higher b_{bp}/MFLH combined with yellowish-green water discolouration, and blooms of filamentous cyanobacterium *Trichodesmium* were characterized by high b_{bp}/MFLH and tinted the sea surface to pale brown colour (Zhao et al. 2015).

Similar approach for HABs monitoring in the Gulf was recently used by Ghanea and colleagues (2016). The newly described in this study method was based on normalized FLH (nFLH) and diffuse attenuation coefficient at 490 nm (Kd_{490}) derived from MODIS remote sensing reflectance data. The authors have proposed Hybrid Ocean Colour Index (HOCI) as a new measure discriminating bloom areas from non-bloom under high concentrations of CDOM and suspended solids. The combination of the HOCI and ratios of the remote sensing reflectance measured at different bands has been found to be effective to discriminate between different bloom-forming phytoplankton species and allows to quantify the bloom intensity. For example, $R_{rs}(667)/R_{rs}(443) > 1$ is a statement of *C. polykrikoides* bloom identification (Ghanea et al. 2016). Additionally, MODIS data were used to study *Cochlodinium* bloom in November-December 2008 in the eastern part of the Gulf, the Strait of Hormuz and the western Gulf of Oman to test FHL data against satellite-derived chlorophyll (Moradi and Kabiri 2012). In this study, FLH data were found to be a better index than the satellite-derived chlorophyll to differentiate phytoplankton blooms due to less sensitive of FLH to interference from CDOM, shallow bottom and atmosphere.

4 Current Issue and Challenges

Compared with conventional field-based methods, satellite remote sensing can provide synoptic views of the marine environment over large temporal and spatial scales. During the past years, satellite-based remote sensing approach has been widely involved for characterization of Gulf's marine environment using a broad range of satellite platforms, sensors and retrieval algorithms to assess the monthly/seasonal/yearly/interannual chlorophyll variability and to show the chlorophyll climatologies, anomalies and long-term trends, to detect, map and monitoring of the phytoplankton blooms, and to identify the species-specific spectral signatures for common bloom-forming species/groups in the Gulf area and nearby waters.

Recent satellite-based studies of bio-optical properties of the Gulf have largely contributed to the current knowledge of its marine ecosystem status in synoptic scale, however, the optical and hydrographic complexity inherent to the Gulf's waters poses a range of challenges to the accurate retrieval of remotely-sensed ocean-colour products, in particularly in case of shallow coastal areas.

4.1 Chlorophyll Algorithms for the Gulf

Satellite ocean colour data imagery may represent an efficient and reliable tool to synoptically monitoring of spatial distribution and variability of chlorophyll concentrations throughout the Gulf, and therefore can be useful for express assessment of the environmental conditions. However, the specific detection of chlorophyll-*a* by satellite ocean colour sensors using blue-green spectral bands may yield large uncertainties in shallow optically complex waters of the Gulf, where often CDOM rather than phytoplankton dominates blue-light absorption.

Satellite ocean colour data should be considered with care when interpreting the inshore, in particular estuarine chlorophyll patterns that may be influenced by strong land-derived CDOM and particulate matter loads, which do not co-vary with chlorophyll-*a*, resulting in overestimation of satellite-retrieved chlorophyll concentrations (e.g. Chang and Gould 2006).

Another important factor leading to the overestimation of the remotely-sensed chlorophyll in shallow waters is bottom reflectance, which depends on bathymetric depth and water transparency (Maritorena et al. 1994). The bottom reflectance could be significant in the larger part of the Gulf and, therefore, chlorophyll concentrations might be considerable, but irregularly, overestimated (Nezlin et al. 2010). For these reasons, in "optically shallow" Gulf's waters, ocean colour satellite data should be processed using special algorithms adjusted for bottom reflectance correction and taking into account significant turbidity signal.

To derive ocean colour products accurately over strongly absorbing aerosols such as dust, the development of region-specific dust-correction algorithms represented the range of local dust aerosol properties has been recommended (e.g. Banzon et al. 2009) due to significant variability in the dust optical properties among the different source regions (Dubovik et al. 2002). Despite the Arabian Peninsula is ranked amongst the major sources of dust storms (e.g. Hamza and Munawar 2009), the development of region-specific dust-correction algorithm for remotely-sensed ocean colour is presently limited due to the scarce dataset on dust storm frequency, mineralogy, sources, and deposition. Region-specific atmospheric correction for strongly dust-affected Gulf area is evidently required to incorporate the site-specific dust properties for Kuwait area (Al-Dousari and Al-Awadhi 2012), UAE (Hamza et al. 2011) and Qatar Peninsula (Al-Ansari et al. 2015).

In optically complex Case 2 waters, blue-green reflectance band-ratios become less sensitive to changes in chlorophyll concentrations, requiring the use of other spectral bands located in the red and near-infrared part of the spectrum (e.g. Xing et al. 2007; Blondeau-Patissier et al. 2014). Recently developed fluorescence-based algorithms have huge advantages in retrieval of phytoplankton chlorophyll-*a* concentration in eutrophic coastal waters because firstly, the fluorescence signal is uniquely attributed by chlorophyll-*a* and, therefore, is unlikely to be confused with non-phytoplankton suspended matters; secondly, the smaller atmospheric influence in the red part of the spectrum than in the blue-green requires a simple atmospheric correction; and thirdly, absorption-based signals are a decreasing function of chlorophyll-*a*, whereas fluorescence contrastively increases with it (e.g. Gower et al. 2005; Xing et al. 2007).

On the other hand, a range of challenges has to overcome for the fluorescence-based algorithms to be applied properly, including influence of fluorescence quantum yield, physiological status of the phytoplankton algae, substances with related optical property in the ocean, atmospheric absorption, and submerged inshore vegetation (Xing et al. 2007; Blondeau-Patissier et al. 2014). Additionally, fluorescence-based chlorophyll assessments in Gulf may be limited to the less penetration depth for the fluorescence wavelength (1.96 m) compared to blue light (more than 50 m for clear ocean waters) (Xing et al. 2007). Consequently, these algorithms are well suited for phytoplankton blooms with surface expression, but may be less efficient to detect the typical for Gulf's waters pronounced subsurface chlorophyll distribution (Subba Rao and Al-Yamani 1999).

5 Conclusions

The reviewed here factors affecting the chlorophyll variability throughout the Gulf, and different approaches for interpretation of remotely-sensed data highlight a several promising applications useful for HABs detection and monitoring within the Gulf area and the adjacent waters. However, development of extensive spatial and temporal data set of field-based regional-specific observation on chlorophyll concentrations is vital to validate the satellite-retrieved chlorophyll assessments within optically complex coastal Gulf's waters. Unfortunately, in situ oceanographic measurements within the Gulf area are limited in time and space and too scarce to be sufficient to develop and validate the regional-specific atmospheric and in-water algorithms for satellite-derived ocean colour data so far.

Remote sensing of the phytoplankton distribution still remains a tool poorly used routinely for HAB detection and monitoring in the Gulf. Accordingly, the potential of modern technologies for detection and monitoring of algal blooms in the Gulf area is not achieved yet. One of the first priorities remains the detection and discrimination of HAB events from non-harmful high biomass phytoplankton blooms in shallow and optically complex Case 2 waters of the Gulf, representing an important contribution to the ecosystem-based management of marine resources in the broadest sense. Recently developed fluorescence-based algorithms regionally tuned to the Gulf appear to be largely beneficial to study of the phytoplankton variability, in particular in coastal areas.

Acknowledgements We would like to thank Kuwait Institute for Scientific Research (Kuwait) for funding. Authors are grateful to the editorial team, especially to Martin Gade and Vittorio Barale and for their patience support. We greatly appreciate the two anonymous reviewers for their helpful comments and suggestions.

References

Acker JG, Leptoukh G (2007) Online analysis enhances use of NASA earth science data. EOS Trans Am Geophys Union 88(2):14–17

Al-Ansari ES, Abdel-Moati MA, Al-Ansi MA et al (2015) Geochemical composition of dust from Qatar peninsula. In: Qatar university life science symposium 2015. http://dx.doi.org/10.5339/qproc.2015.qulss2015. Accessed 25 Jan 2017

Al-Azri AR, Piontkovski SA, Al-Hashmi KA et al (2014) Mesoscale and nutrient conditions associated with the massive 2008 *Cochlodinium polykrikoides* bloom in the Sea of Oman/Arabian Gulf. Estuar Coasts 37:325–338

Al-Dousari AM, Al-Awadhi J (2012) Dust fallout in northern Kuwait, major sources and characteristics. Kuwait J Sci 39(2A):171–187

Al Gheilani HM, Matsuoka K, AlKindi AY et al (2012) Fish kill incidents and harmful algal blooms in Omani waters. J Agric Mar Sci 16:23–33

Al-Yamani FY, Bishop J, Ramadhan E et al (2004) Oceanographic atlas of Kuwait's waters. Kuwait Institute for Scientific Research, Kuwait

Al-Yamani F, Saburova M, Polikarpov I (2012) A preliminary assessment of harmful algal blooms in Kuwait's marine environment. Aquat Ecosyst Health Manag 15(Suppl 1):64–72

Anderson DM, Cembella AD, Hallegraeff GM (2012) Progress in understanding harmful algal blooms: paradigm shifts and new technologies for research, monitoring, and management. Annu Rev Mar Sci 4:143–176

Antoine D, André J-M, Morel A (1996) Oceanic primary production 2. Estimation at global scale from satellite (coastal zone color scanner) chlorophyll. Global Biogeochem Cycles 10:57–69

Banzon VF, Gordon HR, Kuchinke CP et al (2009) Validation of a SeaWiFS dust-correction methodology in the Mediterranean Sea: identification of an algorithm-switching criterion. Remote Sens Environ 113:2689–2700

Blondeau-Patissier D, Gower JFR, Dekker AG et al (2014) A review of ocean color remote sensing methods and statistical techniques for the detection, mapping and analysis of phytoplankton blooms in coastal and open oceans. Prog Oceanogr 123:123–144

Chang GC, Gould RW (2006) Comparisons of optical properties of the coastal ocean derived from satellite ocean color and in situ measurements. Opt Express 14:10149–10163

Chiffings AW (1995) Marine Region II: Arabian Seas. In: Kelleher G, Bleakley C, Wells SM (eds) A global representative system of marine protected areas, vol IV. IUCN. Gland, Switzerland, pp 40–71

Davidson K, Gowen RJ, Harrison PJ et al (2014) Anthropogenic nutrients and harmful algae in coastal waters. J Environ Manag 146:206–216

Dubovik O, Holben BN, Eck TF et al (2002) Variability of absorption and optical properties of key aerosol types observed in worldwide location. J Atmos Sci 59:590–608

Ghanea M, Moradi M, Kabiri K (2016) A novel method for characterizing harmful algal blooms in the Persian Gulf using MODIS measurements. Adv Space Res 58:1348–1361

Gower JFR, King S, Borstad GA et al (2005) Detection of intense plankton blooms using the 709 nm band of the MERIS imaging spectrometer. Int J Remote Sens 26:2005–2012

Hamza W, Munawar M (2009) Protecting and managing the Arabian Gulf: past, present and future. Aquat Ecosyst Health Manag 12:429–439

Hamza W, Enan MR, Al-Hassini H et al (2011) Dust storms over the Arabian Gulf: a possible indicator of climate changes consequences. Aquat Ecosyst Health Manag 14:260–268

Hamzehei S, Bidokhti AA, Mortazavi MS et al (2013) Red tide monitoring in the Persian Gulf and Gulf of Oman using MODIS sensor data. Tech J Eng Appl Sci 3(12):1100–1107

Heil CA, Glibert PM, Al-Sarawi MA et al (2001) First record of a fish-killing *Gymnodinium* sp. bloom in Kuwait Bay, Arabian Sea: chronology and potential causes. Mar Ecol Prog Ser 214:15–23

Hoepffner N, Djavidnia S, Nykjaer L et al (2014) Thermal infrared remote sensing and sea surface temperature of marine and coastal waters around Africa. In: Barale V, Gade M (eds) Remote sensing of the African Seas. Springer, Berlin, pp 55–74

Husar RB, Prospero JM, Stowe LL (1997) Characterization of tropospheric aerosols over the oceans with the NOAA advanced very high resolution radiometer optical thickness operational product. J Geophys Res-Atmos 102(D14):16889–16909

John VC, Coles SL, Abozed AI (1990) Seasonal cycle of temperature, salinity and water masses of the Western Arabian Gulf. Oceanol Acta 13:273–281

Johns WE, Yao F, Olson DB et al (2003) Observations of seasonal exchange through the Straits of Hormuz and the inferred freshwater budgets of the Persian Gulf. J Geophys Res 108 (C12):3391

Jones DA, Price ARG, Al-Yamani F et al (2002) Coastal and marine ecology. In: Khan NY, Munawar M, Price ARG (eds) The Gulf ecosystem: health and sustainability. Backhuys Publishers, Leiden, pp 65–103

Kwarteng AY, Mozumder C (2016) Monitoring chlorophyll-*a* and sea surface temperature variations in SE Arabian Gulf and NW Sea of Oman from MODIS Aqua data. In: Proceedings of 37th Asian conference on remote sensing, Colombo, Sri Lanka, 17–21 Oct 2016, p Ab0012

Lunde P (2005) The Seas of Sindbad. Saudi Aramco World 56(4):20–29

Mahowald NM, Baker AR, Bergametti G et al (2005). Atmospheric global dust cycle and iron inputs to the ocean. Global Biogeochem Cycles 19(4):CB4025

Mallet M, Chami M, Gentili B et al (2009) Impact of sea-surface dust radiative forcing on the oceanic primary production: A 1D modeling approach applied to the West African coastal waters. Geophys Res Lett 36:L15828

Maritorena S, Morel AY, Gentili B (1994) Diffuse reflectance of oceanic shallow waters: influence of water depth and bottom albedo. Limnol Oceanogr 39(7):1689–1703

Meskhidze N, Chameides WL, Nenes A (2005) Dust and pollution: a recipe for enhanced ocean fertilization? J Geophys Res-Atmos 110(D3):D03301

Moradi M, Kabiri K (2012) Red tide detection in the Strait of Hormuz (east of the Persian Gulf) using MODIS fluorescence data. Int J Remote Sens 33:1015–1028

Nezlin NP, Polikarpov IG, Al-Yamani FY et al (2010) Satellite monitoring of climatic factors regulating phytoplankton variability in the Arabian (Persian) Gulf. J Mar Syst 82:47–60

Nobileau D, Antoine D (2005) Detection of blue-absorbing aerosols using near infrared and visible (ocean color) remote sensing observations. Remote Sens Environ 95:368–387

O'Reilly JE, Maritorena S, Mitchell BG et al (1998) Ocean color chlorophyll algorithms for SeaWiFS. J Geophys Res 103(C11):24937–24953

Olalekan AA, Malik K (2015) Application of Giovanni for rapid assessment of harmful algal blooms in the Arabian Gulf. Arab J Geosci 8(10):8767–8775

Patra PK, Kumar MD, Mahowald N et al (2007) Atmospheric deposition and surface stratification as controls of contrasting chlorophyll abundance in the North Indian Ocean. J Geophys Res 112(C5):C05029

Polikarpov I, Saburova M, Al-Yamani F (2016) Diversity and distribution of winter phytoplankton in the Arabian Gulf and the Sea of Oman. Cont Shelf Res 199:85–99

Reynolds RM (1993) Physical oceanography of the Gulf, Strait of Hormuz, and the Gulf of Oman —results from the Mt. Mitchell expedition. Mar Pollut Bull 27:35–59

Richlen ML, Morton SL, Jamali EA et al (2010) The catastrophic 2008–2009 red tide in the Arabian Gulf region, with observations on the identification and phylogeny of the fish-killing dinoflagellate *Cochlodinium polykrikoides*. Harmful Algae 9:163–172

Shanmugam P, Ahn YH (2007) New atmospheric correction technique to retrieve the ocean colour from SeaWiFS imagery in complex coastal waters. J Opt A Pure Appl Opt 9:511–530

Sharifinia M, Penchah MM, Mahmoudifard A et al (2015) Monthly variability of chlorophyll-*a* concentration in Persian Gulf using remote sensing techniques. Sains Malaysiana 44(3): 387–397

Sheppard C, Al-Husiani M, Al-Jamali F et al (2010) The Gulf: a young sea in decline. Mar Pollut Bull 60:13–38

Subba Rao DV, Al-Yamani F (1998) Phytoplankton ecology in the waters between Shatt Al-Arab and Straits of Hormuz, Arabian Gulf: a review. Plankton Biol Ecol 45(2):106–116

Subba Rao DV, Al-Yamani F (1999) Analysis of the relationship between phytoplankton biomass and the euphotic layer off Kuwait. Arab Gulf. Indian J Mar Sci 28(4):416–423

Subba Rao DV, Al-Yamani F, Lennox A et al (1999) Biomass and production characteristics of the first red-tide noticed in Kuwait Bay, Arabian Gulf. J Plankton Res 22(4):805–810

Subba Rao DV, Al-Hassan JM, Al-Yamani F et al (2003) The elusive red tides in the arid zone sea, off Kuwait, the Arabian Gulf. Harmful Algae News 24:10–13

Tomczak M, Godfrey JS (2003) Regional oceanography: an introduction, 2nd edn. Daya Publishing House, New Delhi

Xing XG, Zhao DZ, Liu YG et al (2007) An overview of remote sensing of chlorophyll fluorescence. Ocean Sci J 42:49–59

Zhao J, Ghedira H (2014) Monitoring red tide with satellite imagery and numerical models: a case study in the Arabian Gulf. Mar Pollut Bull 79:305–313

Zhao J, Temimi M, Ghedira H (2015) Characterization of harmful algal blooms (HABs) in the Arabian Gulf and the Sea of Oman using MERIS fluorescence data. ISPRS J Photogramm Remote Sens 101:125–136

Part V
Enclosed Seas and Lakes

Comprehensive Satellite Monitoring of Caspian Sea Conditions

Andrey G. Kostianoy, Anna I. Ginzburg, Olga Yu. Lavrova, Sergey A. Lebedev, Marina I. Mityagina, Nickolay A. Sheremet and Dmitry M. Soloviev

Abstract Satellite data on sea surface temperature (SST), visible and infrared imagery as well as altimetry and radar measurements were used to monitor different parameters of the Caspian Sea. Interannual variability of sea level (1993–2015) and SST (1982–2015) is considered. The elements of mesoscale water dynamics (vortices, jet-like flows, fine structure of upwelling frontal zone, internal waves, etc.) are discussed and examples of oil pollution are shown.

Keywords The Caspian Sea · Kara-Bogaz-Gol Bay · Satellite monitoring Sea level · Sea surface temperature · Water dynamics · Oil pollution

1 Introduction

The Caspian Sea is the largest enclosed water basin of the world that stretches about 1160 km from North to South. In view of the physico-geographical conditions, the character of bottom topography and marked dependence of sea temperature on

A. G. Kostianoy (✉) · A. I. Ginzburg · N. A. Sheremet · D. M. Soloviev
Shirshov Institute of Oceanology, Russian Academy of Sciences,
Moscow, Russia
e-mail: kostianoy@gmail.com

A. G. Kostianoy
S.Yu. Witte Moscow University, Moscow, Russia

A. G. Kostianoy
Interfacultary Center for Marine Research (MARE) and Modelling for Aquatic Systems (MAST), University of Liège, Liège, Belgium

O. Yu. Lavrova · S. A. Lebedev · M. I. Mityagina
Space Research Institute, Russian Academy of Sciences, Moscow, Russia

S. A. Lebedev
Geophysical Center, Russian Academy of Sciences, Moscow, Russia

D. M. Soloviev
Marine Hydrophysical Institute, Russian Academy of Sciences, Sevastopol, Russia

V. Barale and M. Gade (eds.), *Remote Sensing of the Asian Seas*,
https://doi.org/10.1007/978-3-319-94067-0_28

latitude, the sea is subdivided into the Northern, Middle, and Southern Caspian (Fig. 1); east part of the Middle Caspian is adjoined by the shallow (8–10 m deep) Kara-Bogaz-Gol Bay, the largest salt-generating lagoon (Kosarev 1975; Terziev et al. 1992; Kosarev 2005). The Caspian Sea is characterized by considerable variability of its hydrometeorological and hydrodynamic regimes which are influenced by both climate (global warming) and anthropogenous (e.g., variations of the Volga River discharge) factors. This long-term variability of Caspian Sea regimes, as well as need of control of ecological state of the sea, necessitates comprehensive monitoring of changes in its main parameters [sea level and sea surface temperature (SST)].

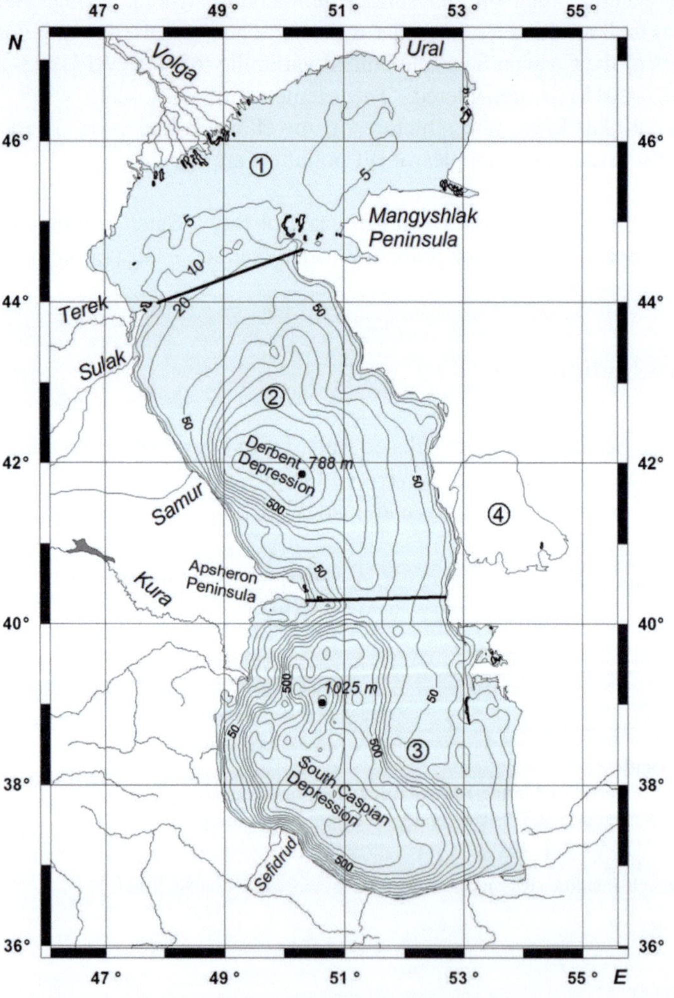

Fig. 1 The Caspian Sea bottom topography (m). (1), (2), (3), (4)—the Northern, Middle, Southern Caspian Sea, and Kara-Bogaz-Gol Bay, respectively

In the condition of rare in situ measurements, appropriate satellite information with good spatial-temporal resolution becomes the base of reliable and regular information on various parameters of the Caspian Sea.

Multi-Channel Sea Surface Temperature (MCSST) data, available since November 1981, the data of the Pathfinder Project (since 1985) and/or GODAE (Global Ocean Data Assimilation Project) High Resolution SST (GHRSST) (since 1982) can be used to track the long-term variability of SST in various parts of the Caspian Sea. These databases with high spatial and temporal resolution (up to 1 km, 1 day) and temperature resolution of 0.1 °C are based on measurements of Advanced Very High Resolution Radiometer (AVHRR) of the NOAA satellites. Altimeters of the TOPEX/Poseidon (T/P) and Jason-1,-2 (J1, J2) satellites provide reliable, regular, quite often (every 5–10 days) and independent on weather information on variability of sea level since October 1992. Images of infra-red (IR) and visible bands acquired from AVHRR radiometer (1 km resolution) of the NOAA satellites, MODIS (Moderate Resolution Imaging Spectroradiometer, resolution of 250 m) aboard the Terra and Aqua satellites, etc. (see Lavrova et al. 2011a) allow us to reveal the poorly known or unknown features of the sea water mesoscale circulation, which provide horizontal water exchange and transfer of various admixtures (in particular, oil pollution) in the sea. Besides, synthetic aperture radar (SAR) images with high spatial resolution (25–100 m) (Lavrova et al. 2011a) give a possibility to detect oil pollution on the sea surface.

Below long-term changes in sea level (1993–2015) and SST (1982–2015) of the Caspian Sea revealed on the basis of satellite data are discussed, as well as features of water dynamics (such as large-scale cyclonic gyre, nonstationary mesoscale vortices in the sea, fine structure of upwelling frontal zone, internal waves, etc.). Manifestation of oil pollution on the sea surface is also considered.

2 Sea Level

Instrumental measurements of the level of the enclosed Caspian Sea, which is conditioned mainly by changes in water balance (Kosarev 1975, 2005; Terziev et al. 1992), are carried out since 1837. In the long-term course (1837–2009) of these measurements of sea level, a number of characteristic periods may be distinguished: relatively stable state with changes within 1 m from −25.6 to −26.6 m (1837–1930), sharp level decrease with a minimal value of −29 m in 1977, increase of the level up to −26.7 m in 1995, and its further gradual decrease (Lavrova et al. 2011a; Kostianoy et al. 2014a). (Sea level is measured in the Baltic system of heights (BS). Counting of absolute heights in meters of BS is conducted from zero of tide gauge in Kronstadt.) Since October 1992, seasonal and interannual changes in sea level are well traced with altimetry data from the T/P and J1, J2 satellites (Lebedev and Kostianoy 2005; Lavrova et al. 2011a; Kostianoy et al. 2014a, b; 2018). Satellite altimeters measure the sea surface height (SSH) relative to a reference ellipsoid (or the gravity center) that allows us to eliminate vertical Earth's crust shifts from interannual level variations. Thus, satellite altimetry has advantages over the old Caspian coastal gauge stations which were not calibrated against the ocean level for several decades, and had no precise 3D GPS stations. An example of Caspian Sea level variations with consecutive alternations of growth and falling of the level from January 1993 to December 2015 is given in Fig. 2a.

It may be seen that from winter of 1992/1993 to summer of 1995 sea level grew to a mark of −26.4 m with a rate of +20.3 cm/year. Then till winter of 2001/2002 it

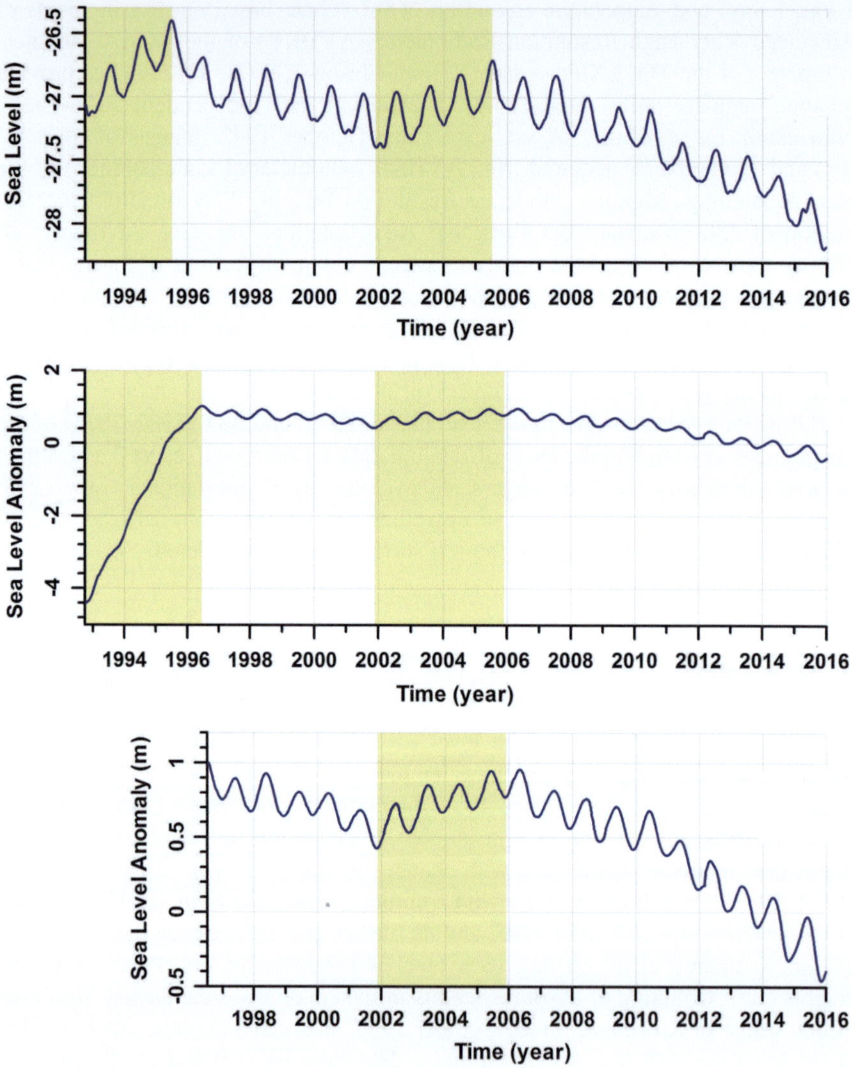

Fig. 2 Variability of water levels of the Caspian Sea (**a**) and Kara-Bogaz-Gol Bay (**b**, **c**) from 1993 to 2015 according to altimeter measurements of T/P and J 1/2. Yellow-shaded areas indicate periods of rising water level

decreased to a mark of −27.4 m with a mean rate of −7.2 cm/year. The subsequent moderate growth to a mark of −26.7 m with a rate of +10.5 cm/year occurred till summer of 2005. Then it went down with a mean rate of −10.3 cm/year. By the end of 2015, sea level has fallen to −28.2 m. Average rate of the Caspian Sea level decrease during the 23-year period 1993–2015 has appeared to be about −5 cm/year. Amplitude of seasonal variations of the sea level is about 30–40 cm.

Satellite altimeter data demonstrate also variations of the level of the Kara-Bogaz-Gol Bay since October 1992 (Lebedev and Kostianoy 2005; Kosarev and Kostianoy 2005; Kosarev et al. 2009; Kostianoy et al. 2011, 2014b). To prevent further fall of Caspian Sea level, which in 1977 had been at its lowest for the past 400–500 years (−29 m), the Kara-Bogaz-Gol Bay was separated from the sea in March 1980 by a sandy dam and by the middle of 1984, the bay has almost completely dried. Since June 1992, when the dam has been destroyed, the natural flow of sea water was restored and level of the bay has begun to increase (Fig. 2b). During 1993–1996, in the process of refilling of the bay, its level increased with a rate of 168 cm/year and by the middle of 1996 has reached a point of about −27 m. Thus, the difference of levels between the sea and the bay has decreased from 6.9 m in June 1992 to 0.2–0.6 m in 1996 (depending on a season) (Kosarev and Kostianoy 2005; Kostianoy et al. 2014b). Further, as appears from comparison of Fig. 2a and c, changes in the level of the bay corresponded to changes in sea level.

3 Sea Surface Temperature

Seasonal and interannual variability of SST was traced in three regions of the sea (Northern, Middle and Southern Caspian) and in the Kara-Bogaz-Gol Bay individually. For this purpose, weekly data of NASA JPL PO.DAAC AVHRR MCSST (Ginzburg et al. 2004, 2005a) for the period 1982–2000 and of the project NASA JPL PO.DAAC AVHRR-Pathfinder (1982–2009) (Ginzburg et al. 2012) were used. For the Kara-Bogaz-Gol Bay, the study was carried out only with MCSST data for 1994–2000 (Ginzburg et al. 2004, 2005a) because the Pathfinder SST values for this bay did not correspond to those known from the literature and characteristic of this region (see Ginzburg et al. 2004): they practically coincided with SSTs for the Middle Caspian that, apparently, was a consequence of an algorithm of the Pathfinder project to process the global AVHRR data without taking into account the isolation of this small basin (Ginzburg et al. 2012).

GHRSST data were used for the 1982–2015 period. It was shown that, on the background of a significant interannual variability of SST, there has been a warming of the Middle and Southern Caspian, on average (Fig. 3). The warming of the waters took place in all seasons (Ginzburg et al. 2012). Trends of mean annual SSTs in the Middle and Southern Caspian, and Kara-Bogaz-Gol Bay in 1982–2015 were equal to +0.05, +0.04 and +0.04 °C/year, respectively, which is several times higher than in the previous at least 20-year period (0.01 °C/year) (Potaichuk 1975; Terziev et al. 1992; Ginzburg et al. 2004). For the Northern Caspian, waters of

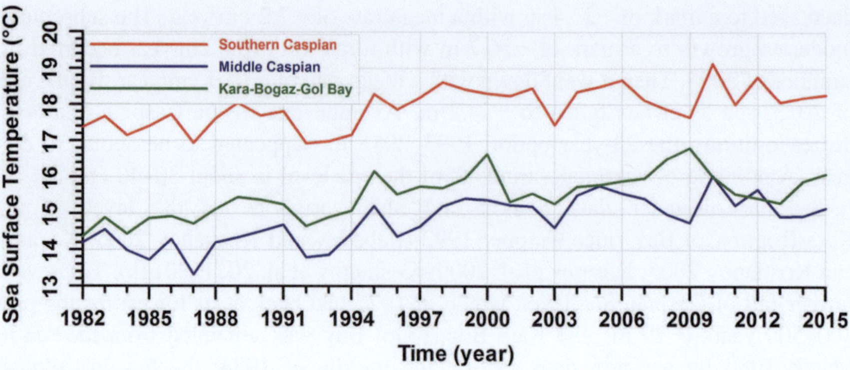

Fig. 3 Variability of SST in the Middle and Southern Caspian Sea, and Kara-Bogaz-Gol Bay (1982–2015)

which freeze in winter, similar study of SST seasonal variability with satellite data was limited to the warm season. However, the known correlation of temperature changes in the Caspian Sea basin (Rodionov 1994) allows us to extend this conclusion on the Northern Caspian Sea as well.

For the period 1982–2015, mean annual SSTs of the Middle and Southern Caspian appeared to be 14.8 and 17.9 °C, respectively, whereas they were correspondingly 13.8 and 17.1 °C in the preceding period (Ginzburg et al. 2004). Thus, mean annual temperatures of these regions in the modern period increased by about 1 °C. In the shallow Kara-Bogaz-Gol Bay, mean annual SST (15.5 °C) in 1982–2015 became 1.4 °C higher, than during the previous period (14.1 °C, see Ginzburg et al. 2004). The revealed warming of the Caspian Sea was a result of global warming in the period under consideration (see Ginzburg et al. 2005a).

4 Water Dynamics

Circulation of the Northern Caspian Sea, which depths do not exceed 20 m, is conditioned generally by wind (Kosarev 1975; Kosarev and Yablonskaya 1994; Terziev et al. 1992). Dynamics of large-scale and mesoscale currents of the Middle and Southern Caspian caused by joint influence of various factors (wind forcing, features of the coastline and bottom topography, baroclinicity of the sea) is studied insufficiently today. The schemes of surface currents in these deep-water parts of the Caspian Sea made by different authors in different years on the basis of in situ measurements (Terziev et al. 1992), calculation by a dynamic method (Kosarev 1975) and modeling (see references in (Lavrova et al. 2011a)) considerably differ. Below we consider the characteristic features of large-scale and mesoscale circulation revealed by the analysis of satellite information. A scheme of nonstationary mesoscale features, frequently observed on satellite imagery during the summer season, is shown in Fig. 4.

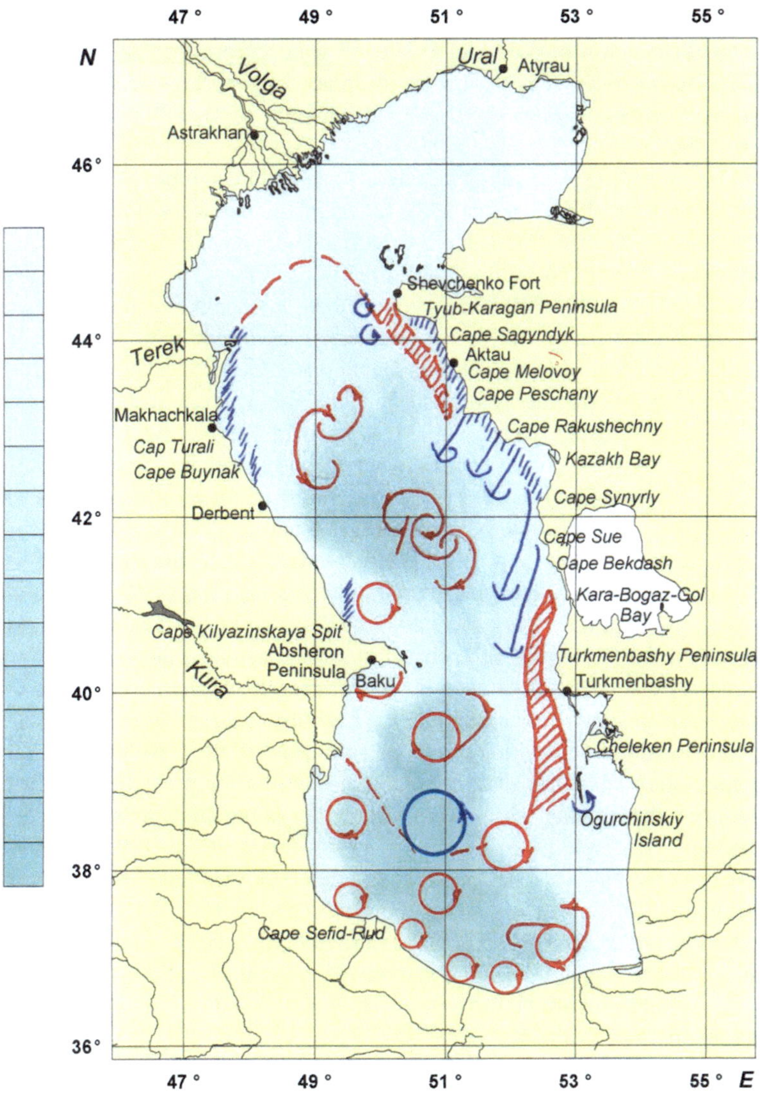

Fig. 4 A scheme of characteristic elements of mesoscale circulation in the Caspian Sea in the warm season

Cyclonic gyre in the Middle Caspian. This feature of large-scale water circulation, characteristic for the cold season, distinctly manifests itself on satellite IR and visible images in March–June and October (see IR satellite images in Sur et al. 2000; Lavrova et al. 2011a) (in winter, informative images are rare because of cloudiness). Involvement of warm waters from the Southern Caspian on eastern periphery of this gyre over the Derbent Depression provides their propagation in the cold period to the north to 43–44°N while southward flow on its western periphery promotes distribution of cold and freshened waters from the Northern to Middle Caspian. During the warm season, at the place of this large-scale cyclonic gyre mesoscale vortices, vortex dipoles and filaments of upwelling are sometimes observed (Fig. 4).

Coastal upwelling. Coastal upwelling is observed in summer along both western and eastern coasts of the Caspian Sea, however the nature of this phenomenon at both coasts significantly differs that is associated with a character of the coastal line/bottom topography, the prevailing local directions of wind, presence of an intensive southeastward current at the western coast. Local incidental (synoptic) rises of cold waters are characteristic of the western coast of the Middle Caspian at wind directions, favorable for upwelling. Generally they are, according to general direction of the coastal line, southeasterly winds on the space from Makhachkala to the Absheron Peninsula (Kosarev 1975; Arkhipkin 1990; Terziev et al. 1992). On quasi-zonal sites of the coast (Makhachkala—Cape Turali, regions of capes Buynak and Kilyazinskaya Spit), cold jet-like structures of the northward direction are formed at the winds of easterly points (Figs. 4 and 5a). These jets are 10–20 km wide and up to $\sim$60 km long, velocity of advance of their front is up to 140 cm/s, temperature drop relative to the surrounding waters—up to 11–12 °C, life time may be up to a few days (Ginzburg et al. 2005b). Time lag of onset of a local upwelling relative to the northern part of the coast (near Makhachkala) increases to the South. Frequency of manifestation of an upwelling at/near the western coast of the Caspian Sea has essential interannual variability (Ginzburg et al. 2005b).

Unlike a synoptic upwelling at the western coast of the Caspian Sea, upwelling along its eastern coast has seasonal nature caused by the prevailing northwesterly winds during the summer season, which coincides in the direction with the coastline (on average) (Kosarev 1975; Arkhipkin 1990; Terziev et al. 1992). Upwelling is observed along the eastern coast from the Tyub-Karagan Peninsula to the Turkmenbashy Peninsula (at strong northerly and northwesterly winds—to the Ogurchinsky Island, see Fig. 6.22b in Lavrova et al. 2011a) from the mid-May to the mid-October. Jagged coastline and variety of wind conditions along the coast with considerable meridional extent cause essential spatio-temporal variability of the upwelling zone (Ginzburg et al. 2006; Lavrova et al. 2011a), and proximity of the 50-m isobath to the coast on the most part of the coast [at this depth there are deep waters with the temperature of 10–12 °C (Kosarev and Tuzhilkin 1995)] is favorable for bright manifestation of upwelling at rather light winds.

Satellite images show (Figs. 4 and 5b, see also satellite images in Sur et al. 2000; Ginzburg et al. 2006; Lavrova et al. 2011a) that the upwelling zone is characterized by the complex multifrontal structure. It includes the alongshore fronts connected with rising of underlying waters to the surface and surfacing seasonal thermocline, fronts on the borders of transformed upwelling waters as a result of their warming up and mixing, and also fronts on the boundaries of the coherent structures accompanying upwelling—filaments (jets) and vortices. Note that these ordered elements of circulation are typical for other zones of coastal

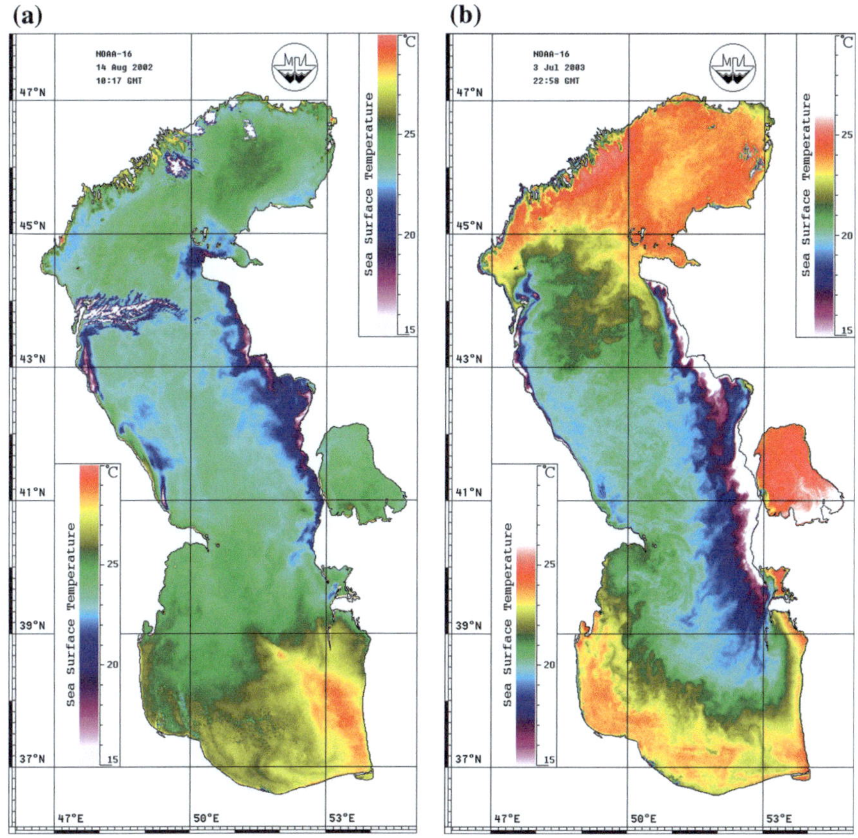

Fig. 5 Coastal upwelling in the Caspian Sea. IR images from the NOAA-16 satellite acquired on August 14, 2002 (**a**) and July 3, 2003 (**b**)

upwelling in the World Ocean and inland seas (Fedorov and Ginsburg 1992; Ginzburg et al. 2006).

Local width of the upwelling zone (distance from the coast, within which the transformed upwelling waters are traced) depends on strength, direction and duration of wind forcing, as well as from horizontal scale of coherent structures accompanying the upwelling. For example, on July 3, 2003 (Fig. 5b), width of the upwelling zone changed from ~25 km to the north of Cape Peschany to ~150 km in the South.

At the same time the extent of coastal areas of lifting of the deep waters with a temperature of 10–12 °C (the upwelling centers) at the southern coast of the Tyub-Karagan Peninsula, behind capes at the eastern coast did not exceed 5–10 km. Filaments of the transformed upwelling waters propagate transverse or inclined to the upwelling front (and to isobaths) to a distance which changes over a wide range: from 20–30 km to ~150 km in Fig. 5b. The distance between the neighboring structures also changes over a wide range—from 20–30 to ~100 km, and their width varies within 8–25 km. Temperature contrast of water in cold upwelling filaments relative to the surrounding waters is usually of 1–1.5 °C. Quite often, they have the T-shaped (mushroom-like) form. Typical lifetime of such filaments is from one to a few days, however sometimes they were traced on IR images within two weeks.

Other characteristic mesoscale elements of the upwelling zone are cyclonic vortices at its boundary (Ginzburg et al. 2006). In particular, frequent formation and separation from the upwelling front of cyclonic vortices about 10–20 km in diameter is observed at the western coast of the Tyub-Karagan Peninsula (Figs. 4 and 5b; see also Fig. 3 in Ginzburg et al. 2006). Their generation and separation from the shore under northerly and northeasterly winds in this zone are likely related to a character of the local shoreline and bottom topography.

Warm jet-like flows from the Northern and Southern Caspian into the Middle Caspian. Synoptic situation with northerly winds along the eastern coast of the sea and anticyclonic circulation in the Middle Caspian can result in propagation of warm jet-like flows to the south from the Northern Caspian along the western coast of the Tyub-Karagan Peninsula to ~43°N (Figs. 4 and 5b; Ginzburg et al. 2006; Lavrova et al. 2011a). Width of such streams of warm waters for July 3, 2003 (Fig. 5b) was ~20 km, and the velocity of advance of their fronts to the southeast according to consecutive satellite images was equal to about 5 cm/s. As it follows from analysis of MODIS-Aqua images, such warm jets from the Northern Caspian are chlorophyll-rich (see Fig. 6.26 in Lavrova et al. 2011a).

Winds of east points, more frequent in August–October, promote propagation of warm jet-like flows from the Southern into the Middle Caspian from 39–40°N to 41° 30′–42°N within the 50-m isobaths (Fig. 4). Velocity of such northward flows estimated on the shift of their head parts between consecutive IR images was about 20 cm/s and duration of their manifestation on the images was about one-two weeks (Ginzburg et al. 2006; Lavrova et al. 2011a).

Mesoscale vortices. Mesoscale vortices (eddies) in the Middle and Southern Caspian are revealed on satellite images of IR or visible bands in the presence of temperature or an admixtures contrasts at the sea surface. Anticyclonic vortex 30–70 km in diameter with the center at 41°–41° 30′N, 50–50° 30′E (between the Derbent Depression and the Absheron Peninsula, Fig. 4) is found on many satellite images. This vortex and associated elements of mesoscale circulation provide propagation of the transformed upwelling waters on a distance to ~150 km from the western coast (Ginzburg et al. 2005b; Lavrova et al. 2011a). Sometimes vortices of approximately the same scale and vortex dipoles (or vortex packing) are observed also in the central part of the Middle Caspian (Fig. 4).

Mesoscale circulation in the Southern Caspian is rather various. Mesoscale vortices are often observed both over the slope of the Southern Caspian Depression (from the Absheron Peninsula to the southeastern "corner" of the sea) and in the deep-water basin, with the centers mainly to the west of 51°E (Fig. 4). In the majority of cases they are anticyclonic vortices or dipoles/tripoles on the base of anticyclones. The centers of frequently observed anticyclones with a diameter

30–50 km are located directly to the South of the Absheron Peninsula, to the South of the Kura River mouth and in southwestern part of the sea. Vortices with a diameter of 20–40 km are found along the southern coast from Cape Sefid-Rud to the southernmost point of the sea (Lavrova et al. 2011a; Katunin and Sapozhnikov 1997; Shipilova 2000).

Anticyclonic and cyclonic vortices of larger sizes, and also vortex dipoles and their combinations are characteristic of deep-water part of the Southern Caspian. A cyclonic vortex up to 70–80 km in diameter sometimes distinctly manifests itself in the western part of the Southern Depression, with the center at about 38° 10′–38° 30′N, 50°–50° 30′E (Figs. 4 and 6a; see also Lavrova et al. 2011a). Its lifetime (manifestation in the field of chlorophyll) was nearly a month.

Cold water patches in the Southern Caspian. Interesting features of water circulation in the southern part of the Southern Caspian are incidentally appearing cold water areas with temperature contrast up to 4 °C relative to the surrounding

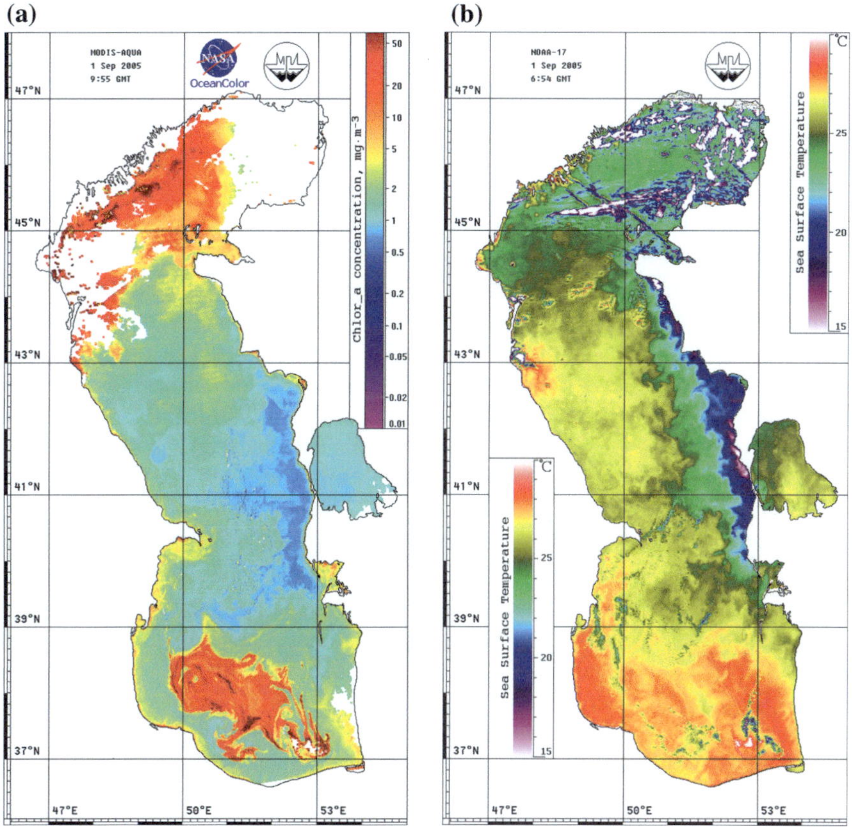

Fig. 6 Chlorophyll-a concentration from MODIS-Aqua (**a**) and IR image from the satellite NOAA-17 (**b**) acquired on September 1, 2005

waters. Such cold water spots with a size about one hundred kilometers in diameter (and more because of involvement of cold water by local vortices) were observed on December 13, 2003 and October 24, 2005 (Lavrova et al. 2011a). Physical nature of these cold water patches is not established yet. Interestingly, that in both known cases the patches were detected in autumn and had a decreased concentration of chlorophyll-a in their waters.

Features of spatial distribution of chlorophyll in connection with coastal upwelling and mesoscale vortices. Except for areas of the increased concentration of chlorophyll/suspended matter evidently associated with river runoff (the Northern Caspian, estuarial zones of the Kura River and the rivers of the Iranian coast) there are zones in the Caspian Sea of the lowered/increased concentration of admixtures caused by the processes of coastal upwelling and vortex dynamics.

Low values of concentration of chlorophyll-a are characteristic of the upwelling zone at the eastern coast of the Caspian Sea (compare Fig. 6a and b), unlike upwelling zones in other seas and oceans. It is determined, apparently, by hydrochemical features of the waters lifted from underlying layers to the sea surface. High concentration of chlorophyll in the southern part of the Southern Caspian in Fig. 6a, comparable with those in the Northern Caspian, had obviously regional character, which was not connected with intake of chlorophyll-rich waters from the Northern Caspian propagating along the western coast of the sea or with river runoff in the Southern Caspian. Note that in two weeks (from August 16 to September 1, 2005), the size of this zone of the increased concentrations of chlorophyll (it was *Cyanobacteria Nodularia,* in accordance with Kopelevich et al. 2008) and its concentration in the zone have increased, during which the greatest concentration of chlorophyll was in the area of a cyclonic vortex (see above).

Internal waves. Regular satellite monitoring of the sea surface since 2009 in the Caspian Sea has allowed to detect surface manifestations of internal waves (SMIW) in radar and optical images of this practically tideless sea, and to reconstruct the pattern of their spatial and temporal variability. SMIW in the Caspian Sea are very numerous (see Figs. 5.12, 5.13, and 5.14 in Lavrova et al. 2011a). In one image of 400 km $\times$ 400 km in size, several IW trains are usually detected and the number of trains reaches six. The length of the leading wave in a train can reach 1500 m, and prolongation of fronts are up to 50 km. Fronts in a train are often concentric arcs; this allows the region of train generation to be detected with a certain accuracy. There are usually heterogeneity of a relief, sandbanks, and sharp rises that are typical of the eastern slope of the Middle Caspian. SMIWs in the Caspian Sea have the form of classical solitons. The seasonal inhomogeneity in the IW surface manifestations is observed; it might well be connected with the position of the density jump layer. A comparison was made between the results derived from satellite data and in situ measurements. The maximal SMIW appearance corresponds to the presence of a sharp and shallow pycnocline. Such conditions are favorable for both origination of internal solitons and manifestation of these solitons at the sea surface, because such conditions provide development of strong orbital currents in the near-surface layer that leads to modulation of the wind-wave spectrum and to manifestation of IW in radar images of the sea surface. A good agreement is observed between the propagation velocity of revealed nonlinear soliton trains estimated from both satellite images and data of hydrological stations: about 15 cm/s in both cases, which indicates the ability with high accuracy to recover the velocity of propagation of packets of internal waves on the radar images. Three main regions of IW manifestation have been distinguished: the southwestern part of the Caspian Sea, region near Cape Limir (end of May); the eastern part of the Caspian Sea from Cape Bekdash to the Cheleken Peninsula (June–July), and the eastern part of the sea from Aktau to the Kazakh Bay (August) (see Fig. 5.15 in Lavrova et al. 2011a).

The tide height in the Caspian Sea is lower than 12 cm on the coast and is 2 cm in the open sea. IW origination in seas with weak tides is possible under active dynamic processes induced by origination and relief of coastal upwelling, vortices of different scales, surging, oscillations of hydrological fronts with an inertial frequency, seiches, and internal bores (Arkhipkin 1990; Lavrova et al. 2011b, c; Mityagina and Lavrova 2016a). In our opinion, seiches are the most probable cause of generation of IWs revealed in satellite images. As follows from observations in Baku and Makhachkala, the amplitude of seiche oscillations can attain 70 cm. It was shown that independent seiche oscillations originate not only in the Middle and Southern Caspian, but also in large bays. Dominating periods of 8.5–8.7 and 4.2–4.6 h are distinguished, caused by the presence of longitudinal seiches of the whole Caspian Sea (uni- and binodal, respectively). One may assume that intense seiche currents meet the scarp and generate thermocline oscillations decaying to soliton trains, as occurs under the action of tidal currents above the edges of ocean shelves.

5 Oil Pollution

The satellite monitoring of the Caspian Sea surface oil pollution was based on a combined analysis of satellite imagery obtained by various sensors that operate in different bands of the electromagnetic spectrum. The main data source was high-resolution Advanced Synthetic Aperture Radar (ASAR) Envisat (till the spring of 2012) and Sentinel-1 SAR imagery (starting from October 2014). The basic SAR data were complemented by other satellite information on sea surface and water condition in order to facilitate the differentiation between various types of surface films, reveal meteorological and hydrodynamic processes, and determine factors governing the spreading and evolution of pollutants.

A comprehensive map of sea surface pollution by oil films was produced based on the satellite remotely sensed data (Fig. 7a). According to this map, two areas of the heaviest pollution are identified. Those are the area of Absheron and Baku Archipelagoes (I) as well as the Western edge of South-Caspian Depression (II). The main source of the open sea surface pollution in the area of Absheron and Baku Archipelagoes is the ingress of the oil caused by the oil production and oil-well drilling, by oil outflow during pipeline breaks as well as by oil outflow from natural seepages at the sea bottom. The most important factors which influence the size,

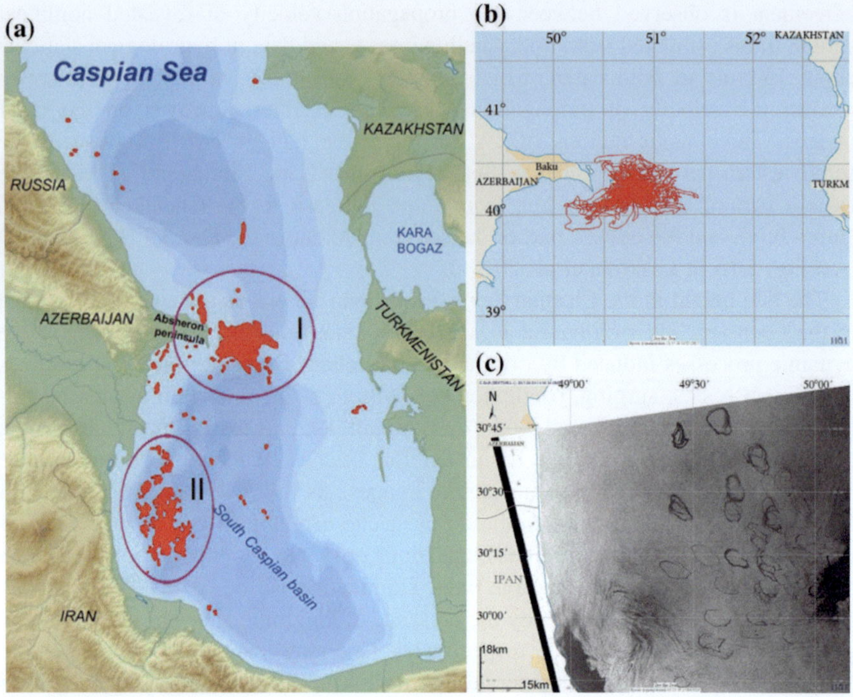

Fig. 7 **a** Map of oil spills revealed from satellite radar imagery in the Middle and Southwestern Caspian in 2010. **b** Consolidated map of oil patches revealed in radar images near the oil-producing platform *Oil Rocks* in 2010. **c** Manifestations of natural hydrocarbon showings from the sea floor via surfacing oil slicks (Sentinel-1A SAR image taken on June 4, 2015 at 14:36 UTC)

extent and spreading of oil films on the sea surface are local winds and surface currents. The system of surface currents in the area near the *Oil Rocks* platforms is rather complicated and unstable. Consolidated map of oil patches revealed in radar imagery in the vicinity of *Oil Rocks* is shown in Fig. 7b. The calculated total surface area exposed to oil contamination in the *Oil Rocks* oil-producing area reaches approximately 6260 km^2.

Surface manifestations of mud volcanoes and seeps at the Caspian Sea bottom are known for a long time. The largest number of mud volcanoes (more than 300) and the biggest ones are concentrated at the north-western edge of the South-Caspian Depression. The majority of the South-Caspian Depression mud volcanoes is in the seepage stage of their lifecycle and is discharging mud, water, gas and oil. Oil slicks are frequently seen in radar images taken over the Southern Caspian. Geographical distribution of these surface films is characterized by their persistent locations and strongly correlates with geographical locations of mud volcanoes and natural hydrocarbons showings at the sea bottom. A characteristic example of oil film pollution detected in the sea surface radar imagery taken in the south-western part of the Caspian Sea is shown in Fig. 7c.

The anthropogenic pollution caused by oil-containing waste water discharges from ships is also detected in SAR images of the Caspian Sea surface. The radar data analysis showed that they are concentrated along the main routes of the oil transportation in the Caspian Sea, namely in the directions of Aktau-Makhachkala, Aktau-Neka and Aktau-Baku. As opposed to what was observed in other enclosed seas, illegal discharges in the Caspian Sea are not the main source of sea surface film pollution, although unfortunately their amount grows year-to-year. Additional examples of oil pollution in the Caspian Sea can be found in (Kostianoy et al. 2011, 2014b; Mityagina and Lavrova 2016b).

6 Conclusions

Satellite remote sensing allows us to trace long-term changes in sea surface temperature and sea level in the Caspian Sea. It was revealed with altimeter data that during the 23-year period 1993–2015 sea level decreased with a rate of about −5 cm/year (on average). By the end of 2015, Caspian Sea level has fallen to −28.2 m. Changes in the level of the Kara-Bogaz-Gol Bay from the beginning of its refilling after destroying of dam (since October 1992) were also traced. The difference of levels between the Caspian Sea and the bay has decreased from 6.9 m in June 1992 to 0.2–0.6 m in 1996. Processing and analysis of GHRSST data have shown that mean positive trend of SST in the Caspian Sea during 1982–2015 was about 0.04–0.05 °C/year, that is several times higher than the trend (0.01 °C/year) in the previous 20-year period. In this case mean annual temperatures of Caspian regions in the modern period increased by about 1 °C. Mean annual SST in the shallow Kara-Bogaz-Gol Bay in 1982–2015 became 1.4 °C higher, than in the 1950s.

Satellite images of IR and visible band acquired from the AVHRR NOAA, MODIS/Aqua, etc., have given us an idea on complicated and variable mesoscale dynamics of the Middle and Southern Caspian, fine structure of the summer coastal upwelling at the eastern coast of the Middle Caspian, whereas SAR images allowed us to trace spots of oil pollution on the sea surface. The results of satellite monitoring of various hydrophysical and hydrobiological parameters are very important for design, construction, and operation of hydrotechnical constructions in coastal and shelf zones of the sea, forecast of evolution of an ecosystem, fishery, tracing regional climate change, ensuring ecological safety of economic activity on the Caspian Sea.

Acknowledgements The research done by A. G. Kostianoy (upwelling), A. I. Ginzburg (water dynamics), N. A. Sheremet (SST), D. M. Soloviev (satellite imagery) was supported by the Russian Science Foundation under the Project N 14-50-00095. The research done by O. Yu. Lavrova (oil pollution), M. I. Mityagina (internal waves), S. A. Lebedev (sea level) was supported by the Russian Science Foundation under the Project N 14-17-00555. The research was partially done in the framework of collaboration between Prof. Andrey Kostianoy and Dr. Marilaure Grégoire from the Interfacultary Center for Marine Research (MARE) and Modelling for Aquatic Systems (MAST), University of Liège, Belgium.

References

Arkhipkin VS (1990) Characteristic features of the structure and dynamics of coastal upwelling in the Caspian Sea. In: The Caspian Sea. Structure and dynamics of waters. Science, Moscow, pp 61–74 (in Russian)

Fedorov KN, Ginsburg AI (1992) The near-surface layer of the ocean. VSP, Utrecht, The Netherlands, 259 p

Ginzburg AI, Kostianoy AG, Sheremet NA (2004) Seasonal and interannual variability of the surface temperature in the Caspian Sea. Oceanology 44(5):605–618

Ginzburg AI, Kostianoy AG, Sheremet NA (2005a) Sea surface temperature variability. In: Kostianoy AG, Kosarev AN (eds) The Caspian Sea Environment. The Handbook of Environmental Chemistry, Springer, Heidelberg, pp 59–81

Ginzburg AI, Kostianoy AG, Solovyov DM, Sheremet NA (2005b) Structure of upwelling at the western coast of the Middle Caspian as inferred from satellite data. Issled Zemli iz kosm 4:76–85 (in Russian)

Ginzburg AI, Kostianoy AG, Solovyov DM, Sheremet NA (2006) Frontal zone of upwelling near the eastern coast of the Caspian Sea as inferred from satellite data. Issled Zemli iz kosm 4:3–12 (in Russian)

Ginzburg AI, Kostianoy AG, Sheremet NA (2012) Long-term variability of the Caspian Sea surface temperature (1982–2009). Sovrem probl distantsionnogo zondirovaniya Zemli iz Kosm 9(2):262–269 (in Russian)

Katunin DN, Sapozhnikov VV (1997) Comprehensive studies of the Southern Caspian Sea ecosystem: seasonal surveys aboard the R/V Gilian of the Russia-Iranian expedition, August–September 1994 to February 1996. Oceanology 37(1):140–142

Kopelevich OV, Burenkov VI, Sheberstov SV (2008) Case studies of optical remote sensing in the Barents Sea, Black Sea, and Caspian Sea. In: Barale V, Gade M (eds) Remote sensing of the european seas. Springer, Heidelberg, pp 53–66. https://doi.org/10.1007/978-1-4020-6772-3_4

Kosarev AN (1975) Hydrology of the Caspian and Aral Seas. Moscow University, Moscow. 272 p (in Russian)

Kosarev AN, Tuzhilkin VS (1995) Climatic thermohaline fields of the Caspian Sea. In: Druzhin SN, Terziev FS (eds). GOIN, Moscow, 96 p (in Russian)

Kosarev AN (2005) Physico-geographical conditions of the Caspian Sea. In: Kostianoy A, Kosarev A (eds) The Caspian Sea Environment. The Handbook of Environmental Chemistry, Springer, pp 5–31. https://doi.org/10.1007/698_5_011

Kosarev AN, Kostianoy AG (2005) Kara-Bogaz-Gol Bay. In: Kostianoy A, Kosarev A (eds) The Caspian Sea Environment. The Handbook of Environmental Chemistry, Springer, pp 211–221. https://doi.org/10.1007/698_5_011

Kosarev AN, Kostianoy AG, Zonn IS (2009) Kara-Bogaz-Gol Bay: physical and chemical evolution. Aquat Geochem 15(1–2):223–236 (Special Issue: Saline Lakes and Global Change). https://doi.org/10.1007/s10498-008-9054-z

Kosarev AN, Yablonskaya EA (1994) The Caspian Sea. SPB Academic Publishing, The Hague, 259 p

Kostianoy AG, Lebedev SA, Solovyov DM (2011) Satellite monitoring of water resources in Turkmenistan. Int Water Technol J 1(1):4–15

Kostianoy AG, Ginzburg AI, Lebedev SA (2014a) Climatic variability of hydrometeorological parameters of the Russian seas (1979–2011). Tr GGO 570:50–87 (in Russian)

Kostianoy AG, Lebedev SA, Solovyov DM (2014b) Satellite monitoring of the Caspian Sea, Kara-Bogaz-Gol Bay, Sarykamysh and Altyn Asyr Lakes, and Amu Darya River. In: Zonn IS, Kostianoy AG (eds) The Turkmen Lake Altyn Asyr and water resources in Turkmenistan, Springer, pp 197–231. https://doi.org/10.1007/698_2013_237

Kostianoy AG, Ginzburg AI, Kopelevich OV, Kudryavtsev VN, Lavrova OYu, Lebedev SA, Mitnik LM, Mityagina MI, Smirnov VG, Stanichny SV, Troitskaya YuI (2018) Ocean remote sensing in Russia. In: Liang S (ed) Compr Remote Sens, Oxford, Elsevier, vol 8, pp 284–325

Lavrova OYu, Kostianoy AG, Lebedev SA, Mityagina MI, Ginzburg AI, Sheremet NA (2011a) Complex satellite monitoring of the Russian Seas. IKI RAS, Moscow, 424 p (in Russian)

Lavrova OYu, Mityagina MI, Sabinin KD (2011b) Study of internal wave generation and propagation features in non-tidal seas based on satellite Synthetic Aperture Radar data. Dokl Earth Sci 436(1):165–169

Lavrova OYu, Mityagina MI, Sabinin KD, Serebryany AN (2011c) Satellite observations of surface manifestations of internal waves in the Caspian Sea. Izv Atmos Ocean Phy 47(9): 1119–1126

Lebedev SA, Kostianoy AG (2005) Satellite altimetry of the Caspian Sea. Sea, Moscow, 366 p (in Russian)

Mityagina MI, Lavrova OYu (2016a) Results of the Caspian Sea satellite survey: internal wave climate. In: Bostater CR, Neyt X, Nichol C, Aldred O (eds) Remote sensing of the ocean. Sea ice, coastal waters, and large water regions 2016. Proceedings of SPIE, vol 9999, pp 99991B. https://doi.org/10.1117/12.2241656

Mityagina M, Lavrova O (2016b) Satellite survey of inner seas: oil pollution in the Black and Caspian Seas. Remote Sens 8(10):875–898. https://doi.org/10.3390/rs8100875

Potaichuk MS (1975) Long-term changes in hydrometeorological regime of the Caspian Sea. Tr GOIN 25:95–123 (in Russian)

Rodionov SN (1994) Global and regional climate interaction: the Caspian Sea experience. Kluwer, Dordrecht, 242 p

Shipilova LM (2000) Eddy formation in the Caspian Sea. In: Lulla KP, Dessinov LV (eds), Dynamic earth environments/remote sensing observations from Shuttle-Mir Missions, Wiley, pp 211–219

Sur HI, Ozsoy E, Ibrayev R (2000) Satellite-derived flow characteristics of the Caspian Sea. In: Halpern D (ed), Satellites, oceanography and society. Elsevier oceanography series, 63. Elsevier, Amsterdam, pp 289–297

Terziev FS, Kosarev AN, Aliev AA (eds) (1992) Hydrometeorology and hydrochemistry of the seas, vol VI. The Caspian Sea. Vyp. 1. Hydrometeorological conditions. Gidrometeoizdat, St.-Petersburg, 359 p (in Russian)

Present-Day Water Balance of the Aral Sea Seen from Satellite

J.-F. Cretaux, A. Kostianoy, M. Bergé-Nguyen and A. Kouraev

Abstract The Aral Sea shrank drastically over the past 50 years, largely due to water withdrawal from Amu Darya and Syr Darya rivers for land irrigation. This has led to the separation of Aral Sea into two (in 1986–1987) and then four (in approximately 2010) water bodies. Lakes and enclosed inland seas are integrators of environmental and climate changes occurring at regional to global scale and present a high variety of behaviors on a variety of time scales (from seasonal to decadal) depending on many factors, natural and anthropogenic. In addition, their crucial importance as water stocks has increased the necessity of monitoring all of their morphodynamics characteristics, such as water level, surface (water contour) and volume. The satellite altimetry and satellite high resolution optical imagery together are now widely used for the calculation of lakes and reservoirs water storage changes worldwide. Based on these different techniques we can determine the water extent within the Aral Sea basin since 1993, as well as volume variations, which is key parameter in the understanding of hydrological regime at time scales ranging from months to decades in this largely ungauged basin. Remote sensing

J.-F. Cretaux (✉) · M. Bergé-Nguyen · A. Kouraev
LEGOS, Université de Toulouse, CNES, CNRS, IRD,
UPS Toulouse, Toulouse, France
e-mail: Jean-Francois.Cretaux@legos.obs-mip.fr

A. Kostianoy
P.P. Shirshov Institute of Oceanology, Russian Academy of Sciences,
Moscow, Russia

A. Kostianoy
S.Yu. Witte Moscow University, Moscow, Russia

A. Kostianoy
Interfacultary Center for Marine Research (MARE) and Modelling
for Aquatic Systems (MAST), University of Liège, Liège, Belgium

A. Kouraev
State Oceanography Institute, Saint Petersburg Branch,
Saint Petersburg, Russia

A. Kouraev
Tomsk State University, Tomsk, Russia

V. Barale and M. Gade (eds.), *Remote Sensing of the Asian Seas*,
https://doi.org/10.1007/978-3-319-94067-0_29

techniques coupled with complementary in situ data have allowed precisely quantifying the water balance of the Aral Sea since 1993 and to understand the recent desiccation of this inland sea. Moreover, unprecedented information can be obtained by coupling models and surface observations with data from space, which offers global geographical coverage, good spatial-temporal sampling, continuous monitoring over time, and the capability of estimating water mass change.

Keywords The Aral Sea · Central Asia · Hydrology · Satellite altimetry Satellite data · Optical imagery · Morphodynamic characteristics

1 Modern Aral Sea Crisis

1.1 Aral Sea Water Balance Over the Last 60 Years

The Aral Sea is a closed lake located in an active graben structure in Central Asia to the south of the Ural Mountains, between the Usturt Plateau to the West, the Karakum Desert to the South, and the Kyzyl Kum Desert to the East (Fig. 1). Two main rivers feed it: the Syr Darya in the north east and the Amu Darya in the south. The climate of the Aral Sea Basin, which encompasses more than 2 million km2, is of arid/semi-arid type and characterized by instability over various timescales ranging from years to millennia (Boomer et al. 2009).

The irrigation in Central Asia exists for centuries, and for example in the mid XIX century, in the delta of Amu Darya, the Karakalpaks (people living in the southern part of the Aral Sea) have maintained sustainable systems of channels for agriculture that insured prosperity of the inhabitants of the delta (Sarybaev and Akimova 2002). However, in 1962, the Aral Sea has started to drastically shrink due to growth of water intake for irrigation and construction of water reservoirs along both Syr Darya and Amu Darya rivers (Micklin 1988; Bortnik 1999). In such arid zone, the irrigation provided the means to reach the planned agricultural objectives of the Soviet Union government. Large-scale development of ground infrastructure (irrigation channels, reservoirs) began and the extent of the irrigated area increased from $\sim$4 billion of hectares in 1960 to $\sim$8 billion ha in 1990.

For an arid water body like the Aral Sea, the water balance which determines the equilibrium level of water is strongly forced by the surface inflow from rivers discharge. Even small changes in this component of the water balance will thus affect significantly the water level since evaporation remains constant for a given climate condition (about 1 m per year) and precipitation is generally very low and not sufficient to compensate for the evaporation. Moreover, the morphology of the Aral Sea (shallow inland sea) induces a very rapid adaptation of the lake surface to any small change in the water balance component (Mason et al. 1994; Cretaux et al. 2016).

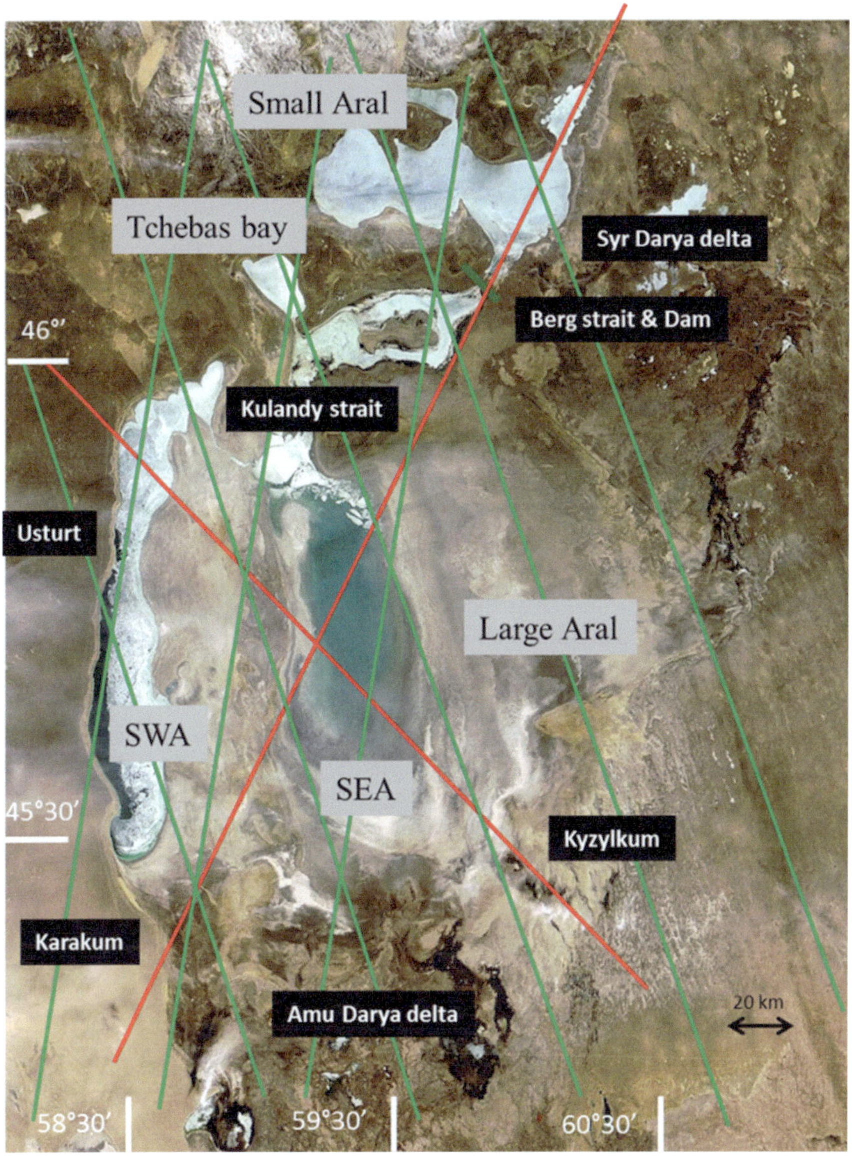

Fig. 1 Aral Sea and its sub-basins: SEA: South East Aral, SWA: South West Aral. Red line represents the Topex/Poseidon (T/P) and Jason-1 (J1), Jason-2 (J2) and Jason-3 (J3) ground tracks. Green lines represent the Enivsat and Saral/Altika ground tracks. The image is from the MERIS instrument onboard the Envisat satellite and was taken on March 6 2008

The balance equation is given by the following equation

$$\frac{dV}{dt} = R + P - E \tag{1}$$

where V is the volume of the lake, R the river runoff, P the precipitation as rainfalls, and E the evaporation. In the case of Aral Sea we can consider that the underground inflow is negligible as shown in Cretaux et al. (2013).

The equivalent response time for a lake to reach 66% of a new equilibrium is given by:

$$\tau_E = \frac{1}{dA/dV(E-P)} \tag{2}$$

where A is the areal extent of the lake.

This equation shows the high dependency of this response time to the bathymetry and for the Small Aral it has a value of 17 years, while for the Large Aral it has a value of 23 years [calculated used a precise Digitized Bathymetry Map (DBM) of the whole the Aral Sea Basin (Cretaux et al. 2005, 2013)].

During the years 1961–1970 the first consequences of reduced river runoff (Fig. 2), particularly along the Amu Darya, was to decrease the Aral Sea level of about 20 cm/year (Bortnik 1999). The following decade (1970s) was characterized by enormous amplification of water intake, with only 17 km^3/year reaching the Aral Sea, instead of 56 km^3 before 1960 (Fig. 2). During the decade between 1975 and 1985 the release of Syr Darya River into the Aral Sea was even close to zero (Fig. 2). Consequently, the rate of decreasing of the water level was about 58 cm/year.

During the 1980s, for some dry years, the Amu Darya could not reach the Aral Sea itself. The average water discharge during this period was about 4 km^3/year (Fig. 2). As a consequence, in the years 1986–1987 the Aral Sea was separated into

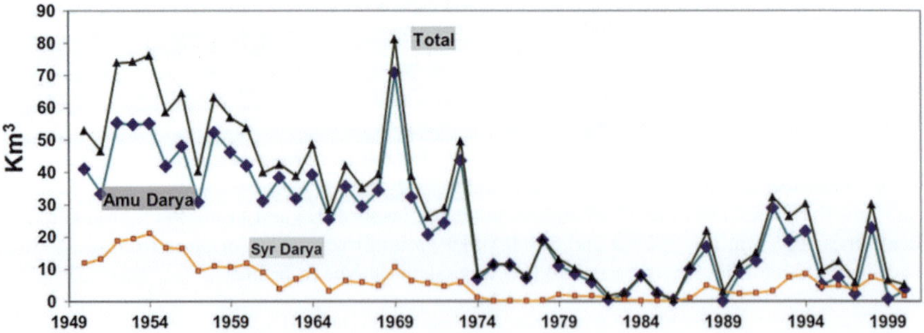

Fig. 2 Annual river discharge from Amu Darya (blue diamonds), Syr Darya (orange square), and total (black triangle) river discharge into the Aral Sea, based on in situ data (www.cawater.info.net)

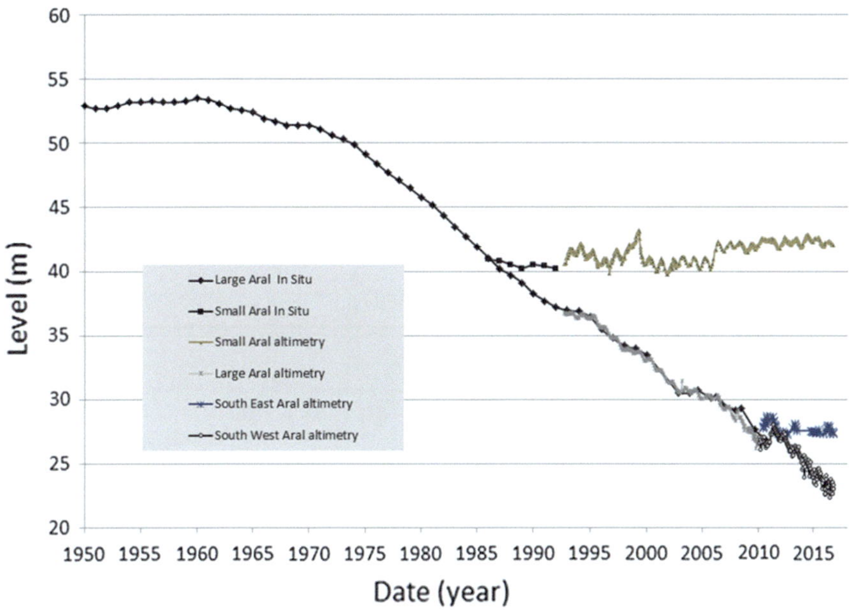

Fig. 3 Variation of the Aral Sea absolute level from 1950 to 2016, using several sets of data: in situ and radar altimetry from T/P, Jason-1, Jason-2, Jason-3, Envisat, & Saral satellites

two distinct water bodies, the North Aral (or Small Aral) and the South Aral (or Large Aral). Small Aral is fed by Syr Darya River, while Large Aral is fed by Amu Darya River. At the time of separation, the level of the Aral Sea was about 40 m above absolute sea (ocean) level (a.s.l.) (Aladin et al. 2005, and Fig. 3).

Since then the Small and Large Aral have evolved in different ways. In the beginning of the 1990s, the Amu Darya still supply around 15 km3 of water per year to the Large Aral due to several years of intense precipitation in the Pamir Mountains (Fig. 2). In the mid-1990s, the water runoff decreased again, and the level of the Large Aral in 2002 was 7 m lower than in the Small Aral. The Large Aral has continued to shrink with an average rate of 80 cm/year until very recently. This rate of level decrease is however not constant from year to year since the discharge of the Amu Darya is highly variable from year to year (Fig. 4).

In wet years of the 2000s, the Amu Darya waters flowed from over-flowing reservoirs artificially created in the Delta into the Large Aral Sea (Zavialov et al. 2012). Note that, unlike the situation with over-flow of waters from reservoirs in Amu Darya Delta into the Aral, which was observed on 13 October 2005 (Ginzburg et al. 2010), the opposite process took place on 30 September and 1 October 2005 under a northeasterly wind. In this case, cold waters from the Aral Sea flowed in artificial reservoirs. However, nowadays similar water exchange between reservoirs of Amu Darya Delta and the eastern part of the Aral Sea is only of historical interest as shrinking of the Eastern Large Aral Sea continued after 2005. This led to a

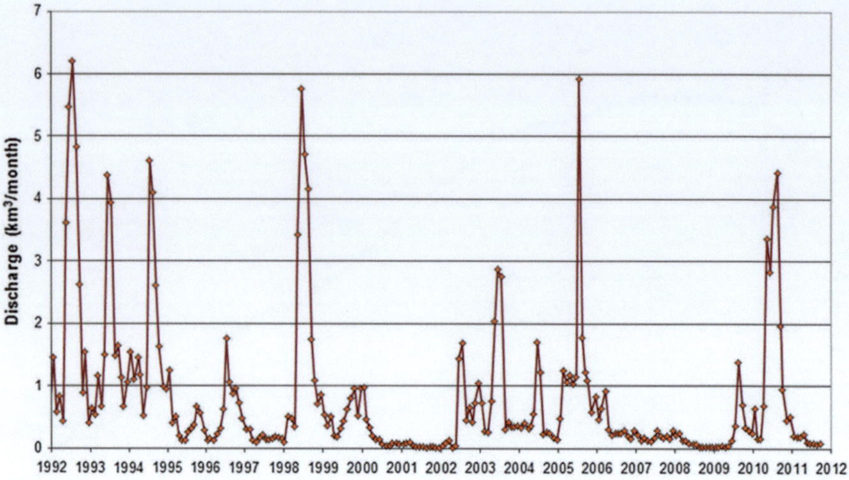

Fig. 4 Monthly discharge of Amu Darya River into the Large Aral Sea from in situ data (www.cawater.info.net)

significant shift of its southern shoreline to the north and as a result to the actual cessation of the existence of this water body.

Since the Small Aral received run-off of the Syr Darya and because of smaller surface of this lake, the evaporation was quite balanced by precipitation and river runoff. The level of the Small Aral has been hence more or less stabilized to an averaged value of 42–43 m a.s.l., with however large fluctuations due to seasonal and inter-annual climate changes, and to the construction of a dam in the Berg Strait at different periods (Figs. 3 and 5). This dam was destroyed (and rebuilt) 3 times during the last 20 years. It has been demonstrated in Aladin et al. (2005) that during the period 1993–1999 the existence of the dam allowed some restoration of the Small Aral. They calculated from water balance that during periods of dam absence, only 20% of the river runoff entering in the delta has reached the sea. The rest was lost to evaporation in the delta and in the desert, and probably some inflow to the Large Aral through the Berg Strait. They also showed that when the dam is in place, it allows water retention of 80% of the river runoff that enters via the Syr Darya Delta. This computation determined the correlation between the amount of water entering into the Syr Darya Delta and the level of the Small Aral. The Small Aral is therefore since 2005 fully controlled by the dam in the Berg Strait, since the discharge from the Syr Darya has remained quasi constant over the last 10 years (Fig. 6).

The differences in the hydrological regimes of the two lakes have thus led to the stabilization of the Small Aral Sea level and continued desiccation and salinization of the Large Aral. All the above has been widely documented in several articles: (Letolle and Chesterikoff 1999; Kostianoy et al. 2004; Cretaux et al. 2005, 2013; Zavialov 2005, 2010; Kouraev et al., 2009; Ginzburg et al. 2010; Zavialov et al. 2012).

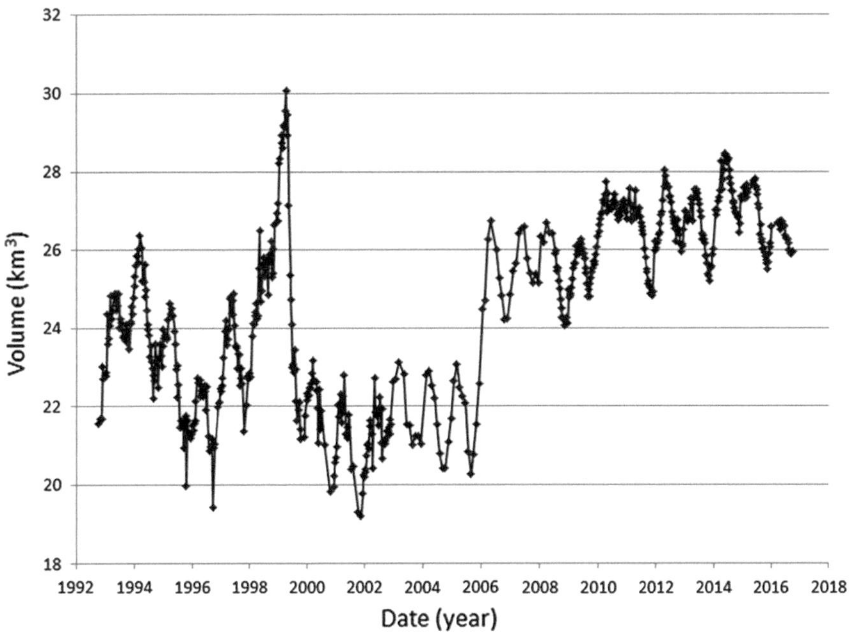

Fig. 5 Volume variation of the Small Aral measured from a combination of altimetry and DBM

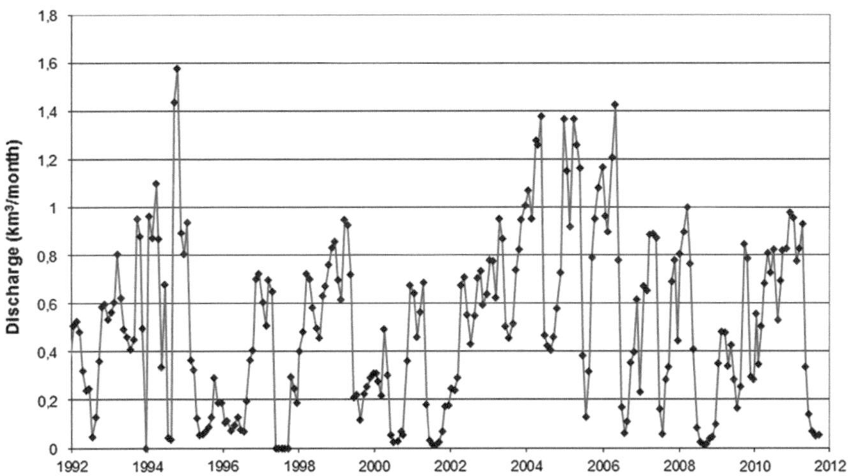

Fig. 6 Monthly discharge of the Syr Darya into the Small Aral from in situ data (www.cawater.info.net)

1.2 Relative Role of Climate Change and Irrigation in the Shrinkage of Aral Sea

It is considered by Bortnik (1999) that after the start of the desiccation in 1960, climate change, as deduced from changes in precipitation and evaporation would have provoked a decline of the sea of 2.3 m in 20 years. The real observe decline was even much higher than 8 m. This seems to slightly over-estimate the impact of climate change with respect to irrigation when we compared with results obtained by Destouni et al. (2010); Small et al. (2001) and Obershänli et al. (2011) as explained below.

Destouni et al. (2010) revealed that global climate change has only slightly modified runoff and evaporation from the Aral Sea, and that in 2002 runoff decreased by 83% due to water diversion for irrigation with respect to pre-1960s conditions and increased only by 3% due to climate change. This result seems to be in good agreement with those of Small et al. (2001) who have shown that the contribution of climate change to Aral Sea desiccation (mainly through increasing direct evaporation) is twice as high from regional manifestation of global climate change than that from feedback from Aral Sea shrinkage. However, Destouni et al. (2010), accounting for the cooling impact of irrigation, conclude that the total increase of temperature over the Aral Sea is likely to be completely due to global climate change, and that the total impact of climate change on Aral Sea water level changes given in Small et al. (2001) are probably a little overestimated. Nezlin et al. (2004) based on the analysis of satellite GPCP (the Global Precipitation Climatology Project) data for 1979–2001 and GPCC (the Global Precipitation Climatology Centre) rain gauges data for 1986–2001 found a significant negative trend in precipitation (decrease of 50%) for the catchment area of the Amu Darya, whereas only a slight decrease in it occurred in the Syr Darya catchment area. These observations show that climate change at regional scale is very variable, hence provoking high decreasing runoff over the Amu Darya and a negligible one over the Syr Darya.

Other studies focus on the contribution of snow to river runoff into the Aral Sea, and also on observation of changes of snow cover and glaciers over the last several decades. In particular, the glaciers of Tien Shan and Pamir decreased up to 28% from 1960 to 2000 (Kutuzov and Shahgedanova 2009). The direct effect of thawing glaciers is the increasing river runoff in the Aral Sea Basin. However, rising temperatures simultaneously lead to a decrease of snow cover (Oberhänsli et al. 2011) with a consequent decrease of river runoff. Oberhänsli et al. (2011) estimated that since 1970 the mean annual snow cover extent was reduced by about 10%, and has therefore helped accelerate the desiccation of the Aral Sea.

This issue is still very much an open debate and needs more accurate quantification of total water volume variation, particularly in the mountain areas, even though it is clear that since 1960 a major part of the Aral Sea desiccation has been coupled with a significant increase in irrigation (almost double since 1960) for both Kazakhstan and Uzbekistan in its catchment area. At the beginning of the 20th century the first systematic measurements of the Aral Sea level started and in 1940 they were already 6–10 ground gauge stations (Bortnik 1999). However, desiccation of the Aral Sea in the anthropogenic period has led not only to significant changes of its morphometric, physical, chemical, biological and other parameters, but also to the destruction of infrastructure in the coastal zone. In particular, since the 1990s, most of meteorological and sea level gauge stations stopped working. Regular monitoring of ice cover on the meteorological stations and with the help of ice air reconnaissance, which was carried out from 1941 until the mid-1980s, also stopped (Bortnik and Chistyaeva 1990).

Since the beginning of 1990s the Aral Sea level was calculated from satellite measurements, through radar altimeters instruments (Cretaux et al. 2005), or via optical satellite imagery of the sea surface (Ginzburg et al. 2010). There is now more than two decades of monitoring of the Aral Sea desiccation using satellite remote sensing data. It has allowed demonstrating the high accuracy of satellite data to measure water storage changes of lakes and reservoirs. Altimetry and multispectral imagery is commonly used in operational system (for example under European the Copernicus program of the sentinel satellites) for continental water study. Over the Aral Sea, due to desiccation, this is not anymore possible to install perennial gauges to measure water level changes. Satellite data offer therefore a suitable tool to replace them in order to insure continuous survey of the Aral Sea surface waters.

2 The Aral Sea Water Level Surface and Volume Changes from Satellite Data

2.1 State of the Art in Altimetry for Inland Seas

Since the mid-nineties, satellite radar altimetry has been a successful technique for monitoring height variations of continental surface water, such as lakes and rivers (Birkett 1995; Alsdorf et al. 2001; Calmant et al. 2008; Silva et al. 2014; Tarpanelli et al. 2015 and many others). The surface water level is measured within a terrestrial reference frame. An altimeter is a nadir pointing system and therefore does not have a global view of the planet. However, the altimetry technique permits systematic continental scale lake and river monitoring and can provide water level estimations in remote areas where in situ data is not available. Another advantage of satellite altimetry is its ability to monitor lakes and rivers over decades with a revisit on each target ranging from a few days to a few months depending on the orbit. For example, the T/P, J1, J2, and J3 satellites have a cycle of 10 days, GFO a cycle of 17 days, Envisat and Satral/Altika 35 days while for Cryosat-2 the revisit cycle is 369 days allowing a very dense coverage of the Earth's surface at a yearly revisit scale. For large lakes, many satellites can pass over them providing continuous surveillance over decades.

Satellite altimetry has some limitation for small lakes since lake-shore topography may affect elevation calculations from the altimeter, but for large lakes, it has been shown (Ričko et al. 2012; Cretaux et al. 2011, 2016) that the accuracy is sub-decimeter with such system.

Consequently, altimeters Topex/Poseidon (T/P), Jason-1, Jason-2 and Jason-3 (J1, J2, J3), Envisat and Saral satellites provide reliable, regular, quite often and weather-independent information on variability of sea level in the Large and Small Aral seas since September 1992.

2.2 State of the Art in Imagery for Inland Seas

Satellite altimetry is used to calculate water level variations over a lake. This is however not sufficient when the objective is to calculate the water balance of a lake which is preferentially represented by the water storage changes. Determination of the absolute water volume of a lake is impossible without precise bathymetry (Crétaux et al. 2005; Lei et al. 2014). For the Aral Sea we have calculate absolute volume change using a DBM. However we also have used the images obtained using the moderate resolution spectro-radiometer MODIS instrument onboard the Terra satellite from 2000 to 2017 to estimate the areal extent of the Aral Sea (Fig. 7). MODIS provides Earth surface reflectance on thirty-six narrow bands of frequency; seven among them are from the visible to mid infrared part of the spectrum with a spatial resolution of 250 (two first bands) and 500 m (all bands):

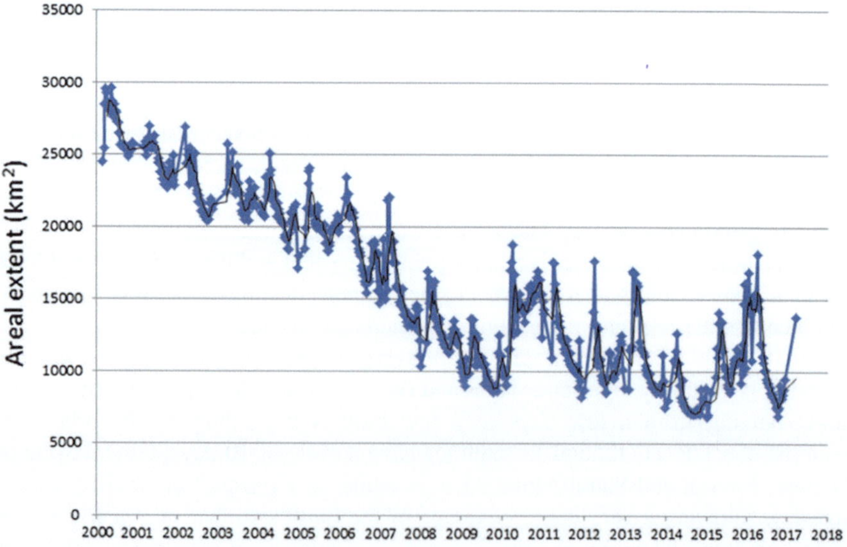

Fig. 7 Areal extent of the Aral Sea (Small and Large Aral together) estimated from MODIS data analysis

- Band 1: 620–670 nm
- Band 2: 841–876 nm
- Band 3: 459–479 nm
- Band 4: 545–565 nm
- Band 5: 1230–1250 nm
- Band 6: 1628–1652 nm
- Band 7: 2107–2155 nm

The methodology to extract water mask from satellite imagery is not straightforward and are generally divided into single-band and multi-band methods. Several methods exist of combination of multi-band ratio to extract water surface from multi-spectral imagery. See McFeeters (1996) for more details on such methods.

A limitation of optical remote sensing data for flood monitoring is cloud cover. Over the Aral Sea regions this principally limits the use of MODIS during winter time, but the use of MODIS 8-days products, which keeps only cloud free pixels allows avoiding generally this problem. the method of land surface classification for hydrology based on a single band approach developed and described in Bergé-Nguyen and Cretaux (2015) has been used on the Aral Sea.

2.3 The Aral Sea Recent Results from Radar Altimetry and MODIS Images Showing the Last Separation of the Large Aral in 2010

After the separation of Small and Large Aral in the end of the 1980s, the inflow to Large Aral has not allowed compensating the evaporation over the Large Aral. It has consequently continued to decline to a certain point where Large Aral itself was separated into two new distinct water bodies, in the years 2010–2012, the South East and the South West Aral (SEA and SWA). What happened then may be understood using the satellite data. Figure 7 shows that after 2010 the areal extent of Small and Large Aral together started to decline at a much lower rate.

From Fig. 8 we moreover can see that after 2010, the water mask performed using the 500 m MODIS images didn't allow detection of water in the Kulandy Strait, in the north of Large Aral, which until 2010 has linked SEA and SWA. This is not possible from MODIS images to determine if this channel disappeared totally. However what is now evident is that water comes from East to West (The delta of Amu Darya is physically linked to the SEA) and during the dry season, quite all water on SEA evaporates. Only a small part may still reach episodically the SWA through the Kulandy Strait. The SEA is indeed much shallower than the SWA. From Fig. 6 we can see the high variability from Spring to Autumn. However what is now happening in the SEA, is that each year, an amount of a few km^3 enter into

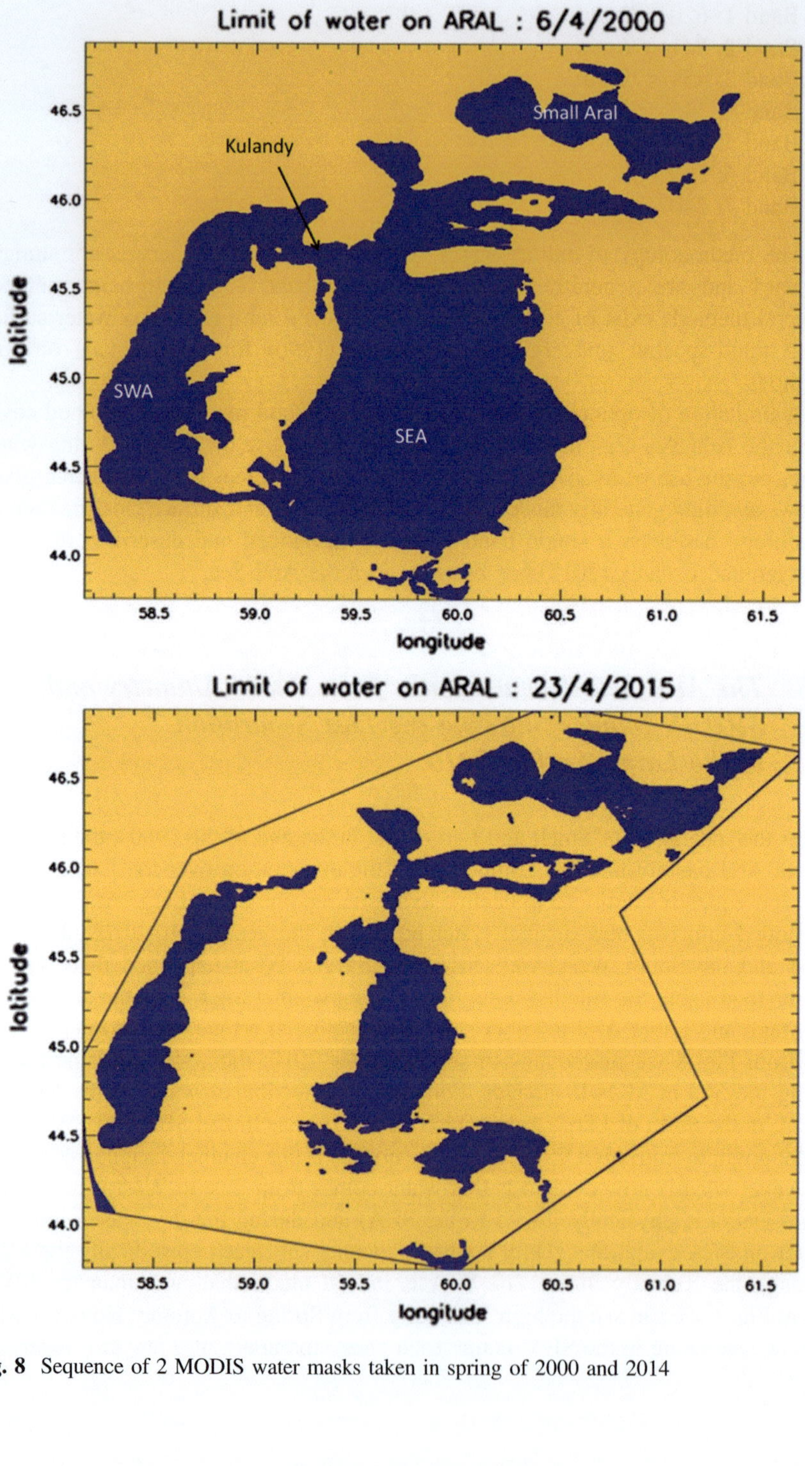

Fig. 8 Sequence of 2 MODIS water masks taken in spring of 2000 and 2014

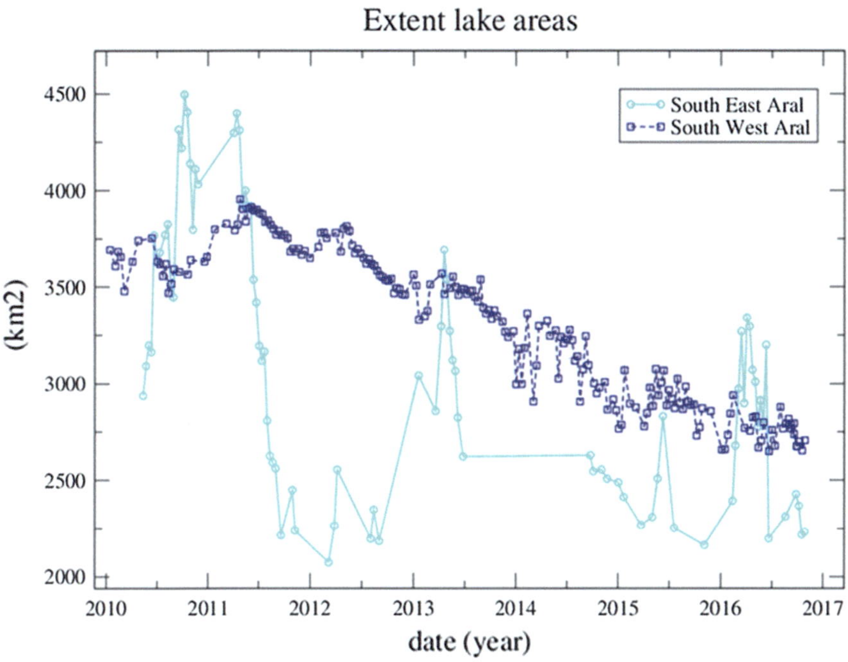

Fig. 9 Variation of the South East and South West Aral Sea areas from satellite altimetry combined to a DBM

the SEA through the Amu Darya delta. Indeed, after 2010, we observe an ephemeral inundation in the SEA from satellite altimetry (Figs. 3 and 9). A water level yearly cycle of amplitude ranging from 20–30 cm to more than 1 m is measured by satellite altimetry and areal extent which is measured using DBM varied from zero (autumn 2014) to 3000–4000 km^2 at maximum as observed in 2010 when a high amount of water was filled from Amu Darya.

On the SWA, the level of water has continued to decline with more or less a constant rate (Fig. 3) which clearly indicates that it is now fully disconnected from SEA. The areal extent of SWA however does not vary as observed in SEA, which is the results of the bathymetry in this part of Aral Sea, much steeper and deeper in SWA than in SEA. We can therefore consider that since around 2010–2012 the process of separation of Large Aral into two new water bodies has occurred and that, without a significant increase of inflow from river runoff, the desiccation of SWA will continue. The SEA has reached a stable situation with seasonal filling and desiccation, from Spring to Autumn. In six years, from 2011 to 2017, the SWA has lost about 15 km^3 (Fig. 10). With a surface of 3000 km^2 in average over this period of time, and a yearly net evaporation of approximately 1 m, the total evaporation over this 6 years period was 18 km^3. This confirms that the current transfer of water from SEA to SWA is likely to be now very small, if not fully stopped. We can also see from Fig. 5 that every year, the Small Aral Sea is

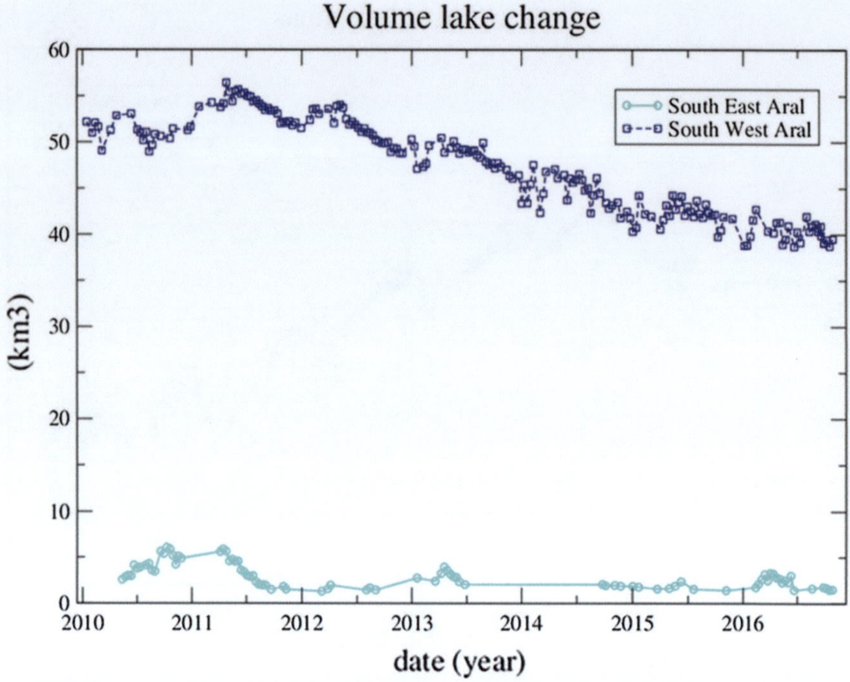

Fig. 10 Variation of the South East and South West Aral Sea volume from altimetry combined to a DBM

controlled by the Dam's infrastructure with some release into the Berg strait. Some small extent of water from Small Aral can reach the SWA through the Kulandy Strait which may explain the small positive balance of 3 km^3 during the period 2011–2017.

It was shown (Nezlin et al. 2004) that the amount of precipitation supplying the Amu Darya and Syr Darya rivers varied with a period of 5–9 years. A significant negative trend in precipitation (decrease of 50%) for the catchment area of the Amu Darya was observed, whereas only a slight decrease in it (the complete lack since 1985) occurred in the Syr Darya catchment area. Identified trends are in good agreement with trends of levels of two separated basins of the Aral Sea. This suggests that not only the anthropogenic impact, but also regional climate change (interannual changes in precipitation in the form of rain and snow over the rivers' catchment areas) determine the levels of the Large and Small Aral seas (Nezlin et al. 2004). Note that seasonal changes in precipitation had little effect on the Large Aral Sea level until the 2000s when it was still a large water body. In contrast, the Small Aral Sea, which has a much smaller volume, is sensitive to seasonal changes in precipitation as now it is observed over SEA and SWA (Figs. 8, 9, 10). Seasonal maximum of atmospheric precipitation occurs in the winter-spring, seasonal minimum–in summer and autumn that is in agreement with seasonal changes in the areas of the Aral Sea basins (Fig. 9 and Kravtsova and Tarasenko 2010; Ginzburg et al. 2010; Zavialov et al. 2012).

3 Conclusions

The desiccation of Aral Sea since 1960s is well documented in the scientific literature and is known to be due to the large increase of irrigation within the Syr Darya and Amu Darya watersheds. Part of the decline in Aral Sea levels is also due to climate change but in a much smaller extent and is the matter of some debate in the scientific community. Since the beginning of the shrinkage of Aral Sea it has separated in the end of the 1980s into two new distinct water bodies, the Small and the Large Aral respectively in the north and the south parts of the basin. During the 1990s they have evolved independently, with the Small Aral being rehabilitated by construction of dam in the Berg Strait, since the Large Aral has continued to decline rapidly. After few years, approximately in 2010–2012 the Large Aral was also separated into two new water bodies, the SEA and the SWA. They are now temporarily linked by a thin channel (the Kulandy strait). In the north, the Small Aral has been stabilized to a level of ~43 m above sea level with seasonal variations due to spring flow from Syr Darya and summer release of water from the Dam in the Berg strait. A very little amount of water (~0.5 km^3) has been transferred each year from SEA to SWA but we can consider that both water bodies have now distinct behaviors. The SWA level declined by approximately 15 km^3 between 2011 and 2017. The level has continued to decrease with a rate of few decimeters per year. SEA is now significantly under climate control: wet year leads to some large inundation in this shallow basin. In contrast, dry years lead to very small inundation like in 2014 when we observed from MODIS images that it has fully dry out. In the SEA we have observed yearly inundation extended over 2–4 thousands of square kilometers.

As the former Aral Sea has shrunk in such a rapid and high magnitude, at the end of the 1990s it was impossible to maintain perennial ground network of Aral Sea level monitoring. However, thanks to satellite altimetry, which provides regular and precise water level since more than 25 years, on a temporal scale of few days, it has been possible to monitor the Aral Sea level decrease. Satellite imagery's system like MODIS has also allowed precisely mapping the water extent overall the Small Aral, the Large Aral, the SEA and the SWA with a temporal sampling of 8 days over the period 2000–2017. Using satellite altimetry together with MODIS and a DBM, it has been possible to measure the past 25 years evolution of the Aral Sea water level, extent and volume, and to describe the different phases of dispersal of former Aral Sea into what is now many small water bodies, permanent or ephemeral.

Acknowledgements The authors acknowledge the Centre National d'Etudes Spatiales (CNES) for financing our project through different TOSCA projects. This work was also supported by the ERA.NET RUS Plus S&T #226 project called ERALECC. A.G. Kostianoy was partially supported in the framework of the P.P. Shirshov Institute of Oceanology RAS budgetary financing (Project N 149-2018-0003). The authors are also grateful to the Centre for Topographic Studies of the Oceans and Hydrosphere (CTOH, www.legos.obs-mip.fr/observations/ctoh) at LEGOS (Toulouse, France) for providing altimetry data in a standard and useful form. The MODIS data was downloaded from the National Aeronautics and Space Administration (NASA) and is courtesy of the online Earth da-ta web site named: Earth Observing System Data and Information System EOSDIS, URL: https://earthdata.nasa.gov/.

References

Aladin NV, Crétaux J-F, Plotnikov IS, Kouraev AV, Smurov AO, Cazenave A, Egorov AN, Papa F (2005) Modern hydro-biological state of the small Aral Sea. Environmetric 6(4): 375–392

Bergé-Nguyen M, Cretaux JF (2015) Inundations in the inner Niger Delta: monitoring and analysis using MODIS and global precipitation datasets. Remote Sens 7:2127–2151. https://doi.org/10.3390/rs70202127

Birkett CM (1995) Contribution of TOPEX/POSEIDON to the global monitoring of climatically sensitive lakes. J Geophys Res 100(C12):25179–25204

Boomer I, Wünnermann B, Mackay AW, Austin P, Sorrel P, Rein-hardt C, Kayser D, Guichard F, Fontugne M (2009) Advances in understanding the late Holocene history of Aral Sea region. Quatern Int 194:79–90

Bortnik VN (1999) Alteration of water level and salinity of the Aral Sea. Cambridge University Press, Creeping environmental problems and sustainable development in the Aral Sea basin, pp 47–65

Bortnik VN, Chistyaeva SP (eds) (1990) Hydrometeorology and hydrochemistry of seas of the USSR. Project Seas of the USSR VII . The Aral Sea. Gidrometeoizdat, Leningrad, p 195

Calmant S, Seyler F, Cretaux JF (2008) Monitoring Continental sur-face waters by satellite altimetry. Surv Geophys. Hydrology from Space. 29(4–5) pp 247–269 https://doi.org/10.1007/s10712-008-9051-1

Cretaux JF, Letolle R, Bergé-Nguyen M (2013) History of Aral Sea level variability and current scientific debates. Global Planet Change 11:99–113

Cretaux J-F, Abarca Del Rio R, Berge-Nguyen M, Arsen A, Drolon V, Clos G, Maisongrande P (2016) Lake volume monitoring from Space. Surv Geophys 37:269–305. https://doi.org/10.1007/s10712-016-9362-6

Crétaux JF, Kouraev AK, Papa F, Bergé-Nguyen M, Cazenave A, Aladin NV, Plotnikov IS (2005) Water balance of the Big Aral Sea from satellite remote sensing and in situ observations. J Great Lakes Res 31(4):520–534

Crétaux J-F, Calmant S, Romanovski V, Perosanz F, Tashbaeva S, Bonnefond P, Moreira D, Shum CK, Nino F, Bergé-Nguyen M, Fleury S, Gegout P, Abarca Del Rio R, Maisongrande P (2011) Absolute calibration of Jason radar altimeters from GPS kinematic cam-paigns over Lake Issykkul. Mar Geodesy 34(3–4):291–318. https://doi.org/10.1080/01490419.2011.585110

Destouni G, Asokan SM, Jarsjö J (2010) Inland hydro-climatic inter-action: effects of human water use on regional climate. Geophys Res Lett 37:L18402. https://doi.org/10.1029/2010GL044153

Ginzburg AI, Kostianoy AG, Sheremet NA (2010) Satellite monitoring of the Aral Sea Region. In: Kostianoy AG, Kosarev AN (eds) The Aral Sea Environment. The handbook of environmental chemistry, vol 7. Springer, Heidelberg, pp 147–179

Kostianoy AG, Zavialov PO, Lebedev SA (2004) What do we know about dead, dying and endangered lakes and seas? In: Nihoul JCJ, Zavialov PO, Micklin PHP (eds.) Dying and dead seas. Climatic versus anthropic causes. NATO ARW/ASI Series, Kluwer Academic Publishers, Dordrecht, pp 1–48

Kouraev AV, Kostianoy AG, Lebedev SA (2009) Recent changes of sea level and ice cover in the Aral Sea derived from satellite data (1992–2006). J Mar Systems 76(3):272–286. https://doi.org/10.1016/j.jmarsys.2008.03.016

Kravtsova VI, Tarasenko TV (2010) Space monitoring of the degradation of the Aral Sea. Vodnye resursy 37(3): 292–303

Kutuzov S, Shahgedanova M (2009) Glacier retreat and climatic vari-ability in the eastern Terskey-Alatoo, inner Tian Shan between the middle of the 19th century and the beginning of the 21st century. Glob Planet Change 69:59–70

Lei Y, Yang K, Wang B, Sheng Y, Bird BW, Zhang G, Tian L (2014) Response of inland lake dynamics over the Tibetan Plateau to climate change. Clim Change 125(2):281–290. https://doi.org/10.1007/s10584-014-1175-3

Letolle R, Chesterikoff A (1999) Salinity of surface waters in the Aral sea region. Int J Salt Lake Res 8:293–306

Mason IM, Guzkowska MAJ, Rapley CG, Street-Perrot FA (1994) The response of lake levels and areas to climate change. Climate Change 27(2):161–197

McFeeters SK (1996) The use of Normalized Difference Water In-dex (NDWI) in the delineation of open water features. Int J Remote Sens 17(7):1425–1432

Micklin PP (1988) Desiccation of the Aral Sea, a water management disaster in the Soviet Union. Science 241:1170–1176

Nezlin NP, Kostianoy AG, Lebedev SA (2004) Interannual variations of the discharge of Amu Darya and Syr Darya estimated from global atmospheric precipitation. J Mar Syst 47:67–75

Oberhänsli H, Novotná K, Píšková A, Chabrillat S, Nourgaliev DK, Kurbaniyazov AK, Grygar TM (2011) Variability in precipitation, temperature and river runoff in W Central Asia during the past ~2000 yrs. Global Planet Change 76:95–104

Ričko M, Birkett CM, Carton JA, Cretaux JF (2012) Intercomparison and validation of continental water level products derived from satellite radar altimetry. J Applied Rem Sensing 6 (1):061710. https://doi.org/10.1117/1.jrs.6.061710

Sarybaev K, Akimova A (2002) L'agriculture irriguée dans le delta de l'Amou Darya à la fin du XIXe et au début du XXe siècle, Cahier d'Asie Centrale, N° 10, Karakalpaks et autres gens de l'Aral, entre rivages et déserts

Silva JS, Calmant S, Seyler F, Moreira DM, Oliveira D, Monteiro A (2014) Radar altimetry aids managing gauge networks. Water Resour Manage 28–3:587–603. https://doi.org/10.1007/s11269-013-0484-z

Small EE, Giorgi F, Sloan LC, Hostetler S (2001) The effects of desiccation and climate change on the hydrology of the Aral Sea. J Clim 14:300–322

Tarpanelli A, Brocca L, Barbetta S, Faruolo M, Lacava T, Mora-marco T (2015) Coupling MODIS and radar altimetry data for discharge estimation in poorly gauged river basins. IEEE transactions on Geosciences and Remote Sensing 8(1):141–149. https://doi.org/10.1109/jstars.2014.2320582

Zavialov P (2005) Physical Oceanography of the dying Aral Sea. Springer-Praxis Publishing, Chichester, UK, p 146

Zavialov PO (2010) Physical oceanography of the Large Aral Sea. In: Kostianoy AG, Kosarev AN (eds) The Aral Sea environment. The handbook of environmental chemistry, vol 7. Springer, Heidelberg, pp 123–145

Zavialov PO, Arashkevich EG, Bastida I, Ginzburg AI, Dikarev SN, Zhitina LS, Izhitsky AS, Ishniyazov DP, Kostianoy AG, Kravtsova VI, Kudyshkin TV, Kurbaniyazov AK, Ni AA, Nikishina AB, Petrov MA, Sazhin AF, Sapozhnikov FV, Solovyov DM, Khan VM, Sheremet NA (2012) The Large Aral Sea in the beginning of century 21. Physics, Biology, Chemistry. Moscow, Nauka, p 231

Ice Cover and Associated Water Structure in Lakes Baikal and Hovsgol from Satellite Observations and Field Studies

Alexei V. Kouraev, Elena A. Zakharova, Frédérique Rémy, Andrey G. Kostianoy, Michail N. Shimaraev, Nicholas M. J. Hall and Andrey Ya. Suknev

Abstract Lakes Baikal (Russia) and Hovsgol (Mongolia) are covered every year by ice for several months. The distribution and state of the ice cover and snow on the ice affect the hydrophysical structure, spring bloom of diatoms and primary productivity, as well as transport on ice. In this respect combination of satellite remote sensing data with dedicated field measurements provide a unique tool for investigating ice cover state and development. Comparison of ENVISAT and SARAL radar altimetry missions data for Lake Baikal shows that SARAL's AltiKa instrument can be successfully used for ice discrimination. We observe large decrease of radar return echo in late spring for both ENVISAT and SARAL and discuss it in the context of ice metamorphism. We then address an interesting

A. V. Kouraev (✉) · F. Rémy · N. M. J. Hall
LEGOS, Université de Toulouse, CNES, CNRS, IRD, UPS, Toulouse, France
e-mail: kouraev@legos.obs-mip.fr

A. V. Kouraev
Tomsk State University, Tomsk, Russia

E. A. Zakharova
Institute of Water Problems RAS, Moscow, Russia

A. G. Kostianoy
P.P. Shirshov Institute of Oceanology, Russian Academy of Sciences, Moscow, Russia

A. G. Kostianoy
S. Yu. Witte Moscow University, Moscow, Russia

A. G. Kostianoy
Interfacultary Center for Marine Research (MARE) and Modelling for Aquatic Systems (MAST), University of Liège, Liège, Belgium

M. N. Shimaraev
Limnological Institute, Siberian Branch of Russian Academy of Sciences, Irkutsk, Russia

A. Ya.Suknev
Great Baikal Trail (GBT) Buryatiya, Ulan-Ude, Russia

V. Barale and M. Gade (eds.), *Remote Sensing of the Asian Seas*,
https://doi.org/10.1007/978-3-319-94067-0_30

natural phenomenon—giant ice rings on lakes Baikal and Hovsgol. Using satellite imagery and photography for 1974–2015 we have identified 45 rings on Lake Baikal (compared to 13 previously known) and also for the first time 4 rings on Lake Hovsgol. The results of our hydrographic surveys beneath the ice rings show the presence of warm double-convex lens-like eddies before and during manifestation of ice rings. These eddies are the driving factor for the formation of ice rings in these lakes. We reassess the existing hypotheses of ice ring formation and discuss the potential mechanisms of eddy formation.

Keywords Lake Baikal · Lake Hovsgol · Ice cover · Satellite remote sensing Radar altimetry · Giant ice rings · Lens-like eddies · Water structure

1 Introduction

Lake Baikal is the deepest lake in the world and contains 20% of the world's unfrozen surface freshwater. The lake bottom is separated into three parts (Fig. 1): deep depressions in the southern and central parts (maximum depths 1461 and 1642 m, respectively), and a less deep region (maximum depth 904 m) in the northern part. High-altitude Lake Hovsgol (also known as Khubsugul or Khuvsgul) in north-western Mongolia is located 250 km south-west of Lake Baikal at an altitude of 1645 m above sea level (1200 m higher than Lake Baikal). It is part of the Baikal rift system with depths of more than 250 m.

One of the typical features for these lakes is the presence of ice cover. Continental climate conditions with long cold winters result in annual complete freeze-up for 5–6 months (December–January to May) for Lake Baikal (Verbolov et al. 1965; Galaziy 1993; Kouraev et al. 2007a, b) and for 6–7 months (November–May–June) for Lake Hovsgol.

The distribution and state of the ice cover and snow on the ice affect the hydrophysical structure, spring bloom of diatoms and primary productivity (Granin et al. 1999; Semovski et al. 2000; Mackay et al. 2003, 2005; Moore et al. 2009), as well as the habitat of the only Baikal mammal—the endemic Baikal seal (*Pusa Sibirica, Phoca Sibirica*). For both lakes, ice is also important for establishing transport on ice, for fishing activities and tourism. In the winter the ice is strong enough (more than one meter thick) to support motorcycles, cars and even heavy vehicles.

Lake Baikal ice cover has been studied extensively for more than a century using observations at coastal stations, as well as (during USSR time) aerial surveys and field expeditions. The number of observations has sharply decreased since the mid-1980s, and observations at coastal stations do not reliably represent ice conditions at a large scale. Thus, different types of satellite remote sensing observations could give an additional and more comprehensive information to study ice cover evolution and properties (ice presence, ice phenology, snow depth, potentially ice thickness etc.).

However, satellite observations alone have limitations, and only their combination with dedicated field observations can really provide insights on ice cover

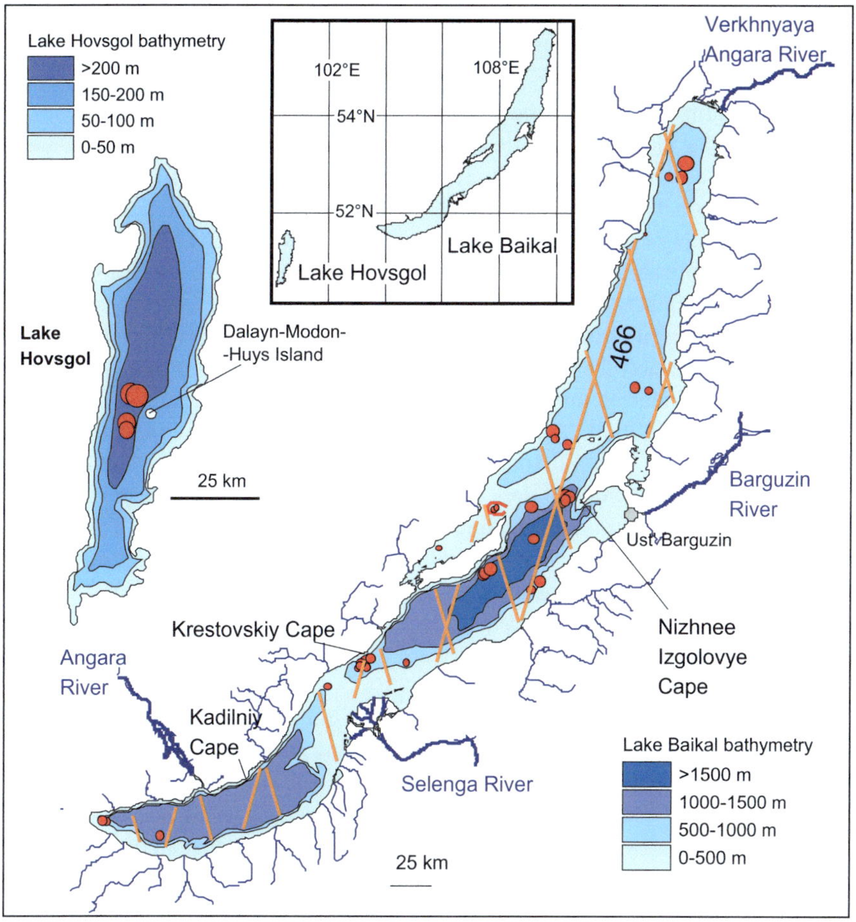

Fig. 1 Overview map of lakes Baikal and Hovsgol. Lake bathymetry (m), ENVISAT and SARAL tracks coverage (orange lines) and location and size of ice rings (red circles) are shown. The scale is different for the two lakes and is indicated on the maps

formation, development and decay. In the framework of French–Russian–Mongolian cooperation every spring (March–April) since 2010 we conduct yearly field observations of ice cover in the central part of Lake Baikal and since 2014 in Lake Hovsgol. In situ measurements include observations of ice thickness and snow depth, as well as ice structure and roughness (we have currently 289 stations for Lake Baikal and 40 stations for Lake Hovsgol). Since 2012 using the YSI CastAway CTD (conductivity, temperature and depth) probe operating down to 100 m depth these observations have been complemented by vertical profiles of temperature and conductivity. Since 2014 we have also performed estimations of currents by defining direction and current strength from CTD cable inclination. The existing dataset contains more than 250 stations located along the tracks of radar altimetry missions (Kouraev et al. 2015), in the regions of known ice ring observations (Kouraev et al. 2016) and other regions with interesting features. These field observations represent an unique database, as no other institutions are currently carrying such observations in the central part of Lake Baikal.

Satellite radar altimetry provides continuous weather-independent observations along the tracks with spatial resolution of 200–400 m. All parts of Lake Baikal are well covered with 17 ENVISAT/SARAL tracks (see Fig. 1). In Sect. 2 we analyze data from the ENVISAT and recently launched (2013) SARAL/AltiKa missions. We estimate applicability of SARAL data for ice discrimination methodology and discuss differences between SARAL and ENVISAT. Then we analyze spatial and temporal evolution of altimetric signal over the ice along the 250 km long track in late spring and discuss them in the context of ice structure evolution.

In Sect. 3 we address an interesting natural phenomenon related to ice cover: giant ice rings (Hsu 2009; "Circles in thin ice…" 2009; Granin 2009; Balkhanov et al. 2010; Kouraev et al. 2010a, b, 2016; Bordonskiy and Krylov 2014; Granin et al. 2015). These are rings of dark (thinner) ice with a typical diameter of 5–7 km and width of 0.9–1.3 km, surrounding and surrounded by white ice. The circular form of these rings and the unclear origins of this natural phenomenon represent a puzzle and a challenge for scientific research.

The size of the ice rings makes them impossible to observe from the frozen lake or from the shore. At the same time they are well visible on the satellite imagery in the visible range due to lower reflectance.

Potential explanations of the ice rings' origin include a wide variety of hypotheses ranging from high-intensity gas venting (methane) from bottom sediments, to atmospheric influence and biological activity in the upper water layer, to UFO activity and hoaxes (crop circles etc.). However, as will be seen later on (Sect. 3), all these hypotheses are tentative, and do not adequately explain the formation of the ice rings.

In order to better understand the phenomena responsible for the formation, development and disappearance of ice rings, we have processed satellite imagery and photography archives to get as full as possible an inventory of the ice rings and their parameters and conducted field hydrographic surveys (2010–2015) in the regions of the ice ring formation that led us to reassess existing hypotheses of the origins of ice rings.

2 Radar Altimetry and Radiometry Satellite Data for Ice Discrimination

2.1 Ice Discrimination Methodology

Since the launch of a series of radar altimeter satellites, the scientific community has now continuous and long time series of both active and passive simultaneous microwave observations, from the same platform and with the same incidence angle (nadir-looking). Active observations are provided by radar altimeters, which are the main instrument onboard altimetric satellites. To get wet tropospheric delay correction to the altimeter range measurements, each altimetric satellite has onboard a microwave radiometer with two or three frequencies, thus providing simultaneous passive microwave observations.

Although the primary mission of satellite altimetry is the study of sea surface topography over the open ocean, this technique has been successfully applied to continental surfaces, including ice cover. Formation of ice cover and its development change the surface properties and emissivity, that for satellite observations in the microwave range mean changes in the backscatter coefficient (active range) and brightness temperature (passive range). We have developed and validated a dedicated methodology for ice discrimination that has been applied for saltwater seas (Caspian and Aral) and freshwater lakes (Ladoga, Onega and Baikal) to estimate dates of ice formation and break-up (Kouraev et al. 2004, 2007a, b, 2008a, b, 2015). This methodology uses the synergy of passive and active microwave satellite data —simultaneous active (radar altimeter) and passive (radiometer) observations available from various satellite altimetry missions (TOPEX/Poseidon, Jason-1,-2, ENVISAT and Geosat Follow-On), as well as passive data from SSM/I (to increase temporal and spatial resolution). This methodology could also be applied for the SARAL/AltiKa data (Kouraev et al. 2015).

ENVISAT (2002–2012) and SARAL (since 2013) missions operate at the same orbit with a 35-days repeat period and have two main nadir-looking instruments: radar altimeter and passive microwave radiometer. ENVISAT RA2 radar altimeter operates in Ku (13.6 GHz, wavelength 2.2 cm) and S (3.2 GHz, wavelength 9.4 cm), and SARAL AltiKa radar altimeter in Ka (35.75 GHz, wavelength 0.8 cm) bands. Radiometers operate in two frequencies: 23.8 and 36.5 (for ENVISAT MWR) or 37 (SARAL) GHz and we use a TB/2 parameter (average value of brightness temperature at two frequencies). The fact that these are simultaneous active and passive observations from the same platform significantly enhances the data analysis capability. We have used 18 Hz sampling frequency data for ENVISAT (cycles 9–83, 2002–2008) and 40 Hz data for AltiKa (cycles 1–12, 2013–2014) with Ice2 altimetric retracker algorithm which is common for both missions.

Comparison of annual TB/2-backscatter plots for ENVISAT and SARAL (Fig. 2) shows that the open water (lower left) and ice (upper right) clusters have very similar shape, thus confirming that by visually defining and then applying a threshold line specific for each dataset (red dotted lines on Fig. 2b) we can reliably discriminate open water from ice also using SARAL/AltiKa data. However, there are some differences in the absolute values of both parameters. Typical values for open water (point A) are lower for SARAL for backscatter (7 vs. 9 dB) as well as TB/2 (152 vs. 156 K compared to that of ENVISAT). Note that the difference between backscatter in both bands over ocean, as over ice sheets is also 2.5 dB

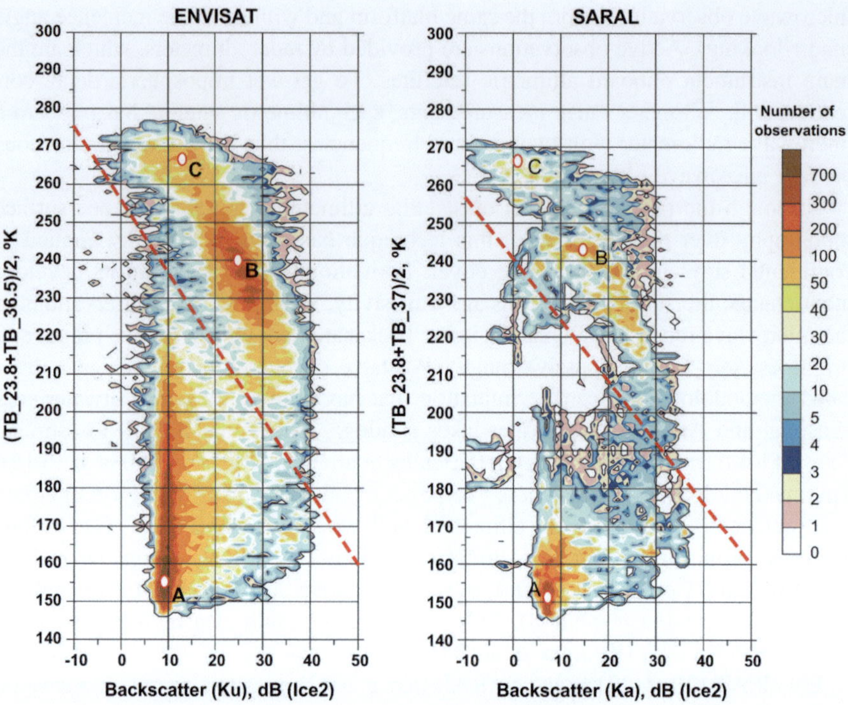

Fig. 2 Two-dimensional histograms (total number of observations) of ENVISAT (2002–2008) and SARAL (2013–2014) data for Lake Baikal in the space of backscatter coefficient (Ka or Ku bands) versus the TB/2 (average value of brightness temperature at two frequencies). A—open water, B—developed ice, C—ice by the end of the winter. After Kouraev et al. (2015)

(Rémy et al. 2015). Such difference could be attributed to the surface component and a difference in perception of surface roughness by altimeter. There are also much more centered values for TB/2 for SARAL (values vary from 145 to just 170 K) than for ENVISAT (much higher variability – from 147 to 220 K) for open water conditions. Values typical for developed ice (point B) show lower backscatter for SARAL (14 vs. 24 dB) and slightly higher TB/2 (243 vs. 240 K). Ice cover observed by the end of winter (Point C) has lower backscatter for SARAL (2 vs. 12 dB) and similar values for TB/2. One of the possible reasons for these changes is the different frequency (Ka vs. Ku band) that could lead to lower backscatter values for all three points for SARAL, and potentially difference in radiometer frequency (37 vs. 36.5 GHz). However, one should also be aware of the different time span of observations and their duration (a bit more than one year for SARAL and 9 years for ENVISAT).

2.2 Temporal Evolution of SARAL/AltiKa Backscatter in Late Spring

We have analyzed temporal evolution of ENVISAT RA2 and SARAL/AltiKa backscatter values along the track 466 (see Fig. 1) for several years and compared cycles dating mid-March (relatively cold air temperatures, thick ice) with consequent ones in Mid-April or early May (thin ice, usually not long before the final ice break-up). Both ENVISAT and SARAL data show an interesting feature—dramatic decrease of backscatter during the period of just one cycle (35 days). These differences reach 20 and sometimes 25 dB for ENVISAT and 30 dB and more for AltiKa (Fig. 3 for examples of 2009 and 2014). For each winter an early cycle has

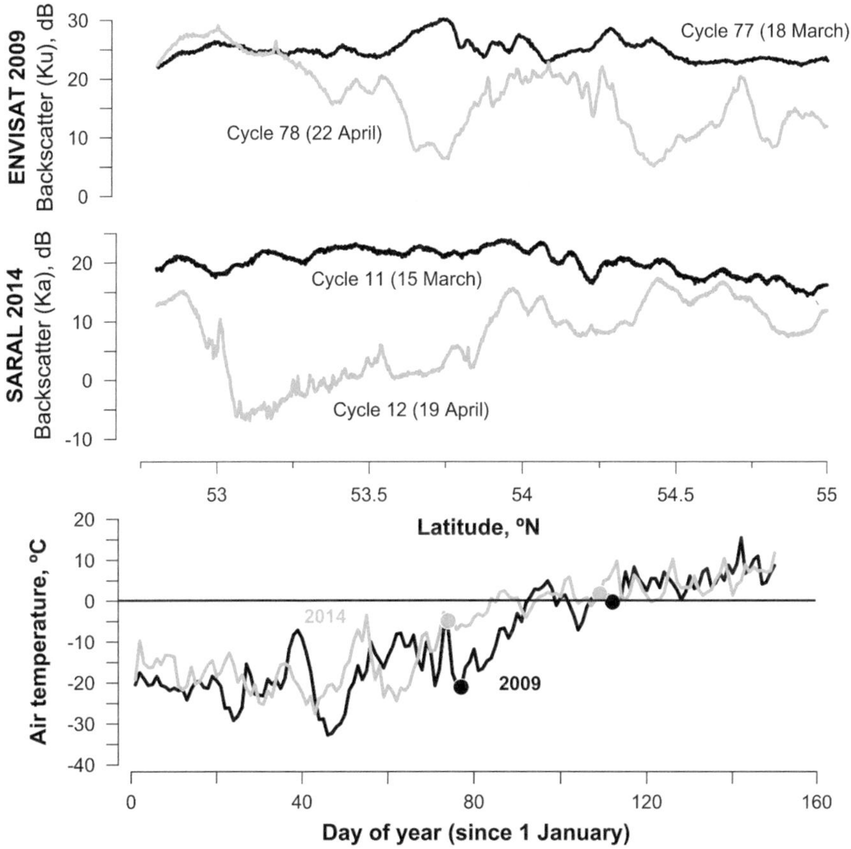

Fig. 3 Spatial variability of Ice2 backscatter values in late spring along 466 track for ENVISAT (in 2009) and SARAL (in 2014) (after Kouraev et al. 2015, please also see this reference for examples for other years), and air temperature variability at the Ust' Barguzin meteorological station in 2009 and 2014. Temperature at dates of satellite overpasses is shown as black (2009) and grey (2014) dots

smooth spatial distribution and a late one shows much more heterogeneity. Not only the absolute values are different, but even the spatial pattern changes quite a lot and it seems that there is only slight memory effect (similar spatial distribution of backscatter) or not at all from one cycle to the next.

Such rapid and significant changes of radar echo are closely related with processes governing evolution and ultimately destruction of ice cover in spring. Increased solar radiation and warmer air temperatures (changes from −21 to −0.25 °C in 2009 and from −5 to +1.7 °C in 2014) in late spring lead to snowmelt, as well as to fracturing and metamorphism of ice cover. Warm air temperature and consequent expansion of ice cover results in hummocking and formation of main cracks. These cracks extend for tens to hundreds of meters and more, forming a complex network and can represent serious difficulties while navigating on ice. Large leads formed during ice formation that have been frozen later on, produce a specular altimeter echo with sharp spike in backscatter values (Zakharova et al. 2015). However, the main cracks and leads represent only a small part of the ice surface and have mostly local impact on altimetric signal.

Much more important for understanding the evolution of backscatter are processes that take place almost everywhere on the ice surface and lead to elevated surface roughness and scattering and thus to lower backscatter. One is the formation of bubbles and air channels on ice surface as a result of intense solar radiation (Fig. 4). They have a typical width of 2–3 mm and are located just below ice surface, giving a whitish snow-like aspect to the ice cover. Their presence not only changes albedo (thus reducing impact of solar radiation and delaying melting) but also significantly affects ice radiometric properties in the microwave range. As noted in (Duguay et al. 2014), radar signal is very sensitive to the size and volume of scatterers in lake ice, especially air bubbles. AltiKa with shorter wavelength is even more sensible to small scatterers than ENVISAT.

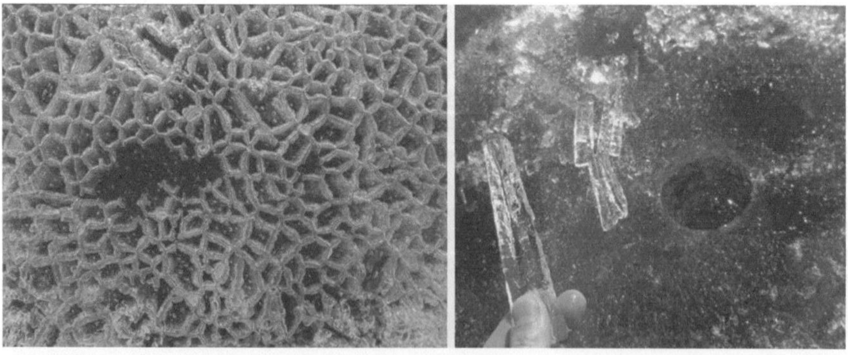

Fig. 4 Left panel: air channels under ice surface during a sunny day (6 April 2012, view from above). Right panel: needle ice crystals from ice bottom (hole diameter—11 cm, crystals size—10–12 cm long, 2–3 cm thick) at the northern part of the ice ring 3 April 2014 (photos by A. Kouraev)

Next step in the ice evolution is the formation of ice needles. When ice is still 50–60 cm thick, under the influence of solar radiation it starts to dissociate onto vertically-oriented ice crystals that are very weakly laterally bound (Fig. 4). Water can easily infiltrate such ice and despite its thickness the ice is no longer solid—even transport by foot becomes dangerous. Gradually ice needles melt down and ice disappears. Needle ice also have quite a different microwave signature than solid ice.

3 Giant Ice Rings on Lakes Baikal and Hovsgol

Changes in ice cover properties could be studied not only in the microwave range. In late winter—early spring ice metamorphism could also be observed as changes in reflectance in the visible and near-infrared ranges. Use of optical remote sensing data together with in situ measures of associated water structure and dynamics makes it possible to investigate the formation and development of giant ice rings on lakes Baikal and Hovsgol ice cover and provides insight into the mechanisms responsible for the ice rings' formation.

3.1 Ice Ring Detection and Inventory

By processing and visually analyzing optical and infra-red satellite imagery from MODIS (2002–2015), Landsat (1972–2015), NOAA AVHRR and satellite photography (with different spatial resolution and time coverage, 1983–2009), from the International Space Station (ISS) and US Space Shuttle (Kouraev et al. 2016), we have been able to identify 45 ice rings (including the 13 previously known) for Lake Baikal, and also—for the first time—4 ice rings for Lake Hovsgol (Fig. 5, see also Fig. 1 for ice rings locations). Although recently documented, the ice rings are not a recent phenomenon, as they have been observed as early as 1974–1975. Observation of ice rings for Lake Hovsgol demonstrates that this phenomenon is not limited to Lake Baikal and that there may be similar physical phenomena leading to ice ring formation elsewhere.

Our inventory shows that ice rings are also observed in regions that are shallower than depth of occurrence of gas hydrates (380 m, Granin et al. 2010), such as Maloye More in Lake Baikal—300–400 m or Lake Hovsgol—200–250 m, or in regions without known gas emission sources. This rules out gas release as a universal explanation of ice ring formation.

Figure 1 shows that ice rings have been observed in all parts of Lake Baikal from its southern to its northern part. Their location and timing of appearance changes from year to year and are hard to predict. The regions with the highest number of ice ring observations are the Krestovskiy and Nizhneye Izgolovye

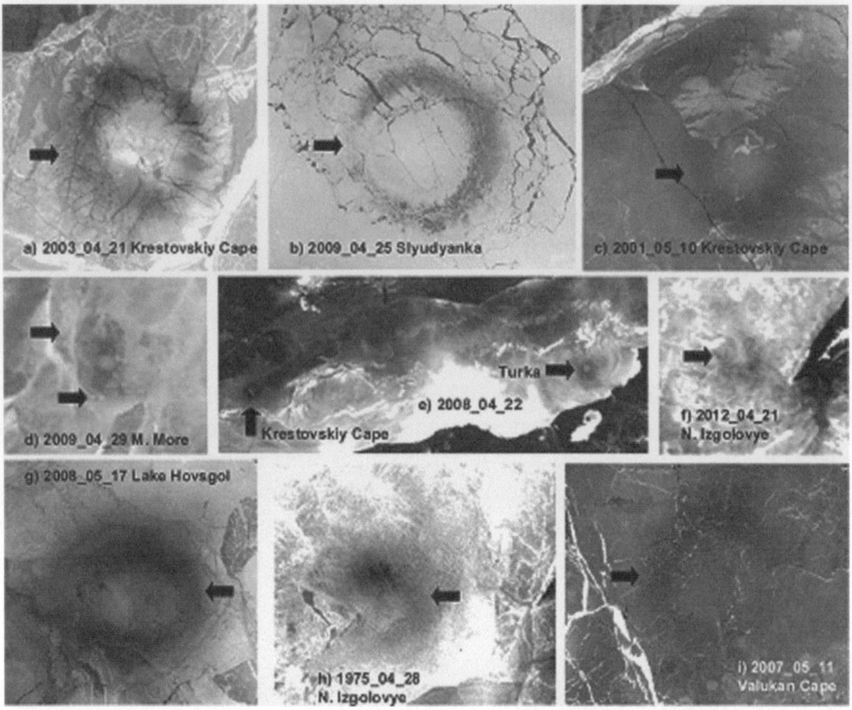

Fig. 5 Examples of ice rings for Lakes Baikal and Hovsgol. Data are from Landsat 1–7 (**a**, **c**, **g**–**i**), MODIS/Terra (**d**–**f**) and ISS (**b**) imagery. All images except (e) are aligned to North. After Kouraev et al. (2016). Diameter of the rings (km): **a**, **b** 5.2, **c** 4.4, **d** 3.8, **e** Kresovskiy Cape—5.4, Turka—4.6, **f** 6, **g** 2.2, **h**, **i** 5.4

Capes. For Lake Hovsgol ice rings are observed near the Dalayn-Modon-Huys Island in the middle of the lake.

The duration of ring observations (defined as difference between the first and the last cloud-free observations) in our dataset varies from 1 day (due to availability of satellite images and/or cloud cover) to 126 days; the typical duration is 5–10 days. Most frequently ice rings are observed in the second half of April (when ice become thinner and snow cover is absent or minimal) before the ice break-up, but they have been observed as early as 31 January and as late as 26 May.

3.2 Water Structure and Ice Surface Characteristics Associated with Ice Rings

As a result of our field observations, there are now four datasets of field hydrographic measurements in the regions where ice rings has been observed the same

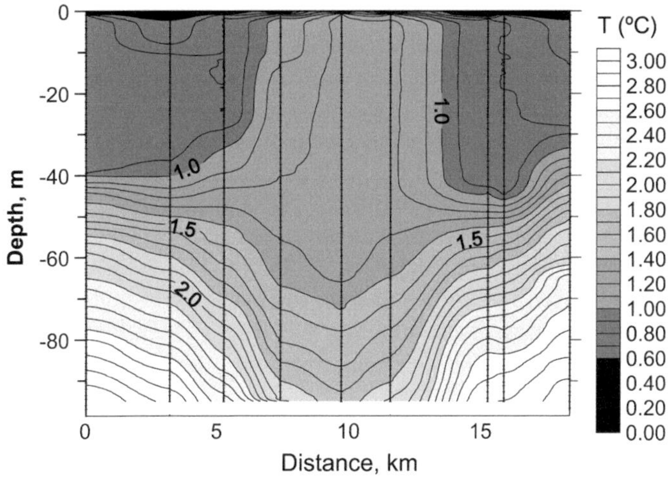

Fig. 6 Vertical profile of temperature (°C) along the transect across the ice rings on 6 April 2012 in the region of the Nizhneye Izgolovye Cape, Lake Baikal. After Kouraev et al. (2016)

winter. For the region of the Nizhneye Izgolovye Cape, measurements in 2012 and 2014 represent situations when an ice ring was already visible on satellite images. For two other cases (Lake Hovsgol in 2015 and the Shartlay Cape in Lake Baikal in 2013) observations have been done prior to ice rings manifestation.

All these hydrographical sections reveal lens-like structures (double-convex form) in temperature (Fig. 6), specific conductance and density fields. These lens-like structures are located at depths (defined as position of undisturbed isopicnal level) of 45–55 m for Lake Baikal and 30 m for Lake Hovsgol and correspond to the location of ice rings (already existing or yet to be manifested). The core of the lens-like structures has fairly uniform values of temperature (1–1.4 °C, which at depth 0–30 m is 0.1–0.6 °C warmer and at depths 80–90 m 0.4–1.3 °C colder than surrounding water).

This three-dimensional spatial distribution of temperature, specific conductance and density is typical for oceanic anticyclonic lens-like eddies (Dugan et al. 1982; Armi and Zenk 1984; McWilliams 1985; Kostianoy and Belkin 1989).

Our observations of water structure for different transects along the ring regions in 2012 and 2014 show that the observed lens-like eddies have an isolated circular form and radial symmetry. The observed anomalous water structure both before (2013 and 2015) and during (2012 and 2014) ring appearance shows that (a) such lens-like structures are directly associated with ice rings, (b) they exist before and continue to exist during ice ring appearance and development, and (c) an ice ring is a surface manifestation of dynamical processes going on below the ice cover.

Ice thickness measurements in the region of the ice ring in Southern Baikal in 2009 (Granin 2009; Granin et al. 2015) and our datasets for the four ice rings in

2012–2014 (Kouraev et al. 2016) show increased ice thickness in the ring center and outside of the ring, and much thinner ice in the ring itself.

For these four observations in ice rings regions ice was black and crystalline (with some ice crust and/or snow on the surface) and without impurities. No water was observed on ice and no gas bubbles were presented in the water below the ice. In 2014 at the stations located in the northern part of the ring we observed large needle ice crystals (up to 12 cm long and 2–3 cm thick, Fig. 4, right panel) at the bottom of the ice. Two neighboring stations presented no needle ice on the bottom, but ice was water-laden from a depth of 30–35 cm. This northern part of the ring represents an advanced stage of ice melting and metamorphism that will later on be typical for other regions of the ring. Further ice ring development will lead to thinner ice in the ring, metamorphism, fracturing, the appearance of leads and finally break-up of the ice.

When travelling on the ice, the main cracks are relatively easy to detect, and ice thickness is similar to other neighboring areas (except in the direct vicinity of the cracks). Ice rings, on the contrary, are not easily visible for travelers, at least not in the final stage of ice ring development. Ice cover here has the same visual aspect as the surrounding areas, but the ice is much thinner. In this respect, early detection and potential forecasting of location, size and properties of zones of weak ice cover associated with ice rings is of paramount importance for ensuring security of transportation and fisheries on lake ice in winter and spring. We have created a web site (www.icerings.org) dedicated to explaining ice rings phenomenon and necessary precautions for travelling in ice rings regions, as well as early warnings in the cases when our field data show presence of lens-like eddies and thus possibility of the ice ring formation later on.

The investigation of the currents field is also important for understanding why eddies didn't create circles, but ice rings with thicker and whiter ice in the central part. Currents observations in 2014 and 2015 confirm typical anticyclonic (clockwise) direction of eddies rotation. Eddies below ice rings are characterized by weak or moderate currents in the central part, and strong currents in the ring region (oriented at about 30°–45° to the left relative to the eddy boundary). This indicates that current speed (and thus intense heat exchange between ice and water), rather than presence of warmer water in the ring center, is the main driver related to ice melting and metamorphism. These regions of thinner and thus darker ice are seen on satellite images as ice rings.

Ice rings are a surface manifestation of these lens-like eddies, however the proposed mechanism of ice ring formation raises a new set of questions related to formation and development of these eddies.

The timing of eddy formation is the most elusive parameter to estimate with field campaigns as it requires long-term monitoring with high spatial and temporal resolution. Eddies may form under the ice when ice cover is stable (influence of seiches, internal waves and surges), or could be formed at some time between the autumnal vertical overturning and ice formation.

As for generation mechanisms there are several candidates: instability of a coastal current, wind forcing before ice formation, seiches induced by the influence

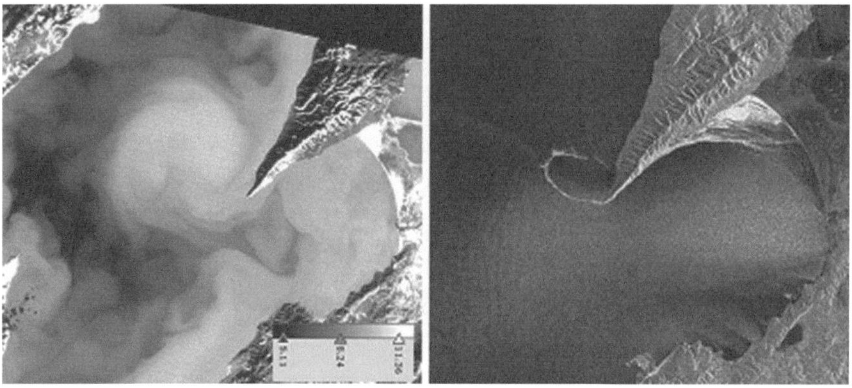

Fig. 7 Water dynamics in the region of the Nizhneye Izgolovye Cape. Left panel: Landsat 7 thermal (°C) image of water surface temperature from 26 September 2002, showing formation of an anticyclonic warm (10.1–10.8°C) eddy. Right panel: Sentinel-1 SAR (IW mode) image from 8 December 2014 showing strong wind blowing out of Barguzin Bay and a patch of new ice following an anticyclonic movement

of atmospheric pressure on the ice sheet, or river inflow and interaction of a coastal current with the coastline, for example near the Nizhneye Izgolovye Cape. In this region, where we found ice rings and lens-like eddies in 2012 and 2014, it is not uncommon to observe clockwise water currents (anticyclonic eddies) in the autumn or during ice formation (Fig. 7).

Detection of lens-like eddies requires dedicated and well-timed field campaigns with very fine spatial resolution between CTD stations of the order of hundreds of meters. In the case of lakes Baikal and Hovsgol, the identification and study of eddies is facilitated by the presence of stable ice cover that makes it possible to observe their manifestation as giant circular ice rings.

4 Conclusions

Our results show that active and passive satellite remote sensing data in different part of electromagnetic spectrum together with field measurements provide new insights into ice cover structure and development, and associated water dynamics. There is a clear need to continue and expand these studies for both ice-covered and ice-free conditions. Additional information could be also gained by numerical modelling approaches (modelling of active and passive microwave return signal, modelling of water dynamics etc.).

Acknowledgements This research is supported by the ERA.NET RUS Plus S&T #226 "ERALECC", CNES TOSCA "Lakes" and "LakeIce", CNRS PICS "BaLaLaICA", RFBR 13-05-91051, IDEX Transversalité 2013 InHERA and by GDRI "CAR-WET-SIB" projects.

A.G. Kostianoy was partially supported in the framework of the P.P. Shirshov Institute of Oceanology RAS budgetary financing (Project N 149-2018-0003). We would like to thank A. Beketov (Ust' Barguzin, Russia), T. Tivikova (Turka, Russia), S. Sushkeev (Tunka, Russia), Hurga and Byamba (Hovsgol National Park, Mongolia), A. Laletin and many others who participated directly or helped to organise field surveys.

References

Armi L, Zenk W (1984) Large lenses of highly saline Mediterranean water. J Phys Oceanogr 14:1560–1576

Balkhanov VK, Bashkuev YuB, Khaptanov VB (2010) Formation of circular rings on the snow covered ice field of Lake Baikal. Tech Phys 55(9):1266–1269

Bordonskiy GS, Krylov SD (2014) On the nature of ring-like structures on satellites images of Lake Baikal ice cover. Issledovaniye Zemly iz Kosmosa N4:27–31 (In Russian)

Circles in thin ice, Lake Baikal, Russia. Online at http://earthobservatory.nasa.gov/IOTD/view.php?id=38721, 25 May 2009

Dugan JP, Mied RR, Mignerey PC, Schuetz AF (1982) Compact, intrathermocline eddies in the Sargasso Sea. J Geophys Res 87:385–393

Duguay CR, Bernier M, Gauthier Y, Kouraev AV (2014) Remote sensing of lake and river ice. In: Tedesco M (ed) Remote sensing of the cryosphere. Elsevier, Amsterdam, pp 273–306. ISBN-10: 1118368851; ISBN-13: 978-1118368855

Galaziy GI (ed) (1993) Baikal. Atlas. Russian Academy of Sciences, Siberian branch. Moscow, Federal Service of geodesy and cartography (In Russian)

Granin NG (2009) The ringed Baikal (in Russian). Sci First Hands (Nauka iz pervykh ruk) 3 (27):22–23 (in Russian)

Granin NG, Jewson DH, Grachev MA, Levin LA, Zhdanov AA, Averin, Gnatovsky RY, Gorbunova LA, Tcekhanovsky VV, Doroshenko LM, Minko NP (1999) Turbulent mixing in the water layer just below the ice and its role in development of diatomic algae in Lake Baikal. Doklady Akademii Nauk 366:835–839 (in Russian)

Granin NG, Makarov MM, Kucher KM, Gnatovsky RY (2010) Gas seeps in Lake Baikal—detection, distribution, and implications for water column mixing. Geo-Mar Lett 30:399–409. https://doi.org/10.1007/s00367-010-0201-3

Granin NG, Kozlov VV, Tsvetova EA, Gnatovsky RY (2015) Field studies and some results of numerical modeling of a ring structure on Baikal ice. Doklady Earth Sci 461(1):316–320

Hsu J (2009) Mystery of giant ice circles resolved. Online at http://www.livescience.com/strangenews/090601-ice-circles.html, 01 June 2009

Kostianoy AG, Belkin IM (1989) A survey of observations on intrathermocline eddies in the World Ocean. In: Nihoul JCJ, Jamart BM (eds) Proceedings of 20-th international Liege Colloquium of Ocean Hydrodynamics. Mesoscale/synoptic coherent structures in geophysical turbulence. Elsevier, Amsterdam, pp 821–841

Kouraev AV, Papa F, Mognard NM, Buharizin PI, Cazenave A, Crétaux J-F, Dozortseva J, Remy F (2004) Sea ice cover in the Caspian and Aral seas from historical and satellite data. J Mar Syst 47:89–100

Kouraev AV, Semovski SV, Shimaraev MN, Mognard NM, Legresy B, Remy F (2007a) Observations of lake Baikal ice from satellite altimetry and radiometry. Rem Sens Environ 108(3):240–253

Kouraev AV, Semovski SV, Shimaraev MN, Mognard NM, Legresy B, Remy F (2007b) Ice regime of lake Baikal from historical and satellite data: Influence of thermal and dynamic factors. Limnol Oceanogr 52(3):1268–1286

Kouraev AV, Shimaraev MN, Buharizin PI, Naumenko MA, Crétaux, J-F, Mognard NM, Legrésy B, Rémy F (2008a) Ice and snow cover of continental water bodies from simultaneous radar altimetry and radiometry observations. Surv Geophys. https://doi.org/10.1007/s10712-008-9042-2 (Thematic issue "Hydrology from space")

Kouraev AV, Kostianoy AG, Lebedev SA (2008b) Recent changes of sea level and ice cover in the Aral Sea derived from satellite data (1992–2006). J Mar Syst. https://doi.org/10.1016/j.jmarsys.2008.03.016

Kouraev AV, Shimaraev MN, Rémy F, Ivanov AY, Golubov BN (2010a) An interesting natural phenomenon—giant rings on Lake Baikal ice. In: EGU (European Geophysical Union) 2010 proceedings, Vienna

Kouraev AV, Shimaraev MN, Rémy F, Ivanov AY, Golubov BN (2010b) Giant rings on ice of the Lake Baikal: a look from space and from beneath. In: Lacoste-Francis H (ed) ESA living planet symposium, Bergen 2010. SP-686 ESA living planet symposium, digital media. ISBN: 978-92-9221-250-6

Kouraev AV, Zakharova EA, Rémy F, Suknev AY (2015) Study of Lake Baikal ice cover from radar altimetry and in situ observations. Mar Geodesy 38: 477–486. https://doi.org/10.1080/01490419.2015.1008155 (Special issue on SARAL/AltiKa)

Kouraev AV, Zakharova EA, Rémy F, Kostianoy AG, Shimaraev MN, Hall NMJ, Suknev AY (2016) Giant ice rings on Lakes Baikal and Hovsgol: inventory, associated water structure and potential formation mechanism. Limnol Oceanogr. https://doi.org/10.1002/lno.10268

Mackay AW, Battarbee RW, Flower RJ, Granin NG, Jewson DH, Ryves DB, Sturm M (2003) Assessing the potential for developing internal diatom-based inference models in Lake Baikal. Limnol Oceanogr 48:1183–1192

Mackay AW, Ryves DB, Battarbee RW, Flower RJ, Jewson D, Rioual P, Sturm M (2005) 1000 years of climate variability in central Asia: assessing the evidence using Lake Baikal diatom assemblages and the application of a diatom-inferred model of snow thickness. Glob Planet Change 46:281–297

McWilliams JC (1985) Submesoscale, coherent vortices in the Ocean. Rev Geophys 23:165–182

Moore MV, Hampton SE, Izmest'eva LR, Silow EA, Peshkova EV, Pavlov B (2009) Climate change and the world's "Sacred Sea"—Lake Baikal, Siberia. BioScience 59:405–417

Remy F, Flament T, Michel A, Blusmstein D (2015) Envisat and SARAL/AltiKa observations of the Antarctic Ice Sheet: a comparison between the Ku-band and Ka-band. Mar Geodesy 38(sup1):510–521. https://doi.org/10.1080/01490419.2014.985347

Semovski SV, Mogilev NYu, Sherstyankin PP (2000) Lake Baikal ice: analysis of AVHRR imagery and simulation of under-ice phytoplankton bloom. J Mar Sys 27:117–130

The Gateway to Astronaut Photography of Earth (NASA Crew Earth Observations) http://eol.jsc.nasa.gov/. Last accessed Nov 2013

Verbolov VI, Sokol'nikov VM, Shimaraev MN (1965) Hydrometeorological regime and heat balance of Lake Baikal. (Gidrometeorologicheskiy regim i teplovoy balans oz. Baikal, in Russian). (Moscow-Leningrad. "Nauka", 1965)

Zakharova EA, Fleury S, Guerreiro K, Willmes S, Rémy F, Kouraev AV, Heinemann G (2015) Sea ice leads detection using SARAL/AltiKa altimeter 2015. Mar Geodes 38:522–533. https://doi.org/10.1080/01490419.2015.1019655 (Special issue on SARAL/AltiKa)

Index

V. Barale and M. Gade (eds.), *Remote Sensing of the Asian Seas*,
https://doi.org/10.1007/978-3-319-94067-0

MIX
Papier aus verantwortungsvollen Quellen
Paper from responsible sources
FSC® C105338

If you have any concerns about our products, you can contact us on
ProductSafety@springernature.com

In case Publisher is established outside the EU, the EU authorized representative is:
Springer Nature Customer Service Center GmbH
Europaplatz 3, 69115 Heidelberg, Germany

Printed by Libri Plureos GmbH
in Hamburg, Germany